POPULATION IN THE HUMAN SCIENCES

POPULATION IN THE HUMAN SCIENCES

CONCEPTS, MODELS, EVIDENCE

Edited by
PHILIP KREAGER
BRUCE WINNEY
STANLEY ULIJASZEK
and
CRISTIAN CAPELLI

OXFORD
UNIVERSITY PRESS

OXFORD
UNIVERSITY PRESS

Great Clarendon Street, Oxford, OX2 6DP,
United Kingdom

Oxford University Press is a department of the University of Oxford.
It furthers the University's objective of excellence in research, scholarship,
and education by publishing worldwide. Oxford is a registered trade mark of
Oxford University Press in the UK and in certain other countries

First Edition published in 2015

Published in the United States of America by Oxford University Press
198 Madison Avenue, New York, NY 10016, United States of America

British Library Cataloguing in Publication Data
Data available

Library of Congress Control Number: 2014956895

ISBN 978-0-19-968820-3

Preface

At the inception of the Human Sciences Programme at Oxford 40 years ago, population was considered the conceptual centre of its several disciplines, and marked by a symposium funded by the Wenner-Gren Foundation, published by OUP as *The Structure of Human Populations*. Human Sciences is a unique interdisciplinary undergraduate teaching programme which unites genetics, evolutionary biology, social and historical demography, social and cultural anthropology, sociology, animal behaviour, and ecology, together with options across a range of related subjects like epidemiology, linguistics, psychology, and archaeology. The programme has been very successful in achieving its goal of preparing students to go on to advanced research and policy applications requiring interdisciplinary knowledge and methods in areas like population ageing, infectious disease, inequality, ethnicity, nutrition, and a wide range of environmental issues.

With the benefit of hindsight, we can say that the founding of the programme took place at a transitional moment in the development of modern population thinking. Evolutionary biology and demography possessed recently established paradigms—the evolutionary synthesis and demographic transition—that were supposed to provide comprehensive frameworks for method and theory. By the end of the 1960s, however, the rise of molecular biology, and the implications of the Princeton European Fertility Project, had indicated fundamental patterns and processes of variation in human populations that the frameworks did not anticipate and could not explain. In 1970, when the symposium was held, fundamental paths of inquiry that were only emerging could not be fully appreciated. New genetic developments (like DNA sequencing or molecular developmental biology) and critical alternatives to classic demographic transition and modernization approaches (arising, for example, from anthropological and historical demography and social network analysis) soon emerged as necessary building blocks of inquiry.

These developments have been joined subsequently by new approaches arising, for example, from biodemography, niche construction, evolutionary anthropology, and the exploration of links between evolutionary and developmental biology. A key element in all of these later developments is consideration of gene–culture–environmental interactions of many kinds. The older frameworks and concepts of population on which they rest, while not displaced by newer approaches, often rest uneasily with them. Meanwhile it has become apparent that the founders and students of the two mid-twentieth-century paradigms made little attempt to clarify relationships between them or the common ground of problems they were encountering. The phrase 'population structure', for example, continues to mean rather different things as used by geneticists and mainstream demographers. Not surprisingly, the broad domain of population-based inquiry into nature and society which current approaches are entering thus appears seriously fragmented. There can be little question that this state of affairs is not a helpful one for scientific research. Even for research that remains grounded in problems of strictly biological or strictly social kinds, there can be little advantage in proceeding oblivious of conceptual, empirical, and modelling developments, or of how new kinds of database can be constructed, that are well advanced in other areas of population research and of wider applicability.

This book provides the reader with revised papers presented at the workshop addressed to these issues, held at the Pauling Centre for Human Sciences, Oxford University, on 7–9 September 2011. The aims of the workshop were to provide a critical overview of major developments in population thinking; to explore analogous practices and problems that have emerged in the several recent disciplinary developments; to consider similarities and differences amongst methods and concepts of population which typify the current state of population thinking; and to consider whether a new and more cohesive framework is emerging.

The editors owe particular thanks to all of the contributors, who entered readily into this collaborative project, and who developed and revised their work to help bring out common themes and clarify outstanding issues. Their interest and enthusiasm made for a very stimulating meeting indeed. We are very grateful to the Wenner-Gren Foundation for once again generously supporting a Human Sciences meeting, together with the John Fell Fund of the University of Oxford. Practicalities of the workshop itself were smoothly facilitated by the team of Human Science students, under the direction of Sarah-Jane White and Amy McLennan. Oxford University has provided us with excellent support in producing the book, and our thanks in particular to Aimee Wright, Jess Smith, Jim Driscoll, and Kumar Athiappan.

Philip Kreager
Bruce Winney
Stanley Ulijaszek
Cristian Capelli

Contents

List of Figures

List of Tables

List of Contributors

Jere R. Behrman W. R. Kenan, Jr Professor of Economics and Sociology, University of Pennsylvania

Walter Bodmer Head, Cancer and Immunogenetics Laboratory, Weatherall Institute of Molecular Medicine, University of Oxford

Melissa J. Brown Frieda L. Miller Fellow in Anthropology, Minnesota Population Center

Francesc Calafell Institut de Biologia Evolutiva (CSIC-Universitat Pompeu Fabra), Barcelona, Catalonia, Spain

Cristian Capelli Associate Professor of Human Evolution, Department of Zoology; Tutorial Fellow, St. Hugh's College, University of Oxford

David Comas Institut de Biologia Evolutiva (CSIC-Universitat Pompeu Fabra), Barcelona, Catalonia, Spain

Jason Dunn Centre for Anatomical and Human Sciences, Hull York Medical School, University of Hull

Sarah Elton Reader, Department of Anthropology, Durham University

Mark Elvin St Antony's College, University of Oxford

Simon Gregson Professor in Demography and Behavioural Science, Faculty of Medicine, School of Public Health, Imperial College, London

Michael Gurven Professor, Department of Anthropology, University of California, Santa Barbara

Timothy B. Hallett Professor of Global Health, Faculty of Medicine, School of Public Health, Imperial College, London

Stéphane Helleringer Assistant Professor of Population and Family Health, Mailman School of Public Health, Columbia University, New York

Paul L. Hooper Omidyar Fellow, Santa Fe Institute, Sante Fe, New Mexico

Graeme Hugo ARC Professorial Fellow and Professor of Geography; Director, Australian Population and Migration Research Centre, University of Adelaide, Adelaide, Australia

Jennifer A. Johnson-Hanks Associate Professor, Departments of Demography and Sociology, University of California, Berkeley

Hillard Kaplan Professor, Department of Anthropology, University of New Mexico

Hans-Peter Kohler Frederick J. Warren Professor of Demography, Department of Sociology and Population Studies Center, University of Pennsylvania

Philip Kreager Senior Research Fellow in Human Sciences, Somerville College; Lecturer and Tutor in Demography, Institute of Human Sciences; Director, Fertility and Reproductive Studies Group, School of Anthropology, University of Oxford

Stephen J. Kunitz Professor Emeritus, Division of Social and Behavioral Medicine, Department of Community and Preventive Medicine, University of Rochester, Rochester, New York

Daniel John Lawson Heilbronn Institute for Mathematical Research, University of Bristol

Klim McPherson Visiting Professor in Epidemiology, Nuffield Department of Obstetrics and Gynaecology, University of Oxford

John Odling-Smee Mansfield College and School of Anthropology, University of Oxford

Elisabeth Schröder-Butterfill Lecturer in Social Gerontology, Centre for Research on Ageing, Faculty of Human and Social Sciences, University of Southampton

Chris Spencer Fellow, Wellcome Trust Centre for Human Genetics, Nuffield Department of Medicine, University of Oxford

Jonathan Stieglitz Research Assistant Professor, Department of Anthropology, University of New Mexico

Mikołaj Szołtysek Research Scientist/Deputy Head, Laboratory of Historical Demography, Max Planck Institute for Demographic Research, Rostock, Germany

Simon Szreter Professor of History and Public Policy, St John's College, University of Cambridge

Stanley Ulijaszek Professor of Human Ecology, Institute of Social and Cultural Anthropology, University of Oxford

Kenneth W. Wachter Professor, Departments of Demography and Sociology, University of California, Berkeley

Susan C. Watkins Professor of Sociology, Department of Sociology and Population Studies Center, University of Pennsylvania

Bruce Winney Senior Postdoctoral Researcher, Department of Clinical Pharmacology, University of Oxford

PART I

POPULATION IN THE HUMAN SCIENCES

AN INTRODUCTION

PHILIP KREAGER, BRUCE WINNEY, STANLEY ULIJASZEK, AND CRISTIAN CAPELLI

> The diversification of models and their means of investigation does not mean that every case is unique and that no generalizations are possible. Such generalizations, however, will turn out not to be of universal application but will be properties that apply to local regions of the model space. The task of population modelling is to map these regions and find their properties.
>
> Richard Lewontin

1 Introduction

For two centuries population has been a highly charged locus of scientific debate and wider social controversy. In the modern view, variations in human numbers shape and strategically limit individual and group capacities. How a population's size, growth, and memberships are structured and managed is a critical factor in its development, the causes and consequences of which can be observed at several levels, from genes and microorganisms to individuals, networks, communities, and wider regional, national, and global entities. The potential for population structures at one level to condition phenomena at others has become an increasing focus of attention, as for example in the way social networks can shape genetic evolution and the spread of microorganisms, or the influence of different intergenerational family structures on the size of the wider population needing state welfare provision. Populations respond to needs and opportunities by altering their size and memberships via many compositional adjustments, and by varying their relationships with other groups. Consideration of intra- and interpopulation dynamics greatly enriches, but can also complicate, our understanding.

As it happens, distinctive and conveniently discrete populations are only rarely found 'waiting' for us to study them in nature. Likewise, the political boundaries and administrative districts that provide units and categories for census and survey tabulations were designed for very different purposes than scientific research; as they commonly cut across environmental, economic, social, and cultural processes, they often omit or arbitrarily carve up important subpopulations, and treat relationships between them very selectively. Hence, deciding units and parameters that enable us to define a population is not just a routine methodological preliminary. These choices determine how and whether we can study the way populations have been formed, adapt, and are (or are not) sustained. Establishing the identity of populations is itself an important objective of research, and one likely to have decisive influence on other research outcomes.

The potential combinations of units and parameters important for specifying populations have, meanwhile, become effectively infinite. For six decades demographers have struggled to narrow down the vast array of social, biological, economic, and other influences that shape fertility and mortality trends. Their aim as it emerged in economic and social demography of the post-war era was to arrive at a general theory of demographic transition: a limited set of deterministic relationships between small subsets of variables that would explain the principal causes and consequences of modern population increase and decrease everywhere. Seeking a universal model of demographic modernization, they instead uncovered a remarkable heterogeneity of pathways and outcomes, the causes of which have resisted reduction to uniform deterministic formulae. In the 1950s, at the same time that demographers were laying down their approach to demographic transition, evolutionary biologists were taking a very different course. They entered the post-war era on the basis of the 'evolutionary synthesis', a framework uniting previously disparate approaches in population genetics, systematics, and laboratory research. Within this framework, biologists actively pursued approaches which, beginning with the discovery of DNA as the information-holding molecule, led to a genomic model of inheritance which intentionally multiplies possible combinations of units, characteristics, and relationships into the billions. Of course, evolution is not by genes alone: understanding why certain genetic changes endure to differentiate species and subpopulations within them requires study of a further diversity of ecological, demographic, and cultural influences which shape inheritance in local populations.

Possible applications of research to topical problems of health and public policy have, of course, been an important driver of both approaches. A common ground of pressing contemporary problems—like AIDS, cancer, ageing, reproductive health, poverty, and networks that act as vectors of genetic, epidemiological, demographic, and social change—call for approaches that build simultaneously on knowledge of biology and of social and cultural formations. The development of approaches able to draw on research at multiple levels—from genes and microorganisms up through the many levels of social complexity—is clearly necessary if issues like these are to be resolved. In other words, when we look more closely at current problems of health and society, we find that not one, but several sorts of population require observation, measurement, and modelling as appropriate to the different levels that make up a given phenomenon, or problems concerning it. An important question thus arises: is this increasing appreciation of the heterogeneity and interrelatedness of population processes, and of the need for collaborative approaches to them, confined only to approaches to topical human concerns? Is there possibly a deeper convergence or complementarity of population concepts that is now emerging as a common ground of the human sciences?

The papers collected together in this book provide evidence and an exploration of this possibility. The principal approaches, while developed in the context of research in different disciplines and subject matters, reveal a common general framework in which methods and models arising from the two main traditions of population thought are allocated as appropriate to the differing levels on which research is being conducted. This framework, which has evolved pragmatically, rather than as a product of formal theory, suggests a much greater underlying coherence of approaches to population than is apparent when, as is common, the fields of evolutionary biology, genetics, history, ecology, demography, and the social sciences are viewed separately. It reveals a common historical ground of population thought preceding the modern habit of separating knowledge into academic disciplines, a ground upon which all of these fields draw.

The aim of this book is thus to clarify and illustrate the concept of population in the wide range of its current scientific and historical uses. This helps, in turn, to shed light on the relation between concept and model developments and the need and demand that science address problems of health and society. The application of specialized methods and knowledge of population to these problems is not only a response to urgent social ills; often it is simply necessary because greater clarity on a given problem is required than public discourse can provide. Fortunately, in recent decades there has also been a growing popular appreciation of the diversity and interrelatedness of populations, notably in respect of the environment, disease, and human rights. This introduction therefore begins by sketching briefly the wider contemporary context.

Population research has responded not only by addressing contemporary issues, but by taking a different path, which is to go back to basics. Asking elementary questions about population definition is critical: it forces us to step back and ask which criteria are needed to specify evidence empirically and comparatively. Being able to tie down particular genetic, environmental, cultural, and other processes to specific aggregates and subgroups in which they are occurring—in contrast or in relation to other populations in which they do not occur or are not significant—is a critical stage in our ability to identify and understand a given phenomenon. Modelling enters this process not only as a tool for interpreting and analysing data, but also as a way of formulating hypotheses about the identity of subpopulations and their characteristics. The interaction of model and evidence may require us to rethink, and thus reshape, what we understand as the very phenomena under investigation.[1] The conceptual formation of a population and its constitution in datasets arises from an interplay of observation, model, and

[1] Inevitably, not all fields for which population definition is a critical issue (whether of research or policy) can be covered in a single book, and of course some fields have advanced much further in assessing the adequacy of conventional population units and relationships than others. For instance, recent research linking evolutionary and developmental biology carries potentially important implications for revising the Evolutionary Synthesis—but it has been noted that, as yet, these developments often do not generate data that can be used to assess population relationships (Müller 2007: 6). How to specify subpopulations for purposes of identifying differential environmental impacts is also an important area to which much more attention could be given (Kreager 2011).

measurement, and only by viewing this process as a whole can we assess how tenable our models and other assumptions are.

1.1 A Tangle of Populations Awaits Sorting

The findings of biologists and demographers often attract wider discussion since they may be potential steps toward hoped-for remedies to human ills, or call attention to potentially dire environmental, economic, and social problems. Populations and their management consequently provide ready subjects for public debate, which often continues for many years without resolution. That Malthus's principle of population still arouses contention two centuries after its first publication is a famous case in point. Uncertainty in the public sphere doubtless owes something to the complexity of defining populations, just noted. Or perhaps more to the point, inability to reach public and scientific consensus is likely to arise when only some levels of population phenomena have been adequately defined.

Debates, for example in the social sciences, have long tended to prioritize the macro level in which generalization normally proceeds on the basis of national and global trends. Nation states are, of course, composite populations made up of myriad cultural, occupational, status, political, age, health, and other groups. The size and compositions of these subpopulations differ, and beneath and amongst them lie possible genetic, physiological, environmental, and other influences. It is clear that a given health or environmental policy, technological development, change of government, spread of pathogens, shift in market sentiment, and so forth, may not affect all subpopulations in the same way, and can alter the behaviour, memberships, and even very existence of groups. Popular recognition has grown in recent years that census, vital registration, and other documentation of constituent subpopulations can be crucial to addressing vulnerability and respecting human rights. Coupled with this is recognition that impacts on one subpopulation carry implications, in turn, for others. What used to exist in the singular—'the population problem'—has shifted to awareness that multiple, interconnected populations are at issue whenever policies and events, such as the contemporary issues just listed, come to the fore. Hence defining characteristics like composition, structure, and relatedness to other populations need to be established if the workableness of policies, and the differing agency or vulnerability of groups, are to be taken into account.

Probably the most singularly vague and controversial proposition about population in recent times was the so-called 'population bomb' widely canvassed in the post-war era, a kind of demographic 'nuclear option' of rapidly increasing birth rates in developing nations that was supposed to preclude their economic development, implying famines and environmental disasters for these countries, with major impacts on the whole world. As the most populous countries, China and India, have emerged as economic giants, and as fertility declines have spread in Europe and most of Asia and Latin America, this intentionally simplified prognosis has gradually receded. In its place a host of new population-related threats have emerged—'cluster bombs', so to speak, both large and small—but still lending themselves to extrapolation across the world, where they are seen as carrying potentially huge and alarming impacts.

Thus, rapid shifts from fertility increase to decline in many countries, both rich and poor, have created marked age-structural imbalances known as population ageing. The general scenario is of shrinking workforces facing the demand to finance both themselves and proportionately larger older cohorts—at a time when the world's population size has more than doubled, escalating economic and social competition and hence the cost of resources. Competition in turn has encouraged ever greater environmental exploitation and pollution, entailing destruction of habitats, the dispersal of local populations, and the extinction of cultures and species. Capital and credit flows in the context of economic globalization forcibly link otherwise disparate populations into this competition, groups that may have little or no contact or conception of each other, from transnational financial elites and laboratory scientists to peasants and hunter-gatherers. Radical and growing disparities between rich and poor translate into conflicts between less and more advantaged subpopulations, while encouraging unsustainable consumption patterns—both of which feed into ever heavier resource use and environmental deterioration. Adjacent populations long characterized by susceptibility to conflict—such as ethnicities, faiths, classes, generations, castes, and migrant/host populations—in consequence find all sorts of reasons to perpetuate their differences. The phrase 'global warming' effectively summarizes the potential of all of these disparities to combine to form the biggest and most interconnected of 'cluster bombs': we face a world with so many people, who live in so many different population groups that are connected in so many ways, and whose impacts on the environment are fuelled by the many differences and connections between them.

The old 'population bomb' theorists, in light of this multiplication, growing scale, and complexity of problems, might feel their warnings were justified. Yet the idea that all of these issues might have been avoided if we could disseminate birth control rapidly enough was clearly an insufficient nostrum in the first place. The large birth cohorts of the 1950s to 1970s were already happening when the 'bomb' was publicized, and now that they, their children, and grandchildren are here, we must expect continued population growth over the twenty-first century. Where younger cohorts are larger than older ones, growth is inevitable if each couple merely replaces themselves. The 'cluster bombs', like ageing, environmental impacts, and economic and health inequalities, still foster macro-level scare scenarios no less vividly than the old 'bomb'. What is clear is that these trends are interconnected and are not affecting all population groups, or arising from them, in the same way. That is the reality that needs disentangling. Even where we can see large-scale implications, the entailments for the behaviour of different populations at different levels will vary. Hence the idea that description and analysis can be confined to two stylized 'universal' levels—macro and micro—has become seriously untenable. It is the heterogeneity of populations, their evolving interrelations, and the processes that renew heterogeneity, that we need to understand.

1.2 Opening Up the 'Local Model Spaces'

Which populations to study, and in *what* networks or set of aggregates they are located, thus become primary questions. From the ancient Greeks to Darwin, the identity and relatedness of populations were matters for careful observation and reflection, whether the issue was membership of the citizen body or characteristics defining a species. Answering the question *What is a population?* (or, *What is the population we need to understand?*) comes down to how it is constituted, whether this means its processes of formation in nature and society, or the classifications and measurement techniques that research uses to specify its limits, composition, and structural characteristics. A population is not a given, but a construct. Early stages of definition commonly entail sorting things that we don't know very well: how to deal with aspects of phenomena that are as yet incompletely observed or simply not observable; possible confounding influences due to external or historical factors; characteristics that seem to stand out, but which may in fact be a function of our own biases; and so forth. Whether the events and relations under investigation are best understood in terms of one or perhaps several populations, or subpopulations, requires clarification. The experience both of evolutionary biology and recent combined methodologies in demography, as we shall see, is that the articulation of multiple subpopulations in which modelling, quantitative, and qualitative observation all have critical roles to play, is likely to be the norm. It is this combination of methodologies that comprises the 'mapping' of local model spaces to which Lewontin's remarks, cited at the beginning of this introduction, refer (2004: 17).

Of course, the fact that populations come to be identified as a function of choices in which researchers and policymakers usually play major roles does not mean that populations are mere scientific or administrative playthings. Populations identified in scientific and administrative discourses readily 'become real', and become objects of popular contest, as the history of economic, political, racial, and other conflicts shows. Differentiating populations and subpopulations creates opportunities for targeting. Research and administrative agendas inevitably mean that some subpopulations are recorded, and others not.[2] The sophisticated apparatus of statistical checks that developed steadily from the nineteenth century can test for formal consistency, but provides no guarantee against the reification of populations conceived for specific scientific purposes coming to stand as social facts. The importance of the 'What is a population?' question lies particularly in requiring observation and analysis always to clarify how population memberships have been arrived at, and in what senses they are artificial.

Two main orientations to observing and modelling aggregate phenomena have emerged over more than two millennia of interest in human populations and their

[2] A substantial literature has come to surround the ways in which census, registration, survey, and the many official datasets have been used to characterize, and in many cases stigmatize, racial, ethnic, national, class, and other populations. Examples include: Cohn (1987); Kreager (1992); Kertzer and Arel (2002); Szreter, Sholkamy, and Dharmalingam (2004); Breckenridge and Szreter (2012).

strategic uses. A convenient way of summarizing these approaches is to refer to them simply as 'top-down' and 'bottom-up'. The fundamental differences between them, and also their potentially complementary roles in description and analysis, were clearly recognized by students of human evolution (notably, Ernst Mayr, 1959) by the middle decades of the twentieth century. Top-down approaches inevitably have a strong element of closure, to be discussed in the next section; bottom-up population thinking may observe open (or not formally bounded) populations that exist in nature and society, but approaches to open population phenomena often draw on closed population models for several purposes, as we shall see. The need for both approaches has come to be recognized much more slowly in demography, history, and social scientific population studies, but has in recent years gathered considerable momentum.

1.2.1 *Closed Populations*

Top-down population thinking is familiar in the principal data sources of demography and public statistics (e.g. censuses, surveys), and of population genetics, medicine, and ecology (e.g. genomics, clinical trials, metapopulations). These compilations may be called *closed populations*, since established units and classifications are usually employed, and any further variables decided before data collection; aggregates are then compiled using comprehensive enumeration or systematic sampling so that datasets may be treated as complete or representative wholes. The way closed population data have been classified for purposes of collection usually dictates the levels of aggregation possible, and there may be a strong element of convention. Thus, demographic data are, as noted, commonly analysed at micro and macro levels; within each, subpopulations at several levels are possible (e.g. provinces within nations, social and economic classes, etc.); and at the micro level, individuals, couples, or households may likewise be used as constructs. Of course, not only convention but circumstance may be at issue: historical demography and archaeology commonly rely on 'opportunistic' samples, i.e. those individuals and groups that happen to survive in archival sources and archaeological remains. Population change within these various closed universes tracks the movement of individuals, genes, or other discrete entities between one existing state or class (for example, age group) and another, which may then be correlated with increases or decreases of given characteristics. While closed population monitoring is necessary to record long-term trends consistently, and to test hypotheses, there is an inevitable tension between retaining units and classifications for such purposes and altering data systems to admit phenomena previously not recognized or included. As several chapters in this volume remark, subpopulations critical at levels not included in closed classification schemes—networks of migration, disease transmission, and intergenerational support, for example—generally cannot be constructed subsequently from most conventional closed data sources.

Closed methodologies are designed not just for monitoring purposes, but with an eye to mathematical statistics, since large-scale enumeration, where possible, enables checks on data consistency that are not possible in small aggregates. Closed population thinking also opens up simulation and other modelling approaches as methods that can assist in identifying and constructing populations. Where unobservable and potentially confounding factors are thought to be important, a model can be built to estimate their possible influence by including parameters for them, and then varying the weight or range of their impacts hypothetically. From this we can sometimes decide whether and to what extent a given factor could significantly influence population structure and behaviour, even if data are seriously incomplete. Of course, the processes giving rise to such factors are not actually observed and documented in this approach.

'Top-down' is an apt way of describing closed methodologies, in two respects. First, and obviously, data compilation and modelling are designed primarily with an eye to phenomena on a large scale. The standardization of units necessary to comprehensive and identical measurement of all subclasses in very large populations (as employed, for example, in national data series or clinical trials), usually has the effect of reducing the categories of population and attributes under study, thus eliminating much variation at lower levels.[3] Second, the approach is 'top-down' because some scientific or administrative body assumes the authority to specify what defines a standard population and its members, which variables are included and which are not, and what relationships between variables (or perhaps proxies for them) can be inferred where they have not actually been observed. Top-down approaches, in other words, permit considerable leeway in deciding which characteristics should be used to define a population. The obvious danger is that populations may be heavily defined by parameters that reflect simplified formal premises, or by agendas of particular research schools and policy interests. They may also simply reflect long-standing social and political norms, as is typical of census and survey compilations. Where population units and definitions are given by convention rather than being principal and open questions to which inquiry is addressed, there is less scope for reducing biases.

As Harrison and Boyce (1972: 3–4) note, demographic or any systematically compiled samples may serve as the basis for subsequent closed population analysis. Units and parameters devised for the purpose of one study often get carried over into further research, and this may include not only assumptions but possible biases they entail. Subsequent analyses can proceed whether or not analysts have in-depth knowledge of the population to which the data source refers, of local contingent factors shaping the original study, or of later changes to the population that may be relevant. Replication of standardized units and classifications in this way is commonly 'applied to many situations which are not primarily concerned with population definition', i.e. where existing or conventional units are simply accepted as real. Social and economic demographers, for instance, frequently employ existing survey tabulations, and while they may apply statistical checks and models to control for possible confounding factors, the use of

[3] Genomics, as we shall see, has opened up the capacity to get around this, since it multiplies the sequences of genetic characteristics into the billions; this enables much finer specification of genotypes, or subsets of genetic traits, and hence of the haplogroups or subpopulations defined by them. However, the number of subsets that are potentially determinant, for example, of the spread of a given disease, also multiplies.

'off-the-shelf' data often means not that such factors are not observable, but that they simply were not addressed in the compilation. In such cases, as Harrison and Boyce remark, whether principal variables and adequate data on them are or are not being included in research becomes arbitrary.

1.2.2 *Open Populations*

There is a need, therefore, for approaches that observe processes by which population memberships in nature and society are actually formed, and how they adapt and change. Instead of inferring relationships amongst subpopulations given by prior classification, the processes that compose relationships and groups based on them are studied directly. Methods used to identify subpopulations obviously vary according to the phenomena under study. Observation, whether in the laboratory or natural and ethnographic contexts, leads to the compilation of events and relationships to which simulation and other closed modelling techniques can be applied to develop hypotheses and test relationships. This combination of open and closed approaches commonly feeds back into deciding how subpopulations may best be specified. Combined open and closed approaches are thus much more flexible than relying solely on closed population units and classifications decided in advance. This flexibility is often necessary for deciding criteria that define *which* population or subpopulation captures processes and trends we wish to understand, and which levels and subpopulation relationships therefore need to be included in analysis.

What counts as the 'bottom' level for purposes of description and analysis consequently varies. For Darwin, writing as a natural historian, the bottom-up perspective implied genetic characteristics that could at the time only be hypothesized. The bottom line of evidence therefore consisted of individual organisms and subpopulations of organisms. Darwin, in proposing that the origin of species arises from the capacity of mating relationships within and between subpopulations to produce members with hitherto unrecognized characteristics, gave the first scientific formulation of open population processes, and made them central to the wider theory of evolution (Darwin 1859; Mayr 1982; Gould 2002). The phrase 'origin of species', in other words, describes a theory of evolution based on subpopulation interaction. While conceived with reference to open populations existing in nature, this approach has proven very apt for bringing insights from closed population data collection and modelling to bear on open population phenomena. The combined perspectives and methods of open and closed population thinking became fundamental to the framework known as the evolutionary synthesis (Mayr and Provine 1980), on which subsequent developments in genetics, ecological heritability, and evolution rest.

Understanding how populations are formed, how memberships vary, and how variation can lead to new group formation is, of course, also recognizable as a normal and fundamental element of social life, in which critical understanding often does not entail either biology or demography as they are now understood. Sophisticated understanding of the open nature of population processes long preceded Darwin. The workings of kin systems, patronage and migration networks, marriage markets, political parties, and many other interest, professional, and factional groups have long involved their members in making strategic assessments of relative group size, composition, and position, and of whether extending, curtailing, or forming new alliances that change the inclusiveness of a population's membership carry individual and collective advantages. The communities, networks, and other aggregates involved in this behaviour are, like those Darwin had in mind, *open populations*, since the boundaries or criteria of membership are subject to change, and people commonly belong to, and change their participation in, several networks. Open population thinking is by nature strategic. It is concerned primarily with *relationships*, whether between individuals or groups, and with how these affect memberships and thus the population composition, size, and structure of states and their capacities. The closed datasets and measures that demographers, population geneticists, and others use to track births, deaths, genotypes, and so forth in effect fix a subset of outcomes of this wider compositional demography at particular points in time. As will be outlined in Chapter 1, open population thinking has a distinguished history integral to the rise of science and the state, beginning from models provided initially by classical antiquity and elaborated in the course of modern European history to shape our basic ideas of how the individual, society, the state, and species in the natural world are related.

While the long history of open population thinking undoubtedly shaped the formation of Darwin's theory, there is also no doubt that our ability to conceptualize open population processes was gradually, but radically, transformed in consequence of his work. The logic of variation that Darwin identified acquired only later an empirical, genetic basis with Mendel's discovery of the way specific characteristics in an open population come to be distributed regularly in the course of mating. This opened a door to increasingly complex population-genetic models that enable the logic of relatively small Mendelian groups to be employed in generating hypotheses for wider aggregates.

Membership of a subpopulation in genetics is defined in terms of networks of mating individuals, and more particularly by molecular elements that individuals come to share. Mating sustains but also may vary their characteristics, creating the possibility of new types of individuals, new demes, and possibly even new species. A membership and its characteristics may change due to genetic mutation and drift, but the continuity of these alterations is subject to members' interaction with the environment. Not all changes that shape evolution are genetic: members inherit ecological adaptations as part of the groups to which they belong, and engage in activities which create and alter the environmental and social niches they inhabit; members may, moreover, migrate, and make use of the phenotypic plasticity that, as individual organisms, enables them to adjust to changing conditions. At any given point in time a number of subpopulations are usually identifiable in a given species; while each may have some potential to separate into distinct species, this is less likely than continuing compositional evolution by which demes may gradually appear and disappear, increase or decrease, within a species.

An obvious but very important pay-off of this interplay of observation and modelling is that what we learn is not confined to the data categories we begin with. Changes that effect population composition may not just alter trends in known types of individuals and events as in closed population thinking: they can give rise over time to new and different populations or subpopulations in which processes and characteristics not recognized in existing classifications have become, or are becoming, primary. Heterogeneity and its renewal are thus canonical to natural selection. And, as already noted, the continuing heterogeneity of modern reproductive and other population trends makes the issue of *what sustains heterogeneity* no less a fundamental issue for demography and related social science disciplines.

The discovery of the role played by DNA and the rise of genomics became possible in part because an immense amount of direct observation in laboratory and natural settings enabled Mendel's work and its development by population geneticists in the 1920s and 1930s to be integrated into Darwin's theory, while also significantly correcting and revising it. Models may now be used to infer the existence of subpopulations and to estimate possible influences on their evolution. Genomics, in radically increasing documentation of sequences of variation, has greatly enlarged the possibility of identifying subpopulation memberships, since (with the availability of billions of combinations of Single Nucleotide Polymorphisms (SNPs)) much finer identification of local memberships or model spaces becomes possible. With the rise of evolutionary biology and the study of heredity, Mendelian populations (i.e. those determined by degrees of relatedness established through mating), have become the most extensively documented populations established by combined open and closed population thinking.

The 'bottom' level has thus descended from Darwin's populations of organisms and their environments to ever more micro levels of phenomena. As Kaplan and Gurven (2008) note, bottom-up research now refers to a continuum of levels from microscopic intracellular processes to populations of human and other organisms. Bottom-up explanations proceed in terms of processes at and between several levels (genes, proteins, subcellular pathways, cells, tissues, organs); relationships within and between these levels are, so to speak, the mechanisms which bring about an organism's development, and may thus influence the composition and development of whole populations. For a time, the idea that genes could determine higher processes at all of these levels even suggested to some authors that organisms are little more than passive vehicles shaped by genes. The emerging complexity of the human genome and the importance of feedbacks from the environment, however, have led to a more cautious view: documenting the complete composition of the human genome still leaves open the fundamental question of what triggers the expression of some genes rather than others. Not only the potential combinatorial complexity of some twenty-three thousand genes for each individual, but their relationships—the multifunctionality of genes, the influence of different cell environments, the role of epigenetic factors, etc.—provide multiple pathways of development. Feedbacks from several levels—such as economic and social constraints on embryological and physiological development; the ways in which different marriage systems shape mating; advantages and disadvantages of a given niche; the differing structure of proteins; the influence of maternal genetics inherited with the egg; and so forth—can determine which combinations of genetic material are activated. Genetic material may thus come to be expressed via different pathways connecting the several population levels:

this 'redundancy', or capacity for determination to occur via multiple routes, is often a key reason why modelling is necessary to disentangle the possible determinants of observed population characteristics. It is common now to speak of *coevolution*, in which several genetic, ecological, and social organizations evolve in tandem.[4]

1.2.3 *The Emergence of Complementary Approaches in the Study of Human Society*

Prevailing attitudes to the two principal approaches to population in the social sciences have, until recently, been very different than in evolutionary biology. Over the course of the nineteenth and twentieth centuries, the institutionalization of vital social and economic statistics in government, academic, and other public inquiries has given closed and top-down population methods powerful predominance, to the almost complete exclusion of open population thinking as an explicit methodological and theoretical perspective. Although the global proliferation of survey and census systems in the twentieth century has enabled populations to be studied at many levels from the state to the household and individual, their use has been guided by economic and social theories that radically reduce attention to the heterogeneity of subpopulation composition and feedbacks between multiple population levels as major factors in human adaptation and change. Three well-known and closely linked theoretical assumptions may be summarized, very briefly.

One is the overriding attention given to analysing and interpreting variables at just two levels: the highest or macro level (typically represented by national, regional, or global datasets), and the lowest or micro level (typically individual and household data). Population structure in this approach usually refers to the age/sex composition of a national population pyramid that results from mortality and fertility rates; explaining why rates vary then proceeds via correlation with individual and household characteristics, occupational groups, geographical regions, and other closed units and series as given in survey and census schedules. To the extent that units and classifications capture all relevant constituent subpopulations accurately, this approach is comparable to the concept of population structure that prevails more widely in population research, i.e. of an evolving set of relationships that constitute and change memberships of subpopulations.

[4] Geneticists, epidemiologists, and other population biologists have in recent years written very stimulating and readable introductions to current developments, of which notable instances are Lewontin (2000) and Noble (2006). Use of 'level' here follows these texts, and it is important to note that the term is often one of convenience both in the biological and social sciences, rather than an agreed canon. In closed population data and models, different levels like socio-economic classes or feedbacks between genetic and physiological processes, can be strictly defined. Where research combines open and closed population approaches, as, for example, in exploring how information and mating networks are related in the transmission of HIV/AIDS, 'level' refers to a number of subpopulations (e.g. pathways of migration and community networks) that are incompletely mapped and are still in the process of exploration and definition. The long-standing predominance of closed population description and analysis in demography and the social sciences appears to have had the consequence that there is no general agreement on what the principal boundaries are between subpopulations identifiable across the range from macro to micro levels.

Macro-level analysis reveals an impressive convergence in fertility and mortality trends over the last century and a half (Wilson 2011), yet the diversity of the timing, duration, and trends of declines (Chesnais 1992; Watkins 1987), continuing variations at levels at or below population replacement (Kohler et al. 2002), and more particularly the many possible mechanisms that might explain this diversity, remain objects of much speculation.

However, as is well known, there are many intermediate subpopulations (social and economic networks and hierarchies, ethnicities, migration networks, emerging economic sectors, religious groups, political factions and patronage networks, patterns of morbidity and health-seeking behaviour, etc.), as well as individuals' management of their several group identities, that are commonly excluded from, or can be resistant to, census and survey procedure. The memberships and dynamics of these subpopulations frequently cut across and subdivide standard census and survey units. Demography's preoccupation with determinants of fertility and mortality trends measured only in standard units over most of the twentieth century has led, unfortunately, to much less attention being given to the way people use network and other compositional adjustments of their several population memberships to address their needs and problems. The conventional macro/micro focus in consequence lacks flexibility to observe and track important demographic phenomena, especially how subpopulations are formed, the ways group composition can be varied to achieve changing needs and ends, and how such adjustments within and between subpopulations combine to influence trends and processes at the macro level. Reliance on census and survey formats further restricts evidence and explanation to events and relations that people are willing to report, and even then reports must be assigned only to those classifications recognized prior to data collection. Survey and census data are commonly not underpinned by sustained direct observation of processes of variation and change that shape memberships, such as the study of natural history, laboratory research, and ecology provide in evolutionary biology.

A second assumption is commonly employed to stand in lieu of direct observation, and of such observation informed by modelling, in defining subpopulations such as communities, networks, and other local units. The study of population dynamics at macro and micro levels has been guided by a general framework of national economic development and modernization which stipulates the main variables relevant to population changes. Particularly in the early post-war era, measures of urbanization, income, education, and so forth tended to preclude interest in exploring the way local histories of institutional and resource development, of ecological changes, of political and social hierarchies, and of family and kin systems, have constituted local subpopulations and their networks.[5] Of course, such histories are likely to play a significant role in shaping how and whether hypothesized general forces of industrial, market, financial, and other modernization actually influence population sizes, trends, and compositions.

[5] Hauser and Duncan's *The Study of Population* (1959) is commonly cited as a landmark statement of post-war population thinking in the social sciences. Relations between formal demography, economics, and sociology predominate in their account of demography as a body of knowledge. Ecology and psychology are discussed as potentially important areas in need of development, but chapters on history

Where data and analysis are confined to idealized macro and micro levels, development models assume perforce that intermediate population structures, and local and regional trends in them, are conforming automatically to motives and outcomes hypothesized in development theory. Variation, of course, can arise from local priorities that differ significantly from model assumptions. Unfortunately, where only macro and micro levels are considered, and standard household, income, education, and other classifications are used, then feedbacks involving the agency of networks, the influence of historical institutions, and of ecological pathways that facilitate adaptations within and between subpopulations, can only be captured incompletely and incidently.

A third and related theoretical assumption is that people's motivations underlying population change can be modelled realistically in terms of idealized rational choices amongst a limited and discrete set of short-term demographic means and ends. To take a common instance, analysis may consider theorized macro-level market forces in the economy (e.g. maximization of income) and general social norms (e.g. higher education facilitating maximization of income) as determining individual demographic behaviour (fertility reduction). Or, they may treat such hypothesized macro-level forces as aggregate outcomes of ideal-typical rational choices made by individuals or household units at the micro level. While choice models can be subtilized by many further assumptions, their aim is to reduce heterogeneity to a limited, uniform set of means and ends, rather than to observe and conceptualize the renewal of population heterogeneity as a mechanism of adaptation and change.

Taken together, these assumptions inevitably narrow the kinds of populations and behaviour that can be considered. Relations between subpopulations, historical and ecological factors shaping group memberships, individuals' differing positions in networks and their agency in restructuring them, become secondary influences, rather than population phenomena in their own right. Indeed, key intermediate subpopulations and network processes may simply never appear in the empirical record. With important exceptions (e.g. Bongaarts 1978), feedbacks between biological population systems (ecological, physiological, genetic) and social organizations are not standard components of economic development and modernization models. The problem of arbitrariness, noted by Harrison and Boyce (see section 1.2.1), becomes particularly relevant here. Populations selected for analysis because they are found in conventional census and survey tabulations will, for the reasons noted earlier, effectively deselect other subpopulations that do not fit readily into closed, top-down institutional data systems. In consequence, investigators can only infer, or may simply be unaware of, their influence.

and anthropology are notably absent. Major recent and contemporaneous developments in population biology were not reviewed, nor considered significant sources of conceptual and theoretical development of population in the social sciences. The demographic transition is viewed with some scepticism as a potentially special case of growth theory or, alternatively, as 'population history' unsuitable for theorizing in social science (since there would be so many differing cases that generalization would become impossible) (1959: 95). Demographers, of course, were not deterred from attempting to generalize the transition and apply it to policy (Hodgson 1983; Szreter 1993).

This influence as a factor shaping the behaviour of groups that do make it into conventional closed data sources becomes especially important when working in non-Western cultures and historical periods for which standard sources are often incomplete. As we shall see, potential biases of this kind have become a major conceptual and methodological concern in evolutionary biology, epidemiology, and ecology.

Reaction against the prevailing economic development and macro–micro framework began to emerge early in the post-war era, but has never displaced its dominance over economic and social demography. 'Middle-range' sociological theory pointed to the existence of other levels of social action, and together with community studies and ethnography, stimulated the now rapidly expanding field of social network research (Coleman et al. 1966; Mitchell 1969; Granovetter 1973; Hedström and Swedberg 1998). Recognition of the importance of clarifying relationships between networks at different levels, and the basic distinction between closed populations and the functioning of networks with open or unfixed boundaries (usually studied as 'ego-centred' networks) resurfaced in this literature (Marsden 1990), apparently without awareness that this kind of conceptual distinction was already mainstream to the evolutionary synthesis.

Independently of these sociological developments, historical demographers, social anthropologists, and social historians began to reconstruct the population history of communities, applying family reconstitution and ethnographic methods which enabled them to track family and kin networks, community welfare, and other institutions, from the sixteenth century and sometimes earlier (e.g. Wrigley 1966; Wrightson and Levine 1979; Kertzer 1984; Smith 1984; Smith and Pelling 2003). The conjunction of historical demographic and other archival sources with the ethnography and oral history of contemporary communities has provided hypotheses regarding local population formation processes, enabling linkage of demographic, legal, religious, labour, health, and other records at parish and other intermediate levels of aggregation (e.g. Viazzo 1989; Segalen 1991; Schneider and Schneider 1996; Fisher 2006). Reconstituting populations from the bottom up has enabled aggregated local-level data to be developed for closed population modelling (Wrigley and Schofield 1993), and macro-models of age-structural transition to be checked against local population heterogeneity (Kreager and Schröder-Butterfill 2010). This emergence of approaches combining bottom-up and top-down perspectives, and closed and open population thinking, is analogous to evolutionary biology, yielding a much more nuanced picture than is possible from analysis of closed data on their own. A kind of renaissance in population history and its methodology has emerged in consequence (Bonfield et al. 1986; Kertzer and Barbagli 2001–2003). The fundamentally open nature of historical communities evident, for example, in patterns of marriage and migration, stands in marked contrast to early post-war ideas of local populations in pre- and early modern societies as traditional, static, and isolated. What was supposed to be an unavoidable bias of historiography—that it would favour elite values and activities simply because these predominated in the record—has given way to extensive use of commercial, personal, legal, church, and other records on different subpopulations. This research has done much to revise the notion that social and economic change is simply a linear 'trickle-down' process in which change occurs as people imitate their betters.[6]

In recent decades, multidisciplinary demographic exploration of network populations in contemporary societies has accelerated, in which processes of communication within and between groups are increasingly placed in the context of wider social structural change. Notable examples include studies of the evolution of inequality in colonial and postcolonial societies (e.g. Schnegg 2007), of family and community support networks for older people (e.g. Schröder-Butterfill 2005), of the diffusion of contraceptive methods (e.g. Watkins and Warriner 2003), and of sexual and related networks in the spread of AIDS (e.g. Weinreb 2006). A convergence on approaches that combine top-down and bottom-up perspectives is also evident in recent attempts to develop shared disciplinary vocabularies of conjunctural action (Johnson-Hanks et al. 2011), compositional demography (Szreter and Garrett 2000), bio-demography (Wachter and Finch 1997), and gene–culture coevolution (Laland et al. 2010).

1.2.4 *A Case in Point: Tracking HIV/AIDS*

A more detailed example can help to illustrate how problem solving in the human sciences relies on combined top-down and bottom-up approaches, and brings together insights from several disciplines working at differing levels of observation and analysis. As remarked in section 1.1, combined approaches are necessary for specifying and differentiating the several kinds of population that contribute to current health and environmental crises; one crucial role they play is to provide methods and hypotheses that narrow down the range of populations and relationships that are critical.

Thus, a complete empirical tracking of how and why HIV/AIDS has spread so extensively in sub-Saharan Africa would be a vast undertaking, requiring extensive network data on sexual contacts, coupled with clinical data on all members of these networks. Even if people were not generally hostile to the idea of close surveillance of intimate relations, and if health infrastructures in the region were up to the task, such comprehensive, bottom-up tracking of the population would pose all but insurmountable problems given the scale of the epidemic, the continually changing nature of networks, and the fact that populations defined by their risk to HIV/AIDS are bound up with other populations that can also often be sensitive and difficult to monitor. The important influence of these other populations on the epidemic is widely acknowledged. Migration streams, for example, and memberships in ethnic, class, and other status subpopulations shape mating and the spread of disease, while being subject to conflict and competition due to factors largely separate from disease transmission. Poverty, an economic and social classification describing populations that cut across most of these groups, is a major factor

[6] Szreter (1996: 9–21, 238–282) provides a detailed case study for England of how the idea developed and became conventional that lower classes undergo fertility transition via downward diffusion of values and practices from higher strata—together with an evaluation of evidence as to whether this adequately explains fertility declines. Schneider and Schneider (1996) provide a parallel account in the context of an evolving rural society.

limiting people's access to treatment and hence our awareness of the extent of the epidemic. Poverty and status differences are also important in shaping the flow of communication: subpopulations with adequate information about HIV/AIDS and how to limit it, and those without, are shaped by the fact that such information may readily permeate some group boundaries, but not all. And, of course, the symptomatology of HIV/AIDS means that a potentially large subpopulation of people spread across all of these groups can have the disease for some years without knowing it. Epidemic HIV/AIDS, in short, is an exemplar of population phenomena that are incompletely observable and subject to potentially myriad influential biological and social factors.

In consequence, conventional health data, like clinic attendance records and prevalence surveys, are likely to give at best very incomplete and hard-to-interpret pictures of who is at risk and who is not. Where ticking the same box may fail to discriminate between different background factors (that influence, for example, not only how a question is answered but who is there to answer it), population analysis based solely on conventional closed data sources will be inadequate. Transmission and prevalence reflect the way local sexual, community, and information networks function, so it is the relationships composing these networks that need to be understood, and incorporated in models built to identify which subpopulations are critical. More accurate top-down monitoring using conventional closed systems may then be facilitated as the nature of subpopulation compositions and structures shaping constituent subpopulations are understood. Some variables, of course, will continue to remain incompletely observed, hence models will continue to be essential for estimates of some parameters, both in local populations and at higher levels of aggregation.

Some of the subpopulations that shape HIV/AIDS risk already have established local identities. Kin, patronage, religious, and other community structures are known by their members to shape marriage patterns and who talks to whom, and about what. Group norms which allow multiple partners, whether within or without marriage, or which effectively enjoin stable monogamy, have an obvious bearing on risk. Mating, in short, is wider than marriage. While the composition and structure of mating networks and awareness of HIV/AIDS in such subpopulations may overlap with groups having established norms and identities, they are unlikely to be identical with them. In other words, the key component populations we need to specify in order to understand disease transmission and, hopefully, reduce it, can be said to involve two sorts of network population. On one hand, there are subpopulations, such as the pool of existing and potentially suitable marriage partners, which are recognized in local cultures (even if the boundaries of such populations are open and changing, and require ethnography to delineate accurately). On the other, there are network populations which overlap and interact with these populations—actual patterns of sexual mating—that are defined by behaviour as well as group norms, and commonly do not conform strictly to norms. Within the latter there may be subpopulations of particular importance, both local and outside of a locality, notably groups with higher levels of HIV/AIDS. Local individuals may mate, for example, with sex workers during migration, and their mating on return to a community is then likely to expand the pool of HIV-positive members both inside and outside of marriage. Subpopulations without established cultural identities are also likely to influence the spread of pathogens. Thus, whether all or some mating subpopulations have ready access to condoms, and circumcision practices which may be common to otherwise differing populations, also have a bearing on transmission and risk.

In circumstances such as these, it is easy to see why deciding the units and variables appropriate to define subpopulations is not just a methodological preliminary: relationships and differences that constitute subpopulations will influence members' sexuality and the spread of information, and are thus vectors or mechanisms of the epidemic. Programmes aimed at preventing AIDS are, in effect, trying to bring the second sort of network population closer to the first: the wider range of mating behaviour would be better protected if people develop, as part of their network linkages, recognized group norms that reduce risk. At this point it is important to note that research activities cannot be fully separated from the information flows that shape network behaviour. Because research that discusses HIV/AIDS with people who belong to relevant networks may influence their knowledge, it is part of changes that affect behaviour, and may alter the bottom-up processes that form and change network subpopulations.

Bottom-up approaches also enter into understanding biological components of transmission and how they relate to social and sexual networks. The genetics of HIV/AIDS depends on observation to specify subpopulation differences: for example, by tracking polymorphisms in host–pathogen relationships, and seeking to identify why some subpopulations have greater resistance than others. At the cellular level, the nine genes that transmit information-forming HIV-related proteins, and which govern their replication, may be differentially expressed across members of different groups as the result of different socio-environmental conditions. Subpopulations of people are here defined by populations of molecular characteristics and their interactions over time. Once a person is infected, there is a natural history of HIV development in their body which carries important implications for subsequent transmission: sex with a currently asymptomatic carrier, for example, entails different risks of partner infection depending on how far the disease has developed in the carrier. These risks interact, in turn, with social population structures, especially where there is mating with multiple partners. Transmission via breastfeeding, rather than sex, is also a possible avenue of risk. Thus, overlapping but not coterminous with social and sexual network populations are gene and other cellular populations with their own networks of communication.

Understanding the nature of such bottom-up processes of transmission as they spread in localized open populations provides a crucial foundation for understanding trends at higher levels of aggregation. Over the duration of an epidemic, the accumulation of bottom-up processes creates a characteristic demographic profile that can be expressed in conventional closed models, like life tables. Put very simply: network members who take more risks tend to die sooner. The deaths of greater risk takers may lessen subsequent spread of the disease, at least for a time. This is because, where subpopulations at greatest risk disappear, the incidence of disease in the population as a whole can decline, whether or not mating behaviour in the rest of the population has changed to incorporate marital and contraceptive norms that lessen risk. New high-risk takers may, of course, emerge to counter such trends. Finally, the arrival of anti-retroviral treatment (usually distributed unevenly in society) creates a subgroup of HIV-positive members able to live longer, and this can have the opposite effect to deaths of greater risk takers, since their survival increases the prevalence of the disease in a population even where rates of incidence are constant or beginning to decline.

Epidemic HIV/AIDS in Africa and elsewhere thus presents a complex compositional demography. Clearly there is a need to identify which subpopulations are the important ones, whether for understanding the nature of transmission or effecting prevention, but for this there needs to be some standard of comparison. Population models have played an important role in cutting the Gordian knot of multiple, influential, and interrelated subpopulations that may or may not shape the course of the epidemic. As Hallett et al. (2009) show, normal features of the natural history of epidemics—notably, that riskier subpopulations tend to die out first—can be used to create a model that simulates subpopulation differentials underlying changes in the population of HIV-positive people and, more particularly, to provide a standard that can be used to assess whether recent declines in prevalence show the influence of shifts to less risky behaviour. The model allocates individuals to three subpopulations (those having sex with spouses only, those in long-term casual partnerships, and those with high numbers of sexual partners), then introduces demographic parameters such as group size and frequency of partner change by gender, which may be varied for each subpopulation. The epidemic mortality pattern of high-risk takers will have different implications for each of these subpopulations, and for the population as a whole, depending on the relative size and other parameters of the respective groups. Variant subpopulation patterns as shown in the model can then be used to guide interpretation of prevalence survey data for Zimbabwe: findings suggest that current HIV/AIDS trends reflect not only the normal decline in prevalence consequent on deaths in the high-risk group, but also behavioural changes toward reduced risk.

These encouraging results then turn attention to identifying which subpopulations and characteristics are influencing behavioural and cultural change. The model, as noted, recognizes only subpopulations defined as closed groups and characterized by crude mating patterns. As information and access to prophylaxis are conditioned not only by AIDS prevention and treatment programmes, research needs to understand how social subpopulation networks existing normally in communities actually channel communication, sexuality, and collective behaviour, and how this may assist, shape, or provide alternatives to programme activities. Again, models provide a helpful conceptual tool—for example, by distinguishing network structures that assist social learning about HIV/AIDS from those that actually influence sexual behaviour. The distinction between social learning and social influence emerged in related demographic research on networks shaping contraceptive practice (Montgomery and Casterline 1996). This approach is helpful because it recognizes that local networks have an agency that functions independently of intervention programmes, and is an example of research that has begun to address intermediate populations and processes of subpopulation formation and change that are inaccessible to exclusively top-down approaches. As indigenous networks shape people's several needs and resources, they give rise to differing health and other priorities. Models may thus assist our understanding of the relative roles and importance of different sorts of network in a given community, of relationships between networks, and how they may influence sexual and other behaviour. We are no longer in the conceptual trap that attributes all change directly to external programme influences, or which assumes that subpopulation structures all function in the same way and may simply be lumped together as a single population.

Models of social learning and influence need, of course, to be given empirical referents. We need to know, first, the composition and structure of local networks, how they differentiate subpopulations in a community, and with what implications for HIV/AIDS incidence and risk. The spread of information about HIV/AIDS and its influence on sexual networks does not occur in isolation: it is shaped by and will very likely be an aspect or extension of relations within and between extant networks and the many social functions and subpopulations they define. The spread of information about HIV/AIDS, and the discussion of alternative courses of action, in effect 'piggy-back' on existing social networks. Second, understanding local priorities and risk concerns—from circumcision to witchcraft—requires detailed contextual knowledge about the content of communication for which the inevitably simplified formats of prevalence surveys have neither space nor sufficient subtlety (e.g. Beckmann 2012; Niehas 2012).

2 The Chapters

At the start of this introduction we noted what might be seen as a paradoxical feature of contemporary population research: its emphasis both on population heterogeneity and on the interrelatedness of populations. How can populations be different if their relationships mean that they share members and major characteristics? The paradox is more apparent than real: as more and more populations have come to be recognized as material to explaining human society and evolution, their differentiation requires us to appreciate and disentangle all sorts of interdependencies and overlapping memberships amongst them. Populations are open phenomena composed by relationships between subpopulations and their members. Methods that combine measurement, modelling, and sustained observation of processes of group formation and change are necessary to decide which constituent units we may consider separable, and why.

Some further at first glance contradictory features have also been noted. For example, explanation has continued to rely on observation and modelling of local populations to identify core processes that give populations their structure, even as the scale of variables and populations that need to be considered has radically expanded. As we have seen, capturing local population processes is not just a local matter. Formal modelling and simulation at higher levels of aggregation can play a major role in hypothesizing and documenting the presence of features of potential local population structures we need to observe and clarify. Evolutionary biology's massive genomic programme of research is doubtless the best known of these developments. The last three decades have also seen the gradual opening up of demography to the complexity of social networks and to historical and cultural variation at differing levels of social organization. While models at the many possible social levels may not have the fine structure that millions of SNPs give to genomics, their role has likewise been directed to patterns and sources of variation in local population structures that are the fundamental ground of explanation. Specialist fields and subdisciplines have proliferated, even as the need for collaboration has grown.

The common ground shared by these developments can now be summarized. Human populations are composites studied at several levels, including genes, cells, organisms, individuals, networks, communities, and regional, national, and global entities. Human groups or subpopulations on any given level have an internal structure, which reflects the ways in which members' relationships have shaped group formation, and their active and continuing construction of a given environmental and cultural setting or niche. While, for the purposes of modelling and the comparative analysis of members' characteristics, it is necessary to control for some characteristics and features of the environment (thus treating populations as closed and their members as identical), the openness of populations to compositional changes both from within and without remains the fundamental ground of the processes of evolution and social change we seek to understand. Subpopulations interact not only with each other but activate feedbacks involving populations at other levels, and also interact with other species. As population structures at one level may define and redefine parameters of behaviour at others, both bottom-up and top-down perspectives linking different levels are necessary.

Recognition of this shared framework across the many fields of population inquiry has not arisen in consequence of a new or emerging 'grand theory' (cf. Skinner 1985). Indeed, differing interpretations of formative influences, such as Darwin's theory of natural selection, have continually provoked debate, and required scepticism and revision where they tend to confine research only to certain levels, or block attention to phenomena that renew population heterogeneity. Since the evolutionary synthesis, biologists have had a significant advantage of working within a framework that requires both open and closed methods, explanation requiring both local observation and aggregate modelling of sources of variation and their evolution. In demography, however, disagreement over the adequacy of economic development and modernization as an explanatory framework for population history has proven harder to overcome. The heterogeneity of demographic transitions has confounded explanations confined to idealized macro/micro levels of economic-driven development and modernization, and no alternative universalist framework has emerged.

On the one hand, this absence is not surprising from the standpoint of the human sciences. As Lewontin's observation, cited at the beginning of this introduction, testifies, universals have given way to 'local regions of the model space' as a basis for evolutionary theory. The ways in which populations develop and change reflect variations built up by relationships between members and subpopulations on several levels, and for which general formulae that privilege any one discipline or level are incomplete. One of the purposes of several chapters in this book is to chart the growing influence of this perspective, which has made cultural and social formations one key to tracking ecological and genetic inheritance. On the other hand, as other chapters note, major shifts in sociological theory and historiography have also moved strongly against older ideas of generalization, notably the artificiality and often politicized agendas that excessive reliance on theories of modernization and idealized macro or micro levels entails. As we shall note later in more detail, compositional demographic approaches in social and demographic history, social network analysis, and anthropological demography have been able to draw both on evolutionary and critical sociological theory, their research providing a substantive ground on which these theoretical developments can, in turn, be tested and extended.

Each section of this book draws on contributions from several disciplinary perspectives across history and the biological and social sciences, following the organization of the Oxford meeting at which the papers were first presented. Chapters in Part II outline the history of the main conceptual orientations to population, noting the emergence of analogous approaches and problems shared by disciplines, and tracking recent conceptual developments, particularly implications of evolutionary genetics. Chapters in Part III present a set of conceptual frameworks for the study of intra- and interpopulation dynamics which together define an emerging agenda for research. Chapters in Part IV show how one of the key components of such frameworks—differentiation of subpopulations at local levels—opens up mechanisms of demographic and evolutionary change, bringing the heterogeneity of intergenerational networks, ethnicity, and migration more fully into explanations of variability at higher levels of aggregation, for example, in fertility and mortality transitions and in impacts of population ageing. Chapters in Part V provide examples of how the growing convergence of human science approaches to population informs and opens up key problems in long-term population history. Part VI, in returning to several of the current and pressing epidemiological and health problems noted earlier in this introduction, shows how fundamental issues in the way populations are specified, including the modelling of long-term population processes, play a critical role in designing and assessing health interventions. There is active dialogue and discussion between chapters in all of these sections, with authors taking up and further developing themes raised by others. An outline of the five sections, noting some main links between chapters, completes this introduction.

2.1 What Is a Population?

Population as a term for aggregate properties of society and nature enters the English language in the early seventeenth century. Its main reference at that time was to people, understood as a body of inhabitants; this usage was not actually coined *de novo*, as the opening chapter on the history of population concepts, by Philip Kreager, makes clear. The scientists and philosophers who created what we now call the scientific revolution, and who wrote the founding texts of European democratic government, were in close dialogue with texts of classical antiquity. This is evident in the most basic terms of reference. *Demos*, the Greek root of 'demography', and *populus*, the Latin root of 'population', refer generally to people residing in a state, but more particularly to citizens. States and the subpopulations that composed them were understood as complex and shifting hierarchies of group memberships, and as formed out of relationships between such subpopulations. Populations were understood as simultaneously natural and political phenomena. This was the source of open population thinking described in preceding parts of this introduction. Classical open population thinking, while an apt and subtle basis for describing and analysing sixteenth-, seventeenth-, and eighteenth-century republics and monarchies, proved increasingly inadequate to the complexity of governmental, economic, and other problems they faced. Enumeration, calculation, and simple closed models drawing on the calculus of probability were introduced into open population thinking in this period to better define, and hopefully manage, complexity. Potential applications to problems like mortality and public revenues were gradually explored. Although methods remained tentative up to the end of the eighteenth century, their application in local institutions provided the template for the idea that whole national populations might be compiled and analysed as closed units.

This latter approach was then generalized as 'statistics' in the course of the nineteenth century, radically transforming population thinking at the social level. Pride of place now went to compiling closed, top-down, macro-level datasets (like censuses) supposed to provide the universal standard upon which health, social, and political reforms would be based, and upon which applications of advanced mathematical statistics would discover general social scientific laws. Major processes of population composition and change, like migration, ethnicity, and inequalities, however, were not effectively encompassed by closed methodologies, and universal social laws did not emerge from the programme. Biology here remained a major exception to the nineteenth-century prioritization of top-down population thinking, as Darwin's theory of natural selection remained strongly shaped by older bottom-up and open population approaches, notably natural history and the continuing influence of eighteenth-century political economy. Over the twentieth century, Darwin's theory, as with the development of closed population models within seventeenth- and eighteenth-century open population thinking, came to admit both bottom-up and top-down perspectives, which enabled the evolutionary synthesis and molecular genetics then to develop successfully. In the last decades of the twentieth century, demography has begun to reintegrate approaches to open population phenomena, notably networks, migration, ethnic and cultural variation, more fully into prevailing closed population approaches.

Population, embracing open and closed perspectives, is thus a central concept in European intellectual history within which ideas of scientific method and government have been framed. The chapter by Walter Bodmer and Bruce Winney carries this historical perspective forward over the twentieth century, as population genetics has developed progressively finer subpopulation specifications. Early in the century, biologists began to differentiate human populations into Mendelian groups according to blood types. Where blood groups of people related by place of birth and residence over more than one generation in particular places were used as a proxy for gene frequencies, differing frequencies provided a first geographic and evolutionary map of subpopulations. As variation in blood grouping began to be studied in terms of constituent proteins, and other genetic marker systems became available, further subpopulation differentiation was possible, so that the spatial distribution came to identify broadly defined continental groups across the world. Following the discovery of DNA structure, molecular genetics has steadily expanded the range of subpopulations that can be identified, opening up further sources of population variation. Some markers of genetic inheritance enable us to trace major historical population movements still visible in the genetic make-up of regional populations. Bodmer and Winney show how this approach illuminates the historical peopling of Britain. Related approaches have subtilized our understanding of susceptibility to disease: it is not just the presence of an allele or genetic marker, but observed relationships between markers across the genome that identify key epidemiological and other subpopulations.

Human populations in this way came to be defined and modelled in terms of a wide range of genetic, cellular, and physiological components. In genomics, individuals in society are, so to speak, reduced to members of subpopulations defined by molecular elements that specify only certain features of shared ancestry or disease inheritance. Mechanisms of population evolution can then be identified by genomic signatures in pathways such as immune response, DNA replication and repair, protein synthesis, and gene regulation that are created as molecular information is transferred down generations. This procedure is a very sophisticated example of how closed models linked to empirical tests are constructed in order to isolate evolutionary and epidemiological processes occurring in actual open populations. However, as noted in preceding sections, closed models select only some features or parameters for membership, and the empirical tests that are then carried out rely on samples designed to help isolate these features. This has caused some concern in recent biological literature, which has come to highlight the 'What is a population?' question both in population ecology (Waples and Gaggiotti 2006) and in epidemiology and issues of social and health inequality (Krieger 2012). These articles address the potential arbitrariness of closed population modelling, noted earlier, in much greater detail for those fields of research. One issue concerns implications of so-called 'ghost' populations: i.e. related but unsampled populations that may affect model estimates of key population parameters such as migration rates, population sizes, and genetic diversity (Beerli 2004; Slatkin 2005). Second, there is a lack of consensus regarding quantitative population definitions (Waples and Gaggiotti, for example, list some 18 definitions in recent biological literature). As we shall see, the influence of related but unobserved subpopulations has also come to occupy a central place in recent historical and social demography.

These issues form an important subject of generative and statistical models of genetic inheritance, discussed in Dan Lawson's chapter. Generative approaches, such as demic diffusion models, are simulations in which members of a population are defined by shared SNP alleles. Models function to trace divergent demes or subpopulations back to their separation from a hypothetically 'pure' ancestral population on the assumption that there has been no significant mating between subpopulations, there is only local migration, and that any subsequent evolution of characteristics owes solely to genetic drift. These models rely on sparse data, i.e. a small set or sample of individuals whose ancestry can be identified with contrasting geographical, linguistic, or other characteristics (see also, for example, the studies discussed by Francesc Calafell and David Comas in Chapter 13). Statistical models, like STRUCTURE and ADMIXTURE, cluster individuals on the basis of their allelic profile using allelic frequencies in such ideal ancestral populations. In contrast, other methodologies, such as the one presented by Lawson, entail the full set of SNPs for a population, and infer subpopulations in terms of recurring clusters or 'chunks' of SNPs, rather than frequencies of selected genes. Inclusion of the full range of individual genetic characteristics enables much finer analysis of genetic variation, opening up the possibility of studying admixture (i.e. where mating between demes has occurred), and allowing for linkage disequilibrium.

However, given the scale of variables making up the human genome, the number of clusters or subpopulations that can be inferred expands rapidly—from an analytical point of view, as Lawson says, 'the method almost has *too much* power'. Hence data or hypotheses about particular networks of population variation as functioning in open populations, such as kin structures, are needed for fine-tuning the resolution of such bottom-up approaches. Put another way, models do not just sift and structure congeries of variables. They have an important role in telling a story: more limited observational data, of which kin structures are one instance, imply a narrative of possible relatedness or separation of subpopulations, enabling analogies to models that have been developed for better-known subpopulations to provide new models and further empirical testing. Such 'stories' are not necessarily a check on the arbitrariness of model assumptions, but they can play an important role in clarifying the plausibility and limits of assumptions made.

In both of the approaches Lawson describes, a population is defined at a minimum as a unit of statistically equivalent individuals. This minimum describes one sort of reality: single organisms carrying certain genetic material in common with other members of an observed or sampled mating group. In studying human populations, the organism is of course also an individual person and a member of social populations and subpopulations. As we saw above in discussion of HIV/AIDS, further questions about subpopulations that combine genetic and social mechanisms are bound to arise: How can we formulate the overlap or simultaneous action of genetic and social influences? Which social characteristics or relationships in a subpopulation act as biological triggers affecting the incidence of gene expression, and how may this vary between cultures and over time and space? The local model space becomes one in which social and biological mechanisms interact, and in which what we commonly understand as different kinds of mechanism combine or may turn out to be effectively the same. Some local mechanisms will be critical variables influencing progressively wider populations of which subpopulation memberships are a part.

This meeting ground, on which fields like evolutionary biology, epidemiology, and social demography are converging, can alter what is at issue in formal demographic models. As Kenneth Wachter remarks in the concluding chapter of this section, taking population heterogeneity seriously means that the central and, indeed, oldest demographic model of risk—the life table, used to calculate individual longevity since the seventeenth century—encompasses variations that would now more aptly be described as *biodemography*. In effect, tables for whole populations need to be broken down into more specific subpopulations experiencing different mortality risks. As Wachter shows, analysis then needs to become much clearer about the different components of risk that individuals and groups actually face. His example of longevity is, of course, a central topic in studies of population ageing. It is also a good instance of the tendency of population inquiry to go 'back to basics' when confronted with the definitional issues posed by the increasing number and complexity of variables.

As previously unobservable processes gradually open to scientific observation, different model assumptions become possible that enable analysis to reflect actual natural and social processes more closely. The stories our models tell begin to change. In traditional life table analysis, as in population genetics, population members begin as statistically equivalent individuals. However, what such tables show, by tracking their deaths at different ages, is that members are not in fact the same. As Wachter says, for modelling purposes they differ by successive 'rolls of the dice', i.e. by all of those unobserved events and processes over a person's life course that influence individual health and susceptibility to specific causes of death. The observed presence of certain alleles (for example, indicating sufferers of Huntington's disease) enables us to identify Mendelian subpopulations whose members have different survival chances than the general population, and thus require their own distinctive model of longevity. For most subpopulations, however, we do not have the certainty of single gene causation. Exploring possible mechanisms often then pursues cross-tabulations of variables (say, for smoking together with dietary, income, ethnicity, or genetic characteristics). The danger here is of infinite regress, since combining social and genomic variants produces many more variables than individuals. Social as well as genetic research thus confronts a problem of 'too much' analytical power. Even where a significant statistical association is found, the mechanism can only be inferred; and of course other unobserved 'ghost' populations may be influencing results. The way out of infinite regress depends, as in the role of kin structures in population genetics noted already, on the interplay of models and observation, and more particularly on how this interplay suggests hypotheses or 'stories' in which population differences owe to potentially observable processes that shape open populations. As Wachter shows, in the biodemography of ageing, hypotheses draw on likely life course processes that lie behind the life table, like long-term physiological consequences of early life diet and infant care, or occupational and other life course patterns, that facilitate resilience in old age.

2.2 Rethinking Intra- and Interpopulation Dynamics

The composite structure of human populations and the complexity of mechanisms shaping them have evidently stimulated new collaboration between disciplines and encouraged human scientists to think cohesively. New frameworks that enable us to draw on observation and modelling in different fields, and apply them together in a given setting, are one fruit of this process. John Odling-Smee's chapter on niche construction theory outlines one such framework in which genetic, ecological, and cultural mechanisms shaping population structures are brought together. Study at the molecular level of gene replication, mutation, and expression has not lessened the importance of the environment in evolution, but increased it. Environments influence genetic processes not only by regulating mating, but by acting, for example, on gene expression (by shaping embryological and neuronal development) and via natural selection. Inheritance is thus not only via genetic but ecological characteristics, including all of those adaptations passed on through social learning and technological development that members of society inherit with cultural and social organization. The overlap between biodemography and ecological inheritance is likely to be considerable. Ecological inheritance over the long term reflects increasingly complex social organizations that have shaped diet, settlement patterns, and agricultural systems, fostering both genetic advantages (e.g. adjustments to disease, inhospitable climates, and famine conditions) as well as liabilities (e.g. the spread of malaria, obesity) that shape population memberships and structures.

As the phrase 'niche construction' indicates, species actively create and reshape their environments, and pass this experience on to later generations. Just as the transfer of information is not passive at the level of genes so it is of critical importance that ecological inheritance at the level of individuals and groups is understood as more than behavioural accommodation to fixed environmental constraints. Population structures are more than mating patterns that determine the genetic make-up, fertility, and mortality of subgroups. The structure of subpopulations, for example, includes status, capital, and age criteria such that senior members of a community influence younger members' production and reproduction opportunities, and shape in other ways the transmission of experience that sustains development of a group's niche and relations to other subpopulations. Grandparenting, for example, provides support to younger, more active adults, assisting their production and reproduction; over time, the advantages that elders' active longevity gives the group may help to encourage values and social organizations that enhance, in turn, the genetics and physiology of members' longevity. This aspect of ecological inheritance may thus help us to explain why the human lifespan extends well beyond what is necessary for individual reproductive fitness at younger ages (Wachter and Finch 1997).

As Odling-Smee remarks, understanding niche construction has the effect of altering how we formulate some classic and long-unresolved demographic problems. Since Malthus, debates over population and resources have relied on definitions of carrying capacity that we can now see are incomplete. Malthus (1890) famously regarded limits on population in terms of exogenous constraints on food production inherent in land availability and extant technology. Boserup (1981) countered that population growth stimulates technological innovation that can alter land use and production levels, thus potentially solving carrying capacity issues. Their economic demography, however, gave no primary place to changing relations within and between groups (beyond possible technology transfer, or the enforcement of bourgeois reproductive values on the poor), nor to the differential access of individuals and groups to means and fruits of production. Relative carrying capacity depends on relationships between the several populations living in a given environmental setting that are not decided by resource limitations alone. 'Spillover' effects of ecological adaptations (like collaborative nurturance, just noted) give some groups advantages over others in the division of labour, fertility, and access to technology. Limits to carrying capacity are thus altered not only by Malthusian 'checks' on population growth or technological developments (such as neo-Malthusian contraception), but by the distribution of social learning: not all technologies are equally accessible to everyone, and this depends both on the nature of the technology and how the internal composition and structure of groups and their relations with others shape learning. As several papers in this volume (Kohler and his colleagues; Schröder-Butterfill; Gregson and Hallett) show, such differential access requires us to document the networks that constitute different groups in society and their advantages and disadvantages. Put another way, population structures and compositions, and the niches they occupy, are built in important respects by communication: the sharing (or not) of knowledge and experience that develops capacities and sustains collective organizations.

Communication as a mechanism that builds and differentiates population structures and memberships is the subject of a second integrative framework, described in the chapter by Simon Szreter. If access to productive and reproductive advantage is mediated by relations within and between groups, and depends particularly on how these relationships shape flows of information, then the 'local model spaces' we need to identify and understand are, as Szreter says, 'communication communities'. The theory of compositional change he outlines arises from a substantial body of historical demographic, social historical, and anthropological demographic research which, analogous to evolutionary biology, has focused on how compositional adjustments in the populations of local cultures and environments serve as mechanisms of wider demographic variation and change. One of Szreter's principal concerns is to show that the way we identify and compare the constituent subpopulations that together make up fertility trends at the level of the nation-state leads to a very different explanation of how and why population change is taking place than consideration of that macro level without such decomposition. In the period since 1870, fertility declines have swept across many states, often in a matter of a few decades, yet the timing, trend, and resulting lower reproductive levels show considerable and continuing variation. Far from being a welter of incommensurable local studies, compositional demography, by addressing the different scales or levels of change which together produce macro-level demographic trends, and by observing where local mechanisms do and do not operate, opens the door to a cohesive account of sources of fertility variation at macro as well as lower levels of aggregation.

Early in the post-war era, demographers reasoned that fertility declines according to a single recipe: societies would adopt low ideal family size norms, and once couples reached their desired number of children, they would take up contraception and stop childbearing. Subsequent research in Europe, Africa, and elsewhere, however, has revealed other established strategies that function via shifts in household composition and labour arrangements, enabling members to postpone childbearing either via marriage rules or birth spacing (e.g. Hajnal 1982; Das Gupta 2005; Bledsoe 1999). Compositional adjustments give members a range of alternatives to meeting labour, production, care, and other needs, which help to explain the heterogeneity of fertility before and during secular declines. Szreter calls attention, moreover, to a key factor underlying those strategies and people's need for alternatives: the variability of fecundity. The ability to have children is circumscribed by sterility, which increases with age and may be increased radically by pathologies. In English history before 1840, for example, with widespread delayed marriage and one in five women not marrying over extended periods (Schofield 1985), childlessness was a known and dreaded reality. Parish data show that over long periods 20% of couples had between zero and three children—in a period in which infant and child mortality further removed approximately two of every five births.[7] In short, people understood very well that it is untenable simply to assume that you can have the number of children you want.

Not only having children, but fitting childrearing into one's life (for example, for the many women in historical Europe who were servants or factory employees working very long hours) posed challenges that restricted reproductive opportunities—challenges that remain no less fundamental to fertility differentials in the modern world (cf. Johnson-Hanks' chapter, discussed later). As Szreter remarks, this changes the central problem requiring explanation: the issue is not, as post-war fertility transition theorists supposed, the spread of a simple mechanism underlying secular declines in *childbearing* (i.e. leading to a convergence in relatively low numbers of children per woman), but how the heterogeneity of transitions reflects the negotiation of *childrearing norms across the life course in different communication communities*, resulting in a range of fertility outcomes in a given population. Responses to industrialization and changes in welfare in historical Europe, as Szreter describes, led to variations between regions, local cultures, and classes. Different occupational groups and gender relations led to not one but at least twenty distinctive fertility regimes in a country like Britain at the time of its 1911 census (Szreter 1996: 363). Explanation thus begins to address not only processes that induce heterogeneity of transitions, but factors that continually renew the heterogeneity of reproductive change.

The question of how social learning varies across communication communities, and with what consequences for different subpopulations, is the subject of a third integrative framework, social network analysis, demonstrated in the chapter by Hans-Peter Kohler

[7] Recognition that being able to have children and rear them is not a safe assumption is not just a historical phenomenon. A 20% childless rate has been reached again in Britain by the cohort of women born 1960–1964, reflecting postponement (supported by contraception), late marriage, rising levels of divorce, and career conflicts (Evandrou and Falkingham 2004). Where pathological sterility is widespread, as in Central Africa, up to 40% of women may have between zero and two children, in a context in which five or more children are desired (Larsen 1994). Childlessness thus becomes a central theme of population history (Kreager 2004).

and his colleagues. Networks are obviously instrumental to forming subpopulations and their cultural, economic, and other characteristics, since it is through networks of marriage, labour, production, religious and ethnic affiliation, and so forth that population memberships are built up and change. The way networks select individuals, and individuals select the networks with which they identify, is a key dynamic underlying compositional demography: networks have the capacity to structure the communication of information and practices to which members have access; to reinforce or alter values; and to shape how external influences on a community are experienced differently in different subpopulations. Tracking individuals' memberships and positions in networks reveals general network properties—like the density, content, and symmetry or asymmetry of relationships—which facilitates comparison and the identification of key sources of variation and change.

Communication is not only about the spread of identities, information, and technologies, but of pathologies and the perception of risks that all of these may entail. Such diffusion, which was once thought to be understandable as entirely innovative behaviour (e.g. in the adoption of modern contraception), is now recognized as shaped in important respects by existing networks which serve functions normal to different parts of a community. Just as genomic research may be guided by knowledge of kin and mating structures, so Kohler and his team note that the differential spread of HIV/AIDS appears to be shaped by the influence on mating of local markets and the networks they create. Networks are by nature selective: they are processes that *form* subpopulations that differ from others in systematic ways, reflecting institutional and other preferences that people have which shape their choices and exposure to risks. Awareness of extant networks and their influence is thus key to identifying unobserved influences—the problem of selection bias or 'ghost populations' remarked earlier in genetic and ecological studies (see sections 1.2.1; 2.1). As Kohler et al. note, how networks are formed and sustained is integral to social learning and is thus crucial to distinguishing processes of epidemiological and social change, and to designing health interventions to check the diffusion of HIV/AIDS. Their research, in bringing together qualitative and quantitative methods, shows how the modelling assumption of random mating which is normal to closed demographic data collection systems and their analysis, may be replaced via ethnographic observation of open population processes, which can then serve as an empirical basis for developing causal inferences and models for interpreting survey and other sources. Their account of the Likoma Network Study shows how this logic informs comprehensive, or sociocentric, network research on a large scale.

The frameworks put forward in the first three chapters in Part III move toward a progressively finer focus: from populations at species and subspecies levels, to national and regional populations and subpopulations, to the network phenomena that build and change subpopulations from the bottom up. Darwin, of course, emphasized that it is individual variation—the fact that no two individuals and the events they face over their life course are identical—that provides the raw material for selection and thus for evolution. With Jenna Johnson-Hanks' paper, aptly titled 'Populations Are Composed One Event at a Time', we reach the fundamental social stratum of how populations are actually composed. Individual choices about reproduction—whether to have a child or children, when, with whom, in what sort of relationship or group a child should be raised, whether to terminate a pregnancy, and so forth—define a nexus at which mechanisms of population membership and natural selection are simultaneous. It is noteworthy that trajectories of modernization or socio-economic development at this point fall away from discussion: Johnson-Hanks' examples are drawn interchangeably from Cameroonian and American settings.

Johnson-Hanks first sets out a simplified framework for comparing processes by which people select, or are selected into, a population membership. The framework leads to a very clear account from a sociological perspective of the problem of unobserved variables, or 'ghost' populations, raised in earlier chapters on genetics and evolutionary biology. In a social context, selection is unlikely to occur as a single event or choice, and is better understood as a series of *conjunctures* in which people negotiate and construe the implications of alternative courses of action and the subpopulations that they thereby come to be associated with. Sometimes this process depends entirely on person-specific characteristics, but often self-selection follows wider institutional and cultural identities that lead people to associate with others in similar social positions to themselves. As Johnson-Hanks notes, this creates a space for network approaches; as already noted, subpopulations are communication communities that differentially facilitate social learning and social influence. The central part of her paper shows how this compositional demography enables phenomena of selection to be grounded in major trends in sociological theory, as represented in the works of Bourdieu, Giddens, Sewell, and Blau. This framework is particularly important not only because it integrates reproductive behaviour with other aspects of population composition that shape selection, but also because it shows how it is possible for demographic research to avoid the tendency to collapse back into pat oppositions between the traditional and modern, cultural versus economic rationalities, ideational versus material causes, and less developed versus developed societies.

2.3 Mechanisms of Local-Level Variation and Change of State

Many anthropological demographers and migration specialists have for some time worked within a basically compositional approach to population. A population's members make use of several subpopulations to facilitate things they need and want to do: recruit labour for different modes of production; arrange marriage and political alliances; exploit new economic opportunities; secure or improve their status in local hierarchies; negotiate access to health care and state services; and so forth (e.g. Goody 1972; Hugo 1982; Massey 1990; Kertzer and Laslett 1995; Bochow and Kreager 2015). Subpopulations in which shared activities and the interests behind them take place may be characterized as, for example, kin networks, economic sectors, marriage markets, socio-economic strata, castes, and so forth. They are not usually closed groups, even when their members consider them to be. It is in the context of such subpopulations that we find the conjunctures described by Johnson-Hanks, in which individuals make choices that shape fertility, mortality, and migration trends. Variables that mainstream demography has long emphasized, like the timing of marriage, birth control, maternal and child health, and intergenerational relationships, are thus not only factors that shape vital rates: they are evidence of how, and how well, members' strategies addressed to other aspects of subpopulation composition, are working.

Subpopulations at the local level therefore need to be understood in 'bottom-up' terms (i.e. of how they are formed and sustained by compositional strategies) as well as 'top-down' (e.g. 'determinants' of fertility and mortality prescribed by demographic transition and modernization theories). In this way we see how population structures inscribed in institutional arrangements, network relationships, and other elements of local population composition act directly and indirectly to regulate overall subpopulation sizes and differential growth. Aggregate outcomes, in other words, reflect the agency of individuals and subpopulations, whose priorities of relevance to demography are not confined to fertility or reaction to external interventions. Indeed, subpopulations arriving at similar reproductive rates often do not rely on the same repertoire of marriage, family formation, contraceptive, and migration strategies, and can vary preferences within their repertoire as circumstances change.[8]

An analogy can be made here to evolutionary biologists' emphasis on 'redundancy'. As noted in section 1.2.2, relations within and between subpopulations provide multiple routes of genetic and ecological inheritance; population heterogeneity is renewed because sets of potential determinants of gene expression depend on the several ways in which genetic and environmental influences may combine. As with differing routes to gene expression, a given fertility or other vital rate may be the outcome of processes that link groups in different ways, with not always the same set of determinants or variables that characterize these subpopulations being required to reach the same trend.

[8] For example, Monica Das Gupta (2005) notes how the study of joint and stem family systems has dispelled an assumption commonly made by early advocates of demographic transition theory, viz. that only small family systems typical of North-West European history exercised effective homeostatic controls on fertility and group size. Her account of the history of family groups in India shows (anticipating Szreter's argument in Part III) that members were concerned with a recurring problem of insufficient numbers (reflecting impacts of mortality and infertility) which shaped their ability to distribute members effectively in relation to resources. An extensive range of alternative strategies were employed, involving several levels (community, lineage, and household), since controls were exercised collectively by adult males at one or more of these levels. Adjustments included: adoption; regrouping households, either by division or to include other families; flexible land use strategies, like sharecropping; and a number of marriage strategies which impact on childrearing and fertility (selective non-marriage of men, informal fraternal polyandry, selective female infanticide, importing brides). Knowledge of this homeostatic system is important for understanding responses to the mortality declines that began in the 1920s. For example, some practices (like fraternal polyandry and adoption) were abandoned, while others (like male non-marriage) soared. The joint family has thus remained the principal logic which has enabled groups to adjust in several ways to changing Indian society and economy, including taking up contraceptive technologies to achieve traditional ends.

As with biologists' concept of redundancy, there are at least three aspects of variation that require attention. One is multifunctionality: a given institution, say, marriage, is an active factor which may be shaping a group's demography in several ways simultaneously. The second is the existence of multiple pathways or networks that link subpopulations at different levels, along which the influence of a multifunctional institution, like marriage, exercises its influence in different ways. Third, local and external agencies may not simply alter trends and individual or group characteristics, but lead to major changes of state, for example, repositioning and empowering some groups, and opening or terminating access to key resources, thus radically altering subpopulation identity and even existence. Chapters in Part IV provide examples of all three forms of variation.

Establishing mechanisms that renew heterogeneity, as noted in section 1.2, is fundamental, and depends on identifying '*which* populations to study, and in *what* networks or set of aggregates they are located'. Elisabeth Schröder-Butterfill's chapter demonstrates how combined open and closed population methodologies can be used to identify a range of subpopulations critical to the study of population ageing, revealing both the processes that have given rise to their demographic characteristics (like age-structural imbalances), and the several routes that coexist in a given community for dealing with vulnerability in old age. The subpopulations that emerge from her research—the community or village; socio-economic strata; and family and neighbourhood support networks—are different levels of social and demographic interaction, and variations can be traced both within levels and between them. An elder's location or situation at each of these levels is defined by the multifunctional roles they play; for example, as grandparents and economically active family members, their actions can expand or reduce the range of useful kin ties available to them; these ties may open up opportunities for access to people in local government or community institutions, i.e. people of higher strata. Inferior social status or position in a network, however, may permanently restrict access to influential kin and others. Over the life course, ties that do or do not permit recurring exchanges with others impact in turn on individual and family reputation and participation in the wider community. As Schröder-Butterfill shows, a typology of support networks and their distribution in relation to socio-economic structure thus becomes possible, and these network and socio-economic structures may be compared for different communities to explain their differing demography. Focusing on one of the three Indonesian communities in her research, she shows how vulnerability late in life is something that affects rich as well as poor, and that the differing nature of that vulnerability is a consequence of the series of conjunctures across individual elders' life courses that have shaped their networks and the resources available to them. Clearly, some subpopulation networks expand and thrive, whilst others die out.

These Indonesian communities have familiar features of social and economic diversity that mainstream modernization and demographic transition approaches often pass over: a considerable variety and fluidity of family and household arrangements; differing patterns of internal and international migration; dependency ratios that vary markedly between communities and regions; different ethnicities and languages; and multiple socio-economic strata with differing combinations of prestige, employment, incomes, and assets. Not surprisingly, we see striking variations between these communities that are not what standard assumptions of modernization and economic development research would lead us to expect. One, which echoes Szreter's chapter on the importance of the range of low fertility in understanding the behaviour of communities, is the presence of completed childbearing at levels of two children or less *before* fertility transition that is strongly associated with greater poverty and vulnerability. Another, echoing Kohler and his colleagues' emphasis on the importance of networks in local problem solving, is the presence of key support arrangements between households that are not visible in survey data without the evidence of network ethnography to interpret them. Schröder-Butterfill's approach relies on sustained ethnography prior to panel surveys which are then informed by and couched in local idioms. This approach allows attention to open population processes, notably variations in network relationships and the qualitative content and structure of ties, thus opening up many issues of multifunctionality: the different sorts of problems that network ties are being used simultaneously to solve, and the several ways in which a person's position in networks can become useful. Schröder-Butterfill also confirms a basic theme of Graeme Hugo's chapter, to which we now turn: people's problems and problem solving—such as their needs for intergenerational support, and how they are resolved—commonly arise not in conjunctures confined to explicit planning of fertility, but from opportunities made possible elsewhere in their networks. The long-standing capacity of migration both to remove and replace members is, of course, a primary example.

The difficulty, however, of bringing migration fully into mainstream demographic method and theory remains one of the major legacies of the promulgation of statistics in the nineteenth century, noted in section 2.1. Put very simply, census and survey data systems designed to capture population changes in series of discrete and unrepeatable events cannot cope with the most common forms of human movement. This situation stands in sharp contrast to evolutionary biology, as several chapters in this volume show: migration and its impact on the composition and relations between populations is a major factor in genetic and ecological inheritance, which models of population structure cannot credibly ignore. Graeme Hugo's chapter provides a succinct account of the conceptual shortcomings of conventional demographic data on migration, then drawing on two unique data sources for the movement of South-East Asian populations to highlight the importance of these movement patterns and their diversity for understanding the economic and social history and development of the region. We see both what is left out in conventional data sources, and what we gain in understanding when it is put back in.

In place of the usual dichotomies that separate permanent moves from circulation, Hugo outlines a framework in which there is a continuum of forms of movement, with multiple criteria differentiating them. Rather than data that prioritize permanent international moves, as in current systems, emphasis is placed on the much more common forms of temporary internal migration and international movement, including the rise of transnational migration strategies. The 1930 Dutch colonial census of Indonesia, while not comprehensive, provides an extensive picture of internal movement, making clear the long historical importance of non-permanent migration in local population adaptation. Indonesian migration, as a response to environmental crises, as a means employed by some ethnic groups to establish major new economic niches, and as a varied response to colonial development of the plantation system, are all evident. Hugo's own research in West Java reveals the continuing rise in the scale and importance of non-permanent movement in that region after independence. He then turns to the extensive data collected by the Australian government on international migration flows, including circular movements, long- and short-term patterns, and counterflows involving the several South-East Asian states, which enable several subpopulations with distinctive movement patterns to be distinguished. It is clear that the effort needed to redesign data systems along the lines Hugo describes would carry major pay-offs in opening up the role of networks and compositional adjustments as central mechanisms of contemporary population and development.

Melissa Brown's chapter addresses a second major population phenomenon that has resisted the top-down, closed population methods that demography inherited from the nineteenth century: changing ethnic identities. The several Chinese and Taiwanese populations she discusses have undergone repeated identity changes over several centuries in consequence of combined pressures on several levels: major regime change and mass migration at the level of states; the impact of in-migrants on community hierarchies; interventionist colonial policies that promote certain ethnic groups at the expense of others; and consequent changes in family, marriage, and locally recognized ethnic markers. Whether in Hubei or Sichuan on the mainland, or in Taiwan, these changes centred on areas with high population densities, poverty, and material resource pressures. Confirming John Odling-Smee's observation that old debates over population and resources need to be rethought where relations between subpopulations are a major influence on their ability to create and utilize economic niches, Brown shows that access to resources, productive capacity, material well-being, and survival vary with the changing structure of local hierarchies in which ethnic groups are ranked. Groups adjust identities actively via demographic behaviour, such as adopting new marriage patterns, recruitment to different labour sectors, and reconfiguring relationships to other subpopulations. Passive acceptance occurs when groups become powerless to oppose ethnic redefinitions imposed by state authorities, although Brown shows how different accommodations and identities can nonetheless be achieved.

The changing composition of subpopulations means that an ethnic label does not refer to the same demographic entity at successive points in time. To take a Taiwanese instance: Japanese colonial regulations that ended footbinding were intended to change the position of women as economic actors, opening up the agricultural sector to their employment. However, because footbinding had been a key marker distinguishing ethnic groups, it was multifunctional, opening up changes in the marriage market and the structure of kin and affinal ties, resulting in changes in ethnic identification. Brown, echoing a point made by Szreter, shows that, in both of these cases, an apparently externally driven or top-down stimulus to change is actually effected—as it spreads across a wider population—by multiple routes or feedbacks initiated at local levels.

Group redefinitions are thus not simply imposed top-down, nor are they simply a kind of 'noise' muddying the demographic record, but active demographic, economic, and evolutionary behaviour that reflects evolving relations between populations.

Without doubt the most widely remarked aspect of population heterogeneity as a central demographic phenomenon is the diversity of secular fertility and mortality declines, and of the historical and contemporary settings in which they begin. In the early post-war era, fertility transitions were assumed to be directly consequent everywhere on a key precedent, i.e. significant earlier falls in mortality. However, as Kaplan and his colleagues remark, early post-war evolutionary biology, unlike demography, considered the bidirectionality of relationships between fertility and mortality as fundamental (for example, infant mortality may be associated either with increased or decreased life course fertility, depending on variations in child and adult health care, breastfeeding, diet, and so forth). Sustained declines in fertility, from this perspective, turn out not to be a more or less automatic demographic change of state that accompanies increased modern influence. Rather, sustained *increases* in fertility occurring before or alongside major falls in mortality, and before fertility transition, have proven to be important. First detected in local studies (Nag 1980; Romaniuk 1980), this pattern has been confirmed at national levels in many cases (Dyson and Murphy 1985).

Kaplan and his colleagues show how prospective studies can now be addressed to mechanisms underlying such early stages of fertility and mortality transition. Their chapter addresses variations between individuals and across multiple communities, taking account of both demographic and evolutionary implications. They address demographic sources of heterogeneity (e.g. parity, mortality, and delayed childbearing) and contextual variations (e.g. geography, body mass, childcare, and educational, linguistic, health, and other factors). Indigenous populations of lowland Latin America, such as the Tsimane studied by Kaplan and his team, make clear the composite nature of total fertility rates as usually calculated for populations as a whole; important variation exists beneath such rates, and the authors track social and environmental factors that appear to combine in different ways to produce different individual and subpopulation reproductive outcomes. As they note, their study also points toward the need for further research clarifying unobserved population variables, notably the way social networks differentiate the behaviour of subpopulations and possibly influence factors like childcare practices that have multifunctional demographic and evolutionary consequences.

2.4 Constructing Populations in the Long Term

The developments traced in preceding sections have helped to open up major new lines of research in long-term human history. Evidence over very long durations—millions of years for hominins, hundreds of thousands in the study of modern humans, and hundreds in the case of modern demographics—is limited inevitably by what happens to survive in the archaeological or archival record. Technical advances in population modelling and the construction of datasets, however, are now able to draw on combined insights of ecological, anthropological, linguistic, and genetic research into current subpopulation processes, both to formulate hypotheses and interpret results. Where genetic data on current populations can be linked to archaeological remains, it becomes possible to compare and possibly link temporally and spatially different sites to each other. Tracking early human and hominin evolution relies on hypotheses, as noted in the preceding sections, in which myriad possible population relationships are delimited by drawing on the 'stories' that environmental, social, and cultural evidence provides about which factors are likely to have shaped mating and other behaviour in the past. Likewise, archival information once considered by demographers and social historians as too fragmentary to be of interest may now be modelled and linked to other records in light of what anthropological, linguistic, and related research tells us about community, kin, and family processes. Together these sources enable long-term demographic and genetic relationships to be understood in the context of a much fuller array of environmental and institutional relationships.

Limitations of the historical record, however, mean that the populations constructed and analysed are inevitably 'opportunistic'. As Calafell and Comas note in the first chapter in Part V, what we can know about the emergence of early human species in Africa depends on the relatively few archaeological remains that have survived over millennia, and also on small numbers of individuals thus far chosen for genomic analysis. Subpopulations constructed by using the latter to interpret the former stand as proxies for actual populations in history for which the full demography needed for a truly representative sample simply is not possible. In consequence, the problems discussed above—of unobserved events and processes, and of whole 'ghost' populations absent from the record whose possible influence we are not able to document—loom very large. Both population genetics and historical demography are, perforce, reliant on populations that have to be treated as closed. The open population processes that shape variation are inferred and, as Calafell and Comas remark, these include principal forces in population history, like natural selection, mutation, migration, and differences in population size in key periods that appear in the archaeological record to have acted as founding events, bottlenecks, or expansions.

The population genetics of Africa as the 'cradle' of humankind is a topic of wide popular interest—an interest often not critically informed by awareness of the limiting assumptions under which past populations have to be constructed. Calafell and Comas provide an authoritative review of this literature, showing how current modelling of sparse data applies approaches outlined in earlier chapters by Bodmer and Winney, and Lawson. Populations are at base idealized mating groups which can be differentiated regionally by combinations of geographical and linguistic criteria together with genetic data. STRUCTURE and ADMIXTURE approaches are also used to construct populations, the results again standing in for actual populations. Despite limitations, these methods have greatly extended our knowledge, for example by characterizing African foraging populations (like the Khoisan, Pygmy, and Hazda, including some subpopulations within them), and then differentiating these groups from neighbouring agro-pastoralist groups arriving in more recent times, and specifying the degree of gene flow amongst them. The precision of these identities, however, comes at the cost of reification: much subpopulation variation remains unobserved, including local environmental adaptations and influences of natural selection. In consequence, adjudication of competing claims for different African regions as the locus of the earliest emergence of anatomically modern humans remains unresolved.

Hominin archaeological data for Africa extend much further into the past, up to some four million years, and pose comparable issues of interpretation. Tracking variation over time in order to distinguish subpopulations below species level has been fundamental to evolutionary theory ever since Darwin identified such variation as a key mechanism in the origin of species. However, as Elton and Dunn remark in the second chapter of this section, identification of distinctive criteria that successfully differentiate subpopulations does not necessarily indicate possible speciation. More likely, this points to demes or subpopulations within them. Given that hominin species are spatially structured, a necessary first step is to identify the geographic range of species currently recognized in the archaeological record. Elton and Dunn's chapter applies GIS modelling to the fossil record of three African hominins in order to calculate their distribution and range, which in turn provides a framework for exploring the likelihood of subspecies diversity. The archaeological record also provides evidence of changing environmental features like rainfall, forestation, and topography; modern primate data can then be compared to these data, suggesting possible adaptive characteristics and variation in features like body mass, arboreal and terrestrial habitats, and of how subpopulations are distributed in relation to these features within an overall geographical range. There is no necessary relationship between range and subspecies variation: some species inhabiting extensive areas show very little variation, while local environmental factors (like island habitats and topological barriers) can result in marked diversity in a limited region. The model shows that two of the hominin species survived for a million years or more over a very extensive geographical range. Evidence of the diversity of habitats across this range, of factors like individual dietary variability, and of morphological changes, all suggest that these hominin populations provide likely instances of Sewall Wright's (1932) classic framework of intraspecific population variation.

With Mikołaj Szołtysek's chapter we turn to a much shorter 'long' duration of a century and a half in Central and Eastern European household systems. His study sharply demonstrates the dangers of opportunistic sampling, while showing how improvements based on closed data sources becomes possible. Historical demography's prevailing image of household and family formation systems up to 1900 was until recently based on Hajnal's Europe-wide synthesis (Hajnal 1965, 1982), although his evidence for Central and Eastern Europe was powerfully shaped by a single community study. Hajnal's 1965 paper famously drew an imaginary line from St Petersburg to Trieste, arguing that each side of the line was characterized by a different marriage pattern; in 1982 he further suggested how this pattern existed in conjunction with contrasting household and family arrangements. Subsequent authors, while suggesting some modifications, for a long time accepted his account, despite the appearance of more systematic comparative analysis of the region for the relatively late date of 1900, and further local studies, that showed marriage and household patterns not conforming to the 'Hajnal line', along with some important variations within the region. The 'ghost' populations not included by Hajnal emerge as real with Szołtysek's construction of an extensive database of 21,000 households for an extensive region of Poland and Lithuania (which would now include Belarus and parts of the Ukraine) from the mid-seventeenth to late eighteenth century. His discussion additionally draws on the construction of comparable datasets for some other parts of Central and Eastern Europe. Even when simplified to three principal subpopulations, Szołtysek's data reveal patterns and subpatterns that do not conform to Hajnal's monolithic East versus West depiction, and that accounting for this heterogeneity requires consideration of several levels of compositional demography, taking in variations in systems of land tenure, labour practices, agricultural technology, ethnic and religious communities, and kinship organization.

Szołtysek's comparative database, of course, remains inevitably an opportunistic sample insofar as it depends on those areas for which records survive. Other (currently 'ghost') populations may yet be identified. Yet the remarkable change of perspective marked by his research owes not simply to building new databases, since their construction is informed historically by knowledge of the ecological, cultural, and economic factors which Szołtysek uses to hypothesize what shaped relationships between households and their members, and which forms of household labour and other relationships promoted the contrasting compositional structures that emerge from the archival record. To this extent, Szołtysek appears not as a critic, but as following on from Hajnal, who emphasized the importance of examining populations in terms of their compositional patterns, and produced principal findings for North-West Europe that ran against ideas of modernization and economic development prevailing at the time of his writing.[9]

As in Part III, the chapters in Part V move progressively from a wider, evolutionary perspective toward a focus on local human populations and subpopulations, that is, the level at which processes give rise to heterogeneity and of the several mechanisms of adaptation that exist in any group, enabling it to alter and renew itself. For explanatory purposes, both local variation and longer-term continuities need to be demonstrated. Mark Elvin's chapter tracks the continuity of a mortality regime in rural and small town areas of central China, specifically in the township of Xiaoji. He compares two periods in the early nineteenth and twentieth centuries on either side of the considerable upheaval known as the Taiping Uprising. Elvin's focus is on specifically demographic measurement: life expectation in relation to parallel shifts in age-specific mortality, fertility, and

[9] For example modernization and demographic transition in the 1960s to 1980s were widely thought to require a shift from rural societies having complex and extended family forms to industrial and increasingly urbanized societies with nuclear family systems. Hajnal showed that, in contrast, small family systems in Western Europe, whether nuclear or stem, prevailed in rural areas, facilitated homeostatic reproductive controls, and *preceded* industrialization and rapid urban movement by two or more centuries. His 1965 paper notes that, when fertility declines occurred around 1900 in Western Europe, this involved a re-endorsement of pre-industrial norms such as high percentages of non-marriage and delayed childbearing, not simply the adoption of birth control.

marriage patterns. The chapter exemplifies two key characteristics of research based on opportunistic information sources: imaginative use of unique records that reflect, and require understanding of, local cultural priorities and environmental constraints; and, in view of the specificity and inevitable limitations in the records, critical assessment of which measures and models can appropriately be applied to them.

Elvin brings out vividly the extreme hardship that Xiaoji's inhabitants experienced: shortage of land, and intense labour required to enhance its productiveness; chronic levels of indebtedness; recurring epidemics (cholera, measles, malaria); out-migration of surplus labour; the fluctuating and often seasonal or temporary nature of migration, both because secure economic niches were not available elsewhere and in consequence of local productive shortfalls and crises; and corresponding fluctuations in fertility that appear to reflect local mechanisms for keeping household membership in line with resources (non-marriage, alternative marriage types, and significant levels of unreported female infanticide that tend to disguise actual reproductive levels). Data from the 1930s reveal marked annual variations in births, deaths, and marriages; the difficulties faced by households and the community resulted in seemingly contradictory patterns, in which an overall rate of natural increase coexisted with population decline. It thus becomes necessary to look beneath the aggregate picture. Elvin compares the 1930s data to his construction of life tables for thirteen local sites in the early nineteenth century, which relies on extensive records of 'faithful widows' and their families, compiled as part of the role of institutions of state over many centuries in support of Confucian ideals of family virtue (Elvin 1984). Elvin uses this richer compositional evidence to consider variations in marriage and migration in relation to the alternative strategies these institutions entailed. The pay-off for interpreting the demography of both periods is both methodological and conceptual. Comparison reveals, first, why the timing and organization of a modern survey would be likely to produce a systematic undercount of infant mortality captured in locally kept records that are informed by local historical processes; second, and continuing a major theme of Parts III and IV, we see why the continuity of mechanisms ensuring population renewal depends on a wider set of alternative courses of action than supposed by Malthusian models confined to fertility and mortality control and material resources. There was a long and widespread tradition of faithful widows, a minority of them even taking their own lives to safeguard their purity, despite what was often severe impoverishment, and unrelenting pressure from their families to remarry. The virtue of loyalty to a dead husband contrasted radically to the impurity of return to sexuality and fertility.

2.5 Identifying Subpopulations for Disease Treatment and Control

Successful design and implementation of health interventions is normally a learning process in which populations at risk gradually come to be differentiated and defined more and more accurately. Risks, as we have seen, reflect subpopulation relationships and events at several levels. Estimates of risk are likely to be based only on some selection of these relationships, as appropriate to the disease and contexts in which it becomes a problem. To begin with, we need to specify the subpopulations or groups in society in which members participate, and the implications and outcomes of differing negotiations they have with each other and with disease and other environmental conditions. These negotiations impact over time on the demography of population and subpopulation compositions, sizes, and renewal. Differing outcomes at social levels impact on other population levels, notably genetic and cellular relationships; there are then feedbacks from lower to higher levels, notably variation in resistance and susceptibility to disease vectors. Finally, there are the often unobservable individual life course processes which, as Wachter notes in Chapter 4, can only be inferred as random 'frailty' and stochastic 'lifeluck'. Improved estimates depend on a balance of direct observation, modelling, and conceptual refinement with which all the chapters have been concerned. Getting this balance right is obviously pressing for health interventions, and the key issues are the same: assessing the influence of unobserved subpopulations likely to be related to those to which interventions are addressed; bringing out the multifunctionality of relationships and of feedbacks within subpopulations and with the environment; and the need to place current events and networks in the context of the history of subpopulation relationships.

Chris Spencer's chapter continues the historical theme of the preceding section, since models of genetic relatedness in the study of differential disease susceptibility often depend on events long in the past which led to the separation of contemporary subpopulations. Separation allows different patterns of natural selection and genetic drift to evolve. In consequence, genetic and physiological characteristics (for instance, sodium uptake, a key correlate of heart disease) may be used to define populations with significantly differing vulnerabilities. In giving an admirably clear exposition of the role of genomics in disease research, Spencer naturally turns to a key problem raised in Dan Lawson's chapter: how to isolate subsets of alleles—amongst the millions of SNPs of the genome—that are relevant to a particular disease. Because the constituency of groups genetically susceptible to a given health problem is an unknown whose membership needs to be established, genomics has had to go beyond conventional closed models in which subpopulations are specified a priori. As Spencer asks, provocatively, what do we do when a population is not a population? The solution, made possible by the extreme detail of the genome, is to begin by treating each individual as its own population, then modelling covariance in alleles between individuals to establish the structures of population relatedness characterized by particular sets of alleles shared by individuals across the genome.

There is an echo here of papers in Part III, but at a very different level of population composition. By analogy, Jenna Johnson-Hanks' phrase 'populations are composed one event at a time' here refers to alleles shared by pairs of individuals, reflecting some past event which gave these individuals a common genetic characteristic. We might easily imagine that the same event was also, at the individual and social population levels Johnson-Hanks is considering, a choice of partner, or the decision to keep a pregnancy and bear the child. Or, to apply Simon Szreter's phrase 'communication communities' at a micro level, Spencer's correlation matrices are communities of individuals built up by shared genetic information—a sharing which is likely also to be a function of shared economic, cultural, or other information in social networks. As Spencer notes, a matrix of correlation that links the alleles of individuals is a model space that lends itself to defining principal genetic components shared by large populations, and where it can explain most of the variability in a dataset, it may then correspond to geographic clines, i.e. long-term population continuities of the kind described by Bodmer and Winney in Part II. We thus return to narratives of long-term population history, building upwards from subsets of alleles and individuals at the micro level.

Gregson and Hallett, in their chapter, are directly concerned with the simultaneity of biological and sociocultural mechanisms of disease transmission, the key events being the communication (or not) of HIV/AIDS, which entails variable flows of genetic information (in the form of infections) and of public and private knowledge and behaviour (not only regarding infection and prevention, but including social factors and networks that shape sexual mating). As noted in section 1.2.4, the history of the epidemic in Africa reveals remarkable heterogeneity, and provides important clues as to why there has been so much variation between subpopulations: widely differing mating patterns; selective migration; varying prevalence of other STDs; differences in circumcision practices; access to and use of condoms; transmission via breastfeeding; cultural institutions, geographical and economic differences that shape flows of information; and differing public health approaches and access to retroviral treatment. People's awareness of HIV/AIDS often, of course, alters mating preferences and behaviour: the population at risk in an epidemic is a moving target.

Gregson and Hallett, in reviewing clinical surveillance and prevalence survey approaches, note that these studies provided a helpful basis for early models, suggesting the importance of networks in which a core subpopulation (often female sex workers) infected wider groups via a 'bridging' subpopulation (their male clients). Randomized control trials were designed to compare particular intervention strategies across multiple communities, but could not produce successful tests as too many 'real-world' factors apparently intruded—it proved impossible, in short, to create a closed population for research which randomizes all the sources of variation already noted. More generally, interventions that have been successful in one area have proven difficult to replicate in other populations. Population networks outside the remit of subpopulations specified for trials or prevalence surveys—for example, membership in community organizations present in some places and not others—appear to be acting as adjacent and overlapping subpopulations that shape communication and sexual behaviour, and account for an important part of temporal variation in the incidence and spread of HIV/AIDS. Mortality due to AIDS obviously reduces those whose sexual behaviour puts other people at risk, and knowledge of their deaths spreads rapidly across social networks (cf. Beckmann et al. 2014). Local processes thus impact on incidence and spread, together with improved information, contraception, and treatment provided by national intervention programmes. As Gregson and Hallett conclude, knowledge of the composition of differing subpopulations and their interactions lies at the core of understanding and addressing the spread of HIV/AIDS.

With Stephen Kunitz's chapter we step back from the current crisis of HIV/AIDS to recognize that contemporary issues of health intervention in Africa take their place in an established but evolving framework of public health programmes that are based on certain assumptions about the identity and nature of populations. Kunitz outlines a general typology of interventions, illustrated by the history of health programmes for native American Indian populations and their long interaction with surrounding American society. In general, public health approaches have favoured a combination of three types of intervention: *definitive* procedures, like immunization, which do not entail ongoing behavioural change; and either *population-based* procedures that can be applied uniformly across all individuals and groups, or *clinical* interventions that focus on high-risk individuals. These strategies, of course, reflect top-down and closed population approaches in which interventions are identical for all people within the population specified for treatment, and comprehensive monitoring provides systematic means of assessing impacts. In contrast to definitive procedures are *halfway* technologies and approaches, necessary where health problems can be ameliorated but not fully resolved—of which retroviral treatment of AIDS would be an example—and which also involve *pervasive* intervention, since they characteristically involve significant, general changes of behaviour. As Kunitz notes, for American Indian populations, as with much of the world since the epidemiological transition, problems requiring halfway and pervasive approaches—like diabetes, heart disease, cancers—have now become much more prevalent. They also pose significant problems for conventional statistics and closed population models, as both the nature of halfway interventions and of pervasive behavioural adjustments entail continuing social evolution and repeated interventions of several kinds, the consequences of which do not fit easily into closed datasets composed of discrete classifications and unrepeatable events.

Kunitz provides several cases in point: for example, diabetes and heart disease as exacerbated by dietary issues and historical changes leading to group identity differences. Thus, the Dakota Sioux, earlier in the twentieth century, were faced with the loss of traditional livelihoods and came to depend on government rations that shifted their primary diet toward rich stews and fry bread; gradually these foods came to be absorbed as traditionally Indian, playing a central part of shared meals and household organization. With time, government assistance priorities have changed, reflecting health authorities' concern about high rates of diabetes and heart disease, but these groups now resist attempts to lessen their susceptibility via consumption of more vegetables and fruits. Diet, in short, is multifunctional: it belongs to political and cultural norms in the northern and mid-western Indian reservations with which the Sioux identify, becoming bound up in a much longer history of disadvantage these groups have suffered from surrounding American society—with consequent scepticism of health interventions, amongst the many government assistance programmes of which they are suspicious. Indian groups in this area have, not surprisingly, begun to take over management of their public health provision. This, in turn, has become part of what differentiates them, for example, from south-western Indian populations in which there is greater response to external public health service interventions to prevent and treat ischemic heart disease, and consequently lower rates of mortality. Which health, cultural, and other population markers are needed to guide biomedical interventions is thus a question that requires bottom-up as well as top-down approaches. As Kunitz remarks, new clinical interventions addressed to high-risk individuals, and the potential of genetic research to enable treatment tailored to recognized individual health risks, would certainly be relevant to improving Indian health; yet, as the majority of deaths to causes like diabetes and heart disease affect the larger part of the Indian population (who are lower risk), these developments may not address the main populations in need.

Klim McPherson's chapter tells the story of a half-century of research and debate over the synthesis of exogenous hormones and their applications in oral contraception and replacement therapy, and associated risks of heart disease and cancers. The field is a case in point of the gradual nature of risk clarification, the crucial role that different ways of specifying populations at risk plays in clarification, and the constraints on research that may keep influences that are difficult to observe from being taken fully into account. As in Stephen Kunitz's chapter, populations relevant to biomedical interventions employing exogenous hormones are also populations subject to cultural, economic, and political definitions that reveal the limitations of conventional top-down population methodologies used on their own. The immense costs of testing hormonal products, disagreement within the scientific community regarding the results of research, and commercial and legal pressures surrounding the pharmaceutical industry, are all well known. McPherson, however, calls our attention to a fundamental and recurring 'disconnect' which is arguably the central problem confronting current methods: a recurring failure to specify the population at risk accurately, or at least to agree on adequate procedures for doing so.

As McPherson notes, two mainstream methodologies are employed. The first is controlled randomized trials designed to pick up short-term effects of exogenous hormone preparations by comparing a representative sample population taking a new drug with samples taking a placebo. The second is observational epidemiology which tracks large cohorts over longer periods, in which comparison is of those taking a medication with those not. The latter method has the advantages of being of much larger scale and of taking place in normal social conditions of use; it is, however, like all closed population methodologies, subject to problems of unobserved influences, such as discussed in the preceding chapters by Gregson and Hallett, and Kunitz. The duration and large scale necessary for cohort studies usually mean that few are carried out. Randomized trials are preferred in company product development, as shorter durations coupled with controlled sample selection enable quicker and more economical research. Such trials, however, cannot tell us about longer-term physiological impacts, nor the influence of variations in actual social conditions of use. The element of controlled comparison on which trials are based nonetheless leads results to be generalized as if relative risks are the same in any population: top-down approaches, as remarked in section 1.2.1, are susceptible to biases disguised by formal conventions where they are not checked by bottom-up and open population thinking. This assumption becomes problematic in hormone studies, as McPherson notes, since endogenous hormone levels vary between people, notably for women according to age and time of the month, as well as diet and other aspects of culture and the environment. Combined methodologies which include sustained ecological, dietary, anthropological, and other observation of behavioural patterns and constraints over time, as shown necessary for accurate research in several preceding chapters, might allow such variation to be considered. They are not, however, normal requirements of either of the two mainstream approaches that McPherson addresses.

Historians of science have shown that controlled trials, like many scientific methods that have attained widespread application, were developed in the context of competing interests and new public regulatory institutions subject to those interests.[10] McPherson's four case studies document how several of the kinds of problems central to this literature remain unresolved. Two may be noted briefly here, one in which prior population specification resulted in the high-risk subpopulation being largely excluded from the sample; in the other, differing academic and institutional interests have led to different base populations being used for estimates and to continuing disagreements over appropriate trial criteria. At base, the question is whether current procedures ensure that the use of closed statistical samples and models in fact captures accurately the populations at risk, and does so in a timely manner. McPherson's first example concerns tests of early oral contraceptive users, from the 1960s and 1970s, and associated risks of breast cancer. Large cohort studies were established to assess long-term effects of the pill, based on samples primarily of married women who had already commenced childbearing; subsequent clinical studies indicated, however, that risks were much higher for women who started the pill prior to their first childbirth. Although the mismatch is evident, large case control studies on contraceptive histories and outcomes of women in the latter subpopulation have not been undertaken. The second instance concerns hormone replacement therapy and its possible associations with cardiovascular disease and breast cancer. Results both of large-scale trials and of findings synthesized from multiple local trials have shown significantly increased risks, yet have been contested. The regulatory framework, in short, does not resolve what historians of science call the social construction of evidence, which is an inevitable influence on the design and interpretation of research, and hence on which treatments are available. What is clear from these examples is that exclusive reliance on closed population models allows biomedical products with significant counter-indications to go on being prescribed despite potentially significant evidence of risk to individual women.

[10] Edwards' (2007) history of the development of the controlled trial framework notes that Bradford Hill and others, in promoting their method, frequently used the term 'controlled' in ways that were not justified by its applications, and that this played an important role in the method's wider acceptance. Debate over current uses of the method in the development of new pharmaceutical products is discussed in Light (2010). More generally on the element of social construction in shaping modern statistical methodologies, see Porter (1990), who is followed on the subject of trials and medical statistics generally by Mathews (1995) and Will and Moreira (2010).

References

Beckmann, N. (2012). 'Medicines of Hope? Antiretroviral Treatment in Tanzania: A Critical Analysis of Large-Scale Biomedical Health Intervention', *Journal of East African Studies* 6/4: 690–708.

Beckmann, N., A. Gusman, and C. Shroff (eds) (2014). *Strings Attached: AIDS and the Rise of Transnational Connections in Africa*. Oxford: British Academy/Oxford University Press.

Beerli, P. (2004). 'Effect of Unsampled Populations on the Estimation of Population Sizes as Migration Rates between Sampled Populations', *Molecular Ecology* 13: 827–836.

Bledsoe, C. (1999). *Contingent Lives*. Chicago: Chicago University Press.

Bochow, A. and P. Kreager (2015). *Fertility, Conjuncture, Difference: Anthropological Approaches to the Heterogeneity of Modern Fertility Declines*. Oxford: Berghahn.

Bonfield, L., R. M. Smith, and K. Wrightson (eds) (1986). *The World We Have Gained*. Oxford: Basil Blackwell.

Bongaarts, J. (1978). 'A Framework for Analyzing the Proximate Determinants of Fertility', *Population and Development Review* 4/1: 105–132.

Boserup, E. (1981). *Population and Technology*. Oxford: Blackwell.

Breckenridge, K. and S. Szreter (2012). *Registration and Recognition: Documenting the Person in World History*. Oxford: Oxford University Press.

Chesnais, J.-C. (1992). *The Demographic Transition*. Oxford: Clarendon Press.

Cohn, B. (1987). 'The Census, Social Structure and Objectification in South Asia', in *An Anthropologist Amongst the Historians and Other Essays*. Delhi: Oxford University Press, 224–254.

Coleman, J. S., E. Katz, and H. Menzel (1966). *Medical Innovation*. Indianapolis: Bobbs-Merril.

Darwin, C. (1859). *On the Origin of Species*. London: Murray.

Das Gupta, M. (2005). 'Strategies for Managing Household Resources in Rural North India', in T. Engelen and A. P. Wolf (eds), *Marriage and the Family in Eurasia: Perspectives on the Hajnal Hypothesis*. Amsterdam: Aksant, 195–214.

Dyson, T. and M. Murphy (1985). 'The Onset of Fertility Transition', *Population and Development Review* 11/3: 399–440.

Edwards, M. (2007). *Control and the Therapeutic Trial: Rhetoric and Experimentation in Britain, 1918–1948*. Amsterdam: Rodopi.

Elvin, M. (1984). 'Female Virtue and the State in China', *Past and Present* 104: 111–152.

Evandrou, M. and J. Falkingham (2004). 'Demographic Change in Europe: Implications for Future Family Support for Older People', in P. Kreager and E. Schröder-Butterfill (eds), *Ageing Without Children: European and Asian Perspectives*. Oxford: Berghahn, 175–197.

Fisher, K. (2006). *Birth Control, Sex, and Marriage in Britain, 1918–1960*. Oxford: Oxford University Press.

Goody, J. (1972). 'The Evolution of the Family', in P. Laslett (ed.), *Household and Family in Past Time*. Cambridge: Cambridge University Press, 103–124.

Gould, S. J. (2002). *The Structure of Evolutionary Theory*. Cambridge, MA: Belknap Press.

Granovetter, M. S. (1973). 'The Strength of Weak Ties', *American Journal of Sociology* 78/6: 1360–1380.

Hajnal, J. (1965). 'European Marriage Patterns in Perspective', in D. Glass and D. E. C. Eversley (eds), *Population in History*. London: Edward Arnold, 101–143.

Hajnal, J. (1982). 'Two Kinds of Pre-Industrial Household Formation System', *Population and Development Review* 8/3: 449–494.

Hallett, T. B., S. Gregson, O. Mugurungi, E. Gonese, and G. P. Garnett (2009). 'Assessing Evidence for Behaviour Change Affecting the Course of HIV Epidemics: A New Mathematical Modelling Approach and Application to Data from Zimbabwe', *Epidemics* 1: 108–117.

Hauser, P. M. and O. D. Duncan (1959). *The Study of Population: An Inventory and Appraisal*. Chicago: University of Chicago Press.

Harrison, G. A. and A. J. Boyce (1972). *The Structure of Human Populations*. Oxford: Clarendon Press.

Hedström, P. and G. Swedberg (eds) (1998). *Social Mechanisms: An Analytical Approach to Social Theory*. Cambridge: Cambridge University Press.

Hodgson, D. (1983). 'Demography as Social Science and Policy Science', *Population and Development Review* 9/1: 1–34.

Hugo, J. G. (1982). 'Circular Migration in Indonesia', *Population and Development Review* 8/1: 59–84.

Johnson-Hanks, J. A., C. A. Bachrach, S. P. Morgan, and H.-P. Kohler (2011). *Understanding Family Change and Variation: Toward a Theory of Conjunctural Action*. Dordrecht: Springer.

Kaplan, H. and M. Gurven (2008). 'Top-Down and Bottom-Up Research in Biodemography', *Demographic Research* 19/44: 1587–1602. Available online at <http://www.demographic-research.org/volumes/vol19/44/19-44.pdf>.

Kertzer, D. I. (1984). *Family Life in Central Italy, 1880–1910*. Brunswick, NJ: Rutgers University Press.

Kertzer, D. I. and D. Arel (2002). *Census and Identity: The Politics of Race, Ethnicity, and Language in National Censuses*. Cambridge: Cambridge University Press.

Kertzer, D. I. and M. Barbagli (eds) (2001–2003). *The History of the European Family*, 3 vols. New Haven Yale University Press.

Kertzer, D. I. and P. Laslett (eds) (1995). *Aging in the Past: Demography, Society, and Old Age*. Berkeley: University of California Press.

Kohler, H.-P., F. C. Billari, and J. A. Ortega (2002). 'The Emergence of Lowest-Low Fertility in Europe during the 1990s', *Population and Development Review* 28/4: 641–680.

Kreager, P. (1992). 'Quand une population est-elle une nation? Quand une nation est-elle un état? La démographie et l'émergence d'un dilemme moderne, 1770–1870', *Population* 6: 1639–1656.

Kreager, P. (2004). 'Where Are the Children?', in P. Kreager and E. Schröder-Butterfill (eds), *Ageing Without Children: European and Asian Perspectives*. Oxford: Berghahn, 1–48.

Kreager, P. (2011). 'The Challenge of Compositional Demography', *Asian Population Studies* 7/3: 85–88.

Kreager, P. and E. Schröder-Butterfill (2010). 'Age-Structural Transition in Indonesia: A Comparison of Macro- and Micro-Level Evidence', *Asian Population Studies* 6/1: 25–45.

Krieger, N. (2012). 'Who and What Is a "Population"? Historical Debates, Current Controversies, and Implications for Understanding "Population Health" and Rectifying Health Inequities', *The Milbank Quarterly* 90/4: 634–681.

Laland, K. N., J. Odling-Smee, and S. Myles (2010). 'How Culture Shaped the Human Genome: Bringing Genetics and the Human Sciences Together', *Nature Reviews Genetics* 11: 137–149. Available online at <http://www.nature.com/nrg/journal/v11/n2/full/nrg2734.html>.

Larsen, U. (1994). 'Sterility in Sub-Saharan Africa', *Population Studies* 48: 459–474.

Lewontin, R. (2000). *The Triple Helix: Gene, Organism, and Environment*. Cambridge, MA: Harvard University Press.

Lewontin, R. (2004). 'Building a Science of Population Biology', in R. S. Singh and M. K. Uyenoyama (eds), *The Evolution of Population Biology*. Cambridge: Cambridge University Press, 7–20.

Light, D. W. (ed.) (2010). *The Risks of Prescription Drugs*. New York: Columbia University Press.

Malthus, T. R. (1890). *An Essay on the Principle of Population*. London: Ward Lock & Co.

Marsden, P. V. (1990). 'Network Data and Measurement', *Annual Review of Sociology* 16: 435–463.

Massey, D. (1990). 'Social Structure, Household Strategies, and the Cumulative Causation of Migration', *Population Index* 56/1: 3–26.

Mathews, J. R. (1995). *Quantification and the Quest for Medical Certainty*. Princeton: Princeton University Press.

Mayr. E. ([1959] 1976). 'Typological versus Population Thinking', in *Evolution and the Diversity of Life: Selected Essays*. Cambridge, MA: Belknap Press, 26–29.

Mayr, E. (1982). *The Growth of Biological Thought*. Cambridge, MA: Belknap Press.

Mayr, E. and Provine, W. B. (eds) (1980). *The Evolutionary Synthesis: Perspectives on the Unification of Biology*. Cambridge, MA: Harvard University Press.

Mitchell, J. C. (1969). *Social Networks in Urban Situations*. Manchester: Manchester University Press.

Montgomery, M. R. and J. B. Casterline (1996). 'Social Learning, Social Influence, and New Models of Fertility', *Population and Development Review* 22 (Supplement), 151–175.

Müller, G. B. (2007). 'Evo-Devo: Extending the Evolutionary Synthesis', *Nature Reviews Genetics* 8: 943–949. Available online at <http://www.nature.com/nrg/journal/v8/n12/full/nrg2219.html>.

Nag, M. (1980). 'How Modernization Can Also Increase Fertility', *Current Anthropology* 21/5: 571–587.

Niehas, I. (2012). *Witchcraft and Life in the New South Africa*. Cambridge: Cambridge University Press.

Noble, D. (2006). *The Music of Life: Biology Beyond Genes*. Oxford: Oxford University Press.

Porter, T. M. (1990). 'The Quantification of Uncertainty after 1700: Statistics Socially Constructed', in G. M. von Furstenberg (ed.), *Acting Under Uncertainty: Multidiscplinary Conceptions*. Dordrecht: Kluwer, 45–75.

Romaniuk, A. (1980). 'Increase in Natural Fertility during Earlier Stages of Modernization: Evidence from African Case Study, Zaire', *Population Studies* 34/2: 293–310.

Schnegg, M. (2007). 'Open Boundaries: The Long-Term Development of a Peasant Community in Rural Mexico', *Journal of Anthropological Research* 63/1: 5–31.

Schneider, J. C. and P. T. Schneider (1996). *Festival of the Poor: Fertility Decline and the Ideology of Class in Sicily, 1860–1980* Tucson: University of Arizona Press.

Schofield, R. (1985). 'English Marriage Patterns Revisited', *Journal of Family History* 10: 2–20.

Schröder-Butterfill, E. (2005). 'The Impact of Kinship Networks on Old-Age Vulnerability in Indonesia', *Annales de Démographie Historique* 2005/2: 139–163.

Segalen, M. (1991). *Fifteen Generations of Bretons: Kinship and Society in Lower Brittany, 1720–1980*. Cambridge: Cambridge Universitiy Press.

Skinner, Q. (ed.) (1985). *The Return of Grand Theory in the Human Sciences*. Cambridge: Cambridge University Press.

Slatkin, M. (2005). 'Seeing Ghosts: The Effect of Unsampled Populations on Migration Rates Estimated for Sampled Populations', *Molecular Ecology* 14: 67–73.

Smith, R. (ed.) (1984). *Land, Kinship, and Life-Cycle*. Cambridge: Cambridge University Press.

Smith, R. and M. Pelling (eds) (2003). *Life, Death, and the Elderly: Historical Perspectives*. London: Routledge.

Szreter, S. (1993). 'The Idea of Demographic Transition and the Study of Fertility Change', *Population and Development Review* 19/4: 659–702.

Szreter, S. (1996). *Fertility, Class and Gender in Britain 1860–1940*. Cambridge: Cambridge University Press.

Szreter, S. and E. Garrett (2000). 'Reproduction, Compositional Demography, and Economic Growth: Family Planning in England Long Before the Fertility Decline', *Population and Development Review* 26/1: 45–80.

Szreter, S., H. Sholkamy, and A. Dharmalingam (eds) (2004). *Categories and Contexts: Anthropological and Historical Studies in Critical Demography*. Oxford: Oxford University Press.

Viazzo, P. P. (1989). *Upland Communities: Environment, Population, and Social Structure in the Alps since the Sixteenth Century*. Cambridge: Cambridge University Press.

Wachter, K. W. and C. E. Finch (eds) (1997). *Between Zeus and the Salmon: The Biodemography of Longevity*. Washington, DC: National Academy Press.

Waples, R. S. and O. Gaggiotti (2006). 'What is a Population? An Empirical Evaluation of Some Genetic Methods for Identifying the Number of Gene Pools and their Degree of Connectivity', *Molecular Ecology* 15: 1419–1439.

Watkins, S. C. (1987). 'The Fertility Transition: Europe and the Third World Compared', *Sociological Forum* 2/4: 645–673.

Watkins, S. C. and I. Warriner (2003). 'How Do We Know We Need to Control for Selectivity?', *Demographic Research* Special Collection 1, Article 4: 109–142. Available online at <http://www.demographic-research.org/special/1/4/s1-4.pdf>.

Weinreb, A. A. (2006). 'Substitution and Substitutability: The Effects of Kin Availability on Intergenerational Transfers in Malawi', in A. H. Gauthier, C. Chu, and S. Tuljapurkar (eds), *Allocating Private and Public Resources Across Generations*. Dordrecht: Springer.

Will, C. and T. Moreira (2010). *Medical Proofs/Social Experiments: Clinical Trials in Shifting Contexts*. Farnham: Ashgate.

Wilson, C. (2011). 'Understanding Global Demographic Convergence 1950–2010', *Population and Development Review* 37/2: 375–388.

Wright, S. (1932). 'The Roles of Mutation, Inbreeding, Crossbreeding, and Selection in Evolution', in Donald F. Jones (ed.), *Proceedings of the Sixth International Congress on Genetics*. Austin, TX: Genetics Society of America, i.356–366.

Wrightson, K. and D. Levine (1979). *Poverty and Piety in an English Village: Terling 1525–1700*. Oxford: Clarendon Press.

Wrigley, E. A. (ed.) (1966). *An Introduction to English Historical Demography*. New York: Basic Books.

Wrigley, E. A. and R. S. Schofield (1993). *The Population History of England, 1541–1871*. Cambridge: Cambridge University Press.

PART II

WHAT IS A POPULATION?

CHAPTER 1

POPULATION AND THE MAKING OF THE HUMAN SCIENCES

a historical outline

PHILIP KREAGER

AGGREGATE properties of human groups have been fundamental to observation and reasoning about society and the state at least since Aristotle's *Politics*. The framework which he and his contemporaries put forward long remained influential, spreading via humanism into the seventeenth and eighteenth centuries, when it shaped the emergence of the term 'population' and its measurement as a fundamental ground of scientific inquiry into human society. The development and use of mathematical models in population thinking began under the influence of this early modern context, giving rise very gradually to modern demographic approaches, chiefly from the nineteenth century. The long gestation of population thinking is important because it has left questions as to whether demographic models fulfil criteria for scientific reasoning. At the heart of these questions are two different ideas of population that came to coexist during the seventeenth and eighteenth centuries, in both of which aggregate relationships between people are understood as integral to the natural world. One is addressed primarily to processes by which population and subpopulation memberships are formed; the other to demographic, epidemiological, and evolutionary outcomes of these processes, usually at higher levels of aggregation.

This chapter sketches the evolution of the two concepts and their relation to each other. This perspective is useful, first, for understanding why population thinking has remained fragmented for so long. Not only did biological and social approaches to population tend to go their separate ways, a recurring theme from the nineteenth century onwards is one of antagonistic debates arising *within* biology and *within* the study of society, in which only one or the other concept was recognized as primary or even legitimate. In consequence, investigation of some fundamental dimensions of population change was marginalized over long periods of modern history. In recent years, the ground occupied by what may be termed the 'classical model', from Aristotle through its early modern development, has had to be largely reinvented in the social sciences and history. Old antagonisms can still exercise a powerful grip on our thinking.

Historical overview of the two concepts carries a second advantage: it calls attention to three main turning points in the history of these concepts which may help us to overcome the old antagonisms and to recapture fundamental continuities in modern population thought. Each point reflects a change in how the two concepts are related. The first concerns elementary methodological choices made in the rapid rise of statistics in the early nineteenth century—i.e. before the human sciences of genetics, demography, and anthropology had even emerged. At that time, processes of population formation previously taken as fundamental were marginalized in data collection because they appeared to conflict with contemporary social and political reforms which the statistical movement was promoting. This marginalization has subsequently made many aspects of inter- and intragroup population dynamics difficult to integrate into social science approaches.

The second turning point concerns a puzzle that confounded one of the most eminent historians of population thought, Ernst Mayr. Clarification of Darwin's central contribution to population thinking is one of Mayr's major contributions, yet in the 960 pages of his *Growth of Biological Thought* (1982) he was unable to identify the specific intellectual sources that enabled Darwin to formulate his distinctive account of the fundamental role of population in natural selection. The answer lies in recent historical research which has shown Darwin's reliance on eighteenth-century writings in which the classical model remained normative, and which have come by his influence to inform the predominant post-war biological paradigm, the evolutionary synthesis. This point is important on two counts. One is that Darwin provided a framework incorporating *both* concepts of population in his theory. Their joint power has become evident in the immense strides that evolutionary biology has achieved beginning from the Synthesis, in a range of fields from molecular biology to ecological inheritance and environmental research. The other is that the importance of classically informed influences on Darwin's theory—notably, the work of Adam Smith—reveals that the two concepts are a fundamental conceptual ground shared by the biological and social sciences. There is no priority of one over the other.

A potential third turning point is posed by demography's hesitant rapprochement in recent decades with population as a subject of social and cultural history and of evolutionary biology. There has long been a striking imbalance in post-war demographic theory. As is well known, the era of mass declines in fertility and mortality from 1870, popularly known as demographic transition, is a highly exceptional era in population history. Whether an era of exception provides a realistic general basis for explaining the nature of population adjustment and change can certainly be questioned. Moreover, in its single-minded focus on providing a unitary account of the transition, demography privileged lines of argument confined to the second of the population concepts. Priority was given to fertility and mortality over other elements of population composition and structure characteristic of the many subpopulations that make up a society or state. Aggregate reproductive and mortality trends were analysed in terms of models of economic and social modernization, giving a secondary role to the distinctive historical trajectories of different societies and cultures. Biomedical innovations in fertility and mortality control were emphasized, not advances in methods and concepts of evolutionary biology. While this approach has proven very fruitful in demonstrating the heterogeneity of declines and continued variation after transition, an agreed unified and deterministic explanation of this diversity has not been forthcoming. Since the 1980s, different approaches emanating from historical demography, social history, anthropological demography, biodemography, population ecology, and social network research have shown that populations and subpopulations employ a much wider range of mechanisms to adjust their composition and size than the more or less exclusive reliance on mortality and fertility controls supposed by transition theory. As understanding these mechanisms depends in important respects on the nature and history of group formation, and on how networks and other compositional processes shape relations between subpopulations, it appears that approaches drawing on both population concepts are gradually coming to be recognized as necessary. From this perspective, demographic transition may in fact simply describe one phase of a longer evolutionary process, an era in which certain compositional adjustments were more visible than others.

1.1 Early Modern Population Thinking: A Short History

The human sciences are disciplines in which people and their actions form the object of study. While the nineteenth century gave us most of the conventional rubrics into which the human sciences can be divided (human biology, demography, anthropology, sociology, genetics, psychology, archaeology, epidemiology, statistics, and so forth), our interest is in the common ground of concepts and problems they share in population. This common interest, as historians of science often remark, arose as part of the seventeenth- and eighteenth-century aim to develop methods for observing and quantifying human relationships, a movement which embraced the natural sciences together with the humane subjects of politics, morals, and philosophy. The *OED* neatly encapsulates this by noting that the earliest modern usage of the term 'population' owes to Francis Bacon (Bacon 1612). Bacon, of course, wrote his seminal texts on scientific method chiefly in Latin, reminding us that the 'human' in human sciences is cognate with humanism, the diffusion of Greek and Roman texts and creative development of their themes from the fourteenth to seventeenth centuries. Hence the *OED*, in its entry for 'human science', gives the following usage as indicative:

> It is not too much to say that the great contribution of modern specialists [in] the human sciences is less in the realm of general theory than in the perfecting of method by which ancient speculations can be confirmed, modified, or rejected.[1]

[1] The quotation is from Lasswell and McDougal (1943: 214).

This observation may, however, come as a surprise to an audience concerned with population today. In modern society, populations are understood as essentially statistical, and it may seem almost nonsensical to say that systematic reasoning about population could, as in the ancient world (and, indeed, up to the eighteenth century), proceed without regular enumeration or at least careful estimation. Can there be systematic population thought with little reference to number? Was classical and humanist understanding of human populations simply swept away by the rise of statistics in the nineteenth century? Answering these questions is important for more than historical reasons: the answers enable us to see how approaches to population following Darwin, and recent compositional demography, have begun to reinstate a fundamental dimension of population thought.

1.1.1 The Classical Model and Its Influence

Population emerges as a distinct and sustained subject of discussion in the political writings of Aristotle and Plato in the third century BCE (Kreager 2008; Charbit 2010). Its fundamental importance could scarcely have begun more emphatically: the composition and size of constituent subpopulations, and more particularly the patterns of association between them, define and determine the size, stability, relative growth, and development of states. Differences in population composition were used to characterize the main types of polity (monarchy, aristocracy, republics), the characteristic problems of each, and the principal explanations of why and how these types of society and government evolve. Down to the eighteenth century, the diverse composition of people making up the state, and the artful management of relations of power and influence within and between the several groups and interests they formed, remained a fundamental ground of writings on human government. When population arithmetic emerged in the seventeenth century, authors following the lead of Graunt's seminal work (1662) devised their measures as a way to better define constituent populations and to show which properties of their composition made these groups more or less effective for purposes of state. That is: quantitative approaches to human population emerged in the context of a conceptual structure that differs markedly in several respects from demography, or at least from the way demographers and social scientists from the nineteenth century came to define and analyse populations. In the classical model, fertility and mortality are not the primary determinants of a state's population size, structure, and growth. Rather, population structure, as the configuration of subpopulations making up a state, shapes levels of reproduction differentially in these groups, as well as other mechanisms of population change. Fertility, mortality, and their impact on age/sex structures are, from this point of view, only secondary aspects of patterns of human association that determine the balance of power and human resources amongst subpopulations.

The motivations underlying this classical logic were eminently pragmatic. Society was considered a natural phenomenon arising from humans' innate needs for association.

Aristotle sums this up pithily in his well-known remark that 'man is a political animal' (Aristotle 1995: 3). The key question in the collective life of people (*politicos bios*) is how to coordinate the constituent populations making up the state so that they answer to the good life for its citizens (i.e. their natural, political, and social needs). This is not easy: because individuals and groups are not alike, association gives rise necessarily and perpetually to differences, and hence to competition and conflict. Differences begin in patterns of cohabitation, family, and household formation. Via networks of descent, marriage, alliance, patronage, and community ties, they then become bound up in population recruitment to wider political, social, and economic organization. The latter, in turn, feed back into the differing composition of family, marriage, patronage, and other patterns of association. This perspective has some similarities to the bottom-up, family reconstitution approach of historical demography—but at base there are three critical, conceptual differences that give classical reasoning about populations its distinctive character.

Population Memberships Are Not Principally Numerical Aggregates

Association by its nature entails multiple memberships: people participate simultaneously and over time in changing and frequently overlapping sets of people, or subpopulations. The units making up a population's structure are thus defined primarily by relatedness that involves both how people are linked to each other and relations between groups. A membership may, in short, be called a 'network population'. This is where the pragmatism of the classical model is most apparent in contrast to modern demography: fertility is a slow way to build up a network; much quicker and more effective is to draw on actual and potential ties between groups that enable a body of people to be assembled who have common interests and are able to act concertedly. Different combinations of people may be drawn on simultaneously for a number of purposes, thus enhancing agency and flexibility. The political import of population composition and structure follows directly: competition between members and their networks was strategic to sustaining privileged subpopulations in a hierarchy of groups. Whether for political control, warfare, labour, social security, making suitable marriages, or other recurring needs, being able to call on and adjust human resources via network memberships opens up many opportunities. It also stimulates competition for people and resources: from classical sources down to the late eighteenth century, what may be called the political demography of factionalism was of much greater concern than the use of enumerations. From Aristotle to Rousseau, the familiar refrain is that it is not numbers or the size of a population that is critical, but how well its constituent units are balanced and can work together in a coordinated way. The defeat of the mighty Persian Empire by the tiny Greek city states was a staple case in point.

The several subpopulation memberships making up a hierarchy, and the groups formed by ties between them or by shared differences to other subpopulations, could, of course, be classified according to a few important attributes of individual members, and the population as a whole then enumerated. This approach, which seems obvious to a modern reader, is one in which classical authors were noticeably uninterested. Regular censuses and registrations never seem to have occurred.[2] It is not difficult to understand why. Standardized classifications would, of course, radically simplify the many purposes and alternative arrangements that the mutability of ties make possible, not to mention revealing potentially sensitive issues amongst competing groups. Those higher up the hierarchy did not see themselves as countable in the same way as their clients (see next section). Putting numbers on entities, even if perhaps useful for identifying a current state of play, would still require interpretation resting on a good knowledge of events and processes of group formation and reformation that determined members' behaviour. The reliability, value, and durability of enumerations, in short, were not obvious, not to mention the trouble and expense of maintaining them. In sum, a network population is an aggregate, but in classical population thinking its quantitative properties are secondary and often incidental outcomes of the way relationships between members are formed, maintained, or changed.

Heterogeneity versus Homogeneity

One consequence of the mutability of memberships and of competition amongst them is that the renewal of heterogeneity in a state's population was recognized as a fundamental and unending process. Assumed homogeneity of the kind that has prevailed in modern demographic thinking (for example, where models assume that individual behaviour responds uniformly to macro-level economic trends) is replaced with the view that large-scale aggregates are composites of subpopulations behaving in decidedly different ways. Classical focus was more on what would now be called relations between local and meso-level populations. More than this, the classical view emphasized that certain sorts of population are incommensurate, such that it is nonsense to compare and count them together. Incommensurability is fundamental to why fertility takes a secondary role. The citizens of ancient republics, like Athens for example, practised birth control and selective migration of family members. This reflected a society in which many citizens were agricultural smallholders needing to restrict the subdivision of property. This group, however, was also of immense political importance as they formed the main body of citizens whose concerted action could keep in check the oligarchic aims of wealthier proprietors. Birth control and migration as mechanisms ensuring the continuity and stability of this middle group (i.e. between rich and poor) defined the political balance amongst competing groups in the state. The birth control of middle strata was, however, of scant importance for regulating the population of the city state as a whole. Citizens were a minority (about one-fourth of the whole population, by later estimates; Aristotle 1997: 79). The majority was made up of guest workers and slaves. Population control of these groups was achieved chiefly via migration: where employment opportunities for guest workers declined, many inevitably left; excess slaves were sold. Hence the practical common sense of the classical point of view: to calculate a fertility rate that combined all of these groups, while possible, was of no interest since it ignored the intrinsically

[2] Even in ancient Rome, in which 'censors' were principal public officials charged with carrying out censuses, these records were not conducted regularly and were in any case confined to a citizen minority.

different natures and political statuses of subpopulations and the differing processes of group formation and change that regulated their memberships.

Open versus Closed Populations

As both preceding points make clear, memberships, and the flexibility with which they can be altered, mean that units of population are by nature *open*. A population here is defined as a limited but unbounded network or set of networks in which patterns of association between members enable the addition or subtraction of ties as appropriate. Openness is critical both internally to the state (e.g. as competing factions realign members, power, and patronage, thus changing the composition and balance of subpopulations) and in its external relations (internal factional realignments, by altering capacities for trade, warfare, and migration, thereby change the state's population and its pattern of growth as a whole). Interestingly, openness does not preclude members from believing they are a closed group. Citizenship was regarded by Athenians as closed: no guest worker or slave could marry a citizen or otherwise gain access to citizenship; yet incorporation of small numbers of successful guest workers did nonetheless occur. More important, the status of citizenship and the size of the citizen population were not fixed. If one faction succeeded in seizing power and establishing an oligarchy or tyranny, then it was not a good idea to be of the opposing faction: your status as a citizen would probably cease, your property would very likely be confiscated, and your death or banishment (and likewise that of subpopulations allied to you) were likely. Regime change was a major demographic force by which memberships were radically altered, with implications for the respective fertility, mortality, and migration of competing groups. (This is, of course, no less true for a number of places in the world today.) All of this is, of course, very different to modern demographic population thinking, in which a state's population is ostensibly compiled comprehensively by its census, which thus fixes it, usually for a ten-year period, and thence enables closed models like representative samples, life tables, and so forth to be compiled and calculated. Regime change, rather than a major mechanism of population change and a central problem in its analysis, is in modern demography a black hole, since in these contexts censuses and representative surveys cannot be conducted.

I hope this sketch, however rapid, conveys something of the sophisticated and practical character of early population thought.[3] We can stylize classical thinking as a 'model' on at least two counts. First, Aristotle's *Politics* was written as a critical and pragmatic synthesis—a model or paradigm intended to systematize contemporary understanding of aggregate human relations and its problems. Second, this schema provided a template on which classical and early modern discussion was based. We turn now to the influence of this model, briefly outlining its absorption into influential theories of state and society in the seventeenth and eighteenth centuries, and how this context shaped the emergence of population measurement. While early modern population arithmetic was conceived

[3] A more detailed account is given in Kreager (2008).

as an aid to contemporary statecraft, and therefore as consistent with existing political norms and ideologies, applications of descriptive numerical analysis, and increasing recourse to concepts drawn from the calculus of probability, did more than simply reinforce the status quo. The early history tells us, as we shall see, that what became demography and statistics in the nineteenth century was born out of open population thinking in the preceding era, but that issues in that era led to attempts to put quantitative reasoning about nature and society on a radically different—i.e. closed, statistical—footing. In the second part of this chapter we note how, from the mid-nineteenth century, a succession of population problems emerged that require a combination of open and closed population thinking.

1.1.2 The Emergence of 'Population'

Bacon's early use of 'population' occurs in his essays 'Of the True Greatness of Kingdoms and States' and 'Of Seditions and Troubles' (1612). As these titles suggest, the first is about getting balance right in the body politic: there should not be too many nobles; immigration is key to providing manpower proportionate to an expanding state; relative numbers of 'middle people'—i.e. having many small landowners—are necessary to balanced government and defence; and so forth. The latter essay is about checking the growth of factions: policies that reduce envy, inequitable taxation, religious discontent, etc. will keep factional subpopulations from growing, thus enabling stability and good government. Ancient Greek and Roman texts provided Bacon both with a logic and specific examples to explain events in contemporary Europe. A central theme of these essays is that small, well-coordinated populations are superior to strength in sheer numbers for purposes of economic growth, political stability, and sometimes defence; but Bacon also notes the disadvantages of too much influence wielded by a small, tightly connected group, i.e. oligarchy.[4]

As political historians have remarked, Bacon's *Essays* drew closely on Machiavelli's (1517) vastly influential extended commentary on classical texts, which circulated widely in the sixteenth to eighteenth centuries (Skinner 1980). From Machiavelli to Rousseau, the classical model dominated: each type of political regime continued to be characterized by the relative size of its ruling membership and how it exercised authority over other subpopulations. Monarchies (rule by one person) are for a time able to amass much greater territory and population, accruing economic and military advantages accordingly; they tend to collapse into tyranny, however, because their severely imbalanced distribution of power leads kings to become tyrants, fostering factional conflict between subpopulations. Unequal taxation, political favouritism, sectarian conflict, and general oppression all flourish; in consequence, individuals and groups are

[4] 'A few that are more Stiff (hold together best) do tire out a greater Number' (Bacon 1931: 155); 'Neither is the Population to be reckoned only by number; for a smaller number, that spend more and earn less, do wear out an estate sooner than a greater number that live more frugally' (1931: 42).

exiled or emigrate, they reduce their fertility and die in conflicts. Aristocracies (rule by a few and, ostensibly, 'the best' citizens) suffer similar problems, drifting into oligarchy. Republics (rule by the citizen body as a whole) are likely to work only in city states where rich or aristocratic elements can be incorporated in balance with significant numbers of middle strata actively engaged in government. Republics have to be small, because members of the citizen body need to know each other to ensure collective responsibility. Even so, these small states are highly vulnerable: first, because tyrants in neighbouring monarchies may amass greater military forces; and second, because factionalism (usually reflecting an imbalance between rich and poor) can push republics into monarchy, oligarchy, or tyranny of the majority.

Although theories of the state are not our direct concern here, it is important to note that population thought evolved because it was integral to conceptualizing human government, both from an individual's point of view and collectively. From a modern perspective, this may seem surprising: we are accustomed to see population as a routine object of a state's administrative functions, not as a primary concept arbitrating the forms governance can take and their legitimacy. In the early modern era, as we have noted, the several aggregate associations composing the population of a state as a whole were important on two principal counts. First, successful management of constituent groups and their resources, including factional interests, determined the success or failure of regimes, whatever their type. The conventional formula, carried over from classical sources, describes the 'balance' or 'good proportion' amongst the 'members' of the political body these members should be well distributed in respect of its several tasks (production, trade, defence, etc.) just as the members of a well-proportioned human body exhibit strength, grace, and capability. The second importance of population followed directly: a balanced or imbalanced population was evidence of the relative legitimacy of a government's policies and of the relative virtues of different types of regime. Imbalance meant a greater likelihood of bad government, and declining population (consequent on the dire effects of factionalism). Political arithmetic, when it emerged as an intended scientific activity in the mid-seventeenth century, addressed these issues of legitimacy directly; in contrast, its potential administrative uses (e.g. regular census-taking) could only be tentatively sketched, and were largely ignored by governments for another century and a half.

The absorption of classical population thinking into political and social discourse of the *ancien régime* helps us to make sense of the simple quantitative practices that were used to compile populations over the pre- and early modern era. From the ancient civilizations onwards, complex state organizations mounted, from time to time, 'musters', '*catastos*', or censuses, usually for tax purposes. There is no suggestion that these headcounts were designed to assist analysis of quantitative composition and structure of a population, or systematic comparison of states (Hecht 1980; Dupâquier and Dupâquier 1985; Gehrman 2009). Bacon comments that, while 'the population may *appear* by musters, . . . there is not any thing amongst civil affairs more subject to error than the right valuation [of] the power and force of a state' (1931: 80, emphasis added). What musters counted was geared not to this issue but to short-term administrative agendas.

That is: what we do *not* find are records that count people as if they were of the same importance, count them exhaustively, and do so on a regular basis. Only certain members were worth bothering with, and then only when a need arose. This varied, but counts were always partial and selective. Fiscal administrations counted politically significant members of the body politic, i.e. propertied household heads. Their properties and trades which helped to sustain the body politic might also sometimes be counted and, as part of this, those able-bodied household members and clients material to taxation (wives, adult children, servants, slaves). Military musters also included age groups of young men. The Church, as the pre-eminent international institution of the early modern era, was likewise selective. Parish registers in England, for example, counted politically and spiritually relevant persons—Anglicans—not Catholics, Dissenters, or Jews. In short, where population structure is understood primarily as a hierarchy of known groups, priority information reflects the power structure of those groups. To erect standardized, regular, and comprehensive methods of data collection, as censuses have aimed to do since the nineteenth century, seemed a massive exercise in counting the unimportant.

1.1.3 Enter Population Arithmetic

The *ancien régime* was not static, and limitations of the classical model were widely recognized. A concept of population intended for ancient city states was hardly adequate for big growing polities like France, Spain, or England. From the sixteenth to the eighteenth century, the more or less permanent instability described by the classical model for monarchies and aristocracies, and the inescapable vulnerability which appeared to doom all republics, meant that it accounted for contemporary problems of government accurately, but provided no way of resolving them. Republics like Florence, mid-seventeenth-century England, and the Netherlands did collapse, and were forced to restore a pre-eminent princely component. The egregious mismanagement of resources, sectarian conflicts, and the endless wars that characterized French, English, and Spanish monarchies, made the classical model very apt at describing tyranny and the critical destabilizing role of competing royal, aristocratic, and popular factions. The major line of Enlightenment political and economic writings treating population, from Montesquieu and Rousseau to Turgot, Condorcet, Smith, Hume, and Steuart, continued to see population problems as a product of bad government, especially observed monarchical policies. But the model proposed no escape.

Population arithmetic, as is well known, originated in the mid-seventeenth-century work of a London merchant, John Graunt, and, as we would expect from the preceding, immediately became known as 'political arithmetic'. The object of his *Natural and Political Observations* (1662) was to use simple numerical measures as a way of demonstrating instances of good and bad government, and to open up potential improvements to scientific assessment and discussion (Kreager 2005). The history of seventeenth- and eighteenth-century political arithmetic, however, has unfortunately been written largely as a prehistory of statistics and demography, i.e. with no reference to its classical sources (e.g. Westergaard 1932; Lorimer 1959; Glass 1973; Dupâquier and Dupâquier 1985). As Innes's (2009) recent survey shows, this focus on proto-demography has meant that such histories take into account only a small subset of the development of quantitative approaches to population, society, and government in this era. This is doubly unfortunate. First, in the absence of attention to classical and humanist sources, proto-histories struggle to show population arithmetic as a sophisticated body of thought. The common view has been, rather, that Graunt and his successors may have produced ingenious ideas, sometimes hitting brilliantly on elements of much later mathematical statistics, yet they never achieved an integrated science of population. As political arithmetic in this view remained a set of flawed attempts to invent statistics, its development is said to have 'stagnated' (Westergaard 1932: 44) by the mid-eighteenth century.

A second consequence of this anachronistic view is that the possibility of early modern population thought exerting any significant influence on later research across the human sciences has been marginalized. Whether in mainstream demography, population biology, or in early modern historical research more generally, the view of early population arithmetic as no more than a set of ad hoc anticipations of later statistical approaches implies that it can have nothing new to tell us. Looking at early modern population thinking in light of its classical sources, however, reveals a very different picture: seventeenth- and eighteenth-century rethinking of the classical model redefined how and why it is important to quantify population, and in so doing, played a central role in reshaping the whole theory and institutional structure of government and its relation to science. Population arithmetic opened a door that was hoped would be instrumental to escaping the dilemmas of factional subpopulations in the *ancien régime*.

Graunt is a key transitional figure who worked within the classical concept of population, yet produced ways of thinking that could no longer be contained by it. His central insight was simple but far-reaching. Given that balanced proportions of people are necessary to the good functioning of the body politic, and enable us to assess the legitimacy of governments and their policies, a way needed to be found to express these proportions in succinct ratios, and to track their movement over time. Graunt effectively proposed applying empirical tests to the classical model. Using bookkeeping arithmetic, he applied this strategy to mortality records kept in London, and also to some parish records, giving rise to vital rates which he said showed the body politic of England to be well balanced. Graunt clearly saw the political import of all of his method, saying 'a clear knowledge of all these particulars, and many more, is necessary in order to good, certain, and easie Government, and even to balance parties and Factions, both in *Church* and *State*' (1899: 397). While Graunt, writing at the end of the Civil War, was understandably chary of applying his method directly to factions, he had another motive for avoiding controversy, which was that such methods should be independent of party. His ratios might be of use to the king and his ministers, but his method and results should be assessed by scientists, represented by the Royal Society (1899: 323, 397). In this Graunt explicitly followed Bacon, modelling his method on the latter's instructions for compiling natural histories. By calling his work *The Natural and Political Observations*, Graunt also signalled his allegiance to the classical view of man as a political animal. However, he broke with Bacon and Aristotle in developing an explicitly quantitative method for this purpose. His work showed that there were many more issues that needed to be counted for purposes of state than those dictated by networks of power and connection. For the first time, quantitative information on human society, which mixed the 'incommensurate' higher and lower strata of society, was seen explicitly as a means of analysing natural regularities beyond the control of princely power or factional interests.

The *Observations* created a sensation and, as Innes (2009) shows, opened up a much wider set of possible quantitative issues that were increasingly taken up in the eighteenth century. The proliferation of population arithmetics saw most calculations move away from ambitious assessment of the legitimacy of government to addressing local concerns, from smallpox inoculation to improving agricultural production. Amongst the several issues that continued to emphasize the role of arithmetic in assessing national policy, like the implications of rapidly rising national debt, or the adequacy of relief provision for the poor, one continuing issue—the use of number in political evaluation of constituent subpopulations—deserves particular note here. Two broad approaches may be recognized. One was the move by established monarchies in Sweden and Prussia to begin regular compilations of the nation's population, or at least much more extensive one-off compilations, as in France and England. Access to these records, however, was closed, so this information could not become a major venue for science or debate over public policy.[5] Numerical compilations in monarchies were confined chiefly to the classical formula of showing the legitimacy of kings by using ratios and other measures, indicating that population and economy were growing.[6]

The other approach arose in discussion of the role of parliament in mixed monarchies, notably in England. In addressing the question of numerical balance between significant groups in the body politic, population arithmetic led logically to questions of political representation. How should the main subpopulations be represented? What criteria should be used to define those members who qualify for political representation, and those who do not? A long line of political arithmeticians, including Petty, King, Davenant, Hume, Price, and Condorcet, experimented with different proportional models to define and represent citizens. To begin with, the assumption of all analysts up to Price and Condorcet continued to follow the classical model: only property owners count as citizens. As Pole (1966) shows, in the American colonies this assumption proved difficult to apply, and the several colonies proposed different arguments and experiments in legislative organization. This contest came to a head in debates over the national constitution. As the central text in the debate, Hamilton, Jay, and Madison's *The Federalist* (1788), shows, the classical model continued to provide the crux of the argument and its principal examples: How could the fledgling

[5] Two exceptions might be noted: Gregory King's unpublished work was based on private access to government sources, and the data on which eighteenth-century Swedish life tables were based was drawn from government compilations.

[6] J. P. Süssmilch's *Die Göttliche Ordnung* (1741) is the best known, of which there is a French edition and translation, J. Hecht, *L'Ordre Divin* (Paris: INED, 1979).

small republic regulate internal factions, particularly by balancing wealthier with common interests? How could it protect itself from the predations of more powerful European monarchies? How could the excesses of mass democracy be checked? A working answer, promulgated by Madison, took a numerical form, apparently cutting through this long insoluble dilemma: all adult males would be citizens, signified by their equal vote in parliamentary elections. Certain subpopulations, however, remained incommensurate and were therefore excluded, notably women and slaves. Each major subpopulation—i.e. the thirteen states of the new republic—would have numerical representation in the parliament or congress, proportional to their population sizes. A population census thus emerged as a central constitutional criterion. A senior body—the Senate—would provide a check on factions and potential tyranny by the majority; its members would, it was assumed, be drawn from more established and distinguished groups in the population, although no constitutional formula could be devised to ensure this.

1.1.4 The Emergence of Closed Population Thinking

The idea that a State should count all its citizens, and count them equally, suggests a modern census. Madison's proposal, however, like the several French attempts to establish a *Statistique Générale* in the aftermath of the Revolution, stopped well short of this. In the American case, the role of the census was conceived initially with reference solely to political representation; only later, under the influence of the statistical movement in Europe, did the census move to become the basis of a comprehensive national accounting system (Anderson 1988). The French, as it happens, also lagged behind: Napoleon tried to turn the *Statistique Générale* back into a monarch's private data source (Perrot 1984). The reality of a modern census was a nineteenth-century achievement that would require major changes in the concept of population, departing substantially from the classical model. These changes took shape gradually in the early decades of the century in the Low Countries, German principalities, and England.

The break from the classical model was achieved via a host of procedural developments intended to free the census, as a system of classification and enumeration, from *ancien régime* interests. First, the census would constitute populations without reference to existing power networks amongst influential subgroups. Second, it would employ an objective or politically neutral system of classification and enumeration that was comprehensive and rigorously precise. Everyone was supposed to be counted equivalently. Within the period and space of its coverage, there could be no events or statuses—births, deaths, marriages, occupations, etc.—that existed outside its purview. Or, put negatively, groups and events other than those recognized in its system of classification were excluded. Third, the census's authority would therefore rest on analysis based entirely on numerical relationships between the classes of events it recognized. Finally, the founding of a census bureau would ensure that a public institution above the level of local factionalism was charged with defining and tracking the national population.

Under this idealized system, for which actuarial methods provided an acknowledged model, population became not a flux of subgroups with shifting boundaries constituted by changing networks, but a *closed* or containerized entity. The population as compiled by a census was then used for purposes of state—regardless of inevitable incompleteness and biases—as a standard, empirical referent. This approach to population contributed, in turn, to wider shifts in the idea of the state and its inhabitants. Instead of territory and population membership responding to the prerogatives of elite networks, the state was identified with a whole national population. Membership, although now defined primarily by classification and enumeration, not networks of association, nonetheless carried over many features of traditional ties between people that were difficult to reconcile with the State's new uniformist statistical identity. Normally, a State was deemed to be a nation, that is, a people identified with a shared cultural and historical identity, occupying the given territory now charted by the census. However, as most states included several cultural and historically distinctive groups, some identities were inevitably marginalized. In consequence, the census, although intended to be above partisan interests, could not avoid interpretations that gave formidable support to tenets of nineteenth-century nationalism, racial stereotypes, and social Darwinism (Kreager 1992). Finally, the emergence of the conventional modern view of population as containerized would alter fundamentally the interpretations of economic and environmental constraints on the state that had prevailed in the classical model and its early modern continuation. As an essentially closed entity, a nation state could have too many people—an interpretation that arrived forcefully with the works of Malthus and Fichte (Kreager 1997).

1.1.5 Enter Formal Population Modelling

To understand the emergence of this containerized view, we need to look at a lineage of population thought that evolved in the context of open population thinking under the *ancien régime*. The origins and early development of this line have been considered for the most part by historians of technical developments in statistics, probability, and demography (e.g. Hacking 1975; Krüger et al. 1987; Hald 1990). A main focus has been the first applied mathematical model of population—the life table—which by the later eighteenth century was being used successfully in limited institutional settings (life insurance companies) in which exhaustive classification and counting of mortality could actually be achieved. The wider influence of the life table followed from this development, which provided stringent methodological criteria seen as ideal for registration and census data collection in the nineteenth century. In other words, it was not so much the life table as a model of mortality that was crucial, as the wider influence it came to exercise on how to construct rigorous classification and enumeration systems for health, economic, and social monitoring. The influence of early vital statistics thus lay in the far-reaching implications of its closed population approach as a basis of government and scientific method.

Because the claim of statistics as it emerged in the nineteenth century was to constitute the state by identifying and compiling all citizens on an equal basis, it also embodied a political ideal, offering a potentially crucial instrument for addressing the dilemma that the classical model could not resolve: How could people escape the endless cycle of vulnerable republics and tyrannical monarchies? As a movement claiming to be based in scientific methods, however, the nineteenth-century promotion of statistical, sanitary, and social reforms eschewed a direct approach to the thorny problem of political representation. As social and medical historians have shown, it focused instead on health and environmental issues that plagued the great mass of labouring poor (Eyler 1979). Greater health and social equality would become possible only if the ground of statistical data documenting the population as a whole, and its problems, was established. The scientific basis for this approach was argued on two grounds. One was that statistical compilations were objective, comprehensive, and purely descriptive accounts of material facts of demography, economy, and society. The other was that advanced probabilistic methods, when applied to these data, were grounded in purely formal, universal regularities, the so-called 'laws of large numbers' (Desrosières 1993). These claims were not realized in practice: censuses often did not achieve completeness or uniformity, and they were subject to party interests and the systematic neglect of major groups and trends. As the many other compilations that nineteenth-century statisticians pursued also generally fell short of the universalist ideals of the movement, the claim that statistical procedures directly establish a value-neutral science came to be questioned. Despite these major shortcomings, the practice of compiling large, quantitative datasets was widely established as necessary to public policy, and the development and application of formal methods like the life table to such closed populations was *de rigueur*.

Graunt is often said to have introduced the first life table, but this is not strictly speaking true. He composed a hypothetical population structured by age that allowed for deaths over the life course, which he used to make his calculation of 'fighting men'.[7] There is no doubt, however, that mathematicians who pioneered the calculus of probability immediately saw a different relevance to Graunt's table: the calculation of life expectation. Within five years of the appearance of the *Observations*, Huygens, Hudde, and de Witt (and subsequently a long further line of distinguished mathematicians, including Leibniz, the Bernoullis, Halley, DeMoivre, Euler, Price, and Gomperz) redeveloped Graunt's table as the first potential real-world application of the calculus (Hacking 1975; Hald 1990). By the mid-eighteenth century, attempts had been made to use life table models to calculate government annuities and private life assurance, to rationalize land rents, to solve inheritance or other legal problems, and to evaluate the efficacy of smallpox inoculation. The mathematics of probability, in conjunction with population arithmetic, appeared to promise ways of using science to improve society and government. Graunt was right, therefore, that his arithmetic would attract scientific

[7] Graunt's own calculation of life expectation occurs elsewhere in his text, and follows classical principles, i.e. propensity to death reflects subpopulations defined by the environment of specific localities (Kreager 1988).

interest: awareness of his work spread widely across the humanist network of European natural scientists, mathematicians, and literati, who were in regular and frequent correspondence and often in the service of heads of state.[8] While the life table became the founding example of how mathematically informed science might reform human society and government, its uses remained speculative until the 1780s, and no wider institutional or social impact was achieved.

The situation changed with the work of the noted mathematician and radical republican, Richard Price (1792). Seeking finally to develop the life table as a model that would allow working men's friendly societies, and life assurance societies, to avoid bankruptcy, Price turned the hitherto impossible data demands made by life table techniques into a virtue. His work on a number of issues, like the national debt, had drawn on the growing interest and experience of local systems of data collection that, by the middle years of the eighteenth century, characterized a range of institutions (hospitals, the military, the board of trade, prisons). While these systems were not designed for mathematical analysis (usually records were used to demonstrate the volume of services provided and to attract charity), this procedure for classifying and enumerating clients had the important feature of treating populations as closed. That is: all population characteristics were reduced to a limited and fixed classification of events; the entry, exit, and all changes of a member's status were exhaustively recorded so that any event merely tracked a member's movement from one class to another. Only events in a fixed period of membership were recorded. Working by analogy to such institutions, Price located an unusually complete and accurately compiled set of parish mortality records for a community whose members could be taken to resemble those typically seeking life insurance from an early company, the Equitable Society. The 'Northampton Table' which he constructed was, as with life tables demographers now construct, not a real population—even though based on meticulous data—but a 'stationary' population frozen in space and time. The Equitable, in adopting the table, in effect reversed the hitherto normal relationship between ratios and society. Since Graunt, ratios had been interpreted as indicative of real relationships amongst people. Now, the ratios came to dictate what the behaviour of those people, and their relation to the Equitable, would be: entering the society meant that actuaries would assess a person's life and death according to the pattern of mortality in the table. Not only did this prescribe when they were expected to die, but the actual course of events: the flow of premiums they would pay up to a given age, the payouts they might then receive, and indeed whether their expected mortality allowed them entry into the society at all. For the first time, a population model dictated the limits of real behaviour.

[8] Hacking (1975) remarks that Graunt's reasoning places him at the heart of the 'emergence of probability'—yet Graunt alone of the analysts just cited had no knowledge of advanced mathematics. The answer to this puzzle, as Garber and Zabell (1979) show, returns us once again to the classical model: the basic ideas of mathematical probability from which the calculus of chance was born in this era were already part of qualitative Greek and Roman logics that formed the basis of contemporary schooling. Graunt's detailed use of these logics in writing his book is discussed in Kreager (2002).

The immense financial success of the Equitable soon produced imitators, the insurance industry expanded as a major source of national economic investment, and the professional role of the actuary was born. More important for what became vital statistics and demography in the nineteenth century, the life table provided early advocates of national data systems, like William Farr, Adolph Quetelet, and Jean-Baptiste Fourier, with formal criteria specifying that numerical information be collected and compiled according to precise actuarial standards. These were laid out by Farr (1837) in his prolegomena to vital statistics, which led to his appointment as Compiler of Abstracts at the Registrar-General's Office, and to his construction of the first national life table in 1843. The era began in which the ideal was to summarize an entire national population in a single formal model.

1.2 Turning Points

The later salience of the life table and the formal data collection procedures it helped to promote need little introduction here. The life table is the daddy of mathematical population models. Fisher's *Genetical Theory* (1930), which came to play a key role in the evolutionary synthesis, relied on conventional life table analysis combined with tables of reproduction by age and genetic variation. Lotka's stable population theory (1939) used the life table as the starting point for demonstrating how inertial principles of thermodynamics could be applied systematically to fertility and mortality, his synthesis subsequently becoming fundamental both to population ecology and mathematical demography. Of course, Fisher's and Lotka's work, as foundational to population genetics and demography, rested on the remarkable nineteenth-century ascendance of statistics, both descriptive and mathematical. Historians and philosophers of science refer to the former as 'the avalanche of numbers' and the latter as 'the probabilistic revolution' (Hacking 1991; Krüger et al. 1987).

The foundation provided by statistics, however, is not all there is to the study of population composition and change. Other nineteenth-century developments, particularly in biology, continued to emphasize key aspects of the classical model, such as the way populations are formed out of relationships between subpopulations. Observation and knowledge of intra- and interpopulation dynamics remained crucial to the study of speciation and variation in the natural world. Closed systems of classification were evidently inappropriate to observing the emergence of new species and forms of variation, since these systems require units and types of population to be decided in advance. Change in a closed framework, in other words, is confined to trends and shifts from a given category, state, or type of population to another, rather than observing the capacity of networks to mutate, blend, or carve up existing entities into new sorts of individuals, groups, and relationships. Where the linear continuity of existing memberships and their characteristics cannot be assumed, understanding the principles of formation and successive reformation of groups becomes necessary. Darwin's importance is of note here because his concept of natural selection made understanding processes of new population and subpopulation formation central to establishing mechanisms of genetic and ecological inheritance, and thus to the dynamics of population variation more generally. With Darwin, observing processes of variation within and among groups on a small or local scale again became fundamental, providing the basis on which Mendelian and later population genetic models could be developed. Darwin's approach to population thus created the possibility of a synthesis uniting the study of open population processes with closed population modelling.

The history of statistics and of evolutionary theory are now established academic subjects, and by drawing on this substantial body of research,[9] three key turning points in modern population thought may be recognized. The first was a consequence of the remarkable spread of descriptive population statistics in nineteenth-century health and social sciences, which had the effect of marginalizing key population processes concerned with subpopulation relationships and change. In the second, debates surrounding Darwin's theory of natural selection in the early and mid-twentieth century reasserted these processes for purposes of biological research. Third, a look at current developments in demography suggests that similar steps are being taken to reintegrate these processes into the study of human societies.

1.2.1 What Got Lost in the 'Avalanche of Numbers'

The continuity of basic vital measures, from early modern political arithmetic to nineteenth-century vital statistics and demography, has often been remarked. The ratios and tabulations of the former passed seamlessly into the latter in the representation of age and sex structures, life expectation, crude and age-specific mortality and reproductive rates, urban/rural differentials, and recognition of the importance of cause of death data. However, as national statistical offices proliferated, these measures were not only refined, but put on a radically different footing. National bureaux spread as foundational to the modern nation state in a period of social and political reform that saw even established monarchies promulgate modernizing bureaucracies. Following France (1800), Prussia (1805), and Bavaria (1808), national offices were formed in Tuscany (1818), in several other German states (beginning with Wurtemburg in 1820 and ending in Saxony in 1831), in the Netherlands (1828), Austria (1829), Belgium (1831), Sicily (1832), Denmark (1833), Norway (1837), and England (1837). From early in the century, official statisticians and members of the newly founded London Statistical Society and other such societies argued that systematic classification based on discrete material events (like death and age) and precisely definable statuses (like marriage and occupation) defined an objective, empirical stratum applicable uniformly to all aggregate social phenomena. The aim

[9] For the purposes of this paper I rely on: Provine (1971), Mayr (1982), Porter (1986), Bourguet (1989), Hacking (1991), and Gould (2002).

of standardized and complete population accounting for entire states became the norm. There was no longer room for the particularistic reasoning fundamental to the classical model, in which the several populations making up a state are incommensurate because defined by the distinctive histories of relationships, or by their place in a hierarchy. While national statistical offices had descriptive aims, from the mid-nineteenth century, Quetelet and others argued that probability as applied in astronomy to correct for observational error could be used to establish statistical normality in human population data.

The impact of bringing applications of the probability calculus to bear on human populations was threefold. First, allying population statistics with the physical sciences greatly enhanced the argument that statistics could provide a universal method. Second, and more particularly, Quetelet and others claimed that, as long as sufficiently large numbers of events and statuses were rigorously collected, the regularities constituted evidence of deterministic relationships in social life, a 'social physics'. Most official statisticians, while sometimes drawing on his argument, were more sceptical, and were content to use their compilations to substantiate or refute deductive 'laws' of the kind that political economists, like Malthus, had proposed. By end of the century, nonetheless, Quetelet's determinism was expanded by Durkheim (1897) into a general sociological method in which social phenomena would be characterized as *sui generis* aggregate regularities that determine the actions of individuals and groups without their awareness. The particularism of the classical model was apparently swept away. If population statistics was about large, national aggregates, then the diversity at local levels of vital events, economic relationships, and social norms could be regarded as mere micro-level noise. Of course, as in the eighteenth century, a great deal of local, problem-oriented, quantitative social inquiry continued without reference to grand notions of statistical laws, or reference to the calculus of probability. Yet there can be little doubt that the rise of descriptive statistics, its probabilistic rationale, and the ubiquitous acceptance of its relevance at the national level, mark a major turning point in the conceptualization of population.

A third development, however, effectively undermined these assertions even as they were being made. Critical revision of determinism emphasized that central tendencies like norms were not evidence of natural and social laws, and that the phenomena of interest lay, rather, in characterizing the principles of variation that typified different population groups. There was, however, no consensus amongst the critics. Approaches like the work of German probabilist, Wilhelm Lexis, were based on a social historical tradition that denied determinism and demonstrated its limits (Porter 1987). Others, like Galton and Pearson, attempted to refine determinism by developing powerful new statistical techniques; their attempt to apply these methods to crude eugenic classifications of population was to discredit the collaboration of evolutionary biology and social sciences for several decades. In a third, more careful, development, Lotka took up Quetelet's advocacy of physical models, but sensibly removed their applicability to a purely formal mathematical plane (Lotka 1939). Meanwhile, Quetelet's expected roster of deterministic social-cum-physical laws never emerged, and historians of this era now speak of the gradual *erosion*, not the triumph, of determinism (Hacking 1991).

One reason for all of this scepticism is that many of the problems that concerned analysts of the nineteenth and early twentieth century did not fit closed population models, and that attempts to make them fit have come to be recognized as problematic. It is without doubt significant that human sciences like anthropology, genetics, and ecology emerged in an era in which the incompleteness of closed population models was beginning to raise serious questions.

A good example, alluded to already, was the difficulty of reconciling the multiplicity of groups residing within the boundaries of states with the idea that a state's population should be treated as a unified whole. Subpopulations have agency too—networks based for example on ethnicities, communication communities, local economies—which commonly do not agree with assumptions coming from a state's administrative centre. Over the latter half of the nineteenth century, eight international statistical congresses were convened for the purpose of promoting uniform data collection, and the question of how to treat distinctive subpopulations was raised repeatedly. The report of the first congress noted as fundamental the principle that actual populations ('population de fait') should be studied, not merely populations as defined by administrative boundaries ('population de droit') (Congrès Général de Statistique 1872: 35). The difficulty of carrying this out became clear four years later in the meeting at Vienna, which proposed that all states compile a 'statistique éthnographique' in which the territory, customs, language, physical characteristics, and several relations between constituent national groups would be recorded (Congrès Général de Statistique 1872: 42). Again, in the meeting in Florence in 1867, an even more elaborate 'statistique communale' was proposed, acknowledging the need to understand hierarchical relationships between groups, and the capacity of such groups to stimulate action and adopt varying forms of solidarity (Congrès Général de Statistique 1872: 265–266). Discussions at other meetings of the congress could easily be cited, but the stated need for methods that track the agency of differing constituent populations and their varying solidarity makes clear that the central concerns of the classical model remained, and were not addressed by the new population statistics. However, only a few states in which claims of constituent groups were powerful—Austria-Hungary, Russia, British India—paid attention to these protocols. The absence of long-term data series on constituent ethnic, linguistic, and religious groups has continued to characterize most of Europe until the present day. And, of course, the absence of evidence of the kinds proposed in Florence turned out to be precisely what, a century later, the Princeton European Fertility Project found it needed—but lacked—to explain the diversity of reproductive change in nineteenth- and twentieth-century Europe (Coale 1969).

The containerized census concept of population has suited discrete, unrepeatable, and irreversible events like births and deaths, and statuses like age and sex. The considerable effort and ingenuity needed to reduce complex natural and institutional processes (like multiple causes of death; differing kinds of partnership; shifting household and family types; class and caste formations) to classifications suitable to this schema has, over the last century and a half, been extended gradually to survey methodologies and their analysis. Reification was inevitable in such a strongly typological approach and, as Wilson and Oeppen (2003) pointedly remark, remains a common problem in demography today. The heuristic premise that population units, events, statuses, and influences on them can be treated exclusively as if they are actually discrete or containerized is now a normative premise of much more elaborate analyses, so that any social, economic, or environmental variable thought to influence reproductive behaviour or death may be taken to act independently of all others. By a long-established convention, population variables (usually those describing fertility and mortality) are 'dependent' variables that can change only as acted upon by 'independent' variables located in the economy and society. The strong tendency of post-war demography to treat fertility and mortality as if they are the only primary population factors has marginalized the rest of compositional demography, notably internal and external group relationships that adjust memberships and subpopulation structures, and which thereby function as mechanisms of agency and solidarity—as if these factors were always secondary, or not even fundamental demographic phenomena. Downgrading subpopulation relationships has the important consequence of restricting attention to potentially major sources of population heterogeneity, and stands in sharp contrast to approaches that have transformed genetics and evolutionary biology during the twentieth century.

1.2.2 Enter Darwin

Darwin (1859), writing in the midst of the 'avalanche', conceived his theory of natural selection using a concept of population that ran counter to then prevailing closed statistical approaches, in three key respects.[10] The first is a matter of scale. Darwin returns from the macro to the local as the primary level of observation and analysis: the population of a species as a whole may be large, but understanding selection processes that sustain or alter that population requires us to look at relationships within and between local subpopulations (demes) of which that species is composed. Demes are memberships defined by individual mating, the variation in genetic material transmitted by mating, and the aptness of these processes to local environments that individuals and demes may inhabit. Thus, at any given point in time, a species is a composite, shaped by the intra- and intergroup dynamics of its subpopulations that are anything but closed by nature. A second difference to closed population thinking follows directly: the statistical convention which classifies all types of members and their characteristics at the start of research is insufficient where genetic and environmental population changes are occurring, because mating and environmental adaptation are capable of producing individuals, groups, and adaptations not envisaged in any existing set of characteristics. Demes are open populations, not only because they admit some mating with other demes, but because of changes in the genetic and ecological inheritance of individuals and groups introduced by mutation, drift, and other evolutionary processes. The third difference

[10] This section draws on the extended account of Darwinian population thinking in Kreager (2009).

between Darwin's population thinking and closed data systems as employed in demography is thus that the primary analytical and substantive concern goes beyond the renewal of a population to *the renewal of its heterogeneity*. Aggregate change is brought about not only by trends that vary statuses and numbers of extant types of individuals, but also by variations leading to new types of individual and subpopulations that can, in turn, produce further new variants.

Evidently, with Darwin we are back to a kind of population reasoning that is much like the open population thinking that was conventional before the nineteenth-century rise of statistics. While there are basic similarities between Darwinian population thinking and the classical model, there are also major differences. What is at issue is not a return to that earlier framework, but a substantially new articulation of open population thinking. The most obvious change is that open populations are now recognized as fundamental to species and their evolution in the natural world, as well as in human political and social relations. The second is that Darwinian population thinking proved to be much more open to formal quantitative development than the classical model.

Ernst Mayr, to whom we owe clarification of Darwin's thought on these issues, remained puzzled about how Darwin came to conceive populations as open systems. Mayr was conversant with several major sources of the classical model, for example, repeatedly expressing his own admiration for Aristotle and noting Darwin's similar estimation (1982: 87–91, 149–154, 305–307). However, it was Aristotle's keen powers of observation, not his classificatory approach to natural history, which they chiefly admired. For Aristotle, all species were fixed, and there can be no new ones. Open population thinking, in other words, owes to a different part of Aristotle's *oeuvre* than his natural history. It is as if Darwin took the population thinking in Aristotle's *Politics* and applied it to the natural world. While such a phrasing is clearly a *reductio ad absurdum*, clarification of the classical model in preceding sections enables us to show how the link occurred.

First, as we have seen, the open population approach was common sense in Europe when thinking about population for most of history up to Darwin; and second, as recent historical scholarship (Schweber 1980, 1985; Gould 2002: 230) has shown, Darwin's own reading in the period in which he formulated his ideas on natural selection drew on eighteenth-century political economy, in which the classical model was customary. As is well known, Darwin's notebooks attribute a key stage in his thinking about species competition to his reading of Malthus: the superfecundity of populations gives mortality a key role in natural selection by weeding out less competitive members of a population (Mayr 1982: 477–478). As Schweber and Gould note, Darwin also reasoned by analogy to political economy in making use of the concept of the division of labour, drawn from Adam Smith's *Wealth of Nations* (1776). Population was important in Darwin's thought not only because mortality served as a mechanism of adaptation, but because he needed to explain continued heterogeneity—why species go on developing and diverging. As taxa specialize increasingly into definite and restricted roles in local ecologies, more species can be supported, and Darwin drew a direct parallel between this diversification in nature and the role of the division of labour, in which people's self-sufficiency and production evolve via the diversification of the labouring population according to their differing tasks and the relation of these populations to each other. Natural selection, like the division of labour, articulates an open conception of population in which the primary issue is not simply how groups survive, increase, or reduce their numbers, but how this process continually opens up new, variant forms of adaptation and populations arising from them.

Schweber's and Gould's detailed reading of Darwin's notebooks thus confirms a key instance of the major influence of eighteenth-century and earlier population thinking on his work. Darwin was able to rely on a general and long-standing fabric of European thought in which populations consist of memberships built and changed via patterns of association between groups and their members. Darwin's work, however, did not leave this fabric as it was. At the heart of his theory there remained unanswered questions about mechanisms of natural selection. Subsequent research and debate inspired by his methods, but going beyond them, succeeded in developing sophisticated closed population models within his overall logic of open population reasoning, which have greatly expanded our knowledge of evolutionary mechanisms.

This rearticulation of the two concepts was without precedent. As we have seen, in the seventeenth and eighteenth centuries, closed population analysis of the life table could grow up with population arithmetic in the context of open population thinking assumed by the classical model—but without altering the latter as a normative political approach to populations, and without being applied successfully to key issues like political representation or life insurance until the very end of the period. Natural historians of that era also experimented with how to use number and probability to overcome arbitrary hierarchical criteria then used to distinguish species, but did not see such approaches as means of reconceptualizing open population thinking.[11] Only with Darwin's theory was the natural history of species reformulated in terms of populations, and the classical view reasserted that human society must be seen as fundamentally part of the natural world and its explanation. While Darwin articulated the logic of natural selection, he remained unable to identify and quantify what are now known to be genetic mechanisms of selection. This central challenge led subsequent research and debate to the rediscovery of Mendel, the development of laboratory genetic research, the confirmation of natural historical evidence by Fisher's *Genetical Theory*, and Dobzhansky and Wright's (Lewontin 2003) adaptation of statistical models to field conditions. Together these developments have come to be known as the evolutionary synthesis (Mayr and Provine 1998).

This remarkable rearticulation of the relationship between open and closed population thinking can, perhaps, be summarized very simply as follows. Evolutionary population thinking needs to reason both from the bottom-up and the top-down. In the former,

[11] Lamarck, for instance, attempted to develop a measure of degrees of similarity between individual members of a species (Stevens 1994: 18), which Adanson then sought to improve by altering closed classifications to polythetic ones (Sneath 1964).

observation of local mating networks as essential to establishing and tracking mechanisms of genetic change was established by Mendel (Mayr 1982: 720–722). Mendelian populations are loci of intra- and interpopulation change, and the identification of alleles transmitted via mating networks provides biomarkers that may then be tracked and modelled in carefully selected large aggregates. Of course, hypothesized genetic links and biomarkers intended to track them don't always prove to be correct. Hence there is also an important role for top-down population reasoning, in which models and analysis at higher levels of aggregation may reveal problems and relationships, and give rise to further hypotheses requiring more local observation and analysis. Participants in the evolutionary synthesis took sharply opposed views of this complementarity, as historians of biology have documented (Provine 1998: 51–58). Natural historians, paleontologists, and some experimental biologists, argued that statistical modelling becomes critical only after observation at local levels has identified primary mechanisms of genetic and environmental change. Mayr went so far as to imply that statistical methods are not really population thinking at all (1959). Geneticists trained in Fisher's methods, in contrast, saw laboratory work and Dobzhansky and Wright's synthesis of field observation and statistical analysis as chiefly documenting relationships established by mathematical population genetics. The contrasting positions reflect differing attitudes to what makes natural selection a scientific theory: whether extensive observation of local processes following Darwin's methods is or is not a sufficient basis for a general theory; and whether quantitative relationships demonstrated in large aggregates nonetheless necessitate assumptions about genetic and other processes that require documentation at local levels.[12] The relative priority of the two approaches depends, of course, on the diverse nature and differing scale of the problems at hand. Studies of genetic and of ecological inheritance, for example, address different aspects of natural selection, and face different limitations of available evidence that require different modelling assumptions. The extent and way in which they rely on both concepts of population may differ accordingly, and a number of chapters in this book (Odling-Smee, Lawson, Bodmer and Winney, Calafell and Comas, Elton and Dunn, and Spencer) provide the reader with examples of several approaches possible.

1.2.3 Compositional Demography

By the mid-twentieth century, anyone looking to further the conceptual development of population thinking was in a position to draw on four remarkable legacies. First, Darwin had freed open population thinking of the negative political associations of the *ancien régime*. The general importance of intra- and interpopulation networks and relationships now extended far beyond the parochial interests of oligarchies and other political elites: they were fundamental elements of the natural as well as social world.

[12] Different approaches to the adequacy of the evolutionary synthesis as a scientific framework are reviewed by Bowler (1984: 327–334).

Second, statistics had meanwhile transformed and generalized population arithmetic as a descriptive and formal system of closed population analysis. Third, the limitations of the overly ambitious deterministic expectations of social physics and eugenics had been demonstrated in the critique of Quetelet and the alarming consequences of eugenics under National Socialism. Finally, the evolutionary synthesis had shown that both population concepts are necessary to understanding population heterogeneity and change, and could play complementary roles at a number of levels of analysis.

Unfortunately, the immediate context of post-war demographic research was not conducive to drawing on all of these legacies. Mayr's (1959, 1982) careful clarification of Darwinian population thinking as extended by the evolutionary synthesis only began to appear after the main strands of post-war demographic approaches had come to be established.[13] Most demographers were, in any case, unaware of developments in evolutionary biology, the relevance of which remained for them under the cloud of eugenics. Moreover, demography as fostered by the nineteenth-century statistical movement was first and foremost a field of macro-level population statistics. As such, it was bound to have other pressing priorities in the post-war era. The prospect of rapid population growth in the developing world, with its implications for health, resources, and political stability in a Cold War context, had tremendous impetus in giving demography a global and institutional remit, both to help identify fertility and health problems, and to help design policies addressing them. This impetus was directed by the idea of 'demographic transition' enunciated by Notestein (1945), which became the prevailing framework for social population research over the remaining half of the twentieth century.

The approach to transition was one of which nineteenth-century determinists, like Quetelet, would have approved. The methodology of statistical determinism was revived and applied to the transition in a way that privileged European macroeconomic and social norms as determinants of fertility and mortality declines for the whole world (Szreter 1993). The historical and cultural relationships of local and national subpopulations, their formation and reformation, were displaced by the standard gradient of economic development and modernization theory. Widely influential syntheses (Hauser and Duncan 1959) characterized demography as allied intellectually and methodologically to physics; in this context, population thinking before 1800 was of interest only to the extent that it anticipated nineteenth- and twentieth-century statistics (Lorimer 1959); developments in evolutionary theory were not considered an important intellectual resource for population theory and research (Kreager 2009); and the importance of biology under the transition was confined to issues of public health and contraceptive technology, while social science research became subservient to their promotion (Demeny 1983). The priority of macro-level economic and population relationships in Notestein's framework pushed key aspects of population variation and change to the margins, and these were again those dimensions of population thought and

[13] Major critiques of the nineteenth and twentieth-century ideal of physics as a model for biological and social inquiry also took place in this period (e.g. Canguilhem 1977), but were not remarked by demographers and social scientists at the time.

phenomena that had proven impossible to address in the deterministic programme of the nineteenth-century statistical movement: intra- and interpopulation dynamics of family and kin systems, of ethnic and community organizations, of migration, and of the unequal distribution of status and power. A large critical literature surrounds the premises and development of Notestein's original proposal and problems it subsequently encountered (e.g. McNicoll 1992; Szreter 1996; Greenhalgh 1996).

As is well known, comparative historical and contemporary research on demographic transitions confounded the macro-synthesis Notestein had put forward. European transitions were remarkably diverse, and this diversity did not correlate strongly or uniformly with standard indices of economic growth or with most expected modernization variables; similar findings were soon confirmed in the developing world (Chesnais 1992). The transition framework, however, led helpfully to increasing documentation of local and regional variation in European history and elsewhere. This proved to be immensely fruitful in stimulating methodological innovations, for example, in the study of the heterogeneity of pre- and early transitional fertility, and more recently in analysing demographic change across generations. By the 1970s and 1980s, widely read and influential syntheses pointed to phenomena bypassed in Notestein's early transition theory as at least potentially of greater explanatory value (Anderson 1986; Cleland and Wilson 1987; Watkins 1987). These again called attention to phenomena that had been central to the classical model and were marginalized during the nineteenth-century statistical movement.

Although the significance of the classical model was unrecognized at the time, this questioning of demographic transition as a scientific theory and attempts to revise it have led population studies increasingly to rethink the influence that local communities and networks of communication play in demographic change (see Szreter, and Kohler et al., both this volume). Historians have for some time questioned the idea that modern economic and social development is a universal trajectory to be followed by all nations, noting that distinctive environmental and cultural constraints existing prior to industrialization have continued to shape population and resource development (Wrigley 1972). The inability of standard socio-economic classification schemes to distinguish the reproductive patterns of distinctive regional, socio-economic, and cultural subpopulations was documented (Szreter 1996). Continuing variation of vital trends in regional, local, cultural, and other subpopulations have suggested that central measures, like total fertility rates of national populations, are composites that disguise significantly different conjunctures (Garrett et al. 2001; Lesthaeghe and Niedert 2006). The role of internal and external migration as normal adaptations by local population groups was recognized as a fundamental and neglected issue (see Hugo, this volume). The fact that members of different groups in society find their fertility constrained by factors that reflect local norms of childrearing, health, family, and gender relations, as well as economic and educational opportunities, has led to a rich literature on the cultural variability of reproduction (e.g. Kertzer and Fricke 1997; Bledsoe 2002; Johnson-Hanks 2006), as well as to the concerns about reification noted earlier.

The need to consider fertility and mortality trends more systematically in relation to other adjustments of population composition is apparent. The renewal of populations is not simply a matter of the fertility and mortality of extant groups if those groups are changing under other compositional pressures. The diversity of demographic transitions, and the variability of birth rates after them, make understanding mechanisms that continue to renew population heterogeneity fundamental to demography. Examination of intra- and interpopulation relationships may provide key data on mechanisms of population adjustment and change analogous to study of their role as mechanisms of natural selection. The scientific status of demographic theory thus faces the same issues raised by Darwinian population thinking: both observed population processes at a local scale and testable models at higher levels of aggregation are necessary, and theoretical formulations confined to one or the other are incomplete. Not only differential recourse to contraception, but study of issues like the differing availability of intergenerational support for older people (Schröder-Butterfill, this volume), the differential incidence of chronic causes of death (Kunitz, this volume), the differential impact of environmental change and policies on different subpopulations (Kreager 2011), the uneven spread of epidemic diseases like HIV-AIDS (Gregson and Hallett, and Kohler et al., this volume), and the diversity of local reproductive regimes before transition (Kaplan et al., this volume), are all cases in point. Evolutionary biology's reliance on both closed and open population thinking at this point becomes potentially much more than a framework in which demographers may find many fruitful analogies. Social and cultural approaches to population can make potentially crucial contributions to evolutionary theory, and identify evolutionary factors that continue to differentiate subgroups in society (Kendal et al. 2011).

1.3 Conclusion

One advantage of taking a long-term historical perspective is that it readily throws up different ways of thinking about things that may now be seen as obvious and settled. Population, understood as the aggregate properties of human groups, usually coupled with ideas about how they are regulated, has been fundamental to European conceptualization of nature and society since the Greeks. Since the nineteenth century, the main approach to population has appeared blindingly obvious: count! (And then apply whatever numerical and algebraic methods are appropriate to identifying quantitative traits, trends, and differentials.) As we have seen, this approach was not obvious or favoured during the great majority of European history, in which direct observation of patterns of association within and between local populations was the priority. Despite the great progress of quantification and modelling, our modern number-minded attitude to population remains profoundly influenced by the legacy of population thought in preceding centuries: people evidently go on experiencing, planning, and adapting to change in terms of the sets of social relationships in which they live, and the subpopulation memberships and networks by which they recognize and alter these relationships. The way groups are built up and varied by such associations is fundamental to natural and social adaptation. Application of increasingly complex quantitative methods and models to society and nature during the last two centuries has helped to transform the technical and material capacities of society, and thus the wider contexts in which these associations are lived. It has not, however, been able to resolve the issues of human government and welfare that gave rise to people's open population behaviour and thinking long before that.

It is possible that we have reached an important stage in the long history of the two concepts of population. Renewed recognition of the usefulness of combining open and closed approaches to population heterogeneity and change is a major legacy of Darwinian population thinking. It does not, however, seem any more advisable to try to reinvent Darwin as an infallible guide to population for demographers than it was for the social sciences to chase Quetelet's dream of a social physics. Darwin's own reliance on ideas basic to open population thinking, as well as the emergence of mathematical population models in the seventeenth and eighteenth centuries, owed much to the earlier classical model—revealing that the two main conceptual orientations to population are rooted in deeper sources of European culture. Both open and closed population reasoning are fundamental to explanation, and how we decide their relation to each other is bound to shape the evidence we collect on any given problem and the models with which we analyse it. The proper relation between the two is an open question that needs discussion. Their respective roles are likely to vary with the specific substantive nature of each problem we are trying to solve in the human sciences.

References

Anderson, B. (1986). 'Regional and Cultural Factors in the Decline of Marital Fertility in Western Europe', in A. J. Coale and S. C. Watkins (eds), *The Decline of Fertility in Europe*. Princeton: Princeton University Press, 293–313.

Anderson, M. J. (1988). *The American Census: A Social History*. New Haven: Yale University Press.

Aristotle (1995). *Politics: Books I and II*, trans. and commentary T. J. Saunders. Oxford: Clarendon Press.

Aristotle (1997). *Politics: Books VII and VIII*, trans. and commentary R. Kraut. Oxford: Clarendon Press.

Bacon, Francis ([1612] 1931). *Essays*, ed. A. S. West. Cambridge: Cambridge University Press.

Bledsoe, C. (2002). *Contingent Lives*. Chicago: Chicago University Press.

Bowler, P. J. (1984). *Evolution: The History of an Idea*. Berkeley: University of California Press.

Bourguet, M.-N. (1989). *Déchiffrer la France*. Paris: Editions des Archives Contemporaines.

Canguilhem, G. (1977). *Idéologie et Rationalité dans l'historire des sciences de la vie*. Paris: Librairie J. Vrin.

Charbit, Y. (2010). 'History and Utopia: The Platonic City', in *The Classical Foundations of Population Thought: From Plato to Quesnay*. Dordrecht: Springer, 13–42.

Chesnais, J.-C. (1992). *The Demographic Transition*. Oxford: Oxford University Press.

Cleland, J. and C. Wilson (1987). 'Demand Theories of Fertility Transition: An Iconoclastic View', *Population Studies* 41/1: 5–30.

Coale, A. J. (1969). 'The Decline of Fertility in Europe from the French Revolution to World War II', in S. J. Behrmann, L. Corsa, and R. Freedman (eds), *Fertility and Family Planning*. Ann Arbor: University of Michigan Press, 3–24.

Darwin, C. (1859). *On the Origin of Species*. London: Murray.

Congrès Général de Statistique (1872). *Compte Rendu Général*. St Petersburg: Congrès Général de Statistique.

Demeny, P. (1988). 'Social Science and Population Policy', *Population and Development Review* 14/3: 451–479.

Desrosières, A. (1993). *La Politique des Grands Nombres*. Paris: Editions la Découverte.

Dupâquier, J. and M. Dupâquier (1985). *Histoire de la Démographie*. Paris: Perrin.

Durkheim, E. (1897). *Le Suicide: Etude de Sociologie*. Paris: Alcan.

Eyler, J. (1979). *Victorian Social Medicine: The Ideas of William Farr*. Baltimore: Johns Hopkins University Press.

Farr, W. (1837). 'Vital Statistics, or the Statistics of Health, Sickness, Disease and Death', in J. R. M'Culloch, (ed.), *A Statistical Account of the British Empire*. London: C. Knight & Co., ii.567–601.

Fisher, R. A. (1930). *The Genetical Theory of Natural Selection*. Oxford: Clarendon Press.

Garber, D. and D. Zabell (1979). 'On the Emergence of Probability', *Archives of the History of Experimental Science* 21: 35–53.

Garrett, E., A. Reid, K. Schürer, and S. Szreter (2001). *Changing Family Size in England and Wales, 1891–1911*. Cambridge: Cambridge University Press.

Gehrman, R. (2009). 'German Census-Taking before 1871', Working Paper 2009-023, Max Planck Institute for Demographic Research.

Glass, D.V. (1973). *Numbering the People*. Farnborough: Saxon House.

Gould, S. J. (2002). *The Structure of Evolutionary Theory*. Cambridge, MA: Belknap Press.

Graunt, J. ([1662] 1899). *Natural and Political Observations on the Bills of Mortality*, in C. H. Hull (ed.), *The Economic Writings of Sir William Petty*. Cambridge: Cambridge University Press, ii.314–436.

Greenhalgh, S. (1996). 'The Social Construction of Population Science: An Intellectual, Institutional, and Political History of Twentieth-Century Demography', *Comparative Studies in Society and History* 38: 26–66.

Hacking, I. (1975). *The Emergence of Probability*. Cambridge: Cambridge University Press.

Hacking, I. (1991). *The Taming of Chance*. Cambridge: Cambridge University Press.

Hald, A. (1990). *A History of Probability and Statistics and their Applications before 1750*. New York: John Wiley and Sons.

Hamilton, A., J. Jay, and J. Madison ([1788]1961). *The Federalist*, ed. J. E. Cooke. Middletown, CT: Wesleyan University Press.

Hauser, P. M. and O. D. Duncan (1959). 'Overview and Conclusions' and 'Part I. Demography as a Science', in P. M. Hauser and O. D. Duncan (eds), *The Study of Population: An Inventory and Appraisal*. Chicago: University of Chicago Press, 1–117.

Hecht, J. (1980). 'L'idée de dénombrement jusquà la Révolution', in *Pour une Histoire de la Statistique*. Paris: INED, 21–82.

Innes, J. (2009). 'Power and Happiness: Empirical Social Enquiry in Britain, from "Political Arithmetic" to "Moral Statistics"', in J. Innes, *Inferior Politics: Social Problems and Social Policies in Eighteenth-Century Britain*. Oxford: Oxford University Press, 109–178.

Johnson-Hanks, J. (2006). *Uncertain Honor: Modern Motherhood in an African Crisis.* Chicago: University of Chicago Press.

Kendal, J., J. J. Tehrani, and J. Odling-Smee (2011). 'Human Niche Construction in Interdisciplinary Focus', *Philosophical Transactions of the Royal Society B: Biological Sciences* 366/1566: 785–792.

Kertzer, D. I. and T. Fricke (eds) (1997). *Anthropological Demography.* Chicago: University of Chicago Press.

Kreager, P. (1988). 'New Light on Graunt', *Population Studies* 42/1: 129–140.

Kreager, P. (1992). 'Quand une population est-elle une nation? Quand une nation est-elle un État? La démographie en l'émergence d'un dilemme moderne, 1770–1870', *Population* 47/6: 1639–1656.

Kreager, P. (1997). 'Population and Identity', in D. Kertzer and T. Fricke (eds), *Anthropological Demography: Towards a New Synthesis.* Chicago: University of Chicago Press, 139–174.

Kreager, P. (2002). 'Death and Method: The Rhetorical Space of Seventeenth-Century Vital Measurement', in E. Magnello and A. Hardy (eds), *The Road to Medical Statistics.* Amsterdam: Rodopi, 1–36.

Kreager, P. (2005). 'John Graunt', *Encyclopaedia of Social Measurement*, ed. K. Kempf-Leonard. San Diego: Elsevier, ii.161–166.

Kreager, P. (2008). 'Aristotle and Open Population Thinking', *Population and Development Review* 34/4: 599–629.

Kreager, P. (2009). 'Darwin and Lotka: Two Concepts of Population', *Demographic Research* 21/16: 469–502.

Kreager, P. (2011). 'The Challenge of Compositional Demography', *Asian Population Studies* 7/3: 85–88.

Krüger, L., L. Daston, and M. Heidelberger (eds) (1987). *The Probabilistic Revolution*, 2 vols. Cambridge, MA: MIT Press.

Lasswell, H. D. and M. S. McDougal (1943). 'Legal Education and Public Policy', *Yale Law Journal* 522: 203–295.

Lesthaeghe, R. and L. Niedert (2006). 'The Second Demographic Transition in the United States: Exception or Textbook Example?' *Population and Development Review* 32/4: 669–698.

Lewontin, R. (2003). 'Introduction: The Scientific Work of Theodore Dobzhansky', in R. C. Lewontin et al. (eds), *Dobzhansky's Genetics of Natural Populations: I–XLIII.* New York: Columbia University Press, 93–115.

Lorimer, F. (1959). 'The Development of Demography', in P. M. Hauser and O. D. Duncan (eds), *The Study of Population: An Inventory and Appraisal.* Chicago: University of Chicago Press, 124–179.

Lotka, A. J. (1939). *Théorie Analytique des Associations Biologiques*, ii. Paris: Herman et Cie.

McNicoll, G. (1992). 'The Agenda of Population Studies: A Commentary and Complaint', *Population and Development Review* 18/1: 399–420.

Machiavelli, N. ([1517] 1950). *The Discourses of Niccolo Machiavelli*, 2 vols, ed. and trans. L. J. Walker. London: Routledge and Kegan Paul.

Mayr, E. ([1959] 1976). 'Typological versus Population Thinking', in *Evolution and the Diversity of Life: Selected Essays.* Cambridge, MA: Belknap Press, 26–29.

Mayr, E. (1982). *The Growth of Biological Thought.* Cambridge, MA: Belknap Press.

Mayr, E. and W. B. Provine (eds) (1998). *The Evolutionary Synthesis: Perspectives on the Unification of Biology.* Cambridge, MA: Harvard University Press.

Notestein, F. (1945). 'Population: The Long View', in T. W. Schulz (ed.), *Food for the World.* Chicago: Chicago University Press, 36–57.

Perrot, J.-C. (1984). 'The Golden Age of Regional Statistics', in J.-C. Perrot and N. Woolf, *State and Statistics in France, 1789–1815*. London: Harwood Academic, 81–170.
Pole, J. R. (1966). *Political Representation in England and the Origins of the American Republic*. London: Macmillan.
Porter, T. M. (1986). *The Rise of Statistical Thinking, 1820–1900*. Princeton: Princeton University Press.
Porter, T. M. (1987). 'Lawless Society: Social Science and the Reinterpretation of Statistics in Germany, 1850–1880', in L. Krüger, L. Daston, and M. Heidelberger (eds), *The Probabilistic Revolution: Ideas in History*. Cambridge, MA: Bradford Books/MIT Press, i.351–375.
Price, R. (1792) *Observations on Reversionary Payments*, 5th edn, 2 vols. London.
Provine, W. B. (1971). *The Origins of Theoretical Population Genetics*. Chicago: University of Chicago Press.
Provine, W. B. (1998). 'Genetics', in E. Mayr and W. B. Provine (eds), *The Evolutionary Synthesis: Perspectives on the Unification of Biology*, 2nd edn. Cambridge, MA: Harvard University Press, 51–58.
Quetelet, A. (1869). *Physique Sociale*, 2 vols. Brussels: C. Muquardt.
Schweber, S. S. (1980). 'Darwin and the Political Economists: Divergence of Character', *Journal of the History of Biology* 13: 195–289.
Schweber, S. S. (1985). 'The Wider British Context in Darwin's Theorizing', in D. Kohn (ed.), *The Darwinian Heritage*. Princeton: Princeton University Press, 35–69.
Skinner, Q. (1980). *The Foundations of Modern Political Thought*. Cambridge: Cambridge University Press.
Smith, A. (1776). *An Inquiry into the Nature and Causes of the Wealth of Nations*. London: Strahan and Cadell.
Sneath, P. H. A. (1964). 'Mathematics and Classification, from Adanson to the Present', in G. H. M. Lawrence (ed.), *Adanson: The Bicentennial of Michel Adanson's 'Familles des plantes'*. Pittsburg: Carnegie Institute of Technology, 471–498.
Stevens, P. F. (1994). *The Development of Biological Systematics*. New York: Columbia University Press.
Szreter, S. (1993). 'The Idea of Demographic Transition and the Study of Fertility Change', *Population and Development Review* 19/4: 659–702.
Szreter, S. (1996). *Fertility, Class and Gender in Britain, 1860–1940*. Cambridge: Cambridge University Press.
Süssmilch, J. P. (1741). *Die Göttliche Ordnung*. Berlin: Realschule; ed. and trans. J. Hecht (1979), *L'Ordre Divin*. Paris: INED.
Watkins, S. C. (1987). 'The Fertility Transition: Europe and the Third World Compared', *Sociological Forum* 24: 645–673.
Westergaard, H. (1932). *Contributions to the History of Statistics*. London: King and Son.
Wilson, C. and J. Oeppen (2003). 'On Reification in Demography', in J. Fleischhacker, H. de Gans, and R. K. Burch (eds), *Populations, Projections, Politics: Critical and Historical Essays on Early Twentieth Century Population Forecasting*. Amsterdam: Rozenberg, 113–130.
Wrigley, E. A. (1972). 'The Process of Modernisation and the Industrial Revolution in England', *Journal of Interdisciplinary History* 3: 225–259.

CHAPTER 2

POPULATION GENETICS

the study of the genetic structure of human populations

WALTER BODMER AND BRUCE WINNEY

2.1 Introduction

The aim of this chapter is to introduce the concepts of population genetics and the way they can be interpreted in a population sense, followed by a description of some of the latest work that has been undertaken on a study of the population of the British Isles.

Human population genetics really began with the discovery by Karl Landsteiner of the *ABO* blood groups in 1900. These basically consist of four simple types of blood group, *A, B, AB*, and *O. A* and *B* were initially identified by simply combining blood from different people, with the observation that when blood from different groups is combined, the red cells agglutinate, forming large visible clumps. Although Landsteiner did that work in 1900, the year of the rediscovery of Mendel's work, it was not until ten years later in 1910 that von Dungern and Hirszfeld interpreted the *ABO* blood groups in terms of Mendelian simply inherited variations. However, at that time it was assumed that the *A* and *B* groups were determined by separate genes, which later turned out not to be correct. The *ABO* blood groups were thus amongst the first Mendelian inherited variations to be studied in human populations. It was the Hirszfelds who realized that this could be applied to human population studies and soon after the First World War, in 1919, published a paper in *The Lancet* showing the frequencies of *A* and *B* in different populations. They pointed out that there was a huge variation in the frequencies of the *A* and the *B* types between major population groups. Thus, the frequency of *B* as compared to *A* was much higher in Asian than in European populations. Indeed the Hirszfelds proposed that humanity was divided into the *A*s and the *B*s and hybrids between them. Their definitions of populations were broad (e.g. English, Greek, Turk, Indian) and largely based on geography, but with an obvious cultural component, especially when it came to Jews. This is obviously a very much oversimplified view of the composition of human populations. Nevertheless, the Hirszfelds' work was the first application of polymorphic variation to human population studies, namely finding different versions of genes at different frequencies in different populations.

Later detailed studies have largely confirmed what the Hirzfelds originally found. Figure 2.1 (from Bodmer and Cavalli-Sforza 1976) is a representation of the *B* type throughout the world, excluding the Americas. The shadings (darker for higher frequency) represent different frequencies interpreted in terms of a continuous model, namely no discrete sharp borders between populations. Some places have relatively high *B* blood group frequencies, for example in central and southern Asia, but there is substantial variation in the *B* frequencies elsewhere, especially in Africa and Europe.

Watkin's (1956) research on *ABO* frequencies in Wales is an example of a more detailed local study. Morgan Watkin was very keen on the proper definition of the population he was studying, and so was careful to sample only people whom he could say were of a certain origin both by place of birth and by surname. By that time, many people had been typed for *ABO* because of the importance of blood transfusion, which developed in the UK especially during the Second World War. His results (Figure 2.2) show that there is

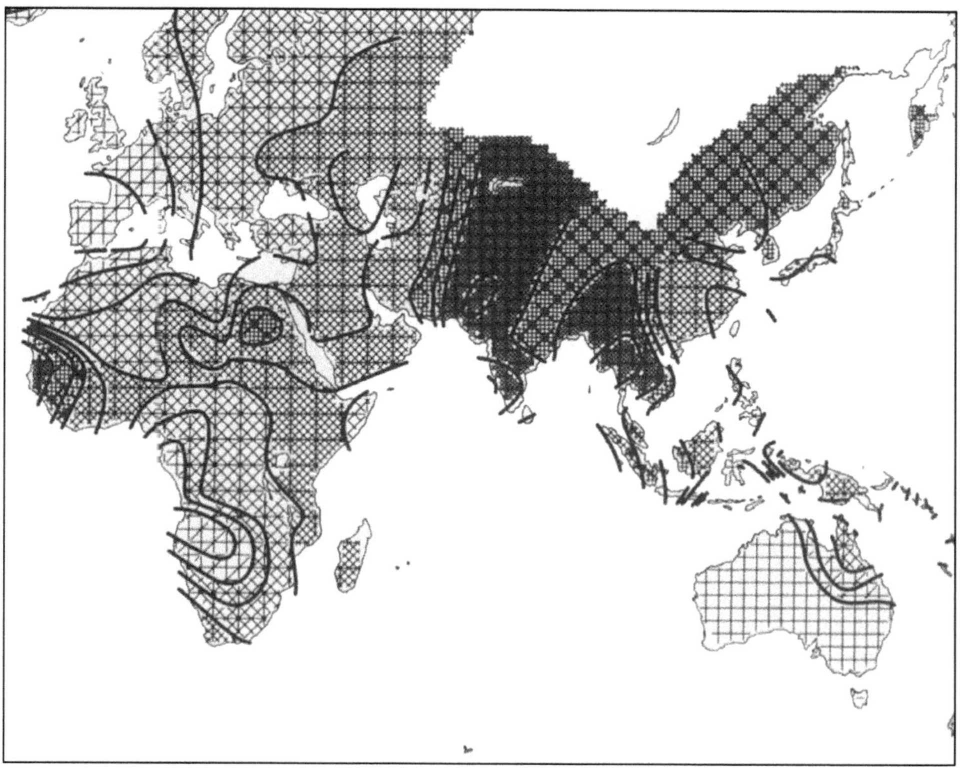

FIG. 2.1 Worldwide (excluding the Americas) *B* blood group frequencies, from Bodmer and Cavalli-Sforza (1976). The shadings are different frequencies interpreted in terms of a continuous model, with the darker the colour, the higher the frequency. Whilst some places have high frequencies, there is also a substantial amount of variation.

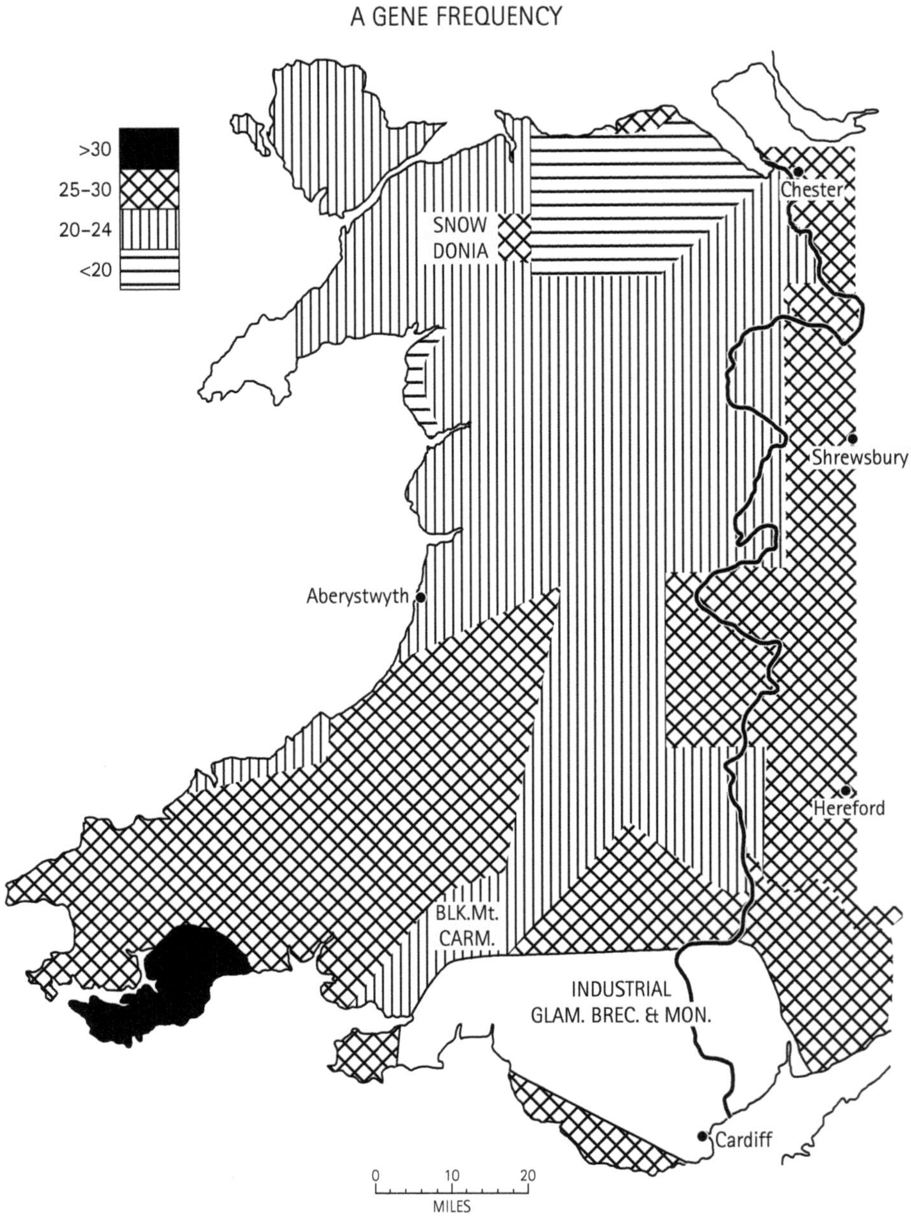

FIG. 2.2 Distribution of *A* blood group frequencies in Wales, from Watkin (1956), with the highest frequency observed in south Pembrokeshire.

a small pocket of people with a very high frequency of type *A* in south Pembrokeshire. This represents what is often called 'Little England', beyond Wales. Flemish farmers were settled there in the twelfth century by King Henry II, and they presumably had a different frequency of type *A* than the local Welsh population. Because these settlers more or less isolated themselves from the rest of the surrounding population, this difference in *A* frequency has been maintained so that a little residue of history is still reflected in the *A* gene frequency today. So, with careful population sampling, we can begin to say something about the relationship between the genetics and the history of a population. The subsequent discovery of many other polymorphic markers—initially further blood groups, and then enzyme variants detected by starch gel electrophoresis—made the notion of humanity divided into just *A*s and *B*s totally unsustainable, and led to the development of approaches to the genetic characterization of human and other populations using combinations of large numbers of polymorphic markers.

2.2 Some Theory

Before going much further, it is necessary to describe some of the basic mathematical ideas involved in population genetics. A very important part of what underlies population genetics and the way we interpret populations comes from a very simple binomial theorem, known as the Hardy–Weinberg law. The result of the Hardy–Weinberg (HW) law was extremely important in the early days of the interface between Mendelian genetics and Darwinian thought. Assume that individuals mate at random in a population and consider the distribution of two alternative versions (alleles) of a gene (labelled *m* and *n*). Then it can be shown with some simple algebra that random mating in the population with respect to the three genotypes, *mm*, *mn*, and *nn*, is equivalent to combining eggs and sperm at random. The result of this can be represented by a simple binomial formula. Thus if p is the frequency of *m* genes and q the frequency of *n* genes in the population, so that $p + q = 1$, you get the famous HW law that the frequencies of the three possible genotypes *mm*, *mn*, and *nn* can be represented as $p^2(mm) + 2pq(mn) + q^2(nn)$, where $p^2 + 2pq + q^2 = (p + q)^2 = 1$ (($p + q$) in males multiplied by ($p + q$) in females). Since in humans genes are carried in pairs, one from the mother and one from the father, the two types of homozygotes, *mm* and *nn*, arise when the versions of the gene received from the mother and father are both the same, the chance of this being p^2 and q^2 respectively. For the heterozygotes (*mn*), one parent passes on *m* and the other *n* and this can happen in two ways, hence the frequency $2pq$. This simple result is a very powerful description of the genetic composition of a population, which, in the absence of selection and ignoring chance variation, can be shown to stay the same from generation to generation. Thus, Mendelian inheritance at its simplest ensures that genetic variation is not lost from the population, a key requirement for Darwinian evolution by natural selection. Notwithstanding all the complexities of what one can do now in terms of investigating much more complex models of population behaviour, the Hardy–Weinberg law still fundamentally remains valid as a description of the genetic constitution of a human population that can easily be generalized to multiple loci and multiple alleles at each locus.

It is important to bring into this discussion the whole notion of what happens if a population is subdivided, in other words what happens if there is population structure.

Instead of thinking of the single population as panmitic (i.e. random mating throughout), the population can be thought of as structured genetically so that it comprises a series of subpopulations within each of which mating is at random, but between which there is some migration leading to gene flow between them. In actual human populations this structure is likely to correlate with geography to a considerable extent, rather than with the age–sex structure usually analysed by demographers. However, it is eventually, as we shall see, the genetics that should be used to define the subpopulations in the first instance, not the geography.

For simplicity let us assume that the subpopulations are all the same size and that there is random mating within each population, but no matings between the populations. Consider, as before, just two alleles, *m* and *n*. Then within each population the two alleles will distributed according to the Hardy–Weinberg law, but each population will be expected to have a different pair of allele frequencies.

What is the consequence of considering the overall population simply as a pool of the constituent populations, ignoring the subdivisions between them? The frequencies of alleles *m* and *n* in this pooled population will just be the average of all the frequencies in the different constituent populations, say p and q, and so, ignoring the subdivision, and assuming random mating in the population as whole with these frequencies, the expected frequencies of the homozygotes and heterozygotes would just be $p^2(mm)$, $2pq(mn)$, and $q^2(nn)$, as shown in Figure 2.3. The effect of treating the population as a pool, and averaging the allele frequencies, can be seen by comparing the genotype frequencies averaged over the separate populations with the expected frequencies for the pooled population. The pooled population has lower homozygote frequencies by an amount σ^2 and higher heterozygote frequencies by an amount $2\sigma^2$, where σ^2 is the variance of the different allele frequencies in the constituent populations. Thus, the effect

		H-W Genotype Frequencies		
Population	Gene Frequency	AA	Aa	aa
1	$p_1 = (1 - q_1)$	p_1^2	$2p_1q_1$	q_1^2
2	$p_2 = (1 - q_2)$	p_2^2	$2p_2q_2$	qp_2^2
...	...	...	...	...
k	$p_k = (1 - q_k)$	p_k^2	$2p_kq_k$	q_k^2
Pooled Population	$\bar{p} = (1 - \bar{q})$	$\bar{p}^2$	$2\bar{p}\bar{q}$	$\bar{q}^2$
Average of *k* populations		$\bar{p}^2 + \sigma^2$	$2\bar{p}\bar{q}\{1 - \sigma^2/\bar{p}\bar{q}\}$	$\bar{q}^2 + \sigma^2$

Where mean gene frequency is: $\bar{p} = \Sigma p_i/k$

And the variance of the gene frequencies: $\sigma^2 = (\Sigma p_i^2/k) - \bar{p}^2$

FIG. 2.3 Effect of population substructuring on Hardy–Weinberg equilibrium (Wahlund's formula).

of subdividing the population is to increase the relative homozygosity and decrease the relative heterozygosity (the Wahlund effect). The amount by which this is done depends on the variance of the gene frequencies divided by the product of the average gene frequencies, namely, σ^2/pq. This is a measure of the extent to which the subdivision leads to a departure from the Hardy–Weinberg law expectation on the assumption of no substructure. This quantity is essentially equivalent to assuming a level of inbreeding (mating between relatives rather than at random) in the pooled population of $F = \sigma^2/pq$ where F is called the inbreeding coefficient. The greater the variation between the subpopulations, the greater the effect of the subdivision. So, the HW law, and its generalizations to multiple loci and multiple alleles, effectively provides the basis for modelling the genetic structure of human populations.

To allow for the inevitable variation in actual data, we must use a statistical test for assessing the significance of a departure from the Hardy–Weinberg law. It turns out that there is a simple relationship between chi-square (the statistical measure of departure from what is expected) and the inbreeding coefficient F, and this relationship can be used to test the significance of the departure from Hardy–Weinberg expectation association, namely whether the departure is more than would be expected just by chance. The chi-square (χ^2) is equal to the total population size (N) times the inbreeding coefficient, namely $\chi^2 = NF$ (for one degree of freedom). We can now ask: what size of population would you need to detect a certain level of departure from the Hardy–Weinberg expectation? And the answer is that it is huge for reasonable values of F. The size of the inbreeding coefficient, F, as measured in most human populations is at most about 1–2%. To detect significantly a 1% departure from HW expectation would require a population of 40,000, while to detect a 2% departure would require a population of 10,000. What this implies is that, while the HW law has very specific assumptions, and can be used as a test for departure from random mating, it is actually extraordinarily robust to variation in its basic assumptions and it takes a large amount of data to see any effects of substructure. The extent to which the detailed structure of a given set of populations is detected depends substantially on how wide-ranging a study of different populations it is. For a quite wide-ranging study, there may not be too much concern about substructure in each constituent population. However, over a much narrower range (for example, in the UK as will be described later), the effect of population substructure becomes more important (for further background on this aspect of population genetics see e.g. Bodmer and Cavalli-Sforza 1976).

2.3 Some Early Studies

An early study on human population relationships using classical blood group and enzyme markers was done by Luca Cavalli-Sforza, a notable figure in the history of human genetics and a student of Fisher's in the late 1950s, and Anthony Edwards, another but later Fisher student. They devised the first procedure for using gene frequency data to construct a phylogeny of populations, which remains the basis for most subsequent population phylogenetic analysis. The starting point is to devise a measure of the extent to which any two populations differ in their gene frequencies for a defined set of variants, the genetic distance. Simple measures of the genetic distance between two populations are, for example, the average variance of gene frequency differences or, when gene frequencies are plotted as a single point in a multidimensional space, the Euclidian or angular distances between the populations. Cavalli-Sforza and Edwards (1963) used such distances to estimate the predicted genetic relationships amongst a set of populations in terms of a phylogenetic tree, which is an objective evolutionary tree that puts together those populations that are most similar in terms of patterns of their genetic frequencies. Figure 2.4 is the tree that they constructed based on a fairly limited amount of blood group data. It is quite remarkable that already at that time they were beginning to be able to say something at a broad level about relationships between populations using not just one marker but a range of different markers. Even these early analyses showed a remarkable agreement between the phylogenetic tree and what might be expected from historical and archaeological records. Clearly, the bigger the genetic distances, on the whole the more accurate the resulting tree, so that relationships between major ethnic groups are reasonably well defined, while those between more closely related populations, such as within the UK, require a much more sophisticated analysis.

Gradually over time more polymorphisms came to be discovered, and it is worth mentioning one that Bodmer has been involved with for many years and that is the HLA

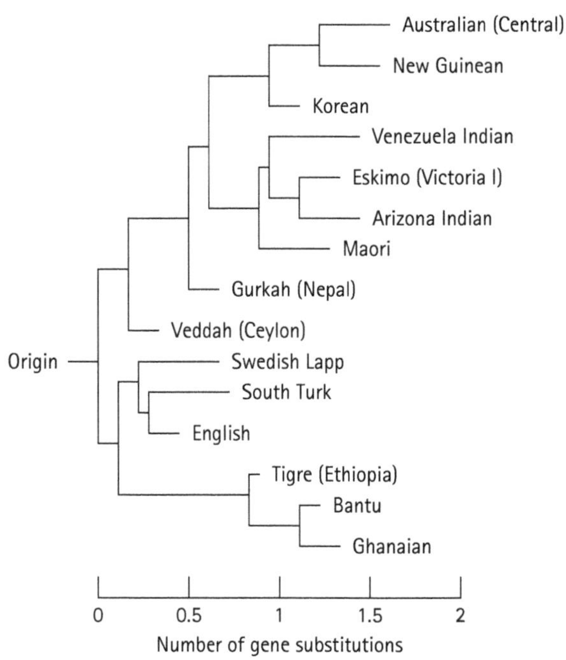

FIG. 2.4 Evolutionary tree computed from blood group frequencies, redrawn from Cavalli-Sforza and Edwards (1963).

system. The HLA system is like a very complex blood group system, where what you are detecting are different types of proteins expressed on the surface of the white blood cells and most tissues in the body. The system was discovered through the aim of finding out what the differences were between individuals that led to rejection of kidney grafts or bone marrow transplants. Originally, the HLA types were identified serologically. They can be divided essentially into six groups, each of which is now known to be determined by an individual gene. There are now thousands of alleles of these genes that can be objectively identified at the DNA level. They are closely linked, and occur together in a cluster on chromosome 6. So looking at a family makes it is possible to determine the haplotypes (the set that goes together on one chromosome) and work out what the six-fold combinations are that are inherited.

This HLA system, even in the days before direct DNA typing, provided a huge increase in the extent of variation that could be used for analysing populations. One example is the application that Luca Cavalli-Sforza and his colleagues (Menozzi et al. 1978) used to interpret patterns of gene frequency change across Europe. What they found was that, when gene frequencies of many different blood groups and enzyme differences across Europe were investigated, they seemed to follow a gradient that was in common between the different genes. That is very important because, when investigating population structure using genetics, it is no good looking at only one marker as that might have its own idiosyncrasies. Although analysis of mitochondrial DNA and *Y*-chromosome variation, each of which is a single marker, has helped to distinguish patterns of female versus male variation, especially with respect to migration, single markers have been subject to a unique evolutionary history that may reflect that specific region of the genome. So, for example, a marker might have been subject to a particular selective event that has not affected the rest of the genome. This means that it is important to look at what is in common between different markers to get an overall genome-wide view of a population's genetic structure.

Cavalli-Sforza and his colleagues pointed out that one can see a common gradient of gene frequencies that goes from somewhere in the Middle East out to the northern periphery of Europe. They suggested that this was correlated with the advance of agriculture from the Fertile Crescent. On this basis, they put forward the demic diffusion model that agriculture was not merely transmitted culturally, but that its spread involved a gradual migration of the people taking agriculture to their neighbours, intermarrying, and this process took place relatively slowly, around 1 km a year. So if agriculture started in the Fertile Crescent about 10,000 years ago, it got to the far north of the British Isles about 5,000 or 6,000 years ago.

This analysis is based on a model that Fisher put forward in 1937, the wave of advance of advantageous genes, and it is often misinterpreted. The whole idea behind it is a gradual dilution out of whatever genetic differences there might be between the originating population and where you actually end up. So by the time that the British Isles were reached, the influence of whatever existed in the original agricultural people genetically had been almost totally diluted out. The demic diffusion model is not, and never has been, a notion that the agriculturalists advanced and completely replaced the populations into which they had advanced. This is one of the many examples where, if that latter assumption is made, the effects of migration are hugely exaggerated. The development of the demic diffusion model of agriculture was a very early and telling application of the combined use of many genetic markers. There have subsequently been many discoveries using combinations of genetic markers, including most notably the notion that modern humans came out of Africa some 60,000 to 100,000 years ago. This was initially put forward on the basis of an analysis of mitochondrial DNA variation by Wilson and Cann (1992) and supported since then by many other studies.

Returning to the HLA system, it turns out to not be created in some way to frustrate transplant surgeons but to have a major function in controlling our immune response to foreign pathogens, and that is probably why it is so polymorphic. One consequence of this, which has turned out to be very important, is the association between HLA types and auto-immune diseases, where the immune system inappropriately recognizes self-determinants and this leads to significant disease. Ankylosing spondylitis, a disease where there is painful fusion of adjacent vertebrae, coeliac disease (reaction to gluten in cereals), and rheumatoid arthritis are good examples. The simple notion to test this was to see whether there are any HLA types that are more frequent in people with a given disease than in controls. For example in the early 1970s it was found that almost everyone with ankylosing spondylitis has the genetic marker *HLA-B*27* (Brewerton et al. 1973; Schlosstein et al. 1973), which in European populations occurs at a frequency somewhere between 5–10%. There are some populations, notably native northern Americans, where the frequency is much higher (Khan 1997), and they also have a higher frequency of ankylosing spondylitis. It is important to emphasize that having *HLA-B*27* still only means that your chance of getting ankylosing spondylitis is a few per cent, but this is compared to a virtually zero chance if you do not have *HLA-B*27*. This is a classical example of an inherited tendency for a disease.

The idea that what is being looked at may not be the effect of *HLA-B*27* itself but another variation at a close-by gene is important. This is because *HLA-B*27* may be a marker for the causative variation at a closely linked locus, and the two variants are held together on the same chromosome because they are closely linked, a phenomenon called 'linkage disequilibrium' (see later for an explanation of this). This linkage-disequilibrium-based explanation for how a genetic marker can indicate the presence of a causative genetic disease susceptibility variant at a closely linked locus is the fundamental underlying idea behind what are now called genome-wide association studies (GWAS). The whole idea of looking for effects of genetic variation on diseases, by comparing frequencies in cases and controls, comes from the HLA studies, and this can now be done with very large numbers of genetic markers covering the whole of the human genome. It is striking that, given the huge amount of GWAS data on common chronic diseases that has been obtained over the last 5–10 years, the HLA associations are by still by far the strongest.

The ability to detect polymorphism at the DNA level has been the most important advance in human population genetics following these earlier investigations. This was first achieved through the use of restriction enzymes that cut DNA at particular nucleotide pair sequences. Then came more and more efficient techniques for detecting single base pair differences, now called SNPs (for single nucleotide polymorphism), and finally, just round the corner, highly efficient 'next generation' DNA sequencing technologies that are beginning to be applied to population studies of whole human genomes. The DNA techniques have the huge advantage of using a common technology and source of material for the detection of all polymorphisms, allowing very high throughput automated genotyping. The ability to type large numbers of people for one million or more SNPs has generated huge bodies of data on different human populations, and stimulated a variety of approaches to the analysis of these data, both for disease association studies, and also for genetic population structure analysis (see, for example, chapters in this volume by Lawson, Calafell and Comas, and Spencer).

A point of note here is that SNP is often used inappropriately to refer to a DNA sequence variant that is presumed to be non-functional, namely selectively neutral in its effect, without evidence that the variation is actually polymorphic, namely that the less frequent variant occurs with a population frequency of, say, at least 1%. Such variants should be called SNVs, where V stands for variant rather than P for polymorphism.

2.4 The People of the British Isles Project

An example of the application of our current ability to genotype millions of SNPs, and so to be able look at population subdivision on a much finer scale than before, is our Wellcome Trust supported study on the UK population (People of the British Isles, PoBI, <www.peopleofthebritishisles.org>). From a medical point of view the aim was to provide the controls for GWAS case control studies on relatively common chronic diseases. As a fortunate by-product, however, such a study can tell us a lot about how to relate the history of British populations to their genetic fine structure.

The PoBI goal was to collect blood samples (as a source of DNA) from throughout the UK. But we did this in a very specific way in order to tell us about the original inputs into the British population and its structure before the major admixture that came most obviously with the Industrial Revolution which stimulated movement from rural areas into cities. We therefore sampled volunteers who were not closely related in rural areas. There can be an argument about how to define rural areas, but there are objective ways of doing this (Winney et al. 2012). Furthermore, we only sampled individuals whose four grandparents came from roughly the same area, to exclude recent immigrants. The whole aim of this approach to sampling was to try to get people who were, as far as possible, characteristic of the geographical area from which they came. The definition of a population is strictly by the geography, but is limited in this very important way in order to get some idea of the major inputs into the British population. The target was to collect

3,500 to 4,000 samples from throughout the UK and so far we have about 4,000 samples. In a sense, our aim has been to create a genetic map of the people of the British Isles.

It is clear that, in order to get a good picture of the distribution of genetic variation throughout the UK that reflects its earlier history, the sampling procedure and quality of the data collected are critical. We were able to collect basic demographic information (with appropriate ethical approval) from all our volunteers, including sex, age, surnames, and, very importantly, the parental and grandparental places of birth, where known, which enabled fine geographic localization. An individual was thus located at the mean place where their grandparents were born. Our sampling criteria applied, we estimate, to probably no more than 5% at most of the population in the urban areas we went to. Figure 2.5 shows the distribution of the samples collected. Scotland was represented relatively sparsely, but there are fewer people in Scotland. The project was not supported for collections in the Republic of Ireland, so Ireland is represented only by Northern Ireland.

Before discussing the genetic results, let us consider briefly a history of the peopling of the British Isles. The first modern humans who settled permanently in the British Isles came after the end of the last Ice Age, about 12,000 years ago. The development of agriculture then came about 5,000 or 6,000 years later. Although the Roman invasion started around AD 50, this probably did not involve a large number of immigrants relative to the size of the population as it was then. There were no further known major migrations into the British Isles until the Anglo-Saxon invasions that came from the Danish peninsular and north Germany around AD 400 to 500. Next came the Viking invasions from Denmark, Norway, and Sweden around AD 800 and finally the Norman conquest in 1066. These historical events clearly must have important implications for understanding the genetic make-up of the present-day British populations.

The *Y*-chromosome carries genetic variation that can tell us something about male as opposed to female migration. Figure 2.6 is a characteristic representation of frequencies where the size of the sector in each circle is proportional to the frequency of different *Y* types at that location. Of special interest is the sector labelled *R1a1*. It is found at a particularly high frequency in Norway and across northern Russia, and so indicates that Norway has contributed particularly to the Orkney population, presumably through the obvious influence of the Norse Vikings who had a kingdom centred on Orkney from about the ninth to the fifteenth century AD.

Surnames can be considered as a form of surrogate for the *Y*-chromosome and so can provide their own information about people's origins through their male lineages. We have analysed the distribution of surnames collected from our samples in collaboration with Paul Longley and his group at University College, London, who have excellent data on the distribution of surnames throughout the UK at different times. The stability of geographic distributions for some surnames is quite striking. An example of this stability is that for the surname Rendall, a very characteristic Orkney name. The distribution of Rendalls in the 1881 census compared with that in the 1998 electoral roll has hardly changed (Figure 2.7). Although you can find Rendalls all over the world, they still have their highest density in Orkney and this stability is quite characteristic of many surnames.

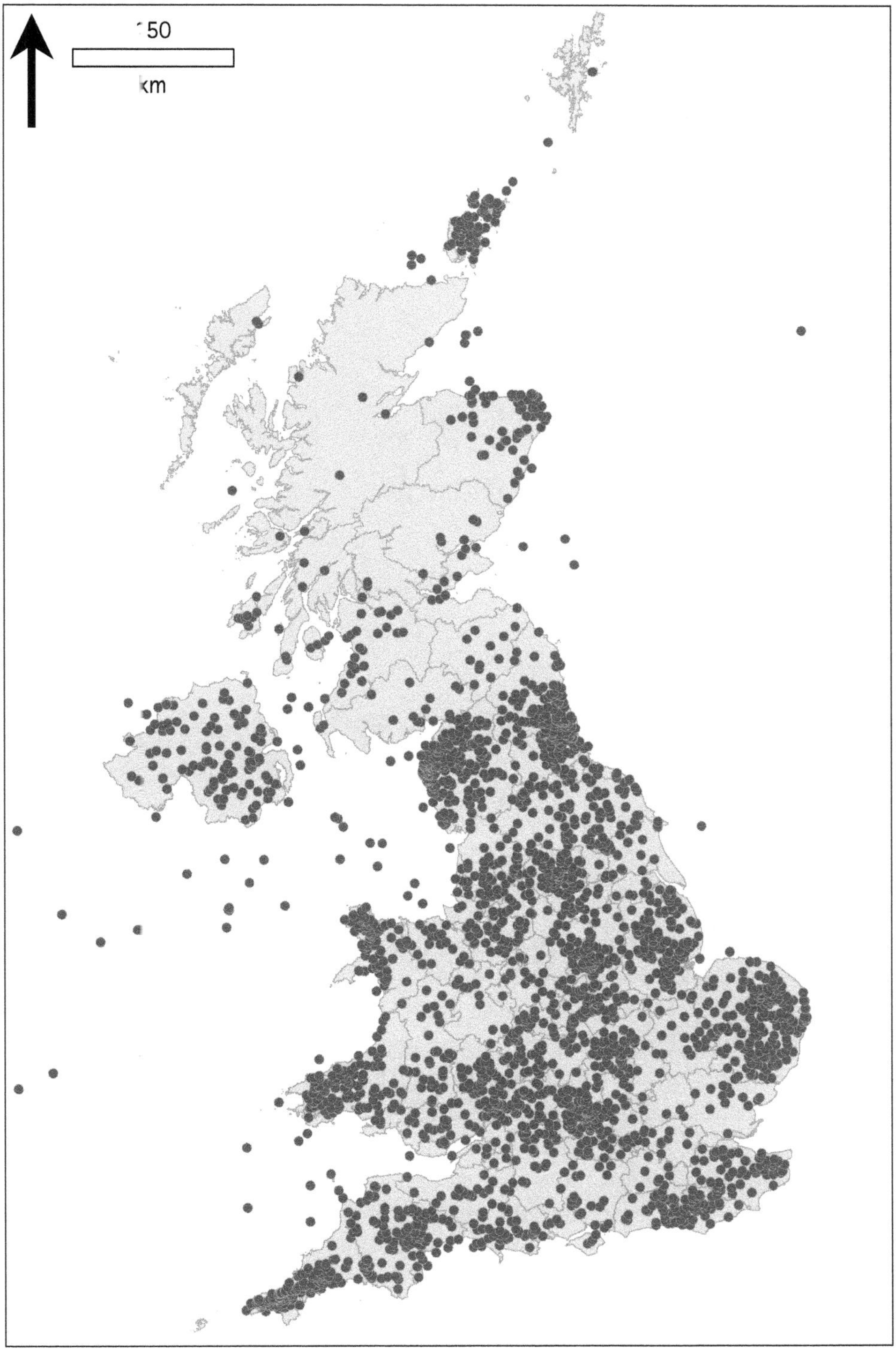

FIG. 2.5 Distribution of blood samples collected, adapted from Winney et al. (2012).

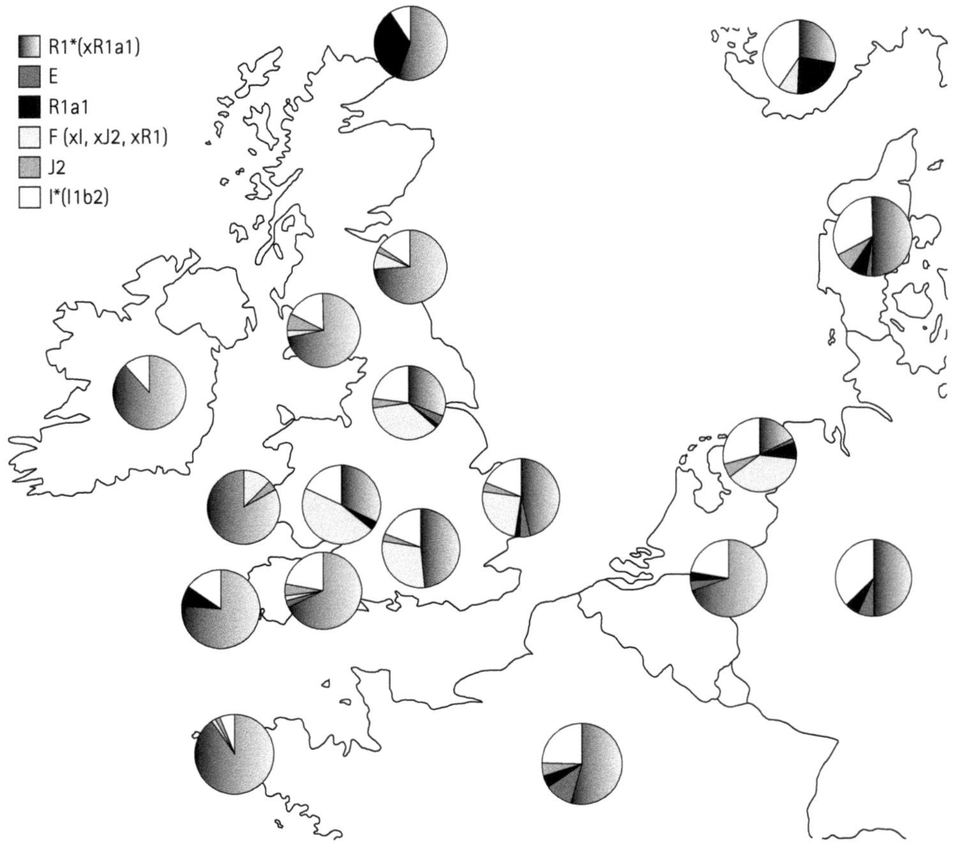

FIG. 2.6 *Y*-chromosome haplogroup frequencies across Europe. Each shade of the pie represents a different group. Data from Winney et al. (2012) and literature. Map adapted from <http://d-maps.com/m/europa/europemin/europemin10.gif>.

One way to think about the distribution of a given genetic combination in different populations is to ask, what is the relative probability that particular genetic make-up would be found in one area rather than another? One individual we looked at, for example, was much more likely overall to have an ancient British than a Norse origin, and yet he had the Norse *R1a1* *Y*-chromosome type. This highlights the sort of mixed background that can be found in a single individual.

As already mentioned, using a large number of genetic markers spread throughout the genome gives much more information about population origins than using just one or a few markers, such as the mitochondrial and *Y*-chromosome, at a time. Using the DNA-based techniques now available, as well as 500,000 or one million markers, means that there may be several markers quite close together in any one region on a chromosome. These markers, as was discussed when describing the HLA and disease associations, tend to stay together on the same chromosome through the phenomenon of linkage disequilibrium. Figure 2.8 illustrates this for two variants, one at the *HLA-A* and the other at the *HLA-B* locus, respectively, *A*01* and *B*08*.

FIG. 2.7 Distribution of 'Rendall' in 1881 and 1998 from <gbnames.publicprofiler.org>. The depth of shade represents the relative density of the surname.

FIG. 2.7 Continued.

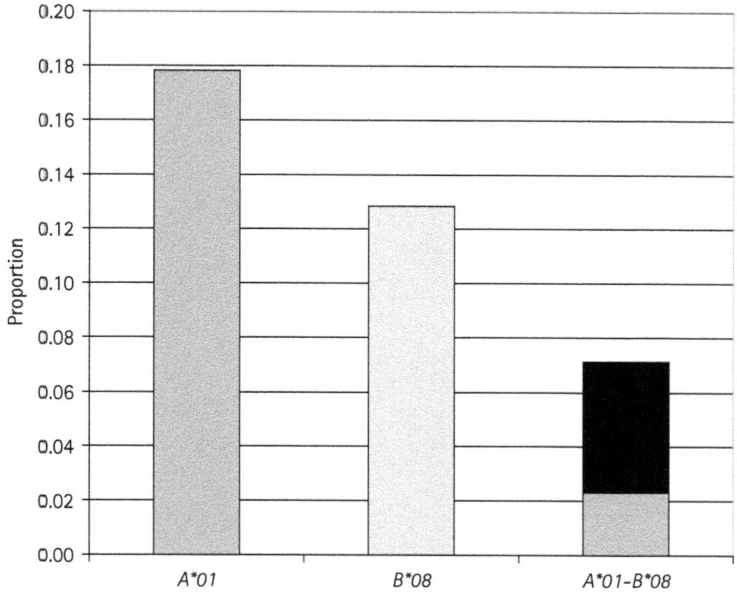

FIG. 2.8 Linkage disequilibrium (LD). The frequencies of *A*01* and *B*08* in the population are shown, along with the frequency of *A*01-B*08* haplotypes. The light shading in *A*01-B*08* is the expected frequency if *A*01* and *B*08* are independent (namely the product of the separate frequencies of *A*01* and *B*08*), whilst the dark shading is the contribution caused by LD. Data from Winney et al. (2012).

Thus the frequency in the UK population of the combination of *A*01* and *B*08* on the same chromosome, which is called the *A*01-B*08* haplotype, is much more than the frequency that would be expected if they were independent of each other, which is the product of the individual frequencies of *A*01* and *B*08*. The difference between the observed and the expected frequencies is a measure of the extent of linkage disequilibrium between these two markers in the UK population. It has been shown, initially using data on HLA frequencies, that such haplotype frequencies can give a much better resolution of differences between populations than just using each individual marker frequency on its own and ignoring the extent of linkage disequilibrium between them.

About 3,000 of the PoBI samples have recently been typed for about a million SNPs, of which 2,500 have been subject to a very detailed genetic analysis. The approach has been to cluster people according to how similar they are genetically, using the method described elsewhere in this volume by Dan Lawson, which takes into account the information provided by linkage disequilibrium. Individuals from each of the derived clusters were placed on the map of the British Isles entirely based on their genetic similarities without any reference to where the samples were collected, as will be described. The FineSTRUCTURE analysis initially gave rise to 54 clusters, not all of which were meaningful. These clusters were then grouped hierarchically by combining the two most similar clusters so that there were at first only

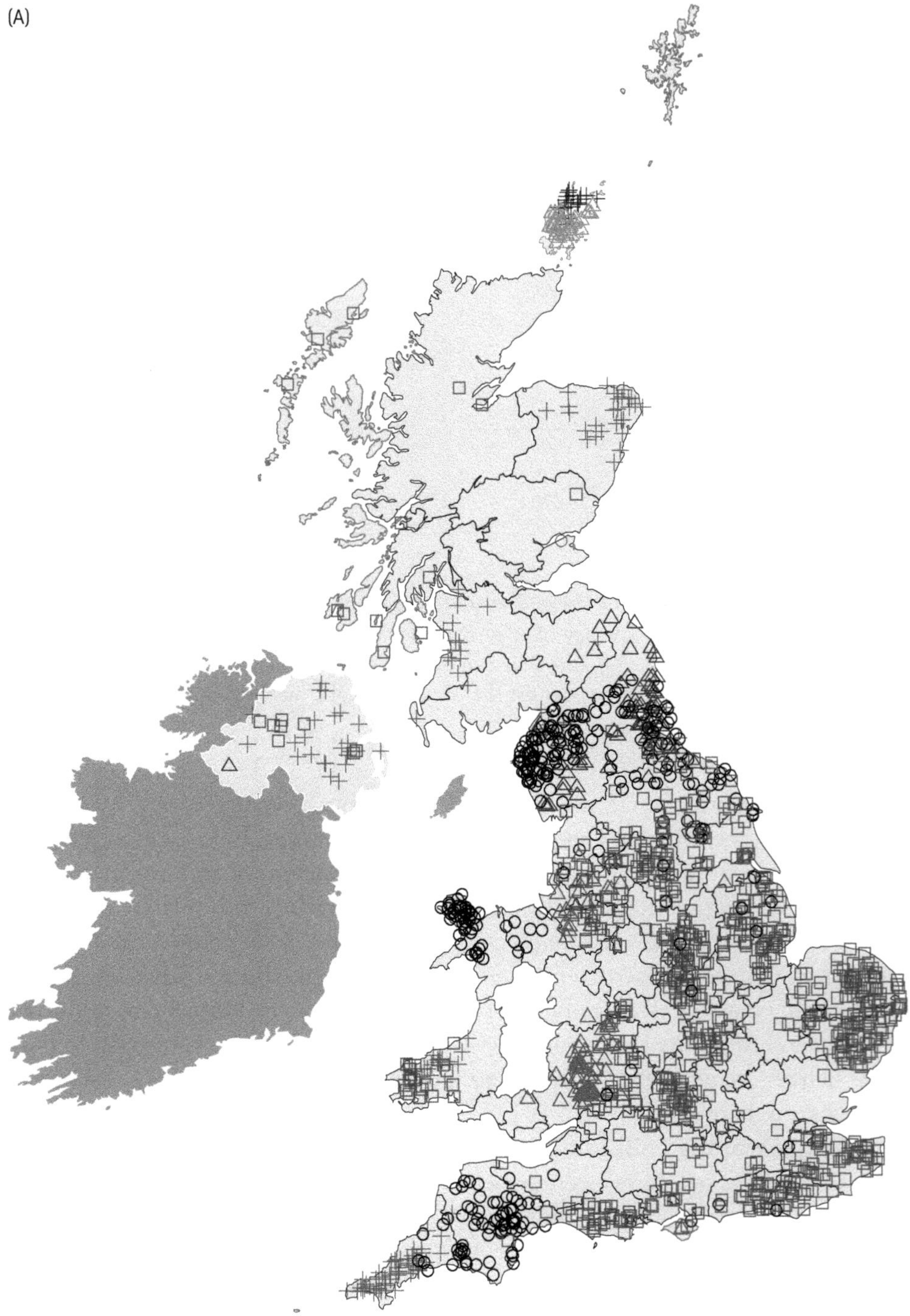

FIG. 2.9 UK map showing the fineSTRUCTURE clustering of the PoBI samples. (a–c) Each sample is plotted at the centroid of its grandparental places of birth. Shade and plotting symbol indicate the inferred cluster. (a) At this level in the hierarchical tree of cluster relationships, the geographic distribution of 17 clusters is shown. (b) Detail of the two clusters observed in Orkney. (c) Detail of Wales and the south-west. For colour maps, see Leslie et al. (in press).

(B)

FIG. 2.9 Continued.

53 clusters. This was repeated until there were only 17 clusters. These clusters, it must again be emphasized, were constructed entirely independently of any knowledge of where each sample came from. Then each member of each of these groups was plotted on a map of the UK in the position from where they originated, as defined by the mid position of the grandparental places of birth, with a quite striking result (Figure 2.9a). The groups defined by genetics were colour- and shape-coded and when plotted on a map, individuals from the same group came together in a most extraordinary way. The genetics thus fits with the geography, although they were obtained totally independently. At this level of resolution there are intriguing differences, for example, between Cornwall and Devon. Further distinct groups are: Orkney; northern England and Scotland; North and South Wales; the Welsh borders; central and southern England. The Cornish group is very tight and the difference between Devon and Cornwall is quite unexpected. This is a much greater difference than has ever been described before. This fine analysis of the UK genetic population structure was possible because of a combination of careful attention to the way the samples were collected and genotyped, together with the FineSTRUCTURE approach to the clustering that took into account the linkage disequilibrium between closely linked markers.

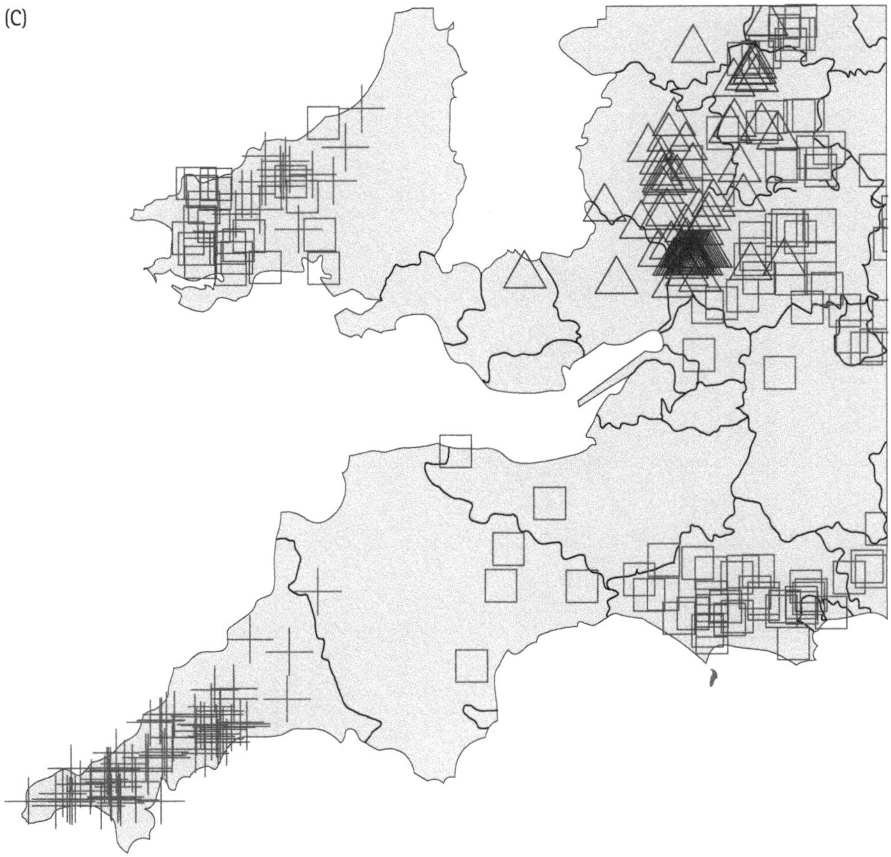

FIG. 2.9 Continued.

Figures 2.9b and 2.9c show parts of the map in more detail. Figure 2.9b shows the separation of Orkney into two groups. As one moves down the hierarchical tree, the northern England/Scotland population is separated into at least three components. Two obvious Scottish groups appear, which clearly overlap with Ireland. There is another Scottish group in Aberdeenshire and several groups throughout northern England. Figure 2.9c shows in more detail the extraordinary resolution of the analysis and its ability to define different geographically cohesive groups. The fine level of resolution of the genetically defined population groups was quite unexpected and shows what can be done with careful sampling, using many markers and an appropriate method of analysis.

The plot of Anglo-Saxon settlements shown in Figure 2.10 clearly suggests that the large southern and central England group may represent the result of Anglo-Saxon invasions. However, a steady earlier influx to the British Isles from modern-day Denmark, the Netherlands, and northern Germany, even extending to Doggerland, which formed a land bridge between Britain and the European mainland after the last Ice Age about 10,000 years ago, is likely to have been an important influence

FIG. 2.10 Early Anglo-Saxon colonization, from Roberts and Sunderland (1973).

on the composition of this major and remarkably homogeneous British population subgroup.

2.5 Final Remarks

The attempt to define populations that are the result of restricted gene flow between subgroups of individuals has a history spanning back nearly a hundred years. However, the practical tools and analytical processes available for doing this have developed remarkably during that time. Most human population genetic studies have used as a starting point a combination of geographical and cultural criteria to define the populations to be studied. Thus in the UK, for example, the definition of population can range from the whole of the UK to one small village in Cornwall, depending on the fineness of scale of the intended analysis. It is relatively easy to differentiate populations over large geographic distances, but only recently have the tools become available to discriminate between very closely related groups, such as those we have studied in the UK. To accomplish this requires, as we have emphasized, both a carefully thought-out sampling strategy and a sophisticated analysis based on using a large number of genetic markers spaced throughout the human genome. Our work on the UK populations shows that it is possible to establish the relationship between genetics and fine-scale population structure. The results suggest that populations tend to be much more stable than has so far been thought to be the case. It is only because of this stability that we have been able to see such a good correlation between the geographic and genetic definitions of the subpopulations of the British Isles. History tends to deal with leaders and those who change culture, whereas the genetics reflects the composition of the whole population and not just the ruling elite. This is a very important distinction that must always be borne in mind when doing fine-scale population genetic studies, especially given the evidence we have found in the UK for a remarkable stability of the population as a whole.

References

Bodmer, W. F. and L. L. Cavalli-Sforza (1976). *Genetics, Evolution and Man*. San Francisco: W. H. Freeman.

Brewerton, D. A., F. D. Hart, A. Nicholls, M. Cafferey, D. C. O. James, and R. D. Sturrock (1973). 'Ankylosing Spondylitis and *HLA-27*', *The Lancet* 1: 904–907.

Cavalli-Sforza, L. L. and A. W. F. Edwards (1963). 'Analysis of Human Evolution', in S. J. Geerts (ed.), *Genetics Today: Proceedings of the XI International Congress of Genetics, The Hague, The Netherlands, September, 1963*. New York: Pergamon Press, iii.923–933.

Dungern, E. von and L. Hirszfeld (1910). 'Über vererbung gruppenspezifischer Strukturen des Blutes', *Zeitschrift für Immunitätsforsch* 6: 284–292.

Fisher, R. A. (1937). 'The Wave of Advance of Advantageous Genes', *Annals of Human Genetics* 7: 355–369.

Hirszfeld, L. and H. Hirszfeld (1919). 'Serological Differences between the Blood of Different Races: The Result of Researches on the Macedonian Front', *Lancet* 2: 505–537.

Khan M. A. (1997). 'Prevalence of *HLA-B27* in World Populations', in C. López-Larrea (ed.), *HLA-B27 in the Development of Spondyloarthropathies*. Austin, TX: Landes/Springer, 1–16.

Landsteiner, R. K. (1900). 'Zur Kenntnis der antifermentative, lytischen und agglutinierenden Wirkungen des Blutserums und der Lymphe', *Zentralblatt für Bakteriologie* 27: 357–362.

Leslie, S., B. Winney, G. Hellenthal, D. Davison, A. Boumertit, T. Day, K. Hutnik, E. C. Royrvik, B. Cunliffe, Wellcome Trust Case Control Consortium 2, International Multiple Sclerosis Genetics Consortium, D. J. Lawson, D. Falush, C. Freeman, M. Pirinen, S. Myers, M. Robinson, P. Donnelly, and W. Bodmer (in press). 'Fine-Scale Genetic Structure of the British Population', *Nature*.

Menozzi, P., A. Piazza, and L. Cavalli-Sforza (1978). 'Synthetic Maps of Human Gene Frequencies in Europeans', *Science* 201: 786–792.

Roberts, D. F. and E. Sunderland (eds) (1973). *Genetic Variation in Britain*. London: Taylor and Francis.

Schlosstein, L., P. I. Terasaki, R. Bluestone, and C. M. Pearson (1973). 'High Association of an *HLA* Antigen, *W27*, with Ankylosing Spondylitis', *New England Journal of Medicine* 288: 704–705.

Watkin, I. M. (1956). '*ABO* Blood Groups and Racial Characteristics in Rural Wales', *Heredity* 10: 161–193.

Wilson, A. C. and R. L. Cann (1992). 'The Recent African Genesis of Humans', *Scientific American* 266: 68–73.

Winney, B., A. Boumertit, T. Day, D. Davison, C. Echeta, I. Evseeva, K. Hutnik, S. Leslie, K. K. Nicodemus, E. C. Royrvik, S. Tonks, X. Yang, J. A. Cheshire, P. A. Longley, P. Mateos, A. Groom, C. Relton, D. T. Bishop, K. Black, E. Northwood, L. Parkinson, T. D. Frayling, A. Steele, J. R. Sampson, T. E. King, R. Dixon, D. Middleton, B. Jennings, R. Bowden, P. Donnelly, and W. F. Bodmer (2012). 'People of the British Isles: Preliminary Analysis of Genotypes and Surnames in a UK-Control Population', *European Journal of Human Genetics* 20: 203–210.

CHAPTER 3

POPULATIONS IN STATISTICAL GENETIC MODELLING AND INFERENCE

DANIEL JOHN LAWSON

3.1 Introduction

VARIATION within biological species has long been described by grouping individuals into 'populations'. However, it is not clear to what degree populations are real biological entities, and to what degree they are a social and linguistic convenience imposed by the human desire for classification. In sexually reproducing organisms, most individuals are genetically unique in some way. Typical organisms undergo numerous mutations per generation (Lynch 2010) and sex acts to reassort the genome of both parents (Creighton and McClintock 1931), leading to these mutations being found in complex patterns in later generations (Hudson 1983). Although full pedigree[1] reconstruction would be the ideal description of populations, this is not possible in general. We argue that we can rigorously use the population concept to understand genetic variation—a very clear and unambiguous definition of a population is available using the well-understood statistical properties of inheritance. But there are limitations, as applying this population concept to real organisms cannot be easily done in all cases. Despite this, the statistical concept of a population is extremely useful, both for conceptual understanding and detailed modelling of genetic variation.

There are two main approaches we can take to defining a population. The first of these is a *generative* modelling approach, in which populations are a component in an explicit model for how genetic variation arises. The second approach uses *statistical* modelling, in which populations are inferred from genetic data. In both approaches, there is in fact a generative model for the genetic structure within a population—they differ on how

[1] A pedigree is the full description of who was the parent of whom, until all individuals are traced back to the most-recent common ancestor of the whole sample.

differences between populations are treated. A statistical approach, which superficially may appear naive, is to define a population as a group of individuals that cannot be distinguished on the available data. This is a less precise definition than the alternative of providing an explicit generative model. The appeal is that under certain ideal—but reasonable—conditions that we will discuss later, the statistical and generative definitions overlap, allowing a very clear definition of what we mean by 'population'. Under the statistical framework, current tools allow for significant progress in understanding many features of genetic structure. Of course, reality can be more complicated and there are at least some cases for which the specifics of genetic inheritance and demography can make even sophisticated attempts at classification into populations seem clumsy and arbitrary.

Central to the majority of theoretical approaches is the idea that, over relevant timescales, all individuals within a single population can be viewed as mating at random within that population (see e.g. Wright 1984). This timescale caveat is extremely important, as we know that real individuals of any species do not randomly mate. Yet mating decisions—based on chance factors such as geographical location and certain individual preferences—are, in most circumstances, 'self-averaging'[2] such that over many generations, random mating is an appropriate approximation. Random mating is a useful assumption as it describes the simplest unit of population structure, on which complexity can be built. But—particularly with modern advances in data acquisition (Schuster 2008; Mardis 2011)—detection of structure within populations is possible (Lawson et al. 2012). Were sufficiently detailed data available, every individual would be a 'population'. This occurs because we can detect some familial relationships (Pemberton et al. 2010; Kyriazopoulou-Panagiotopoulou et al. 2011; Stevens et al. 2011) such as distant cousins, violating the random mating assumption and identifying each individual as unique. We will discuss relaxations to allow for this type of relatedness, which requires an appropriate description of the sort of deviation expected within a population.

Population concepts transcend species boundaries and we aim to be as general as possible. The assumption of random mating is the starting point for the theoretical justification of the approach, and applies with varying degrees of accuracy depending on the genetic and life cycle details of organisms. Generally speaking, the approach describes sexual species well, varying from a near-perfect description of Drosophila (Henderson and Lambert 1982), an adequate description for humans (e.g. Rosenberg et al. 2002, discussed later), to a difficult fit for fungi (Milgroom 1996) and plants (Platt et al. 2010).

Not all species reproduce sexually. Sexual species (including most animals) have two (or possibly more in the case of plants) copies of each chromosome, one of which is inherited from each parent. In contrast, single-celled organisms (prokaryotes such as bacteria and archaea) have a very different genetic structure with only a single copy of each chromosome. Standard sexual reproduction is therefore not possible and it might

[2] A self-averaging process (Lifshitz 1952; Kadanoff 2000) is one for which averaging over a single realization results in the same answer as averaging over many realizations, and occurs most simply when random fluctuations are *additive*, i.e. not state-dependent.

be assumed that random mating is not an appropriate description at all. However, many such species can be represented using this approach if their recombination rate is sufficiently high (Falush et al. 2003)—bacteria can still exchange DNA via non-sexual reproduction, by taking a small fragment of DNA from another organism either accidentally from the environment or via various deliberate (evolved) mechanisms. The amount of recombination bacteria engage in is highly species-dependent.

When recombination is rare or absent, a different population concept is needed. Individuals that occupy the same ecological niche compete neutrally, in which case individuals are sampled under an important genetic distribution called the 'coalescent' (Kingman 1982), discussed in detail later. The genetic distribution within a non-recombining population is very different to that for a recombining population. Rather than variation around a single genetic type being associable with a population, very distinct genetic subtypes can arise due to random fluctuations (since individuals do not exchange genetic information). The coalescent fully describes how genetic variation is distributed, but this typically splits up into multiple separate subtypes when variation is viewed in few dimensions (Lawson and Jensen 2007). Therefore, for non-recombining organisms, a population concept exists as more of an ecological constraint on evolution. Although this population concept is interesting, it is very different from that which is seen in our focus of recombining organisms.

Data collection is predominantly based around sequencing or genotyping many individuals from a small set of sample locations (e.g. Rosenberg et al. 2002; International HapMap Consortium 2007). This naturally leads to finding population-like structures in the data because individuals found at the same location tend to have a similar genetic history. However, as sampling becomes more dense, and genomic techniques more precise, a more continuous picture arises. Strong substructure can be found within single sampling locations (e.g. Lawson et al. 2012) and yet no clear demarcation may exist between samples. Whilst the concept of a randomly mating population is helpful for understanding the genetic structure, we will be increasingly required to work with more complex descriptions. Although there is no clear solution to these issues, we will discuss some possible approaches and how they relate to standard methods.

3.2 Generative Models

Mathematical or computer models that can simulate artificial datasets are called generative models. The concept of a population and the generation of the features of the population are explicitly defined in the model, and individuals can be sampled from them. Generative models typically allow for very complex features to be included, but are hard to perform inference with. The concept of a population used by such models can be very flexible and cannot be generally defined. Typically, however, populations appear as randomly mating groups of individuals, with additional and often complex modelling for how individuals in different populations can interact.

Many population modelling approaches involve generative models, and we focus on two strands here. The first grew from 'statistical phylogeography' which used 'multiple locus coalescent models' (Knowles and Maddison 2002; Chikhi et al. 2002) that consider several 'independent' genomic regions (loci) within which recombination is absent. On a single locus, a 'population' has the same meaning as for non-recombining species described earlier. Loci are then combined assuming a randomly mating population. Modern incarnations (e.g. Thornton and Andolfatto 2006) consider fully sexual populations in a historical scenario in which an ancestral population experiences separation and subsequent genetic drift (i.e. populations are related via a tree, possibly with migration). Throughout this history, individuals randomly mate. Inference of the full demographic history requires complex inference tools such as Approximate Bayesian Computation (Beaumont 2010). However, some historical hypotheses, such as the presence of gene flow between populations (i.e. admixture), can be tested for directly (e.g. Durand et al. 2011) in certain restricted ways. Model organisms such as the fruit fly Drosophila (Thornton and Andolfatto 2006) and humans (e.g. Cornuet et al. 2008 and references within) have been treated to the full modelling approach. However, the multiple locus approaches are widely applicable to non-model organisms (Knowles 2009) from plants (Herrera and Bazaga 2008) to frogs (Bonin et al. 2006) as costly data collection is not needed.

A lot is understood about the genetic structure produced under a tree model for population history, i.e. if individuals mate randomly within populations that have been subjected to arbitrary genetic drift without admixture. The observed genetic variation typically takes the form of Single Nucleotide Polymorphisms (SNPs, which are 0 or 1, either present or not). While the raw distribution is difficult to visualize, differences between individuals (and their populations) can still be plotted. McVean (2009) describes how simple generative models project into their principle components (PCs) based on decomposing the genetic structure into independent directions of variation.[3] These directions are ordered so that the direction of largest variation appears first; see Figure 3.1C–D. Individuals that share many SNPs tend to be close, and SNPs that vary consistently between individuals tend to dominate the higher PCs (those that are typically plotted). The lower PCs describe SNPs that show no particular pattern and are usually considered to be 'noise', e.g. random sampling. Individuals within a single population are characterized by having the same SNP frequencies, and hence approximately share the same location in PC space.

Differences between populations in a PC analysis can be interpreted. Genetic drift affects the PCs by separating the mean position of populations, as shown in Figure 3.1A,C with distances $d1$ and $d2$ becoming separations $f(d1)$ and $f(d2)$.

[3] We use a Singular Value Decomposition (SVD), a form of PC analysis. The SNP data matrix X containing l rows (SNPs) and n columns (individuals) is decomposed into UDV^T where U is an l by n matrix, D is a rectangular diagonal matrix of the eigenvalues, and V is the n by n eigenvector matrix relating individuals. Columns of V are the PCs (which are usually what is plotted), and each row is an individual. Lawson and Falush (2012) has more details about the interpretation.

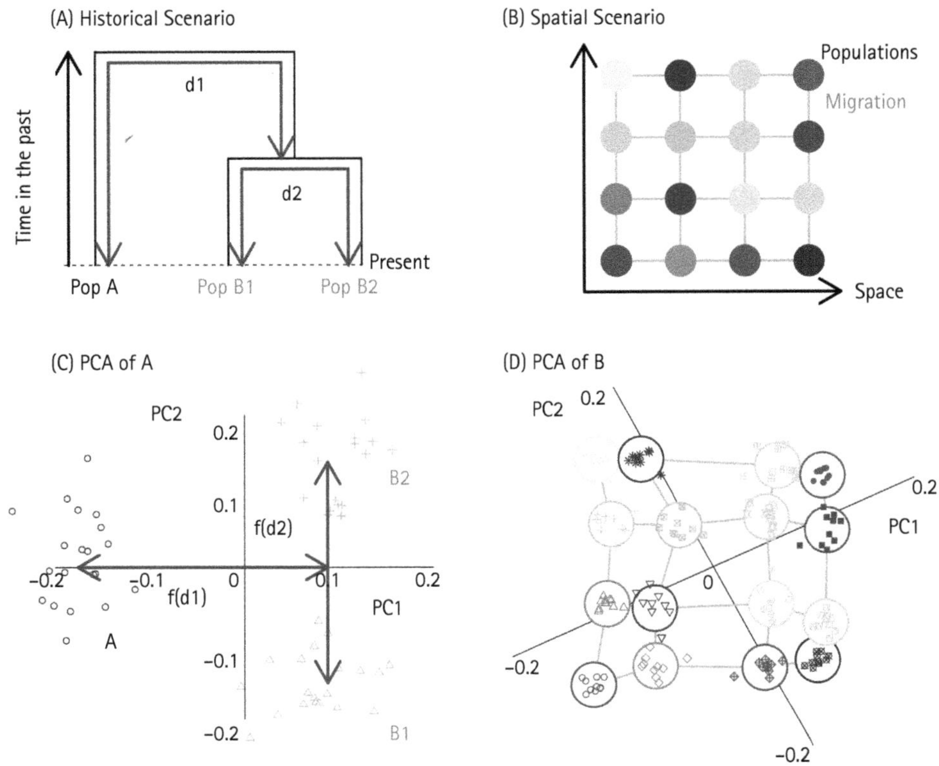

FIG. 3.1 Simple population scenarios* and the resulting PC decomposition. (A) A simple hierarchical tree model for population splitting. d1 and d2 represent genetic drift times. (B) A deme-based spatial migration scenario in which migration occurs at constant rate along edges and each node contains a population. (C) The PC decomposition of the scenario in A. Each PC is associated with an independent drift component, with genetic drift time affecting both the eigenvalue and the distance relative to the within-population variation (i.e. f(d1) and f(d2)). (D) the PC decomposition of C recapturing the spatial structure in this case. Were sample sizes not even, this may not have occurred, and higher PCs contain a complex pattern created by local drift and historical admixture.

* For A, 1000 unlinked SNPs simulated as in Lawson et al. (2012). For B, simulated using the software MSMS (Ewing and Hermisson 2010) sampling 20 individuals per population with the parameters '–r 500 –t 50' resulting in 6754 linked SNPs, and recent migrants removed to reduce the dataset to 10 individuals per population.

Additionally, an individual who is admixed between two populations is located between them in PC space (section 3.4). Unfortunately, sample size strongly affects both the location and order of drift effects in the PCs, and interacts with both drift and admixture in a non-trivial way to determine the location of populations in PC space. Even worse, although an admixture scenario has a predictable effect on the PCs, this is not always inferable (see section 3.3).

Another generative model is demic diffusion (Ammerman and Cavalli-Sforza 1984; Sokal et al. 1991) which captures complex geographical structure. Waples and Gaggiotti

(2006) perform a thorough review of this approach, with specific emphasis on what a population means in this context. Again, populations are randomly mating groups of individuals living in demes at a specific geographic location, and migration occurs between nearby demes (see Figure 3.1B,D). Like the explicit historical model already mentioned, complex histories can be specified. However, the full history of a population is only implicitly described in this approach, as individuals migrate between demes. Recent migrations (e.g. within the last few generations) do not fit neatly into a population description.

Demic diffusion models and PC analysis are usually used together to make inferences about the nature of historical and geographical processes shaping evolution. Unfortunately, there are many historical scenarios consistent with the major PCs (François et al. 2010) which casts doubt on the inferences that can be directly made. In principle, there is more information contained in the lower PCs, describing e.g. genetic drift unique to each population, but this is difficult to extract. This is in contrast to simple statistical populations (section 3.3) which can be inferred using PC analysis (Lawson and Falush 2012). The inference problems are related to the presence of strong admixture in these models, which is discussed in section 3.4. The utility of PC analysis for inference about population structure is therefore a contentious topic.

There is more that can be said about PC analysis; in particular, methods accounting for linkage (section 3.3, details in Lawson and Falush 2012, and Lawson et al. 2012) can be used to improve PC decomposition. However, complex genomic scale datasets are not completely described by the top few PCs and so we view PC analysis as a useful visualization tool rather than a primary analysis technique for understanding populations.

3.3 Statistical Definitions of a Population

Statistical approaches such as STRUCTURE (Pritchard et al. 2000) define a population as a cluster of statistically equivalent individuals. For SNP data taking the value of either 0 or 1, this means that there is a single frequency for each SNP within each population, and a population is defined by the full set of SNP frequencies. Although a single SNP is not very informative (unless it perfectly segregates populations), multiple SNPs can be combined to identify fine population differences.

Each SNP in an individual in the STRUCTURE model is sampled 'independently and identically distributed' (IID) from the population SNP frequency. This means that SNPs are both independent and all drawn with the same frequency, so there can be no correlation between the SNPs from e.g. related individuals. More generally, we can define a population as a collection of individuals who are genetically *exchangeable* (Bernardo 1996). Exchangeability within a population is less restrictive than IID sampling—the

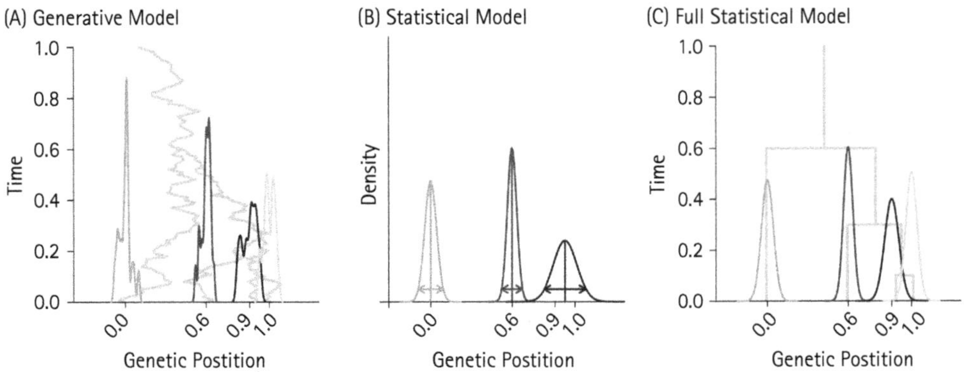

FIG. 3.2 Populations as they may hypothetically appear in different types of model. (A) A fully generative historical model describing the genetic drift process (grey tree) differentiating populations ('peaked' curves centred on marked locations on the x-axis) in genetic space (i.e. a direction in PC space). The details of the drift process must be specified. (B) A simple statistical model, in which only the within-population model (here the means and variances) must be correctly described, and population-level parameters are inferred. Populations may be merged (rightmost populations) if they are not sufficiently distinct. (C) A sufficiently flexible statistical model for the population-level parameters can allow these to be inferred more accurately; here a hierarchical model matches the data better and therefore infers the subtle population differences correctly.

SNPs between individuals may be correlated, providing that the order that individuals are placed within the population does not matter.[4]

In the STRUCTURE model, related individuals are assigned their own population which has a different SNP frequency from the population the individuals were sampled from (and may be close to 0 or 1). More general definitions of a population based around exchangeability would allow individuals to be sampled as e.g. a mixture between a random individual (i.e. using the population frequency) and a previously sampled individual. The number of related individuals within a population and the degree of relatedness would be additional parameters to be inferred. Such models are more complex to implement (and hence are not considered in any major inference algorithm), but should remain part of our conceptual understanding of a population.

To say that individuals belong to different populations, we only need a model of how individuals within a single population vary, as STRUCTURE does via the SNP frequency. In a simple statistical approach, a correct model for how differences between populations arise is not needed (Figure 3.2B). This contrasts with generative models (Figure 3.2A) which require an explicit description of how the current genetic

[4] A classical example is the Pólya urn (Johnson and Kotz 1977) in which a red and blue ball are placed in an urn, and then a ball is repeatedly removed at random and replaced by two balls of the same colour. The probability of drawing a red ball depends on the history of the process but knowing only the number of each type of ball in the urn is enough—the order does not matter. The coalescent process in genetics (Kingman 1982) is but one very important example of exchangeability in action.

distribution was reached. Despite this, a model for population differences is very helpful as it increases statistical power to separate populations (Figure 3.2C). In this example, the underlying generative model (Figure 3.2A) is diffusion along a tree. Because the underlying population separation method was hierarchical, a tree-based statistical model can capture the main differences between populations. Models of population differences are sometimes omitted because (a) sufficiently flexible models should all tend to the truth as the amount of data gets large, even if they ignore between-population differences, and (b) if the model is incorrect, it can lead to problems.

The genetic structure within a randomly mating population is well understood, and is characterized by the mathematical model known as the Ancestral Recombination Graph (ARG) (Griffiths 1981; Hudson 1983) (Figure 3.3A). This describes the genetic history of all individuals using three basic ingredients. Firstly, exactly when their ancestors were shared with ancestors of other individuals in the sample (coalescent events; times T2–4). Secondly, when ancestors mated with ancestors of other individuals (i.e. recombination events which split the genome between these two parents, time T1). Thirdly, when mutations arose (although for our purposes mutation times need not be inferred as these can be integrated out). As such, the ARG is a complete probability model of a closed population.[5] Note that an ARG without recombination reduces to the coalescent model, which is a tree with a probability model for the coalescent times. Although the ARG is a statistical model, it is equivalent to a generative model of a neutrally competing population with fixed recombination rate, mutation rate, and population size (Hein et al. 2005), and a wide variety of different modelling assumptions all lead to the ARG under reasonable limiting conditions, such as large population sizes, so that mating can be considered to occur randomly.

Unfortunately, it is not practical to perform inference with the ARG, and so most approaches are based around approximations to it. Almost all population inference methods start by assuming that genetic drift is weak, because this is relatively harmless as we can always detect the difference between populations when drift is strong. Another common approximation is that the recombination rate is high enough to treat SNPs independently. The most widely used tools, including STRUCTURE (Pritchard et al. 2000), STRUCTURAMA (Huelsenbeck and Andolfatto 2007), PARTITION (Dawson and Belkhir 2001), and ADMIXTURE (Alexander et al. 2009) start with this assumption, which leads to modelling SNPs only via their SNP frequency as described earlier.

The SNP frequency approach works well when SNPs are approximately independent, but becomes increasingly inaccurate in the presence of linkage disequilibrium (LD). LD is the correlation between SNPs along the genome due to sharing the same genetic history; for example, in Figure 3.3A, SNPs 1–2 are in LD because they are each described

[5] The ARG specifies the probability of a particular set of recombination and coalescent events. We start with $k = N$ lineages, one for each genetic sample. Coalescence between lineages occurs at rate 1 between all $k(k-1)/2$ pairs of lineages, and recombination occurs on each lineage at rate ρ. Therefore the time between events is exponential with rate $k(k-1)/2 + k\rho$. The coalescent model is the ARG with $\rho = 0$.

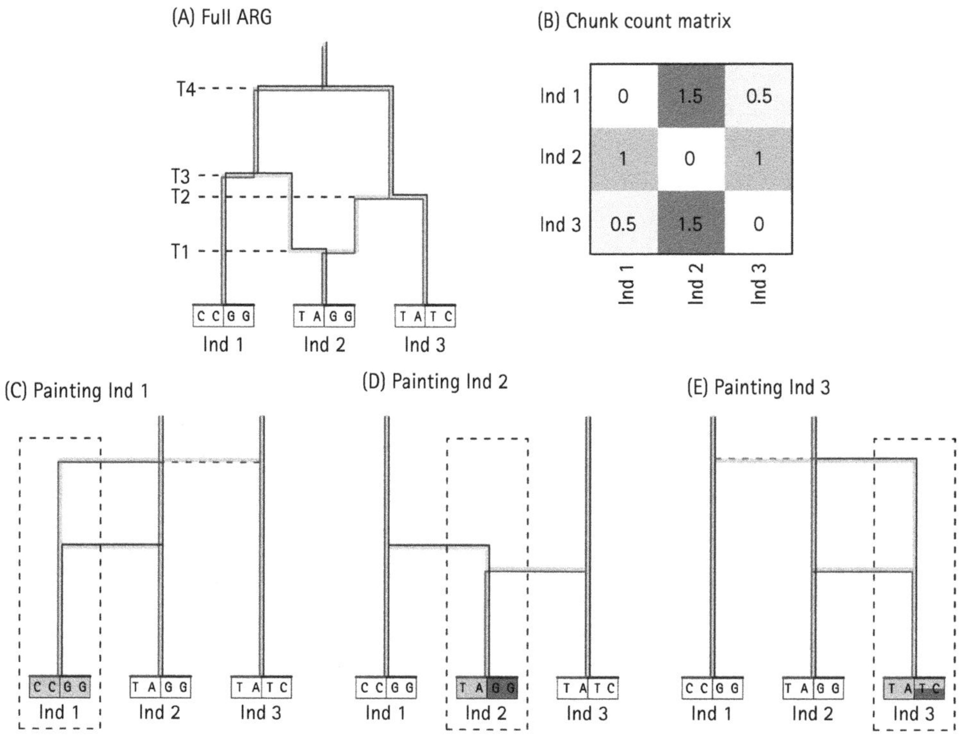

FIG. 3.3 Cartoon illustrating the ChromoPainter model as an approximation to the ARG on a simple sequence. (A) The ARG describes the full history of recombination and coalescence between individuals, terminating with a Most Recent Common Ancestor. The ARG consists of a series of coalescence events (times T2–4), and recombination events (time T1). Recombinations split the genome into segments that have (possibly) distinct tree relationships between individuals. (B) The ChromoPainter chunk count matrix counting the number of haplotype chunks shared between each pair of individuals (which is asymmetric in general). It counts the *expected* number of haplotype blocks used by the painting (C–E), rather than the number of SNPs. (C–E) The chunk count matrix approximates the ARG by describing only the most recent relationship for each individual separately at each genomic location. Coalescence and recombination events are forbidden between other individuals. Where multiple individuals are equally close (dashed lines in C and E), these are averaged and fractions of the chunk are assigned from each. For example, in Painting Ind 1 (Figure C), SNPs 1–2 are on the same haplotype which is uncertain between Inds 2–3, using 0.5 chunks from each. Due to requiring fewer mutations, SNPs 3–4 can be inferred as from Ind 2 (using a single chunk), leading to the top row of (B).

by similar (and in this case, the same) trees. As more recombination events split the genome between two SNPs they become less correlated. Therefore, LD is stronger in genomic-scale sequencing datasets produced by modern sequencing technologies than in older techniques that sampled fewer loci. Modelling LD is becoming more important, and although STRUCTURE has an option (Falush et al. 2003) for this, the approach is too computationally costly for most applications.

A powerful approach for handling LD, which has an additional benefit of speeding the computation, is that of ChromoPainter (Lawson et al. 2012). This uses the 'painting' algorithm of Li and Stephens (2003) to identify long tracts of genetic material that have been inherited as a single contiguous chunk, which can be seen as an approximation to the ARG (Figure 3.3). Each haplotype[6] is described at each genomic location by finding the haplotype from a pool of potential donors that most closely matches it. The donor at a particular genomic location has to be inferred, and there is a constant probability per unit recombination distance[7] that the donor will change to any of the other possible donors. The result is that 'chunks' of DNA are found in which relatively few SNPs differ between the donor and the haplotype being painted. The model takes a computationally convenient form called a Hidden Markov Model,[8] which allows for a fast inference of a wide range of useful summaries with complete accounting for statistical uncertainty. We focus on the 'chunk count' matrix X_{ij}—the (expected) number of chunks donated to individual i from individual j.

Under reasonable assumptions (details in Lawson et al. 2012), a shared haplotype chunk can be viewed as a single observation of coinheritance between two populations in a similar way to the sharing of a SNP. The FineSTRUCTURE model combines this information over all individuals by treating a population as having a characteristic rate of donating chunks to, and receiving from, each other population. The only information required is the number of chunks donated between each pair of individuals,[9] which is a matrix of dimension $N \times N$—a significant reduction in the amount of data that must be processed as there is no dependence on the number of SNPs. Additionally, there are no population-level parameters to be inferred, which makes inference both very fast and very reliable, as the parameter sampling does not tend to get stuck in local modes. SNP frequency-based approaches (e.g. STRUCTURE and ADMIXTURE) must infer the underlying SNP frequencies which can be multimodal. FineSTRUCTURE can therefore handle large sample sizes; thousands using MCMC,[10] and potentially many more using

[6] A haplotype is a particular copy of a chromosome; recall that humans and other animals have two copies of each chromosome, which are called haplotypes. Current sequencing technology cannot tell which copy of DNA a particular mutation is on, and therefore we observed the 'genotype', i.e. the mixture of the two. Haplotypes are inferred using a technique called phasing to separate out which mutations appear on which copy of the chromosome.

[7] Recombination does not occur uniformly over the genome but most frequently occurs at 'recombination hotspots' (Myers et al. 2010). This can be accounted for in the model by measuring distance along the genome in recombination distance.

[8] In our context, the presence of a mutation at a genomic location i depends only on who the donor haplotype is (which is 'hidden'). In turn, the probability of the donor haplotype x_i depends only on the donor haplotype at the previous location x_{i-1} and not on any other genomic positions (this is what is meant by 'Markov').

[9] This is a sufficient statistic for the model likelihood.

[10] Markov Chain Monte Carlo (e.g. Gamerman 1997) is a very general and powerful inference technique based around sampling parameters near to previously sampled parameters, as these are likely to be 'good' at describing the data (i.e. have a high likelihood). The MCMC samples converge to independent samples from the posterior distribution if the sampler is run for 'long enough', which may take a long time.

greedy optimization approaches. However, whilst fast, greedy methods usually do not find the best solution and may get stuck at suboptimal parameter values that cannot be improved by small changes.

Figure 3.4 shows an example of what a population looks like to FineSTRUCTURE, using 102 African individuals from the Human Genetic Diversity Panel (HGDP)—a collection of 1,043 individuals from throughout the world (Pickrell et al. 2009). If the 'randomly mating' population model is applicable, all individuals in population *a* will receive the chunks at the same rate from individuals in population *b*—there is no further dependence on who the individual is. This leads to a 'block' structure, clearly visible in the figure, with some labelled populations perfectly described this way (e.g. the Yoruba). However, FineSTRUCTURE has identified substructure within population labels—for example, there are related pairs of individuals. Related individuals are split off in the tree and have high diagonal elements (e.g. within the BantuKenya, the Mandenka, MbutiPygmy, and BiakaPygmy). They can still be identified as being from a particular population because they otherwise share DNA similarly with all of the non-related populations. Admixture-induced substructure can also be seen, e.g. the BantuSouthAfrica consists of two main subpopulations: one related to the BantuKenya, and the other closer to the San (but not the Pygmy populations, to whom the San are also related). In this case the population model is still a useful description, even though it does not directly model the admixture.

The FineSTRUCTURE definition of a population appears very different to that in STRUCTURE—being based on the sharing of haplotypes, rather than the frequency of SNPs. Yet surprisingly, the model of a population in FineSTRUCTURE applied to unlinked (i.e. ignoring LD, although it may be present) ChromoPainter data is formally equivalent to that in STRUCTURE.[11] Despite this strong similarity, there are very important features that differ between the various models available to perform population clustering, which can lead to quite different inferences about populations. The most important is the model for *differences* between populations, which are usually expressed as a model for the population-level parameters. Figure 3.2 shows how these might be important, and Table 3.1 gives some common approaches.

None of the methods discussed makes powerful use of the difference between individuals being exchangeable within a population, and them being IID. The closest is the FineSTRUCTURE approach, in which populations are initially inferred on the basis of individuals being drawn IID from a population, but then familial relationships are immediately merged in a 'tree-building'[12] step, using a heuristic model for kinship. The heuristic model assumes that related individuals will be placed into the same population and will have additional chunks shared with each other, but the same distribution of

[11] Formally, FineSTRUCTURE and STRUCTURE share the same asymptotic likelihood when SNPs are treated as independent, drift is weak, and number of individuals sampled is large. Proof in the appendix of Lawson et al. (2012).

[12] We omit the details of the tree-building, but broadly, the merging of populations that is 'least bad', i.e. causes the lowest increase in the likelihood, is repeatedly applied until only a single population remains.

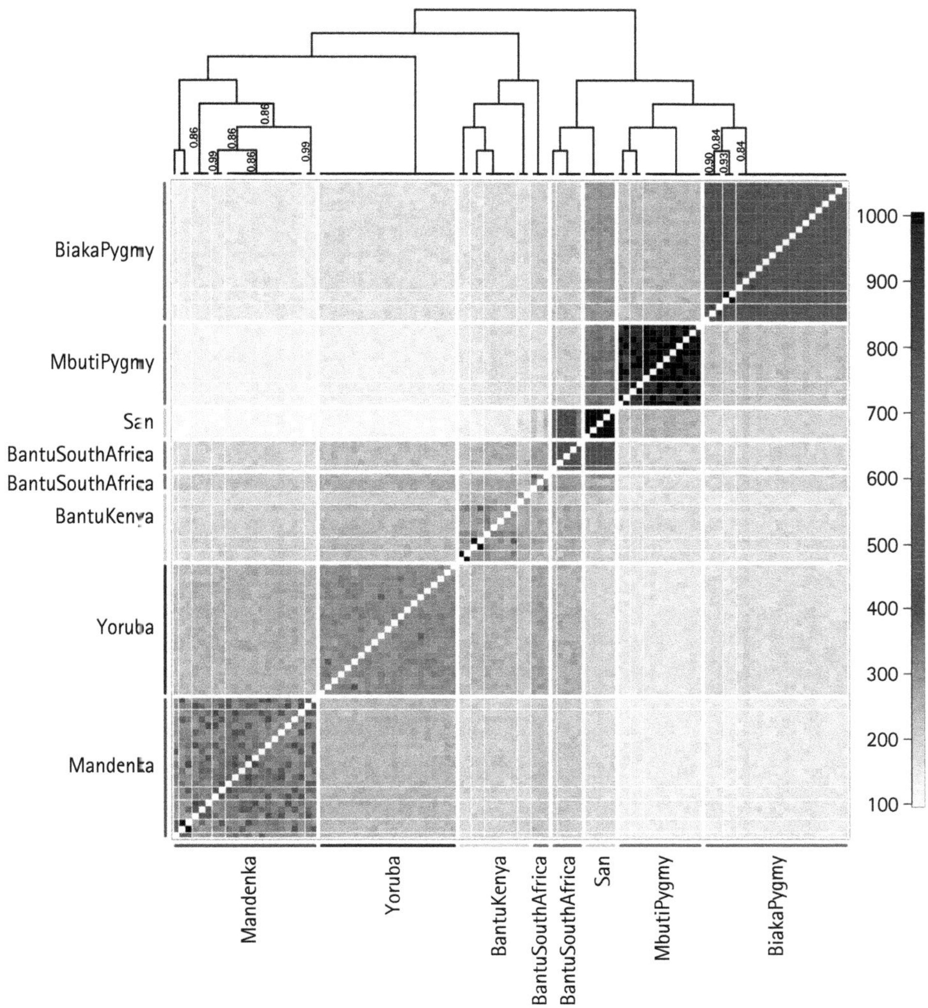

FIG. 3.4 FineSTRUCTURE applied to the African samples of the HGDP dataset.* The main figure shows the chunk count matrix donated between pairs of individuals, as output by ChromoPainter. FineSTRUCTURE generates MCMC samples of the population structure, which for clarity are not shown; instead the maximum a posteriori population assignment (i.e. the best single guess) is shown by the white lines which separate populations, and in the top tree: individuals in the same population are in the same lowest-level branch. Each edge in the tree produced a bipartition; if this is not perfectly observed in all MCMC samples, the proportion is shown on the tree. Thick white lines separate major 'clades'. A 'population' to FineSTRUCTURE is the average of the chunk counts within that population to (columns) and from (rows) all other populations, so within each white square all values should be the same except for noise. Drawn from data described in Lawson et al. (2012).

* ChromoPainter was applied to the whole HGDP dataset as described in Lawson et al. (2012). For simplicity chunks donated to/from other continents are not displayed, although they were used during population assignment.

Table 3.1 Population separation models in popular software

Software	Model
STRUCTURE (Pritchard et al. 2000)	SNP frequencies are correlated due to common descent, meaning that each SNP frequency drifts independently from a single ancestral frequency in a 'star' structure—each population is independent conditional on the ancestral frequency. Because the same number of SNP frequencies as SNPs must be inferred, this prior is quite strong. See Falush et al. (2003) for details.
ADMIXTURE (Alexander et al. 2009)	A frequentist method with no explicit population prior, effectively assuming that each population is sampled IID with a uniform SNP distribution.
STRUCTURAMA (Huelsenbeck and Andolfatto 2007)	As STRUCTURE, but with a model for the *number* of populations (a Dirichlet Process prior; Ferguson 1973) that is convenient for fast inference.
FineSTRUCTURE (Lawson et al. 2012)	Haplotype sharing rates between different populations are IID with a Multinomial prior. Because there are only K haplotype rates to infer, this is a weaker prior than used by STRUCTURE. A Dirichlet Process prior is also used for computational convenience.

other chunks as they would have if the related individual were not in the sample. Related individuals that share a population can then be merged with their underlying population. Although this works well in many examples (e.g. Figure 3.4), the technical details of this two-stage approach are not perfect and new methods are required.

We have described several methods in this section, and claimed that some are 'fast' whilst others are not. However, the speed of these algorithms is strongly dependent on the amount of data available—'fast' here describes algorithms that scale well with the number of genomic loci (and the number of individuals) sampled. Unfortunately, this is difficult to quantify exactly for MCMC algorithms because how quickly the algorithms converge is dependent on the data. Broadly though, the ChromoPainter/FineSTRUCTURE approach is fast because the computational cost is linear in the number of loci L; the STRUCTURE/STRUCTURAMA approach is at least quadratic. All these approaches scale similarly with the number of individuals N. The exception to these issues is ADMIXTURE which is faster (by roughly a factor of N) because it does not perform MCMC. For reference, the absolute computational cost of running genomic-scale datasets (700K SNPs) containing a thousand individuals is that ChromoPainter/FineSTRUCTURE may take a week to run (although much of this cost is parallelizable); STRUCTURE and STRUCTURAMA would take months; and ADMIXTURE may take only hours/days but usually fails to converge to the global solution, and typically a stable approximation at low k is used. All approaches run accurately within minutes or hours on smaller datasets (say, 10,000 SNPs and 100 individuals).

3.4 Admixture Models

Both STRUCTURE and ADMIXTURE have a model permitting an individual to relate to multiple populations, known as admixture. These important models are widely applied in practice as they are typically closer to reality than no-admixture approaches, and—when LD is ignored—little more computationally costly. Some high-profile examples for humans are Rosenberg et al. (2002) which examined the human population genetic structure, and Redon et al. (2006) and Sladek et al. (2007) which analysed functional genetic features, using admixture models to account for the confounding effect of population structure.

The admixture likelihood of a SNP is the same as in the standard no-admixture model, if we knew which population the SNP came from. However, an individual is no longer assigned to a single population—each SNP within an individual is viewed as coming from one of K unobserved, 'pure' populations, and individuals are seen as mixtures of these populations. Therefore the model for 'what a population is' remains the same, only now we do not observe populations directly, but mixtures from them. Figure 3.5 illustrates how this can form standard, inferable populations (even without new genetic material in the form of drift) as well as continuous lines in PC space that do not correspond to a population.

Populations are defined similarly to no-admixture models—as genetically statistically distinct. However, discrete models will use additional populations to describe mixtures of 'real' populations that have undergone genetic drift. As we shall see, if the distribution is complex then admixture models can have multiple interpretations, and a lack of visualization tools makes assessment more difficult. For example, we cannot plot pure populations in PC space. However, PC analysis can still help visualize the results (Figure 3.5).

Part of the difficulty in interpreting mixture models for admixture is that they are an incomplete description of how the present-day genetic distribution arose. Differences between pure populations can be described by complex models such as trees (Falush et al. 2003), but the mixing only happens at the present time. However, admixture may have occurred at any point in history. For SNP frequency models such as ADMIXTURE and STRUCTURE, the mixture approach captures much of the available information. There are currently no approaches that correctly handle LD and admixture in general. A problem is that the length of LD blocks contains strong information about admixture dates, which would ideally be used in an admixture description. Therefore current admixture models need complete revising to utilize this information.

To construct admixture models from LD data, it becomes necessary to form specific hypotheses about historical drift and admixture scenarios. These include: How do populations separate and merge? How is DNA shared between individuals in different populations via migration? Such a model might resemble the generative models described in section 3.2, or might take inferred populations from STRUCTURE/FineSTRUCTURE and explain their relationship *post hoc*. TreeMix (Pickrell and Pritchard 2012) is one method for this; it describes the relationship between populations as a tree, subject to small amounts of historical admixture. However, *post hoc* approaches typically lose

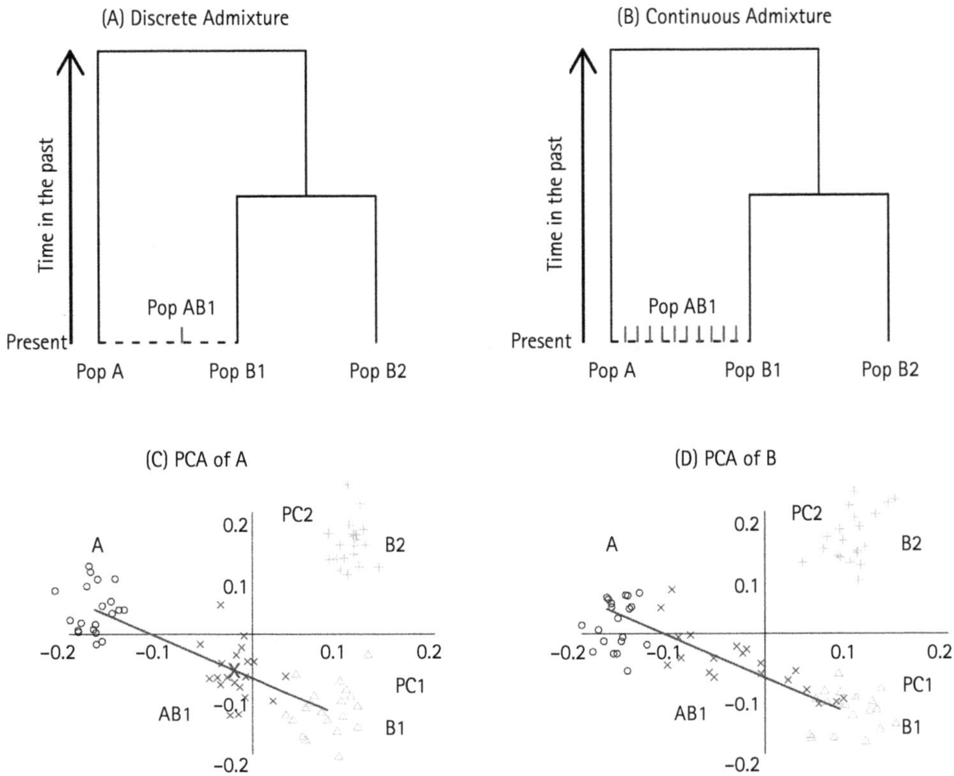

FIG. 3.5 Two simple present-day admixture scenarios for populations split via a tree. (A) Discrete admixture, in which two populations are mixed with some characteristic frequency. Population models can infer an additional population (in this case, AB1) which can either be directly described as a mixture of other populations, or via post hoc methods such as PCA. (B) Continuous admixture, which causes problems for population assignment algorithms. (C–D) PC decompositions of the scenarios from A–B. Admixture fractions will correlate with the position on the admixture line, shown. Note however that in both cases, historically admixed populations may have undergone genetic drift. In this case, there will be additional PC directions in which they are not describable as a mixture.

information, both by discarding information present in the raw data, and by discarding uncertainty present in the first stage of modelling. Conversely, two-stage approaches are usually faster, and in some cases (as discussed for ChromoPainter/FineSTRUCTURE) it can be shown that no information is lost.

3.5 Population-Free Approaches

We may view a ‘population’ as a somewhat arbitrary boundary between local variation and larger-scale differences of relevance to real-world questions. If we are willing to abandon neat segregation of individuals into populations, we can provide a more accurate description of the underlying genetic structure, at the price of interpretability. This trades off one purpose of population modelling—aiding understanding of the underlying variation—with accurately representing the underlying genetic structure in a low-dimensional representation.

STRUCTURE-like models are 'semi-parametric' because they use a parametric description of a population (e.g. a SNP frequency or a haplotype sharing distribution based on results from a generative model of how variation arises) but allow an arbitrary number of populations to describe the sample. However, they penalize the use of too many populations in order to obtain interpretable results, and the description of a population is fully parametric. Non-parametric methods[13] (e.g. Wasserman 2005) can provide a more accurate representation of the underlying genetic structure than parametric methods when modelling assumptions are invalid. For example, real populations are often continuously admixed over many timescales, and have complex substructure. This typically leads to methods such as FineSTRUCTURE inferring a large number of populations that are hard to interpret. We could consider a non-parametric description of the populations themselves, and potentially obtain simpler results.

An interesting feature of admixture models is that they can correct for inadequacies in the conceptual model of a population. By allowing a population to be described as a mixture of (very) many unobserved components, any genetic distribution can be represented. This is essentially a non-parametric statistical view of a population. However, this view is not commonly used for two reasons. Firstly, it confuses the theoretical population concept with the observed populations, so that it is no longer clear what a 'population' is. Secondly, a lot of data is required to infer a large number of mixture components (Kalai et al. 2012).[14] Inference algorithms frequently get trapped at suboptimal population mixtures, leading to repeated runs of the same algorithm finding different solutions. Figure 3.6 illustrates how admixture appears in inference. Admixture components are not always associable with either a cluster, nor a peak in the density in a PCA decomposition (Carreira-Perpinan and Williams 2003) (Figure 3.6D). The drift and admixture fractions in this heavily admixed scenario are very hard to separate and lead to unstable estimates. The end result is that admixture models robustly identify only a small number of populations which may not give a good 'non-parametric' description of the data and can still be hard to interpret. If multiple pure populations are mixed, this becomes even more complex.

Species occupying continuous habitat ranges may require a spatially explicit model. Guillot et al. (2009) reviews isolation-by-distance models, which allow for individuals to be distributed at any point over a spatial region. 'Random fields' can be used to describe

[13] 'Non-parametric' methods describe data using an infinite-dimensional representation in an appropriately chosen basis; of course only a finite number of components are used in practice, which is typically equal to or less than the number of data points.

[14] The number of samples is exponential in the number of mixture components, when they overlap. This makes inferring more than a few components theoretically 'impossible'.

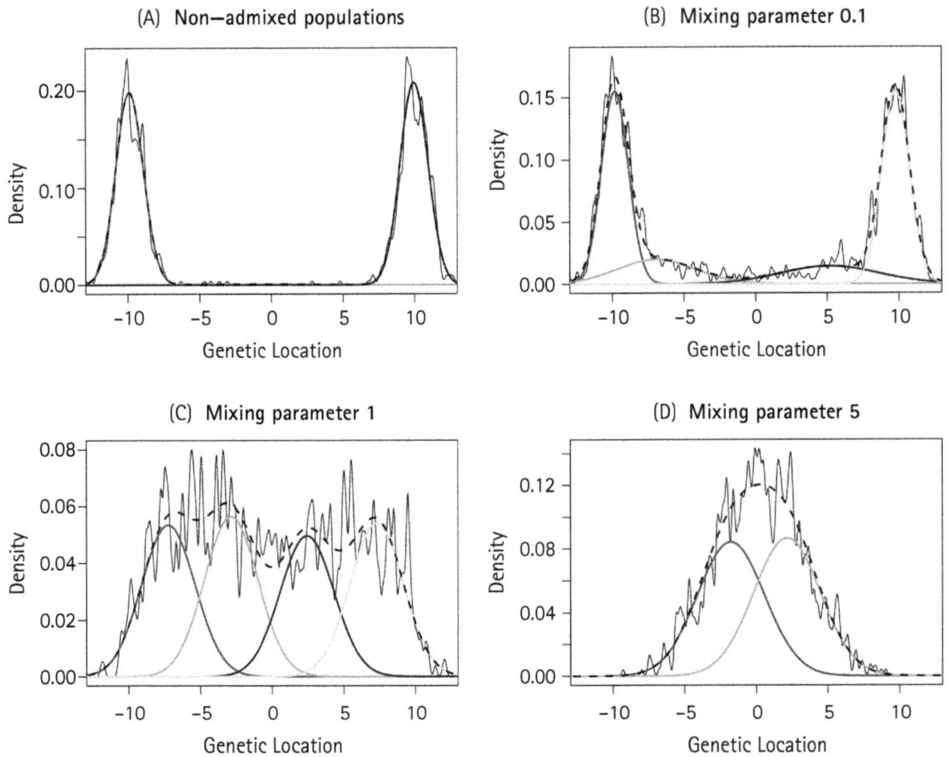

FIG. 3.6 The effect of admixture on population inference in 'genetic space' (equivalent to the first PC of a PC analysis of these two populations). Mixtures between two randomly mating populations were generated (solid wiggly line) by sampling individual's proportions from a Dirichlet (p,p) distribution, for (A) $p = 0.01$, (B) $p = 0.1$, (C) $p = 1$, and (D) $p = 5$. Low values of p mean that individuals predominantly are from a single population; high values mean they are close to equal proportions of both. Also shown is the 'best' no-admixture population assignment* ('bell' curves) and the resulting density this generates (black dashed line). Note that population methods infer 'spurious' populations to explain the intermediate types (B–C). In (D), there is a 'single' population but it has complex structure and is inferred as two populations. This is inferable because although there is a single mode, the distribution is not Normal and the variance is inflated.

* Population assignment performed using MCLUST (Fraley and Raftery 2009) as described in Lawson and Falush (2012). 1000 individuals were sampled; 'populations' here are Normal distributions representing a hypothetical first PC (e.g. McVean 2009).

how populations change over space, either using discrete populations (Guillot et al. 2005; François et al. 2006; Corander et al. 2008), or admixture (Durand et al. 2009). This approach allows for very flexible specification of how spatial correlations arise (a popular choice is the Markov random field; Kindermann and Snell 1980). These models resemble the approaches described already but use geographical distance to help explain differences between populations. Admixture models are more interpretable when using spatial information, and may be more robust (François and Durand 2010), as additional information is available to infer mixture components. However, general application requires care in case spatial correlations are not the primary source of variation. While inference in this framework is very challenging (Guillot 2009) there has been significant recent progress in the technical methods (Grelaud et al. 2009; Everitt 2012).

Although this section has taken a negative view of the population concept, in practice many species can be correctly modelled with it. Fragmented habitats due to man-made or geographical barriers to migration are naturally represented by discrete populations, and spatial structure can be handled either via standard methods or networks of demes (Fortuna et al. 2009). Historical populations are only accessible via rare fossil DNA and are therefore rarely continuous in practice. Population models will therefore coexist with approaches that define genetic distributions implicitly and both may be suitable for different questions even for the same species.

3.6 Discussion

There are many features on which we may wish to group individuals into 'populations'—language, culture, geography, and history will all lead to different groupings suitable for different purposes. Here we focus on population-genetic approaches that group all individuals which cannot be statistically distinguished into the same population. This is an appealing approach since it is not necessary to define how population differences arise—we only need a model for what it means to be in a population. A natural model for a population is a randomly mating group of individuals, which allows for powerful theoretical results to be applied.

There are two main reasons to search for population structure in the genetics of individuals. The first is for understanding; we wish to classify the world around us in the simplest way possible (but no simpler). This is the main motivation behind the generative models from section 3.2, and the statistical approaches from section 3.3 that tie into these. We describe present-day individuals using populations so that we can then reason about the historical forces of separation, migration, admixture, etc. that have influenced the populations we observe. However, the second major application is to remove population structure as a confounder of genetic correlation, for the understanding of genomic function. Because genomic variation occurs at far more loci than the number of individuals we can hope to sample, naive inference is confounded by many factors including population structure. In this case, the non-parametric viewpoint (section 3.5) is appropriate. Major population structure approaches (Tiwari et al. 2008) typically use either PC analysis (Price et al. 2006) or STRUCTURE-like models for this. Neither fully captures complex population structure, particularly when provided with linked data. Although linkage mapping approaches exist (e.g. Redden et al. 2006), these can only account for simple population structure and we expect to see new methods for detecting genetic selection based on linked data.

We have reviewed the main population assignment approaches for weakly linked genetic data (e.g. STRUCTURE; Pritchard et al. 2000) and seen that populations in such models have a clear interpretation in terms of generative models for how genetic variation arises within a population. Models of differences between populations can be used where appropriate to gain further statistical power, and then the model is fully generative.

When data are densely sampled from the genome, handling linkage disequilibrium becomes a necessity. For population identification, our ChromoPainter and FineSTRUCTURE (Lawson et al. 2012) approach is the most powerful currently available. Here, individuals are first summarized in terms of the number of contiguous haplotype blocks they share with each other, after which they can be clustered. This approach theoretically matches previous SNP models but has dramatically improved computational performance, as well as providing a more in-depth description of the data.

Unfortunately, modern methods for population inference have *too much* power to infer randomly mating populations, and can potentially find as many populations as there are individuals. To address this, we have discussed some relaxations from random mating which can be applied when defining a population. Some unambiguous 'within-population' variation is familial structure, which is well understood in principle, and may fit within standard statistical methodology. A more difficult extension is admixture between subpopulations, which is difficult because there is no line between a 'subpopulation' and a 'true population'. Although such deviations may be hard to distinguish either practically or theoretically from differences between populations, we remain positive about the utility for theoretical population concepts in understanding genetic variation.

References

Alexander, D. H., J. Novembre, and K. Lange (2009). 'Fast Model-Based Estimation of Ancestry in Unrelated Individuals', *Genome Research* 19: 1655–1664.

Ammerman, A. and L. L. Cavalli-Sforza (1984). *The Neolithic Transition and the Genetics of Populations in Europe*. Princeton: Princeton University Press.

Beaumont, M. (2010). 'Approximate Bayesian Computation in Evolution and Ecology', *Annual Review of Ecology, Evolution, and Systematics* 41: 379–406.

Bernardo, J. M. (1996). 'The Concept of Exchangeability and its Applications', *Far East Journal of Mathematical Sciences* 4: 111–121.

Bonin, A., P. Taberlet, C. Miaud, and F. Pompanon (2006). 'Explorative Genome Scan to Detect Candidate Loci for Adaptation along a Gradient of Altitude in the Common Frog (Rana temporaria)', *Molecular Biology and Evolution* 23: 773–783.

Carreira-Perpinan, M. and C. Williams (2003). 'On the Number of Modes of a Gaussian Mixture', in Lewis D. Griffin and Martin Lillholm (eds), *Proceedings of the 4th International Conference on Scale Space Methods in Computer Vision*, Lecture Notes in Computer Science 2695. Berlin and Heidelberg: Springer, 625–640.

Chikhi, L., R. A. Nichols, G. Barbujani, and M. A. Beaumont (2002). 'Y Genetic Data Support the Neolithic Demic Diffusion Model', *Proceedings of the National Academy of Sciences of the USA* 99: 11008–11013.

Corander, J., J. Sirén, and E. Arjas (2008). 'Bayesian Spatial Modeling of Genetic Population Structure', *Computational Statistics* 23: 111–129.

Cornuet, J.-M., F. Santos, M. A. Beaumont, C. P. Robert, J.-M. Marin, D. J. Balding, T. Guillemaud, and A. Estoup (2008). 'Inferring Population History with DIY ABC: A User-Friendly Approach to Approximate Bayesian Computation', *Bioinformatics* 24: 2713–2719.
Creighton, H. and B. McClintock (1931). 'A Correlation of Cytological and Genetical Crossing-Over in Zea Mays', *Proceedings of the National Academy of Sciences of the USA* 17: 492–497.
Dawson, K. J. and K. Belkhir (2001). 'A Bayesian Approach to the Identification of Panmictic Populations and the Assignment of Individuals', *Genetical Research* 78: 59–77.
Durand, E. Y., F. Jay, O. Gaggiotti, and O. François (2009). 'Spatial Inference of Admixture Proportions and Secondary Contact Zones', *Molecular Biology and Evolution* 26: 1963–1973.
Durand, E. Y., N. Patterson, D. Reich, and M. Slatkin (2011). 'Testing for Ancient Admixture between Closely Related Populations', *Molecular Biology and Evolution* 28: 2239–2252.
Everitt, R. (2012). 'Bayesian Parameter Estimation for Latent Markov Random Fields and Social Networks', *Journal of Computational and Graphical Statistics* 21/4: 940–960.
Ewing, G. and J. Hermisson (2010). 'MSMS: A Coalescent Simulation Program Including Recombination, Demographic Structure and Selection at a Single Locus', *Bioinformatics* 26: 2064–2065.
Falush, D., M. Stephens, and J. K. Pritchard (2003). 'Inference of Population Structure Using Multilocus Genotype Data: Linked Loci and Correlated Allele Frequencies', *Genetics* 164: 1567–1587.
Ferguson, T. S. (1973). 'A Bayesian Analysis of Some Nonparametric Problems', *Annals of Statistics* 1: 209–230.
Fortuna, M. A., R. G. Albaladejo, L. Fernández, A. Aparicio, and J. Bascompte (2009). 'Networks of Spatial Genetic Variation Across Species', *Proceedings of the National Academy of Sciences of the USA* 106: 19044–19049.
Fraley, C. and A. E. Raftery (2009 [2006]). *MCLUST Version 3 for R: Normal Mixture Modeling and Model-Based Clustering*, Technical Report 504, University of Washington.
François, O., S. Ancelet, and G. Guillot (2006). 'Bayesian Clustering Using Hidden Markov Random Fields in Spatial Population Genetics', *Genetics* 174: 805–816.
François, O., M. Currat, N. Ray, E. Han, L. Excoffier, and J. Novembre (2010). 'Principal Component Analysis under Population Genetic Models of Range Expansion and Admixture', *Molecular Biology and Evolution* 27: 1257–1268.
François, O. and E. Durand (2010). 'Spatially Explicit Bayesian Clustering Models in Population Genetics', *Molecular Ecology Resources* 10: 773–784.
Gamerman, D. (1997). *Markov Chain Monte Carlo: Stochastic Simulation for Bayesian Inference*. London: Chapman and Hall.
Grelaud, A., C. P. Robert, and J.-M. Marin (2009). 'ABC Methods for Model Choice in Gibbs Random Fields', *Comptes Rendus de l'Académie des Sciences Paris* 347: 205–210.
Griffiths, R. C. (1981). 'Neutral Two-Locus Multiple Allele Models with Recombination', *Theoretical Population Biology* 19: 169–186.
Guillot, G. (2009). 'On the Inference of Spatial Structure from Population Genetics Data', *Bioinformatics* 25: 1796–1801.
Guillot, G., A. Estoup, and F. Mortier (2005). 'A Spatial Statistical Model for Landscape Genetics', *Genetics* 170: 1261–1280.
Guillot, G., R. Leblois, A. Coulon, and A. Frantz (2009). 'Statistical Methods in Spatial Genetics', *Molecular Ecology* 18: 4734–4756.

Hein, J., M. H. Schierup, and C. Wiuf (2005). *Gene Genealogies, Variation and Evolution: A Primer in Coalescent Theory*. Oxford: Oxford University Press.

Henderson, N. R. and D. M. Lambert (1982). 'No Significant Deviation from Random Mating of Worldwide Populations of Drosophila melanogaster', *Nature* 300: 437–440.

Herrera, C. M. and P. Bazaga (2008). 'Population-Genomic Approach Reveals Adaptive Floral Divergence in Discrete Populations of a Hawk Moth-Pollinated Violet', *Molecular Ecology* 17: 5378–5390.

Hudson, R. R. (1983). 'Properties of the Neutral Model with Intragenic Recombination', *Theoretical Population Biology* 23: 213–201.

Huelsenbeck, J. P. and P. Andolfatto (2007). 'Inference of Population Structure under a Dirichlet Process Model', *Genetics* 175: 1787–1802.

International HapMap Consortium (2007). 'A Second Generation Human Haplotype Map of over 3.1 Million SNPs', *Nature* 449: 851–861.

Johnson, N. and S. Kotz (1977). *Urn Models and their Application*. New York: John Wiley.

Kadanoff, L. P. (2000). *Statistical Physics: Statics, Dynamics and Renormalization*. Singapore: World Scientific.

Kalai, A. T., A. Moitra, and G. Valiant (2012). 'Disentangling Gaussians', *Communications of the ACM* 55/2: 113–120.

Kindermann, R. and J. L. Snell (1980). *Markov Random Fields and their Applications*. Providence, RI: American Mathematical Society.

Kingman, J. F. C. (1982). 'The Coalescent', *Stochastic Processes and their Applications* 13: 235–248.

Knowles, L. L. (2009). 'Statistical Phylogeography', *Annual Review of Ecology, Evolution, and Systematics* 40: 593–612.

Knowles, L. L. and W. P. Maddison (2002). 'Statistical Phylogeography', *Molecular Ecology* 11: 2623–2635.

Kyriazopoulou-Panagiotopoulou, S., D. K. Haghighi, S. J. Aerni, A. Sundquist, S. Bercovici, and S. Batzoglou (2011). 'Reconstruction of Genealogical Relationships with Applications to Phase III of HapMap', *Bioinformatics* 27: i333–i341.

Lawson, D. J. and D. Falush (2012). 'Population Identification Using Genetic Data', *Annual Review of Genomics and Human Genetics* 13: 337–361.

Lawson, D. J., G. Hellenthal, S. Myers, and D. Falush (2012). 'Inference of Population Structure Using Dense Haplotype Data', *PLoS Genetics* 8: e100245.

Lawson, D. J. and H. J. Jensen (2007). 'Neutral Evolution in a Biological Population as Diffusion in Phenotype Space: Reproduction with Local Mutation but without Selection', *Physical Review Letters* 98: 098102.

Li, N. and M. Stephens (2003). 'Modeling Linkage Disequilibrium and Identifying Recombination Hotspots Using Single-Nucleotide Polymorphism Data', *Genetics* 165: 2213–2233.

Lifshitz, I. M. (1952). 'On a Problem of the Theory of Perturbations Connected with Quantum Statistics', *Uspekhi Matematicheskikh Nauk* 7: 171–180.

Lynch, M. (2010). 'Evolution of the Mutation Rate', *Trends in Genetics* 26: 345–352.

Mardis, E. (2011). 'A Decade's Perspective on DNA Sequencing Technology', *Nature* 470: 198–203.

McVean, G. (2009). 'A Genealogical Interpretation of Principal Components Analysis', *PLoS Genetics* 5/10: e1000686.

Milgroom, M. G. (1996). 'Recombination and the Multilocus Structure of Fungal Populations', *Annual Review of Phytopathology* 34: 457–477.

Myers, S., R. Bowden, A. Tumian, R. E. Bontrop, C. Freeman, T. S. MacFie, G. McVean, and P. Donnelly (2010). 'Drive Against Hotspot Motifs in Primates Implicates the PRDM9 Gene in Meiotic Recombination', *Science* 327: 876–879.

Pemberton, T. J., C. Wang, J. Z. Li, and N. A. Rosenberg (2010). 'Inference of Unexpected Genetic Relatedness among Individuals in HapMap Phase III', *American Journal of Human Genetics* 87: 457–464.

Pickrell, J. K., G. Coop, J. Novembre, S. Kudaravalli, J. Z. Li, D. Absher, B. S. Srinivasan, G. S. Barsh, R. M. Myers, M. W. Feldman, and J. K. Pritchard (2009). 'Signals of Recent Positive Selection in a Worldwide Sample of Human Populations', *Genome Research* 19: 826–837.

Pickrell, J. K. and J. K. Pritchard (2012). 'Inference of Population Splits and Mixtures from Genome-Wide Allele Frequency Data', *PLoS Genetics* 8/11: e1002967.

Platt, A., M. Horton, Y. S. Huang, Y. Li, A. E. Anastasio, N. W. Mulyati, J. Ågren, O. Bossdorf, D. Byers, P. D. Kathleen, M. Dunning, E. B. Holub, A. Hudson, V. L. Corre, O. Loudet, F. Roux, N. Warthmann, D. Weigel, L. Rivero, R. Scholl, M. Nordborg, J. Bergelson, and J. O. Borevitz (2010). 'The Scale of Population Structure in Arabidopsis thaliana', *PLoS Genetics* 6: e1000843.

Price, A. L., N. J. Patterson, R. M. Plenge, M. E. Weinblatt, N. A. Shadick, and D. Reich (2006). 'Principal Components Analysis Corrects for Stratification in Genome-Wide Association Studies', *Nature Genetics* 38: 904–909.

Pritchard, J. K., M. Stephens, and P. Donnelly (2000). 'Inference of Population Structure Using Multilocus Genotype Data', *Genetics* 155: 945–959.

Redden, D. T., J. Divers, L. K. Vaughan, H. K. Tiwari, T. M. Beasley, J. R. Fernández, R. P. Kimberly, R. Feng, M. A. Padilla, N. Liu, M. B. Miller, and D. B. Allison (2006). 'Regional Admixture Mapping and Structured Association Testing: Conceptual Unification and an Extensible General Linear Model', *PLoS Genetics* 2: e137.

Redon, R., S. Ishikawa, K. R. Fitch, L. Feuk, G. H. Perry, T. D. Andrews, H. Fiegler, M. H. Shapero, A. R. Carson, W. Chen, E. K. Cho, S. Dallaire, J. L. Freeman, J. R. Gonzalez, M. Gratacos, J. Huang, D. Kalaitzopoulos, D. Komura, J. R. MacDonald, C. R. Marshall, R. Mei, L. Montgomery, K. Nishimura, K. Okamura, F. Shen, M. J. Somerville, J. Tchinda, A. Valsesia, C. Woodwark, F. Yang, J. Zhang, T. Zerjal, J. Zhang, L. Armengol, D. F. Conrad, X. Estivill, C. Tyler-Smith, N. P. Carter, H. Aburatani, C. Lee, K. W. Jones, S. W. Scherer, and M. E. Hurles (2006). 'Global Variation in Copy Number in the Human Genome', *Nature* 444: 444–454.

Rosenberg, N. J. Pritchard, J. Weber, H. Cann, K. Kidd, L. Zhivotovsky, and M. Feldman (2002). 'The Genetic Structure of Human Populations', *Science* 298: 2381–2385.

Schuster, S. C. (2008). 'Next-Generation Sequencing Transforms Today's Biology', *Nature Methods* 5: 16–18.

Sladek, R., G. Rocheleau, J. Rung, C. Dina, L. Shen, D. Serre, P. Boutin, D. Vincent, A. Belisle, S. Hadjadj, B. Balkau, B. Heude, G. Charpentier, T. Hudson, A. Montpetit, A. Pshezhetsky, M. Prentki, B. Posner, D. Balding, D. Meyre, C. Polychronakos, and P. Froguel (2007). 'A Genome-Wide Association Study Identifies Novel Risk Loci for Type 2 Diabetes', *Nature* 445: 881–885.

Sokal, R. R., N. L. Oden, and C. Wilson (1991). 'Genetic Evidence for the Spread of Agriculture in Europe by Demic Diffusion', *Nature* 351: 143–145.

Stevens, E. L., G. Heckenberg, E. D. O. Roberson, J. D. Baugher, T. J. Downey, and J. Pevsner (2011). 'Inference of Relationships in Population Data Using Identity-by-Descent and Identity-by-State', *PLoS Genetics* 7: e1002287.

Thornton, K. and P. Andolfatto (2006). 'Approximate Bayesian Inference Reveals Evidence for a Recent, Severe Bottleneck in a Netherlands Population of Drosophila melanogaster', *Genetics* 172: 1607–1619.

Tiwari, H., J. Barnholtz-Sloan, N. Wineinger, M. Padilla, L. Vaughan, and D. Allison (2008). 'Review and Evaluation of Methods Correcting for Population Stratification with a Focus on Underlying Statistical Principles', *Human Heredity* 66: 67–86.

Waples, R. S. and O. Gaggiotti (2006). 'What Is a Population? An Empirical Evaluation of Some Genetic Methods for Identifying the Number of Gene Pools and their Degree of Connectivity', *Molecular Ecology* 15: 1419–1439.

Wasserman, L. (2005). *All of Nonparametric Statistics*. New York: Springer.

Wright, S. (1984). *Evolution and the Genetics of Populations*. Chicago: University of Chicago Press.

CHAPTER 4

POPULATION HETEROGENEITY IN THE SPOTLIGHT OF BIODEMOGRAPHY

KENNETH W. WACHTER

A person's survival curve—does such a thing exist? Demographers are heard to use such words, although, on the face of it, vital rates are only definable for population aggregates, not for individuals. Conditioning, in fact or in imagination, on more and more sources of heterogeneity makes constructs like survival curves become more and more nearly personal. How must our concepts respond to the shifting fuzzy boundary between observed and unobserved, observable and unobservable, heterogeneity?

Endeavouring to extend evolutionary theories of senescence and account for commonalities across species in the shapes of demographic schedules, biodemographers have developed distinct specifications of heterogeneity. I discuss the conceptualization of population heterogeneity in current research in the biodemography of longevity.

4.1 Populations of Many and One

I came a while ago on a quotation ascribed to Lord Palmerston. As friends and family gathered round what they took to be his deathbed, he turned to them and said,

> 'Die? That is the last thing I shall do.'

I like this statement for three reasons. First is the insight. It is a gift to be able to speak truth by truism, playing on the fine line between conveying information and affirming features of the nature of the world. My aim in this essay is not to convey information but to bring from the background into the foreground some things that we already know.

Second, I like the recognition of agency. 'Die . . . [a] thing I shall do.' In our decades quantitative social science is becoming more and more the study of full life histories. Longitudinal perspectives invite us to think of outcomes not just as events that happen but as culminations of paths broadly shaped and partly chosen. Even dying. Volition may often be inaccessible to standard sample survey instruments and inconvenient for predictive modelling, but it lurks.

Third, Lord Palmerston's words are about death and survival. Death and survival and mathematical curves that describe them are the examples I shall be using. They provide good illustrations for the conceptual issues that arise with all sorts of outcomes that social scientists measure and model in populations.

Death, survival, and populations: each of us has a death. Does each of us have a survival curve? Demographers certainly do talk about your survival curve and my survival curve. However, when pressed, we tend to fall back on saying that what we mean is not a survival curve for you or for me but for the population or populations to which you and I belong. What is that or what are those? How is it that we define a population when we are trying to define a survival curve?

I shall let this question about populations and survival curves stand in for the question, 'What is a population in general?'

Purists may jump in and say that each of us does have a survival curve. For me, it is the step function that equals 1 for ages up to my age at death and equals 0 beyond my age of death out to infinity. My survival curve is the survival curve for the population of size one consisting of me.

In this sense, collections of people, any collection of people, also have a survival curve, fully observable a hundred and some years from now, built from step functions. The curve for a group of size n has a downward step of size $1/n$ at each age that is the age at death of one person in the group.

All this is true, but when we teach demography we have to finesse a distinction, because the survival curves we teach are not typically full of jerky steps but smooth. They have to be smooth, because we teach how to take the slopes of logarithms of reciprocals of survival curves at every age and call these slopes hazard functions. The jerky curves have no slopes at the downsteps and no well-defined hazard functions. So, at the outset, a subtle move comes into play. In place of thinking about all lives that are, we start to think about lives that might be. We define not in terms of what can be measured but in terms of what can only be imagined.

This move from actual to imagined is a first example for my general theme. Let me go over the storyline. What is a population? Start with an innocent definition:

> A population is a collection of people with certain characteristics.

When demographers say such words, they begin by having in mind real people and measured characteristics, many real people and not too many characteristics. People can be listed on a census and tabulated according to responses. The United States Constitution enshrines the words 'actual enumeration'. Often the people are those specified by characteristics like citizenship or residence. Samples may also supply data, but behind a sample is seen to be a census-like sampling frame. However, demographers come to consort with mathematical statisticians, folks who prefer to say

> A population is a collection of real or imagined people with certain measured or hypothetical characteristics.

I focus on this slide from the measured to the hypothetical. There is an irony about it. The more we are able to measure, the less salient the measures. The more salient the hypothetical becomes.

Elaborating on this irony is my aim in this essay.

The tension between the measured and the hypothetical is at the heart of any answer to the question, 'Does a person have a survival curve?' Within a statistical framework, the difference between survival and death is being made instant by instant, as it were by rolls of dice. The population to which a person belongs is a collection of imaginary people alike in all respects except for different rolls of the dice. One real person, many imaginary people. In this framework, each person does have a survival curve, specifying the distribution of the deaths of the doppelgänger for all countless possible combinations of rolls.

You may worry that we have here a distinction without a difference, because we only learn about imaginary people alike in all respects by comparing actual people alike in some respects. Yes and no. Yes: the empirical underpinning of statistical accounts comes mostly from observations on the populations of real people whom demographers tabulate. No: there is a kind of alternative. If we can understand the mechanisms by which dice rolls are connected to deaths, we can hope to be building a story world which life will imitate.

Philip Kreager has suggested to me an example which carries further what I have in mind. In his words,

> we enter a world in which what common sense understands as real or potentially real is decided or at least strongly led by formalized 'stories' or models....

When I purchase an annuity from an insurance company, the flow of premiums I shall send and the payments I shall receive will be entirely tangible. But they are determined by an actuary who is fitting some set of my characteristics to an imaginary population whose survival is shown in a hypothetical, predictive life table. My life, with its premiums and payments, is already imitating a model. An imaginary population with its hypothetical life table is being used to create an actual population, the members vested in the annuity scheme, with the purpose of solving the problem we face in the workaday world of planning for mortality.

This example brings to mind the parsing of concepts in play in early life tables, as noted by Philip Kreager (this volume). John Graunt's life table was not a tabulation of observations of survival from a delineated group of people but rather a hypothetical construction directed at a practical purpose, as indeed, it seems, was the ancient Roman life table attributed to Ulpian. Life tables were taken up and developed in the 1600s and beyond in tandem with mathematics for predicting outcomes from the throwing of dice, applying the results to eminently human science. My discussion in this essay concentrates on concepts of population and heterogeneity in their present guise and shortchanges their origins. But it is clear that these ideas have deep historical roots.

The purpose for which hypothetical populations, governed by rolls of dice, are being invented is usually prediction. We keep in mind two kinds of prediction, predicting my death or your death and predicting changes in population patterns of death. Progress on the first goal does often bring progress on the second goal, as, for example, from what we have learned about smoking. But the slide from the measured to the hypothetical, bound up initially with the first goal, may be specially relevant for promoting the second goal.

What we have been saying about the meaning of words like 'population' and 'survival curve' brings us directly to the other chief word for this essay, 'heterogeneity'. A population of real people is typically an aggregate of populations of real people each more finely specified by a suite of characteristics the members share. That is to say, the aggregate is heterogeneous. The observed aggregate is often just a starting point, typically coinciding with the collection of people covered by an administrative data system that happens to be in place. Conceptual issues come to the fore with the fine specification of constituent populations. To preview a main example, it is becoming routine to measure millions of genetic markers per individual. What should be meant by survival curves when we condition on vastly more specifiable combinations than there are people alive today or ever?

On the shore, when we see a wave break, we probably do not ask ourselves, 'Why did this wave break just here?' But when we see a loved one die, we so often do ask, 'Why now?'

> Why lung cancer, when she never smoked?
> Why a heart attack, when he ate nothing but health foods?
> Why did she live to 97, when she never exercised and her mother died before 50?

We feel an impulse to find a reason or a handful of reasons for each one person's death. Science has found some factors that on repeated observation have big enough effects on survival to count as reasons for a death. But mostly the scientific picture, at odds with personal impulse, sees cause as a matter of many many factors with mostly small, merely statistically detectable effects, interacting, cumulating, counteracting. As for waves, physicists know a lot about the statistics of wave height, wave speed, wave length, about probabilities for cresting and for water droplets to race on and spill over into foam, and physicists imagine that one could know all the positions and momenta (or the wave function of the wave in its function) and with total information predict each wave's demise. Nonetheless, a Zen master might still ask, what is the cause of one wave breaking?

With people, this metaphor has another kind of force, or at least I fear so. As health scientists are able to gather more and more individualized data, we might imagine that we are proceeding, though far off, in the direction of the physicists' dream of total knowledge for total prediction. However, the kinds of microdata that predominate in demographic studies of mortality do not seem to be endowing demographers with more leverage to predict before the fact or account after the fact for large-scale shifts in trends, for big-picture phenomena. It seems to require different kinds of knowledge to pick out from all the single waves the signal of a sea change.

4.2 Risks, Frailties, and 'Lifeluck'

I now centre in on my main theme. I consider three kinds of heterogeneity and draw on three examples from recent work in biodemography that bring them into focus. These examples epitomize the irony that ever more measurement makes measurement less salient.

In distinguishing kinds of heterogeneity, it is helpful to have a concrete setting. For this purpose I choose a small exercise I use in teaching the demographic method of proportional hazards. The step from general hazards to proportional hazards is itself an example of the slide from the measured to the hypothetical, but I skip over that part of the story. Proportional hazards retain enough complexity to illustrate our main distinctions. For more detail, see Chapter Eight of Wachter (2014).

For my class exercise on proportional hazards, I put each student in possession of three personalized numbers, called 'risk', 'frailty', and 'lifeluck'. I shall describe them shortly as components of explained, unexplained, and—for want of a better word—currently unexplainable heterogeneity. The first of these numbers, 'risk', is a number each student calculates for himself or herself from coefficients I write on the blackboard. The other two numbers for each person are numbers I draw from a random number generator and hand to them. I pass around on graph paper a baseline survival curve. Each student adds his or her three numbers together, applies an exponential function to the sum, and comes up with an answer. The exponential function is a technical step which turns additive effects into proportional factors. The student goes on to find the age on the graph paper at which the height of the survival curve equals his or her own answer. For purposes of the class, this is the student's age at death.

Some students find that they will live beyond 100, and they are generally pleased, perhaps not for good reasons. Occasionally a student finds himself or herself hypothetically facing imminent death, and thinks again.

The student's index of risk is the output of a calculation which, in this context, is taking account of some specified subpopulation to which students can observe themselves to belong. Here the meaning of population comes to the fore. In a simple case, an African-American non-smoking male student receives the risk for African-American non-smoking males. We are interested as well in less simple cases. The index of risk encapsulates the variability in chances of death that we are in a position to explain.

The student's index of frailty is a random number, such as statisticians put into a 'random effects model'. It takes into account differences within a population due to intrinsic susceptibilities that are unobserved at an individual level but which make individuals differ from each other. This component of heterogeneity is inferred but not yet explained. The boundary between explained and unexplained heterogeneity shifts as science progresses. Genetic alleles with which a student is born would have been a prime example of a contributor to frailty a few decades ago, but now some of them can begin to be subsumed into indices of risk.

The third component I call 'lifeluck'. When risk and frailty are given, in effect a student is assigned to a population that is specified as narrowly as can be specified. But, in this way of thinking, what is specified is not enough to determine the exact age at death. Much remains indeterminate or inaccessible to knowledge, the province of residual or inescapable haphazardness. Within a population, ages at death still vary. Lifeluck fixes for you where in the distribution of ages at death in your subpopulation your death falls. Are you at the longest-lived tenth percentile of your peers? At the median? Lifeluck places you, handed down by me in class or by fate in life.

As with the boundary between risk and frailty, the boundary between frailty and lifeluck, between unexplained and 'unexplainable' heterogeneity, is fuzzy and shifting. In practice, the distinction is made by imposing certain technical 'identifiability restrictions' in the style of econometricians, which reflect hypotheses or prejudices about what purified underlying shapes of survival curves should be.

Consider in more detail the index of risk and the role that measured characteristics play in the specification of a population. The coefficients which I write on the blackboard for students' calculations are estimated from large, often longitudinal health surveys, a central enterprise of the scientific infrastructure of public health. With ever larger samples, risks specific to more and more narrowly defined cross-classified populations are being estimated. But reflect on how this vast enterprise has to leave my students unsatisfied:

> You've written down one coefficient for African American. But my genetic readout tells me I'm 59% West-African American, 29% East-African American, and 12% Irish. What population do I belong to? What is my coefficient?
>
> I'm an ardent bungee jumper. Must that enter into my calculation of risk?
>
> I used to be a bungee jumper, but I stopped. Do I penalize my risk score for the impulses that led me to try bungee jumping in the first place, or, having stopped, do I gain full membership in the non-bungee-jumping population?

These questions challenge the conceptualization of population in related ways. Richer datasets and larger samples lure us toward considering calculations of risk specific to populations delineated by finer and finer cross-classifications and by broader classes of determinants. It may seem an easy move from variables for race to variables for mixed-race profiles, from behaviours like smoking to constellations of behaviours full of fine-grained daily choices, from controls for self-selection by, say, income, to controls for innate tendencies, or, inversely to controls for opting to act in accordance with innate tendencies. In practical terms, genetic markers, longitudinal surveys, biometric measurements, continuous monitors, and other advances are providing variables which seem suited to such moves. But going ever further down the path to more and more variables shifts what we mean by population.

The subtleties of volitional self-selection, of ever-shifting populations formed out of the choices and affinities of individual actors, are treated in depth in other contributions to this volume by Jenna Johnson-Hanks, by Simon Szreter, by Philip Kreager, and others. These issues are one central part of what is making definitions of populations in terms of specific dimensions of explained heterogeneity problematic. Here I concentrate on another part, the generic issue of finer and finer specification.

4.3 Risk Groups and Genetic Markers

In the domain of explained heterogeneity and observable contributions to risk, the intersection of biodemography with medical genetics offers readily visualized examples.

Most of us are aware of famous conditions and diseases that are caused, in a fairly direct sense, by the presence of certain alleles or genetic variants in a person's DNA. Haemophilia is one, besetting royal families in Europe. Huntington's disease is another, passed on when a person inherits a stretch of DNA with a pattern that looks like an excessive stutter when the genetic information on the fourth chromosome is written out in customary form with letters from a four-letter alphabet of G, A, T, and C. Such Mendelian diseases are dreadful but rare. In these cases, the population to which a person carrying the mutation can be assigned, when a survival curve is to be computed, can be quite sharply defined. The population is the group of individuals carrying the mutation. It is formed over generations via mating and procreation as a subset of descendants of individuals in whose genomes the offending ancestral mutations first arose.

This picture extends easily to populations defined by a criterion that their members carry some particular combination of variants of several genes, or a dozen genes, or a hundred genes. But the present state of medical genetics pushes further and is straining this picture. Today studies are available which measure combinations of a few million variants for thousands of respondents. Soon, unless derailed by patent claims on the living constituents of our own bodies, it will be feasible to measure all the variants any one of us carries.

Among the datasets available to qualified researchers are genetic markers for some thirteen thousand respondents to the US Health and Retirement Survey (HRS), the premiere study sponsored by the US National Institute on Aging. The data include two-and-a-half million new variables on each respondent. Analyses using data of this kind are called genome-wide association studies (nicknamed GWAS). With the HRS, many years of longitudinal data are also in hand tracking the circumstances, health conditions, self-reported proclivities, behaviours, and survival of these respondents. A number of other leading surveys constructed to be representative of national population aggregates also have or shortly will have such rich data. In prospect are even richer collections, relating to proteomes, metabolomes, and other frontier objects of biomedical investigation, still mysterious to many social scientists.

With some millions of variables per person, what can I stand to learn? Already students in my class can probably buy, and soon cheaply, their personal read-outs on many of these variables. They can know in detail exactly what population each of them belongs to, if by population we mean a group defined by combinations of genetic variants cross-classified by circumstances and behaviour. Each student might be in a position to calculate an intimately personal index of risk, but only if I could find enough chalk in the world and knowledge at my fingertips to put all the relevant coefficients on the blackboard. What population does each of them belong to? Cross-classifying these variables would give more categories than there are protons, not to say people, in the visible universe. Each of my students belongs to a population consisting solely of her or him.

Within a genome-wide association study, if an analyst puts all the variables into a familiar linear regression (for, say, five-year survival), perfect within-sample fit is achieved. But out-of-sample predictive power is negligible. Increasing sample size a hundredfold, stretching any budget, would leave the situation the same. The number of variables outruns the number of respondents.

What do afficionados of such genome-wide studies actually do? They run two million separate regressions, one for each marker. They look for isolated heros, variables with univariate main effects (here, on survival) so large that they leap out above what is to be expected by chance alone. For some outcomes, a handful of such effective predictors have been found, but heros are few and far between. With the most traditionally tractable outcomes, like height, it appears that identifiable genetic markers account for only small portions of the measured heritable variation in height. Most of that variation must be caused by the statistical cumulation of large numbers of small effects, below the threshold of detectability.

In this picture, characterization of the mechanisms of cumulation, modes of independence and interdependence of small contributing effects, is an essential element. Thinking about survival curves in social science starts to be reminiscent of statistical mechanics in the history of physics. For understanding temperature and pressure, physicists put together a theoretical story about the mechanisms of molecular motion. For understanding survival curves, social scientists have a need, not just for more and more variables, but for theoretical stories that give leverage over ways many small biological and social effects can plausibly combine.

I take my illustration from genetic markers. But at this level of generality, much of this discussion applies also to the proliferation of data and categories from longitudinal studies of life histories. Prominent current work includes studies of early-life effects on later-life mortality or on trajectories of life experience that shape resilience at older ages. Human experience is various, and studies easily record so many combinations that, in terms of them, the population to which any individual belongs consists solely of him or her self.

Here, then, we have an example of my general theme. The expansion of measurement shifts practical and conceptual emphasis away from the measured onto the hypothetical. While there are small rosters of measured variables with strong effects, like smoking or cholesterol, on which there can be interventions at societal and individual levels that matter for survival, the preponderance of ever-richer measurements merge into a statistical melange.

4.4 Frailties and Accumulating Mutations

We arrive, then, at the subject of unexplained or unobserved heterogeneity, the insertion of random effects on top of fixed effects in the statistical formulation of survival curves and hazard functions.

The familiar form in demography for a random effect is 'frailty', a random person-specific proportional factor multiplying the hazard function, introduced by Vaupel et al. (1979). My class exercise uses this simple set-up. Each person is assigned a lifelong unchanging random value. I follow the common practice of drawing this multiplier from a statistical gamma distribution and taking its logarithm, so that my students can add it to risk and lifeluck before they undo the logarithm by applying an exponential function.

In this standard setting a population could be regarded as a collection of people sharing both measured characteristics and a hypothetical value of the frailty index. The variability in frailty across an aggregate population shows up through demographic selection. Members with higher levels of frailty are culled from the population by death at earlier ages, leaving an aggregate group of survivors enriched for lower levels of frailty at older ages. A survival curve for the aggregate is forced to flatten out in comparison to survival curves for constituent populations.

As I have been arguing, this standard treatment is intrinsically hypothetical. The constitutent populations are unobservable. Shapes for the hazard curves that are rescaled by the multipliers and statistical distributions for the multipliers themselves have to be posited by an investigator, on the basis of convenience, whimsy, or theoretical speculation about underlying mechanisms. The investigator has wide discretion, but not unlimited freedom. As we come to be able to measure more and more, each shift over time in the boundary between measured and unmeasured variation gives some empirical checks on the consistency and parsimony of assumptions.

From students calculating life chances in my class, the implication of shifting attention from the measured to the hypothetical is that more of what they might expect to put into their index of risk turns up instead in their index of frailty. They know the values of more and more variables about themselves, but I am unable to write coefficients on the blackboard for ever-finer categories. What seems like progress toward more explained heterogeneity turns into greater emphasis on unexplained heterogeneity. It is the statistical properties of the extra variables that are potentially exploitable, not the individual-level values. In the language of statistical modelling, fixed effects are not profitably added *ad infinitem*; random effects take on appeal.

Biodemography supplies a more far-reaching example of the characterization of hazard functions in terms of random effects or unobserved heterogeneity. Such characterizations are coming to the fore in models for the genetic process of mutation accumulation. The idea of mutation accumulation was introduced by Sir Peter Medawar (1952) as a possible evolutionary explanation of senescence. He proposed that there might be large numbers of genetic alleles with small but eminently age-specific deleterious effects on survival (or fertility). Alleles whose negative consequences are felt predominately at young ages have big effects on reproductive fitness and natural selection removes them quickly, generation by generation, from representation in the population. Alleles with late effects are removed more slowly and are found at higher frequencies when an equilibrium is reached between new mutations and removal through natural selection.

This is not the place to go into technical details, which may be found for instance in Wachter et al. (2009) and Evans et al. (2012). What is relevant for our purpose is the treatment of population heterogeneity. The basic ingredient is a hypothetical space of possible (fixed or random) profiles for the age-specific action of alleles. A population for which a survival curve can be specified is defined to be the hypothetical collection of individuals sharing a specific combination of alleles. In most applications it is assumed that there is an infinite variety of possible age-specific profiles, hence an infinite hypothetical aggregate population, from which members of an actual finite population are assumed to be drawn by a process of random sampling.

The description of possible profiles can be informed by the measured age-specific action of the handful of known alleles or combinations of alleles so far discovered whose effects are large enough to be detectable. But alleles with detectable effects may not be entirely informative with regard to larger numbers of alleles with effects below the threshold of detectability. What makes the approach feasible is our theoretical understanding of the mechanism of natural selection, along with the generation-by-generation intervention of genetic recombination. Losses to reproductive fitness depend on age in ways that can be specified in some generality. It turns out that the genetic mechanisms constrain the shapes of survival curves that can be generated within this framework. To a considerable extent, the details of the hypothetical actions of individual alleles do not matter to predicted aggregate survival curves.

Mutation accumulation, then, supplies an example where the ability to measure more and more genetic markers directs attention beyond the markers themselves to mechanisms and constraints which allow hypothetical specifications of population structure to support predictions of aggregate features of survival curves. I suspect that the ability to measure more and more experiential and behavioural markers from longitudinal studies has similar potential to direct attention beyond the markers to theoretical mechanisms and constraints. It also seems likely to me that, in the face of an abundance of small, interacting effects, insight into mechanisms and constraints will be what is most useful for predicting population-level change points in trends and the efficacy of societal interventions.

4.5 Lifeluck and Stochastic Vitality

We have discussed so far two kinds of heterogeneity and their implications for our concept of a population. We come now to a third kind, which I call lifeluck. It registers the residual heterogeneity in ages at death that remains after controlling for observed dimensions of heterogeneity and adjusting for explicitly modelled random effects.

In the familiar setting of proportional hazards, such residual heterogeneity is expressed by the baseline hazard function. We create a baseline to correspond to some standard set of values for measured covariates and to a value of unity for frailty. The hazard function for individuals who share these standard values is the baseline hazard function. These individuals do not all die at the same age. Their ages at death spread out in accordance with the baseline life table calculated from the baseline hazard function. The remaining variation reflects heterogeneity that is currently unexplainable by the fixed and random effects incorporated into models.

Lifeluck specifies where in the distribution of ages at death a person's own death falls, among deaths distributed according to the baseline life table. As mentioned before, it is conveniently measured on a logarithmic scale, for consistency with proportional effects of risk and frailty.

One way to visualize lifeluck is to fall back on the idea of fate and think of a kind of predestination, in which each newborn comes into the world carrying within himself or herself a ticket telling the time of his or her demise. In this essay, I have been relying on a different approach to visualization, in which lifeluck is honed over time by countless rolls of a pair of dice. It turns out that recent work in probability theory by Henning, Ettinger, and Evans (2014) allows a formal elaboration of this picture of random progression. Essentially any life table survivorship function whose hazard rate is smooth and bounded (never going to infinity) can be represented by an elementary kind of random process in which a hidden state, which we may call 'vitality', moves up or down in a continuous version of a random walk called a Brownian motion. Up or down depends in effect on throws of dice. Death occurs, not instantly but at high risk, when a person's vitality drops below a threshold. The threshold changes as a function of age. By adjusting the threshold, almost any baseline life table can be realized as such a dice-driven process.

This representation of life tables through Brownian motions helps us circle back to the conceptual move which took us at the outset from demographers' populations of real people to statisticians' populations of imaginary alter egos, alike in everything except for countless different rolls of dice. In stochastic vitality models, each person's life course is one realization of a random trajectory. Each person has a survival curve of his or her own, formed from the distribution of deaths that would take place, given all the different rolls of dice that there might be.

Biodemography supplies examples of more ramified versions of stochastic vitality models. These models are also called models of changing frailty. They can be regarded as attempts to generalize models with fixed frailty, in which the frailty multiplier for an individual stays the same throughout life, to allow for time-varying suites of random effects. Naturally, full forms of these models are not intended to stand on their own, but, as in our description of lifeluck, to come into action after as much as possible of measured risk and inferred frailty have been taken into account.

As with mutation accumulation models, stochastic vitality models allow predictions of generic features of survival curves from hypothetical descriptions, in this case descriptions of random ups and downs in intrinsic physiological states. Beside a few large measurable effects, large numbers of small cumulating background effects take their place in shaping life history trajectories.

Study of stochastic vitality models has proceeded in two main directions. In one of these directions, researchers have studied models which fit into well-analysed domains of probability theory. (Cf. e.g. Steinsaltz and Evans 2004; Aalen and Gjessing 2001.) Similar to fixed frailty models, these models describe aggregate populations that are subject to demographic selection or culling. But over the life course an aggregate population is not losing to death members with life-compromising attributes. It is losing members who have hypothetically been dealt unfavourable outcomes cumulating in small steps over many random draws.

The other direction in the study of stochastic vitality, often called stochastic risk factor models, has a long history. Over several decades Kenneth Manton and his collaborators, including Anatoly Yashin, have taken the lead in developing models in which identifiable risk factors have transitions in levels across the life course with probabilities estimated from longitudinal surveys. (Cf. e.g. Yashin and Manton 1997.) The cumulative impact of small undetectable effects is subsumed into a random specification which mediates between risk factors and survival outcomes. Over the years most of the attention in development of these models has gone into the identifiable risk factors. Less attention has gone into the random process specification.

What happens as the number of measured variables accessible from longitudinal data grows far beyond our ability to obtain estimates of their joint effects with any practicable sample size? Emphasis necessarily shifts from the identifiable factors in the model to the random specifications. The two directions we have mentioned start aligning with each other. The demand grows for theoretical understanding of mechanisms of combination and for predictions of generic features of survival curves.

Let me list four questions about such generic features:

> Why are measured effects on hazard functions so often proportional? What mechanisms underlie the success of proportional hazards models?
>
> Why are increases from age to age in mortality rates from mid-adulthood to moderate old age so often proportional? Such proportional increases lead to what are called Gompertz mortality models. What accounts for the ubiquity of Gompertz mortality?
>
> Why do these proportional increases with age level out at extreme ages, not only in humans but in many other species? Is the levelling entirely produced by culling in the presence of one or another of the forms that population heterogeneity can take?
>
> Why are many features of aggregate population survival curves found in common across species with widely dissimilar body plans and life history strategies? How important is our evolutionary heritage to human demographic regularities?

It is ironic that the more we are able to measure, the more hypothetical our concept of population becomes. It is also promising because, I believe, the ability to understand and foresee big-picture trends and change points demands big-question answers.

When I first came to the UC Berkeley Statistics Department, Jerzy Neyman, who inadvertently saddled social science with the stars and double-stars of statistical significance testing, was fond of saying

> Life is complicated, but not uninteresting.

But in science, what is too complicated becomes uninteresting. Too many predictors, too many interactions, can mean too much specificity. A population defined as a collection of actual people with certain measured characteristics can carry us a long way, but when it carries us to the point that each person belongs to a population consisting solely of himself or herself, it carries us too far. The more we can measure, the more the hypothetical comes to the fore, or so our illustrations from biodemography suggest to me. We move toward a concept in which a population is the collection of characters in a story, in some story we can begin to tell.

A story is a fiction, but this kind of story is also a prediction. As we live out our lives and march toward our deaths, the characters in the story turn out to be us. In ways that other contributors to this volume help to clarify, the story is not just something that happens but is something partly chosen. The population to which each of us belongs is being built by our decisions and affiliations. The examples in this essay drawn from biodemography largely address opportunities and constraints shaped by physiology, genetic inheritance, and evolution. The kinds of heterogeneity they bring into focus show up across human sciences, as envivornment, economy, society, and will power exercise their own sway over survival curves.

It is doubtful that great precision will ever be achieved in predicting individual death. The residual role of lifeluck is not likely to fade away. But the analysis of modes of heterogeneity serves another purpose, directing attention to mechanisms as well as measurements. Hope persists that trends and turning points in demographic rates, sea changes in the distributions of heterogeneities, can come to be anticipated.

Composing this essay, I was walking on the beach, and I turned to watch a sea-star on a slightly offshore rock. Little waves were dribbling in, some a little more, some a little less, so I was stepping backwards and forwards to keep my feet dry. Then one so little wave swept twenty feet further up the sand, and I backstepped as fast as I could ahead of it. I said to myself, 'The tide has turned. It's coming in.' The following thirty waves looked like any earlier thirty. Hardly a difference to detect. But the tide came in.

ACKNOWLEDGEMENTS

I appreciate support from the Behavioral and Social Research Division of the US National Institute on Aging, Grant P01 AG022500-06A1. I am grateful to Philip Kreager and Jenna Johnson-Hanks for stimulating suggestions.

REFERENCES

Aalen, O. O. and H. K. Gjessing (2001). 'Understanding the Shape of the Hazard Rate: A Process Point of View', *Statistical Science* 16/1: 1–22.

Medawar, P. (1952). *An Unsolved Problem in Biology: An Inaugural Lecture Delivered at University College, London, 6 December, 1951*. London: H. K. Lewis & Co.

Evans, S. N., D. Steinsaltz, and K. W. Wachter (2012). *A Mutation-Selection Model for General Genotypes with Recombination*, American Mathematical Society Memoir 222(1044) (ebook).

Hening, A., B. Ettinger, and S. N. Evans (2014). 'Killed Brownian Motion with a Prescribed Lifetime Distribution and Models of Default', *Annals of Applied Probability* 24: 1–33.

Steinsaltz, D. and S. N. Evans (2004). 'Markov Mortality Models: Implications of Quasi-Stationarity and Varying Initial Conditions', *Theoretical Population Biology* 65/4: 319–337.

Vaupel, J. W., K. G. Manton, and E. Stallard (1979). 'The Impact of Heterogeneity in Individual Frailty on the Dynamics of Mortality', *Demography* 16/3: 439–454.

Wachter, K. W. (2014). *Essential Demographic Methods*. Cambridge, MA: Harvard University Press.

Wachter, K. W., D. R. Steinsaltz, and S. N. Evans (2009). 'Vital Rates and the Action of Mutation Accumulation', *Journal of Population Ageing* 2/1: 5–22.

Yashin, A. I. and K. G. Manton (1997). 'Effects of Unobserved and Partially Observed Covariate Processes on System Failure: A Review of Models and Estimation Strategies', *Statistical Science* 12/1: 20–34.

PART III

RETHINKING INTRA- AND INTERPOPULATION DYNAMICS

CHAPTER 5

NICHE CONSTRUCTION IN HUMAN EVOLUTION AND DEMOGRAPHY

JOHN ODLING-SMEE

5.1 Niche Construction Theory

NICHE construction theory is a branch of evolutionary biology that focuses on the capacity of organisms to *perturb* and modify their environments through their metabolisms, activities, and choices (Odling-Smee et al. 2003). The term also includes migration, dispersal, and habitat selection, when organisms *relocate* in space and encounter different selective environments by doing so. The defining characteristic of *niche construction* is not the modification of environments per se, but changing niche relationships between organisms and their environments, caused by the prior modification of some natural selection pressures by niche-constructing organisms, coupled with the subsequent inheritance of previously modified selection pressures by descendent organisms via a second inheritance system, *ecological inheritance*. Ecological inheritance is the inheritance, through an external environment, of one or more natural selection pressures previously modified by niche construction. Niche construction thus becomes a source of feedback in evolution that, in conjunction with natural selection, co-directs the evolution of organisms by the activities of organisms.

Innumerable examples of niche construction are well known to biologists, even though they are not always described as niche construction. For example, animals manufacture nests, burrows, webs, and pupal cases; plants modify fire regimes, levels of atmospheric gases, and nutrient cycles; fungi decompose organic matter; and bacteria fix nutrients. Niche construction has also been modelled by theorists. Components of niche construction are implicit in many standard evolutionary models of frequency- and density-dependent selection; habitat selection; maternal inheritance; indirect genetic effects; and coevolution. More recently, non-standard models have investigated some general consequences of niche construction not previously investigated (e.g. Laland et al. 1996, 1999, 2001; Schwilk and Ackerly 2001; Ihara and Feldman 2004; Silver and Di Paolo 2006). To date, all the latter models have found niche construction to be consequential. Typically, it changes the dynamics of the evolutionary process, sometimes radically.

5.1.1 Why Niche Construction Theory Is Different

The population geneticist, Richard Lewontin (1983), was one of the earliest advocates of niche construction. He once neatly summarized the principal difference between standard evolutionary theory (henceforth SET) and niche construction theory (henceforth NCT) in terms of two pairs of coupled differential equations. His first pair (1) summarized SET:

$$dO / dt = f(O,E), \tag{1a}$$

$$dE / dt = g(E) \tag{1b}$$

In (1a) evolutionary change in organisms *dO/dt*, depends on both organisms' states, *O*, and environmental states, *E*. In (1b), however, environmental change, *dE/dt*, depends exclusively on environmental states. With many caveats and complications, organisms are not treated as the cause of any evolutionarily significant changes in their environments by SET.

His second pair (2) summarized NCT:

$$dO / dt = f(O,E), \tag{2a}$$

$$dE / dt = g(O,E) \tag{2b}$$

Change in organisms, *dO/dt*, is again assumed to depend on both organisms' states and environmental states (2a), but environmental change, *dE/dt*, is now assumed to depend on both environmental states and the niche-constructing activities of organisms (2b).

Figure 5.1 illustrates the same difference graphically. In SET the evolution of organisms is directed exclusively by sources of natural selection in their environments (Figure 5.1a). Selective environments, E, act on populations of diverse phenotypes, and influence which individuals survive and reproduce, and pass on their genes to the next generation through a single inheritance system, genetic inheritance. In contrast, in NCT (Figure 5.1b) the evolution of organisms is directed by both natural selection and niche construction. The transmission of genes by ancestral organisms to their descendants is influenced by natural selection. However, genetic inheritance is also influenced by niche construction. This is because selected habitats, modified habitats, and modified sources of natural selection in those habitats, are transmitted by niche-constructing organisms to their descendants by *ecological inheritance*, relative to their genetics. In NCT the

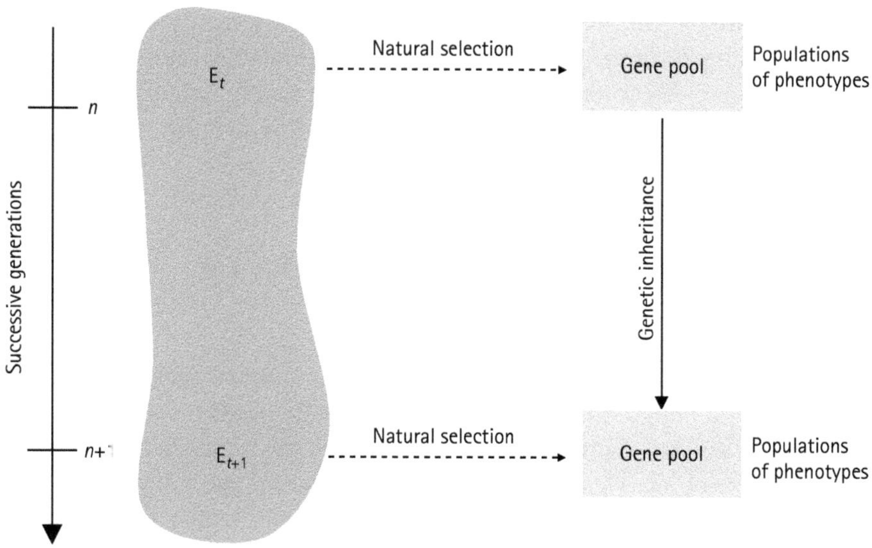

FIG. 5.1a Standard evolutionary theory.

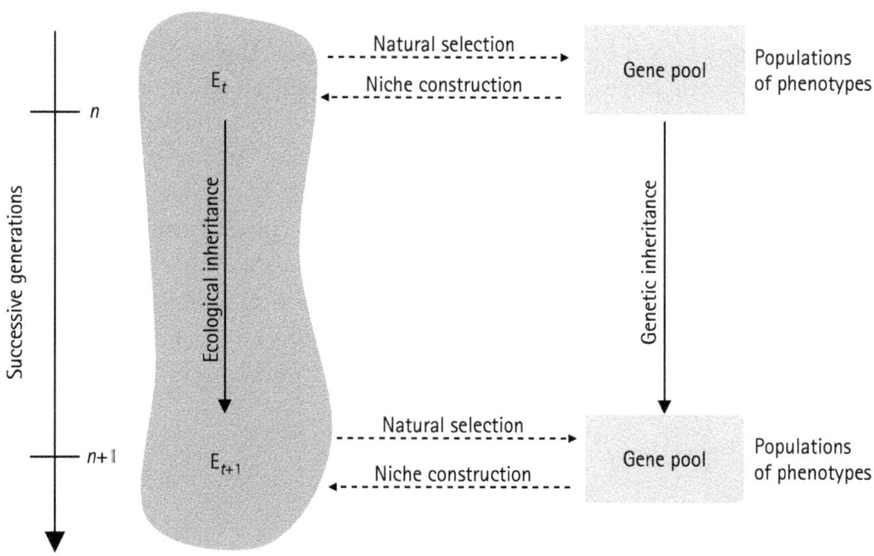

FIG. 5.1b Niche construction theory.

selective environments of organisms are therefore partly determined by independent sources of natural selection—for instance, by climate, or physical and chemical events. But they are also co-determined by what niche-constructing organisms do or previously did to alter some of the natural selection pressures in their own or in each other's environments.

5.2 Human Evolution

How does NCT differ from SET in the way it describes human evolution? The majority of contemporary approaches to human evolution are based on SET. Stemming, in part, from E. O. Wilson's (1975) sociobiology, we now have human behavioural ecology (Smith 2000), evolutionary psychology (Dunbar and Barrett 2007), ecological psychology (Heft 2001), and cognitive evolutionary anthropology (Dunbar and Schultz 2007). Each of these approaches rightfully claims to have contributed some novel insights to our understanding of human evolution. However, because all of them are still based on SET, they all share one key limitation. SET only offers a single inheritance system in evolution, genetic inheritance. Ultimately that forces all these approaches to evaluate or 'cash in' whatever it is they are studying in human evolution, exclusively in terms of the contributions humans make to human genetic inheritance. They are therefore forced to minimize the role of human phenotypes in human evolution whether they want to or not. In SET, phenotypes in all species have only a single job to do in evolution, to survive and reproduce differentially relative to the next generation. That limitation accounts for the reduction of phenotypes, by Dawkins (1976), to just 'survival machines' or 'vehicles' for their genes. Contrary to some of Dawkins' critics, particularly his more vociferous critics in the social sciences, that does not mean human evolution is genetically determined. But it does mean that the only way in which it is possible for any organisms, including humans, to contribute to the evolutionary process, must be by contributing their genes to their descendants because, according to SET, nothing except genes is transmitted by ancestral organisms to descendants.

5.2.1 Pastoralism and Lactose Tolerance

However, it is no longer possible to account for some of the things we know have happened in human evolution in terms of genetic inheritance only. One widely known example of a phenomenon that clearly depends on more than the transmission of genes is lactose tolerance among pastoralists. Because of its relevance, I'll risk a rapid rehearsal.

When our pastoralist ancestors first domesticated cattle and started to drink milk, they eventually affected their own genetic evolution by inadvertently setting up self-induced natural selection pressures in favour of genes that increase lactose absorption in human adults. Why?

All of us are mammals, so we can all digest milk when we are babies. But we lose that facility as adults unless we carry lactase-persistent genes that allow us to go on drinking milk after our childhood is over. Typically, those of us who had pastoralist ancestors can do that because we carry lactase-persistent genes. But people who did not have pastoralist ancestors usually do not carry lactase persistence genes, and usually they do not like milk. It can make them sick. For instance, in Africa, the frequency of lactose-tolerant genes, or to be precise, the frequency of three different genes in different African populations (apparently there has been convergent evolution) is significantly higher among pastoralists than it is among otherwise comparable non-pastoralist people (Gerbault et al. 2011).

In Europe, lactose tolerance depends on a fourth lactase persistence gene, different again from all the African ones. It is called −13.910*T, and it is found on chromosome 2. Roughly speaking, the frequency of −13.910*T increases in European populations the further north you go, that is with increasing latitude. There is also another intriguing point. The key mutation is not found on the lactase gene itself, but on another gene, MCM6, close by, also on chromosome 2. MCM6 apparently regulates −13.910*T.

One question repeatedly asked is what came first? Did these mutations happen first, and by doing so, make it possible for some human populations to become dairy farmers? Or did the dairy farming come first, and did that legate one or more modified natural selection pressures, in favour of lactase persistence, to succeeding generations of pastoralists, which eventually selected for these lactose-tolerant mutations? In Europe, at least, the evidence strongly suggests the latter. For instance, Burger and his colleagues (2007) reasoned that if dairy farming really did come first, were they to look at the genomes of pre-agriculture people, they should not find any lactase-persistent genes. Accordingly, they sampled the ancient DNA of pre-agricultural Neolithic people in Europe, and they found no lactose tolerance genes in Neolithic people. It is hard to prove a null hypothesis, but this result gets close.

5.2.2 Shaping the Human Genome

Lactose tolerance is an excellent example of how human cultural niche construction affected human genetics by modifying natural selection pressures, but it is not the only one. Laland et al. (2010) published a paper provocatively entitled 'How Culture Shaped the Human Genome'. It includes a table (Table 5.1) of more than 100 genes that are strong candidates for either having recently been selected, or for being under current selection by ecologically inherited, culturally modified natural selection pressures in human environments. Table 5.1 is divided into three columns: the left-hand column lists the relevant genes. The middle column lists the known or likely functions of these genes in humans. The right-hand column lists the recent cultural niche-constructing activities that were probably responsible for modifying one or more natural selection pressures in human environments, in favour of the genes on the left.

Table 5.1 Genes identified as having been subject to recent rapid selection with inferred culturally modified selection pressures

Gene	Known or likely function	Known or inferred cultural selection pressure
LCT (C/T-13910, G/C-14010, T/G-13915, C/G-13907)	Digestion of milk and dairy products	Dairy farming and milk usage
G6PD	Resistance to malaria	Farming (e.g. fava beans)
ADH	Alcohol metabolism	Alcohol consumption
FOXP2	Language skills, vocal learning	Language use, vocal learning
Cytochrome P450 genes, esp. CYP3A5, CYP2E1, CYP1A2, and CYP2D6	Detoxification of plant secondary compounds	Domestication of plants
SLC24A5, SLC25A2, EDAR, EDA2R, SLC24A4, KITLG, TYR, 6p25.3, OCA2, MC1R, MYO5A, DTNBP1, TYRP1, RAB27A, MATP, MC2R, ATRN, TRPM1, |SILV, KRTAPs	The externally visible phenotype (skin pigmentation, hair follicles, eye and hair colour, freckles)	Culture-facilitated dispersal and local adaptation, or possibly sexual selection
CDK5RAP2, CENPJ, GABRA4, PSEN1, SYT1, SLC6A4, SNTG1, GRM3, GRM1, GLRA2, OR4C13, OR2B6, RAPSN, ASPM, RNT1, SV2B, SKP1A, DAB1, APPBP2, APBA2, SKP1A, PCDH15, PHACTR1, ALG10, PREP, GPM6A, DGKI, ASPM, MCPH1	Nervous-system genes and genes involved in brain function and development, brain size genes	Selection for complex cognition on which culture is reliant
CD58, APOBEC3F, CD72, FCRL2, TSLP, RAG1, RAG2, CD226, IGJ, TJP1, VPS37C, CSF2, CCNT2, DEFB118, STAB1, SP1, Zap70, BIRC6, CUGBP1, DLG3, HMGCR, STS, XRN2, ATRN	Immunity genes, pathogen response	Dispersal, agriculture, domestication, and aggregation exposed humans to new pathogens
MAN2A1, SI, SLC27A4, PPARD, SLC25A20, NCOA1, LEPR, LEPR, ADAMTS19–20, APEH, PLAU, HDAC8, UBR1, USP26, SCP2, NKX2-2, AMY1	Genes related to the metabolism of carbohydrates, starch, protein, lipids, and phosphates, including genes involved in metabolizing mannose, sucrose, and fatty acids. Carbohydrate metabolism and blood-sugar regulation	Culturally modified dietary preferences
BMP3, BMPR2, BMP5, GDF5	Various genes involved in skeletal development	Culture-facilitated dispersal and local adaptation, sexual selection
MYH16	Gene deletion reduces jaw muscle fibres	Invention of cooking
AHSA1	Heat shock genes	Culture-facilitated dispersal and local adaptation

Table 5.1 includes both pan-human genes, that is genes found in all human beings, as well as regional genes, like the lactase ones. It is important to stress that, as yet, in only a few cases are the feedback loops between culturally modified selection and subsequently selected genes tied down as firmly as they are in the dairy farming case. This is partly because pinning down entire feedback loops typically requires the close interdisciplinary cooperation of evolutionary biologists, geneticists, ecologists, and some social scientists, for example, archaeologists and anthropologists, who are often not used to working with each other, and who initially may have little knowledge of each other's theories or data (Kendal et al. 2011). Table 5.1 may also include some false positives.

Conversely, we know some genes were left out because, since Table 5.1 was published, other candidate genes have been discovered. For example, Barnes, et al. (2011) found the frequency of an allele (SLC 11A1 1729 + 55del4), associated with natural resistance to tuberculosis and leprosy, correlates significantly with the duration of urban settlement. One consequence is that human populations that have a long history of living in towns (ancient urbanization, based on archaeological data) are apparently better adapted to resisting these infections. In another example, Hünemeier et al. (2012) found a robust correlation between maize culture propagation and the frequency of a 230Cys allele, in Mesoamerican agricultural sedentary populations that were heavily dependent on maize consumption, but they did not find the same correlation in South American populations that depended far less on maize consumption (see section 5.3.1).

For several reasons, therefore, Table 5.1 is best regarded as a work in progress. Nevertheless, all the genes in Table 5.1 are strong candidates for having been selected by natural selection pressures that were recently modified, or are currently still being modified, by human cultural niche construction within, and sometimes well within, the last 10,000 years.

But if that is true, none of these findings can be fully explained by SET. SET cannot explain how human cultural niche construction, based on human cultural activities, acquired by individual human phenotypes through cultural inheritance and not by genetic inheritance, can change human genomes. If that has happened, and the data indicate that it has, then human phenotypes must have more than one job to do in evolution. Like all other organisms, human phenotypes not only survive and reproduce differentially under natural selection. They also modify some of the natural selection pressures in their own and in each other's environments. By doing that, we humans too must actively participate in our own evolution. Unlike most other organisms, apparently we do it primarily through cultural niche construction.

5.2.3 Gene–Culture Coevolution

The first biologists to realize this, often working in close association with human scientists, were the gene–culture coevolutionists. They developed dual inheritance models of human evolution that, for the first time, explicitly included non-genetic cultural inheritance systems, as well as genetic inheritance, in models of human evolution (Cavalli-Sforza and Feldman 1981; Boyd and Richerson 1985; Durham 1991). Initially, their models did not

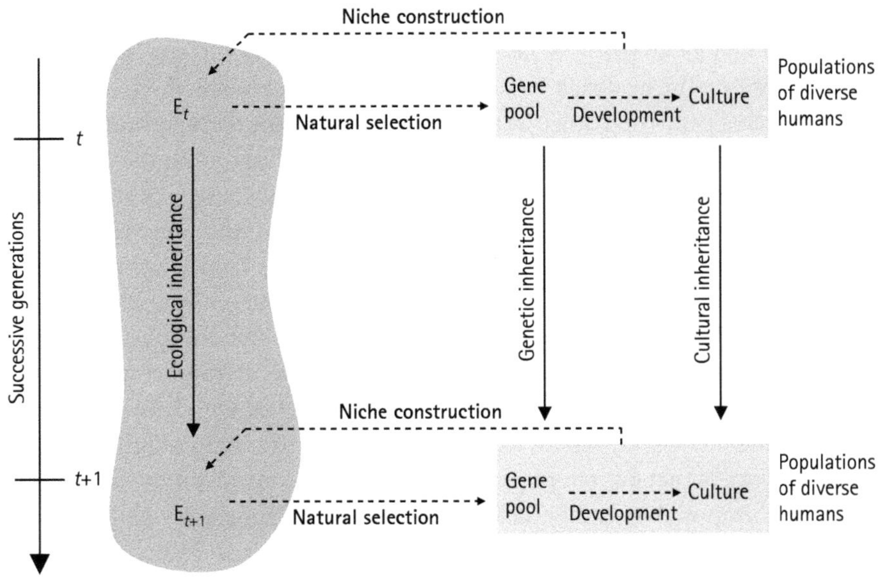

FIG. 5.2 Triple inheritance: Principal inheritance systems in human evolution.

include niche construction or ecological inheritance. However, they were able to account for the evolution of lactose tolerance in pastoralists and other phenomena too (see Table 5.1) in terms of interactions between cultural inheritance and genetic inheritance in human evolution. With the exception of the anthropologist William Durham's exclusively verbal model, their mathematical models defined cultural processes very simply. These models make sense to biologists, but they can appear over-simple to social scientists. Typically they focus on the non-genetic transmission of cultural information, in animals as well as in humans, through social learning, in ways that affect phenotypic variance, beyond genetics (Feldman and Cavalli Sforza 1989). A few years later, Laland et al. (2000) explicitly combined gene–culture coevolutionary approaches with NCT, and by doing so, ended up with the triple inheritance model of human evolution in Figure 5.2. This model includes genetic inheritance, directed by natural selection, as per SET; ecological inheritance, directed by niche construction, as per NCT; and cultural inheritance, directed by various cultural processes, as per the gene–culture coevolutionary models. In Figure 5.2, both genetic inheritance and ecological inheritances are general inheritance systems in evolution that occur in all organisms. Cultural inheritance, however, is almost exclusively human.

5.2.4 Feedback Paths from Cultural Niche Construction

How does human cultural niche construction work in this triple inheritance model? Potentially, human cultural inheritance can influence both human genetic inheritance by affecting human survival and reproduction, and human ecological inheritance by causing changes in natural selection pressures in human environments. These influences are achieved through feedback paths that travel from prior cultural niche construction, to human environments, and back again to subsequent human gene–culture coevolution. Figure 5.3 illustrates the three main paths via which, according to NCT, feedback from prior human cultural niche construction can affect subsequent human evolution. It can do so either by changing subsequent cultural processes only, via Route 1, or by changing human genetics as well as human cultural processes, via Routes 2 and 2a (Creanza et al. 2012).

Feedback from cultural niche construction via Route 1 modifies a human environment, but it does so in a way that only causes further cultural changes. It affects human cultural evolution, but it need not affect human genetic evolution at all. Consequently it may change human history, or prehistory, but it may not have any effect on human genetics. In such cases, social science explanations alone should suffice for the investigation of Route 1 type phenomena. There may be no need for any evolutionary biology here.

That is not true for Route 2 or for Route 2a type phenomena. Route 2 is the more complicated feedback path shown in Figure 5.3, and it does affect human genetics. It is likely to be relevant whenever there is either no culturally adaptive response to a significant human-induced environmental change (via Route 1), or an insufficient one. There are several reasons why this sometimes happens. For example, a human population may be unaware of the ecological or the evolutionary consequences of some of the changes it is causing in its environment by positive or, especially, negative cultural niche construction. Alternatively, a human population may be aware that it is damaging its environment, for instance, by overexploiting it or polluting it, but it may still be unwilling or

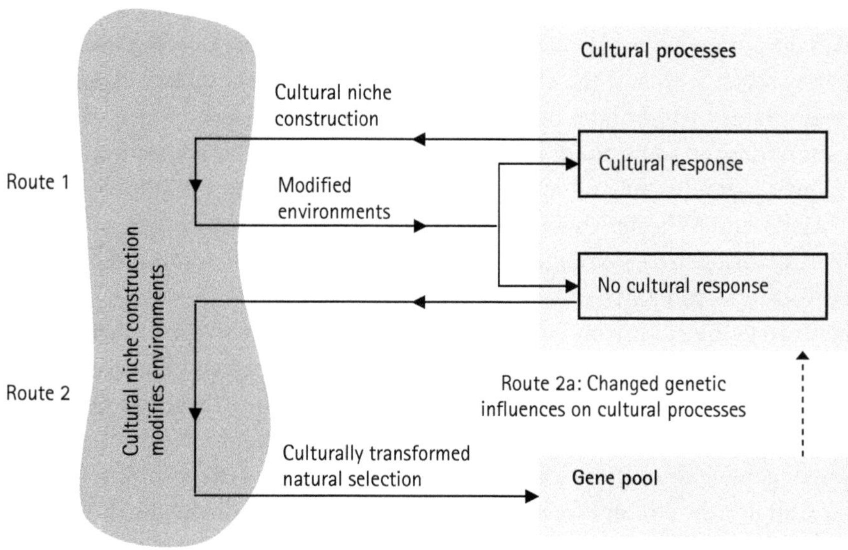

FIG. 5.3 Alternative responses to feedback from cultural niche construction.

unable to pay the full economic or political cost of rectifying the damage it is causing, for example, by investing in additional corrective or *counteractive* niche construction to offset the damage, and thereby converting the threat of a Route 2 phenomenon back to a more tractable Route 1 type phenomenon.

A frequently cited example occurred among Kwa-speaking cultivators in West Africa, from approximately 3000 years ago, who inadvertently caused a genetic change in their population as a consequence of their practice of cutting clearings in tropical rainforests. The clearings enabled the Kwa to grow their crops, but they also increased the amount of standing water in their environments, and thus provided better breeding grounds for malaria-carrying mosquitoes (Durham 1991). That increased the prevalence of malaria, which apparently modified natural selection in favour of an increase in the frequency of the hemoglobin S allele in the Kwa. It did so because, in heterozygotes, the S allele confers a degree of protection against malaria. It is 'the' textbook example of heterozygote advantage. Its disadvantage is that in the homozygote condition the S allele can cause fatal sickle-cell anaemia. If the Kwa had realized what they were doing, they might have alleviated their malaria and sickle-cell problems by further cultural niche construction instead of submitting to a genetic change, by digging better drains to reduce the amount of standing water, for instance, or less probably, by discovering and applying effective medical solutions.

Finally, the Route 2a path in Figure 5.3 is just a special case of Route 2. The same thing happens again, except this time, the genes that are selected by culturally modified natural selection happen to be genes that underpin or influence the evolution of culture itself. For instance, it is likely that some of the nervous system genes, in row 7 of Table 5.1, are candidate Route 2a genes, in contrast to the genes associated with lactose tolerance in the top row of Table 5.1, which are only Route 2 genes. However, the feedback loops associated with Route 2a are often extremely complicated, making it harder to establish the connections between culturally modified natural selection pressures—the alleles they select for—and the changes in the phenotypic traits they express, whether in human brains or in human behaviours. A useful candidate example, that has been in the literature for some time, illustrates this point. It is the probable consequences of food processing by the cultural activity of cooking for the evolution of the human brain (Aiello and Wheeler 1995). Some kind of evolutionarily significant feedback is extremely plausible here (Wrangham et al. 1999), but it is still proving very hard to pin down (Wollstonecroft 2011).

In contrast to the exclusively cultural phenomena that depend on Route 1, any changes in human evolution that depend on either Route 2 or Route 2a in Figure 5.3 must necessarily involve both biological and cultural processes. They therefore need the combined efforts of biologists and social scientists to investigate them. Ideally they require integrated gene–cultural coevolutionary models, combined with interdisciplinary empirical approaches to capture how human genetics and human cultural processes are actually interacting with each other in human evolution (Laland et al. 2010; Kendal et al. 2011; Hünemeier et al. 2012).

5.2.5 Criticisms of Gene–Cultural Coevolution

Gene–culture coevolutionary approaches to the study of human evolution, combined with NCT, are undoubtedly more complicated than any of the approaches based on SET alone. They may be more accurate. They probably describe data, such as the genetic data illustrated in Table 5.1, better than SET can, but at a cost. The added complexity of gene–culture coevolutionary theory raises new problems, and is open to new criticisms.

Proliferating Inheritance Systems

One example is that the gene–culture coevolutionary approach requires a proliferation of inheritance systems in human evolution. It replaces genetic inheritance only in human evolution, as per SET (Figure 5.1a), with the triple inheritance system shown in Figure 5.2. Others have proposed still more elaborate versions of multiple inheritance systems in evolution in general (Bonduriansky 2012) and in human evolution in particular (Danchin et al. 2011; Furrow et al. 2011). Durham (2008), for example, suggests that four distinct inheritance systems are in play in human evolution. A better-known scheme is in Jablonka and Lamb's book, *Evolution in Four Dimensions* (2005), which similarly proposes four inheritance systems in human evolution, although they are not identical to Durham's four.

One potential solution to this proliferation problem is illustrated in Figure 5.4 (Odling-Smee 2007, 2010). NCT differs from SET by proposing that, in all species, evolution always depends on two inheritance systems, genetic inheritance and ecological inheritance, operating relative to each other (Odling-Smee et al. 1996, 2003). Ancestral organisms 'transmit' and descendent organisms 'receive' both kinds of inheritance through two distinct transmission channels. The first channel, Channel 1, is the familiar one that has long been associated with genetic inheritance. It has always been modelled by SET. Ultimately Channel 1 depends on the mechanisms of cell division and cell fusion that were first discovered by Schwann, Schleiden, and Virchow in the nineteenth century, and that eventually gave rise to cell theory (e.g. 'all cells arise from pre-existing cells', etc.). Channel 1 connects the internal or cellular environments of parent organisms to the internal or cellular environments of their offspring, via reproduction. In Figure 5.4, any kind of inheritance that travels directly from cell to cell during reproduction is treated as a Channel 1 type inheritance.

In contrast, Channel 2 is the novel transmission channel introduced by NCT. Channel 2 carries ecological inheritances from ancestral organisms to descendent organisms, through external environments, in the form of legacies of modified natural selection pressures relative to the genetic inheritances of recipient populations. Channel 2 is mediated by external environments. It works by connecting modified natural selection pressures in the external environments of 'transmitting' populations, to modified natural selection pressures in the external environments of 'receiving' populations, in evolution. One complication is that the 'receiving' organisms need not be in the same populations as the 'transmitting' organisms. A niche-constructing

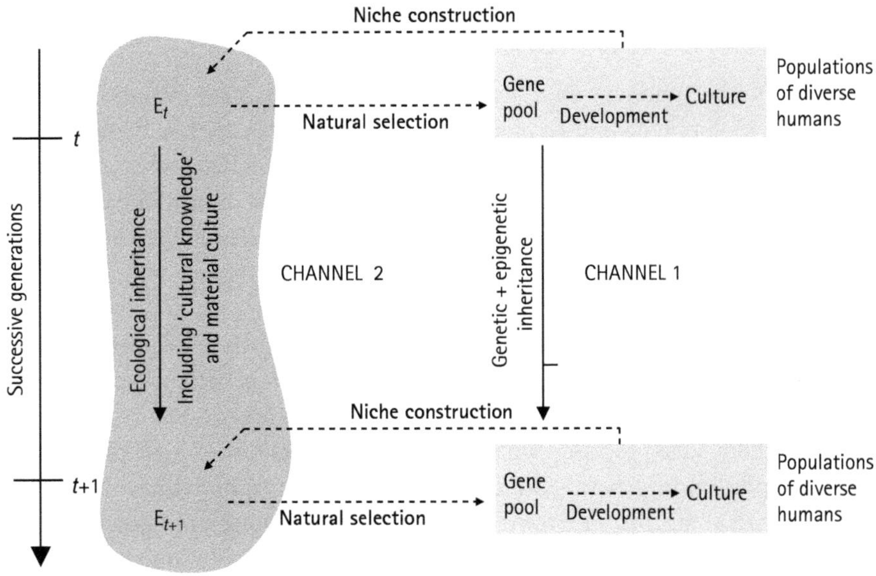

FIG. 5.4 Dual inheritance: Principal inheritance systems in human evolution. Human cultural inheritance is now treated as a component of human ecological inheritance.

population can modify natural selection pressures for one or more other 'recipient' populations, instead of or as well as for itself. Unlike Channel 1, Channel 2 therefore has the potential to connect multiple different populations in coevolutionary relationships, in ecosystems (Odling-Smee et al. 2003; Post and Palkovacs 2009). In human evolution, extensions of Channel 2 probably played a part in the domestication of plants and animals by humans (e.g. Beja-Pereira et al. 2003; Smith 2011), and in the evolution of human diseases (e.g. Barnes et al. 2011).

Given that NCT depends on both Channel 1 and Channel 2, and disregarding, for the moment, any extensions of Channel 2 to other species, it should be possible to assign any supplementary inheritance system in human evolution to either one or other of the two basic inheritance transmission channels in Figure 5.4. Thus, epigenetic inheritance, as described by Jablonka and Lamb (2005), should end up on the right in Figure 5.4, as a supplementary Channel 1 kind of inheritance. Much less obviously, human cultural inheritance, inclusive of both inherited cultural knowledge and inherited material culture, should end up on the left, since all cultural inheritances, including those that depend on social learning, or teaching, or on formal education, are invariably transmitted through external human environments, and therefore by Channel 2. Elsewhere (Odling-Smee 2007, 2010; Odling-Smee and Laland 2012) we have argued that human cultural inheritance is an exceptionally potent component of a more general human ecological inheritance system in human evolution. Significantly, human cultural inheritance works in very similar ways to other ecological inheritance systems, but differs radically from genetic inheritance. For example, cultural inheritance is typically transmitted by innumerable 'cultural parents', rather than by only two biological parents.

It is transmitted continuously, within as well as between generations (Cavalli-Sforza and Feldman 1981), and it does not depend exclusively on the transmission of discrete replicators such as genes or 'memes' (Dawkins 1976; Odling-Smee and Laland 2012).

Different Rates of Genetic and Cultural Evolution

Another frequent criticism of gene–culture coevolutionary theory is that human cultural evolution is fast and it seems to be getting faster. In comparison, human genetic evolution is slow, partly because we are a long-lived species. So for all practical purposes, we can surely afford to forget about human genetic evolution at least during the Holocene (i.e. during approximately the last 11,000 years)? That would allow us to concentrate exclusively on human cultural evolution.

But appearances can be deceptive. Rapid genetic evolution is possible, even in long-lived organisms such as humans. For instance, ecologists have recently begun to realize that evolutionary changes can occur as rapidly as ecological changes, and that 'contemporary evolution' must therefore be included in ecosystem-level ecology (e.g. Hairston et al. 2005; Ellner et al. 2011; Schoener 2011). That insight raises a related question in human evolution. To what extent is contemporary genetic evolution relevant to human gene–culture coevolution today? A preliminary point is that most of the candidate genetic changes in Table 5.1 probably occurred within the Holocene anyway. For example, the changes in the human genome caused by feedback from agricultural niche construction must be relatively recent. They must have occurred since the origin of agriculture about 12,000–10,000 years ago (Smith 2011). Changes due to urbanization (Barnes et al. 2011) must be even more recent.

Second, some genetic evolutionary changes caused by feedback from cultural niche construction are probably still happening. A candidate example concerns the genetic coevolution of hereditary deafness in response to the cultural invention of sign language (Laland et al. 2010). Most inherited forms of human deafness—for instance, connexin deafness—are monogenic. They typically depend on children inheriting a single recessive gene from both their parents. Before the invention of sign language, deaf people had low genetic fitness. They did not survive long or reproduce often. Genes for hereditary deafness were therefore comparatively rare in human populations. After the cultural invention of sign language, however, two changes occurred. First, the genetic fitness of deaf people improved dramatically. Second, deaf people, using sign language to communicate with each other, were highly likely to marry and mate with each other rather than with non-deaf people. In other words, deaf people invested in assortative mating. In combination these two innovations led to a doubling of the incidence of inherited deafness in humans in the USA during the last 200 years (Nance and Kearsey 2004). As far as is known, this example of gene–culture coevolution is still ongoing, although the recent cultural invention of multiple forms of electronic communication permit deaf people today to communicate with non-deaf as well as deaf people without sign language, and that could change the scenario again.

There is also an important caveat. If rapid genetic evolution is possible sometimes, so is its opposite. It is by no means true that all genes in all genomes are equally susceptible to natural selection from contemporary external environments: far from it. Working at an extraordinary level of detail, Davidson (2006) has shown how development in animals is typically controlled by hierarchies of developmental gene regulatory networks, or GRNs, that can be decomposed into different modular subcircuits. A useful rule of thumb is that whatever is expressed earliest in the development of an animal is likely to be controlled by the most evolutionarily conserved subcircuits in its genome, and vice versa for whatever is expressed latest. Davidson and Erwin (2006) then identified a set of highly conserved genetic subcircuits at the cores of these GRNs, called 'kernels'. Kernels are 'extremely conserved, highly refractory and recursively wired suites of genes that are crucial for the specification of many of the characteristic morphologies of major clades' (Erwin et al. 2011: 1094). Thus, kernels are associated with phylum-level body plans: for example, the body plan shared by all vertebrates. Apparently kernels have been conserved in animal genomes since the Cambrian period, for more than 500 million years. Because GRNs are hierarchical in structure, 'features can be added downstream' but 'alteration near the top is often catastrophic, and is thus eliminated by internal selection' (Davidson and Erwin 2010: 185). So kernel genetic subcircuits are all but impervious to natural selection by contemporary external environments. Rather they are extreme examples of how evolution can 'lock in' the products of earlier evolution. The earlier evolution forbids further evolution by imposing developmental constraints. At the opposite extreme, other genes can express phenotypic traits that are highly evolutionarily labile. They are very susceptible to contemporary natural selection by external environments, including selection pressures only recently modified by niche construction. They can and sometimes do demonstrate rapid evolutionary responses (Table 5.1).

For anyone interested in human gene–culture coevolution, the 'message' is that rates of genetic evolution in humans probably vary considerably, from fast, to slow, to completely impervious to contemporary external natural selection, regardless of whether the selection has been modified by human cultural niche construction. Sweeping statements about the role of genes in human evolution are misleading or meaningless. Instead it is necessary to know as much as possible about the functions and the regulation of whatever particular genes are under scrutiny in human genomes. Then it may be possible to demarcate legitimately between cases where we can afford to ignore human genetic evolution in practice, and can concentrate exclusively on human cultural evolution, and other cases where we cannot.

5.3 Some Implications of NCT for Human Demography

Might NCT offer a better basis than SET for connecting human demography to human evolution? I'll consider two points only. The first concerns the definition of populations. NCT requires us to make explicit some factors that contribute to the structure of populations that, more usually, are only treated implicitly. The second concerns the carrying capacities of environments relative to one of the classic problems addressed by biologists and economic demographers. Are limits to population size and growth a stimulus or a check with respect to what happens in ecology or in human economics? NCT has the potential to integrate some divisive issues.

5.3.1 Defining Populations

Evolutionary biologists assume that 'ideally' populations share homogenous gene pools, share the same environments, and that every individual in a population has an equal chance of interacting with every other individual in the same population. Idealized populations like this are called panmictic (Barton et al. 2007). In reality no populations are panmictic. All real populations deviate from the panmictic ideal in some way or other. For example, populations frequently divide into sets of spatially distributed subpopulations across heterogeneous or 'patchy' environments. Insofar as populations deviate from the panmictic ideal, they are called structured. In reality, all populations are structured. How they form and how they interact to form more durable metapopulations is the subject of metapopulation biology in ecology (Hanski 1999) and in evolution (Barton et al. 2007). If NCT is substituted for SET, it must be able to capture some of the ways in which the niche-constructing activities of organisms influence the basic demographic processes of birth, death, immigration, and emigration, and the life history processes responsible for establishing populations, better than SET.

How much difference might NCT make? A plausible initial answer is 'not much' because, although it is seldom explicit, niche construction is already implicit in a lot of standard models. For example, the persistence of metapopulations typically requires a stochastic balance between local extinctions and recolonizations in ecology (Hanski 1999). That balance depends on gene flow between subpopulations caused by active as well as passive dispersal mechanisms, and conversely, sometimes on the prevention of gene flow by self-induced isolating barriers. Insofar as gene flow or barriers to gene flow depend on the non-random choices or migrations of animals, for instance (Piersma and van Gils 2011), or on active mate choice, or assortative mating, then it will depend, at least partly, on niche construction. In that case, so will its demographic consequences. That much is already implicit in many standard models (Hanski 1999; Barton et al. 2007).

However standard models seldom take the interactions between genetic inheritance (Channel 1, Figure 5.4) and ecological inheritance (Channel 2, Figure 5.4) sufficiently into account. If they did, models based on NCT instead of SET probably would lead to a better understanding of the structures of subpopulations and metapopulations than SET. Models that already take NCT into account suggest that in some circumstances this is true. For example, Silver and Di Paolo (2006) showed how spatial clusters of niche-constructing organisms can evolve in unfavourable environments, and how such clusters can expand and establish colonies of niche-constructing organisms in hostile environments. Other provocative results have been reported by Krakauer et al. (2009) and Kylafis and Loreau (2008, 2011).

Defining Human Populations

Human populations do not differ much from other long-lived animal populations with respect to their capacity to respond genetically to changing natural selection pressures in their environments, as described by SET. Where we do differ radically is in our capacity for changing environments, and therefore for modifying natural selection, as described by NCT. The primary reason for this is because most human niche construction is cultural, which makes it exceptionally powerful (Smith 2007; Kendal et al. 2011). Therefore, substituting NCT for SET should be especially relevant to understanding human populations. It implies that human populations should be defined by what they do 'culturally' in and to their environments, inclusive of their social environments, as well as by their responses to natural selection. Rendell et al.'s (2011) model of runaway cultural niche construction, capable of overwhelming strong counter-pressures from natural selection (in a way similar to runaway sexual selection), offers a case in point.

An Empirical Example

Hünemeier et al. (2012) deliberately tested NCT and gene–culture coevolution empirically. They defined 19 Amerindian populations from Meso-, Central, and South America in time (< 15,000 years, because humans were not present in South America before), place (the geographic coordinates of each population), and by their niche-constructing activities (what they did or still do). In the case of the Mesoamerican and Andean populations, niche construction refers to agriculture, particularly maize agriculture, versus hunter-gathering and foraging populations elsewhere in South America. The authors then genotyped a total of 1905 individuals from these different subpopulations. They found a significant correlation between an autochthonous native American allele (ABCA1*230Cys) and the distribution of *Zea* pollen relics (associated with maize domestication and production) in the Mesoamerican agricultural populations, but did not find the same correlation in the Andean or the South American populations that, from the archaeological record, were known to be either less dependent or not dependent on maize. They interpreted these results in terms of the modification of natural selection by the 'agri-cultural' niche-constructing activities of the Mesoamerican maize growers, as per NCT.

But, as Hünemeier et al. pointed out, this interpretation is not straightforward. Apparently, the early Mesoamerican agriculturalists, the people who were most dependent on maize, suffered from periods of food scarcity due to plantation losses, crop diseases, and food storage problems. They suggest that the 230Cys allele probably conferred a selective advantage for these people during periods of food shortage. In interpreting their results they therefore integrated a version of the 'thrifty gene' hypothesis with NCT and gene–culture coevolutionary theory. We can summarize their reasoning approximately by saying that 'thrift' refers to any gene assumed to influence the 'preparation' of organisms for future famines by promoting the deposition of extra fat during periods of abundance.

If they are correct, there is a further twist in this scenario. The natural selection pressures that were actually modified by cultural niche construction, that is, by the Mesoamericans' maize production, cannot have stemmed directly from the production of maize. Instead they must have stemmed from an indirect consequence, namely from the periods of food shortage that the over-reliance of the Mesoamerican populations on their maize crops induced. In contrast, the South American hunter-gathering/foraging subpopulations probably inherited unchanged selection pressures in their ecological inheritances because, counterintuitively, they enjoyed a more varied, more reliable, and nutritionally sounder diet (Poinar et al. 2001; Hünemeier et al. 2012). Thus Hünemeier et al.'s data make better sense in terms of NCT than SET because NCT explicitly recognizes that different ecological inheritances, caused by different niche-constructing activities, should lead to different genetic consequences in populations. SET does not.

One further point in Hünemeier et al.'s (2012) study that is likely to be of interest to demographers studying long-term population change, is that if the agricultural niche-constructing activities of these American subpopulations had involved other crops, for example, beans or squashes instead of maize, and if the production of these other crops had been distributed differently from the production of maize among these subpopulations, it would have been necessary to define the subpopulations differently. So it may be necessary to define subpopulations differently relative to different kinds of niche construction. That re-emphasizes the relativity of structured subpopulations to the questions being asked. It also raises the possibility that individuals might belong to more than one subpopulation simultaneously. However, these problems are not introduced by NCT. They are not new (see Wachter, this volume).

5.3.2. The Carrying Capacity of Environments

A second implication of NCT for human demography concerns the ability of niche construction to alter the carrying capacities of environments. Here, there are some notable differences between how biologists treat carrying capacities in general and how economists treat them in particular. Given our contemporary concerns about the consequences of human-induced climate change or reductions in biodiversity, I will briefly consider how NCT might contribute to debates between biologists and economists.

'K' in Biology

The equation biologists most often use to describe how populations interact with the carrying capacities of their environments is the familiar logistic growth equation:

$$dN/dt = rN(1 - N/K) \tag{3}$$

where N is population size, r is the intrinsic rate of growth of the population, and K is the carrying capacity of the population's environment. K represents both the maximum resources available to a population in its environment and the maximum population size its environment can support (e.g. Gotelli 1995). N is expected to increase, as per r, until the population hits whatever 'Malthusian ceiling' is set for it independently by K. Then it should reach equilibrium and stop. Typically, although not invariably, K is treated as an exogenous variable by biologists (Lehmann and Feldman 2009).

What is wrong with that? Quite a lot, according to the paleobiologist Douglas Erwin (2007, 2012). First, K is not fixed, but is constantly changing, both positively and negatively. Second, estimates of K only apply to particular times and places, and to the particular adaptations of particular organisms (Erwin 2007). Third, the concept of an exogenous carrying capacity is tied to what Erwin calls outmoded equilibrial views of biodiversity (Erwin 2012) as, for instance, in the once-popular concept of the 'balance of nature'. Fourth, treating K as if it were independent of the actions of organisms excludes the possibility that a population can change the value of K for itself, either by evolving new adaptations, or by expressing positive or negative niche construction (Erwin 2007). Fifth, the concept of an exogenous carrying capacity is difficult to apply beyond the level of single populations, i.e. to communities or ecosystems. This is because each niche-constructing population in an ecosystem can potentially alter not only its own K, but also the Ks of other populations in its ecosystem by its environment-altering niche-constructing activities (Odling-Smee et al. 2003). Erwin (2012) calls these kinds of between-population interactions 'ecological spillovers'. Sixth, the concept of an exogenous K entirely fails to explain how, over time, 'ecological spillovers' between populations apparently did open up new 'living spaces' for successions of different species of organisms, by expanding resources in ecosystems.

A shorter-term example of this phenomenon is 'ecological succession', in which successions of engineering or niche-constructing species induce changes in their environments that subsequently feed back to impact negatively on themselves, but positively on different successor species (Odling-Smee et al. 2003). Erwin notes that something similar seems to have happened in 'deep time', over millions or billions of years of life on earth. He cites many examples, including the emblematic one of oxygen production by photosynthesizing microorganisms that 'eventually generated a complete change in the redox state of the Earth from anoxic and euxinic oceans and atmosphere to the well ventilated systems we have today . . . in the process facilitating the generation of far greater biomass and biodiversity than would otherwise have been possible' (Erwin et al. 2011: 4). As Erwin points out, this change had both negative and positive consequences. It had a negative impact on anaerobic microbes, but it was enormously positive for the biosphere as a whole (Erwin 2007).

Erwin's overall point is that the variable K in the logistic equation (3) is never determined by exogenous variables alone. It is always the product of interactions between exogenous and endogenous variables (e.g. Henn et al. 2012).

'K' in Human Economics

Now we can turn to the carrying capacity of human environments and to some of the arguments between economists and biologists, for example, that there are no limits to the growth of national GDPs (from economists) versus that there are such limits to growth (from ecologists).

In a review of the work of the anti-Malthusian agricultural economist, Esther Boserup, the demographer Ronald Lee (1986) supports Boserup's point that 'the power of [human] ingenuity will always outmatch that of demand'. 'Boserup observes quite plausibly that the likelihood of technological advance rises in response to increasing population density' (Lee 1986: 100). Probably most economists would agree. If we assume the power of human ingenuity is equivalent to the power of human cultural niche construction, why should not the rest of us agree too (see Lehmann and Feldman 2009)?

Up to a point, Boserup is obviously right. For example, it is undeniable that: (i) human agriculture has increased the carrying capacities of human environments; (ii) increases in agricultural production frequently do depend on technological advances (Smith 2007, 2011); and (iii) they correlate with the size of human populations. A negative example, from Henrich (2004), illustrates these points. When Tasmania was cut off from Australia between 12,000 and 10,000 years ago by rising sea levels in the Bass Strait, the isolated Tasmanian population (only a few thousand people at the time of contact with Europeans) lost most of their cultural toolkit, apparently because they were not sufficiently numerous to retain the cultural 'know-how' to sustain it. By the time the Europeans arrived, the Tasmanian toolkit was down to about 24 items only, compared to the hundreds of tools still used by similar but more numerous people on the other side of the Bass Strait. In Tasmania fishing, not agriculture, was the principal economic casualty. The archaeological record reveals that the Tasmanians relied heavily on fishing from 8000 to 5000 years ago. Then fishing disappeared. On the Australian side of the Bass Strait fishing never stopped (Henrich 2004).

In spite of this support for the idea that the carrying capacities of human environments depend on human 'ingenuity' and on human population size, others query the hypothesis. For instance, Richerson et al. (2009) claim that population growth is a consequence of technical innovation, not its cause. It does not precede it. Second, technical innovations may increase the carrying capacities of human environments by introducing pulses of increased resources, but subsequently human populations use up the bounty of each resource pulse by growing again, sometimes explosively, until they reach the limits set by a new K. Third, technical innovations are themselves costly, which limits the extent to which innovations actually do increase carrying capacity. For instance, the 'Green Revolution' in agriculture, introduced by Norman Borlaug in the late 1960s, yielded better crops, but only at the price of more artificial fertilizers, pesticides, and water, each of which increases the potential for environmental degradation.

Erwin (2008) draws attention to another consequence of human technical innovation, this time from economics itself. Solow (1956) established that economic growth always requires technological innovation, and that without it economic growth stalls. But later, Romer (1990) showed that the consequences of technical innovation are not simple. Romer distinguished between economic 'goods' that are 'rivalrous', meaning that they can only have one user at a time, a bicycle for instance, versus 'non-rivalrous' goods, for example, the global positioning system (GPS) that can be used by any number of people simultaneously. Romer also distinguished 'excludable' from 'non-excludable' goods with reference to how easy it is for a user to exclude other users from sharing the same good. Romer's crucial insight was that economic growth ultimately depends on technical innovations generating non-rivalrous, non-excludable goods, but not rivalrous, excludable goods (Erwin 2008; Lehmann and Feldman 2009).

'K' in Biology and Economics

Erwin has translated Romer's insights to ecosystem ecology, by proposing that 'the greatest impact of niche construction and ecosystem engineering is when they produce innovations that are analogous to non-rivalrous, non-excludable goods' (2008: 308), and he gave some examples. He again suggested that oxygen production by photosynthesizing organisms is a paradigmatic example of a non-rivalrous, non-excludable good. That probably explains why it did facilitate far greater biomass and diversity. In contrast, living space for organisms that live in intertidal zones (for example, barnacles) is a highly rivalrous, highly excludable resource.

Translated back into human environments, these insights indicate that different kinds of technical innovation may have different consequences for the carrying capacity of human environments. Some will increase K, in the manner described by endogenous growth theory in economics. However, others may do the opposite. They are more likely to generate 'tragedy of the commons' scenarios characterized by too many 'selfish' consumers competing for too few resources, and eventually destroying their common resource as a result (Hardin 1968). Overfishing by competing fishing nations is a contemporary human example.

In sum, both biologists and economists appear to be partly right and partly wrong about the carrying capacities, K, of environments, including human environments, but intriguingly, they are usually right and wrong in opposing ways. The biologists who still accept SET typically underestimate the potency of endogenous variables, including the potency of niche construction, in favour of exogenous variables. Conversely, economists typically emphasize endogenous variables, and claim they trump exogenous variables. For instance, the concept of resource substitutability implies that whenever environmental resources become sufficiently scarce to threaten human growth, human ingenuity should find substitutes. If this diagnosis is correct, what is needed is a theoretical framework that can resolve these disciplinary differences. NCT is a candidate theoretical framework that might do the job because it is a priori based on two reciprocally interacting causal processes in evolution: exogenous natural selection and endogenous niche construction.

5.4 Conclusion

At present, there is an increasing recognition by many biologists of the need for a more inclusive theory of evolution (Pigliucci and Muller 2010; Laland et al. 2011), and NCT is a candidate theory. NCT is more inclusive because it describes four evolutionarily significant processes, natural selection, niche construction, genetic inheritance, and ecological inheritance (Figure 5.1b), instead of only the two classical processes of natural selection and genetic inheritance, described by SET (Figure 5.1a). When NCT is applied to humans, in lieu of SET, it introduces a variety of novel implications, including the two discussed here. First, it assigns a major role to cultural processes in human evolution, based on the human capacity for changing environments by cultural niche construction, and causing different kinds of feedback in human prehistory, history, and evolution (Figure 5.3). In particular, NCT draws attention to data indicating that human cultural niche construction has shaped the human genome by modifying the natural selection of some genes (Table 5.1). Therefore a major implication of NCT for human evolution is that we cannot ignore the consequences of human cultural processes in human evolution. It will be necessary, from hereon, to model human evolution in terms of gene–culture coevolutionary models that explicitly include cultural niche construction.

Second, we discussed two implications of NCT for human demography. One is that human subpopulations need defining in terms of what they do to their environments culturally, as well as in other ways. The other is that it is unlikely to be possible to understand the carrying capacities of human environments, either in biology or in economics, except in terms of the interactions of exogenous variables, represented primarily by natural selection, and endogenous variables, represented primarily by cultural niche construction.

Famously, the idea of natural selection first occurred to Darwin when he 'happened to read for amusement Malthus on Population . . .' (see Darwin's autobiography, 1968 [1902]). Natural selection was initially derived from a way of thinking that prioritized exogenous variables (Figure 5.1a). NCT now adds the modification of natural selection by niche construction (Figure 5.1b). Thus, NCT acknowledges that endogenous variables stemming from niche-constructing organisms, including human technical 'innovations', constantly interact with exogenous 'Malthusian' variables by modifying them. NCT could therefore provide demographers, and maybe economists too, with a better basis for connecting human demography to human evolution than SET can.

References

Aiello, L. C. and Wheeler, P. (1995). 'The Expensive Tissue Hypothesis', *Current Anthropology* 36: 199–221.

Barnes, I., A. Duda, O. Pybus, and M. G. Thomas (2011). 'Ancient Urbanisation Predicts Genetic Resistance to Tuberculosis', *Evolution* 65: 842–848.

Barton, N. H., D. E. G. Briggs, J. A. Eisen, D. B. Goldstein, and N. H. Patel (2007). *Evolution*. Cold Spring Harbor, NY: Cold Spring Harbor Laboratory Press.

Beja-Pereira, A., G. Luikart, P. R. England, D. G. Bradley, O. C. Jann, G. Bertorelle, A. T. Chamberlain, T. P. Nunes, S. Metodiev, N. Ferrand, and G. Erhardt (2003). 'Gene–Culture Coevolution between Cattle Milk Protein Genes and Human Lactase Genes', *Nature Genetics* 35: 311–313.

Bonduriansky, R. (2012). 'Rethinking Hereditary, Again', *Trends in Ecology & Evolution* 27: 330–336.
Boyd, R. and P. J. Richerson (1985). *Culture and the Evolutionary Process*. Chicago: University of Chicago Press.
Burger, J., M. Kirchner, B. Bramanti, W. Haak, and M. G. Thomas (2007). 'Absence of the Lactase-Persistence-Associated Allele in Early Neolithic Europeans', *Proceedings of the National Academy of Sciences of the USA* 104: 3736–3741.
Cavalli-Sforza, L. L. and M. W. Feldman (1981). *Cultural Transmission and Evolution: A Quantitative Approach*. Princeton: Princeton University Press.
Creanza, N., L. Fogarty, and M. W. Feldman (2012). 'Models of Cultural Niche Construction with Selection an Assortative Mating', *PLoS One* 7(8): e42744. doi:10.1371/journal.pone.0042744.
Danchin, E., A. Charmantier, F. A. Champagne, A. Mesoudi, B. Pujoi, and S. Blanchet (2011). 'Beyond DNA: Integrating Inclusive Inheritance into an Extended Theory of Evolution', *Nature Reviews Genetics* 12: 475–486.
Darwin, C. (1968 [1902]). *The Autobiography of Charles Darwin*, ed. F. Darwin. New York: Dover.
Davidson, E. H. (2006). *The Regulatory Genome: Gene Regulatory Networks in Development and Evolution*. New York: Academic Press.
Davidson, E. H. and D. H. Erwin (2006). 'Gene Regulatory Networks and the Evolution of Animal Body Plans', *Science* 311: 796–800.
Davidson, E. H. and D. H. Erwin (2010). 'Evolutionary Innovation and Stability in Animal Gene Networks', *Journal of Experimental Zoology, Part B: Molecular and Developmental Evolution* 314/3: 182–186.
Dawkins, R. (1976). *The Selfish Gene*. Oxford: Oxford University Press.
Dunbar, R. I. M. and L. Barrett (2007). *The Oxford Handbook of Evolutionary Psychology*. Oxford: Oxford University Press.
Dunbar, R. I. M. and S. Schultz (2007). 'Evolution in the Social Brain', *Science* 317: 1344–1347.
Durham, W. H. (1991). *Coevolution: Genes, Culture, and Human Diversity*. Palo Alto: Stanford University Press.
Durham, W. H. (2008). 'When Culture Affects Behavior: A New Look at Kuru', in M. J. Brown (ed.), *Explaining Culture Scientifically*. Seattle: University of Washington Press, 139–161.
Ellner, S. P., M. A. Geber, and N. G. Hairston (2011). 'Does Rapid Evolution Matter? Measuring the Rate of Contemporary Evolution and its Impacts on Ecological Dynamics', *Ecology Letters* 14: 603–614.
Erwin D. H. (2007). 'Increasing Returns, Ecological Feedback and the Early Triassic Recovery', *Palaeoworld* 16: 9–15.
Erwin D. H. (2008). 'Macroevolution of Ecosystem Engineering, Niche Construction and Diversity', *Trends in Ecology & Evolution* 23: 304–310.
Erwin, D. H. (2012). 'Novelties that Change Carrying Capacity', *Journal of Experimental Zoology, Part B: Molecular and Developmental Evolution* 318/6: 460–465.
Erwin D. H., M. Laflamme, S. M. Tweedt, E. A. Sperling, D. Pisani, and K. J. Peterson (2011). 'The Cambrian Conundrum: Early Divergence and Later Ecological Success in the Early History of Animals', *Science* 334: 1091–1097.
Feldman, M. W. and L. L. Cavalli-Sforza (1989). 'On the Theory of Evolution under Genetic and Cultural Transmission with Application to the Lactose Absorption Problem', in M. W. Feldman (ed.), *Mathematical Evolutionary Theory*. Princeton: Princeton University Press, 145–173.

Furrow, R. E., F. B. Christiansen, and M. W. Feldman (2011). 'Environment-Sensitive Epigenetics and the Heritability of Complex Diseases', *Genetics* 189: 1377–1387.

Gerbault, P., A. Liebert, Y. Itan, A. Powell, M. Currat, J. Burger, D. M. Swallow, and M. G. Thomas (2011). 'Evolution of Lactase Persistence: An Example of Human Niche Construction', *Philosophical Transactions of the Royal Society B* 366/1566: 863–877.

Gotelli, N. J. (1995). *A Primer of Ecology*. Sunderland, MA: Sinauer.

Hairston, N. G. J., S. P. Ellner, M. A. Geber, T. Yoshida, and J. A. Fox (2005). 'Rapid Evolution and the Convergence of Ecological and Evolutionary Time', *Ecology Letters* 8: 1114–1127.

Hanski, I. (1999). *Metapopulation Ecology*. Oxford: Oxford University Press.

Hardin, G. (1968). 'The Tragedy of the Commons', *Science* 162: 1243–1248.

Heft, H. (2001). *Ecological Psychology in Context: James Gibson, Roger Baker, and the Legacy of William James's Radical Empiricism*. Hillsdale, NJ: Lawrence Erlbaum.

Henn, B. M., L. L. Cavalli-Sforza, and M. W. Feldman (2012). 'The Great Human Expansion', *PNAS* 109: 17758–17764.

Henrich, J. (2004). 'Demography and Cultural Evolution: How Adaptive Cultural Processes Can Produce Maladapative Losses—the Tasmanian Case', *American Antiquity* 69: 197–214.

Hünemeier, T. C., E. G. Amorim, S. Azevedo, V. Contini, V. Acuna-Alonzo, F. Rothhammer, J. M. Dugoujon, S. Mazieres, R. Barrantes, M. T. Villarreal-Molina, V. R. Palaxao-Cortes, F. M. Salzan, S. Canizales-Quinteros, A. Ruz-Linares, and M. C. Bortolini (2012). 'Evolutionary Responses to a Constructed Niche: Ancient Mesoamericans as a Model of Gene–Culture Coevolution', *PLoS One* 7/6: e38862. doi: 10.1371/journal.pone.0038862.

Ihara, Y. and M. W. Feldman (2004). 'Cultural Niche Construction and the Evolution of Small Family Size', *Theoretical Population Biology* 65: 105–111.

Jablonka, E. and M. J. Lamb (2005). *Evolution in Four Dimensions: Genetic, Epigenetic, Behavioral and Symbolic Variation in Life History*. Cambridge, MA: MIT Press.

Kendal, J., J. J. Tehrani, and J. Odling-Smee (2011). 'Human Niche Construction in Interdisciplinary Focus', *Philosophical Transaction of the Royal Society B* 366/1566: 785–792.

Krakauer, D. C., K. M. Page, and D. H. Erwin (2009). 'Diversity, Dilemmas, and Monopolies of Niche Construction', *American Naturalist* 173: 26–40.

Kylafis, G. and M. Loreau (2008). 'Ecological and Evolutionary Consequences of Niche Construction for its Agent', *Ecology Letters* 11: 1072–1081.

Kylafis, G. and M. Loreau (2011). 'Niche Construction in the Light of Niche Theory', *Ecology Letters* 14: 82–90.

Laland, K. N., F. J. Odling-Smee, and M. W. Feldman (1996). 'The Evolutionary Consequences of Niche Construction', *Journal of Evolutionary Biology* 9: 293–316.

Laland, K. N., F. J. Odling-Smee, and M. W. Feldman (1999). 'Evolutionary Consequences of Niche Construction and their Implications for Ecology', *Proceedings of the National Academy of Sciences of the USA* 96: 10242–10247.

Laland, K. N., F. J. Odling-Smee, and M. W. Feldman (2000). 'Niche Construction, Biological Evolution, and Cultural Change', *Behavioral and Brain Sciences* 23: 131–175.

Laland, K. N., F. J. Odling-Smee, and M. W. Feldman (2001). 'Cultural Niche Construction and Human Evolution', *Journal of Evolutionary Biology* 14: 22–33.

Laland, K. N., F. J. Odling-Smee, and S. Myles (2010). 'How Culture Shaped the Human Genome: Bringing Genetics and the Human Sciences Together', *Nature Reviews Genetics* 11: 137–148.

Laland, K. N., K. Sterelny, F. J. Odling-Smee, W. Hoppitt, and T. Uller (2011). 'Cause and Effect in Biology Revisited: Is Mayr's Proximate–Ultimate Distinction Still Useful?', *Science* 334: 1512–1516.

Lee, R. D. (1986). 'Malthus and Boserup: A Dynamic Synthesis', in D. Coleman and R. Schofield (eds), *The State of Population Theory: Forward from Malthus*. Oxford: Blackwell, 96–130.

Lehmann, L. and M. W. Feldman (2009). 'Co-evolution of Adaptive Technology, Maladaptive Culture and Population Size in a Producer–Scrounger Game', *Proceedings of the Royal Society B* 276/1674: 3853–3862.

Lewontin, R. C. (1983). 'Gene, Organism, and Environment', in D. S. Bendall (ed.), *Evolution from Molecules to Men*. Cambridge: Cambridge University Press, 273–285.

Loreau, M. (2010). *From Populations to Ecosystems: Theoretical Foundations for a New Ecological Synthesis*, Monographs in Population Biology 46. Princeton: Princeton University Press.

Nance, W. E. and M. J. Kearsey (2004). 'Relevance of Connexin Deafness (DFNB1) to Human Evolution', *American Journal of Human Genetics* 74: 1081–1087.

Odling-Smee, F. J. (2007). 'Niche Inheritance: A Possible Basis for Classifying Multiple Inheritance Systems in Evolution', *Biological Theory* 2: 276–289.

Odling-Smee, F. J. (2010). 'Niche Inheritance', in M. Pigliucci and G. B. Muller (eds), *Evolution: The Extended Synthesis*. Cambridge, MA: MIT Press, 175–207.

Odling-Smee, F. J. and K. N. Laland (2012). 'Ecological Inheritance and Cultural Inheritance: What Are They and How Do They Differ?', *Biological Theory* 6/3. DOI: 10.1007/s13752-012-0030-x.

Odling-Smee, F. J., K. N. Laland, and M. W. Feldman (1996). 'Niche Construction', *American Naturalist* 147: 641–648.

Odling-Smee, F. J., K. N. Laland, and M. W. Feldman (2003). *Niche Construction: The Neglected Process in Evolution*, Monographs in Population Biology 37. Princeton: Princeton University Press.

Piersma, T. and J. A. van Gils (2011). *The Flexible Phenotype: A Body-Centred Integration of Ecology, Physiology and Behaviour*. Oxford: Oxford University Press.

Pigliucci, M. and G. B. Muller (2010). *Evolution: The Extended Synthesis*. Cambridge, MA: MIT Press.

Poinar, H. N., M. Kutch, K. D. Sobolik, I. Barnes, A. B. Stankiewicz, et al. (2001). 'A Molecular Analysis of Dietary Diversity for Three Archaic Native Americans', *Proceedings of the National Academy of Sciences of the USA* 98: 4317–4322.

Post, D. M. and E. P. Palkovacs (2009). 'Eco-Evolutionary Feedbacks in Community and Ecosystem Ecology: Interactions between the Ecological Theatre and the Evolutionary Play', *Philosophical Transactions of the Royal Society B* 364/1523: 1629–1640.

Rendell, L., L. Fogarty, and K. N. Laland. (2011). 'Runaway Cultural Niche Construction', *Philosophical Transactions of the Royal Society B* 366/1566: 823–835.

Richerson, P. J., R. Boyd, and R. L. Bettinger (2009). 'Cultural Innovations and Demographic Change', *Human Biology* 81: 211–235.

Romer, P. M. (1990). 'Endogenous Technological Change', *Journal of Political Economy* 98: S71–S102.

Schoener, T. W. (2011). 'The Newest Synthesis: Understanding the Interplay of Evolutionary and Ecological Dynamics', *Science* 331: 426–429.

Schwilk, D. W. and D. D. Ackerly (2001). 'Flammability and Serotiny as Strategies: Correlated Evolution in Pines', *Oikos* 94: 326–336.

Silver, M. and E. Di Paolo (2006). 'Spatial Effects Favour the Evolution of Niche Construction', *Theoretical Population Biology* 20: 387–400.

Smith, B. D. (2007). 'The Ultimate Ecosystem Engineers', *Science* 315: 1797–1798.

Smith, B. D. (2011). 'General Patterns of Niche Construction and the Management of "Wild" Plant and Animal Resources by Small-Scale Pre-industrial Societies', *Philosophical Transaction of the Royal Society B* 366/1566: 836–848.

Smith, E. A. (2000). 'Three Styles in the Evolutionary Analysis of Human Behavior', in L. Cronk, N. Chagnon, and W. Irons (ed.), *Adaptation and Human Behavior: An Anthropological Perspective*. New York: Aldine de Gruyter, 27–46.

Solow, R. M. (1956). 'A Contribution to the Theory of Economic Growth', *Quarterly Journal of Economics* 70: 65–94.

Wilson, E. O. (1975). *Sociobiology: The New Synthesis*. Cambridge MA: Harvard University Press.

Wollstonecroft, M. M. (2011). 'Investigating the Role of Food Processing in Human Evolution: A Niche Construction Approach', *Archaeological and Anthropological Sciences* 3/1: 141–150. DOI: 10.1007/s12520-011-0062-3.

Wrangham, R., J. H. Jones, G. Laden, D. Pilbeam, and N. L. Conklin-Brittain (1999). 'The Raw and the Stolen: Cooking and the Ecology of Human Origins', *Current Anthropology* 40: 567–594

CHAPTER 6

POPULATIONS FOR STUDYING THE CAUSES OF BRITAIN'S FERTILITY DECLINE

communication communities

SIMON SZRETER

6.1 Introduction

In this chapter it is proposed that the concept of communication communities can provide a theoretical formulation of the distinctive form of population which is most relevant for studies of Britain's fertility decline, *c.*1860–1940. It is argued that change in fertility behaviour needs to be contextualized through the study of changes in the perceived relative costs and benefits of childrearing, not parity alone. Attention is additionally drawn to the importance of sterility, which is responsible for the wide degree of variability ubiquitously found in human fertility. This is a key consideration which facilitates the proposition that communication communities can form a foundational theoretical component for both understanding and explaining fertility change, including secular fertility declines in large aggregates, such as national populations.

6.2 Possible Populations for Studying Demographic Change

In his framing chapter for this book, Philip Kreager helpfully points out that while there have always been at least two distinct approaches to the study of populations in the human sciences, in fact only one of these two has tended to predominate during the last century and a half or more. Since at least Aristotle's *Politics* it has always been possible either to study the processes by which the membership of populations and of their defined subpopulations come to exist; or to study at higher levels of aggregation the demographic, epidemiological, and evolutionary outcomes of those processes as 'population change' (Kreager 2008). The latter approach, dominant in demography, assumes in effect that we already know what the population in question is and that it is primarily general change at the aggregate level which is the heart of the scientific matter, rather than delving within the category of 'population' to observe possible forms of 'internal' change in its component parts and their mutual relations as a method for understanding and explaining change in the aggregate (Kreager 2009).

Equally helpfully, Kreager notes that a primary historical influence which has produced the long-standing dominance within demography of the second of these two approaches has been the choices made by nineteenth-century nation-making states and their bureaucracies about the data they wished to collect, collate, and deploy. These were informed by protean practical political and administrative exercises in the formation of populations, in the first sense, through inscribing individuals' membership in a territorially bounded sovereign entity through nationally resourced processes of census taking and civil registration.[1] However, in an age characterized by a positivist understanding of the status and validity of the information collected, having created the data on populations and their changing character through censuses and vital registration, the nationalist (and other sociocultural) assumptions and ideology motivating this process were not themselves subject to critical questioning, only the technical and statistical adequacy of the administrative and analytical procedures used (Eyler 1979; Schweber 2006). Indeed, this now long-standing officially managed process of constructing populations as nations has not, until relatively recently, received much critical scholarly attention. When it has, it has been mostly historians and anthropologists, not demographers, who have begun to open up these questions about the categories and classifications with which demographic statistics are imbued; consequently many students of population change in the human sciences do not refer to this growing literature.[2]

This bias within the historical data legacy since its origins in the early nineteenth century has been amplified by the continued intimate association of demographic social science with nation state policies during the late nineteenth and throughout the twentieth centuries. Alison Bashford (2013) argues that it was the dawning perception in the 1880s and 1890s of a 'closed world' with no further unclaimed territory in either the New World or Africa that resulted in much enhanced attention to global settlement patterns and expansionist claims for 'Lebensraum' for the 'vigorous' European populations during the period

[1] In fact, by contrast with census taking and their classifications (see next note), civil registration and identity registration in general have received virtually no academic attention as a subject for comparative historical research until extremely recently. See Breckenridge and Szreter (2012).

[2] These include for instance: Hacking (1986); Anderson (1988); Desrosières (1993); Caplan and Torpey (2001); Curtis (2002); Kertzer and Arel (2002); Szreter et al. (2004); Hirsch (2005).

*c.*1880–1939 among certain social scientists in the imperialist nation states. Thus, flows of migration—sponsoring, prohibiting, and attempting to direct them in conformity with nationalist policies—became a subject of scientific discourse among students of population problems, at the same time that international migration was becoming the intense focus of nationalist and imperialist administrative practices across the globe. Adam McKeown (2008) has examined the global construction of 'national' borders at this time in the much-enhanced effort to police all individuals' movements in conformity with the notion of their exclusive membership of countable national populations conceived as closed entities (and see also Bashford 2006: part II).

From the 1880s through until the 1940s this nationalist demography was also informed by imperialist and social Darwinist agendas, which entailed race-based, class-based, and gendered typologies. After the 1940s this was superseded, though not entirely replaced, by a nationalist agenda of economic growth and development. This once again emphasized the nation as the fundamental accounting entity with the global application of a national economic accounts framework for managing growth as a national process. Demography, wedded to the modernization and associated demographic and epidemiologic transition theories, found ready policy applications and international demand for its practitioners alongside this growth accounting framework. This was in response to an urgent post-war realpolitik agenda to promote liberal models of national economic growth among the world's newly independent nations as an alternative to the promises of communist ideology and planning (Szreter 1993a; Gilman 2003).

As Kreager observes, consequently the dominant approach within twentieth-century demography, which proceeds on the assumption that the internal structure and dynamics of national populations and their memberships are not a key focus for scientific enquiry, has only begun to be seriously challenged during the last two decades. Barbara Anderson (1986) in the summary volume of the Princeton European Fertility Project drew attention to the issue of linguistic and ethnic regional divisions. These acted as cultural boundaries apparently dividing the experience of national fertility declines in many countries, such as Great Britain, Germany, Austria, Italy, Belgium, Russia, Portugal, and Switzerland. Kreager notes that this basic insight has since been more intensively explored and developed in the work of historical demographers and anthropological demographers, to which one might also add those who have directly studied the intellectual and administrative histories of the data-producing official institutions and their models.[3]

Thus, a number of the contributions in this volume from anthropological and historical demographers demonstrate various ways in which apparently uniform population aggregates, sometimes exhibiting no apparent change in their aggregate demographic variables of mortality or fertility over the course of a century or more,

[3] On the latter, see references in notes 1 and 2 and also Szreter (1984); Boltanski (1987); Cohn (1987); Szreter (1993b); Bayly (1996). For important collections of research in anthropological and historical demography, see Greenhalgh (1995); Kertzer and Fricke (1997); Basu and Aaby (1998); Bledsoe, Lerner, and Guyer (2000); Janssens (2007).

are in fact composed of diverse and changeable subpopulations. This is demonstrated in Mark Elvin's chapter on the Yangzi Valley 1810–1934 and Mikołaj Szołtysek's on Eastern and Central European household demography in the eighteenth century; and on a smaller scale by Kaplan et al.'s on Bolivia. Another important group of chapters demonstrate the importance for students of population change to be willing to adopt a range of investigative methodologies that, *de facto*, make the opposite assumption to that which focuses on the mortality and fertility of aggregates. These are explicitly focused on subpopulations and their processes: specific localities, communities, subgroups or networks of individuals. Both Melissa Brown and Elisabeth Schröder-Butterfill have utilized multigenerational ethnographic interviews, Gregson and Hallett analyse locally contextualized community programmes of intervention, Kohler et al. review studies of networks for research on HIV/AIDs and sexual norms and behaviour. While Johnson-Hanks has pursued a combination of demographically contextualized interviews in Cameroon and survey evidence from Kenya to present the argument that the agency of individuals—in making both apparently mundane and more momentous choices in their lives—contributes to the relative expansion or contraction of the diverse sociodemographic subpopulations that compose a society. Individuals 'select' themselves into and out of affiliation with different social identities and possible futures, thereby gradually changing society's composition.

6.3 Communication Communities

Each of these authors in the following chapters does not claim to be studying something statistically representative of the larger national aggregate. The focus is instead on a contextualized set of motives, choices, and actions which are diverse but which ultimately contribute to the overall sum of demographic change occurring in and through the communication communities they choose to study. The term 'communication community' is a deliberately arresting term (one might say an awkward term, apparently committing the user to a linguistic redundancy). I coined it in *Fertility, Class and Gender* (*FCG*) to denote all those diverse subpopulations within the population of England and Wales *c*.1860–1940, which could be discerned exhibiting distinctive trajectories of fertility during the era of Britain's secular fertility decline, once the official census data was subjected to an independent, secondary analysis (*FCG*: ch. 7 and 546–558). Other researchers, with diverse methodologies addressing demographic change at the subnational level, have used terms such as localities, networks, kinship groups, and communities to describe the various subnational social groups they are studying. I would argue that most of these may be encompassed in the more general concept of communication communities where the study of fertility decline is concerned. A crucial aspect of this term, communication community, is that it focuses attention *both* on the importance of studying the constituent 'subpopulations', the communities which, summed together, compose the conventional larger aggregate, the national population, but also, equally, it insists on the significance of studying the content and forms of communication that occur between constituent members of each such community. It is this which differentiates communication communities from each other, as it is these processes of communication which determine the character of the communities and their influences over the group's demography.

The claim for the formative importance of communication communities when studying the causes of the kind of large-scale observed change in fertility behaviour which occurred during the modern fertility declines, follows from the theoretical proposition in *FCG* that parents' evaluations of their own reproductive behaviour relate to the whole process of childrearing, as they see it, and not simply to decisions about whether or not to have another birth. It was argued that significant, sustained, and voluntary change in reproductive behaviour (observable in frequency or age of commencing childbearing or in age-specific fertility rates) will be the result of modification in the responsible agents' *perceived relative costs and benefits of childrearing* (*FCG*: 443–447). It is important to note of this formulation, firstly, that 'childrearing' is not a simple noun, but itself a radical variable: different cultures and social groups have widely varying and continually renegotiated expectations of what is required to raise the next generation and the age at which childrearing responsibilities are deemed to diminish or formally end (*FCG*: 447–465). Secondly, the responsible agents for upbringing vary legally and normatively between different times and places, typically comprising part of the kinship and socializing group, but varying from a single parent or guardian to a multigenerational extended family or lineage clan, a religious community, or even virtually the entire matriarchy of a poor urban street or neighbourhood. Furthermore, in any polity which provides its citizens with a significant form of collectively funded social services or welfare, there is a sense in which the state itself is also one of the responsible agents. Major changes in its policies affecting families must certainly be a variable factor also influencing the perceived relative costs of childrearing among parents and all others involved (*FCG*: 513–525).

Thirdly, it is to be noted that the 'relative costs and benefits' are 'perceived'. This is crucial. It means that this formulation does not imply that it is an optimal approach to the study of human fertility change that we should follow the economic accounting approach associated with Gary Becker (1960). The latter focuses explanatory effort on the deployment of the marginal utility framework of analysis of microeconomics to convert all known factors to a common arithmetic calculus of household value or utility. Such an approach has a somewhat intellectually reductive goal, which is to demonstrate that if one makes the simplifying assumption that all observable human behaviour is based on a calculus of rational self-interest, then models can be built (highly complex ones if necessary) to reduce all observable behaviour to an apparent statistical consistency. This does not seek to understand human fertility variation as a communicative and causal process but to explain it (in the narrowly statistical sense of this word) in order to appear to demonstrate accounting consistency with a universal, and critically unquestioned, theory (in fact an assertion) about human motivation. Such an approach is rather uninterested in studying perceptions and their communicative formation, except insofar as they may be rendered into mathematically tractable modifications of the utility model's schedule of preferences.

Finally it should be clear that communication communities also strongly mediate involuntary, as well as voluntary, adjustments in fertility behaviour. Communication communities refer to an encompassing sociocultural environment of language, values, and roles in which individuals and families participate and through which they form and negotiate their meanings, goals, and social identities. Communication communities are, in turn, in an inescapable relationship with the material, biophysical, and epidemiological context in which all human societies must exist, and they are endowed with a range of institutions for coping with that relationship. This means that the perceived relative costs of childrearing of a communication community may also alter due to involuntary or seemingly 'unconscious' (in the sense of not conatively processed) causes. Such non-voluntary fertility change would occur in the form of unreflexive, normative responses to alterations in the biophysical, economic, or political environment within which the communities are located in conformity with the accepted behaviour sanctioned as normal—and therefore perceived as merely 'natural'—by the communication community's extant institutions. This could have a direct consequence of changed fertility, though such change would not be the primary function of the institutions nor would individuals have perceived themselves as doing anything except acting naturally, normally, and morally. For example, people may postpone getting married or—in contemporary, routinely contracepting societies—getting pregnant, not because they wish to avoid having children but precisely because in their communication community they believe it is necessary to attain a conventional minimal standard of living in an independent household in order to have children; and yet this postponement may lead to a radically reduced family size—or no children at all—though this was never a desired or designed outcome.

Thus, having introduced the concept of communication communities, the remaining sections of this chapter explore the wider implications of taking this concept seriously for approaching the study of fertility declines. Firstly, it will be argued that this approach foregrounds study of the social and cultural processes of childrearing, in place of a more restricted focus simply on 'decisions' about fertility—in the sense of an individual woman's or couple's sequential parity choices. Consequently, communication communities can be seen to influence rather more than the sphere of conscious fertility decisions. Their functioning is articulated by and integrated with the transgenerational institutions and norms which constrain and frame people's values and motives in any specific time and place. This affects fertility outcomes directly, and also through behavioural responses to challenges and opportunities thrown up by changes in the wider socio-economic, physical, and political environment within which communication communities are lodged.

In the second section to come, an important apparent theoretical objection to the proposed approach for studying large-scale fertility change will be discussed and answered. This could be called the problem of scale. Even supposing that communication communities do exist in the form envisaged, if one is trying to understand the causation involved in such a large-scale change as has repeatedly occurred when national populations have experienced secular fertility declines, should we not be looking for common factors of variation across these different communication communities—even across national aggregates—rather than spending our time and energy on the diverse characters and unique sets of internal relationships of communication communities?

Finally, the chapter will conclude with a brief discussion of how communication communities change. How do the perceptions of the component individuals and their awareness of each other within communication communities translate into secular demographic change? Most of the examples given throughout this chapter will relate to the history of secular fertility decline in England and Wales only. This demonstrates the point that there is a sufficiency of enlightening diversity to be studied among the communication communities within this single 'national' aggregate; however, relevant research on other populations will also be mentioned as appropriate.

6.4 Childrearing, not Parity, as the Focus for Understanding Change in Fertility Behaviour

While the approach advocated here apparently seeks to reduce the scale of the social context for operational study of the processes of fertility decline from the nation to the communication community, it simultaneously proposes to radically expand the socio-cultural scope of the primary subject for research. Demographers have focused fairly exclusively on the observation, precise measurement, and explanation of changes in parity—i.e. the numbers and timing of live born children produced by each married woman. It is proposed here to place changes in parity patterns within the context of a broader focus on historical, contextualized agents' evaluations of changes in the whole process of childrearing, as formulated in *FCG*:

> Perceived relative costs of childrearing shifted primarily because the gendered and generational roles and norms, upon which this variable equation was premised, were themselves altered. It was not really the quantities in the equation which changed (a primarily economic issue) but the qualities: the identities of the terms themselves. The implications of childrearing and the perceptions of parental roles and responsibilities were shifting, because of transpositions in social identities and, therefore, in the information contexts.[4]

With childrearing as the focus, this means that we need to take into account wider and potentially much longer-term considerations by reproducing couples than whether or

[4] (*FCG*: 546). And on the theoretical importance of distinctive information contexts for relativizing the 'rationality' which any individual can be observed to be pursuing, see *FCG*: 39–41.

not to have an additional child at a point in time. It brings to our attention perceptions of intergenerational relationships, duties and obligations. Also how changing economic opportunities, collective institutions, national policies, or local social provisions might be interacting with the shared roles and norms that govern such relationships in each communication community. This may well engender change in the meaning of key social roles and identities in relation to gender and generational relationships: change in what it means to be a father and a husband, a mother and a wife, with entailed implications for change in reproductive behaviour. For instance, John Gillis has stimulatingly suggested that among the Victorian middle class there was a shift from an understanding of motherhood as childbearing to one of childrearing—as something like a lifelong 'career' (Gillis 1992: 32). On a somewhat fuller evidential basis, in his classic study *Prosperity and Parenthood*, J. A. Banks (1954) documented how the perceived paternal financial responsibilities for maintaining a genteel household and for schooling genteel children escalated in Victorian society. At the end of her excellent chapter discussing the ways in which poor British working-class mothers and fathers were subject to ever more stringent checks and demands on their childrearing practices by the public health officials of local government during the first two decades of the twentieth century, Ellen Ross cites a Rotherhithe working woman saying in 1919, 'Sometimes I hanker after another baby, but then I say to myself the pleasure of it would only last a year or two, and afterwards there'd be nothing but worry' (Ross 1992: 63). This testifies to the way in which the greater public accountability for rearing their children for many more years beyond early childhood could be producing a reflective attitude of modification towards their own fertility among members of this communication community. These are all, then, different examples of the challenging and negotiated sociocultural and political processes, whereby perceived relative costs of childrearing may have changed among different communication communities in British society during the era of secular fertility declines.

Of course one might expect periods of dramatic labour mobility and migration to new settlements and the growth of urban communities to be associated with substantial intergenerational change in the meanings of parental roles and of childrearing. This happened in Britain during the period *c.*1730–1830 when marriage became relatively young, universal, and more fertile among the labouring classes. However, crowding into new urban environments was in fact a declining feature of society by the time British marital fertility fell so dramatically, *c.*1875–1935. It did so in mostly well-established urban communities, no longer subject to an influx of rural and Irish newcomers but in fact more likely to be subject to a rising rate of overseas emigration, especially in the 1880s and again in 1901–1931 (Tranter 1996: 20, 38).

During this period of relatively stable and settled town and urban life, communication communities predominantly took the form of neighbourhoods of street communities among the working classes. There was a marked pattern of residential segregation, with most of the middle-class families inhabiting distinct streets and squares, even whole suburbs where their numbers were sufficient, as in Birmingham's Edgbaston. However, among the educated, literate, and aspirational upper and middle classes, participation in their more networked communication communities did not necessarily depend upon residential propinquity. Individuals and couples living in different towns, but striving after similar social and cultural goals and adopting similar gender and work roles and sharing a similar language, could be participants in the same communication community, though they would never meet or know of each other's existence. This could happen most obviously among the educated professional and managerial classes because, with a national daily press delivered by the rail network, they read the same daily and weekly newspapers and journals and the novels serialized in them. They could participate in this whether they chose to reside with their families in one of the more exclusive enclaves of central London, in the many bucolic commuter retreats of the Home Counties, or their equivalents in Leeds, Birmingham, or Manchester (*FCG*: 551–552). Subject to affordability, parents in the upper, middle, and lower middle classes wanted a similar privileged educational formation for their children to ensure a secure and prosperous future. By contrast, for most families in the working classes—the vast majority of the British populace throughout the era of fertility decline—communication communities would be more geographically circumscribed to the town or city they lived and worked in. Here they shared a common dialect, cheap local press, and a similar range of employment and leisure options (*FCG*: 546–555).

Individuals could become members of various communication communities during their lifetimes. They learned their complex codes principally through their childhood and early adulthood upbringing, during which they were socialized into a minimum of two such communities: that of their family (or institutional) home and that of the streets outside it, their neighbourhood (or boarding school if born into the elite). In some cases these two might be culturally almost identical and strongly reinforcing, but for others they were not.[5] For instance, Robert Roberts' memoirs attest to his keen childhood consciousness of a difference between his own family's status code and that of the Salford working-class neighbourhood of streets which his parents' corner shop served (Roberts 1971). Social identities were formed through participation in communication communities. For some, like Roberts, this could represent a selective choice. For most others, especially those born into poorer working-class families, choices were likely to be far more restricted to the encompassing proletarian communication community of their town of birth, unless they took the drastic step of leaving their home town, which might perhaps also require divesting themselves of its spoken accent.

[5] The pioneering work of the educational and linguistic sociologist, Basil Bernstein, focused precisely on this issue in his exposition of the formation of cultural capital, in Bourdieu and Passeron's sense (as expounded by Pierre Bourdieu 1977b). Bernstein (1977a,b) identified the disjuncture between the culture and language of the home and that of the school as a critical source of educational difficulty for working class children, an obstacle which was encountered in a much less extreme form by the children of middle class parents. The latter were, by definition of their parents' occupational roles in society, more assimilated to the distinctive culture and language codes of schools. Assimilation required a record of relative success in the credentialling institutions of the nation's educational system, in order to gain entry to the middle class jobs which they held. For an influential ethnographic study in post-war Britain further illustrating these insights, see Willis (1977).

As Elizabeth Bott's seminal 1950s research found, Roberts' perception of a primary allegiance to the values and goals of his own nuclear family as being individually distinctive from those of other neighbours in the vicinity around him, is itself a selectively 'middle-class' or bourgeois trait (Bott 1957). Leonore Davidoff's most recent study of the bourgeois Victorian family has emphasized how throughout the period 1780–1920 emotional and social reliance on the ties of extensive kinship between siblings, cousins, and their closest friends, across considerable distances at times, became an ever more emphatic element of shared upper- and middle-class culture. Embracing 'close marriage' this was functional to their dynastic success as financial and cultural capital-accumulating lineages (Davidoff 2012). Bott's work suggested that reliance on kin was also a working-class trait. She and other urban anthropologists in the 1950s found this operating primarily through a highly gendered mother–daughter relationship that was strongly dependent on geographical propinquity, while working-class men tended to fraternize primarily with unrelated other men from the workplace or neighbourhood. Davidoff's research emphasizes that bourgeois kin closeness was distinctive in that it was as likely to be bilateral, in gender terms, and intra- as well as intergenerational, among brothers and sisters and their cousins; and that significant effort was made to maintain such relations independently of possible geographical separation.

In *FCG* it was suggested that the differential fertility patterns, according to male occupation of the head of household in the 1911 census, which were there reanalysed in detail, indicated that there was something of a sociological fault-line running between, firstly, proletarian families identifying primarily with the goals and values of most others living in their local communication communities and, secondly, those identifying with more 'virtual', networked communication communities associated with their wider kin contacts and participation in a national literary and scientific media. Thus, *FCG* found that the differential fertility patterns of the working-class mass of the nation reflected a diversity of local industrial regimes (such as the Staffordshire potteries, the West Yorkshire wool and worsted industry, the coal-mining regions, etc.), rather than the neat, socially graded class model of diffusion 'down' a national social hierarchy which the influential Registrar-General's official social classification scheme had been invented to portray (Szreter 1984).[6] By contrast, *FCG* found that the variation in fertility within the numerically smaller part of the nation, who could be termed upper, middle, or lower middle

[6] The fundamental problem with this approach is that to resort to use of the concept of diffusion in order to explain change over time is to do no more than deploy a description as if it were an explanation. See Kreager's (1998) masterly critical essay, 'The Limits of Diffusionism'. In the historiography of the study of fertility declines it is hard to overestimate the importance of the adoption of a notion of 'trickle-down' diffusion by the official statistician who was responsible for presenting the Fertility of Marriage Report on the 1911 census, the seminal enquiry into the nation's fertility decline. As *FCG* (9–21 and ch. 5) shows the theory of diffusion and the model of the nation's social structure and its fertility decline as a neat linear status hierarchy went hand in hand. This diffusion model then became almost ubiquitous in interwar international demographic research on fertility declines and the associated model of social structure, persisting for almost the entire twentieth century in British official publications and derivative social science.

class, represented a more 'national' set of hierarchically graded status groups, less tied to any specific locality (*FCG*: 310–322).

The term 'communication communities' was selected, rather than 'discursive communities', because the latter, though more elegant, would privilege language alone as the shared meaning system of a social group. Of course, non-verbal practices, daily routines, unspoken understandings, and 'body languages' (Bourdieu's fruitful concept of 'habitus', 1977a) were also all intended as aspects of the mutually influencing communication occurring as a result of observing and being observed among those recognizing each other as belonging to a shared communication community. This was the basis on which they differentiated themselves and their choices from the codes and values held by alternative communication communities. Furthermore, it was a crucially important part of the concept that physical communities—places and localities—could be host to a number of such communication communities simultaneously. Although these would be alternatives to each other, they would not necessarily be straightforwardly mutually exclusive alternatives, in that individuals might participate in several communication communities at once, shifting their allegiances and alliances as either context or motive altered. This is exemplified in some of the research which Jennifer Johnson-Hanks' chapter presents. She shows how individuals altered their fertility behaviour by choosing or 'selecting' their social identities and futures, while she also adds an insistence, which follows from this, on the importance of studying the process of selection itself as a key to understanding fertility change.

Individuals may well participate in several communication communities at once, providing them with various resources, but also with potentially conflicting loyalties and values. In any physical community, such as a town or neighbourhood, there may well be a number of communication communities of varying social extents and powers overlaid upon each other. Communication communities form a social palimpsest, since they are essentially complexes of multiplex relationships, not places. In addition to the often gendered educational and employment opportunities that may be available in any specific location, the diversity of religious institutions and ethnic divisions—often related to a history of population movements (as occurred on a substantial scale in Liverpool, Glasgow, Manchester, or in London's East End, for instance)—would further multiply or strongly influence the character of the competing communication communities and available sources of social identities in any town or city. However, where ethnicity and religion are involved, this may render the rivalrous communities more mutually exclusive. This can also carry dangers for individuals attempting to mix their allegiances—a Romeo and Juliet predicament. Excellent studies specifically exploring the relationship between divergent but related forms of fertility change among rival communication communities characterized by religious difference sharing the same physical locality have, for instance, been published in relation to the Jews and Gentiles of Venice, the province of Quebec in Canada, and Catholic and Protestant communities in Switzerland (Derosas 2006; McQuillan 1999; Praz 2005).[7]

[7] See also the various contributions to Derosas and van Poppel (2006).

It is important to emphasize that this view of communication as positively including the non-verbal—the merely observational and the silently reflective upon others can be particularly significant where the question addressed, as in *FCG*, is changing marital fertility as in the late Victorian and Edwardian era. In relation to British society a number of studies have confirmed that the explicit discussion of reproduction, childbearing, and sexuality was a subject rarely raised in public, respectable discourse, including even among neighbours and friends, during the period before the Great War (Porter and Hall 1995: part II; Cook 2004: part II). It was only slightly more subject to explicit acknowledgement in the interwar period—and then mainly in the form of written manuals which tended to circulate only in private and discreetly among a more progressive section of the educated middle classes.[8] Such conditions of public reticence on matters of sex and reproduction have in fact characterized many (though certainly not all) societies in which dramatic fertility reductions have occurred in the past. Absence of a frank, public discourse on sex and reproduction also continues, for instance, to characterize many African and Muslim societies today. Thus, Uganda was virtually unique in Africa in proving capable of reducing its HIV/AIDs infection rate in the 1990s because President Museveni boldly broke public codes of non-discussion of sexual affairs and marital fidelity. His 'zero grazing' government propaganda campaign was shown to have licensed explicit and active discussion and awareness of the problem in civic social networks throughout Ugandan society, quite unlike the silence prevailing elsewhere (Stoneburner and Low-Beer 2004).

As Fisher (2006) has documented in interwar British society through an oral history methodology, parents of both sexes, even of very small families at the end of the process of secular fertility decline in the 1930s, did not necessarily perceive themselves to have taken planned or rational decisions about family size. Rather, they claimed to have moderated their sexual behaviour—preferring the use of the 'natural' and spontaneous method of *coitus interruptus*, which they saw as most 'healthy'—in accordance with what they felt was a normal and sensible family size to maintain a respectable standing among their neighbours and in their wider community.[9] As Garrett et al. (2001: 268–289) found in their in-depth demographic analysis of 13 contrasting localities in England and Wales, 1891–1911, in those communities dominated by working-class family households (defined by paternal occupation), resident middle-class households tended to have higher fertility than middle-class families living elsewhere (in less proletarian neighbourhoods). Conversely, working-class couples residing in predominantly

[8] For evidence on this see Szreter and Fisher (2010: ch. 2). Even taking a wide definition of the middle classes to include the lower middle classes, this comprised a mere 14% of the adult male workforce of 1921, and still only 21% as recently as 1961 (Szreter and Fisher 2010: 21, n. 51). There is also evidence among working-class communities for fairly widespread but interestingly locally specific (again confirming no 'national' explicit discourse) colloquial euphemisms to refer to the practice of *coitus interruptus* (Szreter and Fisher 2010: 239).

[9] Fisher (2006). For a similar preference for 'natural' methods among women in the USA, see Bledsoe (1996); and for similar findings from an interview methodology, of somewhat younger women in four locations in Italy, see Gribaldo, Judd, and Kertzer (2009).

middle-class communities tended to reduce their fertility earlier and more effectively than working-class families living in more proletarian communities, indicating the influence on members of both classes of reflective observation within their local environment.

This evidence demonstrates the potential porosity of social class identities and wealth and income differences between individual households; and also the norm-influencing processes that can occur through daily participation in communication communities, where these are experienced as a locally engulfing cultural environment.

Robert Roberts has been cited earlier, choosing to discuss the differences in values between his own lower-middle-class family and those of the poor, working-class neighbourhood his parents' shop in Salford served. However a key assumption—shared by all those who read *The Classic Slum*—is that he genuinely 'knew' this proletarian community and can report with authority on its mores, and this is testimony to his simultaneous participation in multiple communication communities. Another almost equally well-known memoir is William Woodruff's *Road to Nab End*, which recalls growing up in poverty in interwar Blackburn (Woodruff 2002). Memorably depicting the privations of growing up in the kind of families Roberts' parents' corner shop served, Woodruff himself, nevertheless, through choosing to take opportunities for educational mobility, entered a 'middle-class' milieu in adulthood. That individuals, in association with their adult carers, consciously take such life-defining choices to remain within or pass between communication communities, is a crucial point to emerge from Johnson-Hanks' chapter and is, for instance, confirmed in the Blackburn and Harpenden oral history evidence by various respondents who deliberately did not pursue the kind of educational opportunity which Woodruff grasped by leaving his home town (Szreter and Fisher 2010: 22–23).[10]

6.5 Sterility and Fertility Variation: The Intrinsically Dynamic Properties of Communication Communities

It may be objected to the theoretical position laid out so far that it does not seem to provide for an originating or causal mechanism to account for dynamic change in reproduction at the national scale. If everybody in a communication community is sitting around

[10] After leaving school at 13, like our working-class interviewees, Woodruff then took a very different path by moving to London aged 17, and by age 20 had won an LCC scholarship to enter Oxford University. There were also one or two in our oral history sample who, like Woodruff, had experienced some social mobility from their working-class origins and for whom physical movement out of their working-class communities into more middle-class environments in adulthood had been an integral part of such mobility (Szreter and Fisher 2010: Horace, Hugh, and Norman).

watching and mimicking each other, how does anything ever change? Even if the argument was (which it isn't) that a few middle-class people infiltrate a working-class locality and then begin to exert an influence on the working-class people through their manifest example of lower fertility, that still wouldn't explain how it was that middle-class people began to adopt lower fertility in the first place. While there may be a range of carefully contextualized historical, anthropological, and ethnographic studies which confirm the plausibility of the concept of communication communities, and there may also be evidence that they are associated with distinctive subpopulation fertility regimes, it is not clear how such a framework of thinking about—essentially—cross-sectional diversity can provide a satisfactory account of dynamic, secular change on such a large scale as to encompass ultimately a dramatic reduction, in common, in the fertility of all the different communication communities in a nation. This, after all, is what fertility declines have entailed and why they are so significant demographically.

Isn't this, then, a theoretical mismatch: a theory of comparative statics misapplied to a problem of secular dynamics? Surely this phenomenon of aggregate large-scale change requires giving attention primarily to those factors operating at a trans-community level? Doesn't this require investigating those influences and forces which have the capacity to change behaviour simultaneously throughout the populace, despite the internal socio-cultural and economic diversity of the population, which the concept of communication communities addresses? This, broadly speaking, was the apparently irrefutable common-sense viewpoint which has informed the dominant approach to research on fertility declines within the framework of demographic transition and 'second' demographic transition theorizing. This has produced generations of quantitative studies trying to adjudicate statistically between structural economic or cultural and ideational aspects of 'modernization' as master 'causes' of this transnational phenomenon.[11]

However, as *FCG* (555–558) argued, there is a crucial property of the fertility of most species, and most particularly of humans, which renders this objection to the current approach far less compelling than at first sight appears to be the case. When taken fully into account, this endows the model of communication communities with the power of self-propulsion, in terms of generating dynamic change in reproductive outcomes, even on the scale of the secular national fertility declines. This is the phenomenon of sterility, or, put another way, the natural variability in fecundity of individuals and couples. This results in a wide diversity of fertility rates and family sizes in any society and its component communication communities, even under the conditions of what has been termed 'natural fertility', meaning the known or presumed absence of motives for birth control or the practice of techniques of contraception. Although, interestingly, sterility remains a relatively under-researched area of demography, in their textbook Bongaarts and Potter (1983: 41–42) cite the research of Louis Henry, published in 1965, as, in their

[11] Intriguingly just at the point in time when Coale and Watkins (1986: chs. 6–8, 10) pronounced the funeral rites on the first generation of such research conducted through the series of Princeton studies of national fertility declines over the previous two decades, the approach was reborn as the theory of 'the second demographic transition' (Lesthaeghe and van de Kaa 1986).

opinion, providing 'the best available estimates of couple sterility'. These estimates indicate that 3% of couples will be entirely sterile for involuntary reasons and that by the time a woman is aged 25, this will apply to 5% of her age cohort, rising to 8% by age 30, to 15% by age 35.

More recently the analysis performed by Wrigley et al. (1997) on 25 of the 26 parish family reconstitutions of England in the period 1538–1837, during which very little birth control within marriage is believed to have been practised, showed that among women marrying at the typical age of 20–24, no live births occurred to 3.8% of marriages. (Among the much smaller subset marrying at age 15–19, which was an unusually young age at marriage in England at this time, the corresponding figure was exactly the same as Louis Henry's estimate of 3%.) Among this group marrying at 20–24, a further 3.7% produced only one live birth; an additional 6.8% were limited to two live births; 6.0% produced three births only; a further 12.2% had four only; and 12.6% had five; while the remaining 54.9% produced six or more live births. (Wrigley et al. 1997: table 7.16, 401). So, due to the vagaries of primary and secondary sterility, slightly over 20% of couples throughout this 'pre-fertility-decline' English population were rearing 'modern', post-fertility-decline, small families of zero, one, two, or three children only. Indeed, when one takes into account the combined effects on survivorship to age 5 of prevailing rates of infant and child mortality, ranging in England in the period 1675–1749, for instance, from 170/000 at best (Hartland, Devon), up to 470/000 at worst (March, in the Isle of Ely and almost certainly higher in London at that time), a rather greater proportion, at least one-third of married couples, would in fact only have raised families of three or less children beyond age 5, while others were of course raising larger families of up to a dozen or more.[12] Thus, as can be seen from the parish reconstitution figures, if we are interested in explaining change and the potential for change, in fact the key property of a 'natural' high-fertility society, such as early modern England, is the great variability of individual couples' fertilities. Each parity size of family from zero, through one, two, three, up to about ten or twelve was represented by a substantial fraction of the population.

In a substantially sized parish in early modern England, *c.*1600–1700, containing, say, for convenience about 1,000 married couples, even if everybody in this parish had married relatively young, aged 20–24, about 40 marriages (one in every 25) would have produced no live births; a further 40 would have had only one, 70 would have had just two, and another 70 just three live births. One in four of the households in the parish would have exhibited this extremely low fertility—in the absence of any systematic family limitation. In early modern England there was therefore visible to all living in each parish an enormous variety in fertility and family sizes. Just as many married women, as a proportion of the total in the parish, would have exhibited the lowest fertility of zero or just one birth, as those producing large families of more than a dozen. Of course, as Caroline Bledsoe and colleagues have shown us, the cultural need for women to produce

[12] Wrigley et al. (1997: table 6.16, 270–271). On London see Laxton and Williams (1989), and for the most recent research, Newton (2010).

and rear children can be so strong, as in the African communities she has studied, that a wide range of marriage strategies and fosterage arrangements have developed to satisfy this requirement in the face of the biological facts of human sterility (for instance, Bledsoe and Isugo-Abanihe 1989). In early modern English society the widespread system of service and apprenticeship, whereby the majority of young adults aged 15–24 left home to reside and work in households often in other parishes, has been termed a system of 'ex post facto family planning' (Kussmaul 1981: 26). But it could almost equally be termed a system of ex post facto family building for the many subfertile among the labour-short, servant-receiving households. As the anthropologist Alan Macfarlane noted in his review of Kussmaul's book, 'the total absence of adoption in English history has long been a puzzle as has the absence of pressure towards high fertility' (Macfarlane 1982). For Macfarlane, Kussmaul's classic study of the institution of service solved this puzzle in comparative anthropology.

This state of continuous and radical variability in unplanned fertility throughout world history is of critical importance to the theory of compositional change through communication communities. It provides an inbuilt dynamic mechanism for a theory of the secular fertility decline which proposes to view the process in compositional terms as a set of adjustments occurring within, among, and between the communication communities which form national societies and which inform the norms and norm-referenced choices of individual agents in relation to their fertility behaviour. Given that those participating in any communication community will necessarily be exposed to many examples of a full range of fertility variation outcomes, among those with whom they identify and interact in their parish, community, or network, they will therefore be faced continually with a wide range of practical observations and reflections on the perceived relative costs and benefits of rearing families of all different sizes. This means that when changes are understood to occur in the encompassing economic, cultural, or political contexts of any of the various communication communities in any particular society, there are always practical lessons directly visible to the parents and potential parents—the agents directly responsible for fertility—to show which kinds of fertility outcome benefit families least or most, consequent on these changes in the encompassing environment.

In the variability of human fertility there is thus a continually prevailing situation of 'play' or open experiment, analogous to the variability provided by the phenomenon of mutation in relation to the mechanisms of evolution and the origin of species. Just as in evolutionary theory, the question of which of these continually generated variants will expand or decline proportionately to each other depends on the changing conditions in the 'environment' and the adaptive properties of the 'species' (the communication communities) in question. Just as in natural evolution this process of adaptation does not take place in a naked confrontation between the raw gene mutations—analagous here to the varying family sizes—and the environment. There are important intermediary factors in nature related to how the genetic mutations happen to result in often minute changes in form or function for the macroorganism—the individual member of the species—to which they contribute, and whether this confers any fitness advantage relative to other individuals of the species in the total environment in question. In our analogy to processes of fertility change, the 'mutations' are a far simpler range of variations—being the differing numbers of children born and surviving to adulthood, varying from zero to perhaps fifteen or twenty at the extreme. But where human society is concerned, the 'environment' in which the 'species' (the communication communities and their individual members) must compete is extraordinarily complex, involving not only the economic, epidemiological, and geophysical features, but necessarily also the full range of culture, belief, institutions, technology, and politics. There is also the further complexity that much of the latter also potentially represent parts of the adaptive apparatus available as resources to the individuals as well! Whether, taking all this into account, different forms of human fertility outcome are perceived favourably or not by the observing persons in each communication community depends on the values and norms through which they collectively and individually interpret the meaning and significance of these brute facts of fertility's variability visible before them in their neighbours' family households. The possibilities for adaptability are also complicated because they are strongly influenced by understandings of the coping sufficiency of the specific 'inherited' institutions which their societies and networks have already developed to modify the environment in order to manage this phenomenon of variability (e.g. service and apprenticeship as a leading example of such an adaptive institution in early modern English society, swept away in the course of the period *c.*1730–1830).

6.6 How do Communication Communities Change? How Does Awareness of Variation Translate into Secular Demographic Change?

Short-term and irregular fluctuations, such as in the price of wheat due to vagaries of the harvest, would be unlikely to exert marked effects producing trend changes in people's strategy in relation to their fertility. Thus, while rigorous recent research by Bengtson and Dribe (2006) has established that in early modern southern Sweden the rural poor were deliberately restricting their fertility in response to harvest vagaries, they also show that this short-term adaptation occurred in the context of no overall secular trend of modification in their fertility levels. However, longer-term and persistent changes do appear to have had such effects. Examples include the sustained growth of an accessible centre of new employment opportunities in a nearby prospering town, as happened for many communities near coalfields in the English North and Midlands, and in the Scottish Lowlands from the mid-eighteenth century onwards; or the permanent curtailment by central government of a long-standing general measure of welfare provision, as happened in England in 1834, with the enactment of the New Poor Law. In each of these examples, sustained change in circumstances could, through processes of community members' deliberative reflection, shift the preferences and behaviour of individuals to marry at younger or older ages (Wrigley 1983; Szreter and Garrett 2000: 55–60). Johnson-Hanks' point (cf. her chapter in this volume) would be that, if oral qualitative historical evidence were available on movement to new employment opportunities, the accompanying shifts in fertility behaviour could be shown to have been mediated by individuals selecting themselves into different communication communities as they left rural parishes for factory towns.[13] Or, in the case of the New Poor Law, people faced adjustments, after 1834, to the fact that bearing illegitimate children was becoming a particular source of shame in their communities and among their kin networks. Under the post-1834 Poor Law, parishes still needed to pursue absent fathers to pay back resources the parish was obliged to provide in support of a single mother and her child, but this was now made difficult in law.[14]

These changes would, in turn, result in further modifications that can readily be hypothesized in the character of extant communication communities and their associated norms and institutions. Previously successful social and cultural accommodations to the encompassing political economy might altogether cease to exist along with the customs and ways of thought they entailed, while changed values and practices would emerge as new constellations. Thus, new and expanding communication communities would come to exist, composed of members sharing similar mutual and acknowledged concerns and characteristics, functionally adaptive to the changed circumstances of their lives. Among these different characteristics would inevitably be changed norms and understandings of parental roles, of childhood dependency, of the gendering of work patterns, and associated implications for relations with wider kin and with neighbours and with the civic institutions in the locality and its authorities. For example, we know that the radical cut in the terms and amounts available for familial social security support after 1834, along with the expansion of factory wage labour, led to a revaluation of the wage form in working-class discourse and an adoption of the gender-exclusionary 'male breadwinner' ideology among increasingly organized male-only trade unions (*FCG*: 481–488). This was accompanied also by a dramatic expansion in subscription memberships of mutualist, self-help Friendly Societies to provide a measure of social insurance to replace the lost security of the Poor Law (Neave 1996). These developments established important new, gendered institutions and norms in the practices of many British working-class communities.

Thus, Sian Pooley (2009) has been able to show how the unusually low fertility of the textile mill town of Burnley in 1880s Lancashire was associated with a greater practical equality of income-earning opportunities between husbands and wives than

[13] For some documentation of this in Lancashire during the Industrial Revolution, see Pooley and D'Cruze (1994).

[14] Gillis (1985: 239–241). On the old and new Poor Laws and paternity affiliation orders, see Nutt (2005).

in many other post-1834 working-class communities, such as the proletarian mining community of Bishop Auckland in County Durham, also included in her study. But more importantly, she demonstrates how a public discourse in Burnley's local press, reflected in the policies promoted by its local authority, portrayed the 'private' concerns of working-class family life and childrearing to be an appropriate subject for public discussion of civic ideals and for interventions (e.g. an exceptionally active School Board ensured that children were in school and their school fees paid). This was quite unlike the laissez-faire inactivity and lack of discussion of domestic matters found in the patriarchal mining town of Bishop Auckland. Such community-wide differences would have mediated changing fertility behaviours and may make it meaningful to talk of virtually the whole of Burnley and Bishop Auckland as very different communication communities. These differences between the two towns had a whole range of implications for social practices with fertility effects, influencing several of the variables—or proximate determinants (Bongaarts and Potter 1983: ch. 2)—that demographers use to track variations in reproduction before the rise of modern contraception (such as systematic differences in ages and in proportions marrying and breastfeeding, together with age differences between spouses, coital frequencies, and allocation of food and other resources between family members). The demographic outcome of these changes in fertility behaviour would occur through shifts in the proportions of agents favouring different family sizes. Individuals' deliberative reflections on whether smaller, medium-sized, or large families were most advantageous in the divergent circumstances they were encountering, were referenced both to the extant values and norms in their different towns, with their different and changing economic circumstances, together with the lottery of fecundity, which carried the omnipresent threat that in any communication community there would be those whose childbearing fell significantly short of norms.

6.7 Conclusion

Thus, it has been argued that the study of all forms of fertility change, including secular fertility declines, should give much more attention to the key characteristic of human fertility: that it exhibits continuous and dramatic variability, especially when it is 'natural' in demographers' terminology. Variation in parities should be a much more central focus of study, offering opportunities for productively combining demographic with more qualitative methods. Given an appropriate emphasis on this, it is possible, through the flexible theoretical concept of communication communities—and changes in their character and form by participating reproductive agents—to embrace both a focus on individuals and the perceived relative childrearing costs and benefits which they negotiated, and also a theory of change in reproductive outcomes, which can incorporate long-term change. This could include a change as dramatic as the secular fertility decline. An explanation of macro-scale, national change in fertility indices can be built up from evidence about the micro-scale basics of individuals' motives and values in their households and networks in specific contexts of their communication communities.

We are not, therefore, constrained only to being able to explain large-scale aggregate change in demographic variables with reference to other large-scale aggregate variables. What is needed is a convincing mechanism to explain how the study of locally specific or highly contextualized circumstances, faced by disparate kinds of agents, can sum up to a convincing understanding and explanation of what superficially appears to be a singular national statistical trend, derived from analysis of the output of the state's aggregate data apparatus. The concept of communication communities certainly requires more empirical research to refine it and to explore its properties, but it potentially provides the theoretical basis for a resolution to the impasse of non-communication between studies addressing aggregates and their secular trends and studies exploring contextualized ethnographic detail. There can perhaps be a rapprochement between the two populations and their distinct methods which Kreager's chapter discusses.

The quantitative analysis of aggregated demographic trends and, indeed, subnational statistical indices of marriage rates, marital and non-marital fertility levels, child survival rates, etc. remain absolutely crucial to our capacity to study historic fertility declines. Without such research, founded on precisely defined and categorically classified data, exploiting the assumptions of closed populations that are required for demographic accounting, we cannot know accurately when and where, to what extent and among whom, fertility declines were occurring. However, this essentially descriptive measuring and mapping approach also imposes serious intellectual constraints. Its methodological limitations mean that it needs to be supplemented by a willingness also to investigate using more purely qualitative and ethnographic methods. These are required to explore directly the meanings and significance of the gendered and generational norms, values, roles, and motivations with which agents in the past in particular communication communities, with their specific institutions, understood their own reproduction and sexuality. Only through a preparedness to combine methodologies and to accept that the differences between them are gaps to be bridged, rather than boundaries to be resisted, can we see how change in the perceptions of relative costs and benefits of childrearing might occur—as the content and meanings of the key roles of parenting and childrearing culturally mutated and transformed in response to a changing political economy environment.

It is likely that such momentous change as occurred during secular fertility declines in modern history happened through and between a multiplicity of permeably bounded communicating communities within each national population aggregate. Individuals exercised agency in participating in these communication communities, modifying their values and norms, and in choosing to move between them when they could. To study the causes of such a significant demographic change in the British national population as the modern fertility decline, we need to approach the problem with a willingness to engage with other possible populations than the national aggregate—those subpopulations or communication communities of which it was composed. If it is correct that the crucial problem is to understand changes in perceived relative costs of childrearing, then those collectivities which provided parents and other carers with their cultural reference points when performing a childrearing role need to be identified and studied. It has been argued here that these are best conceptualized as communication communities.

Acknowledgements

I would like to thank Philip Kreager for the great care and attention he devoted to an earlier preliminary version of this chapter, providing a set of extremely stimulating and helpful editorial comments.

References

Anderson, B. A. (1986). 'Regional and Cultural Factors in the Decline of Marital Fertility in Europe', in A. J. Coale and S. C. Watkins (eds), *The Decline of Fertility in Europe*. Princeton: Princeton University Press, 293–313.

Anderson, M. (1988). *The American Census: A Social History*. New Haven: Yale University Press.

Banks, J. A. (1954). *Prosperity and Parenthood: A Study of Family Planning among the Victorian Middle Classes*. London: Routledge and Kegan Paul.

Bashford, A. (ed.) (2006). *Medicine at the Border: Disease, Globalization and Security, 1850 to the Present*. Basingstoke: Palgrave Macmillan.

Bashford, A. (2013). *Life on Earth: Geopolitics and the World Population Problem*. New York: Columbia University Press.

Basu, A. M. and P. Aaby (eds) (1998). *The Methods and Uses of Anthropological Demography*. Oxford: Oxford University Press.

Bayly, C. A. (1996). *Empire and Information: Intelligence Gathering and Social Communication in India, 1780–1870*. Cambridge: Cambridge University Press.

Becker, G. S. (1960). 'An Economic Analysis of Fertility', in National Bureau of Economic Research (ed.), *Demographic and Economic Change in Developed Countries*. Princeton: Princeton University Press, 209–231.

Bengtsson, T. and M. Dribe (2006). 'Deliberate Control in a Natural Fertility Population: Southern Sweden, 1766–1864', *Demography* 43/4: 727–746.

Bernstein, B. (1977a). 'Class and Pedagogies: Visible and Invisible', in J. Karabel and A. H. Halsey (eds), *Power and Ideology in Education*. New York: Oxford University Press, 511–534.

Bernstein, B. (1977b). 'Social Class, Language and Socialisation', in J. Karabel and A. H. Halsey (eds), *Power and Ideology in Education*. New York: Oxford University Press, 473–486.

Bledsoe, C. (1996). 'Contraception and "Natural Fertility" in America', *Population and Development Review* 22: 297–324.

Bledsoe, C. and U. Isugo-Abanihe (1989). 'Strategies of Child-Fosterage among Mende Grannies in Sierra Leone', in R. J. Lesthaeghe (ed.), *Reproduction and Social Organisation in Sub-Saharan Africa*. Berkeley: University of California Press, 442–474.

Bledsoe, C., S. Lerner, and J. Guyer (eds) (2000). *Fertility and the Male Life-Cycle in the Era of the Fertility Decline*. Oxford: Oxford University Press.

Boltanski, L. (1987). *The Making of a Class: Cadres in French Society*, trans. A. Goldhammer. Cambridge: Cambridge University Press (first published in French 1982).

Bongaarts, J. and R. G. Potter (1983). *Fertility, Biology and Behavior: An Analysis of the Proximate Determinants*. New York: Academic Press.

Bott, E. (1957). *Family and Social Network: Roles, Norms and External Relationships in Ordinary Urban Families*. London: Tavistock.

Bourdieu, P. (1977a). *Outline of a Theory of Practice*. Cambridge: Cambridge University Press.

Bourdieu, P. (1977b). 'Cultural Reproduction and Social Reproduction', in J. Karabel and A. H. Halsey (eds), *Power and Ideology in Education*. New York: Oxford University Press, 486–510.

Breckenridge, K. and S. Szreter (eds) (2012). *Registration and Recognition: Documenting the Person in World History*, Proceedings of the British Academy 182. Oxford: Oxford University Press.

Caplan, J. and J. C. Torpey (eds) (2001). *Documenting Individual Identity: The Development of State Practices in the Modern World*. Princeton: Princeton University Press.

Coale, A. J. and S. C. Watkins (eds) (1986). *The Decline of Fertility in Europe*. Princeton: Princeton University Press.

Cohn, B. S. (1987). 'The Census, Social Structure and Objectification in South Asia', in *An Anthropologist among the Historians and Other Essays*. Oxford: Oxford University Press, 224–254.

Cook, H. (2004). *The Long Sexual Revolution: English Women, Sex, and Contraception 1800–1975*. Oxford: Oxford University Press.

Curtis, B. (2002). *The Politics of Population: State Formation, Statistics, and the Census of Canada, 1840–1875*. Toronto: University of Toronto Press

Davidoff, L. (2012). *Thicker than Water: Siblings and their Relations, 1780–1920*. Oxford: Oxford University Press.

Derosas, R. (2006). 'Between Identity and Assimilation: Jewish Fertility in Nineteenth Century Venice', in R. Derosas and F. van Poppel (eds), *Religion and the Decline of Fertility*. Dordrecht: Springer, 177–205.

Derosas, R. and F. van Poppel (eds) (2006). *Religion and the Decline of Fertility in the Western World*. Dordrecht: Springer.

Desrosières, A. (1993). *La politique des grands nombres: Histoire de la raison statistique*. Paris: Editions la découverte.

Eyler, J. M. (1979). *Victorian Social Medicine: The Ideas and Methods of William Farr*. Baltimore: Johns Hopkins University Press.

Fisher, K. (2006). *Birth Control, Sex and Marriage in Britain, 1918–1960*. Oxford: Oxford University Press.

Garrett, E., A. Reid, K. Schurer, and S. Szreter (2001). *Changing Family Size in England and Wales 1891–1911: Place, Class and Demography*. Cambridge: Cambridge University Press.

Gillis, John R. (1985). *For Better, for Worse: British Marriages, 1600 to the Present*. New York: Oxford University Press.

Gillis, John R. (1992). 'Gender and Fertility Decline among the British Middle Classes', in John R. Gillis, L. A. Tilly, and D. Levine (eds), *The European Experience of Declining Fertility: A Quiet Revolution, 1850–1970*. Oxford: Blackwell, 31–47.

Gilman, N. (2003). *Mandarins of the Future: Modernization Theory in Cold War America*. Baltimore: Johns Hopkins University Press.

Greenhalgh, S. (ed.) (1995). *Situating Fertility: Anthropology and Demographic Inquiry*. Cambridge: Cambridge University Press.

Gribaldo, A., M. D. Judd, and D. I. Kertzer (2009). 'An Imperfect Contraceptive Society: Fertility and Contraception in Italy', *Population and Development Review* 35: 551–584.
Hacking, I. (1986). 'Making Up People', in T. Heller, M. Sosna, and D. Wellberry (eds), *Reconstructing Individualism*. Stanford: Stanford University Press, 222–236.
Hirsch, F. (2005). *Empire of Nations: Ethnographic Knowledge and the Making of the Soviet Union*. Ithaca: Cornell University Press.
Janssens, A. (ed.) (2007). *Gendering the Fertility Decline in the Western World*. Bern: Peter Lang.
Kertzer, D. and D. Arel (eds) (2002). *Census and Identity: The Politics of Race, Ethnicity and Language in National Censuses*. Cambridge: Cambridge University Press.
Kertzer, D. and T. Fricke (eds) (1997). *Anthropological Demography: Toward a New Synthesis*. Chicago: Chicago University Press.
Kreager, P. (1998). 'The Limits of Diffusionism', in A. M. Basu and P. Aaby (eds), *The Methods and Uses of Anthropological Demography*. Oxford: Oxford University Press, 298–322.
Kreager, P. (2008). 'Aristotle and Open Population Thinking', *Population and Development Review* 34/4: 599–629.
Kreager, P. (2009). 'Darwin and Lotka: Two Concepts of Population', *Demographic Research* 21: 469–502.
Kussmaul, A. (1981). *Servants in Husbandry in Early Modern England*. Cambridge: Cambridge University Press.
Laxton, P. and N. Williams (1989). 'Urbanisation and Infant Mortality in England: A Long-Term Perspective and Review', in M. Nelson and J. Rogers (eds), *Urbanisation and the Epidemiologic Transition*. Uppsala: Uppsala University, 109–131.
Lesthaeghe, R. and D. van de Kaa (1986). 'Twee demografische transities?' [Two demographic transitions?], in R. Lesthaeghe and D. van de Kaa (eds), *Bevolking: Groei en Krimp, Mens en Maatschappij*. Deventer: Van Loghum Slaterus, 9–24.
Macfarlane, A. (1982). 'Review of Ann Kussmaul, *Servants in Husbandry in Early Modern England*', *Economic History Review* 35/3: 459–460.
McKeown, A. (2008). *Melancholy Order: Asian Migration and the Globalization of Borders*. New York: Columbia University Press.
McQuillan, P. (1999). *Culture, Religion and Demographic Behaviour: Catholics and Lutherans in Alsace, 1750–1870*. Montreal: McGill-Queen's University Press.
Neave, D. (1996). 'Friendly Societies in Great Britain', in M. van der Linden, M. Dreyfus, B. Gibaud, and J. Lucassen (eds), *Social Security Mutualism: The Comparative History of Mutual Benefit Societies*. Berne: Peter Lang, 41–64.
Newton, G. (2010). 'Infant Mortality Variations, Feeding Practices and Social Status in London between 1550 and 1750', *Social History of Medicine* 24/2: 260–280.
Nutt, T. (2005). 'The Paradox and Problems of Illegitimate Paternity in Old Poor Law Essex', in A. S. Levene, T. W. Nutt, and S. K. Williams (eds), *Illegitimacy in Britain 1700–1920*. Basingstoke: Macmillan, 101–121.
Pooley, C. and S. D'Cruze (1994). 'Migration and Urbanisation in North-West England circa 1760–1830', *Social History* 19: 339–358.
Pooley, S. (2009). 'Parenthood and Child-Rearing in England, *c.*1860–1910', unpublished PhD thesis, University of Cambridge.
Porter, R. and L. Hall (1995). *The Facts of Life: The Creation of Sexual Knowledge in Britain, 1650–1950*. New Haven: Yale University Press.
Praz, A.-F. (2005). *De L'Enfant Utile A L'Enfant Precieux: Filles et Garcons dans les cantons de Vaud et Fribourg (1866–1930)*. Lausanne: Editions Antipodes.

Roberts, R. (1971). *The Classic Slum: Salford Life in the First Quarter of the Century*. Manchester: Manchester University Press.

Ross, E. (1992). 'Mothers and the State in Britain', in J. R. Gillis, L. A. Tilly, and D. Levine (eds), *The European Experience of Declining Fertility: A Quiet Revolution 1850–1970*. Oxford: Blackwell, 48–65.

Schweber, L. (2006). *Disciplining Statistics: Demography and Vital Statistics in France and England, 1830–1885*. Durham, NC: Duke University Press.

Stoneburner, R. L. and D. Low-Beer (2004). 'Population-Level HIV Declines and Behavioral Risk Avoidance in Uganda', *Science* 304/5671: 714–718.

Szreter, S. (1984). 'The Genesis of the Registrar-General's Social Classification of Occupations', *British Journal of Sociology* 35: 522–546.

Szreter, S. (1993a). 'The Idea of Demographic Transition: A Critical Intellectual History', *Population and Development Review* 19: 659–701.

Szreter, S. (1993b). 'The Official Representation of Social Classes in Britain, United States and France: The Professional Model and "les cadres"', *Comparative Studies in Society and History* 35: 285–317.

Szreter, S. (1996). *Fertility, Class and Gender in Britain, 1860–1940*. Cambridge and New York: Cambridge University Press.

Szreter, S. and K. Fisher (2010). *Sex before the Sexual Revolution: Intimate Life in England 1918–1963*. Cambridge: Cambridge University Press.

Szreter, S. and E. Garrett (2000). 'Reproduction, Compositional Demography, and Economic Growth in Early Modern England', *Population and Development Review* 26: 45–80.

Szreter, S., H. Sholkamy, and A. Dharmalingam (eds) (2004). *Categories and Contexts: Anthropological and Historical Studies in Critical Demography*. Oxford: Oxford University Press.

Tranter, N. L. (1996). *British Population in the Twentieth Century*. Basingstoke: Macmillan

Willis, P. E. (1977). *Learning to Labour: How Working Class Kids Get Working Class Jobs*. Farnborough: Saxon House.

Woodruff, W. (2002). *The Road to Nab End: An Extraordinary Northern Childhood*. London: Abacus (first published as *Billy Boy* by Ryburn 1993).

Wrigley, E. A. (1983). 'The Growth of Population in Eighteenth-Century England: A Conundrum Resolved', *Past and Present* 98: 121–150.

Wrigley, E. A., R. S. Davies, J. E. Oeppen, and R. S. Schofield (1997). *English Population History from Family Reconstitution, 1580–1837*. Cambridge: Cambridge University Press.

CHAPTER 7

THE SOCIAL AND THE SEXUAL

networks in contemporary demographic research

HANS-PETER KOHLER, STÉPHANE HELLERINGER, JERE R. BEHRMAN, AND SUSAN C. WATKINS

7.1 Introduction

Theories of social interactions in demography rest on the insight that individuals do not make decisions about demographic and other social behaviours in isolation, but rather with others. Rather than being constituted merely by collections of individuals, populations consist of interconnected networks of persons who share information, resources, and often a common understanding of norms. Understanding this interconnectedness is essential for understanding demographic behaviours and the dynamics of demographic processes. While the basic insight about the relevance of social interactions is old, dating back at least to the nineteenth century with the work of the sociologist Simmel (1922), in recent decades researchers have made substantial progress in specifying, measuring, modelling, and understanding the importance of network-based interactions on human behaviours on the micro level, and on societal, economic, and cultural dynamics on the macro level (e.g. Burt 1982; Fischer 1982; Laumann et al. 1983; Granovetter 1973; Marsden and Friedkin 1993; Valente 1994; Montgomery and Casterline 1996). A recent book *Connected: The Surprising Power of our Social Networks and How They Shape our Lives* (Christakis and Fowler 2009) claims that 'how we feel, whom we marry, whether we fall ill, and how much money we make and whether we vote—everything hinges on what others around us are doing, thinking and feeling'. While this specific claim is rather bold and controversial (Kolata 2011), there is indeed growing evidence that a broad range of fertility, health, and related behaviours are associated with, and sometimes even causally affected by, a person's social networks. For example, there is an extensive literature on the importance of social interactions and the diffusion of innovation for the decline in fertility during the demographic transition or in contemporary low-fertility contexts (Behrman et al. 2002; Bernardi 2003; Bongaarts and Watkins 1996; Hensvik and Nilsson 2010; Kohler 1997, 2001; Kohler et al. 2007; Lyngstad and Prskawetz 2010), and recent studies have linked a variety of health outcomes to social network structures, including for instance smoking, alcohol and substance use, obesity, HIV risks, mental health, risk perceptions, and subjective well-being (e.g. Andrews et al. 2002; Bearman et al. 2004; Christakis and Fowler 2007; Fiori et al. 2006; Fowler and Christakis 2008; Helleringer and Kohler 2007; Kaplan et al. 2001; Smith and Christakis 2008).

Several reviews of the network literature have described the origins of network theories and analyses (e.g. Wellman 1988). Network studies were given an early applied focus by European anthropologists studying rapid social change associated with modernization in sub-Saharan Africa, who wrote of the network connections of urban migrants with each other and with the rural communities from which they came (Mitchell 1969), and by anthropologists studying class relationships in Britain (Bott 1971). An interest in social networks then developed among US sociologists. In some cases, the analyses focused more on theories and methods of network analysis (Burt 1982; Marsden 1990; Valente 2005) while others focused more on substantive issues: for example, Fischer (1982) analysed personal networks to investigate the social and psychological consequences of urban life, and Granovetter (1973) emphasized the importance of *weak ties* that transmit unique information across otherwise largely disconnected segments of social networks, thereby facilitating the diffusion of new information. Strong ties and dense networks, on the other hand, are more likely to enforce norms and conventions that represent a proper way to behave. In a similar vein, Burt (1992) pointed to the strategic informational advantage that may be enjoyed by individuals who bridge *structural holes*, that is, those with ties into multiple networks that are largely separated from one another. Social interaction processes and their effects on social dynamics have also been investigated extensively in the context of the diffusion of innovations (e.g. Rogers 2003), social change and collective action (Kim and Bearman 1997; Klandermans 1992), and search or matching processes in the labour market or similar markets (Granovetter 2005, 1973).

In demography, an important stimulus to incorporating social networks in analyses was the attempt to explain patterns of fertility declines in historical Western Europe and the developing world. Although it was expected that fertility declines could be understood as individual responses to structural changes associated with modernization (e.g. urbanization, the transformation of the labour force from agricultural to industrial, declines in infant and child mortality), the associations between measures of these changes and fertility declines were typically modest (Cleland and Wilson 1987; Coale 1986). This led to postulating the importance of interpersonal diffusion (Knodel and van de Walle 1979) and the use of social network theory to structure analyses of diffusion (Committee on Population and Casterline 2001; Montgomery and Casterline 1996; Munshi and Myaux 2006; Watkins 1991). Applications in demography have also included the perception of mortality change (Montgomery 2000; Sandberg et al. 2012), the onset of sexual behaviour among teenagers (e.g. Rodgers and Rowe 1993), and international migration (Massey et al. 1994; Munshi 2003). Conceptually, the analyses of fertility change have focused on two mechanisms through which fertility and contraceptive behaviours might be influenced by social interactions: social learning and social influence (Montgomery and Casterline 1996). The former postulates that decisions about fertility and contraceptive adoption are subject to substantial uncertainty, for example, about the medical side effects and/or the costs and benefits of modern methods of family planning. Learning about other women's experiences through networks may reduce this uncertainty, thus increasing the probability that a risk-averse woman will adopt modern contraception herself. The second aspect, social influence, emphasizes normative influences on behaviour rather than processes of learning about unknown characteristics. Social influence therefore implies that the fertility-related opinions and behaviour of an individual's network partners influence and alter her preferences regarding modern contraception and/or number of children.

Despite evidence collected by demographers interested in understanding incipient fertility declines in the developing world and by advocates interested in promoting the adoption of modern contraception, which showed that developing-country villagers talked to each other about family planning and family size, demographers paid little attention to the role of social networks prior to the 1990s. In part, this was due to the absence of data on networks; that absence, however, was due to a dominant model of social behaviour that privileged individual and family characteristics over social interactions. Only in the 1990s were new data collected in developing countries that permitted detailed descriptions of social networks and rigorous analyses of social network effects: for example, in Ghana (Montgomery et al. 2001), in Thailand (Entwisle et al. 2007), in Kenya (kenya.pop.upenn.edu), and in Malawi (malawi.pop.upenn.edu).

An important limitation of much of the literature on social networks and health has been the use of *egocentric* (or local) network data in the majority of studies on this topic. Egocentric data provide information about the social ties of a survey respondent, but contain no information about the larger network in which the respondent is embedded. Thus, egocentric network studies provide only a very restricted view of a person's social capital and do not allow analyses of how global/local network structures and a person's structural position within a larger community-level social network affect important health outcomes (e.g. Smith and Christakis 2008). Another approach to collecting network data consists of conducting a census of individuals and their relations in a small, well-defined population. This approach, known as sociocentric data collection, has frequently been used in the sociology of organizations, but it is more complicated to employ in the larger populations demographers typically study. Recognizing the limitations of egocentric network data, a small but growing number of researchers have begun to collect *sociocentric network data* (a.k.a. sociometric, complete, or global networks), in which all or nearly all members of a community or group and their linkages to each other are represented as part of *saturation samples*. While egocentric data include only the direct links from the focal individuals (the '*egos*') to other persons, sociocentric networks include both direct and indirect ties and therefore allow community-level networks of social relationships to be mapped. Sociocentric data are less affected by measurement error because all social relationships, and the interactions among network partners, are potentially reported by each of the two members of the relationship. In addition, utilizing the recent advances in network theory (Carrington et al. 2005; Morris 2004), sociocentric network studies—especially when they are longitudinal—can identify the complex patterns of interrelations among persons and their health. Path-breaking results, for example, came from analyses of the Framingham Heart Study (Christakis and Fowler 2008); from studies showing the spread of obesity through social networks (Christakis and Fowler 2007); the sociocentric Adolescent Health Study (AddHealth), which included findings that obese adolescents are often socially isolated in networks of high school students (Strauss and Pollack 2003); and Bearman et al.'s (2004) finding that over an 18-month period in a Midwestern high school, 52% of all romantically involved students were connected so that they were embedded in one very large spanning tree. Recent work on the Framingham study highlighted the role of social networks in the growing obesity epidemic, or in changes in smoking behaviours. Finally, we have recently shown that sexual networks in sub-Saharan Africa can be much more extensive than was previously thought. For example, half of all sexually active respondents on Likoma Island in Malawi were linked together in a *giant network component*, and more than one quarter of these were connected by multiple *independent* chains of sexual relations. Such structural features of sexual networks have been associated with epidemic spread of STIs in high-risk groups (Moody et al. 2003; Newman 2002; Potterat et al. 2002; Rothenberg et al. 1998), but prior to our study (Helleringer and Kohler 2007) had never been documented among a general population in developing countries.

7.2 General Overview of Survey Data and Contexts

Within demography, an important focus has been the empirical measurement and identification of social interaction effects on fertility and health outcomes, relying primarily on sociocentric network data and concerned mostly with the estimation of peer effects on individual behaviours. In a few cases, researchers—including the authors of this paper—have collected longitudinal data that permit careful description and rigorous analyses of the causal impacts of networks. Our work for well over a decade in investigating the roles of interactions in social—and more recently also sexual—networks for important demographic behaviours has been based on data collected in Kenya and Malawi. In our review of this work, we begin with analyses using data from the Kenyan Diffusion and Ideational Change Project (KDICP) and the Malawi Longitudinal Study of Families and Health (MLSFH; formerly Malawi Diffusion and Ideational Change Project, MDICP), which featured egocentric network data. In the second part of this review we turn to sociocentric network data collected on Likoma Island in Lake Malawi, where our analyses have tried to infer the population-level size and structure of sexual networks that are related to the spread of HIV.

The KDICP and the MDICP conducted panel surveys of households and qualitative data. In both cases the initial primary motive for collecting these data was to analyse the role of social networks in the diffusion of innovations to increase the use of modern family planning methods and to reduce fertility, provoked by the limited explanatory power of individual characteristics to explain the European fertility decline (see the introduction). Subsequently the focus of the projects—and therefore of the data collection—shifted increasingly towards HIV/AIDS because of the rapid spread of the epidemic in the populations being studied (Helleringer and Kohler 2005; Kohler et al. 2007). The surveys were conducted in rural areas between 1994–2000 (for Kenya) and 1998–2010 (for Malawi); to compensate for sample attrition in Malawi, a sample of adolescents was added in 2004 (Clark et al. 2009). Longitudinal qualitative data on responses to the epidemic were collected between 1999 and 2010 in Malawi (Watkins 2004; Tawlik and Watkins 2007; Smith and Watkins 2005).

The KDICP and MLSFH data are, to our knowledge, unique in including detailed accounts on women's and men's social interactions about family planning or the HIV/AIDS epidemic.[1] In particular, the data include information on egocentric networks that contain the respondent and those with whom the respondent had chatted about family planning or HIV/AIDS, with detailed information on up to four network partners. The term 'chat' was used in survey questions to indicate informal conversations rather than lectures at clinics. Respondents were first asked about the number of people they had chatted with about family planning or AIDS, followed by a series of questions about these network partners (covering a maximum of four network partners if more than four were identified).[2] The data on the network partners include relationship (e.g. co-wife, sister-in-law, uncle); the degree of closeness (confidant, friend, acquaintance); the network partner's age, sex, and wealth; and the respondent's perception of views and behaviours of the network partner on family planning or their risk of becoming infected with HIV.

In Kenya, the first wave of the longitudinal household survey (KDICP 1) was conducted in December 1994 and January 1995 in South Nyanza District, near the shores of Lake Victoria. The second wave (KDICP 2) reinterviewed these women and men two years later, and a third wave was conducted in January and February 2000 (KDICP 3). Only the second and third waves of the survey addressed AIDS. In total, 545 ever-married women (408 husbands) participated in both of these last two rounds of

[1] Other datasets on AIDS have information on respondents' sexual partners (information that we do not have for our overall samples, though we do have for some subsamples) but not on their social networks in which they discuss HIV/AIDS risks and ways of coping with such risks.

[2] The question about the number of conversations did not have an explicit time reference. A related question in the Kenyan survey about the time of the last conversation shows that many conversations were relatively recent: the last conversation with the network partner occurred within one year prior to the survey in more than 80% of all cases. We expect that this pattern is similar in Malawi.

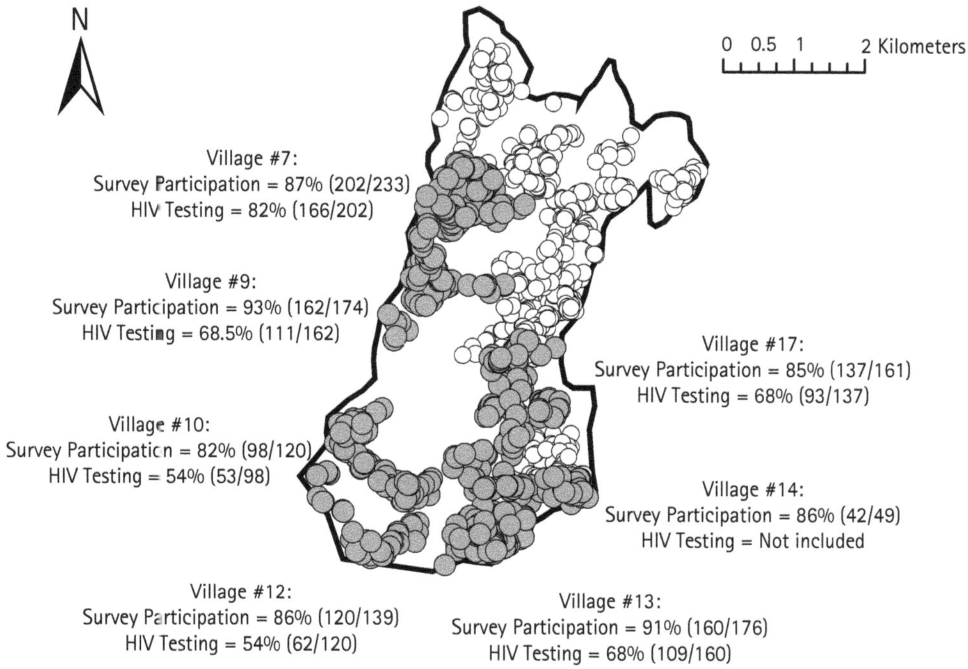

FIG. 7.1 Geographic location of the sampled villages and village-specific participation rates.

the data collection. In Malawi in 1998, the project interviewed 1541 women and 1065 men who were husbands of the women respondents on topics related to AIDS and family planning (MLSFH 1) in the Rumphi (North), Mchinji (Centre), and Balaka (South) regions. Follow-up surveys were conducted in 2001 (MLSFH 2), 2004 (MLSFH 3), 2006 (MLSFH 4), 2008 (MLSFH 5), and 2010 (MLSFH 6). Details of data collection and analyses of attrition and data quality are available at <malawi.pop.upenn.edu> and in a special issue of *Demographic Research* (Watkins et al. 2003).

The second part of our analyses in this paper focuses on the Likoma Network Study (LNS), an innovative sociocentric *sexual network survey* in Likoma, a small island in Lake Malawi with high HIV prevalence (Figure 7.1).[3] The key feature of these data is two rounds of data on sociocentric—rather than egocentric as in most other studies—sexual networks covering the young adult population in seven villages of Likoma. The data were collected in 2005/6 and 2007, and included detailed information on the socio-economic and demographic context of individuals, as well as information about their subjective health, AIDS-related behaviours, attitudes, risk perceptions, and sexual network partners. Studies using a design similar to that of the LNS have been conducted in different contexts (e.g. Bearman et al. 2004; Klovdahl et al. 1994), but there were none for African populations with generalized HIV epidemics.

[3] Likoma extends over only 18 square kilometers, has limited transportation to the mainland, and its population is small with just over 7,000 persons living in a dozen villages.

In order to describe fully the sociocentric sexual networks on Likoma, the data collection occurred in three stages. First, we conducted a census of every individual on Likoma Island to obtain a roster of potential partners. Second, we conducted a sexual network survey with *all* individuals aged 18–35 (LNS 1) or 18–50 (LNS 2) in the study villages (Figure 7.1), asking respondents for information about their romantic and sexual partners. Finally, this saturated sampling strategy permitted us to construct the population-level sexual network by matching the reported sexual partners with the census roster, and then linking the data of all young adults residing in the sample villages. The context and methodology of this survey are summarized below, and additional details are provided in Helleringer, Kohler, and Kalilani-Phiri (2009) and Helleringer et al. (2014).

7.3 Social Networks and the Diffusion of Family Planning

This section summarizes three papers on different dimensions of social networks in the adoption of modern methods of family planning in the high-fertility, poor, developing context of rural Kenya. Specifically, in section 7.3.1 we describe a simple—but nevertheless useful—conceptual model that illustrates the implications of social interactions for the dynamics of fertility decline, and how social interactions affect the impact of policies such as family planning programmes. Section 7.3.2 illustrates how egocentric network data can be used to distinguish between the two primary mechanisms—social learning and social influence—through which social interactions have been hypothesized to influence individual behaviours. These data include information regarding whom a respondent has talked with about family planning and also about the density of the networks of conversational partners. In section 7.3.3, we illustrate the challenges in making inferences about the causal effect of social networks on individual behaviours, and we use longitudinal KDICP and MLSFH data and fixed-effects estimation to show that social interactions about family planning have substantial causal effects on the adoption of family planning methods.

7.3.1 Empirical Assessments of Social Networks, Fertility, and Family Planning Programmes

Important long-standing concerns in demography have been the evaluation of the role of family planning programmes in facilitating the decline of fertility during the demographic transition, and the investigation to which extent diffusion processes have contributed to individual-level and population-level fertility change (e.g. Kohler 2013). This section emphasizes that these two issues cannot be addressed independently, and that there are important implications of social interaction processes for measuring and interpreting family planning programme effects. To provide intuitions for some dimensions of our analysis in this chapter, we first present a simple model to illustrate some of the advantages that our data have for such analysis. The availability of unusual longitudinal data on social networks and the use of statistical methods that may control for unobserved factors provided a unique opportunity to extend the individualistic rational-actor models to incorporate social interaction and to estimate the causal effects of social networks on attitudes and behaviours under certain assumptions about the nature of the unobserved effects.

In particular we use an empirical specification of the relation determining contraceptive behaviours in which there is explicit recognition that, in addition to observed right-side variables (including social networks prior to time *t*), there are unobserved fixed factors that might affect both contraceptive use directly and through social networks. For example, preferences for family orientation or for traditional types of social relations might affect both. A first-order linear approximation to the model for the determinants of contraceptive behaviour is

$$Y_{it} = agN_{it-} + bgX_{it-} + f_i + e_{it},$$

where

Y_{it} is the observed contraceptive behaviour of individual *i* at time *t*;

N_{it-} is the social network for individual *i* prior to time *t* (we use the subscript '$t-$' to emphasize that the variable *N* refers to the time prior to *t*; we use this notation also for other predetermined variables);

X_{it-} is a vector of other state variables for individual *i* determined prior to time *t* (e.g. age, marital status, completed schooling of adults, wealth indicators);

f_i represents unobserved fixed factors that are assumed to affect directly contraceptive behaviours by individual *i* (e.g. the persistent part of preferences, unobserved current community characteristics, expectations regarding future prices, and inter-familial and community resources on which the individual can draw) and which also affect social networks; and

e_{it} is a disturbance term that affects the contraceptive behaviour of individual *i* at time *t* due to, for example, new information on the probability of conceiving children from using traditional contraceptive methods or about the advantages and disadvantages of children in a high HIV/AIDS prevalence context, or price shocks that are deviations from the long-run secular price trends.[4]

[4] The assumption that the disturbance term e_{it} is i.i.d. also excludes autocorrelation; the model therefore assumes that persistent heterogeneity in contraceptive behaviours among individuals with similar observed characteristics is primarily due to heterogeneity in fixed characteristics (captured by the fixed effect f_i) rather than to lasting effects of past '*shocks*' (captured by lagged values of the disturbance term e_{it}).

This modelling approach to social interaction is consistent with Montgomery and Casterline's (1996) social multiplier model of diffusion in which, if b is the direct impact of some change in X_{it} on an individual's behaviour, and Y_{it} and N_{it-} are measured in the same terms (e.g. Y_{it} is the ith individual's contraceptive behaviour and N_{it-} is the average contraceptive behaviour by social network partners), the social multiplier that captures the long-run effect through the network is $1/(1-a)$. Therefore, to estimate this social multiplier, as well as the direct determinants of contraceptive behaviours of an individual, it is important to obtain unbiased estimates of the coefficients a and b.

While the above linear model is conceptually simple, this is not necessarily the case for the estimation of the relevant parameters of this model. In particular, if conventional regression (ordinary least square, or OLS) estimates are made of the above linear equation, inconsistent (biased) estimates are likely to result because the unobserved fixed effects are correlated with the characteristics of social network partners if they indeed affect the choice of network partners, so the OLS-estimated value of a is biased because the social network variable is representing in part the unobserved fixed effects in addition to the effects of social network per se. By virtue of having longitudinal data, however, we are able to control for the individual fixed effects by comparing fixed-effects estimates with the OLS estimates to learn how much difference control for unobserved fixed effects makes in inferences about the magnitudes of network effects.[5] For example, if individuals who are more likely to interact with others about family planning, for instance due to residing closer to a family planning clinic than others, are also more likely to adopt family planning, conventional regressions of the social network on contraceptive use would overestimate the effect of networks on individual behaviour.

Controlling for such unobserved factors that might influence both social networks and contraceptive use is therefore essential for making progress in identifying the causal influences of networks on behaviour. Yet, cross-sectional observational data are often not suitable for doing so, and alternative approaches are required. Before discussing the challenges and potential solutions for estimating the importance of social interactions from empirical data in more detail, we will use the above framework to illustrate how social interactions can change the interpretation of how family planning programmes affect fertility behaviours, and how assessments of the relevance of family planning programmes can depend on whether the analyses account or do not account for the presence of social learning and/or social influence through networks. Specifically, we continue with the assumptions of a linear model and assume that the probability that a woman adopts modern family planning ($y = 1$) is given by $P(y = 1|z, y_c = a * (-.5 + y_c) + b * z + d$ where the term $a * (-.5 + y_c)$ represents the influence of social interaction on a woman's probability of using family planning. The parameter a reflects the 'strength' or relevance of social interaction and determines the extent to which the

[5] Fixed-effects estimates control for unobserved fixed factors, but not for unobserved time-varying factors (see section 7.4.2).

adoption probability is affected by the contraceptive behaviour in the reference group (y_c), which for simplicity we assume is the village in which a woman lives (in our subsequent empirical estimates, this assumption is not required; our analyses later in this section also illustrate that in reality the networks are not necessarily constrained to the village of residence; quite to the contrary, they often include weak ties to relatively distant individuals). If the contraceptive prevalence in the village (y_c) is above 0.5, then social interaction increases the probability of using family planning as compared to the situation when no social interaction is present, and otherwise it decreases the probability. The coefficient b is the direct effect of programme efforts (z), and larger programme efforts increase the probability of using contraception when $b > 0$.[6] The solid line in Figure 7.2a then plots the curve implied by the above linear social interaction model: the vertical axis gives an individual's probability of using contraception as related to the average contraceptive use for the individual's village (y_c, on the horizontal axis) given the programme effort z (e.g. proportion of other villagers who *heard a family planning message on the radio*). The slope of the solid line indicates how the probability of individual use changes when there is a discrepancy between the probability of an individual's use and the average contraceptive use of other women in her village. The linear model in Figure 7.2a exhibits only one equilibrium, the point at which each individual's behaviour mirrors the village average—where the solid line intersects the 45° ray from the origin in Figure 7.2a. This equilibrium therefore satisfies $P(y = 1|z, y_e) = y_e$, where y_e is the equilibrium level of contraceptive use. To the left of it the individual probability of use is above the village average use; therefore the average village use increases because the individual is in the reference group for others in the village, which causes movement to the right towards the equilibrium (and vice versa to the right of the equilibrium).

The above linear framework is useful for illustrating the effects of an increase in family planning programme effort in the presence of social interactions about contraceptive use. The increased family planning effort can, for instance, be due to a new media campaign. This increased programme effort then results in an increase in the probability of using family planning, and this upward shift of the propensity to use family planning before accounting for any additional social interaction effects is indicated in Figure 7.2a as the *direct programme effect*. This direct programme effect is not modulated by social interactions. If, however, the individual adjusts to her reference group through social learning and/or social influence, we get a social multiplier (Montgomery and Casterline 1996). The social multiplier leads to a new and higher equilibrium level of contraceptive use, i.e. where the dashed line intersects the 45° ray. The total increase in the probability of contraceptive use is thus larger and consists of the direct programme effect plus its multiplication by the social interaction term. In Figure 7.2a, therefore, the *total programme effect* is indicated as the sum of both the direct programme effect, resulting

[6] For simplicity, in our discussion of this theoretical model in this section we consider only women who are identical with respect to individual characteristics, which permits us to combine the effect of these characteristics into the constant term d.

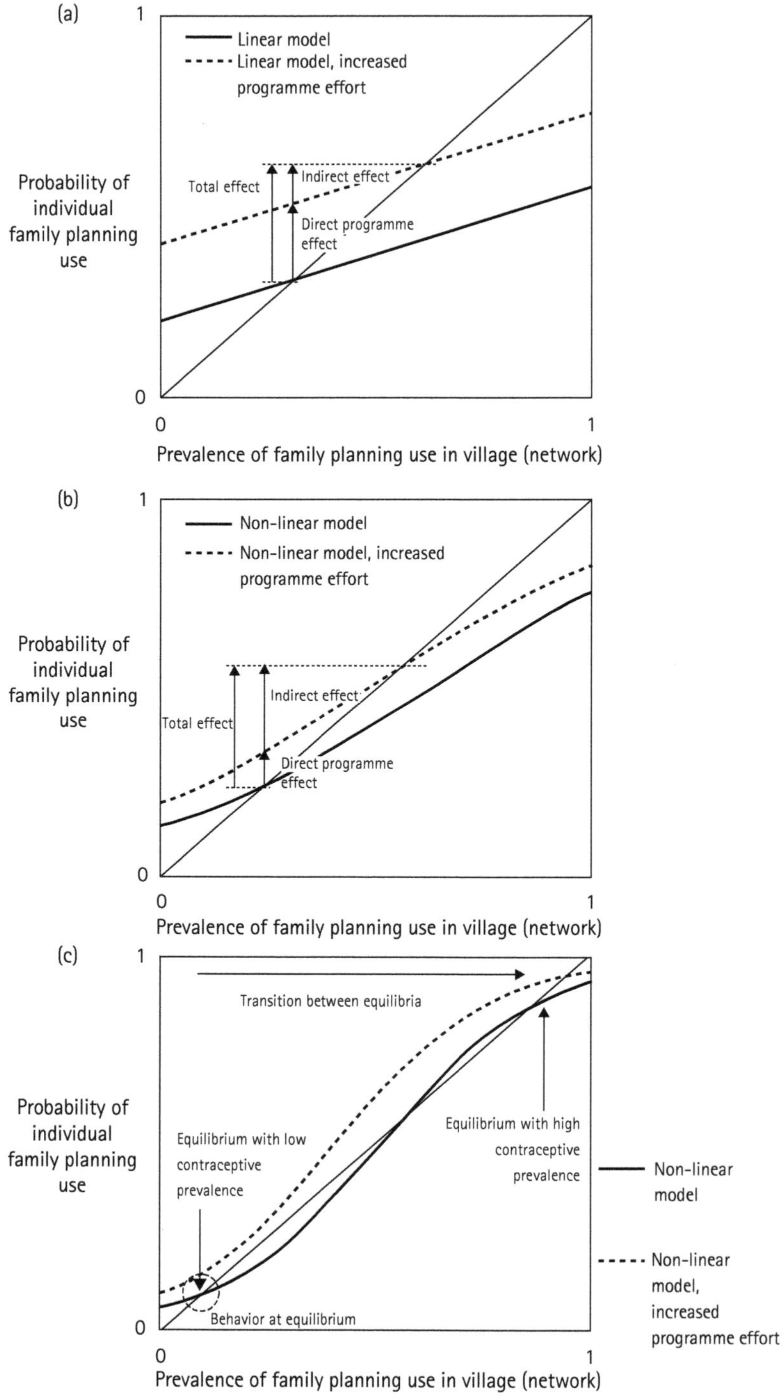

FIG. 7.2 Linear and non-linear model with social interaction.

directly from the policy holding the prevalence of family planning use in the reference group constant, plus the indirect programme effect that occurs more gradually over time and is due to the additional changes in family planning use resulting from social learning and/or social influence occurring in social networks.

In addition to the linear social interaction model in Figure 7.2a, it is also useful to consider a non-linear social interaction model (Figure 7.2b) that is frequently used in theoretical models of social interactions (Durlauf 2001; Kohler 2000; Kohler et al. 2000; Manski 1993) and for empirical estimates (Arends-Kuenning 2001; Godley 1999; Kohler et al. 2001). In this model we assume that the disutility from deviating from the average behaviour of a woman's reference group is related linearly to the difference between an individual's decision to use or not to use and the average reference group behaviour y_c. More specifically, we assume that the social utility term takes the form of $a * (-.5 + y_c)$ where 0.5 is the critical level above which the prevalence of contraceptive use in a woman's reference group has a positive influence on the adoption of family planning, and a is the 'strength' or relevance of this social interaction effect. The standard derivation leads to the probability that a woman uses a modern method of family planning given by a logistic model as $P(y = 1)|z, y_c = F(a * (-.5 + y_c) + b * z + d)$, where d is a constant including the effect of the individual characteristics and F is the cumulative logistic distribution. The total effect of family planning programmes in the presence of social interactions can be characterized in the above non-linear model, as in the linear case, by equilibria in which an individual's choice probability mirrors the reference group average (Figure 7.2b). That is, an equilibrium is a level of contraceptive use that satisfies $P(y = 1|z, y_e) = y_e$, or equivalently, an equilibrium is a fixed point at which $y_e = F(a * (-.5 + Y_e) + b * z + d)$. These equilibria are thus at intersections of the 's-shaped' curve $F(.)$ with the diagonal. The solid line in Figure 7.2b displays a case in which only one such equilibrium exists. The solid line in Figure 7.2c, on the other hand, shows a case with three intersections. The equilibria at low and high levels of contraceptive use are stable for reasons parallel to those discussed with regard to Figure 7.2b. The same reasoning, however, indicates that the centre equilibrium always is unstable. A population converges to one of the two stable equilibria depending on whether it is to the left or right of the unstable equilibrium.

The comparison of the linear and non-linear specification in Kohler et al. (2000), including both the theoretical implications of a linear versus non-linear specification as well as the differences of these models in their empirical estimation using KDICP data, yields the following major results.

First, as noted, we distinguish between the direct effects of a family planning programme on an individual's probability of using family planning and the indirect effects due to social interaction. Our empirical estimates show that the non-linear model of the relations among programme effects, social interaction, and of modern family planning leads to some fairly large differences in the estimates of programme effects from those obtained with the linear model—e.g. with estimated direct programme effects on the ever-use of family planning from 20% lower to 27% higher for the linear than the non-linear model. We then show empirically that in our data as much as 43% of total programme effects are due to social interactions.[7] This social multiplier effect is due to a feedback loop that occurs because social interactions render the family planning decisions of community members interdependent. Because of this social multiplier, attributing all of the total change in contraceptive behaviour to the direct impact of changes in programme effort would be a substantial overestimate of the true direct programme effect.

Second, if the model is non-linear (Figure 2b–c), there may be both a low-level Malthusian equilibrium in which contraceptive use remains relatively low despite ongoing programme efforts as well as an equilibrium in which contraceptive use is high.[8] If a population is at a low-contraceptive-use and high-fertility equilibrium—a situation that may characterize much of sub-Saharan Africa, including places with family planning programmes—small programme changes have relatively small effects. However, large increases in programme efforts—even if transitory—may cause a shift to a high-contraceptive-use and low-fertility equilibrium. In a linear model, in contrast, large programme efforts can lead to high contraceptive use, but the programme efforts must be maintained at high levels to sustain high contraceptive use. Our empirical analyses of data in Malawi and Kenya have not indicated the presence of multiple equilibria in our data. Thus, these estimates suggest that there is little likelihood that a sharp transitory increase in programme activities would lead to a rapid shift to much higher sustained levels of contraceptive use. But such possibilities may exist in other contexts.

Third, one can show formally that intensified social interactions may either increase or decrease the total effect and social multiplier effect resulting from family planning programme efforts, and 'more' social interaction can thus reinforce or retard the diffusion of an innovation. When a non-linear (logistic) model is used, increasing the impact of social interactions is status-quo-reinforcing close to a stable equilibria (whether at low or high contraceptive use) in a multiple-equilibria situation. Therefore, if a new programme effort were to intensify social interactions near the stable equilibria, the total—or long-term—change in contraceptive use resulting from the programme effort is reduced and these more intensive social interactions would retard the diffusion of family planning after the programme interventions. Our non-linear empirical estimates for Nyanza District in Kenya imply that when social interactions are intensified, they reduce the total effect associated with programme interventions, but slightly increase the social multiplier effect. These findings are in contrast to the linear estimates that imply that more intense social interaction leads to a larger social multiplier effect and an increased total effect after the programme interventions.

[7] That is, if the social multiplier is 175%, the proportion of the total effect due to social interaction is 75/175.

[8] Related models of multiple equilibria and path dependency in the context of fertility decline are found in, for example, Becker et al. (1990), Galor and Weil (1996), and Kohler (1997, 2000).

7.3.2 The Density of Social Networks and Fertility Decisions

In this section we turn to the specific mechanisms of how social interactions affect individual and population-level family planning use, and thus alter the potential interpretation of the effects of family planning programmes. We turn to the specific mechanisms through which social interactions may affect fertility behaviours and contraceptive use. Early studies of this issue emphasized the content of social interactions, usually measured by the proportion of contraceptive users in a respondent's network on family planning choices. These studies typically found that the probability of a woman's using contraceptives is related strongly to content, and that the relationship is positive: the more users in a network, the more likely that the woman herself uses family planning. In Kohler et al. (2001), we expanded this approach by proposing that it is not just the number of users: rather, the structure of the network modifies the impact of the number of users in the network. Specifically, in these analyses we have included measures of network density, distinguishing between dense networks, in which all the network partners know each other, and sparse networks, in which the network partners are connected only through their ties to the respondent. The inclusion of network density allowed us to argue that when social learning dominates, network density should not matter. In situations of uncertainty, information is important. Because all members of a dense network are likely to possess the same information, we expect weak, possibly negative effects of density on the adoption decision when the content of the interaction is controlled. If social influence dominates, however, density is expected to be important. In particular, when the normative acceptability of contraceptive use is the issue, dense networks with a low proportion of contraceptive users should reduce the probability of using family planning; dense networks with a high proportion of users should increase that probability; and sparse networks should be relatively neutral.

In Kohler et al. (2001) we used this modelling approach and the KDICP data to estimate the probability of using modern contraception. The key empirical results are reported in Table 7.1. In addition to some standard socio-economic characteristics, the estimated models in Table 7.1 include the proportion of contraceptive users in the social network, the density of the network, and an interaction between these measures of content and structure. We find that both our measure of network content and our measure of network structure are related to the probability that a woman uses family planning. The patterns of the interactions between content and structure in our empirical modelling, however, suggest that context determines whether social learning or social influence dominates. In one of the four regions of our study, Obisa, the probability of a woman's contraceptive use is affected primarily by the measure of the content of the interaction; network structure has little relevance. In Obisa, social learning apparently is the mechanism through which social interaction affects contraceptive decisions. In Owich, Kwadhgone, and Wakula South (OKW),

Table 7.1 Logistic regression of contraceptive use (ever used family planning) on individual and network characteristics. (Sample: currently married women with a family planning network of size three or four.)

	Model 1	Model 2	Model 3
for Owich, Kawadhgo, and Wakula S.			
%users (δ_1)	1.619 (0.442)**	1.596 (0.442)**	−1.712 (1.053)
density (δ_2)		−0.248 (0.531)	−2.756 (0.787)**
density × *%users* (δ_3)			3.867*a* (1.268)**
for Obisa			
%users (δ_1)	2.271 (0.623)**	2.290 (0.542)**	4.495 (2.197)*
density (δ_2)		−1.850 (0.831)*	−0.337 (1.654)
density ×*%users* (δ_3)			−2.814*a,b* (2.614)

Notes: The estimated model is specified as $Pr(y=1|X, \textit{social network}) = X\beta + \delta_1 \cdot (\%users) + \delta_2 \cdot density + \delta_3 \cdot (\%users)\ density$, where *y* equals 1 if the respondent uses (has ever used) family planning, *X* is a set of individual characteristics, *%users* is the percentage of users of modern methods of family planning by the network partners, and *density* is the density of the social relations among the network partners. The individual characteristics included in *X* include age, age^2, number of children ever born, and dummy variables indicating whether the respondent has primary or secondary education. The standard errors, reported in parentheses, are adjusted for the clustering of respondents in villages using the Huber–White estimator of variance. *p-values:* $^*p < 0.05$; $^{**}p < 0.01$. *Additional tests:* (*a*) The linear combination $\delta_1 + \delta_3$ measures the effect on the probability to use family planning due to a change in *%users* in a network with *density* = 1. A Wald test of the null hypothesis $\delta_1 + \delta_3 = 0$ is rejected at the 1% level for OKW, and for Obisa at the 5% level. (*b*) The linear combination $\delta_2 + \delta_3$ measures the effect on the probability to use family planning due to a change in *density* in a network with *%users* = 100%. A Wald test of the null hypothesis $\delta_2 + \delta_3 = 0$ is rejected at the 5% level for Obisa in Panel A.

Source: Kohler et al. (2001).

the other regions, it is not content but social influence that appears to be the primary mechanism through which networks influence individual behaviour. In OKW, the interaction between content and structure is critical: dense networks discourage an individual from using contraception if the network includes few contraceptive users, but dense networks encourage use when contraceptive use in the network is relatively high. Thus, when social learning is the mechanism by which networks affect contraceptive decisions, a comparison across contexts confirms the simple account: the higher the proportion of contraceptive users in a woman's network, the more likely she is to use family planning. Where social influence dominates, however, the influence of networks is ambivalent: they may either facilitate or constrain the adoption of family planning.

These differential implications of social learning and social influence on the probability of using family planning are also depicted in Figure 7.3. Given the same social network, the ever-use of contraception is higher in Obisa than in OKW. If we compare the lines for dense networks and sparse networks in Obisa, we see that a woman is more likely to have ever used modern contraception if she has a sparse network than if she has a dense network, given the same prevalence of family planning in a respondent's social network. Moreover, as the proportion of network partners using family planning increases, the lines diverge. Therefore, when the prevalence of users within the network is low, women with sparse networks are about as likely to use family planning as women with dense networks. When the prevalence of family planning in the network is high, however, women in sparse networks are more likely to use than women in dense networks. These patterns in Figure 7.3 for Obisa thus reflect the implications of social learning. In contrast, the right graph in Figure 7.3 reflects a relation that is typical for social influence. Although the probability of having ever used contraception again increases with the prevalence of use among network partners, the effect is rather minimal and not substantively important for networks with a density of 0.5. Only for relatively dense networks (i.e. density > 0.75) does the proportion of contraceptive users in the network have a relevant influence on the respondent's probability to use family planning. In addition, the lines no longer diverge for increasing levels of contraceptive prevalence in the networks as in the left graph, but rather intersect at a prevalence of about 0.7 that is indicated by the line CC. To the left of the line CC an increasing density of the network reduces the probability of having ever used contraception, holding the prevalence of family planning users in the network constant. To the right of the line CC the social influence is towards modern contraception. In this case, an increasing density of the network, holding the prevalence of contraceptive users in the network constant, increases the probability of using family planning.

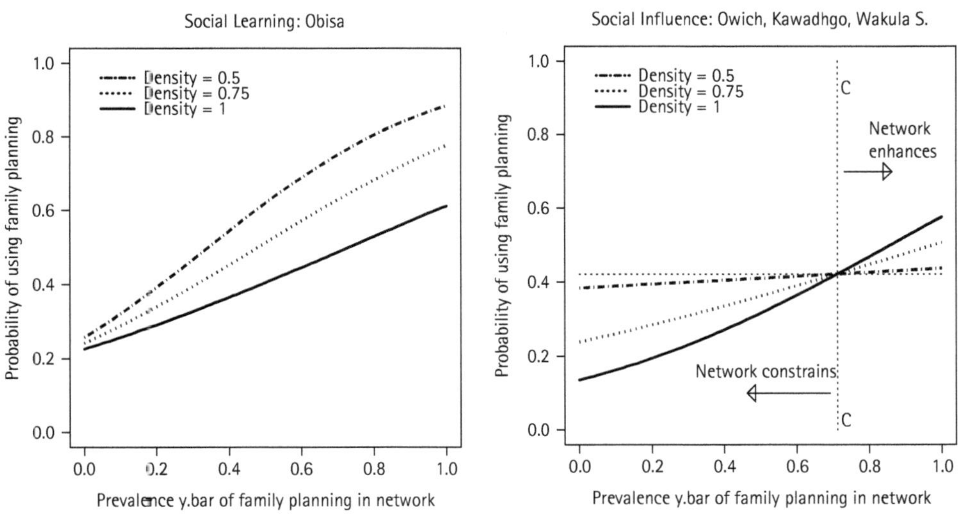

FIG. 7.3 The effect of contraceptive prevalence in the network on the probability of adopting family planning for respondents with networks of different density.

These two regions for which our models are estimated in Table 7.1—Obisa and OKW—are not distinguished by the characteristics of the networks of the respondents who live there, but rather by the extent of market activities: in Obisa, more women are engaged in market activities than in OKW, and they buy and sell at a larger market. We find that social influence is important only where market activity is low. Where market activity is high, social learning dominates. Although the available data do not allow us to investigate in detail the interdependence of social interaction and market activities, and further research on the interaction between networks and market influences is required, the notion that higher market activities favour social learning is plausible. After all, the spread of information is an important aspect of markets, and market participants may focus more strongly on the information provided by their personal contacts than on the social acceptance regarding their family planning behaviour. Our findings about the importance of market activity are consistent with provincial differences in the onset and pace of fertility decline in Kenya: the earliest declines occurred in Nairobi Province and in Central Province. Markets in both of these locations have long been more highly developed than in Nyanza (Bates 1981). This finding also suggests that, even in areas where social interactions currently retard the diffusion of family planning, the dominance of conservative social influence may shift to a dominance of social learning, which will accelerate this diffusion if market development is sufficient.

7.3.3 Social Networks and Changes in Contraceptive Use Over Time

A limitation of the analyses in the previous section, as well as of much of the early demographic literature on social interactions, is that they do not permit confident inferences regarding the causal effects of social networks because unobserved factors that may directly affect attitudes and behaviour may also directly affect choices of the units of social interaction (see previous discussion). Most of the literature on social interactions and demographic behaviours assumes, usually implicitly, that it is acceptable to treat networks as if they were formed randomly. There are at least two reasons to expect that this assumption of random network selection often may be violated. First, empirical studies suggest a non-random selection of network partners. For example, using qualitative data collected as part of the KDICP, Watkins and Warriner (2003) showed that the networks with whom respondents discuss issues of family planning and AIDS are characterized by a tendency to discuss these topics with others who are perceived to be similar ('like me'); in addition, some network partners are deliberately chosen because they are believed to have relevant information or competence. Second, a theoretical consideration of learning under uncertainty suggests that social interactions about family planning are determined by the following factors: (1) the costs and benefits of social learning about family planning and fertility-related issues; (2) the various social constraints imposed on the ability to engage in interactions about family planning due to the availability of suitable network partners and the social acceptability of communications about contraception and fertility reduction within households and communities; and (3) the expected reduction of uncertainty about the benefits and side effects (or other costs) of using family planning through interactions with others, which depends in part on network partners' knowledge, their possibly strategic communication of this knowledge, and the individuals' interpretation of the information they obtain from others. As a result of these processes of how social networks are formed, if the causal direction is unclear, what has been interpreted as the causal effects of social networks may simply be associations that are due to both contraceptive use and network partners' choices being determined, in part, by unobserved factors, such as preferences related to contraceptive use and network partner selection. Therefore we use our longitudinal data with special information on social networks once again to investigate the determinants of contraceptive use in high-fertility rural Kenya.

Four major findings emerge from the analyses in Behrman et al. (2002) (see Table 7.2 for the key regression results). First and foremost, social networks have significant and substantial effects even when we controlled for unobserved factors that may also determine the nature of the social networks. In particular, this study provides what we believe are currently the best available estimates about the effects of social networks on contraceptive use in high-fertility areas. Second, estimates of the effects of social networks that are based on the implicit assumption that they are determined randomly, as in previous studies, may lead to a substantial misunderstanding of the impact of social networks on individual behaviours. With our data, analyses that did not control for the possibility that both contraceptive behaviour and social networks within which this behaviour is discussed are partially determined by unobserved factors, such as preferences, appeared to misestimate the effects of networks. Third, the effects of social networks are not limited to women, even though in local stereotypes women are often characterized as gossiping much more than men. On the contrary, our estimates indicate that, if anything, men are likely to be more influenced by their network partners than are women. This finding may reflect cultural patterns of exogamy and patrilocality that result in men having known their network partners since childhood, whereas women alter their network partners after marriage. Fourth, the effects of social networks found in this study contribute to a better understanding of social change. These effects are generally non-linear and asymmetric. They are particularly large for having at least one network partner who is perceived to be using contraceptives; however, the inclusion of additional network partners with the same characteristic generally has much smaller (and insignificant) effects (for women). This combination of non-linearity and asymmetry suggests that the exchange of information constitutes the primary aspect of social interactions about family planning—social learning, not social influence. In addition, the non-linear and asymmetric pattern of network influences is consistent with stereotypic diffusion models (Rogers 2003). If there are just a few who initially adopt an innovation, they have a relatively large influence because they interact with a relatively large number of individuals who have not yet adopted it; in such cases, they provide

Table 7.2 Females: Fixed-effect and random-effect logit models for currently using family planning with different specifications of network partners' family planning use. Respondent's contraceptive use is measured at K1, K2, and K3

Method	Fixed Effects Logit	Random Effects Logit	Fixed Effects Logit	Random Effects Logit
At least one family planning user in network	0.72 (0.30)*	0.61 (0.25)*	0.69 (0.32)*	0.49 (0.26)+
Number of remaining family planning users in network	0.16 (0.12)	0.49 (0.10)**	0.07 (0.14)	0.49 (0.11)**
At least one non-user in network			0.01 (0.30)	0.27 (0.24)
Number of remaining non-users in network			−0.22 (0.16)	−0.19 (0.13)
Dummy for not married, time *t*–	−0.60 (0.52)	−0.64 (0.41)	−0.59 (0.52)	0.66 (0.41)
Children ever born, time *t*–	0.10 (0.12)	0.06 (0.05)	0.12 (0.12)	0.06 (0.05)
Respondent has radio, time *t*–	0.41 (0.30)	0.38 (0.20)+	0.39 (0.30)	0.37 (0.20)+
Respondent has metal roof, time *t*–	−0.71 (0.37)*	0.08 (0.22)	−0.73 (0.37)*	0.08 (0.22)
Respondent has at least primary schooling		0.83 (0.31)**		0.85 (0.31)**
Respondent has secondary schooling		0.61 (0.28)*		0.61 (0.28)*
Age		0.41 (0.11)**		0.41 (0.11)**
$(Age/10)^2$		−0.59 (0.16)**		−0.59 (0.16)**
Dummy for survey wave Kenya 2	0.35 (0.24)	0.21 (0.22)	0.34 (0.24)	0.21 (0.21)
Dummy for survey wave Kenya 3	0.60 (0.29)*	0.44 (0.22)*	0.63 (0.30)*	0.45 (0.23)*
Constant		−11.36 (1.99)**		−11.35 (1.99)**
N (number of women, each observed at three surveys)	156	497	156	497

Notes: p-values: $^{+}p < 0.1$; $^{*}p < 0.05$; $^{**}p < 0.01$. Fixed-effect logit model is based only on individuals who change their contraceptive behaviour at least once between Kenya 1 and Kenya 3; women with constant contraceptive use in all three survey waves are dropped in the estimation. We use the subscript '*t*–' to emphasize that the variable refers to the time prior to *t*, where *t* refers to the survey wave.

Source: Behrman et al. (2002).

these individuals with at least one adopter, the influence of whom is relatively large. Thus, adoption initially accelerates. As there are more innovators, however, the marginal influence of yet another adopter eventually starts to decline. Interaction processes therefore suggest that social networks are likely to have large effects on behaviour as long as an innovation is not widely disseminated. As innovative behaviour increases, the marginal effect of interactions is likely to be much smaller than in the early phase of the diffusion process.

7.4 Social Networks and Mortality and Death: The Diffusion of Worry about AIDS

Individuals facing the tsunami of the AIDS epidemic in eastern and southern Africa know well that HIV is primarily transmitted in their context by sexual intercourse and that reducing risky sexual interactions can help to protect them from infection and death. Whether correct or incorrect, the subjective perceptions of one's own risk and of one's sexual partner's risk have been shown to be important correlates of whether an individual adopts risk reduction strategies (Cerwonka et al. 2000; Estrin 1999; Weinstein and Nicolich 1993). But what is the process through which these risk perceptions are formed (e.g. Smith 2003)?[9] In Kohler et al. (2007), we therefore investigate the determinants of subjective HIV/AIDS risk assessments, focusing in particular on the hypothesis that individuals assess their risk of infection through interactions with others in their social networks.

7.4.1 Qualitative Evidence on the Content of Conversations about AIDS in Informal Social Networks in Malawi

We begin our discussion by drawing on qualitative data to provide insights into the process through which risk perceptions are formed in social networks. The first round of the MLSFH survey in 1998 had confirmed that respondents talked about AIDS with others in their social networks. These data, however, were inadequate for learning what people were saying to each other. For example, it could have been that there was no uncertainty about AIDS—perhaps all agreed or disagreed about the level of risk that they faced.

[9] For a general discussion of the need to better understand the formation of expectations, including risk perceptions, see Manski (2004). Some of the few studies that explicitly address the determinants of AIDS risk perceptions in sub-Saharan Africa or other developing countries are Bernardi (2002); Bühler and Kohler (2003); Bunnel (1996); Helleringer and Kohler (2005); Kengeya-Kayondo et al. (1999); London and Aroyds (2000); Smith (2003); Smith and Watkins (2005); Watkins (2004).

Or perhaps the conversations did no more than transmit epidemiological information, such that HIV is transmitted sexually and invariably fatal, without comments evaluating the accuracy or appropriateness of the information, or that nothing was said in these conversations to support our assumption that social influence was being exerted in these conversations.

We thus conducted semi-structured interviews and ethnography in order to examine the validity of the assumptions with which we approached our quantitative analyses (see also Watkins 2004; Watkins and Swidler 2009, 2011). Here we draw on the ethnographies, the term we use to describe a large set of field journals collected since 1999. We asked a total of 23 high-school graduates living in or near the MLSFH survey sites to pay attention to public conversations they heard about AIDS during the course of their everyday activities, such as walking to the well for water or having a beer in a local pub. They were then to write as much as they could recollect, and in as much detail as possible, in a field journal, which was then given to an intermediary and sent to us (for more detail see Watkins and Swidler 2009; see also details at <http://investinknowledge.org/projects/research/malawian_journals_project>, for approximately 900 journals).

We begin by summarizing a journal that displays the great diversity in natural conversational settings and in topics covered in the conversations. The journal begins on 14 June 2001, when the journalist, Alice, visits her cousin, who is a nurse at a hospital about an hour's bus ride away. The cousin, who is pregnant, tells Alice that three months after her marriage her husband began coughing, then a headache, then diarrhea, then both diarrhea and shingles, all of which involved stays in the hospital. The cousin herself had become thin. The cousin requested that they both be tested, and both were HIV positive. Later, Alice returns for the husband's funeral, where she talks with her cousin and her cousin's mother. The cousin warns Alice, a widow, to be careful whom she marries and to be sure to have a blood test beforehand. On the way home from the funeral, Alice meets a man at the bus stop who has been to see a brother ill with tuberculosis; he tells her that the TB ward is full—they all have AIDS (presumably including his brother). Another man at the bus stop joins the conversation, asking why it is that women appear to have AIDS more than men. This generates a lengthy discussion about differences in men's and women's behaviour and bodies, whether or not it is possible to use a condom in marriage, medications, and the history of AIDS, with all of these topics introduced not by the journalist but by the men, none of whom Alice knows. Two weeks later Alice returns for the funeral of her cousin's newborn baby. Walking back from the funeral to the bus stop, a neighbour of her cousin asks Alice why her cousin is so thin, and then comments that people are saying she has AIDS because although she herself was innocent, her husband was promiscuous and, as a woman, she could not refuse to have sex with her husband. On the bus, a woman starts a conversation with Alice about AIDS, which is then joined by the third person on their seat, an old man. Again, the others introduce the topics, which cover AIDS as God's punishment, AIDS as witchcraft, AIDS as a governmental plot, and AIDS as a result of youth who disobey the advice of their parents. A few weeks later, Alice goes to the funeral of her cousin, where she overhears others explaining that her cousin was the innocent victim of her husband.

Although funerals are frequent in rural Malawi—the 1998 survey round showed that on average respondents went to 3–4 funerals a month—the scarcity of HIV testing facilities at that time in rural Malawi meant that few would have been tested. Nonetheless, by that time people knew the symptoms of HIV, which they combined with local knowledge about the medical history of the deceased and his or her sexual biography to conduct a 'social inquest' on the cause of death. Seeing someone sicken and then die of something that was believed to be AIDS is likely to have influenced the formation of individuals' own perception of risk, the variable that in our quantitative analysis later in this section is summarized as 'worry' about infection.

In one incident, a young man tells several friends that he has reformed his behaviour. They ask why, and he explains: 'Only because I have seen for myself, some of my friends have died because of this disease AIDS, and I do care for my life. AIDS troubles a lot! I didn't say anything. He kept on, saying, For example, there was a certain army pensioner who was living up there in my village. . . . He was very sick indeed, going to the hospital, no treatment, private hospitals—just wasting money and then he came home and was sick until he became like a very little young child. I was going to see him during the whole course of his suffering. You could liken him to a two-year-old child when he lay down sick. . . . And the way I had seen him suffering, that's when I came to my senses, that indeed AIDS troubles a great deal before one dies.'

The young man attributes his behavioural change to seeing a neighbour whom he knew had many sex partners decline physically, but we know from other journals that this witness had himself been promiscuous. Thus, it is likely that while watching his neighbour waste away he imagined himself as 'a two-year-old child'. We do not know whether the reforms he claimed to have made happened at all, or persisted, or occurred too late. We do know that many in Malawi have had similar experiences watching those with whom they can identify die, as well as hearing about other deaths they did not witness. If we are persuaded by the literature in psychology that 'people disproportionately weight salient, memorable or vivid evidence even when they have better sources of information' (Rabin 1998: 30, citing Kahneman and Tversky 1973), then anecdotes about people who are known in the community are likely to weigh heavily in the process of perceived risk formation, as well as to provide particularly compelling motivations for change.

Rarely is information about HIV or AIDS unaccompanied by comment. The information is often evaluated in terms of its credibility. For example, one conversation turned to the possibility of infection when a man gets his hair cut from a barber. A woman says going to a barber is dangerous, that she heard a radio programme that if a person with HIV is cut, 'the virus sticks to the teeth of the shaver and if other people come to be shaved . . . definitely the other people will contract the virus and start suffering from AIDS'. Another participant, however, evaluates this information by offering a counterfactual: he says 'then if what the government was saying through the radio, that barber shops can also facilitate the spreading of the virus which causes AIDS, was true, a lot of people would have contracted it, almost every man starting from a young boy and men and some of the women and girls'.

People also share their personal worries with others. For example, while walking together to a funeral one woman tells two others that she is worried that her husband will give her AIDS, for he had been having an affair with the deceased, a known prostitute who was believed to have certainly died of AIDS. Now, she says, she doesn't know what to do. The participants discuss the pros and cons of divorce: the conversation ends when one of the participants advises her to have a blood test before she takes a hasty action.

In addition to providing insight into the process by which perceptions of risk are formed in social networks, the above qualitative findings emphasize that many determinants of risk assessment are likely to be unobserved in survey data, including for instance aspects such as an individual's exposure to seeing someone die from AIDS, or being advised to have a blood test before taking the serious step of ending a marriage. These unobserved factors are important because they are prone to affect not only variation in perceptions of risk, but also the size, composition, and selection of individuals' social network partners. Some individuals, for example, are likely to have less tolerance for risk and, because of systematic patterns in the selection of their social networks, are more likely to associate with others who have less tolerance for risk (for a discussion of these aspects of social network selection, see also Behrman et al. 2002; Manski 2000; Watkins and Warriner 2003). Based on considerations about individuals having unobserved characteristics such as those related to risk aversion and social interaction that are likely to affect both their worry about HIV/AIDS and their social networks related to information about HIV/AIDS, parallel to our studies of the impact of social networks on fertility control in section 7.3, we posit that prior social networks are not likely to be random in the sense of being independent of disturbance terms in relations for the estimation of risk perceptions and AIDS-related behaviours at time *t*. Therefore we use an empirical specification of the relation determining risk perceptions and AIDS-related behaviours in which there is explicit recognition that, in addition to observed right-side variables (including social networks prior to time *t*), there are unobserved factors.

7.4.2 Quantitative Evidence on the Impact of Social Networks on Responses to AIDS

Not surprisingly, concerns about the risk of AIDS infection were widespread in both rural Kenya and Malawi from the mid- to late 1990s onward. The MLSFH survey measured this perceived AIDS risk with a question frequently used in research on risk perceptions: 'How worried are you that you might catch AIDS?' Responses to this question ranged from 'not worried at all' to 'worried a lot'. Between 36% and 40% of women in Kenya responded in the 1996/1997 (KDICP 1) and 2000 (KDICP 2) surveys, respectively, that they perceived themselves to have a moderate or high risk of becoming infected with AIDS. For Malawi, 61% and 47% of women perceived a high risk of AIDS in 1998 (MLSFH 1) and 2001 (MLSFH 2), respectively; moreover, their responses are highly and positively correlated with a question about the subjective likelihood that the respondent will become infected with HIV in the future. Respondents were generally also aware of several mechanisms by which HIV/AIDS is transmitted and several means of protection. For instance, in 1996/1997, more than 90% of women in Kenya knew that AIDS can be transmitted by sex, and 48% knew about possible transmission by injections. Similarly high levels of knowledge prevailed in Malawi.

Survey responses in the KDICP and the MLSFH reinforce the perception from the qualitative data that such conversations are frequent. In both the KDICP and the MLSFH, the survey module for collecting the egocentric network data on HIV/AIDS conversations began with a question, 'How many people have you talked with about AIDS?' Very few had talked with no one. The networks are quite dense (most members know each other as well as the respondent) and highly gendered (men talk with men, women with women) (Watkins and Warriner 2003; Zulu and Chepngeno 2003). Responses to other questions provide insight into some topics of their conversations. For example, respondents report on the extramarital partnerships of their network partners and their best friend; a study of a subsample of MLSFH respondents shows that they learn about these relationships directly from one of the couple, indirectly from others who have talked with one of the couple, or from observation ('I saw them coming and going') (Tawfik and Watkins 2007). More than 85% (Kenya) and 87% (Malawi) of women know of at least one recent death that they suspected was caused by AIDS, and more than 30% (Kenya) and 16% (Malawi) know about more than five such cases. The specific question in the KDICP and MLSFH regarding the risk perceptions of the network partners was phrased as 'How worried is name of network partner about getting AIDS?', with the same response categories as for the respondent. Over three-quarters of the women had talked with at least one person about AIDS, and over two-fifths of the women had talked with at least one person who believes that he or she is at moderate or great risk of becoming infected with AIDS. Interestingly, between the 1998 and 2001 rounds of the MDICP, women who were married in the former but divorced or widowed in the latter became less worried about infection, whereas those who were not currently married in 1998 but had married by 2001 were more worried (Smith and Watkins 2005; see also Reniers 2008). In addition to talking with network partners about AIDS, husbands and wives discuss with each other their risks and how they can prevent infection; following HIV testing conducted by the MDICP in 2004, 84.4% of women and 91.9% of men reported having shared their HIV test result with their spouse (Anglewicz and Chinsanya 2011).

On average, women reported in the KDICP 2 and 3 surveys and the MLSFH 1 and 2 surveys that they had talked with 3.9–4.8 network partners about AIDS, and men report slightly more interactions, ranging from close to 4 to about 7 network partners. Detailed information about interactions is available for about 2.4–3.6 network partners. In general, the respondents report more interactions with network partners who perceive high AIDS risks as compared with network partners who assess their risks as low. Neither the size of these networks nor having talked with at least one network partner about AIDS depend strongly on the respondent's risk perception (Kohler et al. 2007: table 3), whereas—as we expect based on the our hypothesis that social interactions are important determinants of risk perceptions—network partner's assessments of HIV/AIDS risks are associated with the respondent's own risk perception. We represent social networks by the extent to which each respondent's network partners are reported to be worried about AIDS. This perception is measured via a categorical variable with four options in Kenya (categories are none (1), some (2), moderate (3), and great (4)) and with three options in Malawi (categories are none (1), moderate (2), and great (3)). The essential variable representing social interactions about HIV/AIDS is therefore the number of network partners with whom the respondent has interacted about HIV/AIDS classified by the respondents' perceived network partners' risk perceptions. Although in what follows we will refer to the network partners' perceptions of risk, these perceptions are reported by the respondent.

The linear social interaction model discussed earlier provides one possible framework for empirically analysing the effect of social interactions on the perceived risk of AIDS, where

Y_{it} is redefined to be the perceived AIDS risk of individual i at time t;

f_I represents the unobserved fixed factors that are assumed to affect risk perceptions and AIDS-related behaviours by individual i (e.g. the persistent part of preferences, unobserved current community characteristics, expectations regarding future prices, and interfamilial and community resources on which the individual can draw); and

e_{it} is a disturbance term that affects the perceived AIDS risk of individual i at time t due to, for example, new information about AIDS prevalence provided by the death of a family/community member from AIDS, new information about the behaviour of the spouse, or price shocks that are deviations from the long-run secular price trends.

As discussed already in the context of estimating social interaction effects for the adoption of family planning, in order to obtain consistent estimates of the coefficient a, which measures the impact of social networks on risk perceptions and AIDS-related behaviours, it is necessary to break the correlation between the term representing social networks (N_{it-}) and the compound disturbance term including both fixed and random elements ($f_i + e_{it}$). However, contrary to our earlier example, a fixed-effect estimation alone may not be fully satisfactory because it relies on the assumption that the social network prior to time t, N_{it-}, does not depend on the lagged disturbance terms $e_{i(t-1)}$ (or higher-order lags). Our estimation strategy in Kohler et al. (2007) allows for such feedback from lagged disturbances affecting HIV/AIDS risk perceptions on the current social network size and composition by combining fixed-effect and instrumental-variable (IV) estimation. Since differencing over time eliminates the individual fixed effect f_i from the estimation relation, variables that are correlated with the fixed effect but uncorrelated with the difference of e_{it} over time can be used as instruments. Specifically, our analyses in Kohler et al. (2007) use the initial 'stock' of social network partners observed in KDICP 2 or MLSFH 1, funerals and other events that led to social gatherings, and other 'stock variables' at the beginning of the panel such as age, education, marital status, and indicators of household wealth.

To demonstrate empirically the relevance of considering the endogeneity of social networks in inferences of social interaction effects, we implemented the following four estimation techniques: (a) standard OLS analyses; (b) fixed-effect estimation; (c) IV fixed-effect estimation that instruments for the change in the social network measures, ΔN_{it-}; and (d) Generalized Methods of Moments IV (GMM-IV) fixed-effect estimation, which uses a more efficient weighting of the moment conditions implied by the IV fixed-effect estimation (Baum et al. 2003; Hayashi 2000).

The regression results for women are shown in Table 7.3, with additional results reported in Kohler et al. (2007). In summary, the key findings of the analyses are as follows. First and foremost, the analyses show that social networks have significant and substantial effects on individuals' AIDS risk perceptions, even when we control for unobserved factors that also may determine the nature of the social networks. Thus, to understand the dynamics and diffusion of behavioural change in response to AIDS, it is essential to incorporate the impact of social networks: the failure to do so may lead to misunderstanding the dynamics of behavioural change. Second, this effect of social networks extends to the area of spousal communication about AIDS risk: interactions with network partners—independent of network partners' risk assessments—tend to increase the probability of husband–wife communication about the disease. Third, the effects of social networks found in this study contribute to a better understanding of diffusion. These effects are generally non-linear and asymmetric. They are particularly large for having at least one network partner who is perceived to have a great deal of concern about AIDS. The inclusion of additional network partners with the same level of concern or with less concern generally has much smaller or insignificant effects. An exception to this asymmetry occurs in the network effects on spousal communication: network partners, independent of their risk perceptions, have strong and significant effects. Fourth, social networks are associated with important social-multiplier effects that reinforce the effects of AIDS prevention programmes. For women, for instance, about one-fifth of the influence of programme efforts on respondents' HIV/AIDS risk perceptions is mediated through social networks.

These findings are of central importance for understanding the spread of AIDS because they document that social interactions constitute important determinants of how individuals and couples develop strategies for coping with the disease. In particular, this study shows that social networks exert systematic and strong influences on risk perceptions and the probability of spousal communication about HIV/AIDS risks in rural areas of two sub-Saharan African countries with high HIV prevalence, and that these influences are in addition to other factors such as programme interventions that disseminate knowledge about the disease, provide access to condoms, and advocate changes in sexual behaviours within and outside marriage. Social networks are also likely to amplify programme efforts aimed at increasing individuals' information about HIV/AIDS and their assessments of their own risks. Thus, social interactions are likely to have a substantial impact on the course of the epidemic and the magnitude of its

Table 7.3 Females: Regression of respondents' risk perceptions on the number of social network partners with high, moderate, and low risk perception and personal characteristics ('nwp(s)' = network partner(s))

	Kenya			Malawi		
	GMM FE-IV	Fixed Effect	OLS	GMM FE-IV	Fixed Effect	OLS
# of nwps with high risk perception, time *t*–	0.2199 (0.0440)**	0.1742 (0.0318)**	0.1618 (0.0237)**	0.1036 (0.0242)**	0.1193 (0.0188)**	0.1549 (0.0131)**
# of nwps with moderate risk perception, time *t*–				−0.1315 (0.0373)**	−0.0639 (0.0246)**	−0.0487 (0.0184)**
# of nwps with low risk perception, time *t*–	−0.0744 (0.0429)+	−0.0448 (0.0286)	−0.0737 (0.0212)**	−0.2189 (0.0366)**	−0.1698 (0.0229)**	−0.1789 (0.0183)**
Children ever born	−0.0907 (0.0902)	−0.0111 (0.0412)	0.0115 (0.0146)	0.0642 (0.0481)	0.0003 (0.0203)	−0.0055 (0.0082)
Dummy for not married, time *t*	0.1913 (0.1903)	0.1725 (0.1814)	0.1894 (0.0976)+	−0.2154 (0.0966)*	−0.1804 (0.0972)+	−0.0942 (0.0543)+
Respondent has radio, time *t*	−0.1456 (0.1005)	−0.1164 (0.1027)	−0.0881 (0.0632)	0.0399 (0.0538)	0.0304 (0.0541)	0.0278 (0.0343)
Respondent has metal roof, time *t*	0.0025 (0.1240)	−0.0020 (0.1276)	0.0430 (0.0670)	0.1011 (0.0957)	0.0867 (0.0927)	0.0295 (0.0587)
AIDS programme effort				0.4442 (0.1733)*	0.4760 (0.1750)**	0.3989 (0.1255)**
Respondent has at least primary education			0.1417 (0.0784)+			0.0853 (0.0368)*
Respondent has secondary education			−0.1180 (0.0876)			0.0721 (0.0692)
Age			0.0216 (0.0274)			0.0130 (0.0102)
(Age/10) squared			−0.0475 (0.0373)			−0.0149 (0.0129)
Dummy for survey wave Kenya 3 or Malawi 2	0.0363 (0.0791)	−0.0190 (0.0617)	−0.0206 (0.0528)	−0.1073 (0.0512)*	−0.1114 (0.0374)**	−0.0964 (0.0333)**
Constant			1.9248 (0.4756)**			1.9322 (0.1774)**
N	545	545	545	1138	1138	1138

Notes: Standard errors in parentheses. *p-values:* + $p \le 0.10$; * $p \le 0.05$; ** $p \le 0.01$.

Source: Kohler et al. (2007).

consequences, and these should be taken into consideration in understanding and predicting behaviours in such high-prevalence contexts and in devising programme interventions with respect to the HIV/AIDS epidemic.

7.5 Sexual Networks and HIV/AIDS: The Likoma Network Study

Sexual networks are the primary mechanism through which HIV is spread and transformed in sub-Saharan Africa (SSA). Theoretical network models have shown that individuals' positions within these sexual networks, and the structural characteristics of the network itself, are important determinants of HIV infection risks and disease dynamics (Ghani and Garnett 2000; Kretzschmar and Morris 1996; Newman 2002). Despite this evidence, empirical network studies of HIV infection risks and disease dynamics in SSA remain very limited. Available data on sexual networks are often based on small populations, frequently restricted to egocentric rather than complete networks, and with the exception of the study described in this paper, not based on an integrated design that includes tracing of sexual networks, HIV testing, and extensive socio-economic data for all members of a population.

In a pioneering illustration of the interactions between networks and health in a poor developing country, the Likoma Network Study (LNS) represents the first complete sexual network data for a large population—that on Likoma Island, Lake Malawi. Combining a census of the population with a sociocentric sexual network survey of *all* adults aged 18–35, Helleringer and Kohler (2007) document the existence of a large and robust sexual network (Figure 7.4). Half of all sexually active respondents were linked together in a *giant network component*, and more than one-quarter were connected together through multiple independent chains of sexual relations. Such structural features of sexual networks have been associated with epidemic spread of STIs in high-risk groups, but prior to this study had never been documented among the general population. This unique design of the LNS has provided important new findings on the role of concurrency of sexual partners and HIV risk, the role of migrants and the contribution to HIV risks within rural populations, the uneven distribution of HIV risks within large sexual networks, data quality and misreporting of sexual relationships, and the determinants of sexual relationships and patterns of homophily and clustering within sexual networks (Helleringer and Kohler 2008; Helleringer et al. 2007; Helleringer, Kohler, and Kalilani-Phiri 2009; Helleringer, Kohler, and Mkandawire 2009; Helleringer, Kohler, Chimbiri et al. 2009).

Epidemics in a mathematical sense are non-linear phenomena. The classical models of mathematical epidemiology rely on the assumption that sexual partners are randomly selected (i.e. the population is assumed to be well-mixed and unstructured) (e.g. Anderson and May 1991; Bailey 1975). In this framework, two key measures to study epidemics are (1) the basic reproduction number, R_0, and (2) the final size of an epidemic s_∞. The basic reproduction number, R_0, is the expected number of secondary infections arising from a single, typical infectious individual in a completely susceptible population (Heesterbeek 2002). In a well-mixed and socially unstructured population (i.e. where individuals randomly select their partners from other members of the population), R_0 is the product of three quantities: the transmissibility of the infection τ, the duration of infectiousness δ, and the rate of contact between susceptible and infectious individuals c. When $R_0 > 1$, an epidemic is certain in a deterministic model and has non-zero probability in a stochastic model. Strategies for disease control and eradication are aimed at bringing R_0 below the threshold of unity, i.e. when the average infection generates fewer secondary infections than necessary for replacement and the epidemic fades. In the well-mixed and unstructured case, the final size of the epidemic is given by the implicit equation $\log(s_\infty) = R_0(s_\infty - 1)$, which has exactly two roots on the interval [0 1] when $R_0 > 1$. The smaller of these roots is the proportion of the population remaining uninfected at the end of an epidemic.

Because HIV is transmitted by intimate sexual contacts between partners, and because people employ varied and elaborate rules to choose their partners (Magruder 2008; Watkins 2004), HIV transmission dynamics in real populations are not well described by the classical epidemiological model. In other terms, c is generally a poor approximation of the patterns of contacts leading to the diffusion of an infection within a population. For instance, while African men (and to a lesser extent, women) do not report having more sexual partners than men elsewhere, they tend to have more than one ongoing long-term relationship at any point in time. Partnerships in SSA can overlap for months, maybe years (Lagarde et al. 2001; Morris and Kretzschmar 2000). This pattern of sexual partnerships that overlap rather than follow each other sequentially, is one of several important characteristics of human sexual networks that violate the classical epidemiological model and importantly affect HIV infection risks and disease dynamics (Kretzschmar and Morris 1996; Moody 2002). *Concurrent partnerships* may thus increase the speed at which HIV spreads through a population, and have probably contributed to the rapid take-off of the HIV epidemic in SSA in the 1980s (Morris and Kretzschmar 2000). Other violations of the classical epidemiological model include: (1) *assortative mixing*, i.e. the selection of sexual partners based on their individual characteristics can structure a network into communities within which the disease spreads rapidly, but across which the spread is slow (Laumann et al. 1994; Laumann and Youm 1999; Morris 1993); (2) *small worlds*, i.e. networks characterized by bridges joining otherwise disjoint clusters (Watts 1999; Watts and Strogatz 1998), can lead to thresholds and rapid disease diffusion to distant subpopulations; (3) *robust networks*, i.e. groups of persons tied together by more than one path in the sexual network, can decrease the ability to control the spread of HIV because redundant connections continue to transmit HIV even after some transmission paths are broken or eliminated (Potterat et al. 2002); (4) *skewed degree distributions*, i.e. networks containing individuals with a very high number of partners (high-degree network members) can result in epidemics driven by promiscuous individuals.[10]

While it is possible to simulate HIV disease dynamics taking these characteristics of human sexual networks into account (Hethcote et al. 1991), only detailed information on the sexual network structures in SSA allows proper calibration of these models. In addition, the investigation of sexual networks has important implications for disease prevention. For instance, in a simple heterogeneous epidemic model structured by degree of sexual activity, the basic reproduction number, and therefore epidemic threshold, is linearly proportional to the variance in partner numbers (e.g. Anderson and May 1991; Bailey 1975). Epidemics will therefore be more difficult to control in populations characterized by behavioural heterogeneity. The structure of sexual relations between the highest-degree individuals and the general population will also determine the effectiveness of control interventions. The design of optimal interventions to control HIV therefore needs to take the prevailing sexual network structure into account. However little is known about these aspects due to the lack of detailed sexual network data.

The literature therefore suggests a considerable potential for important and policy-relevant research based on an empirical investigation of the relationships between sexual networks, HIV infection risks, and HIV/AIDS disease dynamics. While sophisticated analytic methods have recently become available that allow these investigations (e.g. Koehly and Morris 2004; Goodreau et al. 2009), their application to high HIV-prevalence contexts in SSA has been hampered by a lack of suitable data. This is not surprising, as the empirical challenges are formidable: (i) network information needs to extend to (quasi-)complete networks, including information on the structure of the network and the positions of individuals within the network; and (ii) data need to be comprehensive, ranging from sexual networks to risk perceptions, sexual behaviours, and health measures, and be available for *both* respondents and their network partners.

The Likoma Network Study utilizes a sociocentric design to address many of these concerns (see section 7.2 for a discussion of the data collection). Among the 1,858 reports of sexual relationships that were collected during the first round of the LNS, 1,333 (72.7%) were to partners currently residing in Likoma, and 845 (45.5%) were to partners who were also interviewed during the ACASI survey. The reliability of the reports of these within-sample sexual relationships is relatively high as 57.8% are reported jointly by both partners, with marital relationships being substantially more reliable than non-marital relationships. Based on these reports, we reconstructed a large sexual network including 1,803 individuals: 923 respondents of the sexual network survey and 880 sexual partners of respondents who were not interviewed. These network members are connected by 1,614 unique sexual relationships as a total of 244 relationships among

[10] E.g. Liljeros et al. (2001); for a critical perspective, see Handcock and Jones (2004); Jones and Handcock (2003).

survey respondents were concordantly reported by both partners. Several key findings emerge from this study of sexual networks in Likoma (Figure 7.4).

Component size distribution: The sexual network identified by this study contains a total of 256 separate components representing clusters of individuals who are connected to each other through sexual relations that have occurred during a three-year period prior to the survey. The distribution of component sizes is highly skewed: more than 86% of the identified components are of size five or smaller, but include only 34% of all sexually active respondents. On the other hand, two-thirds of network members are embedded in 35 components of size six or larger that are shown in Figure 7.4. Moreover, 883 network members—56% of male and 45.6% of female survey respondents—constitute a single giant component of individuals connected through sexual partnerships having taken place during the three years prior to the survey. Inclusion in this component is not necessarily associated with a large numbers of partners: whereas members of the giant component on average reported a higher number of partners than other respondents (3 vs 1.8, $p < 0.01$), a substantial fraction of members of the giant component (40%) also had at most two partners during the three years prior to the survey. The connectivity of the sexual network therefore occurs not because the distribution of the number of sexual partners is highly skewed (i.e. a small number of individuals are network 'hubs' because they have high rates of multiple partnerships; Liljeros et al. 2001), but as a result

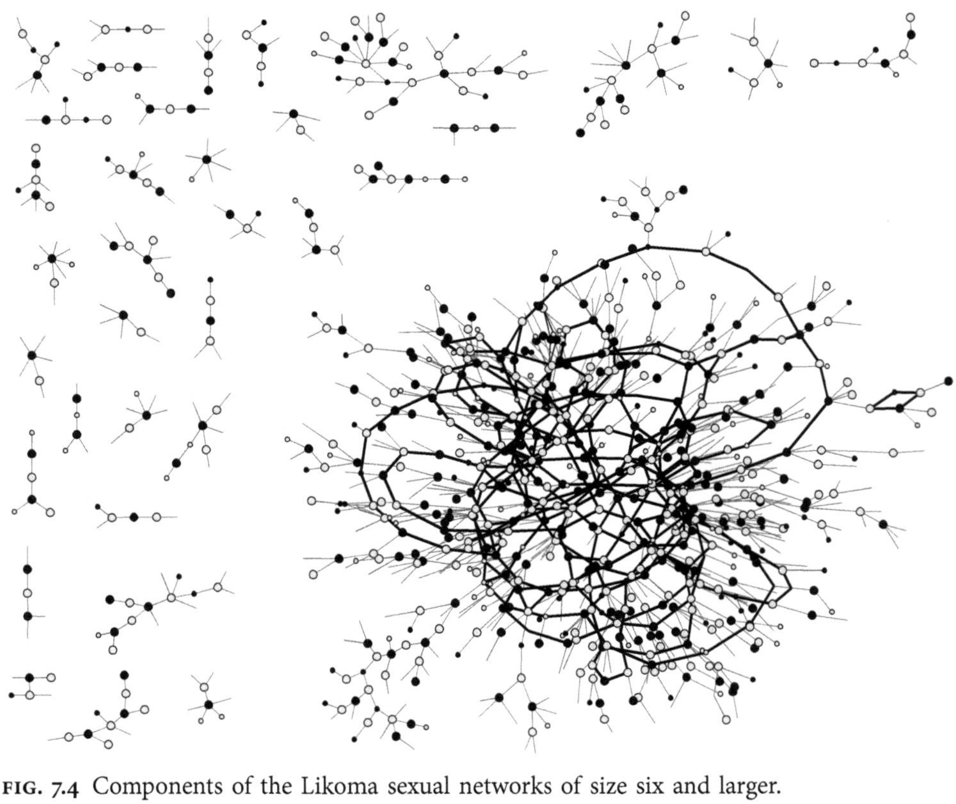

FIG. 7.4 Components of the Likoma sexual networks of size six and larger.

of a generally moderate number of relationships with partners who have (or have had) other partners, who in turn may have had other partners and so on. When we restrict the period of observation to the year prior to the survey, a larger proportion of network members are included in dyads (28.2%) (Figure 7.5). Nevertheless, large connected components still emerge within the network: close to half of all network members belong to components of size 4 and above, and close to one-quarter of all network members are included in structures of size 25 and larger.

High prevalence of cycles in the network: The giant component dominating the sexual network among young adults in the study villages (Figure 7.4) also contains three *bicomponents* that include individuals who are connected by more than one independent path. Two of these bicomponents are simple diamonds connecting four network members in a cycle. The third bicomponent, however, is significantly larger and includes 274 individuals, that is, 15% of all members of the sexual network and close to 25% of survey respondents. Among respondents below age 25, 36% belong to this denser region of the network. Several young adults in our study villages have had more than one partner in common, and a significant proportion of inhabitants of Likoma may be at an increased risk of HIV infection because they are connected by multiple independent chains of sexual relations. Moreover, the occurrence of cyclical structures in the observed sexual network does not depend on the relatively long recall period of three years: the bicomponents in Figure 7.4 contain cycles even if only recent or currently active relationships are considered. For instance, the network of relationships that were active within one year of the survey contains several bicomponents connecting a total of 84 network members (Figure 7.5a). Short-length cycles (i.e. two individuals having two partners in common) are also present within the network of relationships that were ongoing at the time of the survey (Figure 7.5b). For comparison, Bearman et al. (2004) did not find any short-length cycles among students of a US high school over an 18-month period.

Network location and HIV infection: Analysis of the position of HIV-positive individuals within the network indicates that the distribution of HIV is not homogeneous (Table 7.4), ranging from 3% in the bicomponents to 8.9% in the disjoint components and 10.8% in the branches of the giant component. While members of the bicomponents (Figure 7.4) are exposed to multiple pathways of potential HIV infection, HIV prevalence is highest in the sparser regions of the network—i.e. small components and the giant component outside the bicomponents. Several factors contribute to this apparently paradoxical distribution of HIV prevalence. First, the sociodemographic composition of the sparser regions of the network favours an increased prevalence of HIV as some groups who are more likely to be infected—e.g. older respondents, women, and widows—are over-represented in these regions. Second, the prevalence of several risk factors associated with HIV infection also varied significantly within the network (Table 7.4). For instance, the proportion of respondents having engaged in relationships outside of Likoma, or having engaged in a relationship with a partner older than 30 years, were higher in the sparser regions of the network. Relationships—in particular marital relationships—tend to be longer in the smaller components, and as a result concurrent partnerships might actually be more common in these structures. Table 7.4 also shows that the participation in HIV testing

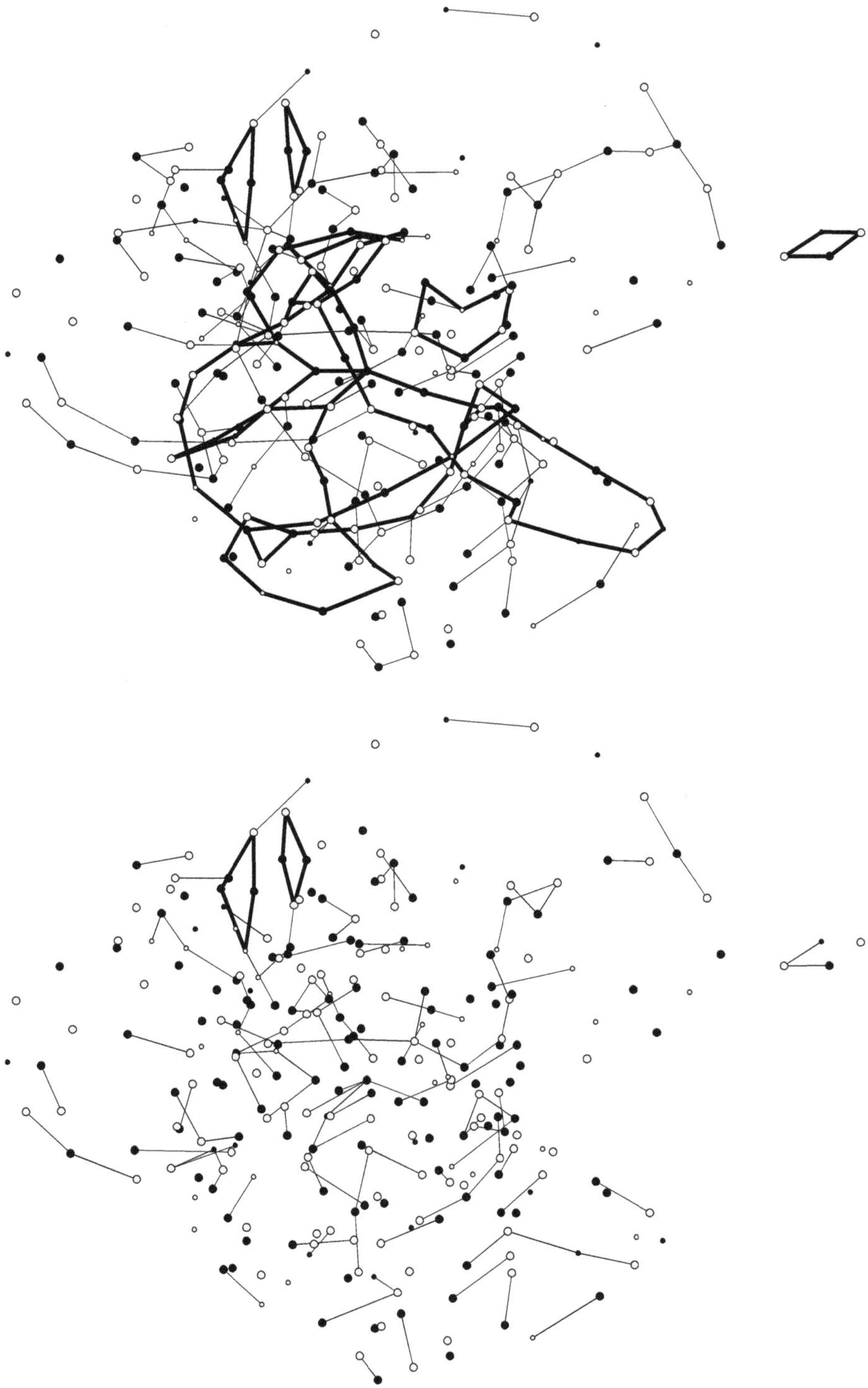

FIG. 7.5 Relationships within the bicomponents of Figure 7.4 that were active within one year prior to the LNS network survey (top panel) and at the time of the survey (bottom panel).

Table 7.4 Prevalence of HIV and risk factors for HIV infection across network locations

	Network Location		
	Small components	Main component	Bicomponents
Mean age (ICR)	27.5 (23,32)	23.4 (19,27)	22.1 (19,25)
Proportion of network members who are			
female	58.9	52.4	46.9[a]
never married	29.1	55.5	63.5[a]
divorced or widowed	10.6	3.9	3.15[a]
Average number of partners			
Males (Std Deviation)	2.1 (1.6)	2.3 (1.2)	3.8[a] (1.4)
Females (Std Deviation)	1.4 (0.8)	2.2 (1.1)	3.7[a] (1.3)
Health variables[b]			
Proportion of respondents who			
were tested for HIV	65.8	72.3	77.0[a]
were infected with HIV	8.9 (5.9,11.8)	10.8 (6.3,15.3)	3.0 (0.6,5.9)
reported symptoms of STIs	16.7 (13.1,20.4)	16.5 (12.1,20.8)	17.5 (12.3,22.6)
received an injection within one year prior to survey	34.2 (29.9,38.5)	40.1 (34.4,45.8)	35.0 (29.3,40.7)
were ever tested for HIV prior to this study	21.9 (18.0,25.9)	21.1 (15.9,26.3)	25.3 (19.7,30.8)
Sexual mixing variables[c]			
Proportion of respondents reporting			
any partner outside of Likoma	40.7 (36.5,44.9)	25.3 (20.6,30.0)	16.9 (14.5,19.3)
any partner above 30 years old	35.8 (32.6,38.9)	23.8 (20.0,27.6)	12.6 (10.7,14.6)

(*continued*)

Table 7.4 Continued

	Network Location		
	Small components	Main component	Bicomponents
consistent condom use	16.7 (12.9,20.6)	16.7 (11.9,21.5)	11.9 (8.8,15.0)
Proportion of relationships that are ongoing at survey among all			
marriages that started more than one year prior to survey	87.2 (83.2,91.2)	93.5 (88.6,98.4)	68.9 (60.2,77.5)
extra-marital relations that started more than one year prior to survey	28.1 (22.0,34.0)	25.6 (20.2,31.0)	21.6 (17.3,26.0)
extra-marital relations that started within one year prior to survey	43.7 (35.1,52.2)	42.7 (34.8,50.6)	40.9 (35.3,46.4)

Network locations: *small components:* members of small components (isolates, dyads, triads, etc.); *main component:* members of the giant component, outside the bicomponents (Figure 7.4a); *bicomponents* members of the bicomponents within the giant component (individuals connected by thick lines in Figure 7.4). 95% Confidence intervals are in parentheses unless otherwise noted.

[a] The difference in proportions between network locations is significant at the .05 level.

[b] Proportions are standardized by age, gender, and marital status.

[c] Proportions are standardized by age, gender, marital status, and number of partners.

Source: Helleringer and Kohler (2007).

differed across regions of the network, but this modest variation is unlikely to explain the differential HIV prevalence across regions of the network.

7.6 Conclusion

Social networks receive an increasing emphasis in theories of fertility change, and sexual networks have emerged as an important research theme for understanding the spread of HIV/AIDS. This interest in social and sexual network studies in demography has a pedigree in anthropology, sociology, economics, and epidemiology, and there is related contemporary work in all of these disciplines.

Our work for well over a decade in investigating the roles of interactions in social—and more recently also sexual—networks for important demographic behaviours has been based on data collected in Kenya and Malawi. This paper began by reviewing the theoretical arguments regarding why an explicit consideration of social/sexual networks is relevant for understanding fertility and health behaviour, and how interactions in social or sexual networks can affect the aggregate dynamics of fertility change or diseases such as HIV/AIDS. We then discussed some of the empirical challenges and approaches to collecting egocentric and sociocentric network data, and we illustrated some of the methodological challenges in using such data to establish causal relationships between social interactions and individual behaviours. And finally, this paper reviewed several analyses that have investigated the role of social networks in the diffusion of fertility control and in shaping perceptions about HIV/AIDS, and we discussed some of the key findings of the Likoma Network Study (LNS) that has provided the first large-scale sociocentric sexual network study in a high HIV prevalence context in sub-Saharan Africa.

References

Anderson, R. M. and R. M. May (1991). *Infectious Diseases of Humans: Dynamics and Control.* Oxford: Oxford University Press.

Andrews, J. A., E. Tildesley, H. Hops, and F. Li (2002). 'The Influence of Peers on Young Adult Substance Use', *Health Psychology* 21/4: 349–357.

Anglewicz, P. and J. Chitsanya (2011). 'Disclosure of HIV Status between Spouses in Malawi', *AIDS Care* 23/8: 998–1005.

Arends-Kuenning, M. (2001). 'How Do Family Planning Worker Visits Affect Women's Behavior in Bangladesh?', *Demography* 38/4: 481–496.

Bailey, N. T. J. (1975). *The Mathematical Theory of Infectious Disease.* New York: Hafner Press.

Bates, R. H. (1981). *Markets and States in Tropical Africa: The Political Basis of Agricultural Politics.* Berkeley: University of California Press.

Baum, C. F., M. E. Schaffer, and S. Stillman (2003). 'Instrumental Variables and GMM: Estimation and Testing', *STATA Journal* 3/1: 1–31.

Bearman, P. S., J. Moody, and K. Stovel (2004). 'Chains of Affection: The Structure of Adolescent Romantic and Sexual Networks', *American Journal of Sociology* 110/1: 44–91.

Becker, G. S., K. M. Murphy, and R. Tamura (1990). 'Human Capital, Fertility, and Economic Growth', *Journal of Political Economy* 98/5 (pt 2): 12–37.

Behrman, J. R., H.-P. Kohler, and S. C. Watkins (2002). 'Social Networks and Changes in Contraceptive Use over Time: Evidence from a Longitudinal Study in Rural Kenya', *Demography* 39/4: 713–736.

Bernardi, L. (2002). 'Determinants of Individual AIDS Risk Perception: Knowledge, Behavioural Control and Social Influence', *African Journal of AIDS Research* 1: 111–124.

Bernardi, L. (2003). 'Channels of Social Influence on Reproduction', *Population Research and Policy Review* 22/5–6: 527–555.

Bongaarts, J. and S. C. Watkins (1996). 'Social Interactions and Contemporary Fertility Transitions', *Population and Development Review* 22/4: 639–682.

Bott, E. (1971). *Family and Network: Roles, Norms and External Relationships in Ordinary Urban Families.* London: Tavistock Publications.

Bühler, C. and H.-P. Kohler (2003). 'Talking about AIDS: The Influence of Communication Networks on Individual Risk Perceptions of HIV/AIDS Infection and Favored Protective

Behaviors in South Nyanza District, Kenya', *Demographic Research* Special Collection 1/13: 398–438.

Bunnel, R. (1996). 'Promoting or Paralyzing Behavioral Change: Understanding Gender and High Levels of Perceived Risk of HIV Infection in Southwestern Uganda', doctoral dissertation, Harvard School of Public Health, Boston, MA.

Burt, R. S. (1982). *Toward a Structural Theory of Action: Network Models of Social Structure, Perception and Action*. London: Academic Press.

Burt, R. S. (1992). *Structural Holes: The Social Structure of Competition*. Cambridge, MA: Harvard University Press.

Carrington, P. J., J. Scott, and S. Wasserman (eds) (2005). *Models and Methods in Social Network Analysis*. Cambridge: Cambridge University Press.

Cerwonka, E. R., T. R. Isbell, and C. E. Hansen (2000). 'Psychosocial Factors as Predictors of Unsafe Sexual Practices among Young Adults', *AIDS Education and Prevention* 12/2: 141–153.

Christakis, N. A. and J. H. Fowler (2007). 'The Spread of Obesity in a Large Social Network over 32 Years', *New England Journal of Medicine* 357: 370–379.

Christakis, N. A. and J. H. Fowler (2009). *Connected: The Surprising Power of our Social Networks and How They Shape our Lives*. New York: Little, Brown & Co.

Clark, S., M. Poulin, and H.-P. Kohler (2009). 'Marital Aspirations, Sexual Behaviors, and HIV/AIDS in Rural Malawi', *Journal of Marriage and the Family* 71/2: 396–416.

Cleland, J. and C. Wilson (1987). 'Demand Theories of the Fertility Transition: An Iconoclastic View', *Population Studies* 41/1: 5–30.

Coale, A. J. (1986). 'The Decline of Fertility in Europe since the Eighteenth Century as a Chapter in Demographic History', in Ansley J. Coale and Susan C. Watkins (eds), *The Decline of Fertility in Europe*. Princeton: Princeton University Press, 1–30.

Committee on Population (National Research Council) and J. B. Casterline (ed.) (2001). *Diffusion Processes and Fertility Transition: Selected Perspectives*. Washington, DC: National Academies Press.

Durlauf, S. N. (2001). 'A Framework for the Study of Individual Behavior and Social Interactions', *Sociological Methodology* 31/1: 47–87.

Entwisle, B., K. Faust, R. R. Rindfuss, and K. Toshiko (2007). 'Networks and Contexts: Variation in the Structure of Social Ties', *American Journal of Sociology* 112/5: 1495–1533.

Estrin, D. (1999). 'In Ghana, Young Men's Condom Use Is Linked to Lack of Barriers: Perceived Susceptibility to HIV Infection', *International Family Planning Perspectives* 25/2: 106–107.

Fiori, K. L., T. Antonucci, and K. S. Cortina (2006). 'Social Network Typologies and Mental Health among Older Adults', *Journals of Gerontology Series B: Psychological Sciences and Social Sciences* 61/1: P25.

Fischer, C. S. (1982). *To Dwell among Friends: Personal Networks in Town and City*. Chicago: University of Chicago Press.

Fowler, J. H. and N. A. Christakis (2008). 'Dynamic Spread of Happiness in a Large Social Network: Longitudinal Analysis over 20 Years in the Framingham Heart Study', *British Medical Journal* 337/2338: 1–9.

Galor, O. and D. N. Weil (1996). 'The Gender Gap, Fertility and Growth', *American Economic Review* 86/3: 374–387.

Ghani, A. C. and G. P. Garnett (2000). 'Risks of Acquiring and Transmitting Sexually Transmitted Diseases in Sexual Partner Networks', *Sexually Transmitted Diseases* 27: 579–587.

Godley, J. (1999). 'The Influence of Sibling Ties on Women's Contraceptive Method Choice in Nang Rong, Thailand', *International Family Planning Perspectives* 27/1: 4–10, 41.

Goodreau, S. M., J. A. Kitts, and M. Morris (2009). 'Birds of a Feather, or Friend of a Friend? Using Exponential Random Graph Models to Investigate Adolescent Social Networks', *Demography* 46/1: 103–125.

Granovetter, M. S. (1973). 'The Strength of Weak Ties', *American Journal of Sociology* 78/6: 1360–1380.

Granovetter, M. S. (2005). 'The Impact of Social Structure on Economic Outcomes', *Journal of Economic Perspectives* 19/1: 33–50.

Handcock, M. S. and J. H. Jones (2004). 'Likelihood-Based Inference for Stochastic Models of Sexual Network Formation', *Theoretical Population Biology* 65/4: 413–422.

Hayashi, F. (2000). *Econometrics*. Princeton: Princeton University Press.

Heesterbeek, J. A. P. (2002). 'A Brief History of Ro and a Recipe for its Calculation', *Acta Biotheoretica* 50/3: 189–204.

Helleringer, S. and H.-P. Kohler (2005). 'Social Networks, Risk Perceptions and Changing Attitudes towards HIV/AIDS: New Evidence from a Longitudinal Study Using Fixed-Effect Estimation', *Population Studies* 59/3: 265–282.

Helleringer, S. and H.-P. Kohler (2007). 'Sexual Network Structure and the Spread of HIV in Africa: Evidence from Likoma Island, Malawi', *AIDS* 21/17: 2323–2332.

Helleringer, S. and H.-P. Kohler (2008). 'Cross-Sectional Research Design and Relatively Low HIV Incidence, Rather than Blood Exposures, Explain the Peripheral Location of HIV Cases within the Sexual Networks Observed on Likoma', *AIDS* 22/11: 1378–1379.

Helleringer, S., H.-P. Kohler, and A. Chimbiri (2007). 'Characteristics of External/Bridge Relationships by Partner Type and Location where Sexual Relationship Took Place', *AIDS* 21/18: 2560–2561.

Helleringer, S., H.-P. Kohler, A. Chimbiri, P. Chatonda, and J. Mkandawire (2009). 'The Likoma Network Study: Context, Data Collection and Initial Results', *Demographic Research* 21/15: 427–468.

Helleringer, S., H.-P. Kohler, and L. Kalilani-Phiri (2009). 'The Association of HIV Serodiscordance and Partnership Concurrency in Likoma Island (Malawi)', *AIDS* 23/10: 1285–1287.

Helleringer, S., H.-P. Kohler, and J. Mkandawire (2009). 'Increasing Uptake of HIV Testing and Counseling among the Poorest in Sub-Saharan Countries through Home-Based Service Provision', *Journal of Acquired Immune Deficiency Syndromes* 51/2: 185–193.

Helleringer, S. J. Mkandawire, L. Kalilani-Phiri, and H.-P. Kohler (2014). 'Cohort Profile: The Likoma Network Study (LNS)', *International Journal of Epidemiology* 43/2: 545–557.

Hensvik, L. and P. Nilsson (2010), 'Businesses, Buddies and Babies: Social Ties and Fertility at Work', Institute for Labour Market Policy Evaluation (IFAU) Working Paper Series 2010:9.

Hethcote, H. W., J. W. van Ark, and I. M. J. Longini (1991). 'A Simulation Model of AIDS in San Francisco: I. Model Formulation and Parameters', *Mathematical Biosciences* 106/2: 203–222.

Jones, J. H. and M. S. Handcock (2003). 'Sexual Contacts and Epidemic Thresholds', *Nature* 423/6940: 605–606.

Kahneman, D. and A. Tversky (1973). 'On the Psychology of Prediction', *Psychological Review* 80/4: 237–251.

Kaplan, C. P., A. Nápoles-Springer, S. L. Stewart, and E. J. Pérez-Stable (2001). 'Smoking Acquisition among Adolescents and Young Latinas: The Role of Socioenvironmental and Personal Factors', *Addictive Behaviors* 26/4: 531–550.

Kengeya-Kayondo, J. F., L. M. Carpenter, P. M. Kintu, J. Nabaitu, R. Pool, R. and J. A. G. Whitworth (1999). 'Risk Perception and HIV-1 Prevalence in 15,000 Adults in Rural South-West Uganda', *AIDS* 13/16: 2295–2302.

Kim, H. and P. S. Bearman (1997). 'The Structure and Dynamic of Movement Participation', *American Sociological Review* 62/1: 70–93.

Klandermans, B. (1992). 'The Social Construction of Protest and Multiorganizational Fields', in Aldon D. Morris and Carol McClurg Mueller (eds), *Frontiers in Social Movement Theory*. New Haven: Yale University Press, 77–103.

Klovdahl, A., J. Potterat, D. Woodhouse, J. Muth, S. Muth, and W. Darrow (1994). 'Social Networks and Infectious Disease: The Colorado Springs Study', *Social Science and Medicine* 38: 79–88.

Knodel, J. and E. van de Walle (1979). 'Lessons from the Past: Policy Implications of Historical Fertility Studies', *Population and Development Review* 5/2: 217–245.

Koehly, L., S. Goodreau, and M. Morris (2004). 'Exponential Family Models for Census and Sampled Network Data', *Sociological Methodology* 34: 241–270.

Kohler, H.-P. (1997). 'Learning in Social Networks and Contraceptive Choice', *Demography* 34/3: 369–383.

Kohler, H.-P. (2000). 'Fertility Decline as a Coordination Problem', *Journal of Development Economics* 63/2: 231–263.

Kohler, H.-P. (2001). *Fertility and Social Interactions: An Economic Perspective*. Oxford: Oxford University Press.

Kohler, H.-P. (2013). 'Population Growth', in Bjørn Lomborg (ed.), *Global Problems, Smart Solutions: Costs and Benefits*. Cambridge: Cambridge University Press, 510–580.

Kohler, H.-P., J. R. Behrman, and S. C. Watkins (2000). 'Empirical Assessments of Social Networks, Fertility and Family Planning Programs: Nonlinearities and their Implications', *Demographic Research* 3/7: 79–126.

Kohler, H.-P., J. R. Behrman, and S. C. Watkins (2001). 'The Density of Social Networks and Fertility Decisions: Evidence from South Nyanza District, Kenya', *Demography* 38/1: 43–58.

Kohler, H.-P., J. R. Behrman, and S. C. Watkins (2007). 'Social Networks and HIV/AIDS Risk Perceptions', *Demography* 44/1: 1–33.

Kolata, G. (2011). 'Catching Obesity from Friends May Not Be So Easy', *New York Times*, 8 August.

Kretzschmar, M. and M. Morris (1996). 'Measures of Concurrency in Networks and the Spread of Infectious Disease', *Mathematical Biosciences* 133/2: 165–195.

Lagarde, E., B. Auvert, M. Caraël, M. Laourou, B. Ferry, E. Akam, T. Sukwa, L. Morison, B. Maury, J. Chege, I. N'Doye, and A. Buvé (2001). 'Concurrent Sexual Partnerships and HIV Prevalence in Five Urban Communities of Sub-Saharan Africa', *AIDS* 15: 877–884.

Laumann, E. O., J. H. Gagnon, T. Michael, and S. Michaels (1994). *The Social Organization of Sexuality: Sexual Practices in the United States*. Chicago: University of Chicago Press.

Laumann, E. O., P. Marsden, and D. Prensky (1983). 'The Boundary Specification Problem in Network Analysis', in R. S. Burt and M. J. Minor (eds), *Applied Network Analysis: A Methodological Introduction*. London: Sage Publications, 18–34.

Laumann, E. O. and Y. Youm (1999). 'Racial/Ethnic Group Differences in the Prevalence of Sexually Transmitted Diseases in the United States: A Network Explanation', *Sexually Transmitted Diseases* 26/5: 250–261.

Liljeros, F., C. R. Edling, L. A. N. Amaral, H. E. Stanley, and Y. Aberg (2001). 'The Web of Human Sexual Contacts', *Nature* 411/6840: 907–908.

London, A. S. and R. Aroyds (2000). 'The Co-occurrence of Correct and Incorrect HIV Transmission Knowledge and Perceived Risk for HIV among Women of Childbearing Age in El Salvador', *Social Science and Medicine* 51: 1267–1278.

Lyngstad, T. H. and A. Prskawetz (2010). 'Do Siblings' Fertility Decisions Influence Each Other?', *Demography* 47/4: 923–934.

Magruder, J. (2008). 'Marital Shopping and Epidemic AIDS', unpublished working paper, Department of Agricultural and Resource Economics, University of California, Berkeley.

Manski, C. F. (1993). 'Identification of Endogenous Effects: The Reflection Problem', *Review of Economic Studies* 60/3: 531–542.

Manski, C. F. (2000). 'Economic Analysis of Social Interaction', *Journal of Economic Perspectives* 14/3: 115–136.

Manski, C. F. (2004). 'Measuring Expectations', *Econometrica* 72/5: 1329–1376.

Marsden, P. V. (1990). 'Network Data and Measurement', *Annual Review of Sociology* 16: 435–463.

Marsden, P. V and N. E. Friedkin (1993). 'Network Studies of Social Influence', *Sociological Methods and Research* 22/1: 127–151.

Massey, D. S., J. Arango, G. Hugo, A. Kouaouci, P. Adela, and J. E. Taylor (1994). 'An Evaluation of International Migration Theory: The North American Case', *Population and Development Review* 20/4: 699–751.

Mitchell, J. C. (1969). *Social Networks in Urban Situations: Analyses of Personal Relationships in Central African Towns*. Manchester: Manchester University Press.

Montgomery, M. R. (2000). 'Perceiving Mortality Decline', *Population and Development Review* 26/4: 795–819.

Montgomery, M. R. and J. B. Casterline (1996). 'Social Learning, Social Influence, and New Models of Fertility', *Population and Development Review* 22 (Supplement): 151–175.

Montgomery, M. R., G.-E. Kiros, D. Agyeman, J. B. Casterline, P. Aglobitse, and P. C. Hewett (2001). 'Social Networks and Contraceptive Dynamics in Southern Ghana', Population Council, Policy Research Division Working Paper 153.

Moody, J. (2002). 'The Importance of Relationship Timing for Diffusion', *Social Forces* 81: 25–46.

Moody, J., M. Morris, J. Adams, and M. Handcock (2003). 'Epidemic Potential in Human Sexual Networks: Connectivity and the Development of STD Cores', unpublished working paper, Department of Sociology, Duke University.

Morris, M. (1993). 'Epidemiology and Social Networks: Modeling Structured Diffusion', *Sociological Methods & Research* 22/1: 99–126.

Morris, M. and M. Kretzschmar (2000). 'A Microsimulation Study of the Effect of Concurrent Partnerships on the Spread of HIV in Uganda', *Mathematical Population Studies* 8/2: 109–133.

Morris, M. (ed.) (2004). *Network Epidemiology*. Oxford: Oxford University Press.

Munshi, K. D. (2003). 'Networks in the Modern Economy: Mexican Migrants in the US Labor Market', *Quarterly Journal of Economics* 118/2: 549–599.

Munshi, K. D and J. Myaux (2006). 'Social Norms and the Fertility Transition', *Journal of Development Economics* 80/1: 1–38.

Newman, M. E. J. (2002). 'Spread of Epidemic Disease on Networks', *Physical Review E* 66/1: 016128-1–016128-11.

Potterat, J. J., L. Phillips-Plummer, S. Q. Muth, R. B. Rothenberg, D. E. Woodhouse, T. S. Maldonado-Long, H. P. Zimmerman, and J. B. Muth (2002). 'Risk Network Structure in the Early Epidemic Phase of HIV Transmission in Colorado Springs', *Sexually Transmitted Infections* 78/1 (Supplement): 159–163.

Rabin, M. (1998). 'Psychology and Economics', *Journal of Economic Literature* 36/1: 11–46.

Reniers, G. (2008). 'Marital Strategies for Regulating Exposure to HIV', *Demography* 45/2: 417–438.

Rodgers, J. and D. Rowe (1993). 'Social Contagion and Adolescent Sexual Behavior: A Developmental EMOSA Model', *Psychological Review* 100/3: 479–510.

Rogers, E. M. (2003). *Diffusion of Innovations*. New York: Free Press.

Rothenberg, R. B., C. Sterk, K. E. Toomey, J. J. Potterat, D. Johnson, M. Schrader, and S. Hatch (1998). 'Using Social Network and Ethnographic Tools to Evaluate Syphilis Transmission', *Sexually Transmitted Diseases* 25: 154–160.

Sandberg, J., S. Rytina, V. Delaunay, and A. Marra (2012). 'Social Learning about Levels of Perinatal and Infant Mortality in Niakhar, Senegal', *Social Networks* 34/2: 264–274.

Simmel, G. (1922). *Conflict and Web of Group Affiliations*. New York: Free Press.

Smith, K. P. (2003). 'Why Are They Worried? Concern about AIDS in Rural Malawi', *Demographic Research* Special Collection 1/9: 279–317.

Smith, K. P. and N. A. Christakis (2008). 'Social Networks and Health', *Annual Review of Sociology* 34: 405–429.

Smith, K. P. and S. C. Watkins (2005). 'Perceptions of Risk and Strategies for Prevention: Responses to HIV/AIDS in Rural Malawi', *Social Science and Medicine* 60/3: 649–660.

Strauss, R. S. and H. A. Pollack (2003). 'Social Marginalization of Overweight Children', *Archives of Pediatrics and Adolescent Medicine* 157/8: 746–752.

Tawfik, L. and S. C. Watkins (2007). 'Sex in Geneva, Sex in Malawi', *Social Science and Medicine* 64/5: 1090–1101.

Valente, T. W. (1994). *Network Models of the Diffusion of Innovations*. Cresskill, NJ: Hampton Press.

Valente, T. W. (2005). 'Network Models and Methods for Studying the Diffusion of Innovations', in P. Carrington, S. Wasserman, and J. Scott (eds), *Recent Advances in Network Analysis*. Cambridge: Cambridge University Press, 98–116.

Watkins, S. C. (1991). *From Provinces into Nations: Demographic Integration in Western Europe, 1870–1960*. Princeton: Princeton University Press.

Watkins, S. C. (2004). 'Navigating the AIDS Epidemic in Rural Malawi', *Population and Development Review* 30/4: 673–705.

Watkins, S. C. and A. Swidler (2009). 'Hearsay Ethnography: Conversational Journals as a Method for Studying Culture in Action', *Poetics* 37/2: 162–184.

Watkins, S. C. and A. Swidler (2011). 'Hearsay Ethnography: A Method for Learning About Responses to Health Interventions', in Bernice Pescosolido, Jan McLeod, Jack Martin, and Anne Rogers (eds), *Handbook of the Sociology of Health, Illness, and Healing*. New York: Springer, 431–445.

Watkins, S. C. and I. Warriner (2003). 'How Do We Know We Need to Control for Selectivity?', *Demographic Research* Special Collection 1/4: 109–142.

Watkins, S. C., J. R. Behrman, H.-P. Kohler, and E. M. Zulu (2003). 'Introduction to "Research on Demographic Aspects of HIV/AIDS in Rural Africa"', *Demographic Research* Special Collection 1/1: 1–30.

Watts, D. J. (1999). *Small Worlds: The Dynamics of Networks between Order and Randomness*, Princeton Studies in Complexity. Princeton: Princeton University Press.

Watts, D. J. and S. Strogatz (1998). 'Collective Dynamics of "Small-World" Networks', *Nature* 393: 440–443.

Weinstein, N. D. and M. Nicolich (1993). 'Correct and Incorrect Interpretations of Correlations Between Risk Perceptions and Risk Behaviors', *Health Psychology* 12/3: 235–245.

Wellman, B. (1988). 'Structural Analysis: Form Method and Metaphor to Theory and Substance', in Barry Wellman and S. D. Berkowitz (eds), *Social Structures: A Network Approach*. Cambridge: Cambridge University Press, 19–61.

Zulu, E. M. and G. Chepngeno (2003). 'Spousal Communication about the Risk of Contracting HIV/AIDS in Rural Malawi', *Demographic Research* Special Collection 1/8: 247–277.

CHAPTER 8

POPULATIONS ARE COMPOSED ONE EVENT AT A TIME

JENNIFER A. JOHNSON-HANKS

8.1 Introduction

On the first day of an undergraduate introduction to demography, I lecture about how population size, structure, composition, and change arise through the interplay of only three processes: fertility, mortality, and migration. True. Even a truism. But at the same time quite untrue. Many of the kinds of populations of contemporary demographic interest arise at least as much out of individual events whereby people are sorted into and out of relevant categories defined by race and ethnicity, religion, class, education, or residence as they do out of dynamics of fertility, mortality, and migration. These subpopulations of demographic interest are thus socially constructed, not (primarily) in the sense of being objects of discourse or cultural reference, but in the literal sense of being built up differentially through sequential, socially meaningful actions of individual agents selecting themselves into or out of given population units through specific events, and in socially structured and meaningful ways. Some of these events are dramatic and visible (Do I carry this pregnancy to term? Do I convert to my husband's religion?). Others are more subtle and quotidian (Do I explain to this shopkeeper that I am not, contrary to his assumption, from the dominant ethnic group? Do I study for that exam tonight or put it off?). In both cases, however, individual actions may alter or reinforce recognized population units and boundaries. The units and boundaries emerge, change, and disappear as an aggregate consequence of often small-scale events. As a result, differences in vital rates between subpopulations are related to the processes through which those subpopulations are put together. This paper attempts to explore these processes and their implications for contemporary demographic theory.

The paper relates to two significant literatures. The first concerns categorization, classification, and social boundaries (see Abbott 1995; Lamont and Molnar 2002; Loveman and Muniz 2007; Szreter, Sholkamy, and Dharmalingam 2004). This elegant literature shows how, at an aggregate level, populations are made as much through political strategy joined to cultural constructions of sameness and difference as through fertility, mortality, and migration. Abbott (1995) argues that groups are constituted by their boundaries, rather than the reverse. In demography, this implies that any time we think about differences in rates between two groups, we should ask to what degree the differences themselves are responsible for constituting the groups. Szreter, Skolkamy, and Dharmalingam (2004) challenge demographers to recognize that our categories do not necessarily correspond to meaningful concepts, schemas, and categories in the populations we study. Categories, Szreter et al. argue, are powerfully influenced by cultural context and history, which matter for what we should measure and how we should measure it. Szreter et al. are particularly attentive to the ways that politics, culture, and history shape the categories themselves. In this paper I am more concerned with the very micro processes through which individuals enter and leave those categories, event by event, in specific conjunctures. Nonetheless, my position is—I believe—fully consistent with their work, as with Abbott (1995).

The second body of literature that is relevant to me here concerns selection bias. Selection in this sense means that initial differences between groups of people are themselves the cause of their group memberships, not the result. People who go to graduate school, are active in political parties, travel internationally, or use payday lenders have specific characteristics that led them to be in those groups. These groups, in turn, become the denominators for all kinds of other measures, and the consequences of being a member of the group get mixed up with the processes that lead certain people to be in the group in the first place. Hundreds if not thousands of papers in our best journals treat the kinds of processes through which individuals are sorted—or sort themselves—into population categories as examples of selection bias, that is, a data problem to be controlled for in statistical analysis, as with fixed-effects models or IV estimates (see, for example, Axinn and Thornton 1992; Berk 1983; Brand and Xie 2010; Chattopadhyay, White, and Debpuur 2006; Hofferth 2006; Western and Pettit 2005). Many of these are excellent, thoughtful papers, which make an important contribution. However, we need to push further on the topic of selection conceptually. In some cases, controlling for selection may obscure many of the key processes of demographic interest. Let us take an example from education, which provides a particularly good example of how attending to selection processes utterly changes our understanding of the nature of population, and to which I will return at some length in the next section. Many young African women seek to become educated partly *in order* to lead 'modern' reproductive lives of limited and late childbearing. In this case, it is a mistake to try to identify the exogenous effect of education on fertility controlling for aspirations. 'Selection' is not something to be excluded from analysis as so much noise—it *is itself the story*.

Some demographic theorists have begun to question whether we need always to control for selection as a matter of course, arguing in the tradition of Freedman (1991) that we should think more and calculate less. For example, Watkins and Warriner (2003) use rich longitudinal data from Malawi to show that women's chatting networks are in fact relatively random—that is, not the product of selection—and therefore statistical corrections for selection bias are not necessary, and may even be worse than nothing, because of the strong assumptions that the models require. The Watkins and Warriner result is significant here for two reasons. First, it shows that my premise has important limits: selection may be pervasive, but it is not universal. Second, it shows the importance of thinking sociologically and looking empirically at the processes through which groups are constituted. On this second point, my position is completely in agreement with that of Watkins and Warriner.

Particularly in complex societies with many subpopulations, populations are constituted processually, and simultaneously at different scales. At the individual level, people follow specific life trajectories, leading them into and out of various populations over time. People may convert to a new religion, get more education, reclassify themselves racially according to whom they marry, or move away from political beliefs that once tied them to a given party. These are the processes that are of most interest to me in this paper, and also the ones that often get called 'selection'. At a more macro level, each individual move incrementally alters both the category that the individual left and the one she entered, often both directly or indirectly: American Episcopalians become more liberal on average with the departure of each conservative parish to the American Anglican Council, but they also—indirectly—come to construe themselves as more liberal since they are now the liberal alternative in a socially meaningful dichotomy (this point echoes Abbott 1995). At a yet more macro level, there are occasional major realignments of population categories, as when whole racial categories on the US census are dropped and others added, or when national borders move as a result of war or treaty. Some 150 years after the Mexican-American war, certain Hispanic political organizers in California hold signs that read: 'We didn't cross the border—The border crossed us!' These large-scale realignments may or may not be the aggregate consequence of smaller-scale events, but either way, what results is a population of which the attributes, attitudes, and vital rates are distributed differently than before. Thus, the question 'who is in which group when?' is a fundamental question for population studies, and a surprisingly theoretical one. At all levels of aggregation, from the most micro of interactional setting to the most macro of global politics, populations are made one event at a time.

8.2 Kinds of Selective Process: An Illustration

Thinking about the events in which individuals are sorted into populations over time requires thinking processually, rather than in terms of static categories. It is noisier and messier. It pushes us away from examining the correlations between attributes, toward examining transitions—and even potential but unrealized transitions. Working with all that mess requires as much simplification as possible. To that end, let us begin with four oversimplified processes. In each case, people are differentially selected into categories A and B, which are in turn associated with outcomes α and β.

1. On the basis of some characteristic C, some people are assigned to category A and others to category B. But characteristic C is also associated with the propensity to outcomes α versus β.
2. Some people choose to do/become/enter A while others choose B. A is associated with α and B with β, but α and β are not relevant to people's choices of A and B.
3. Some people choose to do/become/enter A in order to do/become/enter α, while others choose B for β.

A fourth case is not technically selection, but has similar enough social dynamics that I include it here. Commonly called 'reverse causation', this is the case in which the purported outcome actually drives the categorization:

4. Some people have outcome α, making them eligible for A, while others have outcome β and are thus sorted into category B.

To make sense of these, we need an example. Education is a good category to think with in relation to selection. It is associated with a lot of positive outcomes, many of which (higher income, better health, lower mortality) are quite stable across a wide range of social and economic contexts. It is very complicated, involving a large number of actors and structures. And (those of us who teach hope at least) it is not only a selection machine, but also has content, obfuscating some of the selection processes that we want to look at here. Let us take as our example one of the outcomes stably associated with education: later first births. This one is not necessarily positive, and indeed contributes significantly to unintended childlessness among professional women in rich countries. But it is a relatively clear example with rich data and some interesting social dynamics.

Study after study has shown that women with more education bear fewer children overall and postpone first childbearing relative to women with less education (see Adamchak and Ntseane 1992; Ainsworth et al. 1996; Bledsoe et al. 1999; Castro Martin 1995; Lappegard and Ronsen 2005; Musick et al. 2009). There are a number of ways that this could happen as a direct result of education itself—say, if schools teach girls how to use contraception, or keep them so busy that they have no time to meet boys. As educators, we certainly hope that the education we offer has some substantial direct effects on the life chances of our students. Here, however, my focus is on the processes and events of selection that could also contribute to—or potentially completely account for—the association between more schooling and later entry into motherhood even if education itself is entirely vacuous. The first kind of selection is when people are sorted into categories on the basis of some characteristic that is also associated with the outcome of interest. This form is commonly called 'omitted variable bias'. For example, girls who go to school longer might come disproportionately from families in which late and limited childbearing is already the norm. Similarly, if school performance is correlated with overall competence or confidence, girls who do better in school might both stay in school longer and also be more able negotiate contraceptive use and therefore avert early births. In both cases, there is some characteristic 'C' (for example: family norms, general competence) that accounts both for the sorting into educational category and for the reproductive 'outcome'.

Now that description is fine as far as it goes. It is not hard to imagine how, on average, mothers who give birth later in life might tend to inculcate their daughters with the same inclinations, and might also invest more heavily in keeping their girls in school. But if that is the case, the average arises out of specific events—specific instances in which two courses of action were possible, but only one was taken. In canonical demographic data, we can observe very few of these events, and only the most dramatic. In southern Cameroon in the late 1990s, for example, we can observe that when schoolgirls get pregnant, those with more educated mothers are more likely to abort and return to school than are schoolgirls with less educated mothers (see discussion in Johnson-Hanks 2006). But the events that lead to these different outcomes are far more numerous, and for the most part far less dramatic, than we can capture in survey data. Each time the daughter and mother talk about school; each time they jointly imagine the daughter's future; each time the mother makes a decision about whether to save some money for her daughter's school uniform, or spend it on something else: these are kinds of events that aggregate up to differential selection understood in this first sense.

The next two processes in our list are both driven by the choices of the girls themselves. At young ages, students are not for the most part making their own choices about schooling, but this effect becomes clear if we think about college or especially graduate school. The young women who decide to continue their college education clearly differ in important ways from those who do not (process two). If nothing else, their education is more important to them, and perhaps also their associated careers. Even if they want children just as much as do the women who leave school, they want careers more intently—and for some proportion of those who stay in school, their more intent wanting of careers means that career comes before family formation. Some far smaller number stay in school specifically in order to postpone childbearing (process three), or perhaps more accurately to become the kind of person who tends to end up with postponed childbearing: an economically successful, cosmopolitan, world-wise person. This is the case mentioned in the introduction. While we may have practical reasons for needing to identify the exogenous effect of schooling, controlling for person-specific characteristics, say, in a cost–benefit analysis of building schools compared to wells, as theoretical social science the question becomes nearly nonsensical: it is *precisely* the person-specific characteristics that make it possible for schooling to have the effects it does.

Processes two and three often occur in tandem. This is very clear in the educational aspirations of young Cameroonian women whom I interviewed in the late 1990s and early 2000s. For the most part, they are not going to school in order to have fewer children or have them later, but they are going to school to be the kinds of 'modern', urban, professional women who fit that reproductive profile. For example, in response to the question 'why is it important to you to go to school?' one young woman explained:

> Women who have been to school, they are very different from drop-outs, you see? They know how to carry themselves. They have a certain posture. You will not find an educated woman selling in the street markets. You will not find her with too many children. You will not find her in ragged, ugly clothes. No! An educated woman does not do just anything ('n'importe quoi'). She knows what is what. So that is why it is important to stay in school.

Another made a very similar point. School is important, she explained, because of the kind of future it offers you: legally married, living in the city, enjoying the advantages of a cash income and a nice house. 'We [who go to school] are the modern ones! They [who do not attend high school] don't know anything about what the future can bring.'

Here again, however, the discourse matters for demographic outcomes only insofar as it is translated into concrete events, and these events will often be on a small scale, outside of easy view in the kind of survey data we so often use. What matters is not simply that schoolgirls construe of themselves as more modern (and therefore dignified) than their uneducated counterparts, but that that construal leads them to actions that postpone births. Bourdieu (1984) argued that taste classifies the classifier. Here, I am suggesting that a social actor's construal of the social world locates them in relation to that world, with sometimes important consequences.

The final kind of 'selection' is reverse causation, where the nominal outcome actually drives the categorization in the first place. Reverse causation is common in many domains of social life, and is commonly recognized as bias in quantitative research. In the case of education and fertility, reverse causation happens because continuing in school after having a child is so difficult. Indeed, sometimes just being in the kinds of circumstances that often lead to first children—a committed, particularly cohabiting or married, relationship—is already less compatible with continued education. If many of the women who have early first births drop out of school, then the more educated women will necessarily have later births on average. In many countries in sub-Saharan Africa, formal or informal policies discourage young mothers from continuing in school; reverse causation likely accounts for some fraction of the association between education and reproductive practice (but see Lloyd and Mensch 2008).

Taken together, these four kinds of selective processes define a space of possibility. On one axis, selection can be driven more by the intentional actions of the person in question or more by the other social forces. On the other axis, selection into categories may be blind to the outcome of interest, partially influenced, or even completely driven by it. In most real-world cases, the processes are mixed, with the intentional actions of individuals synchronized with the selective pressures of institutions, and the outcomes sometimes more and sometimes less directly held in view.

8.3 Selection as a Problem in Sociological Theory

Much of the time, different processes of selection work in conjunction. People choose what social structures would in any case suggest, because they choose what they know or like what they are good at. For example, the population who wants to go to college overlaps considerably with the population of people that colleges would be inclined to admit. People's aspirations are often relative to those of their close peers, and so peer groups sometimes act as if in tandem, much to their own surprise. And certain social activities or social positions tend to cluster with others. For example, particularly for whites in the US, the population of people who regularly attend church overlaps far more with the population of people living in the suburbs and exurbs than it does with the population of urban dwellers. The population of people with advanced degrees overlaps considerably with the population of high-income earners. The population of people with school-aged children largely overlaps with the population of people with elderly parents.

Social theory of the last 50 years has regularly drawn attention to social processes and practices that produce these alignments, but the relationships between these theories and quantitative approaches to addressing selection bias are more rarely explored. The mutual alignment of 'choices' or 'preferences' on the one hand and social structures on the other—as when people who excel in school choose to pursue advanced degrees—plays a key role to two important schools of thought: Bourdieu's theory of practice (1977, 2000) and Giddens' theory of structuration (1984), the latter particularly as reformulated by Sewell in his dual theory of structure and 'eventful' history (2005).

8.3.1 Bourdieu: A Theory of Practice

Through his theory of practice, Bourdieu articulates the mutual constitution of a social field, the forms of capital that circulate through that field, and the habitus of the field's participants. A field for Bourdieu is a social arena in which actors manoeuvre, a space of positions and position takings, such as the field of fine art, formal education, or the law. Habitus refers to the patterns of habit and inclination that a person has acquired from repeated interactions in a field; it consists of 'systems of durable, transposable dispositions' (1977: 72). People are inculcated with habitus through their experiences in fields; this habitus in turn inclines them to act in certain ways and not others; these forms of action give life to, and thereby reinforce, the structure of the field. People act effectively when their habitus is aligned to the field in which they are acting. What comes 'naturally' to them is then precisely what is called for; their native intuitions correspond to what their interaction partners both want and expect; they have a 'feel for the game' and can act fluidly without having to explicitly reflect on the rules. Thus the adage, 'l'habitude est une seconde nature'. The form of action that derives from a well-attuned habitus Bourdieu calls 'practice', actions 'which can be adapted to their outcomes without presupposing a conscious aiming at ends or a conscious mastery of the operations necessary to attain them' (1990: 53). Capital for Bourdieu is similar to, but much more extensive than, economic capital as commonly understood: that is, as resources that are unequally distributed and that act as opportunities and constraints for action. What is specific to him is that capital is a characteristic of fields, and its distribution is related to the positions and position takings of the field.

If we think back to the four kinds of selection discussed earlier, it is clear that Bourdieu's practice theory offers an elegant way of explaining why institutional and self-selection so frequently occur in tandem. If the habitus, or predisposition to certain forms of action, is the product of position in an institutional field, then they will almost always align. But practice theory also offers a productive way of thinking about each of the first three kinds of selection independently. For example, the first type of selection, omitted variable bias, is exemplified by the educated mother who both inculcates her daughter with a parenting style that leads her to have few children and also sends her to advanced levels of school. The omitted variable here could be called something like 'family background'. But of course, 'family background' is far more complex than a single variable, even if we were to find a moderately good measure of it that we could use in a quantitative model. Rather, 'family background' refers to position in a social field: the origin of habitus.

8.3.2 Sewell: Dual Structure Theory

Building on the work of Anthony Giddens and Marshall Sahlins, William Sewell proposes a different parsing of the terrain also covered by Bourdieu. Sewell argues that an understanding of action, including demographically relevant action such as getting pregnant or having an abortion, requires us to think about action in social context. Society is made up of 'sites of a multitude of overlapping and interlocking cultural structures' (2005: 209). By 'multitude', Sewell means many—at least hundreds, probably thousands—of structures that make up a society. What is a structure? A durable set of actions, ideas, and things. They vary in scale, from dating and family dinners to systems of social services and economic exchange. They are 'overlapping and interlocking' not only with respect to the domains of action to which they refer, but also in relation to the individuals and groups that populate societies. In addition, Sewell argues that this structure has a dual character.[1] It is 'composed simultaneously of schemas, which are virtual,

[1] Most of what follows Sewell shares with Giddens, but Sewell is both clearer and more careful about his terms. For example, what Sewell calls 'schemas', Giddens calls 'rules', although they are quite un-rule-like.

and of resources, which are actual' (2005: 136). Schematic elements (such as ideas, values, and 'habits of mind') and material elements (such as objects and performances) propel, support, constrain, and transform each other in tangible ways over time.

Schemas are the largely underdetermined and often taken-for-granted ways of perceiving and acting through which we make sense of the world and motivate our actions. They include 'the array of binary oppositions that make a given society's fundamental tools of thought, but also the various conventions, recipes, scenarios, principles of action, and habits of speech and gesture built up with these fundamental tools' (2005: 131). Resources, by contrast, are 'actual', by which Sewell means materially perceptible: for example, buildings, roads, books, boxes, or money. As does Bourdieu, Sewell stresses the mutual constitution of these two key concepts. Schemas are built up through interactions with resources; resources are made meaningful through the schemas that enliven them.

Structures are not fixed for Sewell, but constantly being made and remade in 'conjunctures' or temporary arrangements of circumstance. But conjunctures are not only the loci for the reproduction and transformation of structure, but also the proximal contexts of social action—and therefore of vital events (Sewell 2005: 220–223, cf. Bourdieu 1977: 78). Social actors make sense of what is going on in a given conjuncture by reading them through the schemas that are accessible to them. This process of construal is widely studied in psychology (see, for example, Prasada, Ferenz, and Haskell 2002). It consists of the usually non-volitional perception of a state of affairs as a case of a familiar category. Construals are routinized, but not fixed; we have access to multiple possible construals, and yet structure makes particular construals more accessible than others. In many circumstances of everyday life, a single construal is highly regularized, so that consociates or onlookers would consider any alternative aberrant. When you reach the front of a long checkout line in the drugstore, 'service transaction' is the only viable construal. However, in other circumstances, more than one construal would be possible, opening up uncertainty regarding which schemas social actors will employ in a given situation, and how. For example, is this a collegial coffee break? A chance to discuss my recent performance in the company? A date? Sexual harassment? Because it may be any of these, or several at once, and because both parties may reinterpret the event as it unfolds and even afterwards, some kinds of conjunctures provide likely arenas for structural change.

Here we see how Sewell's approach recasts canonical thinking about selection. What does it mean to be educated? For Sewell, education is a complex structure, associated with specific schemas and resources. How do young people enter this category? They confront a series of conjunctures that they construe according to schemas available to them, deploying resources to advance a line of action. For example, the fourth kind of selection that we considered, usually called reverse causation, is often cited as contributing to the correlation between schooling and fertility in Africa. As many schools either officially or unofficially prohibit attendance by young women who are pregnant or have children, early childbearing may exclude some women from attaining advanced education. But if we think about the unintended pregnancy of an unmarried student in Sewell's terms, we will see that 'reverse causation' is rarely so clear.

Imagine two young women who get pregnant unintentionally while still in school. One has a strong set of schemas that equate life chances with education, and she considers herself the sort of person who can succeed in school, and assumes that she will marry well and live in the city. Her resources are also robust: educated parents with social ties to professionals, including doctors. With her mother's help, she aborts the pregnancy and stays in school. A second young woman has different schemas and resources. She is unfamiliar with people who are highly educated, and construes herself as someone who will marry a farmer and stay in the village, even if she hopes for better. Her parents have little to offer by way of help, but the father of the pregnancy is excited and happy, and offers to marry her. She leaves school to move in with him. Although her leaving school was precipitated by the pregnancy, it would be strange to call the pregnancy the *cause* of her educational trajectory, since the same cause in the first case had no such result. Rather, the outcome is the joint effect of schemas, resources, and the specific conjunctures in which they are enacted.

Although the frameworks of Bourdieu and Sewell have much in common, they do not translate easily one to the other. The habitus is not quite the sum of a person's schemas. Capital is not quite a set of resources. And a field is not merely a structure of virtual and actual components. Each vocabulary makes some phenomena easy to describe and others hard, and each leaves some important questions open. What matters here, however, is that in either framework, we should expect people to select themselves into and out of denominators in systematic and meaningful ways, because their 'choices' are at once voluntary and prestructured by their experiences. Indeed, we might think of the term 'denominator' in a dual sense: the base population for a rate or ratio, of course, but also as a group into which they name themselves, from the transitive verb 'denominate'.

In *Pascalian Meditations*, Bourdieu summarizes how action may correspond to the demands of context without being either rule-bound or conforming to the requirements of utility maximization. (Note that here he refers to habitus as 'systems of schemes', suggesting one way that the translation between the two approaches might begin.) Social agents are endowed with a habitus, inscribed in their bodies by past experiences. These systems of 'schemes of perception, appreciation, and action enable them to perform acts of practical knowledge, based on the identification and recognition of conditional, conventional stimuli to which they are predisposed to react; and, without any explicit definition of ends or rational calculation of means, to generate appropriate and endlessly renewed strategies, but within the limits of the structural constraints of which they are the product and which define them' (Bourdieu 2000: 138). The frameworks of Bourdieu and Giddens/Sewell suggest that what people will want, or try to do, or crave, will be structured by the same forces that establish their access to those things. Thus, what analysts treat separately occurs in tandem, and is experienced simultaneously. People tend to want what they know, like what they are good at, and value the systems of value that value them in turn.

8.3.3 Blau: Structural Consolidation

Bourdieu and Sewell both offer ways for thinking about the mutual alignment of what people choose for themselves and what social structures make accessible, that is, the systematic coordination of self-selection with structural selection. But it is also important to think about the ways that different kinds of structural selection overlap. In statistical terms, overlapping structural selection means a high correlation between many covariates in a model. But sociologically, it is more useful to think in terms of overlapping pools of exposure to risk, that is, when the same people are simultaneously members of differently defined categories, of denominators in rates applying to different social groups. One way of approaching this aspect of selection comes from Peter Blau's work on social structure (1974, 1977), and particularly his concept of structural consolidation. Blau argued that sociology's proper object of study is not the characteristics of individuals, but the ways in which that variation is structured:

> We have been much concerned with the characteristics and behavior of persons, yet little with the forms and degrees of differentiation among them, which constitute the specific structural problems. The subjects of structural inquiry are, for instance, ethnic heterogeneity, not ethnic background; political differentiation, not political opinions; the division of labor, not occupational performance; income inequality, not poverty. (Blau 1974: 616–617)

Blau suggested a narrow definition of social structure, as 'the distributions of a population among different social positions that reflect and affect people's relations with one another' (Blau 1977: 28). Given that people simultaneously inhabit multiple positions (related to religion, education, employment, etc.), societies 'can be defined as a multi-dimensional space of social positions among which people are distributed and which affect their social relations' (Blau 1977: 28)—for example, because people are more likely to associate with others in the same or similar social positions to themselves: that is, homophily. Homophily has become a central concern in the burgeoning literature on social networks (see Kossinets and Watts 2009 for a nice example); network ties are considerably more likely among those similarly positioned in Blau's multidimensional social space.

Blau notes that in a given society, some parameters (or axes in this multidimensional space) are strongly correlated, while others are less so, or even not at all. For our purposes here, it is the correlated parameters that matter: the fact that, whatever might be theoretically possible, in the real world some areas of Blau's multistate space are densely populated and other areas completely void.[2] This phenomenon is called structural consolidation. Sometimes, it occurs because one parameter is functionally related to the other one, such as in the correlation between education and income in most societies. In other cases, the two parameters are related through a shared relationship to values or ethics, as in the case of politics and religion. The nearly universal association between

[2] Abbott (2001) makes a very similar point about this in relation to regression analysis.

women's education and fertility suggests that these axes are relatively consolidated in most societies, although there is considerable debate whether this occurs for the first or second reason. Whatever the reasons for their correlation, once two parameters are associated, it means that the same people are contributing events and exposure in each, and trends in the one will necessarily be echoed by trends in the other.

We can again take an example from education and fertility. In Cameroon, educated women are largely of Bulu-Beti-Fang ethnicity, Catholic religion, live in the south of the country, and marry late. Less educated women are concentrated among Muslim Northerners, who keep cattle and marry young. When we observe distinctive reproductive patterns of educated and uneducated women, we are simultaneously observing differences by religion, ethnicity, region, and so on (Johnson-Hanks 2003). Statistical controls for those other factors make it arithmetically possible to parcel out the associations, but the factors are lived in tandem. Structural consolidation is a very real part of how social life is organized and experienced.

In some cases, the superposition of different subpopulations is not coincidence or artifact, but the product of intentional projects. Through all sorts of mechanisms, people intentionally put themselves into social categories in order to be subject to the different risks that apply there. If you want to stop drinking, you stop hanging around your hard-drinking friends. If you want to find a potential husband, you put yourself in contexts where appropriate single men are likely to go. Blau focuses on the effects of social structure, largely taking that structure as given. Bourdieu and Sewell focus mostly on the ways in which we are unwittingly inculcated with schemas of perception and action by the contexts in which we find ourselves. Those processes are important, and ordinary people know that, and therefore may seek out contexts quite wittingly in order to inculcate themselves with alternative schemas of perception and action.

This intentional self-sorting in order to increase or decrease risk is both common and intuitive. It is also an important tool of most ethics training. When I was young, I was taught that the first step in sin was to be in the wrong place at the wrong time, that is, to put yourself in a context in which sinful actions are more likely. In this same vein, an online catechism explains that, 'Sin is all about motive. Deep down, why were you there, and not somewhere else?' The idea that self-selection into different denominators is an ethical project is also not limited to Christianity. Mahmood (2005) analyses how Muslim women participating in the Cairo mosque movement seek to attain piety through the intentional shaping of their own desires, actions, and wills, which they achieve by repeatedly putting themselves in circumstances that are likely to have the desired effects on their bodies and minds. These women seek to cultivate the inclination for specific emotional states through the structured performance of physical behaviours associated with those states, which Mahmood calls 'rehearsed spontaneity'. In other words, action that appears—and indeed is experienced as—'authentic' or 'spontaneous' is the product of practice and training that begins with the selection of contexts in which to act.

This kind of intentional self-sorting, or self-selection through process three, described at the beginning of the paper, occurs in all kinds of contexts: basically any time a social status is locally construed as being both attainable and associated with desirable outcomes, some people will try to transform themselves into members of the status to achieve the outcomes. Going to school in order to acquire the habits, disposition, and social status of an 'educated woman'—one who lives in the city, works only in clean work, marries well, and has few children—is perhaps the clearest example of this.

8.4 Selection in Time

We have seen so far that people are differentially selected into various denominators both by circumstances and their own actions. Indeed, people sometimes even act specifically in order to put themselves into particular subpopulations associated with specific profiles of risk and rates. So individual characteristics matter for understanding how people are sorted into different denominators. However, selection also works in another way: rather than on the characteristics of people, it can also work on the conjunctures in which people find themselves. Think of the population of people in a hospital waiting room, subway station, funeral home, or grocery store checkout line: in these instances, what is selected for is not so much people with certain traits, habits, inclinations, or skills, but rather people in particular situations. Over time, many different people will move into the relevant situation, and therefore into the population of interest, while others move out. Our category is the tip of a moving wave, and the processes of selection here are temporary and conjunctural.

This notion of temporal selection has been important in thinking about fertility and fertility limitation. If we model a woman's reproductive years as a series of statuses (pregnant, post-partum infecund, fecund), each of which she can enter multiple times and each of which she can leave, then we can think about contraceptive use in a quite different way than if we considered all women together. In particular, it makes the differences in contraceptive use rates between high- and low-fertility countries appear far less dramatic, because in high-fertility countries, a much smaller proportion of women will, at any given time, be in the status 'fecund' and therefore potential contraceptive users. Imagine two populations, one with a total fertility rate of two children and the other with a total fertility rate of six. For simplicity, assume that in both populations, fetal loss is zero (a totally unreasonable assumption that makes the example easier), and that the duration of childbearing lasts from age 20 until age 45 (a less heroic assumption). In the first population, women breastfeed for only three months on average, which—combined with rich diets and limited physical labour—leads to an average experience of only three months of post-partum infecundibility. In the other population, women breastfeed for a year on average, and are infecund the entire time.

It is immediately obvious that in the low-fertility population, women spend a considerable amount of time in the fecund state: any given woman is often in a situation where she is neither pregnant nor infecund, and therefore might contracept. Viewed from the level of the population, at any given time, a large proportion of women are fecund, and therefore potential contraceptive users. In the high-fertility population, the situation is quite different. In order to fit in six pregnancies between age 20 and 45, women experience far fewer durations in which they are fecund and exposed to the risk of pregnancy—or contraception: 174 months as opposed to 276 in the other population. Viewed from the perspective of who is potentially at risk for contracepting, even a twofold difference in total contraceptive prevalence in the two populations is therefore not surprising.[3] The tip of one wave is much larger than the tip of the other.

Although classical demography has attended to temporal selection, as in the preceding example (or Ryder's observation that the relevant fertility behaviour is 'whether to permit the next ovulation to come to fruition'; 1973: 503), more and more social demography focuses on the associations between presumably stable attributes of individuals, rather than on the contexts and conjunctures in which the individuals find themselves (see Johnson-Hanks et al. 2011 for an elaborated critique). This leads us away from understanding some of the most important processes underlying population variation, which concern who is in which groups when. The idea that selection works on people's short-term circumstances as much as it does on their attributes pushes us to think about the multiple kinds of temporalities of social phenomena: populations have specific temporal trajectories, as do the people moving through and across them. Sometimes, the temporal patterns at different levels of aggregation align. Sometimes they do not. But in any case, who is in what population at what time requires far more attention than it has received of late.

8.5 Conclusion

The last several decades have seen demography move further away from formal analysis toward 'social' demography, a much more observational and less model-driven kind of research. New, richer, individual-level data have become increasingly available, along with faster computers and more accessible software. The move toward more 'social' demography has been associated with increasing interest in the differences between subpopulations, as defined for example by race, class, religion, or education, often using regression models to identify individual-level correlates of differential outcomes. However, we have seen much less interest at the theoretical level in the processes through which the central subpopulations are made. In this paper, I have tried to argue that these subpopulations are constructed one event at a time, as individuals differentially select themselves into and out of them, thereby reinforcing or disrupting their boundaries.

[3] Clearly there is a chicken-and-egg problem here. The high-fertility population is able to have high fertility because contraceptive use rates are low, and high fertility in turn makes low contraceptive use inevitable. The problem can be thought in either direction.

Whereas global population is rightly modelled as the product only of natural increase, and some national or continental populations can be treated as explainable through natural increase and net migration, most other populations have boundaries that emerge, change, and disappear as an aggregate consequence of often small-scale events. In many cases, people's motion into and out of subpopulations is either structurally overdetermined or personally meaningful. Pregnant Mexican women crossing the border to give birth to a US citizen; young American families moving out of central cities to suburbs with better public (that is, state-funded) schools; third-generation Mexican Americans declining to categorize themselves as 'Hispanic' once all their children attend school and no longer speak fluent Spanish: even if the numbers of individuals crossing subpopulation boundaries in these ways are small, they are highly consequential for the differential rates we then calculate for the different groups. As a result, differences in vital rates between subpopulations are often related to the processes through which those subpopulations are put together. In many of the cases of contemporary demographic interest, selection should not be treated as a confounder to be controlled in a statistical model. Events of selection, over time and in distinct conjunctures, are instead among the central demographic processes we need to understand.

References

Abbott, A. (1995). 'Things of Boundaries', *Social Research* 62/4: 847–872.

Abbott, A. (2001). *Time Matters: On Theory and Method.* Chicago: University of Chicago Press.

Adamchak, D. and P. G. Ntseane (1992). 'Gender, Education and Fertility: A Cross-National Analysis of Sub-Saharan Nations', *Sociological Spectrum* 12: 167–182.

Ainsworth, M., K. Beegle, and A. Nyamete (1996). 'The Impact of Women's Schooling on Fertility and Contraceptive Use: A Study of Fourteen Sub-Saharan Countries', *World Bank Economic Review* 10/1: 85–122.

Axinn, W. G. and A. Thornton (1992). 'The Relationship between Cohabitation and Divorce: Selectivity or Causal Influence?' *Demography* 29/3: 357–374.

Berk, R. (1983). 'An Introduction to Sample Selection Bias in Sociological Data', *American Sociological Review* 48/3: 386–398.

Blau, P. M. (1974). 'Presidential Address: Parameters of Social Structure', *American Sociological Review* 39/5: 615–635.

Blau, P. M. (1977). 'A Macrosociological Theory of Social Structure', *American Journal of Sociology* 83/1: 26–54.

Bledsoe, C., J. Casterline, J. Johnson-Kuhn, and J. Haaga (1999). *Critical Perspectives on Schooling and Fertility in the Developing World.* Washington, DC: National Academy Press.

Bourdieu, P. (1977). *Outline of a Theory of Practice.* Cambridge: Cambridge University Press.

Bourdieu, P. (1984). *Distinction: A Social Critique of the Judgment of Taste.* Cambridge, MA: Harvard University Press.

Bourdieu, P. (1990). *The Logic of Practice.* Stanford: Stanford University Press.

Bourdieu, P. (2000). *Pascalian Meditations.* Stanford, CA: Stanford University Press.

Brand, J. and Y. Xie. (2010). 'Who Benefits Most from College? Evidence for Negative Selection in Heterogenous Economic Returns to Higher Education', *American Sociological Review* 75/2: 273–302.

Castro Martin, T. (1995). 'Women's Education and Fertility: Results from 26 Demographic and Health Surveys', *Studies in Family Planning* 26/4: 187–202.

Chattopadhyay, A., M. White, and C. Debpuur (2006). 'Migrant Fertility in Ghana: Selection versus Adaptation and Disruption as Causal Mechanisms', *Population Studies* 60/2: 189–203.

Freedman, D. A. (1991). 'Statistical Models and Shoe Leather', *Sociological Methodology* 21: 291–313.

Giddens, A. (1984). *The Constitution of Society*. Berkeley: University of California Press.

Hofferth, S. (2006). 'Residential Father, Family Type, and Child Well-Being: Investment versus Selection', *Demography* 43/1: 53–77.

Johnson-Hanks, J. (2003). 'Ethnicity, Education and Reproductive Practice in Contemporary Cameroon', *Population* 58: 171–200.

Johnson-Hanks, J. (2006). *Uncertain Honor: Modern Motherhood in an African Crisis*. Chicago: University of Chicago Press.

Johnson-Hanks, J., C. Bachrach, S. P. Morgan, and H.-P. Kohler (2011). *Understanding Family Change and Variation: Toward a Theory of Conjunctural Action*. Berlin and New York: Springer.

Kossinets, G. and D. Watts. (2009). 'Origins of Homophily in an Evolving Social Network', *American Journal of Sociology* 115/2: 405–450.

Lamont, M. and V. Molnar (2002). 'The Study of Boundaries in the Social Sciences', *Annual Review of Sociology* 28: 167–195.

Lappegard, T. and M. Ronsen (2005). 'The Multifaceted Impact of Education on Entry into Motherhood', *European Journal of Population* 21/1: 31–49.

Lloyd, C. and B. Mensch (2008). 'Marriage and Childbirth as Factors in Dropping out from School: An Analysis of DHS Data from Sub-Saharan Africa', *Population Studies* 62/1: 1–13.

Loveman, M. and J. O. Muniz (2007). 'How Puerto Rico Became White: Boundary Dynamics and Inter-Census Racial Reclassification', *American Sociological Review* 72/6: 915–939.

Mahmood, S. (2005). *Politics of Piety: The Islamic Revival and the Feminist Subject*. Princeton: Princeton University Press.

Musick, K., P. England, S. Edgington, and N. Kangas (2009). 'Education Differences in Intended and Unintended Fertility', *Social Forces* 88/2: 543–572.

Prasada, S., K. Ferenz, and T. Haskell (2002). 'Conceiving of Entities as Objects and as Stuff', *Cognition* 83/2: 141–165.

Ryder, N. B. (1973). 'Recent Trends and Group Differences in Fertility', in C. F. Westoff (ed.), *Toward the End of Growth: Population in America*. Englewood Cliffs, NJ: Prentice-Hall, 57–68.

Sewell, W. H. (2005). *Logics of History: Social Theory and Social Transformation*. Chicago: University of Chicago Press.

Szreter, S., H. Sholkamy, and A. Dharmalingam (2004). *Categories and Contexts: Anthropological and Historical Studies in Critical Demography*. New York: Oxford University Press.

Watkins, S. and I. Warriner (2003). 'How Do We Know We Need to Control for Selectivity?', *Demographic Research* Special Collection 1, article 4: 109–142.

Western, B. and B. Pettit (2005). 'Black-White Wage Inequality, Employment Rates, and Incarceration', *American Journal of Sociology* 111/2: 553–578.

PART IV

MECHANISMS OF LOCAL-LEVEL VARIATION AND CHANGE OF STATE

CHAPTER 9

NETWORKS, STRATA, AND AGEING

towards a compositional demography of vulnerability

ELISABETH SCHRÖDER-BUTTERFILL

9.1 Introduction

Most thinking on population processes and the implications of population change rests on one of two extreme levels of analysis: the macro level of entire populations or nation states, or the micro level of individuals and the households they belong to. The adoption of one level over another is almost arbitrary, as much of mainstream post-war demographic thinking assumes a close homology between processes and outcomes at individual and aggregate levels (Kreager and Schröder-Butterfill 2010). Preston and Martin (1994: 3) capture this way of thinking when they write:

> Population aging will be one of the most important social phenomena of the next half century.... [I]t is important because changes in the aggregate age structure are mirrored within nearly all social institutions, from firms to families. How these institutions accommodate themselves to impending changes in population age structures will have a significant effect on the quality of life in the twenty-first century.

Several chapters in this volume are addressed to challenging this perspective of homology by highlighting the fact that populations are composed of subpopulations with different characteristics and behaviours. While it is *possible* (and indeed commonplace) to derive an aggregate, average representation from the sum of subpopulations' characteristics and trends, the interesting 'stories' are lost in the conversion. Indeed, as Szreter remarks in his chapter, there may be no 'story' at the aggregate level if different subpopulations encounter notable but counteracting trends. Understanding contemporary population changes and accommodation to them requires us to be able to identify empirically the relevant units of population (regions, communities, classes, networks?) and to examine how these units respond differentially to the mundane challenges of daily life. At base it is an exercise in disaggregating and recombining population units until patterns emerge. As Johnson-Hanks notes in her chapter, the relevant recombinations are likely to differ according to the challenges at hand, just as the composition of constituent population units will change over time.

While studying population composition holds strategic value for understanding vulnerability (among other things), engaging in 'compositional demography' is hampered by demography's preferred approaches and data sources (Kreager 2011). This chapter therefore describes a methodology of successive, exploratory decomposition and recombination developed for studying population ageing and old-age support in Indonesia. The chapter starts by drawing out key dimensions of population heterogeneity as it pertains to population ageing at regional and local levels. Key differentials include the degree of age structure imbalance, the availability of children, and socio-economic position. Combining these dimensions begins to point to significant vulnerable subgroups but falls short of explaining outcomes in later life, because demographic and socio-economic characteristics appear to be mediated by membership of wider networks. Once older people are located within the networks they are part of, patterns begin to emerge. The chapter therefore goes on to describe the nature of older people's networks in rural Indonesia and characterizes a number of indicative network types. The relationship of these networks to each other and to wider social, economic, and religious changes is highlighted. By examining older people's position within their networks and the ways in which different networks accommodate the needs of elders within them, the social, moral, and material significance of networks as relevant population units for understanding experiences in later life become apparent. Networks are of course not bounded, nor static. A key challenge in compositional demography is therefore to understand how membership of networks is established or lost, and how different networks negotiate shifting hierarchies at local and national levels.

9.2 From Nations to Networks

Indonesia, a former Dutch colony which achieved independence in 1945, is the world's fourth largest population, and one of the most rapidly ageing populations worldwide. By international comparisons the proportion of elders is still modest, with currently 7.6% of the population aged 60 and over (Badan Pusat Statistik 2010). However, the size of Indonesia's elderly population is set to treble between 1990 and 2025, and it will take at best two generations before elderly people represent as large a proportion in Indonesia as they do in Western industrialized countries today. The rapidity of population ageing, the sheer numbers involved, and the fact that Indonesia is still firmly a developing country have given rise to a dominant 'crisis scenario' in policy and academic circles (Hugo 2000; Minister for Social Affairs 2002; Chen and Jones 1989). This scenario predicts growing vulnerability among the older population as a result of inadequate state provision and declining family support due to migration, fertility decline, and 'modernization'.

It requires little evidence to begin to unpack this monolithic story of imperilment. Indonesia is a country of unrivalled diversity. Its 240 million inhabitants are spread over 5000 inhabited islands. There are more than 300 ethnic and linguistic groups, and while the majority of the population is Muslim (88%), there are Hindu (1.8%), Christian (8.9%), and Buddhist (0.84%) minorities which reach majority status in some areas, as well as considerable diversity within Islam (see section 9.3.2 and Suryadinata et al. 2003; Beatty 1999). Thus there is not one demographic pattern for Indonesia, but manifold regional and local patterns. These reflect different demographic regimes in the past—notably different levels of pre-transitional fertility, timings of fertility decline, ages at marriage, and patterns of morbidity, mortality, and infertility—as well as different present-day profiles for migration and family formation (Hull and Tukiran 1976; Hugo 1982; Jones 2001). Table 9.1 compares the ageing profiles and population sizes for several of Indonesia's 33 provinces.

Table 9.1 Percentage of population aged 60 and over and total population size (million), selected Indonesian provinces

Province	Percentage 60 and over	Population size
Indonesia Total	7.6	237.6
Yogyakarta	13.0	3.5
East Java	10.4	37.5
West Sumatra	8.1	4.8
East Nusa Tenggara	7.5	4.7
West Java	7.0	43.1
Riau	4.1	5.5
East Kalimantan	4.0	3.6
Riau Islands	3.4	1.7
West Papua	3.1	0.8

Source: Badan Pusat Statistik (2010).

In central Javanese Yogyakarta the percentage of elders is almost double the national average and more than four times greater than in West Papua, Indonesia's 'youngest' province. East Java, the location of one of the study communities reported on here, is the second most 'elderly' province in Indonesia and home to 37 million older people. Several Sumatran provinces, notably Riau and Riau Islands, manifest young age structures as a result of in-migration by young Indonesians seeking work in the national and global businesses located there. Clearly, in some areas the growing significance of the older population is already being felt, in others concentrations of younger people are having to maintain family links over large distances. Where infertility or labour migration have a history, families have long been making adjustments to age structure imbalances at the local level (Kreager and Schröder-Butterfill 2010). Understanding how families and communities respond to population ageing and identifying vulnerable elders necessitate an approach that can get beneath the national and even regional picture to examine demographic change in its local social, cultural, and economic context.

Do data exist to permit such an approach? Like other countries with a nationalist agenda, Indonesia has not escaped the 'avalanche of numbers' that Kreager refers to in his chapter in this volume. Decennial population censuses go back, with some interruptions, to the Dutch era.[1] These censuses are supplemented by inter-censal surveys (SUPAS) and randomized household surveys, most notably the annual National Socio-Economic Survey (SUSENAS) and the Indonesian Family Life Survey (IFLS), conducted since 1993. While the latter explicitly focuses on familial, economic, and demographic processes and relations, it does not allow meaningful breakdown of ageing-related data below the provincial level due to inadequate sample sizes. This means important heterogeneity at a local level cannot be captured. Like most surveys, the IFLS treats households as the relevant unit of analysis. The critical literature on households is vast (e.g. most recently Randall et al. 2011). Three points that are particularly germane to the Indonesian setting may briefly be made. Firstly, there exists significant heterogeneity within household types in terms of underlying resource flows; these are inadequately captured by variables on intra-household exchanges. Thus multigenerational coresidential arrangements may cover net downward flows of support from old to young, upward flows, balanced flows, or even minimal flows between the generations (Schröder-Butterfill 2004b; Kreager and Schröder-Butterfill 2008; Beard and Kunharibowo 2001). Secondly, households in Indonesia are immensely fluid, as manifested in almost constant change in composition. Whole nuclear family units are temporarily incorporated in response to a crisis; elders circulate between the households of different offspring; youngsters join the households of relatives better placed vis-à-vis schools or employment; members depart on migration. In our East Javanese study community, Kidul, more than a

[1] Data on ethnicity were not collected until the 2000 census. Prior to that, researchers had to rely on mother tongue and religion to gain a handle on ethnic differences (Suryadinata et al. 2003).

quarter of households experienced a change in household composition over a period of 18 months. The case of widow Sum, detailed in section 9.4, is merely an extreme instance of residential instability. Thirdly, households simply do not capture all or even the most significant flows of resources: they are *not* the relevant units for understanding old-age support. The most significant patterns of economic vulnerability, as manifested by the one in ten households in our study communities in which consumption exceeds income, are simply not interpretable without considering the resource flows into the household from non-resident family members or charity (cf. the case of Siyati in section 9.4). In short, even realistic assessments of socio-economic position require consideration of wider networks. Existing surveys, with their insistence on random sampling, pluck households out of their contexts and therefore make study of actual networks impossible.

9.2.1 Methodology of 'Ageing in Indonesia'

Since existing surveys are both not fine-grained enough—in virtue of aggregating to units well above the local community—and too fine-grained—in virtue of disaggregating to households disconnected from others and from wider structures—an attempt to capture the subpopulations of relevance requires a different methodological tack. 'Ageing in Indonesia', a project funded by the Wellcome Trust between 1999 and 2006 and directed by Philip Kreager, took a comparative, longitudinal, and mixed-method approach to understanding old-age vulnerability and support.[2] In-depth ethnographic field research of 12 months' duration was conducted in three rural communities selected to capture areas with progressive ageing profiles and some of Indonesia's ethnic diversity. This involved semi-structured interviews with almost all resident elders, followed by repeat interviews, life histories, observations, kin mapping, and interviews with family members among a subset of elders. Interviews were also conducted with key informants, such as heads of village and of neighbourhoods, health professionals, and mosque and temple leaders. Data on networks are thus the outcome of a range of sources and tools employed during ethnography. Towards the end of fieldwork two randomized household surveys were conducted in each community, one on health and health care use, the other on household economy and inter-household exchanges. Samples were drawn from household listings compiled in each community. The surveys provided both independent evidence of network relationships, and the empirical foundations for our socio-economic stratification system, which was further corroborated and interpreted by ethnography. Five years on the communities were revisited for

[2] I am grateful to the Wellcome Trust, the Economic and Social Research Council, and the British Academy for the generous support of this research, and to Philip Kreager, Tengku Syawila Fithry, Edi Indrizal, Vita Priantina Dewi, and Haryono for their contributions to the data, analysis, and discussions resulting in this paper.

qualitative follow-up and a second round of surveys. Table 9.2 provides sample sizes for the different study components in the three locations, and Table 9.3 summarizes key characteristics of the three communities studied. The material presented in this chapter draws most heavily on data from Kidul (East Java), where I conducted doctoral and postdoctoral field research in 1999–2000 and 2004–2005.

Table 9.2 Sample sizes in the project 'Ageing in Indonesia'

	Kidul (East Java)	Citengah (West Java)	Koto Kayo (West Sumatra)
Number of elders interviewed at least once	206	87	101
Number interviewed repeatedly (incl. interviews with family members)	60	25	30
Health survey: number of older individuals	73	50	84
Household survey: number of households	106	101	98
Approximate village size	2,000*	1,350	700*

Note: * 'Village size' refers to the size of the hamlet we primarily worked in.

Table 9.3 Characterizing the three research communities

	Kidul	Citengah	Koto Kayo
District and Province	Malang, East Java	Sumedang, West Java	Tanah Datar, West Sumatra
Main ethnic group	Javanese	Sundanese	Minangkabau
Family system	nuclear and bilateral	nuclear and bilateral	extended and matrilineal
Population aged 60 and over	11%	10%	14%
Elders' children no longer resident in the village	46%	44%	75%
Elders who are childless	26%	7%	7%
Households owning rice land	13%	55%	66%
Workforce employed in . . .			
agriculture	15%	65%	43%
trade	25%	12%	38%
civil service	10%	5%	7%
other occupations	50%	18%	12%
Elderly households in receipt of a pension	20%	31%	3%

Source: Household rosters, fieldnotes, and household survey data, 1999–2000.

9.2.2 Narrowing the Focus to Communities

Indonesian village communities are populations of sociological significance to their members. The communities we studied showed clear geographic boundaries, hierarchical segmentation into hamlets and neighbourhoods, and integration under one village leadership and administration. Villagers displayed a sense of collective identity: for example, contrasting their village with adjacent but less economically developed villages. Village identity finds its most obvious expression in the West Sumatran Minangkabau community of Koto Kayo, where labour outmigration (*rantau*) has a long tradition (Indrizal et al. 2009; Kato 1982). This has resulted in the emergence of migrant associations in major migration destinations on the basis of village of origin. These associations serve as vehicles for mobilizing support for village projects (Indrizal 2004).

Migration—significant in all three communities (see Table 9.3 and Hugo's chapter in this volume)—underlines that villages are open populations. It also identifies a first important type of local subpopulation, namely on the basis of origin. Villagers differentiate individuals who can trace back to original founding families and incomers (who may have arrived as far back as the 1940s). Incomers are hardly a small minority; in Kidul, for example, only just over half of older people had been born in Kidul, although many had moved in from surrounding villages. In Koto Kayo, where village endogamy is preferred, about a quarter of the population are incomers (*pendatang*), many of whom arrived in the post-Independence decade. Incomer status can leave some villagers with very small local kin networks (see Lubis's network, section 9.5). Moreover, in all three communities in-migration resulted in a distinct, spatially concentrated, and less socially integrated subpopulation. In Kidul, Madurese migrants founded a hamlet separated by a river from the rest of the village early in the twentieth century. Inhabitants of this hamlet are supposedly 'less refined' (*kurang halus*). The hamlet is poorer, more agricultural, and more committed to traditional cultural and mystical practices. In Citengah an equivalent hamlet of incomers has built up around the village's upland tea plantations, while in Koto Kayo the incomers from poorer parts of West Sumatra form a disadvantaged subpopulation of agricultural labourers, concentrated in a particular hamlet. As the local economy of Koto Kayo is heavily based on agriculture (Table 9.3), yet rice lands are held inalienably by the matrilines, there is no scope for incomers to improve their lot by acquiring land. In this community, where lineage endogamy is preferred, intermarriage with incomers is virtually non-existent and incomers tend to be excluded from charitable support sent by migrants and distributed by the local mosque. This indicates that subpopulations on the basis of migration status overlap non-randomly with socio-economic subpopulations, to which I now turn.

9.2.3 Population Composition: Socio-economic Strata

Thus far I have narrowed the focus from national and regional-level populations to identify local communities as significant sites of social interaction and differentiation. A further important step in decomposing populations and understanding vulnerability is appreciation of the material realities within which old-age support is negotiated. Relative wealth not only shapes living conditions, consumption, and access to health, but also influences the creation and maintenance of the networks on which people rely for support. These influences are examined in more detail in section 9.3 and in the case studies presented in sections 9.4 and 9.5.

Although Indonesia has witnessed impressive economic growth since the 1970s, studies of rural Indonesia point to growing socio-economic inequalities. Under the New Order regime (1966–1998) the economic and political elite in villages benefited from government subsidies and loans. Entry into the civil service became a profitable strategy for the educated young and contributed to social and economic differentiation (Cederroth 1995; Barnes 2004). Aside from the distinction between landholders and landless (which in places like Kidul is of waning significance; see Table 9.3), there are growing divisions in terms of access to work and income security (Hart 1986). Women, the elderly, and the poor have fared particularly badly under exclusionary labour practices, such as the replacement of sharecropping with the selling of standing crops (Hüsken 2001; White and Wiradi 1989; Breman and Wiradi 2002).

Yet despite widening economic cleavages, an explicit class structure or consciousness is lacking. Different economic positions do not translate into distinct groups or occupations, nor do they segregate into neighbourhoods (with the exception of the incomer hamlets referred to above). While the better-off strive to maintain or improve their position through strategic marriages, and unions across economic extremes are rare, rural Java is not characterized by homogamy or other pronounced forms of social closure (White and Schweizer 1998). Taste, dress, and linguistic etiquette show gradients between the richest and poorest, but all villagers share in a broad set of values or ethos. Moreover, there is considerable transience in people's economic situation: a person can go from 'poor' to 'rich' and back over a lifetime. Economic descent is an ever-present threat that increases with age, declining ability to work, and—for women—widowhood or divorce. This fluidity raises the challenge of empirically capturing economic heterogeneity and identifying economic subgroups without reifying these and treating them as fixed categories or closed populations.

Local observation has to form the starting point. People at both ends of the extreme are most readily distinguished. At the top are a group of rich families (*kaya* or Stratum I), at the bottom families surviving either on an extremely uncertain economic base or on charity. Rather than being called poor (*miskin*), which would be derogatory, they are referred to as being *kurang mampu* ('less capable'; Stratum IV). Distinctions within the families in the middle are less clearly articulated. The largest group of villagers manage on a daily basis without support from others, but they are unable to accumulate wealth. These villagers refer to their economies as *cukup-cukupan* ('just enough', Stratum III), which may be translated as 'getting by'. Some of the middle-range households are doing slightly better than getting by: they lead secure existences, generate surplus, but lack the ostentation of the rich. No specific term was used to refer to these, but people in this group often described their

Table 9.4 Distribution of households by economic strata in the three communities (percentages)

	Rich I	Moderate II	Getting-by III	Poor IV	Total (N)
Kidul (East Java)	16.0	29.2	43.4	11.3	106
Citengah (West Java)	10.9	30.7	49.5	8.9	101
Koto Kayo (West Sumatra)	8.2	36.7	30.6	24.5	98

Source: Household survey data 2000.

situation as *lumayan* ('moderate', Stratum II). None of these groupings is sharply delineated, as underlined by the fact that there was occasional disagreement between informants' evaluations of a particular family's position.

The commonly expressed, informal status distinctions provided a base for a model of social structure in our study communities. By combining household survey data on assets, income, consumption, and savings with knowledge of 'qualitative' status hierarchies in the community, it became possible to assign households to one of four groupings. Table 9.4 shows their distribution in the three villages. The following paragraphs briefly describe the opportunities and constraints underlying this distribution.

*Stratum I: The Rich (*kaya*)*

The small group of rich households, comprising between 8 and 16% of households in the three communities, includes rich landowners, higher-ranking civil servants, members of the armed forces, owners of successful businesses or large shops. Rich households are increasingly able to educate their children to university level. The rich enjoy a high degree of welfare security, as witnessed by their ability to afford hospital treatment: they have guaranteed regular incomes or own valuable saleable assets (e.g. agricultural land), or both. Rich villagers are expected to dispense patronage to poorer kin and contribute generously to village charity.

*Stratum II: The Moderately Well-Off (*lumayan*)*

The moderately well-off, making up around one-third of households, comprise those with modest land-holdings, recipients of small pensions, lower-ranking civil servants, medium-sized business people, shopkeepers, and well-paid factory and transport workers. Economic mobility across the lifecourse means that Stratum II includes a sizeable transient membership. For example, a number of elderly people who used to be rich were merely comfortably off when surveyed. As the case studies later illustrate, downward economic pressure is often exerted by the distribution of assets to the younger generation. Stratum II also recruits by upward mobility, especially from young families enjoying dual incomes. Remittances from international labour migration further fuel this group.

Stratum III: The 'Getting-By' (cukupan)

In 2000, the largest socio-economic grouping in the two Javanese villages was made up of the *cukupan*.[3] Members of this group are typically involved in steady agricultural labour, small-scale trade, work in small factories, transportation, construction work, or food production. Elderly people who no longer work, but either have small assets to fall back on, or receive reliable extra-household support from relatives or patrons, are also included. Households getting by find it difficult, if not impossible, to accumulate wealth; most income goes to covering living costs and participating in community life. Lack of accumulation leaves households highly vulnerable to economic crises and descent into outright poverty, unless wider networks intervene to prevent this. A key distinction between Strata II and III in communities like Kidul is receipt of charity and subsidies (especially subsidized rice): among the 'getting by' two-thirds are entitled to buy subsidized rice and almost half receive *zakat*.[4] Yet far from just depending on village assistance, most villagers in Stratum III actively contribute to local social and ritual life.

Stratum IV: The Poor (kurang mampu)

Around one in ten households in the Javanese communities are poor; in the Sumatran village their ranks are swelled to one quarter by the incomers (*pendatang*). Life histories revealed about half to be lifelong poor; the remainder have descended into poverty. Often it is demographic reasons—widowhood or childlessness—or health crises which precipitate these declines, although gambling, poor business acumen, and excessive generosity towards offspring sometimes play the dominant role. None of the poor households own productive assets or receive a pension, and only slightly over half own the house they live in. Members of these households typically engage in low-paid work, like small-scale sale of food, collection and sale of firewood or herbs, or making traditional medicine. Some describe their work as *seadanya* ('whatever there is'). Only one-quarter of poor household heads have work most days of the week; a quarter are too old or ill to work at all. Household income typically matches or slightly exceeds expenditure on food, making these households unable to survive without at least occasional extra-household support. With one exception (a recent migrant to the village), all poor households in Kidul received both *zakat* and subsidized rice. The vast majority only ever consume meat on ritual occasions; unless helped by better-off kin, they are unable to afford medical treatment. Despite their poverty, the poor attempt to participate in village social life—for example, by conducting small rituals at the appropriate time. This investment in local exchange networks may afford them support in times of need.

[3] By the time of our re-survey in 2005 the socio-economic composition of Kidul had changed slightly. The largest group was then made up by Stratum II (37.6%), while Stratum III comprised only 29.4% of households. The shift reflects the opening up of local job opportunities in new factories. International labour migration was also rising. Households headed by an elderly person have not benefited from these improvements, and the proportion of poor households (Stratum IV) among elderly households has risen from 18 to 24%.

[4] Not all local governments and religious authorities target the poorest in the distribution of subsidies. Some use the welfare measures to win political followers by distributing indiscriminately to rich and poor alike.

Social stratification and its material and social entailments have been described in some detail here, because socio-economic position has an obvious bearing on older people's vulnerability. An elderly person's control over material resources affects diet, housing quality, and access to health care. In a monetized economy such as the rural Indonesian, access to money is also necessary for social participation: attendance at weddings or funerals, the hosting of life-cycle rituals, even religious and social meetings require money (Lont 2002; Nugroho 1996). Yet as consideration of Sum's and Siyati's situations in section 9.4 will reveal, vulnerability or security cannot be deduced simply from elders' household economic status. Strata are significant, but they are not the sole units of population we need to study to understand outcomes in later life. Similar levels of material well-being do not translate into equal sets of opportunities and constraints, because households are embedded in networks of exchange linking households and generations. These networks do not overlap closely with the economic strata described: rural Indonesian kin and social networks are not economically homogenous; they are *not* social classes writ small. Instead, networks cut across economic strata, although their central tendency may lie in one or another broad economic zone. Economic gradients within networks may provide opportunities for patronage, but equally they may represent barriers to interaction and delimit effective networks. These dynamics will be explored in section 9.3.1. Beforehand it is necessary to examine one other key source of local population heterogeneity which is shaped by economic position and tempered by the operation of networks, namely the availability of children.

9.2.4 Population Composition: The Distribution of Children

In terms of the cohorts we studied, only the youngest elders, i.e. those born around 1940 and aged 60 when we first conducted fieldwork, will have been exposed to modern family planning, first implemented in Indonesia in the early 1970s (Niehof and Lubis 2003). All others would have completed childbearing largely or entirely by the 1970s. We therefore expected completed family sizes among villagers aged over 65 to be high, in the region of five to six children on average, and thus in line with the reported Total Fertility Rate of 5.5 for Indonesia prior to fertility transition. In fact, average completed family size among older women in Kidul was only 3.6. We found immense diversity in reproductive outcomes, with large minorities at either extreme of the spectrum. Figure 9.1 presents the evidence for Kidul, the East Javanese community.

The picture suggested by Figure 9.1 is of a strikingly u-shaped, polarized distribution of children. One in five women stated that they had never had a child.[5] By contrast, more

[5] This figure may overestimate primary sterility, as some people without children may not admit that they had a child who died. Data on children surviving are more easily verifiable. The main challenge here is to identify adopted children as adopted rather than own.

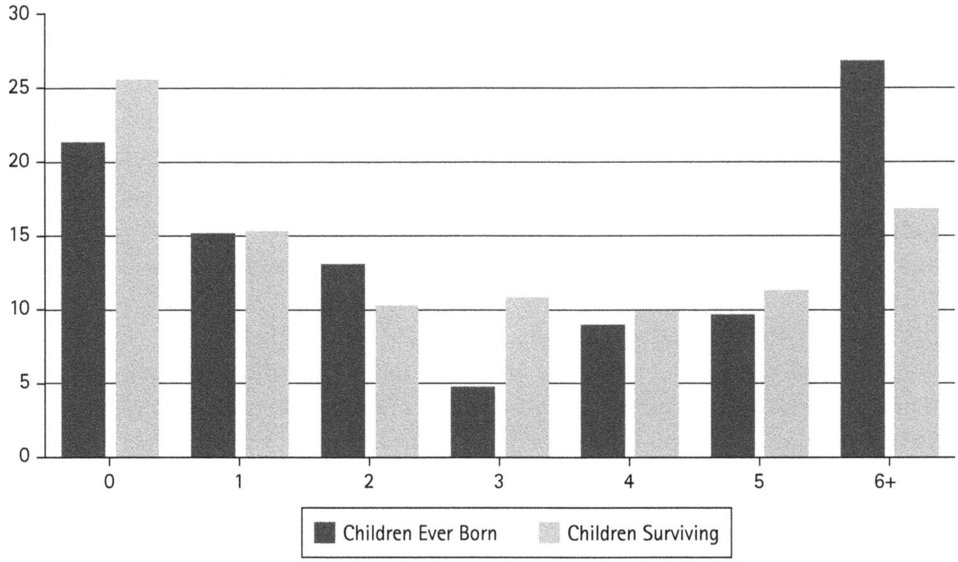

FIG. 9.1 Percentage of older women in Kidul by number of children ever born and number of children surviving.

Source: Fieldwork data 1999–2000. N = 126.

than one quarter had given birth to six or more children. One quarter of elders do not have a surviving child; half of the women have zero, one, or two children, the other half three or more. The data provide an interesting non-European illustration of the point made by Szreter in his chapter, which is that 'due to the vagaries of primary and secondary sterility' members of any given communication community would be exposed to examples of a range of fertility outcomes on the basis of which to judge the merits and disadvantages of small family sizes.

We were able to locate this striking demographic picture within a broader historical and regional pattern and confirm East Java as having experienced pronounced periods of suboptimal fertility. The impression given by national and provincial estimates of completed fertility averaging between four and six births per woman disguises the presence of substantial childless minorities which only emerge when fertility is disaggregated. It is this variation, not central tendency, which is indicative for understanding old-age support. In the 1940s fertility declined and mortality increased, leaving the cohort experiencing its peak childbearing during the 1940s particularly affected. The reasons for the demographic crisis are fairly well established. Major disruptions of economy and society, starting with the world depression and exacerbated during the Japanese occupation of 1942–1945 and the ensuing War of Independence, brought deteriorating health and dietary conditions (Schröder-Butterfill and Kreager 2005). These crises were particularly severe on Java, and led to sharp increases in infant and child deaths (Hugo et al. 1987: 118; Singarimbun and Hull 1977). Marriage and childbearing also suffered. Not all individuals were affected in the same way. Our data confirm a

Table 9.5 Availability of surviving children to older people by membership of upper two and lower two economic strata (percentages)

	Kidul		Citengah		Koto Kayo	
	Upper	Lower	Upper	Lower	Upper	Lower
No children	18.8	32.1	5.0	8.5	5.7	10.7
One or two children	18.8	32.1	0	19.2	15.1	21.4
Three or four children	20.9	19.8	17.5	34.0	24.5	25.0
Five or more children	41.7	16.0	77.5	38.3	54.9	42.8
N =	96	106	40	47	53	28

Source: Fieldwork data 1999–2000.

range of correlations between economic position and demographic outcomes, with the poor more likely to experience divorce and multiple marriages, more likely to lose a spouse to early mortality, and more likely to experience the death of a child. Strikingly, poorer women also experienced lower fertility, an observation census and survey evidence from other parts of Java confirm (Singarimbun and Hull 1977; Hull and Hull 1977; Hull 1976; Gooszen 1980).[6] The reasons for this can be found in the proximate determinants that reduce childbearing in a poverty context, namely divorce, widowhood, malnutrition, and sexually transmitted disease as a result of prostitution or multiple marriages.

The implications of a socio-economic patterning of reproduction are clearly reflected in the evidence from all three study communities, even though East Java has by far the highest levels of childlessness (see Table 9.5).

Inadequate family sizes are not randomly distributed but concentrated among the economically marginal, and they contribute in turn to the persistence of inequalities. Poverty also makes ties to existing children more tenuous. Divorce or widowhood early in life forced many parents to leave their children to be raised by others, and this often weakened their children's sense of responsibility towards them. Similarly, the inability of poor parents to offer assets and an education to their offspring has made these more prone to leave the community and rarely return (Kreager 2006). Under certain circumstances these processes result in parents becoming *de facto* childless. As parents have aged, their lack of adult children on which to fall back has entrenched their material vulnerability and meant that even elders with a favourable economic starting position face economic decline. The networks of Lubis and Rini (section 9.5) illustrate the insecurity and misery that a lack of children in old age can entail.

[6] The observed relationship between fertility and poverty in rural Java contradicts an apparent 'law' in demography: 'The higher fertility of the lower classes has been observed so often and in so many different countries that the existence of a negative correlation between fertility and class . . . has virtually acquired the force of a socio-demographic law' (Wrong 1967, quoted in Hull and Hull 1977: 43).

Table 9.6 Older people's net availability of children by economic strata in Kidul (East Java)

	Rich I	Moderate II	Getting-by III	Poor IV	All
Actually childless	10.0	3.5	15.5	45.7	17.9
Child-poor	27.5	29.9	37.9	30.4	31.9
Child-rich	62.5	66.7	46.5	23.9	50.2
Total (N)	40	57	58	46	201

Note: Own and successfully adopted children are included in the accounting. The 'actually childless' are those with neither own nor acquired children plus the *de facto* childless.
Source: Fieldwork data, 1999–2000.

Of course, environments in which a lack of children is commonplace engender the development of practices through which a more equitable distribution of children may be achieved (Grant and Yeatman 2012). Informal adoption is widespread and acceptable in Indonesia. Kin networks are crucial for the redistribution of children, as it is typically the offspring of siblings, cousins, or other relatives who become *anak angkat* ('raised' children). Thus wider networks may compensate for individual reproductive failure, provided the network's demographic resources are adequate, and the moral and economic standing of the childless person sufficient.[7]

If the impacts of family formation, mortality, *de facto* childlessness, and alternative avenues to parenthood are considered together, it becomes possible to arrive at a preliminary classification of elders' net availability of children and thus a preliminary assessment of the size and composition of vulnerable subpopulations. These subsets then need placing in the context of their wider networks to examine the extent to which kinship and community ties may compensate for, or exacerbate, deficiencies in family networks. Table 9.6 provides this classification for Kidul. It distinguishes between those who are actually childless (i.e. those with neither own nor acquired children as well as those who are *de facto* childless), those who are child-poor (locally perceived as those with only one or two surviving children, own or adopted), and those who are child-rich (with three or more children).

The distribution of types among socio-economic groups is striking. Less than one-quarter of poor elders have a generous supply of children, almost half are actually childless. Only among the upper two strata are large family sizes in the majority. The economically poor and those 'poor in children' are not identical populations, and they each experience different challenges. Yet where the two types of poverty intersect, it

[7] It is worth noting that informal adoption is not found among the matrilineal Minangkabau. According to the logic of matrilineal kinship, a person's sisters' children are like own children. As on Java, children are thus redistributed via the kin network, albeit in different ways (Kreager and Schröder-Butterfill 2007).

becomes possible to identify a subpopulation facing significant vulnerability in later life. This subpopulation combines elders in Strata III and IV who are childless or child-poor and makes up one-third of all elders. However, this preliminary classification, while giving a broad overview of the social and demographic structure of vulnerability and security in rural Indonesia, is inadequate. Wider networks beyond the nuclear family need to be brought into the equation, not least because the availability of children cannot be taken for granted.

9.3 The Nature of Networks in Rural Indonesia

Indonesian villagers are embedded in different, overlapping networks. For old-age security immediate family members, wider kinship links, and neighbours are most relevant, while religious institutions, the village apparatus, and in some cases supra-local institutions also matter. This section outlines Indonesian kinship and community relations and the values guiding the interactions of villagers, as these inform the nature of people's networks. The wealth differences characterized earlier are pertinent to this, not only because material realities shape network members' ability to provide assistance, but also because wealth and status differences influence the readiness with which network members approach each other. How support is negotiated depends on how membership is defined, what expectations adhere to different relationships, and what moral values motivate people's interactions. As will become clear, the term *negotiation* is unusually apt in the Javanese context, because strong normative prescriptions governing intergenerational support—as found perhaps in Confucian systems of filial piety—are lacking. Household constellations are nuclear and fluid, families individualistic, and kinship networks diffuse. This means that exchanges cannot be explained in terms of obligations (cf. Finch and Mason 1993). Instead, flows of support and network members' willingness to intervene rely on the quality of bonds and the reputations of the individuals involved.

9.3.1 Javanese Kinship Networks

Javanese families are not extended but nuclear. In this they resemble other South-East Asian ethnicities (e.g. Malays, Sundanese, or Filipinos) but differ from major Asian family systems, such as the Chinese or many Indian groups. Indeed, in organization and sentiment, Javanese families are closer to northern European families (Goody 1996; Hajnal 1982; Todd 1985). Adult children are expected to set up independent households, although independence is not necessarily immediate upon marriage and may occasionally never be achieved (Jay 1969: 40; Geertz 1961). As in European nuclear family systems, net flows of intergenerational wealth and support are unequivocally downward, from parents or grandparents to children and grandchildren (Schröder-Butterfill 2004b). Independence is considered ideal. Once reliance on others becomes necessary, assistance from children is preferred (Schröder-Butterfill and Fithry 2014). However, childlessness, outmigration, and poverty among children make reliance on wider networks inevitable for many.

Different cultures have different notions about which kin have social significance, how kin should interact, and what roles and expectations attach to different degrees of relatedness (Skinner 1997). Among the matrilineal Minangkabau of West Sumatra, for example, a man will look first to his sister's children for support, rather than to his own, and relatives within the same matriline may routinely be called on for assistance (Kato 1982; Indrizal et al. 2009). Matters are different among the Javanese. Kinship relations on Java are fairly devoid of prescriptions. Who counts, and may be counted on, depends more on social and geographic proximity than on clearly articulated obligations.

The Javanese and Sundanese have a bilateral (or 'cognatic') system of kinship (Jay 1969; Geertz 1961). Kin are traced through both parents, brothers and sisters, and male and female children. As a result, bilateral kinship systems, unlike lineal ones, do not give rise to bounded groups, but to ego-centred networks of kin (or 'kindreds') sharing varying degrees of relatedness. Freeman, in his influential paper on cognatic kinship, writes:

> [B]y the very nature of its composition, the members of a kindred have no collective perception of unity, no persisting common objective and no leader or organization. A kindred, therefore, is not a group in the sociological sense of the term, but rather a category of cognates, a set of persons who have in common the characteristic that they are all related cognatically in varying degrees to the same person.
>
> (1961: 202)

In other words, kindreds are quintessential open, unbounded populations; they are networks, rather than groups. Two apparently contradictory conclusions are generally drawn from this. Firstly, it is argued that wider kin play an unimportant role in structuring social life, and that expectations of, and entitlements to, assistance among kin beyond the nuclear family are limited. The second conclusion is that cognatic systems give rise to a potentially limitless network of ties on which individuals can draw selectively to their advantage (Freeman 1961: 203; Geertz 1961: 25). As we shall see, both statements are borne out in practice, with the significance of kinship links depending on the ability and inclination of actors to differentially manipulate, nourish, or discard ties at their disposal.

Villagers recognize a wide range of bilateral kin. When probed they distinguish relatives on the basis of genealogical distance, although there are neither clear boundaries between close and distant kin, nor terms to denote such groups. A person's spouse (obviously not kin), children, parents, and siblings form the core network. Outside this circle are grandparents and grandchildren, nephews and nieces, uncles, aunts, and cousins, all loosely described as close kin (*saudara dekat*). Interaction with and knowledge about these relatives vary from case to case. Beyond these, people recognize distant relatives (*masih saudara*). Differences between close and distant kin, and distant kin and non-kin are downplayed in public, but in reality genealogical, spatial, and social proximity shape interactions. Where there are significant wealth differences between kin, contact will be avoided. For example, when I asked my elderly neighbour, who belongs to the 'getting by', to identify her relatives in the village, she listed half a dozen names, but then added that so-and-so is 'like not a relative at all, because he is rich'. Her day-to-day interactions are most intense with certain neighbours and relatives who are, like her, not well-off. Javanese people are highly sensitive to status differences and feel awkward (*sungkan*) in the presence of social and economic superiors.[8] Poor villagers will avoid visiting rich kin lest they be accused of pandering for favours. The implications of these patterns of interaction are clear: while from a purely genealogical standpoint a person's kin network might be extensive, the effective network is limited by awareness of status differences. Not only is the effective network only a fraction of the actual kinship network, it often contains some distant kin, while omitting certain close kin. Those relatives best able to provide assistance may fall outside the range of effective kin because of patterns of avoidance. Meanwhile the resources of kin willing to help, and most readily approached, may well be very constrained (Scott 1976: 28).

Clear obligations among distant kin are even more limited than among close kin. The role of kinship is certainly apparent during ceremonial events, when close and distant kin are expected to help with labour and materials. However, outside this domain, obligations among kin are circumscribed and depend both on the closeness of ties and on who else exists in the network. Help among kin outside the nuclear family is normally confined to gifts of food, small sums of money, visits, and practical help. Kin other than children are not expected to cover costly medical fees or make substantial monetary gifts. One elderly widow, whose only son experienced a bad accident, had to sell her house to cover his medical bills. When I asked her siblings, nephews, and nieces whether they had contributed, they admitted freely that they had not. Such instances are neither atypical, nor do they draw critical comment. In this respect Kidul confirms what has been noted in the literature on South East Asia: sharp differences in economic status even among close relatives are accepted, and there is no expectation that better-off kin should try to equalize economic differences (Djamour 1959: 47; Scott 1976: 40). Wealthy kin may step in to prevent absolute destitution—not least to safeguard their own reputations—but they are not expected to substitute for missing or feckless children or spouses. This is manifest in the networks of Siyati, Jasman, and Sunadi (sections 9.4 and 9.5).

[8] Beatty has drawn attention to the importance of emotions in guiding Javanese social interaction: 'one's sense of the invisible boundaries within the village community, and of degrees of solidarity and difference among similar households, is typically expressed in terms of feelings of "reluctance", "embarrassment", "respect" and so on' (2005: 66). These feelings shape who people can and cannot comfortably interact with. In Kidul people used the concepts of *sungkan* and *krasan* to communicate their awareness of social distance or familiarity. *Sungkan* may be approximated as a 'sense of respect mixed with shame, awkwardness and avoidance' (cf. Geertz 1961: 152). People feel *sungkan* when confronted with individuals of appreciably higher status or wealth, and such interactions are consequently avoided where possible. This extends to interaction with relatives who are much better off. The opposite of *sungkan* is *krasan*, which may be roughly translated as feeling at home or at ease. Feeling *krasan* entails regular, informal visiting and an ability to presume on another's assistance (e.g. when needing a small loan or help with preparing a ritual).

My account thus far may seem to suggest that kinship networks beyond the immediate family are of modest significance in rural Java. This would be to forget the sizeable childless subpopulation of elders. Where customary sources of old-age support are lacking, wider kinship networks take on greater significance, and for childless villagers kin represent a primary resource. The best solution to childlessness is to (informally) adopt a child, and the most common source of adoptive children are one's relatives. Through the redistribution of children, kin networks are able to even out inequalities in reproductive outcomes. Kin may also be cast in the role of patron to a poorer relative, allowing relatives of widely differing statuses to engage in mutually beneficial exchanges without upsetting Javanese sensitivities to hierarchy (Schröder-Butterfill 2004a).

9.3.2 Wider Network Links

As we have seen, villagers' networks are composed of kinship ties which operate selectively on the basis of social and spatial proximity. Beyond kin, neighbours and religious networks are most important. Ties with neighbours operate on a similar logic to kin ties. Interactions tend to be most frequent with neighbours who live closest and are most similar in social and economic status and entail the exchange of food, gossip, assistance, and loans. Typically neighbours' supportive role is inversely related to the strength and size of a person's kinship network, with villagers who have adequate local kin networks avoiding heavy reliance on neighbours.

Religious participation is important to most Indonesian villagers, and religious charity can be significant (Benda-Beckmann 1988; Schröder-Butterfill 2006). The West Sumatran and West Javanese study communities were entirely Muslim, and religious diversity confined to degrees of religious zeal. In Kidul the religious landscape is more heterogeneous and shifting, and social interactions consequently more heavily shaped by religious affiliation. Until the mid-1960s many villagers in East Java were followers of Javanese religion (*agami jawi* or *kejawen*) which combines Javanese beliefs and practices with Muslim and Hindu elements (Geertz 1960; Beatty 1999). Following the attempted coup and ensuant communist massacre of the mid-1960s, President Suharto made it compulsory for every citizen to adhere to one of five world religions. Many people in Kidul opted for Hinduism, in part because the then-village head became Hindu. In the 1970s and early 1980s, 40–50% of villagers were Hindu, but the 1990s saw a steep decline. While reflecting broader patterns of religious change across Java, in Kidul the shift can be tied more specifically to the arrival of a Muslim missionary couple in 1989. They targeted women, children, and Hindus and sought to improve the moral and social fabric of the community by condemning drinking, gambling, and prostitution. They also established an outstanding, affordable school in the village. Their approach is a familiar one across Indonesia, namely of combining religious instruction and conversion with the provision of high-quality secular education.

Within Islam a growing number of villagers are becoming visibly devout followers, and every year there are some who make the costly pilgrimage to Mecca. Those who either refrain from religious participation or emphasize mystical, Javanese elements within 'Javanese Islam' are becoming fewer and older. A further recent trend has been the emergence of a fundamentalist stream of Islam within the community. Members of LDII (*Lembaga Dakwah Islam Indonesia*) represent a network most closely approximating a 'group' within Kidul's social landscape, as they emphasize their distinctiveness, worship in a separate mosque, intermarry, refuse to participate in practices which are not in line with Islam proper, and contravene important rural Javanese values, such as easy interaction among the sexes, relaxed attitudes to religious dress, and the use of family planning. Despite other villagers' wariness towards them, their strong ethos of mutual help and welfare provision towards members in misfortune was commended.

The religious changes in Kidul are representative of similar shifts across Indonesia: a move away from syncretic practices and beliefs towards a 'purer' interpretation of Islam, a decline in the importance of Hinduism compared with Islam and Christianity, and the association of devout Islam with modernity, education, and progress (Rinaldo 2008; Hefner 2000). (Fundamentalist Islamic streams are sometimes an exception to the latter point by increasing gender inequality and curtailing reproductive choice.) In Kidul differences in religious adherence continue to be downplayed and are not major determinants of social interaction.[9] A good example of this is people's participation in ritual meals (*slametan*) to mark births or deaths. These include Javanese mystical and syncretic elements not in line with Islam, but most devout Muslim villagers take a pragmatic approach to participation and emphasize social intent over religious content. Nonetheless, the impact of religious changes on the local population's composition has hardly been subtle. By 2005 only 10% of Kidul's population was not Muslim and among children under age 15 it was only 6%, while among the elderly almost a quarter remain Hindu. Over time Hinduism has come to be associated with older, more traditional, less dynamic, and less outward-looking elements within the village, and this is clearly reflected in the socio-economic profile (see Table 9.7).

Households headed by a Hindu are more than four times more likely to be poor than households headed by a Muslim. There are very few notable, wealthy Hindu families left in Kidul, a fact which finds reflection in the dynamics between different networks as described in section 9.5. Long gone are the days in which a Hindu could become village head. That said, as the example of the present village head shows (see the case of Jasman, section 9.5), having a Hindu past is no bar to political success.

This last point brings into the picture village leadership and administration as a potential component of people's networks. Membership of the village apparatus has always been an important avenue to influence and privilege. For example, village office comes with *usufruct* of prime agricultural land. Under the New Order regime, village heads

[9] Significantly, even the distribution of religious charity in the form of Islamic *zakat* does not discriminate between Muslims and Hindus in Kidul. The Muslim missionary couple described as 'fanatical' (in a dig at the LDII mosque which distributes *zakat* only among its followers) the position that *zakat* should be confined to Muslims. Thanks to their effective pro-poor targeting, a higher proportion of Hindu than Muslim households actually received *zakat* in 2000.

Table 9.7 The distribution of households in Kidul by economic strata and religion in 2000 (percentages)

	Rich I	Moderate II	Getting-by III	Poor IV	N
Muslim	17.6	33.0	41.8	7.7	91
Non-Muslim*	5.9	5.9	52.9	35.3	15

Note: *All but one of these are Hindu.
Source: Household survey 2000.

were tasked with implementing national *diktat* and rewarded generously for loyalty. Since democratization and decentralization, village leaders have become more autonomous and are able to raise and control their own revenue (Antlöv 2003). Importantly, it is the responsibility of local government to allocate subsidies, grants, and other welfare measures (e.g. the new social health insurance). This makes good links with village leadership instrumental.

9.4 Two Elderly Women

Thus far the basic parameters of villagers' networks have been described, as has the changing socio-religious context. But what is their significance? How do these play out in the lived experiences of elderly villagers? The cases of two elderly women, Sum and Siyati, will illustrate this. Although Sum and Siyati had started from broadly similar positions in terms of background, wealth, and family building, they found themselves in very different situations at the end of their lives. Sum lived her dying days in pitiful destitution and neglect, while Siyati, economic decline notwithstanding, remains secure and respected.

Sum, born *circa* 1915, descended from one of Kidul's wealthy and respected founding families. Despite making three good marriages (to a Javanese member of the Dutch colonial army, to a railway official, and to a village official), she only succeeded in having two children, a son who died in adulthood and a daughter, now a divorcée in her sixties. Sum used to run a canteen outside a busy factory and was renowned as a traditional healer. Her third husband left her with sizeable plots of land in the village. When I first met her in 1999, she was living in a five-generation household with her widowed daughter, married granddaughter, great-granddaughter, and great-great-granddaughters. Her three widowed sisters lived in the vicinity. Soon after, however, the granddaughter, to whom Sum had sold her house under very favourable terms a few years earlier, ran into debt and sold the house she shared with her mother and grandmother. The granddaughter departed on labour migration, and Sum and her daughter became tolerated guests in the house of Sum's recently deceased sister. That house, too, was sold, and the two elderly women moved into a hastily erected bamboo shack on a tiny plot of land Sum still owned. By 2004, Sum and her daughter had moved again, this time into a lean-to built onto the end of the house of her great-granddaughter, who had in the meantime become the junior wife of a wealthy regional politician. Sum survived on what little her daughter could make selling second-hand clothes, on occasional gifts from her kin, and on charity from neighbours and the mosque. This did not stretch to cover health care, and she died in 2005 after a long period of poor health.

Sum's sorry end was the culmination of a downward trajectory of wealth and status dating from the late 1970s. Following the death of her third husband—a man still remembered for his influence and wealth—Sum repeatedly came under pressure to sell land to cover family needs until none was left: a great-grandchild was hospitalized, circumcision and wedding ceremonies needed paying for, and capital was required to enable her granddaughter to embark on labour migration to Saudi Arabia. Relinquishing ownership of her house proved catastrophic when her granddaughter reneged on her promise to keep Sum. Significant flows of intergenerational wealth to children and grandchildren are common in East Java, and it is particularly elderly widows or divorcées who find it hard to resist expectations from the younger generation (Schröder-Butterfill 2004b).

A contrasting case is provided by Siyati, a divorcée who is about 15 years younger than Sum. Siyati has a son from her first marriage, who was raised by her childless older brother, and two sons and a daughter from her third marriage to a member of the Indonesian army. For health reasons Siyati gave up working as a food-seller some years ago. In old age Siyati has been subject to similar pressures as Sum. She, too, divided up her land among her children, helped her daughter and son-in-law with building a house and setting up a business, and bequeathed her house to her daughter, although she continues living in it. Unlike Sum, however, Siyati has been able to resist her daughter's demands that the house be sold and Siyati move in with her daughter. Instead, Siyati chooses to live alone but in close proximity to her daughter and three siblings. In her opposition to giving up her house she is supported by her sons and her brother, Kolil, who is an influential 'head of neighbourhood' (*ketua RW*). Her resistance on the issue of the house has strained her relationship with her daughter, such that on occasion the two were not on speaking terms. For many years most of Siyati's day-to-day material support derived from her firstborn son, a wealthy retired lieutenant turned businessman, despite him having become her oldest brother's adopted son. His sudden death in 2003 deprived Siyati of her main source of support, but the gap was quickly filled by greater involvement of her two other sons, both of whom do not live in the village, support from her siblings, and gifts from other kin. Thus, although Siyati, like Sum, has no income of her own, she gets by and is assured of assistance with health-related expenses when they arise.

Both women and their families are located within the same context of social, political, and economic change, yet their outcomes in old age are poles apart. The most immediate source of vulnerability for both stems from dependencies and expectations of support in the younger generations. These expectations are not new or unusual but reflect long-standing patterns of gendered and generational access to power and resources. Household composition and household economy cannot explain the women's differing fates. Both women have no income of their own, but it is Sum, living with working-age family members, who is vulnerable, while Siyati, living alone, is secure. Clearly, identities beyond the household are crucial: the women's outcomes can only be understood by placing them in the context of their wider networks and tracing their and their network's trajectories over time. This reveals that Sum's and Siyati's networks occupy starkly different positions vis-à-vis local sites of power, wealth, and status, with direct implications for the elderly women's well-being and security.

Figure 9.2 visualizes Sum's wider kinship network. (A lot of the detail in Generations 2, 3, 4, and 5 has been omitted.) Sum is the oldest of six siblings to survive to adulthood. Her sister, N., had one child which died. She resorted to adopting a son from outside the kindred. The next sister, R., experienced the same fate; she adopted a son and daughter from her husbands' kindred. K., the fourth sister, has one son, who lives locally but is unsuccessful, and additionally raised a cousin's son, who has done well and spent most of his working life on Sumatra. Sum's brother, P., with three children, produced the largest family, but died early. The youngest of the siblings, S., migrated to Kalimantan and died when his only daughter was young. In total, despite marrying at least once, the six siblings have produced only nine offspring between them, of which only six have survived to adulthood. This gives an average completed family size of 1—not even enough for Sum's generation to replace itself. Clearly, the demographic pressures outlined earlier, which influenced childbearing in the 1930s and 1940s, have affected this particular set of siblings heavily. There were several other sibling sets where more than one in three elders remained childless. Such variation directly affects the supply of children for redistribution within a network. Given Sum's and Sum's sisters' wealth and standing when they were younger, they might well have acquired more children through informal adoption, had the preferred supply of such children (relatives via the mother) not been so constrained. The second generation has hardly fared better. If we include adopted children, the ten members of the second generation have had 2.2 children on average. Family planning may have contributed to the curtailment of fertility among the younger members of that generation.

A striking contrast is provided by the kinship network of Siyati, shown in Figure 9.3. She, too, is one of six siblings who cover a slightly later range of birth cohorts than Sum and her siblings (*circa* 1925 to 1940). Like Sum, Siyati descended from an influential village-founding family. All siblings made good marriages, some of them several. Only the oldest of the siblings remained childless, and he was able to adopt two children from among his siblings. Siyati's second-oldest brother had eight children from three marriages; Siyati herself had four children. A younger brother, now deceased, had seven children; her surviving sister has four and her youngest brother nine, of whom two have died. Together this set of six elderly siblings has produced 30 surviving offspring, yielding an average completed family size of 5. This is above the average for that broad cohort for East Java (Schröder-Butterfill and Kreager 2005: 27), and five times greater than that of Sum and her siblings. In the second generation no one has

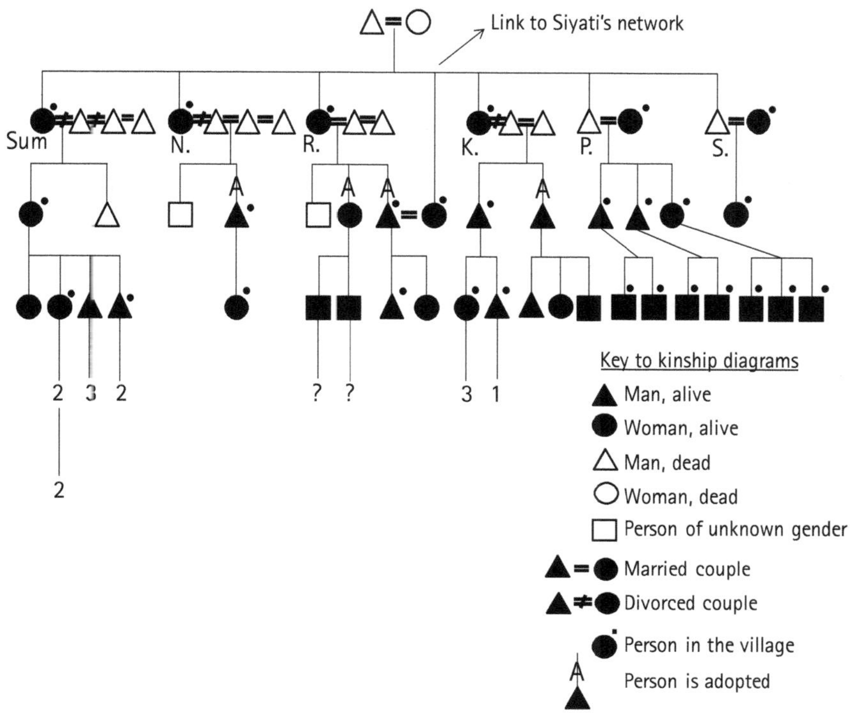

FIG. 9.2 Sum's network.

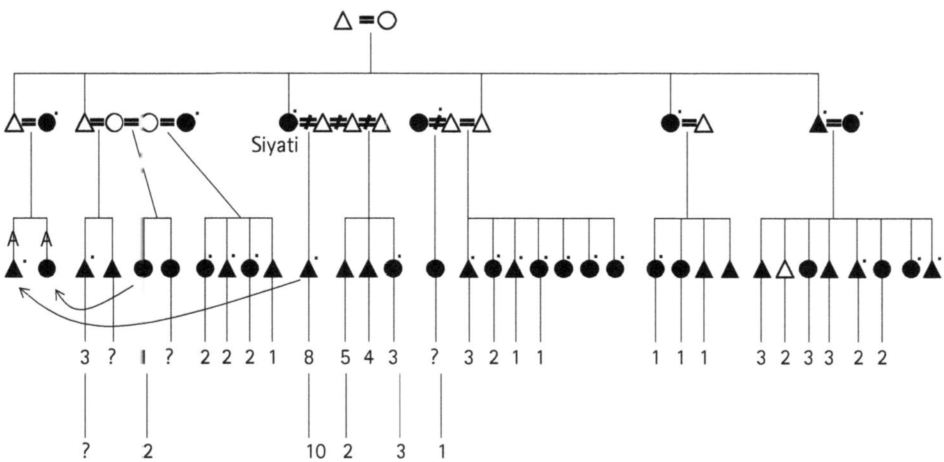

FIG. 9.3 Siyati's network.

remained childless, but with family planning adopted by most members, the average family size is just over two.

Let me use these two similar yet contrasting cases to illustrate the importance of networks in trying to understand vulnerability and security in old age. Networks matter both as sources of material and practical support, and as safeguards of vulnerable members' interests. Sum receives support from only a small handful of kin (chiefly her poor daughter, siblings, and a great-granddaughter), none of it significant, while Siyati reports regular assistance from her sons as well as many of her relatives. Siyati's network is able to cope with the death of her main source of support and with the at best intermittent support from her daughter, while with Sum there is a steady contraction of supportive network and a dramatic worsening of her material situation. In Siyati's case, influential network members (notably her brother and army son) intervened to prevent her from relinquishing her house. The same son had also, a few years earlier, negotiated the redistribution among all elderly siblings of the agricultural land which Siyati's oldest brother had inherited alone. By contrast, Sum was twice made homeless as a result of self-interest in the younger generation, with no one speaking up. When pressed on the issue, relatives argued that it was an 'internal matter' between Sum and her granddaughter.

The reasons for these different network responses have to be sought in the differential composition, wealth, and reputation of the two networks. The most immediate difference between them is, of course, network size. Sum is at a clear disadvantage because her network contains far fewer working-age members who could potentially lend support, and the departure of her granddaughter further depletes the circle of kin she might turn to. Siyati's kin network is vast by comparison, providing a much greater pool of potential supporters. However, network size alone cannot fully account for Sum's vulnerability and Siyati's security. Even Sum has some young family members to hand who might be expected to help, not least in the light of her previous generosity, while Siyati enjoys support from individuals who could easily avoid assisting her, including the son who was raised by her brother and the sons living away from the community. To understand this it is necessary to place the networks within the shifting status hierarchies of the community. Sum's network is not only demographically, but also socially and economically on the decline. Importantly, it lacks the presence of influential men: since the death of Sum's husband in 1979 the wider family has not produced any significant figures in the community. The oldest generation consists entirely of women, and the men in the younger generations are either unsuccessful or absent. It is telling that the only member linking Sum's and Siyati's networks directly (one of Sum's nephews who has married Siyati's niece) expressly aligns himself with his wife's family rather than his own. A number of sexual scandals have damaged the reputation of what had been, in Sum's prime, a leading village family. By the end of her life, Sum was viewed with pity, rather than respect. She had come to epitomize a way of life no longer aspired to, clad in a traditional sarong, an erstwhile Javanese healer and follower of Hindu-Javanese mysticism. Without a 'good family name' to safeguard, any members of Sum's wider network able to intervene have felt little inclination to do so. Moreover, the open disrespect in which Sum is held by her own family makes it difficult for the wider community to continue to hold her in esteem.

Matters are quite different in Siyati's network. Not only is her wider family demographically in ascendancy, it is economically and politically well connected. There is no shortage of influential and ambitious men (retired army members and civil servants, figures in local administration) in whose interest it is to protect the good name of the wider family. Ensuring a basic standard of living among elderly members is part of that reputation. Both families owned land to start off with, and farming and food trade initially formed the economic mainstay of both. However, in the case of Sum's family, land was gradually sold off in response to crises, while in Siyati's family it was for many decades controlled entirely by Siyati's oldest brother. This forced members of Siyati's network to seek alternative economic niches, and placed them in a strategic position vis-à-vis Indonesia's shifting sources of economic, political, and social status. By investing in education early on, members of Siyati's kindred gained entry to urban-based formal employment (army, civil service, factory work) and consolidated their family's economic advantage by providing other kin with access to desirable jobs. Positions of influence in village social and political life have followed from the experience gained in the urban economy and wider Indonesian bureaucracy. Siyati's brother and wife, for example, are responsible for allocating the government rice and health subsidies. With the growing importance of Islam as a sociopolitical force, it comes as no surprise that the vast majority of Siyati's kin are visible followers of Islam. This adds to the esteem in which they are held and ensures the elderly network members' inclusion in the distribution of religious charity.

The two case studies clearly demonstrate the shortcomings of existing methodological approaches to population and population processes. Sum's and Siyati's situations are not explicable by their individual or household characteristics. Instead, it has been necessary to place them in the context of their wider networks and communities. Viewed in isolation, Siyati and Sum have more in common than divides them, although Siyati is child-rich where Sum is child-poor. However, the contrast between their social networks could hardly be starker, and the implications for the two elderly women's welfare follow directly from this contrast.

9.5 Towards a Network Typology

Let me summarize the key variables according to which Sum's and Siyati's networks may be differentiated, so that other networks can be brought into the picture and aligned or contrasted with these networks. Dimensions of apparent importance include: reproductive success, size of local adult kinship network, broad economic position of network, religious affiliation and reputation, the availability of persons of influence within the network, links to key community institutions, and broad livelihood strategy (e.g. in terms of asset ownership, occupational pattern and diversity, investment in education and migration). Obviously, any characterization of an entire network along these dimensions has to be crude, as there is immense variation within any given network. Moreover, no two networks are alike. The aim is to identify a limited number of broad and recurrent network types and derive hypotheses concerning their treatment of elders. The socio-economic and religious patterns and shifts outlined earlier provide axes along which networks may be expected to align themselves differentially.

Table 9.8 summarizes the key characteristics of seven networks in Kidul; these networks will now be described briefly.

Haji Lina's network may be glossed as an *elite network* which has long held privilege and power and has maintained that position by adapting to change. Haji Lina's father was village head under the Dutch and consequently acquired considerable landholdings. Lina's brother enjoyed an unusually good education for his times and was recruited into the Indonesian army. Her husband was also in the army; after retirement he built up a successful local business. The couple remained childless but was given four children to raise from within Lina's kinship network. These children's loyalty to Lina was never in question. Education levels in the second generation have been high, including instances of university education. This has resulted in many of the younger generation leaving Kidul and finding work in the formal sector and civil service. Less economically successful network members have remained local and are able to provide companionship, practical support, and care. Some members of the network have succeeded in marrying into the regional elite, while kin links to local families of poor standing and repute are not actively maintained. Members are devout Muslims, and Haji Lina is revered for being a returned pilgrim from Mecca. When she encountered a care crisis in old age, she was able to co-opt a granddaughter into the role of carer. Other elderly network members are similarly secure and respected.

In the local hierarchy of yesteryear, Sum's and Siyati's networks will have been positioned somewhat below that of Haji Lina's. Both were respected founding families, but Sum's had the edge over Siyati's by marrying into village leadership and the colonial state apparatus. Yet Sum's network is one which today may be caricatured as being *traditional and descending*. The key parameters of Sum's network's decline have already been described. Demographic misfortune, combined with a failure to invest in the education and discipline of younger members (possibly due to a lack of male leadership), have meant that this network has not partaken in Indonesia's take-off. The younger generation is economically unsuccessful and therefore prone to over-reliance on the older generation, leading to a steady dissipation of wealth and an inability to offer much support to the older generation. The only really successful member of the second generation (K's adopted son) has left the community for good. Sum's granddaughter's participation in international domestic labour migration is indicative: it is a strategy which requires limited human capital in terms of education but considerable monetary capital, often, as in this case, provided by the family through sale of land on the promise of substantial returns. However, the profits from this form of labour migration are often invested in short-term conspicuous consumption, rather than becoming the foundation for a more sustainable strategy of investment and accumulation. The elders in this network no longer have power, and the younger generation is not minded to compromise on self-interest and protect them. It is a network which actively exacerbates the vulnerabilities of the old. Significantly, the only one of Sum's sisters who is not destitute has actively curtailed her involvement with her younger relatives and donated her remaining land to the mosque. While Sum's trajectory was extreme, her network's broad experience is

Table 9.8 Key characteristics of seven indicative networks

A	B	C	D	E	F	G	H	I	J
'Name' of network	No. in Gen 1	Average family size	Local kin	Economic status	Land	Religion	Influence	Links	Reputation
'Elite' (Haji Lina)	6	2.3	35 [10]	I	Yes	Islam	Yes (army)	To regional elite	High
Key strategy: early investment in education, formal-sector work, national and international migration with remittances									
'Modernizing, outward-looking' (Siyati)	6	5.0	45 [15]	I-II	Yes	Islam	Yes (village admin, army)	To village admin	High
Key strategy: early investment in education, formal-sector work, national migration with remittances									
'Upwardly mobile' (Jasman)	5	3.0	20 [6]	I-IV	Yes	Javanese/Hindu to Islam	YES (village head)	To regional elite, village admin	High
Key strategy: steady accumulation (shop, land), political ambition, education of selected children, good marriages into state bureaucracy									
'Traditional, localizing' (Sunadi)	3	4.0	30 [15]	II	YES	Islam	Yes (religion, village admin)	To village admin, mosque	High
Key strategy: accumulation of agricultural land, division among children who remain local, little investment in education									
'Locally depleted' (Rini)	7	2.1	15 [2]	III	No	Hindu	No	To neighbours, Hindu temple	Neutral
Key strategy: petty trade, migration without accumulation and remittances									
'Traditional, descending' (Sum)	6	1.0	25 [10]	III-IV	No longer	Javanese, some Islam	No	To neighbours (charity)	Poor
Key strategy: accumulation of land, dispersal of land, lack of investment in education, national and international migration without remittances									
'Inadequate' (Lubis)	2	1.0	5 [0]	IV	No	Non-practising Islam	No	None	Neutral
Key strategy: (failed) attempts to acquire children through adoption and marriage, manual work into old age									

Notes: Column B: Number of siblings in elderly generation (making up Gen 1); not all of these are alive. (In Sunadi's case, Gen 1 comprises three cousins.) Column C: Average number of surviving biological offspring of Gen 1 (making up Gen 2, prior to supplementation through adoption). Column D: Approximate number of adult kin living in the community, with number of close kin (spouse, children, sibs) in brackets. Column E: Central tendency of network's economic status. Column F: Ownership of agricultural land. Column H: Inclusion of influential member(s) in network. Column I: Significant non-kin links. Column J: Indicative reputation of network.

replicated in a number of networks in Kidul where past success in a more traditional Javanese economy and society has not been translated into success in a modernizing environment.

By contrast, Siyati's network has managed to transform itself into a *modernizing, outward-looking* network. A key factor appears to have been the unequal access to land among a large set of siblings with growing families, necessitating pursuit of a non-agricultural strategy for survival.[10] Employment by several network members in a state-owned factory and the army ensured good, reliable income, subsidized housing, and later a small pension. This allowed the second generation to benefit from a relatively good education (though not to university level) and entry into the local and national formal sector. While this has taken a large proportion of the younger generation away from the community, a strong family ethos of mutual, rather than one-way, assistance and interdependence has meant that ties remain strong across distances. Only one network member has participated in international domestic labour migration, and this against the expressed wishes of the parents. The network is devoutly Muslim, has marriage links to other 'good' kin networks (e.g. Haji Lina's), is well-connected to the local mosque and village administration, and has succeeded in avoiding any major scandal. All elderly in this network are well-treated and respected, and all maintain some control over assets, such as house, land, pension, or savings. Relations with neighbours are cordial, but for assistance this is a network that looks to its own ranks first.

An *upwardly mobile* trajectory is also found in the network around the elderly man Jasman. This intensely strategizing network started with none of the privileges of Lina's, Sum's, or Siyati's networks, yet is managing to exploit the opportunities that the changes of the past two decades have created. Jasman and his surviving brother Mis (who is childless) are elderly men directly descended from Madurese migrants, who settled in the incomer's hamlet of Kidul. By cultivating virgin land they acquired small landholdings, and agriculture dominated the livelihoods of the older generation. Jasman is considered a local expert of Javanese mysticism and numerology. Jumadi, the key player in this network, is one of Jasman's five children; he has only basic education. Jumadi was raised a Hindu by his aunt, and to this day a crudely erased swastika tattoo on his arm reminds of this fact, although he converted to Islam upon marriage. Jumadi and his wife (of humble origins but a shrewd penny-capitalist) gradually built up a successful economic base by combining Jumadi's work as a driver with his wife's running a small shop. In the evenings they operated an illegal but hugely profitable beer hall in their living room. When in the early 1990s the Muslim missionaries started gaining followers, Jumadi was offered a post within the village administration, under the condition of 'cleaning up his

[10] According to elderly informants, prior to Independence the oldest child stood to inherit the bulk of parental agricultural land. A change of law (in the 1960s) replaced this with equal inheritance among all children. Apparently families' enactment of this change varied.

act'. His new position entails use rights of prime agricultural land, the profits from which enabled the expansion of the (now entirely respectable) shop. By the time I met them, the couple counted among the rich, but Jumadi's friendly nature and effectiveness as village official meant he remained popular. It came as no surprise that he defeated the incumbent village head in 2001, although a few years earlier such triumph of an upstart over a member of a local elite network would have been unthinkable. Jumadi's success lies in his ability to unite the disparate factions that make up Kidul today by adopting the right tone with everyone and not taking any particular 'sectarian' line. Thus he has skilfully won over both modernist Muslims and villagers for whom Javanese traditions and beliefs remain important, for example by reintroducing the ancient 'village purification ritual' (*slametan bersih desa*). This was traditionally a syncretic, mystical affair, but has been given an Islamic guise through the inclusion of prayers and chants. On his way up the economic and political ladder, Jumadi engaged in generous patronage towards his wider kin, many of whom remain quite poor. His treatment of both his elderly father and childless uncle have been commended as exemplary. He recently arranged very good marriages for his three daughters, thereby forging links directly into the networks connected to the Indonesian bureaucracy and army. Such links are without precedent in this network.

The remaining three networks can be dealt with more succinctly. That of the elderly man Sunadi may be compared with the one just described, because it shares elements of upward mobility, initiative, and gradual consolidation. However, it highlights the drag and inertia that success may acquire. Sunadi's is a *traditional, localizing* network. Sunadi and his wife started off poor but gradually became one of the richest families in Kidul by patiently reinvesting small profits from trade until prized agricultural land could be accumulated. Under the patronage of a previous village head, Sunadi became religious official (*modin*) within the village administration, a role which afforded him respect and further access to land. Yet as in the case of Sum's network, the wealth accumulated by the elderly generation has simply been devolved to the younger generation without a strategy for building on that wealth. All of Sunadi's seven children have been given a house and land in the village. Consequently none of his children has moved away. The younger generation is neither well-educated, nor occupationally particularly successful, although one son has succeeded his father into the role of *modin*. The elders in this network are reasonably secure, but the next generation is poised for decline. Unlike Sum, Sunadi has held onto some land, and where Sum was child-poor, he is child-rich and assured of filial care and support. An elderly childless aunt and childless cousin receive occasional assistance from Sunadi. Sunadi's network type is replicated several times over in the village, namely where the older generation has accumulated wealth and assets (chiefly agricultural land), but then divided up this wealth among children. While this has tied offspring to the community, it has meant missed opportunities in terms of Indonesia's wider socio-economic developments and consequent stagnation.

Opposites of Sunadi's localized, demographically successful network can be found in the last two networks. Rini, the only survivor of seven siblings, is of respectable village stock. She is related to Haji Lina (her mother and Lina's mother were cousins), but in status, size, wealth, and fortunes her network bears no resemblance to Lina's elite network. Contact between these networks is minimal. Three of Rini's sisters remained childless, making this a network with unusually high reproductive failure, and her daughter, Sita, was consequently raised by an aunt. In the late 1960s Rini and several of her network went to Sumatra, following in the wake of poor Javanese migrants sent there as part of Indonesia's transmigration scheme (Rigg 1991). They worked as sellers of Javanese food and did quite well. Sita eventually returned to nurse her ailing adoptive mother and safeguard the small plot of residential land the family owned in Kidul. Rini also returned, but Sita's only son has remained in Sumatra, leaving her *de facto* childless. The elderly mother–daughter duo now has no close family ties in Kidul, merely more distant ties (cousins and beyond). Sita is very active in the Hindu temple and well-respected among the Hindu community. Interactions are chiefly with fellow Hindu kin (none of whom are wealthy), affinal kin who share their economic position, and with a close network of supportive neighbours. Theirs is a *locally depleted* network in which a lack of younger members is compounded by a lack of adequate material resources and non-existent links to positions of local power. This network type is replicated across the community among economically marginal child-poor networks without compensatory connections to wealthy kin or community institutions. At present, the elderly women survive on Sita's income as a petty food seller. Should anything happen to her, there is no safety net and charity would be their lot.

The implications of *inadequate* local networks and reliance on charity are thrown into sharp relief by examining Lubis's as our final network. Lubis moved to Kidul upon marriage to his first wife. This and his subsequent marriage produced no offspring, although both wives have children from a previous marriage. With his second wife Lubis informally adopted two children belonging to a neighbour, but these children's loyalties reverted to their better-off biological parents. Lacking any children or local kin, Lubis worked into deep old age as a generalized labourer, taking whatever work he could get. His second wife left him, and Lubis survived on gifts of cooked food from neighbours. Matters rapidly deteriorated once he required physical care. Initially he depended on a local nephew of his first wife and on a neighbour, but they quickly grew tired of providing intimate care to someone they didn't share blood ties with. Lubis had always been respected for his work ethic. Once physically and materially dependent, he was derided and gossiped about. His carers eventually identified the whereabouts of his closest blood relative (a cousin's son in a nearby town) and left Lubis with him despite the old man's tearful protest. He died soon after. Lubis's fate illustrates the importance of having a local network, both as a source of support and of advocacy. This importance is recognized by villagers, and efforts to build networks through adoption, marriage, or patronage are typically great.

9.6 Conclusion

This chapter set out to present an approach to identifying the subpopulations that are relevant for understanding vulnerability in later life. This has involved starting from local communities which, while by no means bounded, are locally recognized collectivities and socially relevant sites of interaction. In-depth community ethnographies allow the important dimensions of local context, such as economy and stratification, ethos, demographic regime, and religio-political orientation, to be captured and their influence on people's options and actions to be examined. Communities are composed of different subpopulations: for example, in terms of economic position, availability of children, or religious membership. Overlapping memberships to several of these subpopulations may give rise to subgroups which are more or less vulnerable. As the methodology of 'Ageing in Indonesia', outlined earlier in this chapter, brings out, studying vulnerability entails specifying several different subpopulations and relating them analytically to each other so that meaningful aggregates emerge. In the communities we studied, economic disadvantage often went hand-in-hand with demographic disadvantage to produce a significant vulnerable subgroup lacking both material and social resources. In Kidul this group comprised around one in three elders. However, not all vulnerable elders experience a bad outcome, because they are part of larger units which mediate their relative disadvantage. In rural Indonesia the units of greatest social significance for people's lives were shown to be composed of bilateral kinship networks, supplemented by links to neighbours, religious networks, and key community institutions.

Kinship networks are multigenerational, reproductive entities which encompass considerable diversity and cut across socio-economic status and space. They are flexible and dynamic, their membership can be expanded through marriage, adoption, or alliance, or curtailed through divorce, conflict, or avoidance. Their internal diversity allows them to even out and exploit reproductive and economic extremes through the allocation of children, land, labour, economic opportunity, or support. However, as I have tried to show, their relevance for understanding old-age vulnerability and security lies not only in the material flows that occur within networks, but also in their moral force. Although outright poverty affects around one in ten households in contemporary rural Indonesia, virtually none of these households is unconnected to at least some better-off households. Vulnerability or security are in consequence almost always first a question of 'allegiance', by which I mean a question of whether the people in a given network who are *able* to intervene feel a sufficient sense of common interest and common reputation to act on behalf of the vulnerable member. Of secondary (but not negligible) importance are the size, composition, and wealth of a network.

Among the network types characterized, the 'elite', 'modernizing', and 'upwardly mobile' networks appeared to be most successful and to offer the greatest protection to their elderly members. They are networks which in different ways are succeeding in exploiting at a local level the evolving opportunities which contemporary Indonesian society and economy are creating. This means they are doing well economically, but are also aligning themselves favourably with important socio-religious and political trends, thereby mobilizing resources beyond the immediate local network. 'Traditional localizing' networks that have pursued a strategy of investment only in local opportunities and livelihoods are less likely to succeed in the long run, especially in a cultural context in which expectations of downward flows of intergenerational wealth and equal inheritance among children result in the fragmentation and dissipation of wealth. Elders in these kinds of networks may be assured of adequate local support from children, but may preside over their networks' gradual decline in status. Such decline is already realized in so-called 'traditional descending' networks, where failure to hold on to economic privilege has been accompanied by a loss of moral leadership. At best such networks gently unravel, leaving their elderly members surviving on a minimum social and economic base; at worst they disintegrate under the weight of scandal, disharmony, and disrespect, leaving elders completely exposed. 'Locally depleted' and 'inadequate' networks entail obvious vulnerability for elderly members, often through no fault of their own. Sheer numbers matter: networks require a minimum balance of old and young, local and distant members to function adequately. Where childlessness and poverty cluster within a given network, or migration removes network members permanently from potential support roles, then attempts at network building may prove to be no match against demographic disadvantage. Unless there are strategic links to persons or institutions of influence, elderly members of such networks may end their lives in misery.

References

Antlöv, H. (2003). 'Village Government and Rural Development in Indonesia: The New Democratic Framework', *Bulletin of Indonesian Economic Studies* 39/2: 193–214.

Badan Pusat Statistik (2010). 'Sensus Penduduk 2010: Jumlah dan Distribusi Penduduk' [Population Census 2010: Total and Distribution of Population]. Available online at <http://sp2010.bps.go.id/index.php/site/index>, accessed 31 July 2012.

Barnes, R. H. (2004). 'On the Margins of the Middle Class: Becoming Middle Class in Rural Eastern Indonesia', *Asia-Pacific Forum* 25: 33–58.

Beard, V. A. and Y. Kunharibowo (2001). 'Living Arrangements and Support Relationships among Elderly Indonesians: Case Studies from Java and Sumatra', *International Journal of Population Geography* 7/1: 17–33.

Beatty, A. (1999). *Varieties of Javanese Religion: An Anthropological Account.* Cambridge: Cambridge University Press.

Beatty, A. (2005). 'Feeling Your Way in Java: An Essay on Society and Emotion', *Ethnos* 70/1: 53–78.

Benda-Beckmann, F. von (1988). 'Islamic Law and Social Security in an Ambonese Village', in F. von Benda-Beckmann, K. von Benda-Beckmann, B. O. Bryde, and F. Hirtz (eds), *Between Kinship and the State: Social Security and Law in Developing Countries.* Dordrecht: Foris, 339–366.

Breman, J. and G. Wiradi (2002). *Good Times and Bad Times in Rural Java.* Leiden: KITLV Press.

Cederroth, S. (1995). *Survival and Profit in Rural Java: The Case of an East Javanese Village.* Richmond, Surrey: Curzon Press.
Chen, A. J. and G. W. Jones (1989). *Ageing in ASEAN: Its Socio-Economic Consequences.* Singapore: Institute of Southeast Asian Studies.
Djamour, J. (1959). *Malay Kinship and Marriage in Singapore,* LSE Monographs on Social Anthropology. London: Athlone Press (University of London).
Finch, J. and J. Mason (1993). *Negotiating Family Responsibilities.* London and New York: Routledge.
Freeman, J. D. (1961). 'On the Concept of the Kindred', *Journal of the Royal Anthropological Institute* 91/2: 192–220.
Geertz, C. (1960). *The Religion of Java.* Chicago: University of Chicago Press.
Geertz, H. (1961). *The Javanese Family: A Study of Kinship and Socialization.* New York: Free Press of Glencoe.
Goody, J. (1996). 'Comparing Family Systems in Europe and Asia: Are There Different Sets of Rules?', *Population and Development Review* 22/1: 1–20.
Gooszen, A. J. (1980). '(Kinder)rijke en Arme Gezinnen', in R. N. J. Kamerling (ed.), *Indonesië Toen en Nu.* Amsterdam: Intermediar, 53–66.
Grant, M. and S. Yeatman (2012). 'The Relationship between Orphanhood and Child Fostering in Sub-Saharan Africa, 1990s–2000s', *Population Studies* 66/3: 279–295.
Hajnal, J. (1982). 'Two Kinds of Preindustrial Household Formation System', *Population and Development Review* 8/3: 449–494.
Hart, G. (1986). *Power, Labor, and Livelihood: Processes of Change in Rural Java.* Berkeley and Los Angeles: University of California Press.
Hefner, R. W. (2000). *Civil Islam: Muslims and Democratization in Indonesia,* Princeton Studies in Muslim Politics. Princeton: Princeton University Press.
Hugo, G. (1982). 'Circular Migration in Indonesia', *Population and Development Review* 8/1: 59–83.
Hugo, G. (2000). 'Lansia: Elderly People in Indonesia at the Turn of the Century', in D. R. Phillips (ed.), *Ageing in the Asia-Pacific Region: Issues, Policies and Future Trends.* London: Routledge, 299–321.
Hugo, G., T. Hull, V. Hull, and G. Jones (1987). *The Demographic Dimension in Indonesian Development.* Singapore: Oxford University Press.
Hull, T. and V. Hull (1977). 'The Relation of Economic Class and Fertility: An Analysis of Some Indonesian Data', *Population Studies* 31/1: 43–57.
Hull, T. and Tukiran (1976). 'Regional Variations in the Prevalence of Childlessness in Indonesia', *Indonesian Journal of Geography* 6/32: 1–25.
Hull, V. (1976) *The Positive Relation between Economic Class and Family Size in Java: A Case Study of the Intermediate Variables Determining Fertility,* Monograph Series. Yogyakarta: Population Institute, Gadjah Mada University.
Hüsken, F. (2001). *Social Change and Support Systems: Gondosari Revisited,* paper presented at the Third EUROSEAS Conference, SOAS, University of London, 6–9 September.
Indrizal, E. (2004). 'Problems of Elderly without Children: A Case Study of the Matrilineal Minangkabau, West Sumatra', in P. Kreager and E. Schröder-Butterfill (eds), *Ageing Without Children: European and Asian Perspectives.* Oxford: Berghahn, 49–76.
Indrizal, E., P. Kreager, and E. Schröder-Butterfill (2009). 'The Structural Vulnerability of Older People in a Matrilineal Society: The Minangkabau of West Sumatra, Indonesia',

in J. Sokolovsky (ed.), *The Cultural Context of Aging: Worldwide Perspectives*, 3rd edn. Westport: Praeger, 383–394.

Jay, R. (1969). *Javanese Villagers: Social Relations in Rural Modjokuto*. Cambridge, MA: MIT Press.

Jones, G. W. (2001). 'Which Indonesian Women Marry Youngest, and Why?', *Journal of Southeast Asian Studies* 32/1: 67–78.

Kato, T. (1982). *Matriliny and Migration: Evolving Minangkabau Traditions in Indonesia*. Ithaca: Cornell University Press.

Kreager, P. (2006). 'Migration, Social Structure and Old-Age Support Networks: A Comparison of Three Indonesian Communities', *Ageing and Society* 26/1: 37–60.

Kreager, P. (2011). 'The Challenge of Compositional Demography', *Asian Population Studies* 7/2: 85–88.

Kreager, P. and E. Schröder-Butterfill (2007). 'Gaps in the Family Networks of Older People in Three Rural Indonesian Communities', *Journal of Cross-Cultural Gerontology* 22/1: 1–25.

Kreager, P. and E. Schröder-Butterfill (2008). 'Indonesia Against the Trend? Ageing and Inter-generational Wealth Flows in Two Indonesian Communities', *Demographic Research* 19/52: 1781–1810.

Kreager, P. and E. Schröder-Butterfill (2010). 'Age-Structural Transition in Indonesia: A Comparison of Micro- and Macro-Level Evidence', *Asian Population Studies* 6/1: 25–45.

Lont, H. (2002). 'Juggling Money in Yogyakarta: Financial Self-Help Organisations and the Quest for Security', unpublished PhD thesis, University of Amsterdam.

Minister for Social Affairs (2002). 'Indonesia: Statement by Minister for Social Affairs of the Republic of Indonesia at Second World Assembly on Ageing', Madrid, Spain, 9 April.

Niehof, A. and F. Lubis (eds) (2003). *Two Is Enough: Family Planning in Indonesia under the New Order (1968–1998)*. Leiden: KITLV Press.

Nugroho, H. (1996). 'The Social Meaning of Money in Java', *Internationales Asienforum* 27/3–4: 301–326.

Preston, S. H. and L. G. Martin (1994). 'Introduction', in S. H. Preston and L. G. Martin (eds), *Demography of Aging*. Washington DC: National Academy Press, 1–7.

Randall, S., E. Coast, and T. Leone (2011). 'Cultural Constructions of the Concept of Household in Sample Surveys', *Population Studies* 65/2: 217–229.

Rigg, J. (1991). 'Land Settlement in Southeast Asia: The Indonesian Transmigration Programme', in J. Rigg (ed.), *Southeast Asia: A Region in Transition*. London: Allen and Unwin, 80–108.

Rinaldo, R. (2008). 'Muslim Women, Middle Class Habitus, and Modernity in Indonesia', *Contemporary Islam* 2/1: 23–39.

Schröder-Butterfill, E. (2004a). 'Adoption, Patronage, and Charity: Arrangements for the Elderly Without Children in East Java', in P. Kreager and E. Schröder-Butterfill (eds), *Ageing Without Children: European and Asian Perspectives*. Oxford: Berghahn, 106–146.

Schröder-Butterfill, E. (2004b). 'Intergenerational Family Support Provided by Older People in Indonesia', *Ageing and Society* 24/4: 497–530.

Schröder-Butterfill, E. (2006). 'The Role of Secular and Religious Community Institutions for Welfare in Rural Indonesia', paper given at the Cambridge Group for the History of Population and Social Structure.

Schröder-Butterfill, E. and T. S. Fithry (2014). 'Care Dependence in Old Age: Preferences, Practices and Implications in Two Indonesian Communities', *Ageing and Society* 34/3: 361–387.

Schröder-Butterfill, E. and P. Kreager (2005). 'Actual and De Facto Childlessness in Old Age: Evidence and Implications from East Java, Indonesia', *Population and Development Review* 31/1: 19–55.
Scott, J. (1976). *The Moral Economy of the Peasant*. New Haven: Yale University Press.
Singarimbun, M. and T. Hull (1977). 'Social Responses to High Mortality Which Act to Support High Fertility', in *Proceedings of the IUSSP International Population Conference, Mexico*. Liege: IUSSP, 225–239.
Skinner, G. W. (1997). 'Family Systems and Demographic Processes', in D. I. Kertzer and T. Fricke (eds), *Anthropological Demography: Toward a New Synthesis*. Chicago: University of Chicago Press, 53–95.
Suryadinata, L., E. N. Arifin, and A. Ananta (2003). *Indonesia's Population: Ethnicity and Religion in a Changing Political Landscape*. Singapore: Institute of Southeast Asian Studies.
Todd, E. (1985). *The Explanation of Ideology: Family Structures and Social Systems*. Oxford: Basil Blackwell.
White, B. and G. Wiradi (1989). 'Agrarian and Nonagrarian Bases of Inequality in Nine Javanese Villages', in G. Hart, A. Turton, and B. White (eds), *Agrarian Transformations: Local Processes and the State in Southeast Asia*. Berkeley: University of California Press, 266–302.
White, D. and T. Schweizer (1998). 'Kinship, Property Transmission, and Stratification in Javanese Villages', in T. Schweizer and D. White (eds), *Kinship, Networks, and Exchange*. Cambridge: Cambridge University Press, 36–56.

CHAPTER 10

CONSTRUCTING MIGRATION IN SOUTH-EAST ASIA

conceptual, empirical, and policy issues

GRAEME HUGO

10.1 Introduction

Migration has always been an integral part of population change, influencing not only the culture, society, and economy of the communities which people depart from or move to, but also shaping in multiple ways the lives of those who move. Yet while it is one of the trilogy of demographic processes along with fertility and mortality, it has been described as the 'stepchild of demography' (Goldstein 1976) because it is often ignored, misrepresented, or represented only in a partial way in demographic considerations of population change. This demographic neglect has its roots in the conceptual fuzziness surrounding migration and a seemingly intractable measurement problem. Unlike fertility and mortality, measurement of migration is always only going to capture a part of the totality of movement, making comparisons between populations difficult. More importantly, it is the part of migration which is captured in traditional statistical collections which tends to be the focus of research. Yet often the movement that is missed also has considerable social, economic, political, and cultural significance.

The conceptual and measurement problem has bedevilled the study of migration, not only at particular points in time but it has also made it difficult to analyse changes over time. There is no doubt that there are complex and significant interrelationships between mobility on the one hand and economic development and social change processes on the other. Yet whereas the demographic transition has provided a productive theoretical basis for understanding changes in fertility and mortality over time, these barriers have prevented similar success in examining the evolution of migration over time, although there have been some important attempts (e.g. Zelinsky 1971).

This chapter focuses on one of the most diverse and economically and demographically dynamic regions of the world to examine some key conceptual and empirical challenges in migration research from a demographic perspective. It first of all briefly introduces some of the key economic, social, and demographic characteristics of the South-East Asian region. It then addresses some of the challenges which we face in conceptualizing population movement in that region. It is argued that traditional demographic concepts of migration have long failed to capture the diversity of mobility in South-East Asia but this has never been more true than in the contemporary situation. Consideration is given to both internal and international migration, partly because both are of considerable scale and significance but also because, it is argued, the understanding of population movement has been hampered by the imposition of dichotomous categories on migrations when there is often as much in common across as within those categories (King 2002; King and Skeldon 2010). Firstly, some comments are made on historical migration patterns. In the contemporary context it is shown that non-permanent international movement is more significant in scale in South-East Asia than permanent migrations using Australian migration flow data which is unusually comprehensive. The focus is then turned to internal migration in South-East Asia where a similar argument is made. It is demonstrated that non-permanent forms of international and internal migration in the region are of significance because they influence the fundamental population of areas, and also because they are having important developmental impacts not only in destinations but also in origin areas as well. Finally, some of the policy implications of the growing significance of non-traditional forms of mobility are discussed. It is argued that demography's contribution to informing the growing desire of governments in the region to intervene in migration processes is hampered by an adherence to traditional concepts and measures of migration.

10.2 The South-East Asian Region

South-East Asia includes the eleven nations[1] shown in Figure 10.1 which comprise the Association of Southeast Asian Nations (ASEAN). The last two decades have seen unprecedented economic and social transformations in the South-East Asian region and these have been both a cause and a consequence of major demographic shifts. One of the salient features here has been the fast growth of most economies in the region. Rates of economic growth have generally been more than double the global average over recent decades, even during the Global Financial Crisis. South-East Asia contains some of the world's most dynamic economies in Singapore, Malaysia, Thailand, Vietnam, and Indonesia, but also some countries which have generally experienced slow growth (e.g. the Philippines) and others whose lack of development is reflected in their lack of basic economic data. However, while economic growth has been substantial in most countries, there is greater variation between countries in the base on which this growth is occurring and the extent to which the benefits of growth are shared by groups and regions within those countries.

[1] Indonesia, Timor Leste, Malaysia, Singapore, Brunei Darussalam, Thailand, Myanmar (Burma), Cambodia, Laos, Vietnam, and the Philippines.

FIG. 10.1 South-East Asia.

There is enormous variation between and within countries in the region with respect to economic and social development. The World Bank (2011) divides developing countries with more than 1 million inhabitants into low- and middle-income economies (with GNI per person of US $995 being the threshold). By their classification, three South-East Asian countries (Cambodia, Laos, and Myanmar) can be classified as low income. Four countries (Cambodia, Indonesia, Philippines, and Thailand) are rated as 'lower-middle'-income countries while Malaysia is considered an upper-middle-income country. Countries with per capita incomes of US $12,196 or more were considered 'high income' and Singapore and Brunei were in this category in 2009.

Among the most important of these developments is the structural change which has seen the proportion of the workforce engaged in agriculture decline substantially in most countries; significant improvements in participation in education have made primary education almost universal throughout the region; and major changes are occurring in the structure and functioning of the family with a rapid transition from an extended to a nucleated system.

Associated with the massive economic and social shifts in the region have been a number of demographic transitions which in most South-East Asian countries have occurred more rapidly than was the case in Euro-American societies. There are wide variations across the region in population size, growth, and structure but what is common to most is that rapid changes are occurring in the basic demographic processes of fertility, mortality, and migration. Indonesia is by far the largest nation, having over 40% of South-East Asian residents. On the other hand, four nations in the region had less than 10 million residents. All countries in the regions have experienced declines in fertility, although these have been much more substantial in Singapore, Thailand, Vietnam, and Indonesia than elsewhere. Accordingly, countries are situated at different points along the demographic transition, with Singapore having very low fertility and a rapidly ageing society while the Philippines is still growing rapidly and remains quite young.

10.3 Conceptualizing Migration in South-East Asia

There is an increasing appreciation of the significance of place across the social sciences. Issues of place-related identity, the influence of place on social, cultural, and economic behaviour, spatial dimensions of and influences upon economies, labour markets, housing markets, and a myriad of cultural, social, and economic processes are becoming central in social science research. Yet demographic approaches to place remain largely unchanged. Censuses allocate individuals to a point on the earth's surface where they happened to be on the night of the census enumeration, or which is identified as their 'usual place of residence'. Thereafter all of their characteristics are assigned to that location. Yet throughout history people have never been static. Workplace and home place were often separated by some distance.

Hagerstrand (1957) distinguishes migration and circulation as depicted in Figure 10.2. He recognizes that people have a major place of usual residence which they identify as 'home' but they engage in a great deal of circular movement around that place. Migration occurs when an individual or group change this centre of gravity of their mobility. Zelinsky (1971: 225–226) makes the distinction between conventional migration as 'any permanent or semi-permanent change of residence' and circulation as 'a great variety of movements, usually short term, repetitive or cyclical in nature, but all having the lack of any declared intention of a permanent or long-lasting change in residence'.

While conceptually it is possible to make this clear distinction between permanent migration and circulation, in practice it is not as clear-cut as it may seem. The reality is that there is considerable blurring in the permanent/temporary polarization. It could be argued that mobility is more appropriately conceptualized along a continuum. At one end of the continuum is short-term circular movement around the home place such as commuting to a workplace or educational institution. At the other end is a permanent shift in the home place around which short-term circulation takes place. However, there are also a lot of other movements inbetween where the absence from the home place grows longer. In South-East Asia labour migration usually involving an absence of around two years from the home place is the most common form of internal and international migration.

Hugo (1983: 87) has argued that an important concept when considering the difference between permanent and circular migration is that of *commitment*. He argues that all migrants can be located along the continuum depicted in Figure 10.3 according to the degree of commitment that they have to their origin and that which they feel toward the destination. To take an extreme example, an unskilled labourer, circular migrant A, moves from Indonesia to Malaysia with the intention of earning sufficient money to buy a small business in his home community. He leaves his whole family behind in Indonesia, he retains his Indonesian citizenship, and his intention is that once he has reached his

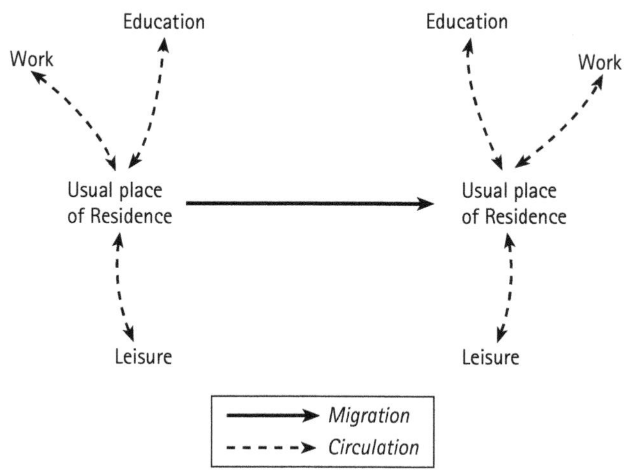

FIG. 10.2 The distinction between migration and circulation.

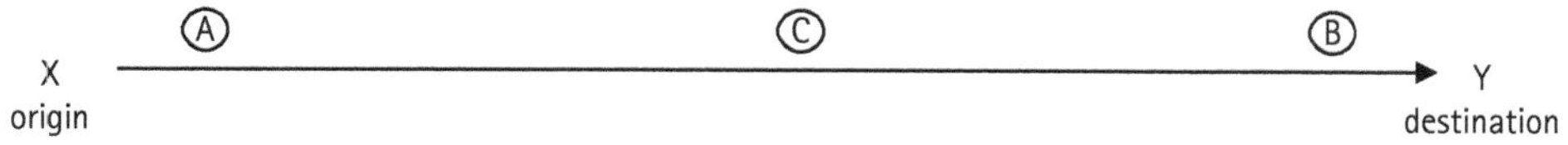

FIG. 10.3 The commitment continuum of migration.

target earnings he will return to Indonesia and stay there. His commitment is almost totally to his Indonesian origin. On the other hand, B in the figure is also Indonesian but she has obtained her high-school and university education in medicine in Adelaide, Australia. During her study she has met a fellow doctor whom she has married and had children with. Her parents have died so she only has distant relatives in Indonesia. She is an Australian citizen and sees her future and that of her family as being totally Australian. Her commitment to her Indonesian home which she left as a child is very limited.

These are cases near the extremes of migrants being almost totally committed to their origin or to their destination. However migrants can be located at any point along the continuum according to the relative mix of their commitments to the origin and their destination. Migrant C, for example, is an ethnic Chinese businessman who left Indonesia during the anti-Chinese riots of 1998 and took his family to Australia where he has set them up in Sydney. His children have gone to Australian schools and one has met and married an Australian partner. However he has retained his business interests in Jakarta and circulates regularly between Australia and Jakarta. He has developed business interests in Australia but still retains his Indonesian enterprise and his parents and siblings still live in Jakarta. His commitment is equally divided between origin and destination. Ley and Kobayashi (2005) show this situation applies to many Hong Kong Chinese in Canada. Moreover, they explain that this group move their place of usual residence between Hong Kong and Canada according to the stage of their life cycle—in Canada when educating their children, returning to Hong Kong in their older years to access family support.

Assimilation models of settlement assume that immigrants forsake their heritage and origins and shift from X to Y quickly. In reality this is not common. Most movers retain a mix of commitment to origin while developing commitments to their destination. Indeed, the concept of transnational families which is increasingly relevant in South-East Asia (Huang, Yeoh, and Lam 2008) involves family members being physically distributed between origin and destination with a mix of commitments to the two. A key distinction between circular migrants and permanent migrants is that the former are located closer to X in Figure 10.3 while permanent migrants are closer to Y.

In most contexts involving both mobility within and between countries, it is possible to develop a number of indicators which could be used to measure the level of commitment to origin and destination for particular migrants. This could include, for example:

- The *location of family members*—leaving nuclear family members in the origin is a strong indication of maintaining a major commitment to the origin community while having only extended family remaining in the origin is also indicative of linkages, albeit less strong.

- In international movement, maintaining full *citizenship* (or even joint citizenship) at the origin is a clear reflection of a strong commitment while temporary resident status at the destination is indicative of a weaker association than with the origin. In some countries this indicator also applies to internal migration since there are registration systems which require movers to nominate their home place. This reaches an extreme situation with the *hukou* system in China (Chan and Zhang 1999), or *Ho Khau* in Vietnam (Hong 2008: 238–243), where there are major differences in entitlements according to where the mover is registered.
- The *location of property* owned by the migrant is also reflective of their relative connections with origin and destination. This is especially the case for businesses, agricultural land, and other fixed assets, but house ownership is also of relevance. Maintenance of substantial location-specific capital (DaVanzo 1981) in origin areas is a clear indicator of commitment.
- The regularity and scale of *remittances* also is indicative of the strength of commitment to the origin community. Circular migrants tend to send more money home and more often than permanent migrants since they are usually supporting the balance of their nuclear family that remains there.
- The balance of locations of *bank accounts* and investments is also an indicator of the degree of commitment to origin and destination.
- The *ethnicity/ancestry of one's partner* can be another indicator of extent of commitment to the origin community. It is likely that where both partners come from the origin country that the family's linkages home will be stronger than if one is a native of the destination.
- *Language* can also be a relevant marker. The extent to which the origin language is spoken at home in day-to-day family communication can reflect the strength of identification with the origin. Similarly the ability of the immigrant with respect to speaking the destination country language can also be an indicator of the extent of commitment to origin and destination. This can apply to internal as well as international migration. For example, in Indonesia each region has its own local language and there is a *lingua franca*, Indonesian, which is most spoken in major cities.
- The extent of *cultural maintenance* can also be a gauge of the degree of commitment to the origin. This may be associated with active membership of migrant associations and organizations based in the destination country.
- *Voting* rights and behaviour may also reflect the relative degree of commitment to origin and destination.

When taken alone and in isolation, each of these *elements* may be fallible as a direct indicator of the relative degrees of commitment to the origin and destination but if they are considered together they should give a clear indication of the relative strength of the migrants' associations with home and destination. The aggregate measure obtained can locate them along the continuum presented in Figure 10.3. Most first-generation movers will not be at either of the extreme poles of total commitment to origin or destination but will have a mix of linkages with both. Circular migrants will be more to the left of the diagram and permanent migrants to the right. Seen in this way, there are clear overlaps between circular and permanent migration.

The study of circular and other temporary migration is hampered by a lack of data relating to its scale, composition, and impact. Standard demographic measurement of migration stocks and flows predominantly captures movement on the far right of Figure 10.3. Accordingly, the scale of non-permanent movement is not appreciated and little evidence is available to policymakers to resolve policy dilemmas regarding it. This applies to both internal and international migration. In the latter, as transnationalism has replaced permanent settlement as the dominant paradigm in international migration research (Glick-Schiller et al. 1992; Portes et al. 1999; Vertovec 2004) there is a growing 'disconnect' between standard migration measurement and the actual nature of population mobility between countries. More than three decades ago, Mitchell (1978: 6–7) pointed out that circular migration within countries has:

> ... remained remarkably intractable to thoroughgoing analysis. Part of this analytical recalcitrance derives from the great difficulties in collecting suitable data to carry out adequate theoretical formulations.

This judgement remains applicable to the study of internal migration but has particular relevance to international migration as it becomes more apparent that the relationships which the linkages established by migration between origin and destination are of major social and economic significance.

The fundamentals of conceptualizing and measuring internal and international migration in standard traditional sources such as population censuses, national labour force surveys, and border control data collections have remained very much as they were articulated in the key United Nations manuals published in 1970 and 1980. Accordingly, the questions which are asked in censuses on internal migration involve detecting permanent changes in residence, usually since the last census. Non-permanent movements are not detected at all, although there is considerable potential in using journey-to-work questions to get some indications. Studies of international migration have likewise focused on permanent relocations, which means that in both areas—internal and international migration—there has been a bias toward what Ley and Kobayashi (2005: 1120) term the 'narrative of departure, arrival and assimilation'. Recommendations on collecting international migration data, even in recent authoritative documents (Center for Global Development 2009), are destination-biased and largely involve people who have made a permanent move. However, increasingly in both internal and international migration much of the social, economic, and political effect of migration is delivered:

- through non-permanent, circular flows;
- by relationships maintained by migrants with their origins;
- in origin communities.

In each of these areas traditional migration data collection is found wanting and this greatly reduces the extent to which we can provide representative, timely information to policymakers on the scale, composition, and effects of mobility on economic and social development.

This disconnect between the nature of contemporary mobility and established ways of measuring them is not new. It was emphasized by scholars working on internal migration decades ago (Chapman 1979; Mitchell 1978; Chapman and Prothero 1985; Bedford 1973; Goldstein 1976; Hugo 1978, 1982). It is interesting, however, that despite the growing dominance of the transnationalism paradigm in international migration research, demographers remain largely wedded to the 'narrative of departure, arrival and assimilation' at least as far as measurement of international migration is concerned. In both internal and international migration, however, it is clear that modern forms of communication and information technology and a massive improvement and cost reduction in transport technology have made it more possible than any time in human history to:

- engage in non-permanent forms of mobility in which individuals and families can have more than a single 'locus of activity' in Hagerstrand's (1957) terms;
- keep immediate, regular, intimate, and cheap contact with relatives, friends, businesses, etc. at very distant locations using modern forms of information and communication technology.

There is then a growing gap between the movement which is being captured by our traditional demographic data collection systems and the mobility which is shaping and being shaped by contemporary economic development and social change.

These shortcomings are increasingly evident in the failure of existing migration data to have a productive input in research into the impact of migration (especially international migration) on development in origin areas (United Nations 2006, 2010; World Bank 2006; Global Commission on Migration 2005). It is apparent that migration can result in positive (as well as negative) impacts in origin communities through remittances, engagement of the diaspora, and return migration. However, the migration datasets which have been assembled to study migration and development by the World Bank, OECD, and United Nations all relate to destinations and largely include stocks of permanent migrants. They provide limited insight for studies seeking to investigate the full impact of mobility on economic development and social change in origin communities.

10.4 Historical Context of Mobility in South-East Asia

The stereotype of Asians being born, living, working and dying in a single rural village without having ventured far beyond it has never been accurate. However, mobility was limited in both pre-colonial and colonial times in the region. The colonial situation was predominantly one in which, as Naim (1974: 3) points out in the Indonesian context:

> with the exception of the privileged native aristocrats and also traders communication systems and travelling facilities could not be enjoyed by the ordinary natives.

The 1930 Dutch East Indies (Indonesian) census gives us one of the earliest accurate quantitative pictures of mobility in the region since it indicated the proportion of 'indigenous' population living outside the district in which they were born. Some 11.5% were in this situation but Table 10.1 shows that this proportion varied between different parts of the archipelago. It indicates that even in colonial times a significant percentage of Indonesians moved.

Hugo (1975) notes in his study of West Java a number of types of internal migration during the colonial period:

- migration in response to crop failures (Kroef 1956: 746);
- flight of inhabitants of regions in which the exactions of the colonial regime were especially severe (Widjojo (1970: 36, 37); Schrieke (1957: 300));
- the movement of some ethnic groups which traditionally had high levels of mobility such as the Minangkabau of West Sumatra and the Bugis of South Sulawesi;
- disaster-related migration such as that associated with the evisceration of Krakatoa in 1883 (Furneaux 1965);
- colonization efforts by the Dutch to resettle people from densely settled areas in new migration areas;
- significant migration to colonial towns and cities (for example, Table 10.2 shows the proportion of the population in major urban areas in 1930 in West Java who were inmigrants).

In addition, although most of the 1930 census data refer to permanent migration, it is clear that the colonial census takers were aware of the existence of significant temporary movements. Enumerators in Java were instructed to determine each

Table 10.1 Indonesia: Percentage of indigenous residents in major regions living outside of their district (if born in Java) or subdivision (if born outside Java) of birth, 1930

Region	Percentage	Region	Percentage
West Java	9.1	Celebes	7.7
Central Java	5.4	Bali-Lombok	2.6
East Java	11.8	Moluccas	8.8
Sumatra	18.5	Outer Islands	2.3
Borneo	14.4	Total	11.5

Source: Volkstelling (1933–1936).

Table 10.2 West Java: Origin of indigenous in-migrants to major cities and towns, 1930

	In-migrants		Origin of In-migrants (percent)			
City	Number	Percent of Total	West Java	Central Java	East Java	Outer Island
Batavia	166,932	51.2	79.4	11.8	2.9	5.9
Meester-Cornelis	34,770	41.6	58.6	27.6	4.2	9.6
Buitenzorg	11,544	21.7	67.5	17.7	4.2	10.6
Bandung	71,766	55.2	70.4	20.4	4.5	4.7
Sukabumi	9,785	36.0	76.4	14.7	5.0	3.9
Cirebon	18,123	42.5	75.7	20.1	2.2	2.0
Serang	3,268	35.4	68.0	17.6	3.2	11.2
Cimahi	9,454	51.6	30.5	41.7	5.3	22.5

Source: Volkstelling (1933).

person's domicile (Volkstelling 1933, I: 93). All persons regarded as temporarily present at their place of enumeration were counted not within that district but within their district of domicile or usual residence. Hence men working at a sugar factory or town who intended to return (even if several months in the future) to the village where they had left their family were considered as having their domicile in that village. The census report further states that when difficult cases were encountered, as happened frequently in large towns, the criterion of a 'fixed domicile' was a minimum period of six months spent in the place of enumeration or intended to be spent there.

No details were published regarding the persons 'temporarily present' apart from the percentages of the total population of each province who were so categorized. These figures, presented in Table 10.3, show that nearly half of all temporarily present persons in Java were enumerated in West Java, which accounted for only 27.32% of the island's total population.

Discussions of population mobility are frequently confined to more permanent types of movement. However, it is clear from the literature that colonialism in Indonesia also had a significant impact on circular or temporary types of population movement. Some temporary migration was associated with much of the labour used on plantations in both Java and the Outer Islands, but this was only one of several important circular population movements in colonial Indonesia. The migration literature written during the colonial period, while meagre, nonetheless reflects the importance of non-permanent forms of mobility. Scheltema (1926: 872–873) and Ranneft (1916), for example, produced a typology of population movement (Table 10.4), the most striking feature of which is its emphasis upon circular movements.

Table 10.3 Java-Madura Provinces: Persons temporarily present, 1930

	Temporarily Present	
Province	Percent	Number
West Java	0.76	83,360
Central Java	0.21	23,297
Yogyakarta	0.18	2,806
Surakarta	0.24	6,156
East Java	0.38	57,212
Java-Madura	0.42	172,831

Source: Volkstelling (1933, III: 32).

Table 10.4 Ranneft's (1916) and Scheltema's (1926: 872–873) classification of population mobility in Java during the colonial period

Type of Movement	Scheltema's Example
(i) Local Movements	Movements to a neighbouring village (e.g. marriage migration)
(ii) Temporary Movements	Movements to an area to construct a railway or irrigation works or to go on *haj* (pilgrimage to Mecca)
(iii) Periodic Movements	Movements caused by seasonal demands for labour such as the Bantamese (people from the northwestern section of West Java) moving to Lampung (Southern Sumatra) to pick pepper
(iv) Semi-Permanent Movements	People who earn their living in one place while their family stays at another, e.g. clerks and servants working in the city while their family stays in their home village
(v) Permanent Movements	a. Within Java b. To other Indonesian islands c. To overseas destinations

Source: Hugo (1975).

Turning to international migration, both permanent and temporary movements were in evidence. Again, drawing on Indonesia as an example, Hugo (1975) notes the following:

- substantial immigration of not only Europeans but also Chinese and Indians to undertake key roles in the colonial economy
- some settlement of Indonesians in Surinam
- deployment of contract labour to Malaya, Fiji, and other areas
- *Haj* pilgrimages to the Middle East.

In short there are strong historical precedents in South-East Asia of population mobility. There is little doubt, however, that the contemporary period has seen a substantial increase in the scale and diversity of movement.

10.5 International Migration in South-East Asia

Of all the demographic, economic, and social changes which have transformed South-East Asia since 1950, few have been so dramatic, or as far-reaching in their impacts, as the increase in personal mobility. Migration, on a permanent or temporary basis, is now within the calculus of choice of most South-East Asians as they weigh up their life chances. This has involved both internal and international migration. Historically South-East Asia has been an important crossroads for migrations from China in the north and India and Europe from the west while its massive internal diversity has resulted in high levels of mobility within the region. In the contemporary world, globalization, the internationalization of capital and labour markets, and the revolution in transport and communications has meant that international population mobility has assumed an even more important role in the region.

However, official data on international migration (United Nations 2013) provides only a limited picture of the significance of international migration in the region. This is partly because the data focus only on destinations and not on origins and the region contains some of the world's major origin countries of migrants. However, it is also because official data, drawn mainly from censuses, is limited in the extent to which it picks up non-permanent migrations. There is also the issue of a failure to include undocumented migrants. Nevertheless, these data indicate that over 7 million people living in South-East Asian countries were born in a different country.

10.5.1 Labour Migration

It is important to appreciate that historically, non-permanent migrations have been important in South-East Asia along with permanent relocations. In fact, almost all of the native populations of the countries of the region except for small Aboriginal populations are descendants of migrants from China and, to a lesser extent, South Asia. However, even during historic times non-permanent migration was also very important. The tradition of sojourners from China who travelled into the region to work or trade while leaving their family behind in China occurred both before and after European colonial domination. Moreover, during colonial times there was significant deployment of 'contract coolies' between colonies which was a precursor to contemporary contract labour migration (Hugo 1975: 135). However, for the quarter-century following World War II there was very little international migration in South-East Asia. Most substantial were the movements associated with decolonization as several of the countries gained their independence from European colonial powers. Since the 1970s, however, international migration of South-East Asians within and out of the region has gathered momentum, increasing not only in scale but also in diversity and economic and social significance.

The largest-scale contemporary international migration in South-East Asia is temporary contract labour migration which involves both low- and high-skilled workers, although the former predominate and face the greatest difficulties. While these movements have their origins in the 'contract coolie' programmes of colonial times, in the contemporary era they began with the flows of Thai and Filipino workers into Vietnam to work for the Americans during the Indo-China War. Then, following the 1973 increase in oil prices which fuelled a building boom in the Middle East, migrant workers from the Philippines and Thailand, and later Indonesia, flowed to Saudi Arabia and other labour-short, oil-producing Middle Eastern countries. Since then, the scale of movement and the number of countries involved in migration has increased. With the emergence of labour shortages in Japan and the 'Asian Tigers' during the 1980s and 1990s, these low-skilled migrant workers increasingly moved to Japan, South Korea, Taiwan, Hong Kong, and Singapore.

Table 10.5 shows that the main origin countries of labour migrants in the region send workers to both the Middle East and other Asian countries. It is important to emphasize that while there is significant movement between ASEAN countries,

Table 10.5 South-East Asian countries: Estimates of stocks of migrant workers in other countries, 2010

Origin Countries	Number	Main Destinations
Brunei	24,300	Australia, Canada, UK, US, Germany, Ireland, Malaysia, Netherlands, NZ
Myanmar	514,200	Thailand, US, India, Malaysia, Australia, UK, Japan, Canada, Korea, Germany
Thailand	810,000	Saudi Arabia, Taiwan, Myanmar, Singapore, Brunei, Malaysia, Cambodia, US, Germany, Japan, Korea, Australia, UK
Laos	366,600	Thailand, US, France, Australia, Canada, Japan, Germany, Belgium, Cambodia, NZ
Cambodia	350,400	Thailand, US, France, Australia, Canada, NZ
Vietnam	2,200,000	Korea, Japan, Malaysia, US, Australia, Canada, Cambodia, Germany, France, Thailand, UK
Philippines	9,400,000	Middle East, Malaysia, Japan, US, Canada, Australia, Italy, UK
Malaysia	1,480,000	Singapore, Australia, Brunei, UK, US, Canada, NZ, India, Japan, Germany
Indonesia	5,500,000	Malaysia, Saudi Arabia, Taiwan, Singapore, South Korea, United Arab Emirates
Total	20,645,500	

Source: Scalabrini Migration Centre (2012).

two-thirds of the stock of ASEAN emigrants shown in Table 10.5 are in destinations outside the region. The Philippines and Indonesia are two of the world's major emigration nations, but Vietnam and, to a lesser extent, Cambodia and Myanmar are of increasing significance. There has been an increase in the scale of outmovement of labour migrants from these ASEAN origin countries and this can be demonstrated with the case of Indonesia. Figure 10.4 shows the number of official labour migrants processed by the Indonesian Ministry of Labour from 1979 to 2011. However, since most workers are on two-year or longer contracts, the actual number of official OCWs overseas in any one year is greater than the numbers deployed in an individual year. In late 2008, it was announced by the Minister for Labour (*Asia Migration News*, December 2008) that there were 5.8 million Indonesians working overseas, including those with official permission as well as undocumented migrant workers. This represents around 5% of the total national workforce. Undocumented labour migration out of Indonesia remains substantial, especially to Malaysia. Accordingly, it is difficult to arrive at an accurate estimate of the numbers of Indonesian labour migrants who are absent at any single point in time.

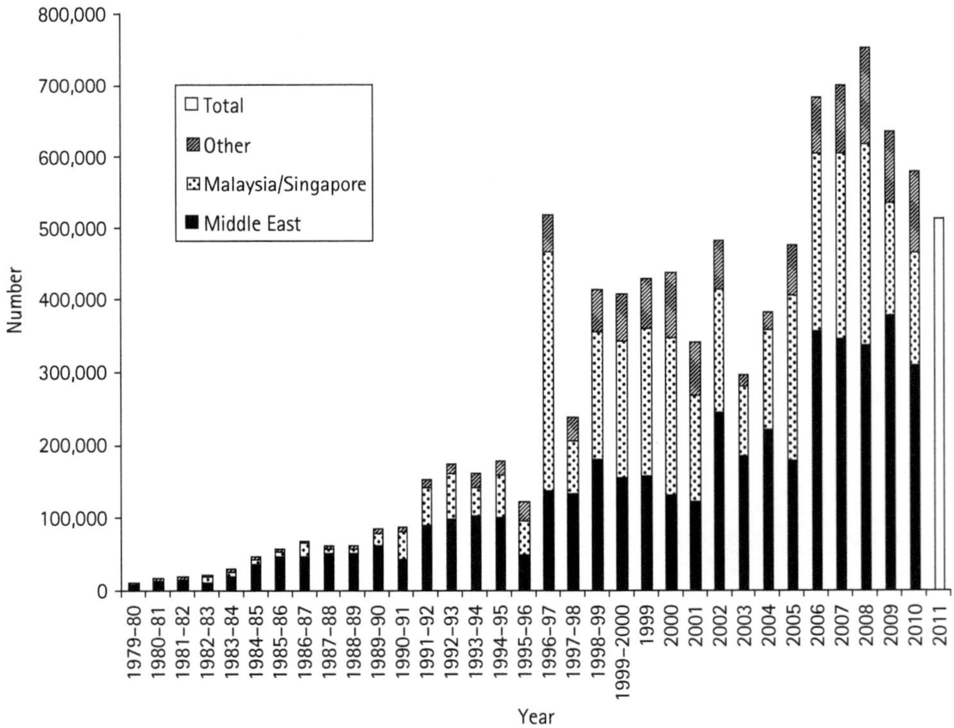

FIG. 10.4 Number of Indonesian overseas workers processed by the Ministry of Manpower, 1979–2011.

Sources: Suyono (1981); Singhanetra-Renard (1986: 52); Pusat Penelitian Kependudukan, University Gadjah Mada (1986: 2); Indonesia, Department of Labour, unpublished data; Soeprobo (2005); Ananta and Arifin (2008); Kim 2011; Data Penempaton Tenaga Kerja Indonesia, unpublished data.

Turning to a destination perspective, Table 10.6 provides estimates of the numbers of foreign workers in ASEAN countries. Although again the data are of limited accuracy, the table shows they are relatively most significant in Singapore and Brunei where they make up more than a third of the workforce. Nevertheless they are also very important in Thailand and Malaysia. In Malaysia, official government figures put the number of migrant workers at 2,235,200 in 2007 when they made up 20.5% of the workforce. Most work in manual occupations in the plantation, forestry, construction, manufacturing, and household domestic sectors. On the other hand, Malaysia sends significant numbers of relatively low-skilled workers to Singapore, Taiwan, and Japan. Similarly, Thailand has more than 1 million workers from neighbouring nations but sends several hundred thousand workers to the Middle East, Taiwan, Japan, and Malaysia. As in the case of destination stocks, there has been a significant increase in recent times in the stocks of migrant workers in South-East Asian destination countries.

A distinctive feature of South-East Asian labour migration has been the increasing significance of women in the movement. Indeed, Table 10.7 shows that women have outnumbered men in many recent years in labour migration out of the two largest sending countries—Indonesia and the Philippines. Women labour migrants are more clustered in a few occupations than is the case for men. Dominant here are household domestic workers, but carers, nurses, entertainers, and, increasingly, factory workers are also important. Several of the occupations taken up by female South-East Asian labour migrants place them at greater risk than their male counterparts of exploitation, partly because they work in non-conventional workplaces.

While the dominant temporary labour migration flow is of low-skilled workers, there are increasing temporary flows of skilled workers within the Asian region.

Table 10.6 Estimated stocks of foreign labour in South-East Asian countries around 2012

Country	Year	Stock	Source
Singapore	2012	1,234,100	Baruah (2013)
Malaysia	2011	1,573,061	Baruah (2013)
Thailand	2012	2,500,000	Baruah (2013)
Brunei	2008	87,867	Scalabrini Migration Center (2012: 104)
Vietnam	2010	80,000	Scalabrini Migration Center (2012: 159)
Indonesia	2012	72,400	Zhengyi (2013)
Philippines	2008	36,150	ILO (2013a)
Laos	2010	200,000	Scalabrini Migration Center (2012: 126)
Total		5,783,578	

Table 10.7 Selected South-East Asian out-migration countries: Proportion of international labour migrants who are women

Country of Origin of International Labour Migrants	Year	Number of Workers Sent	Percent Women	Source
Philippines	2010	1,470,826	54.5	Philippines Overseas Employment Administration
Thailand	2010	143,795	15.7	Huguet, Chamratrithirong, and Richter (2011: 13)
Indonesia	2010	575,803	69.0	Scalabrini Migration Center (2012: 119); data Penempatan Tenaga Kerja Indonesia, unpublished
Vietnam	2012	80,320	33.3	ILO (2013b), <http://www.ilo.org/wcmsp5/groups/public/---asia/---ro-bangkok/documents/presentation/wcms_218167.pdf>

In particular, the large cities of the region have large expatriate Asian populations. The ASEAN Economic Agreement planned for introduction in 2015 contains some provisions for free exchange of some skilled workers between South-East Asian countries. As elsewhere in the world, there is a growing divergence in policy toward high-skilled migration to which countries are welcoming and low-skilled migration against which barriers are strengthening.

10.5.2 Migration to and from OECD Countries

An increasing trend in recent years has been in temporary and permanent migration to OECD countries. The numbers of South-East Asia-born persons living in OECD countries at the time of the 2000 round of population censuses is shown in Figure 10.5 and this indicates that each country, but especially Vietnam and the Philippines, has an extensive diaspora in high-income countries. However, migration data such as those shown in Figure 10.5 fail to indicate the complexity of mobility between South-East Asian countries and high-income nations, and this hampers policy efforts to derive the maximum development dividends from migration. It is possible to illustrate this by focusing on a single destination country, Australia, where the migration data available are more comprehensive than for most other nations.

With respect to migration flows, Australia's geography as an island nation facilitates effective data collection. In effect, *all* moves in and out of the country are captured in the data. All persons entering and *leaving* Australia are asked questions about country of birth, date of birth, gender, occupation, country of origin/destination, intended/actual length of residence in Australia (or in the case of Australians leaving, abroad), and reasons for moving.

In conventional analyses of international migration it is usual to dichotomize countries as being either destination/origin or sending/receiving. However, as King (2002)

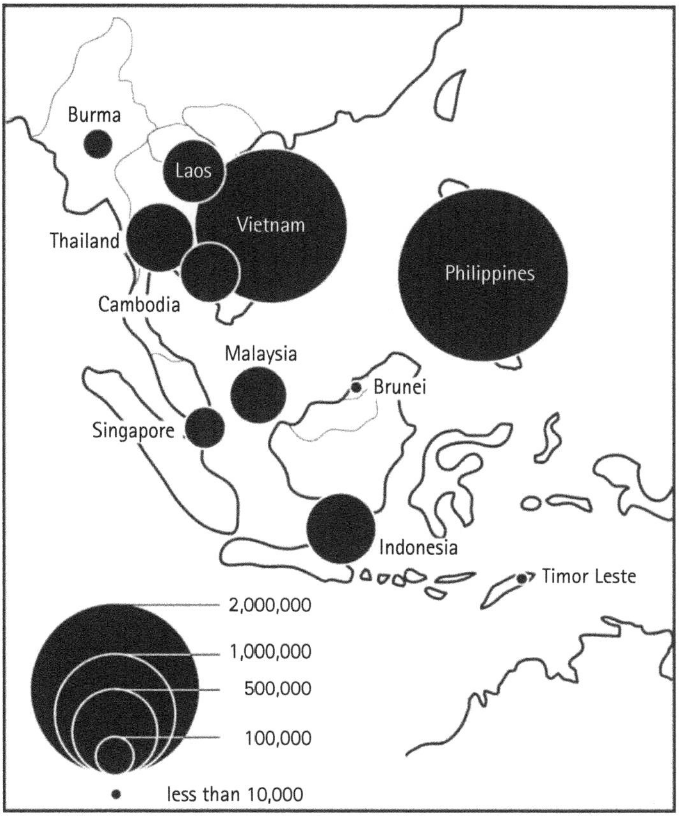

FIG. 10.5 South-East Asian countries: Numbers residing in OECD countries, 2000.

Source: OECD Database on Immigrants and Expatriates.

has pointed out, such conventional dichotomies, while they have always oversimplified more complex situations, are especially inappropriate to contemporary global migration. An analysis of the Australian data suggests that it is more meaningful to consider migration flows or systems involving movers in both directions. Figure 10.6 presents a stylized model of the contemporary migration system between Australia and South-East Asian countries.

Within this system it is possible to recognize a number of distinct structural elements:

(a) *Permanent Settlement of South-East Asians in Australia.* This refers to the traditional, more or less permanent movement of South-East Asians accepted for settlement under the Skill, Family, Refugee-Humanitarian, or Special Eligibility Components of the Australian Immigration Program (DIAC 2008). They take out permanent residence or citizenship in Australia.

(b) *Indirect Settlement Migration to Australia.* Some South-East Asian groups move initially to a third country and subsequently move to Australia. One of the main such avenues is through New Zealand where they can gain citizenship

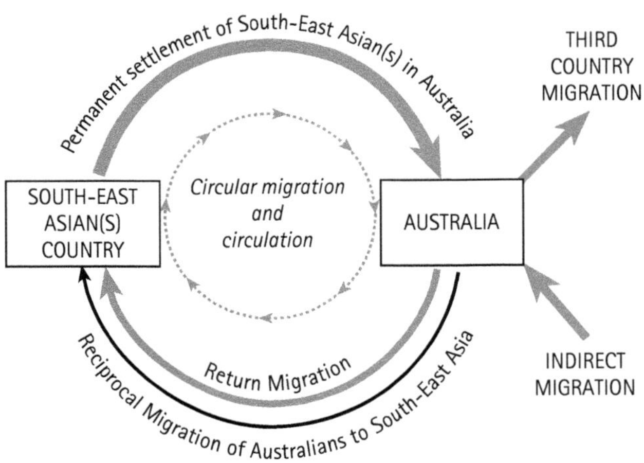

FIG. 10.6 A model of the Australia–South-East Asia migration system.

or residency and then become eligible to move to Australia under the Trans-Tasman Migration Agreement (Bedford et al. 2003).

(c) *Return Migration.* Involving previous settlers in a more or less permanent return to their South-East Asia homeland after a period in Australia. This is sometimes referred to in Australia as settler loss (Hugo 1994).

(d) *Third Country Migration.* Involving a more or less permanent migration of settlers from South-East Asia to a third country destination after a period of residence in Australia. This is referred to in some contexts as remigration.

(e) *Reciprocal Migration.* Involving a more or less permanent relocation of Australians to a South-East Asian country. These flows are usually smaller in size than the flows in the opposite direction.

(f) *Circular Migration.* Involving long-term but temporary migration of South-East Asian people to Australia and Australians to South-East Asia. The main groups are students and long-term temporary business migrants. These are people who will spend more than a year at the destination but always with the intention to return. They take out temporary residency in Australia.

(g) *Circulation.* Involving shorter-term movements from South-East Asian countries to Australia and from Australia to South-East Asian countries.

The picture presented in Figure 10.6 belies the conventional depiction of south–north migration as a more or less permanent flow. It conveys the real complexity of the South-East Asian regional international migration system which can only be captured if comprehensive data are available.

The Australian international migration flow data means that it allows the complexity of the South-East Asia–Australia migration system to be quantified. Since the abolition of the last vestiges of the infamous White Australia Policy in the 1950s, there has been an increase in permanent settlement of South-East Asians in Australia. There has been a permanent settler immigration of South-East Asians to Australia over the 1993–2011 period, comprising

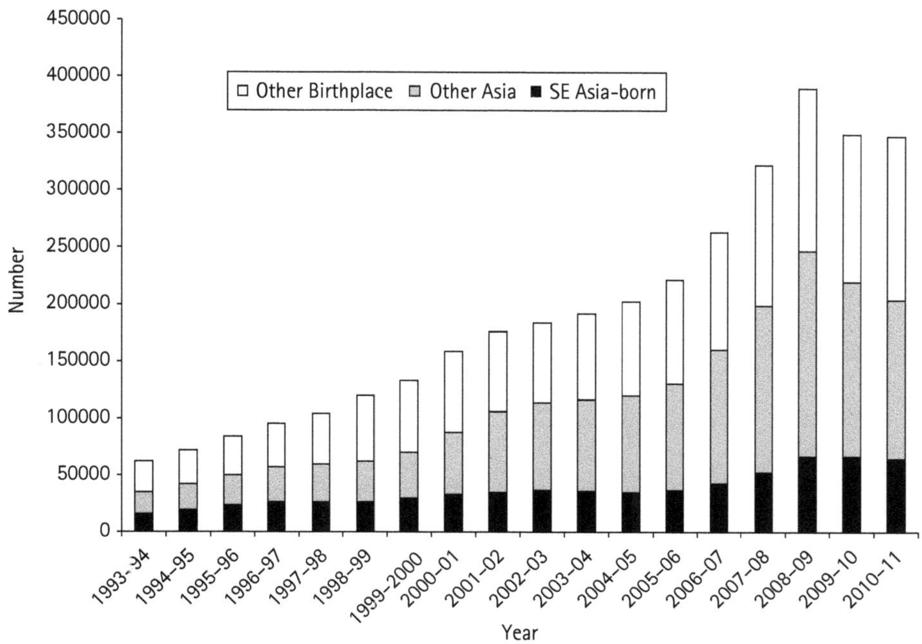

FIG. 10.7 Australia: Long-term visitor arrivals by birthplace, 1993–1994 to 2010–2011.

Source: DIAC unpublished data.

14.3% of the total 1.97 million settler arrivals over those 18 years. The annual numbers fluctuated between 9,700 in 1997–1998 and 21,164 in 2007–2008, but a general upward trajectory is evident in the last few years. The relative significance of the South-East Asian countries has fluctuated over the period and they account for four of the top ten birthplace countries of migrant settlers over the 1993–2011 period, with the largest numbers being from Philippines (74,600), Vietnam (51,285), Malaysia, (40,951), and Indonesia (38,115).

However, the permanent settlement programme only reflects a part of South-East Asian skilled migration to Australia. Perhaps the most striking change in Australian immigration over the last decade has been the increased non-permanent immigration of workers (Hugo 1999). Hitherto, Australia's immigration policy had eschewed temporary worker migration in favour of an overwhelming focus on permanent settlement. This is reflected in the number of long-term arrivals of foreigners, as can be seen in Figure 10.7 which indicates that South-East Asia has been an important source of such arrivals.

One of the largest categories of skilled temporary residents with the right to work are foreign students. In 2010 there were 469,619 people on student visas in Australia and over three-quarters from Asia. Of the largest ten countries of origin, four are South-East Asian—Malaysia (5%), Vietnam and Thailand (4%), and Indonesia (3%).

In December 2010 there were 116,012 persons on 457[2] (Business Long Stay) visas resident in Australia (a 2.6% decrease over the previous year). South-East Asian countries

[2] This visa category is analogous to the HB1 visa in the United States.

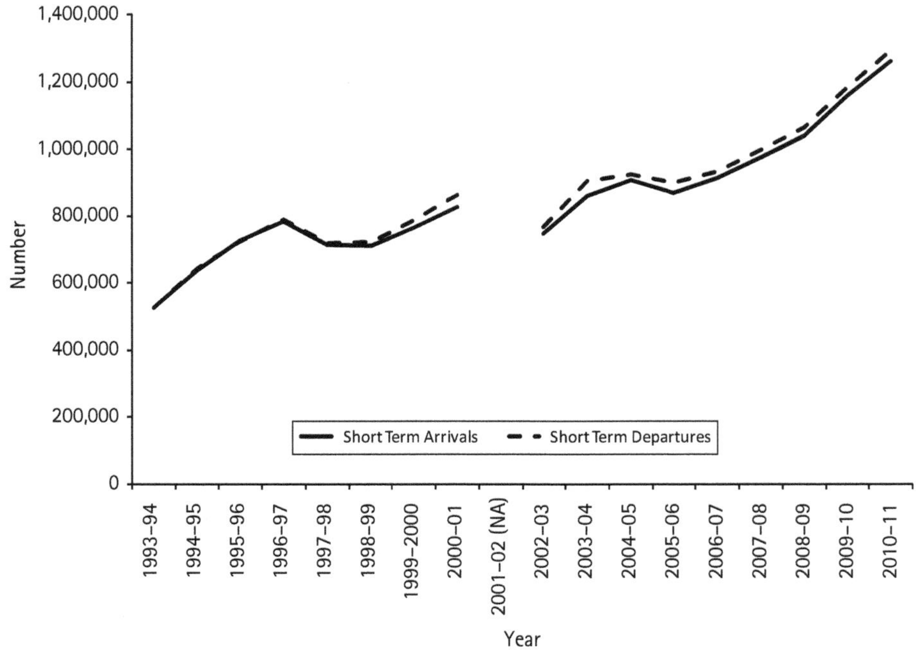

FIG. 10.8 Short-term arrivals and departures of South-East Asia-born, 1993–1994 to 2010–2011.

Source: DIAC unpublished data.

are not as prominent as among students but they account for 14.0% of the 457 population. Of the top ten origin nations, only the Philippines (9.0%) is represented from South-East Asia.

In addition, there has been a significant increase in the number of short-term visitors. Figure 10.8 shows that there has been a steep increase in the number of arrivals of South-East Asia-born people who indicate they are to stay in Australia less than 12 months. This movement is sometimes dismissed as being of little significance but it is apparent that it can have some important economic and social implications. Table 10.8 shows that the reasons given by South-East Asia-born short-term arrivals include economic activity in addition to tourism and visiting of friends and relatives.

Clearly, there has been a significant increase in the inflow of both permanent and temporary migrants to Australia from South-East Asia. However, a major element in the international migration system is the counterflow. As Dumont and Lemaitre (2005) have pointed out, this flow is significantly underestimated in size and influence by analysts and policymakers largely because there is little or no data on its scale. On the one hand, destination nations rarely collect information on who leaves the country and concentrate only on immigration. On the other, source countries have little data on immigration at all, especially immigrants who are citizens returning after a sojourn abroad. Australian data is an exception since information is collected on all who leave the nation

Table 10.8 Australia: South-East Asia-born short-term visitor arrivals from South-East Asia by reason for travel, 2010–2011

Reason for Travel	Number	Percent
Exhibition	1,905	0.3
Convention/Conference	20,675	3.0
Business	57,475	8.4
Visiting Friends/Relatives	133,491	19.6
Holiday	331,649	48.7
Employment	32,275	4.7
Education	69,268	10.2
Other	33,694	5.0
Not Stated	11,902	
Grand Total	692,334	100.0

Source: DIAC unpublished data.

and can provide some insights into this flow. Table 10.9 provides data on permanent departures from Australia to South-East Asian countries over the 1994–2006 period and these are divided into two groups:

(a) The foreign-born who mainly represent *return migration* and *third country migration* of former settlers (61.0% of the migrants from Australia).
(b) The Australia-born who are partly the Australia-born children of those returnees but who are predominantly Australian citizens of long standing, moving for one reason or another to an Asian country (39.0%) as the reciprocal migration referred to earlier.

It is shown in the table that the permanent outflow from Australia over the study period is around 30% the size of the inflow, although its significance varies from nation to nation. It can be seen that there is a very significant backflow to Singapore, Indonesia, Malaysia, and Thailand. Even for Vietnam there has been a substantial return movement. This is a recent phenomenon. Much of the Vietnamese migration to Australia was of refugee-humanitarian migrants and occurred in the first 15 years following reunification in 1975 (Viviani 1996) and was characterized at that time by a very low rate of return migration (Hugo 1994). However, it is apparent that with *doi moi* and the opening of the Vietnamese economy that an increasing number of Vietnamese Australians have returned to their birthplace and taken advantage of the liberalization of the economy to invest and set up businesses.

Table 10.9 Australia: South-East Asian country of origin, permanent arrivals and permanent departures (overseas- and Australia-born), 1994–1995 to 2005–2006

Country of Origin	Settler Arrivals	Permanent Departures Overseas-born	Permanent Departures Australia-born	Permanent Departures	Permanent Departures as percent of Settler Arrivals
South-East Asia					
Mainland					
Burma (Myanmar)	2,027	43	66	109	5.4
Cambodia	5,493	346	219	565	10.3
Laos	478	145	120	265	55.4
Thailand	15,075	3,193	3,341	6,534	43.3
Vietnam	26,946	4,627	1,554	6,181	22.9
Maritime SE Asia					
Brunei	1,378	499	981	1,480	107.4
East Timor	92	80	88	168	182.6
Indonesia	28,686	5,573	4,447	10,020	34.9
Malaysia	29,214	3,085	3,187	6,272	21.5
Philippines	38,639	1,816	1,233	3,049	7.9
Singapore	43,394	9,114	13,547	22,661	52.2
Total South-East Asia	191,422	28,521	28,783	57,304	29.9

Source: DIAC Overseas Arrivals and Departures, unpublished data.

The data hence do show that there is a significant north-to-south flow of skilled workers from Australia to South-East Asia. Hence, while the dominant permanent flow in the South-East Asia–Australia migration system is toward Australia, there is a smaller but nevertheless significant counterflow. Table 10.10 shows that over the 2001–2011 period 194,622 South-East Asians settled in Australia but that 29,095 returned to South-East Asia and there were 51,272 Australia-born persons who moved permanently to South-East Asia. Hence, the net migration gain was 114,255 representing a migration efficiency percentage of 41.5, i.e. it takes more than two immigrants to a get a net migration gain of a single migrant.

Return migration has long been an important theme in the migration literature (King 2000). However, as Conway and Potter (2009: 1) point out:

> To this point, extant scholarship on international return migration has commonly focused on the elderly, first generation retirees who have consummated their long held desires and intentions to return to the global South homelands from which they emigrated in their youth.

However, the return movement from Australia to South-East Asia does not conform to this pattern. Indeed the bulk of returnees are young families who are still available

Table 10.10 Australia: Permanent migration in and out, 2000–2001 to 2010–2011

SE Asia-born moving to Australia	194,622
SE Asia-born moving from Australia to SE Asia	29,095
Australia-born moving from Australia to SE Asia	51,272
Net Migration	114,255

Source: DIAC unpublished data.

for work when they return (Hugo 2008). Moreover, the literature on return migration and its impact on origin countries has almost totally been concerned with permanent return. Yet non-permanent return can also impinge upon development in origin countries. Returnees can not only bring with them money and equipment but also new ideas and new ways of doing things. It is apparent that settlement of South Asian groups has resulted in an upswing of *non-permanent* return migration out of Australia. Figure 10.9 shows how long-term and short-term movement of Australian residents from Australia to South-East Asian countries has increased in recent years. Moreover it is apparent that the South-East Asian-born have been an important component in that temporary emigration of Australian residents. Clearly the permanent settlement of South-East Asians in Australia is creating a significant temporary flow back to countries of origin in which former settlers are an important component. This is the case in both short-term and long-term movement.

A special dataset created using the personal identifiers of all persons moving into and out of Australia allow us to establish the extent to which this temporary movement out of Australia into South-East Asia involves former settlers. Table 10.11 shows the average number of return trips made into and out of Australia over the 1998–2006 period by different South-East Asian birthplace groups. They are differentiated according to their visa status granted by the Department of Immigration and Citizenship.

Some of the key patterns can be summarized as follows:

- The first column shows the average number of trips made home among the settler arrivals from South-East Asian countries who arrived in Australia for the first time in the 1998–2006 period. These ranged from 4.8 for Indonesia to 1 for Burma.
- The second shows that the average number of trips made to Australia by visitors from South-East Asia over the period ranged from 3.2 for Laotians to 7.3 for those from Brunei Darussalam. This indicates that many South-East Asians are coming regularly to Australia.
- The third column is particularly interesting. This shows the average number of temporary trips made by Australian residents who were born in South-East Asia. This shows a high level of travel back to the home country, ranging from an average of 10.3 in Indonesia and 10.2 for Singapore to 4.0 for Cambodia.

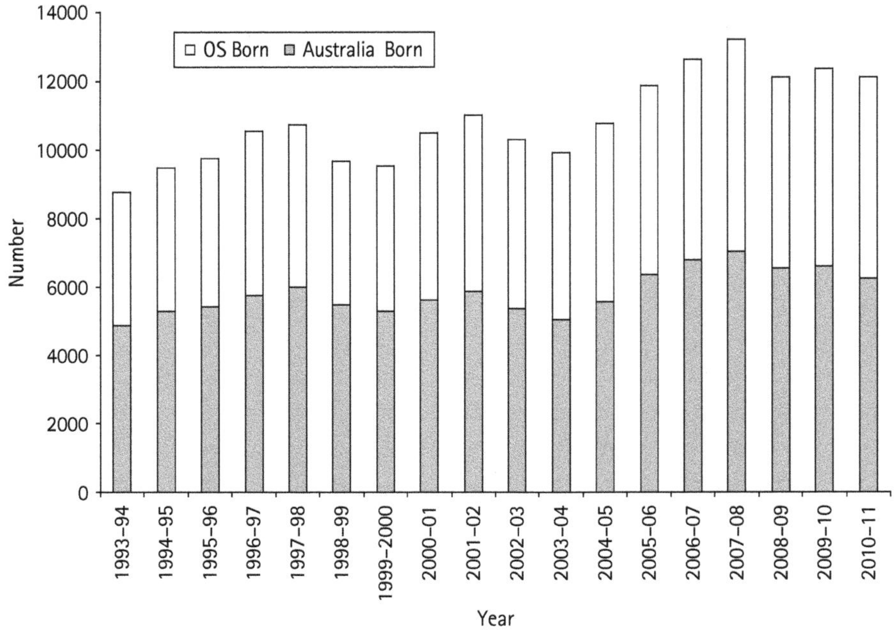

FIG. 10.9 Australia: Long-term departures to South-East Asia of Australia- and overseas-born, 1993–1994 to 2010–2011.

Source: DIAC unpublished data.

Table 10.11 Australia: Country of birth by type of movement: Average number of return trips by individuals, 1998–2006, type of movement

Birthplace of Mover	Settler Arrival	Visitor Arrival	Resident Temporary Departure
Burma (Myanmar)	1.0	3.6	5.9
Cambodia	1.3	4.8	4.0
Laos	1.2	3.2	5.6
Thailand	2.4	4.9	6.2
Vietnam	1.4	3.9	4.8
Brunei Darussalam	4.1	7.3	9.1
Indonesia	4.8	6.1	10.3
Malaysia	4.0	5.9	9.4
Philippines	1.5	3.9	5.0
Singapore	3.7	6.4	10.2

Source: Special dataset received by DIAC.

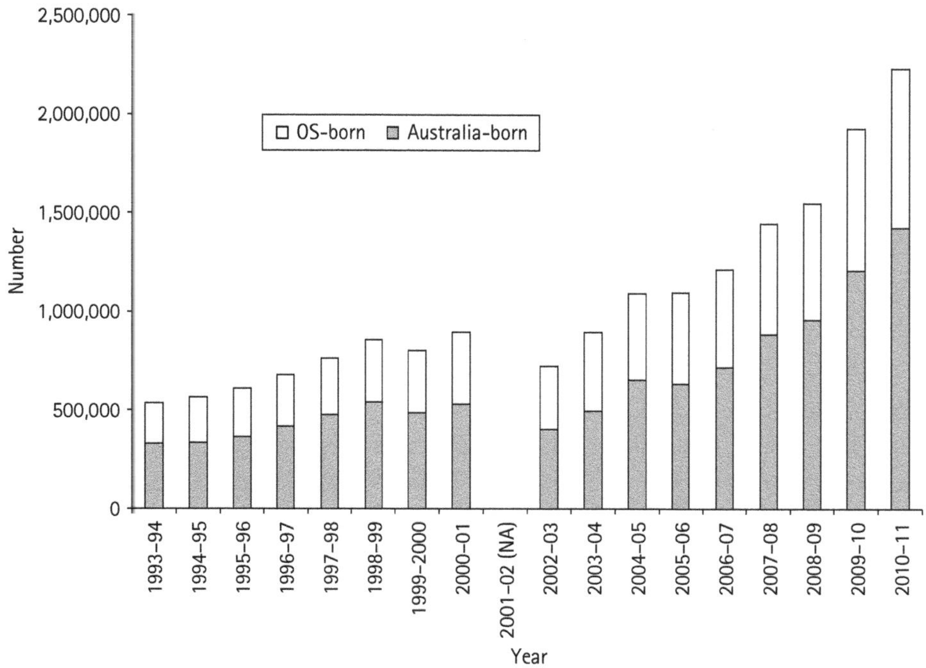

FIG. 10.10 Australia: Short-term resident departures to South-East Asia of Australia- and overseas-born, 1993–1994 to 2010–2011.

Source: DIAC unpublished data.

This clearly shows that there is considerable circulation between Australia and South-East Asian contexts of former settlers and also that there are significant numbers of South-East Asia-born, based in South-East Asia, who circulate frequently to Australia. In both cases the potential for significant development impacts in South-East Asian origins are considerable.

The Australian international migration flow data analysed here have demonstrated conclusively that the South-East Asia–Australia migration system is characterized by a high degree of complexity and circularity. This stands in sharp distinction to the conventional depiction of this system being seen largely as 'south–north migration' where, at least implicitly, it is assumed that the overwhelming dominant pattern is of permanent redistribution of highly skilled people from poorer countries of South-East Asia to better-off Australia. While the explosion of this myth of south–north migration is important for the Australian case, the only difference between it and most other OECD 'destinations' of south–north migration is the fact that it has a more comprehensive data collection system which allows the inherent circularity and complexity in the system to be quantified. Australia is not a special case; such patterns are characteristic of south–north migration in most cases, but the data limitations conceal it. It is argued here that circularity, reciprocity, and complexity are structural features of south–north migration systems—they are not peripheral or ephemeral. The material analysed here point to a pressing need on the one hand to reconceptualize the whole concept of south–north migration so that it recognizes the fundamental complexity of the population flows which are involved. On the other, it also points to the urgency of improving our migration data collection systems which in many countries remain grounded on the outmoded settlement migration model and are biased toward migration-receiving countries and considerations.

10.6 Internal Migration

While international migration has increased greatly in scale and significance in South-East Asia, it is the increase in population mobility *within* nations which has had the greatest impact since it has influenced the majority of the population of the region. While mobility in the past has been underestimated, there have been major extensions of the area which South-East Asians range across in their search for work and to engage in social and economic interaction. Only a minority of these movements are captured in census migration data but they do give some indications of the increase in mobility. For Indonesia, for example, Table 10.12 shows the increase in the percentage of the population who have moved between provinces over the last 30 years. Nevertheless, these data show that only one in ten Indonesians has lived outside of their place of birth. This would suggest that Indonesia has a low level of internal population mobility. The Australian 2006 population census, on the other hand, identified 45.5% of the population as internal migrants. The internal migration data, however, (as is the case for most intercountry comparisons) are not comparable because:

- The Australian data only relate to movers in the five years before the census whereas Indonesians are asked if they had ever lived in another province.
- The Australian data detect all changes in place of residence whereas in the Indonesian case they must cross a provincial boundary.

The Indonesian internal migration data, like that of most censuses, only measures movement across a migration-defining region boundary. Yet Indonesian provinces vary enormously in size and diversity. Some studies (Hugo 1980) have shown that there are around five intraprovincial migrants for every interprovincial move. Moreover, much rural to urban migration in Indonesia which is important from a development perspective occurs within provinces.

A major issue, however, as is the case with international migration, is that the entire focus is on permanent migration yet there are now several decades of field evidence which indicates that non-permanent migration is not only of increasing scale but that it has important development implications.

Table 10.12 Indonesia: Census-based measures of migration, 1971–2000

Year	Migration Measure	Males	Females
1971	Percent ever lived in another province	6.29	5.06
2000	Percent ever lived in another province	10.56	9.57

Source: Indonesian censuses of 1971, 1990, and 2000.

Across the region there has been a massive increase in personal mobility. One indicator of the increased ability of Indonesians to move on a temporary basis is the decreased number of persons per motor vehicles which is shown in Table 10.13. This reflects not only the proliferation of personal motor vehicles but the explosion of public transport of different sizes, costs, and levels of comfort.[3] The impact is that labour markets have been dramatically extended and individual levels of mobility increased considerably, so that South-East Asians are able to move over a wider area in search of work than ever before and average distances between place of residence and place of work have been dramatically extended.

More than a decade ago the World Bank (1984: 20) estimated that at least 25% of rural households on Java have at least one family member working for part of the year in urban areas. This would imply that at least 3.75 million people are involved in this form of migration on Java. Of course, since migrants are only working in the cities for part of the year, the average effect is less than this, but it is not unlikely that about one-sixth of the average daily urban workforce consists of temporary migrants not included in official employment figures. A series of studies sponsored by the Ministry of Population and Environment (Mantra and Molo 1985) examined circular migration and commuting in six Indonesian cities. In addition to establishing how significant and widespread this mobility was in cities of varying sizes in both Inner and Outer Indonesia it was found that the great majority of these movers had only been circulating to those cities since 1980. It would appear that this pattern of an increasing tempo of non-permanent migration has continued over the last decade and became of even greater significance with improvements in transport, advances in education, changes in the roles of women, increased urban and industrial development, etc.

Perhaps the strongest evidence of a pattern of continued increase in the scale and significance of non-permanent migration in Java during the period since the first study in West Java carried out in 1973 (Hugo 1975) is derived from a comprehensive longitudinal study of 37 villages in Java carried out over the period 1967–1991. In that study Collier et al. (n.d.: 1) concluded that:

[3] A further indicator of rapidly increasing circulation in Indonesia lies in the increase in the number of domestic airline passengers, trebling from 8.6 million in 2000 to 25.3 million in 2005 (Ananta and Arifin 2008).

Table 10.13 Indonesia: Number of persons

Year	Persons/Motor Vehicle (including Motor Cycles)	Persons/Motor Vehicle (excluding Motor Cycles)
1950	1,507	1,691
1961	263	447
1971	129	300
1980	38	123
1990	20	64
2000	11	39
2005	6	23
2009	3	18

Source: Biro Pusat Statistik publications on motor vehicles and length of motor vehicle, 1950–2009 roads.

> Twenty five years ago many of the landless labourers on Java had very few sources of income . . . Now most of the landless rural families on Java have at least one person who is working outside of the village, and in a factory or service job.

In all of the villages in the 1992–1993 resurvey, massive migration out of the village to jobs in the larger cities and towns was recorded and only 20% of households depended on agriculture for their total livelihood. The bulk of the movement recorded was on a temporary basis. The fact that those villages were deliberately selected to be representative of villages in highly accessible and low-accessibility areas suggests that the scale of non-permanent rural–urban movement from Javanese villages has increased exponentially in recent years and that such movement still far outweighs permanent relocation from village to city.

While there are many similarities in the contemporary situation with respect to patterns and processes of non-permanent migration compared with the 1970s, there also have been some significant changes. Paramount among these is the increasing involvement of women in both permanent and temporary movements, especially those directed from village to city. Another change is that the increasing size of the formal sector in Indonesian cities, especially Jakarta and other large cities in Java, has led to a growing number of migrants having to be more or less permanently settled in the city and not free to come and go to their village as frequently and readily as was possible when they worked in the informal sector. In many cases, for example with many of the young women working in factories in and around cities like Jakarta and Surabaya, there are intentions to eventually return to settle in their village but the fixed time commitments of their work prevent them circulating to and from the village on a weekly, fortnightly, or monthly basis.

Table 10.14 Population and household heads by residence status in RW 06 Pasawahan Village, 1996

	Household Head			Population
Type of Residence	Male	Female	Total	Total
1. Local	145 (26.3)	11 (6.3)	156 (21.5)	639 (30.1)
2. Permanent	181 (32.8)	28 (16.1)	209 (28.8)	615 (29.0)
3. Temporary	226 (40.9)	135 (77.6)	361 (49.7)	867 (40.9)
Total	552 (100.0)	174 (100.0)	726 (100.0)	2,121 (100.0)

Note: Numbers in brackets are percentages.
Source: Wahyuni (2000).

The widespread nature of circular mobility in Java is evident in a recent study of part of a village in the study of Bandung by Wahyuni (2000). She made a full census of the village by examining family cards (*kartu keluaga*) and checking with local officials. Table 10.14 shows how the household heads and total population are divided between:

- local origin residents
- permanent migrants from elsewhere
- temporary migrants.

Some 40.9% of male household heads and 77.6% of female household heads are temporary migrants in the city and only one-fifth are non-migrant households. This indicates the significance of both permanent and non-permanent migration in Indonesia's large cities.

10.7 Some Implications

The argument presented here is that both internal and international migration have been increasing in scale and complexity in the South-East Asian region but that traditional demographic concepts of that mobility are proving increasingly less adequate to describe and explain that movement. Much of the mobility which is influencing and being influenced by demographic, social, and economic change is non-permanent in nature and does not conform to the narrative of migration as permanent relocation from one place to another which underlies most demographic thinking about migration. This shortcoming is amplified because it is demographic concepts of migration which provide the basis for measurement of internal and international migration in population censuses and other traditional forms of population data collections. In this section we will briefly address some of the important implications which flow from this.

One of the most fundamental pieces of population information relates to the number and characteristics of people living in a particular area. Locating a population in space is crucial to both research and policy. It is basic to all planning of the provision of goods, services, and utilities as well as to much social science research in which it is recognized that individual and group behaviour is influenced by place and context. However, new patterns of population movement compromise the ability of traditional censuses to accurately count the population of particular areas. Censuses are usually collected on a *de jure* (place of usual residence—regardless of location on the night of the census) or *de facto* basis (location on the night of the census). Each person enumerated is assigned to a single point on the earth's surface. However, increasing and widespread patterns of temporary migration, much of it involving people spending considerable periods at different locations, is diluting the strength of the logic which assigns individuals to a single place in censuses.

This concern is applicable at all geographical scales. At a national level one could ask the question of what constitutes the national population. To what extent is it captured in a census? The issue is assuming increasing significance with the growing recognition of the diaspora as an important dimension of nation states. This has long been recognized in emigration nations such as Italy where the diaspora has been represented in parliament. However, the new scale of international migration and modern methods of information and communication technology have given this issue a new significance. Modern diaspora can form a significant part of a national population. Figure 10.11, for example, shows the distribution of Filipinos in foreign nations. In 2009 it was estimated that there were 8.6 million (POEA 2010) Filipinos overseas, equivalent to 9.1% of the total resident population.

Modern forms of communication and the cheapness and speed of international travel make it possible for the diaspora to maintain intimate and active contact with their homeland in a way which was not possible for earlier generations of immigrants. Filipinos are said to use texting more than any other national group. Immigrants and migrant workers are able to maintain social and professional linkages in a personal and intimate way that reinforces their identity with their homeland and their involvement in activities in the homeland. Clearly, diaspora are able to maintain a strong identification with, and carry out a substantial amount of activity with, their homeland even though they are overseas on a permanent or temporary basis. For some purposes it may be useful to be able to include the diaspora in counts of the 'national population'.

A similar argument can be made about counting the population of subnational spatial units. In the contemporary world the home place is only one of the spaces in which individuals and households spend substantial periods of time. In considering the impacts of context and location on the behaviour of people and the impact of people on particular areas, perhaps we also have to consider:

- daytime vs night-time population—this is especially relevant in the central business districts of major cities
- weekday vs weekend populations

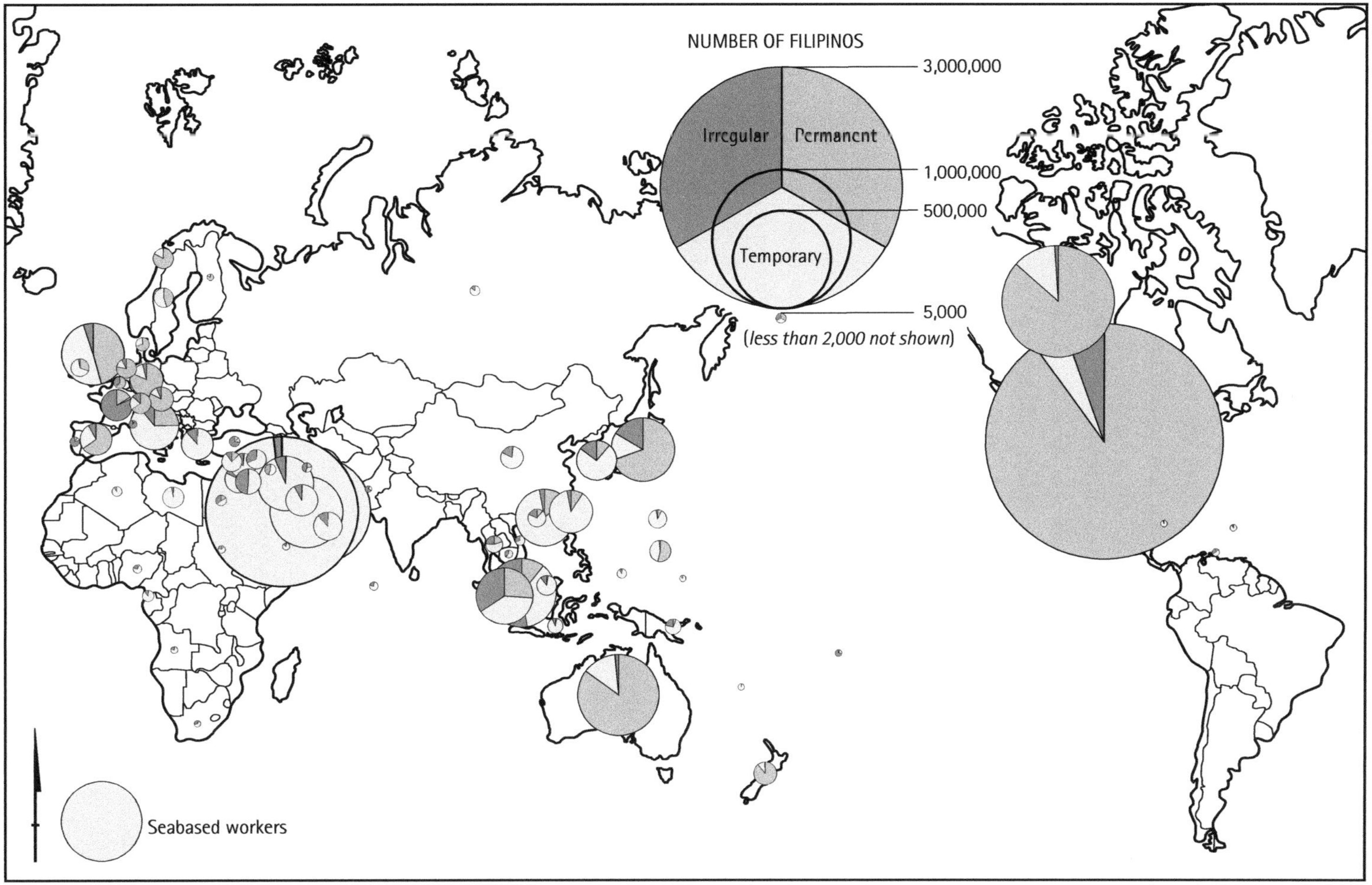

FIG. 10.11 Stock of Filipinos overseas, December 2009.

- seasonal variations in population—in holiday and resort areas
- working vs resident populations—increasingly people are able to live at great distances from where they work.

In South-East Asia, where the scale of temporary movement, especially that between urban and rural areas, is substantial, it is likely that census counts of urban areas really underestimate the populations which functionally are part of those cities. Where commuters and especially circular migrants are counted can have a significant influence on the population of an area and lead to considerable differences between census population and the 'functional' population of localities. Indonesia is a case in point where successive censuses have used a six-month rule whereby people are regarded as being visitors at the destination if they have been there less than six months. Most circular migrants return home at least once every six months so that most should be counted in their home area (Hugo 1982). This issue continues to be a source of confusion in census taking in Indonesia.

A second set of implications relating to the failure to capture non-permanent movement in censuses and other traditional sources concerns the increasingly important issue of migration and development. There have been arguments made that in the case of both internal and international migration, circular migration is better able to deliver development dividends and poverty reduction impacts of south–north migration for low-income origin countries and areas than can permanent relocation. This is predominantly because such migrants are more committed to their home community than their permanent counterparts because they usually leave their nuclear family at home and because they reside part of the time at the origin. Accordingly they are likely to remit a larger proportion of their income and are more likely to eventually return to their homes and contribute to development. This applies to both internal and international migration. What are the types of policies which can maximize the beneficial impacts of circular migration for origin communities?

Remittances are the *raison d'etre* of most circular migration. Best practice in both origin and destination involves educating migrant workers and potential migrant workers about all of the alternatives for sending money home and above all the making available of lower-cost and more secure options. This can be supported by establishing low-cost channels for remittances and setting up establishments in the major destinations and origins to facilitate sending money back. Particular notice should be taken of new low-cost alternatives including cell-telephone-based remittances (World Bank 2006: 150). Familiarizing migrant workers with remittance systems and involving them in them can be their first step toward 'financial literacy' and involvement in the formal financial system which can be of help to them in the future (Terry and Wilson 2005).

One dimension of international temporary labour migration which is often overlooked is the social costs which are endured by the migrant and their family by extended separation (Hugo 2004). There may be a role for governments in ameliorating the social costs on the family that remain behind. At present, it is the extended family, community solidarity, and NGOs that provide support; however, governments may also be able to play a role. In international migration, although sending governments can play a role in the protection and support of circular migrants in other countries, the migrant experience in destinations is influenced more by employers, governments, and society in those destinations. Governments play a central role because they set the conditions under which migrant workers can enter their country and the rights, access to services, and obligations which they have while in the country.

With respect to international migration, Vertovec (2006: 43) has summarized a number of concerns that need to be borne in mind in the rolling-out of circular migration schemes. Prominent in these criticisms is that in practice such schemes lock migrants into dependent and exploitative relationships and offer little opportunity for upward mobility and training. Such programmes often involve a loss of rights among the workers and they suffer lack of integration, exposing them to exclusion. Others suggest that it encourages illegal 'overstaying' and creates substantial compliance problems for destination authorities.

In South-East Asia the temporary labour migration of low-skilled workers has been heavily criticized as being a 'new form of indentured labour'. Some NGOs in the region even equate it with trafficking because of the exploitation of migrants that often characterizes this movement. The undoubted excessive rent taking, social costs of long separation from families, lack of opportunities for social mobility, and lack of opportunity to transfer to permanent residency are also criticized. It is also seen to be associated with undocumented migration. These criticisms can all be sustained by looking at particular migrants in particular flows of temporary low-skilled labour migrants. However the question has to be asked as to whether the problems are intrinsic features of this type of migration or whether they are due to a failure to introduce fair and efficient governance of labour migration systems in the South-East Asia context. There is some evidence that it is the latter rather than the former.

In the region, the potential for this type of migration to have a developmental impact is considerable not only because of their large numbers but also because most retain a strong commitment to their home communities where they leave their families. Accordingly they send back a higher proportion of their income in remittances than is the case for permanent settlers. Moreover they intend to return to their homelands and this is enforced by the migration regimes in destination countries. It has been shown (Hugo 2001) that in many countries (e.g. Indonesia) this type of migrant worker is drawn from some of the poorest areas of the country (e.g. parts of Java, East and West Nusatenggara). In such areas remittances are the only substantial inflow of potential investment money from the outside.

Given this substantial potential for international contract labour migration to deliver financial resources to the grass roots in some of the poorest areas in the region, what are the barriers which are dampening these impacts? One of the main issues here relates to *transaction costs*. For many contract workers the amounts that they have to pay to recruiters, to government officials, to travel providers, for documents, training, etc. are very high and well above what could be considered a reasonable charge for those services. There is too much unproductive rent taking in the burgeoning contract labour migration industry in the region and this is siphoning away money that migrants earn that otherwise would have gone toward development-related activity in home areas. Often migrants have to work several months on arrival at the destination just to pay off the debts incurred by the migration process. If they are duped by recruiters so that the job they were promised is not available, or if they cannot complete their contract for some reason, they and their families have a substantial (and rapidly increasing) debt. Exploitation of migrant workers in the recruitment and preparation for travel processes, en route, at the destination, and on their return home is rife. It should be noted that in some countries it is the documented migrants who have higher transaction costs than undocumented migrants. Indeed one of the reasons why migrants opt to take the undocumented route is to avoid the predations of gatekeepers who extract money, both official and 'unofficial', at every stage of the migration process.

Migrant remittances are a key to labour migration having positive impacts on development and poverty reduction in origin areas. Yet these potential dividends can be reduced firstly by having to pay high rates to send the money home and by the lack of investment opportunities in the home area. It is apparent that despite a range of ingenious methods of sending money home, many systems overcharge migrants to remit money so that the proportion of earnings that eventually get back to the origin is smaller than it could be. In addition, Hugo (2004) found in Eastern Indonesia that the origin area of migrant workers to Malaysia had been so neglected by the central and provincial governments that it lacked the basic infrastructure which would be needed for the successful setting up of new enterprises by returned migrants. There were very few productive channels open to returnees to invest money in productive enterprises other than to purchase agricultural land or buy a passenger motor vehicle.

In short, there is considerable evidence that in South-East Asia the potential for international circular migration to deliver development dividends to origin areas is being compromised by the poor governance of these migration systems which prevails over much of the region. Examples of best practice in circular migration in the region exist but are few and far between. In particular the excessive costs of migration are diverting the earnings of the migrant workers which otherwise could have been directed to development-related expenditure in the origin area. Moreover poor governance results in significant undocumented migration, lack of protection and exploitation of the migrant workers, and destination communities developing anti-migrant attitudes and practices.

10.8 Conclusion

Demographic responses and behaviour are becoming increasingly complex and heterogeneous. This is especially the case with population mobility which is both a cause and consequence of economic development and social change. However, understanding the nature of these complex relationships and translating this knowledge into policy-relevant messages to inform effective and equitable policymaking is being hampered by the inadequacy of traditional demographic conceptualization of internal and international migration. If we are to maximize the benefits and minimize the costs of the new population mobility, it is crucially important that we rethink our concepts of migration and the way in which they are operationalized in standard data collection.

References

Ananta, A. and E. Arifin (2008). 'Demographic and Population Mobility Transitions in Indonesia', paper presented at PECC-ABAC Conference on Demographic Change and International Labour Mobility in the Asia-Pacific Region: Implications for Business and Cooperation, organized by Korea National Committee for Pacific Economic Cooperation (KOPEC) in collaboration with Korea Labour Institute (KLI), held in Seoul, Korea, 25–26 March.

Baruah, N. (2013). 'Trends and Outlook for Labour Migration in Asia', presentation to 3rd ADBI-OECD-ILO Roundtable on Labour Migration in Asia: Assessing Labour Market Requirements for Foreign Workers and Developing Policies for Regional Skills Mobility, Thailand, 23–25 January.

Bedford, R. D. (1973). *New Hebridean Mobility: A Study of Circular Migration*, Publication H9/9, Department of Human Geography, Australian National University, Canberra.

Bedford, R. D , E. Ho, and G. J. Hugo (2003). 'Trans-Tasman Migration in Context: Recent Flows of New Zealanders Revisited', *People and Place* 11/4: 53–62.

Center for Global Development (2009). *Migrants Count: Five Steps Toward Better Migration Data*, Report of the Commission on International Migration Data for Development Research and Policy, May.

Chan, K. W. and L. Zhang (1999). 'The *Hukou* System and Rural–Urban Migration in China: Processes and Changes', *China Quarterly* 160: 818–855.

Chapman, M (1979). 'The Cross-Cultural Study of Circulation', *Current Anthropology* 20/11: 111–114.

Chapman, M. and R. M. Prothero (eds) (1985). *Circulation in Third World Countries*. London, Boston, Melbourne, and Henley: Routledge and Kegan Paul.

Collier, W. L., K. Santoso, K. Soentoro, and R. Wibowo (n.d.). 'New Approach to Rural Development in Java: Twenty Five Years of Village Studies in Java', mimeo.

Conway, D. and R. B. Potter (2009). 'Return of the Next Generations: Transnational Migration and Development in the 21st Century', in D. Conway and R. B. Potter (eds), *Return Migration of the Next Generations: 21st Century Transnational Mobility*. London: Ashgate, 1–18.

DaVanzo, J. (1981). 'Microeconomic Approaches to Studying Migration Decisions', in G. F. De Jong and R. W. Gardner (eds), *Migration Decision Making*. New York: Pergamon Press, 90–129.

Department of Immigration and Citizenship (DIAC) (2008). *Population Flows: Immigration Aspects*. Canberra: AGPS.

Dumont, J. and G. Lemaitre (2005). 'Counting Immigrants and Expatriates in OECD Countries: A New Perspective', Conference on Competing for Global Talent, Singapore Management University, Singapore, 13–14 January.

Furneaux, R. (1965). *Krakatoa*. London: Secker and Warburg.

Glick-Schiller, N., L. Basch, and C. Szanton Blanc (1992). 'Transnationalism: A New Analytic Framework for Understanding Migration', in N. Click Schiller, L. Basch, and C. Szanton

Blanc (eds), *Toward a Transnational Perspective on Migration*. New York: New York Academy of Sciences, 1–24.
Global Commission on International Migration (GCIM) (2005). *Migration in an Interconnected World: New Directions for Action*. Geneva: Global Commission on International Migration.
Goldstein, S. (1976). 'Facets of Redistribution: Research Challenges and Opportunities', *Demography* 13/4: 423–434.
Hagerstrand, T. (1957). 'Migration and Area', *Lund Studies in Geography* B 13: 27–158.
Hong, X. N. T. (2008). 'Migration of Youth to Ho Chi Minh City, Vietnam: Determinants of Mobility and Adjustment Experiences', unpublished PhD thesis, Discipline of Geographical and Environmental Studies, University of Adelaide.
Huang, S., B. Yeoh, and T. Lam (2008). 'Asian Transnational Families in Transition: The Liminality of Simultaneity', *International Migration* 46/4: 3–13.
Huguet, J. W., A. Chamratrithirong, and K. Richter (2011). 'Chapter 1: Thailand Migration Profile', *Thailand Migration Report 2011*. Thailand: IOM (International Organization for Migration).
Hugo, G. J. (1975). 'Population Mobility in West Java, Indonesia', unpublished PhD thesis, Department of Demography, Australian National University, Canberra.
Hugo, G. J. (1978). *Population Mobility in West Java*. Yogyakarta: Gadjah Mada University Press.
Hugo, G. J. (1980). 'Population Movements in Indonesia During the Colonial Period', in J. J. Fox, R. G. Garnaut, P. T. McCawley, and J. A. C. Mackie (eds), *Indonesia: Australian Perspectives*. Canberra: Research School of Pacific Studies, Australian National University, 95–135.
Hugo, G. J. (1982). 'Circular Migration in Indonesia', *Population and Development Review* 8/1: 59–84.
Hugo, G. J. (1983). 'New Conceptual Approaches to Migration in the Context of Urbanization: A Discussion Based on Indonesian Experience', in P. A. Morrison (ed.), *Population Movements: Their Forms and Functions in Urbanization and Development*. Liege: Ordina Editions (for the International Union for the Scientific Study of Population), 69–114.
Hugo, G. J. (1994). *The Economic Implications of Emigration from Australia*. Canberra: AGPS.
Hugo, G. J. (1999). 'A New Paradigm of International Migration in Australia', *New Zealand Population Review* 25/1–2: 1–39.
Hugo, G. J. (2004). 'International Labour Migration and Village Dynamics: A Study of Flores, East Nusa Tenggara', in T. R. Leinbach (ed.), *The Indonesian Rural Economy: Mobility, Work and Enterprise*. Singapore: Institute of Southeast Asian Studies, 103–133.
Hugo, G. J. (2008). 'In and Out of Australia: Rethinking Indian and Chinese Skilled Migration to Australia', *Asian Population Studies* 3/4: 267–291.
International Labour Organization (2013a). Migration for Employment Convention (Revised), 1949 (No. 97)—Philippines (Ratification: 2009). Bangkok: ILO.
International Labour Organization (2013b). 'Vietnam'. Bangkok: ILO.
Kim, K. B. (2011). 'Labour Migration in Asia: Recent Trends and Outlooks', presentation at ADBI-OECD Roundtable on Labor Migration in Asia: Recent Trends and Prospects in the Postcrisis Context, Tokyo, Japan, 18–20 January.
King, R. (2000). 'Generalizations from the History of Return Migration', in B. Ghosh (ed.), *Return Migration: Journey of Hope or Despair?* Geneva: International Organization for Migration, 7–55.

King, R. (2002). 'Towards a New Map of European Migration', *International Journal of Population Geography* 8/2: 89–106.

King, R. and R. Skeldon (2010). '"Mind the Gap!" Integrating Approaches to Internal and International Migration', *Journal of Ethnic and Migration Studies* 36/10: 1619–1646.

Kroef, J. M. van der (1956). 'Population Pressure and Economic Development in Indonesia', in J. J. Spengler and O. D. Duncan (eds), *Demographic Analysis*. Glencoe: The Free Press, 739–754.

Ley, D. and A. Kobayashi (2005). 'Back to Hong Kong: Return Migration or Transnational Sojourn?', *Global Networks* 5: 111–128.

Mantra, I. and M. Molo (1985). *Studi Mobilitas Sirkuler Penduduk ke Enam Kota Besar di Indonesia*, Kerjasama Kantor Menteri Negara Kependudukan den Lingkungan Hidup dengan Pusat Penelitian Kependudukan UGM, Laporan akhir Yogyakarta, Pusat Penelitian Kependudukan.

Mitchell, J. C. (1978). 'Wage Labour Mobility as Circulation: A Sociological Perspective', paper presented to the International Seminar on the Cross-Cultural Study of Circulation, East-West Center, Honolulu, April.

Naim, M. (1972). 'Voluntary Migration in Indonesia', Working Paper 26, Institute of Southeast Asian Studies, Singapore.

Philippine Overseas Employment Administration (POEA) (2010). Commission on Filipinos Overseas: Stock Estimate of Overseas Filipinos as of December 2009.

Portes, A., L. E. Guarnizo, and P. Landolt (1999). 'The Study of Transnationalism: Pitfalls and Promises of an Emergent Research Field', *Ethnic and Racial Studies* 22/2: 217–237.

Pusat Penelitian Kependudukan, Universitas Gadjah Mada (1986). *Mobilitas Angkalan Kerja ke Timur Tengah*. Yogyakarta: Gadjah Mada University.

Ranneft, J. M. (1916). 'Volksverplaatsing op Java' [Population Movements in Java], *Tijdschrift voor het Binnenlandsch Bestuur* 49: 59–87, 165–184.

Scalabrini Migration Center (2012). *Asian Migration Outlook 2011*. Philippines: Scalabrini Migration Center.

Scheltema, A. M. P. A. (1926). 'De Groel van Java's Bevolking', *Koloniale Studien* 10/2: 849–883.

Schrieke, B. (1957). *Indonesian Sociological Studies*. The Hague: van Hoeve.

Soeprobo, T. B. (2005). 'Recent Trends in International Migration in Indonesia', paper presented at the Workshop on International Migration and Labour Market in Asia organized by the Japan Institute for Labour Policy and Training, Japan Institute of Labour, Tokyo, 20–21 January.

Suyono, M. (1981). 'Tenaga Kerja Indonesia di Timur Tengah Makin Mantap', *Suara Karya* h.v.k.: 2–6.

Terry, D. F. and S. R. Wilson (eds) (2005). *Beyond Small Change: Making Migrant Remittances Count*. Washington, DC: Inter-American Development Bank.

United Nations (2006). 'International Migration and Development: Report of the Secretary-General', Sixtieth Session, Globalization and Interdependence: International Migration and Development, 18 May, United Nations.

United Nations (2010). 'International Migration and Development: Report of the Secretary-General', Sixty-Fifth Session, Globalization and Interdependence, 2 August, United Nations, New York.

United Nations (2013). 'International Migration 2013', Department of Economic and Social Affairs Population Division, December, United Nations, New York.

Vertovec, S. (2004). ‘Migrant Transnationalism and Modes of Transformation’, *International Migration Review* 38/3: 970–1001.

Vertovec, S. (2006). ‘Is Circular Migration the Way Forward in Global Policy?’, *Around the Globe* 3/2: 38–44. Melbourne: Monash Institute for the Study of Global Movements.

Viviani, N. (1996). *The Indochinese in Australia 1975-1995: From Burnt Boats to Barbecues*. Melbourne: Oxford University Press.

Volkstelling (Population Census) (1933–1936). *Definitieve Uitkomsten Van de Volkstelling 1930*, 8 vols. Batavia: Department van Landbouw, Nijverheid en Handel.

Wahyuni, E. S. (2000). ‘The Impact of Migration upon Family Structure and Functioning in Java’, unpublished PhD thesis, Department of Geographical and Environmental Studies, University of Adelaide.

Widjojo, N. (1970). *Population Trends in Indonesia*. New York: Cornell University Press.

World Bank (1984). ‘Economic and Social Development: An Overview of Regional Differentials and Related Processes’, main report from *Indonesia: Selected Aspects of Spatial Development*, Report No. 4776-IND. Washington, DC: Country Programs Department, East Asia and Pacific Regional Office.

World Bank (2006). *Global Economic Prospects 2006: Economic Implications of Remittances and Migration*. Washington, DC: World Bank.

World Bank (2011). *World Development Report 2011: Conflict, Security and Development*. Washington, DC: World Bank.

Zelinsky, W. (1971). ‘The Hypothesis of the Mobility Transition’, *The Geographical Review*, 61/2: 219–249.

Zhengyi, Y. (2013). ‘Indonesia Restrict Foreign Labor, The Stock Market Watch’, <http://www.thestockmarketwatch.co/indonesia-restrict-foreign-labor-yin-zhengyi.html>, page no longer accessible.

CHAPTER 11

COLLECTIVE IDENTITIES, SHIFTING POPULATION MEMBERSHIP, AND NICHE CONSTRUCTION THEORY

implications from Taiwanese and Chinese empirical evidence

MELISSA J. BROWN

THERE is still much we need to understand about what empirically constitutes a population, including under what—if any—conditions human populations may be thought closed or fixed in their membership (Kreager et al., this volume). Collective identities, especially ethnic identities, are noteworthy populations because they constitute large potential and actual networks in the contemporary world that people within them recognize. Moreover, politicians and activists generally claim that collective identities represent closed populations, wherein membership is determined by birth, fixed for life, and marked—usually culturally, sometimes physically.

The history of Taiwan and China, however, demonstrate that regime changes and mass migrations constitute macro-level perturbations to historically inherited ethnic relations which can lead to shifts in ethnic membership (Brown 2004, 2008b). Regime changes combined with mass migrations prompted shifts in ethnic population membership in Taiwan in the seventeenth century and again after World War II. Ethnic shifts also derived from mass migrations alone in China over millennia, and from regime changes alone in early twentieth-century Taiwan and Maoist-era China. Although these examples indicate the potential power of states to affect population membership, states' deliberate attempts to shape identity-related population membership are often problematic. Colonial Taiwanese, for example, knew they were not Japanese (Brown 2010).

Causality, however, does not move only from the macro-level structural changes to micropopulations. With appropriate ethnographic data, we can see the dynamics of how individuals construct the social niche in which they live. In early twentieth-century Taiwan, for example, in the aftermath of regime change and colonial development policies, individuals at crucial nodes in hierarchical social networks succeeded in micro-level renegotiation of normative expectations about creating, holding, and performing roles. Their success was not negotiation of a new expectation but negotiation of a different referent group—membership in another ethnic population—with different normative expectations (on referent populations, see Durham 1991: 210–211; Brown 1995, 2008a; Brown and Feldman 2009; Lipatov et al. 2011; and Szreter's 'communication communities' in this volume). Such success served as precedents which facilitated further negotiations that accumulated to allow entire communities to change ethnic identity.

These shifts in the membership of collective identities constitute anomalies which problematize our concept of population, for how can we base discussion on empirical, realistic units of population when ethnic groups—often regarded as quintessentially closed populations—are subject to major changes in membership? A modified niche construction theory approach that views social negotiations as the social-niche-regulating activities of individuals can explain these anomalous dynamics (cf. Brown and Feldman 2009; Lipatov et al. 2011; Odling-Smee 2011, this volume; Odling-Smee et al. 2003). This approach recognizes that social negotiations occur in the context of normative expectations about social roles, semiotic implications of cultural beliefs, and associations across roles and beliefs (cf. Runciman 2009; Johnson-Hanks et al. 2011; Johnson-Hanks, this volume). It models these contextual influences using a network-based concept of society; a population-based concept of culture; distinct processes of social and cultural selection; and associational modifications of these selection pressures, due to epistasis and linkage (Brown and Feldman 2009). This approach to social inheritance also problematizes understandings of culture and society as indistinguishable (e.g. Fox and King 2002; Johnson-Hanks et al. 2011; Johnson-Hanks, this volume); society as reducible to culture (e.g. Odling-Smee 2011, this volume; Odling-Smee et al. 2003); and culture as preceding society (Runciman 2009). It supports the robustness of niche construction theory, at the same time that it suggests a modification of the analogy between ecological inheritance and social inheritance.

11.1 Ethnic Populations

In both China and Taiwan today, Han are the politically dominant ethnic majority, and Mandarin Chinese is the national language (see Figure 11.1, Tables 11.1 and 11.2; Brown 2004, 2010). In the People's Republic of China (PRC), all 'regional varieties' of Han—including speakers of seven mutually unintelligible Chinese 'dialects'—are officially classified as a single ethnic group.

Ethnic identity appears on individuals' national identity cards, based on a government 'ethnic identification' project largely carried out in the 1950s (e.g. Harrell 1995; Mullaney

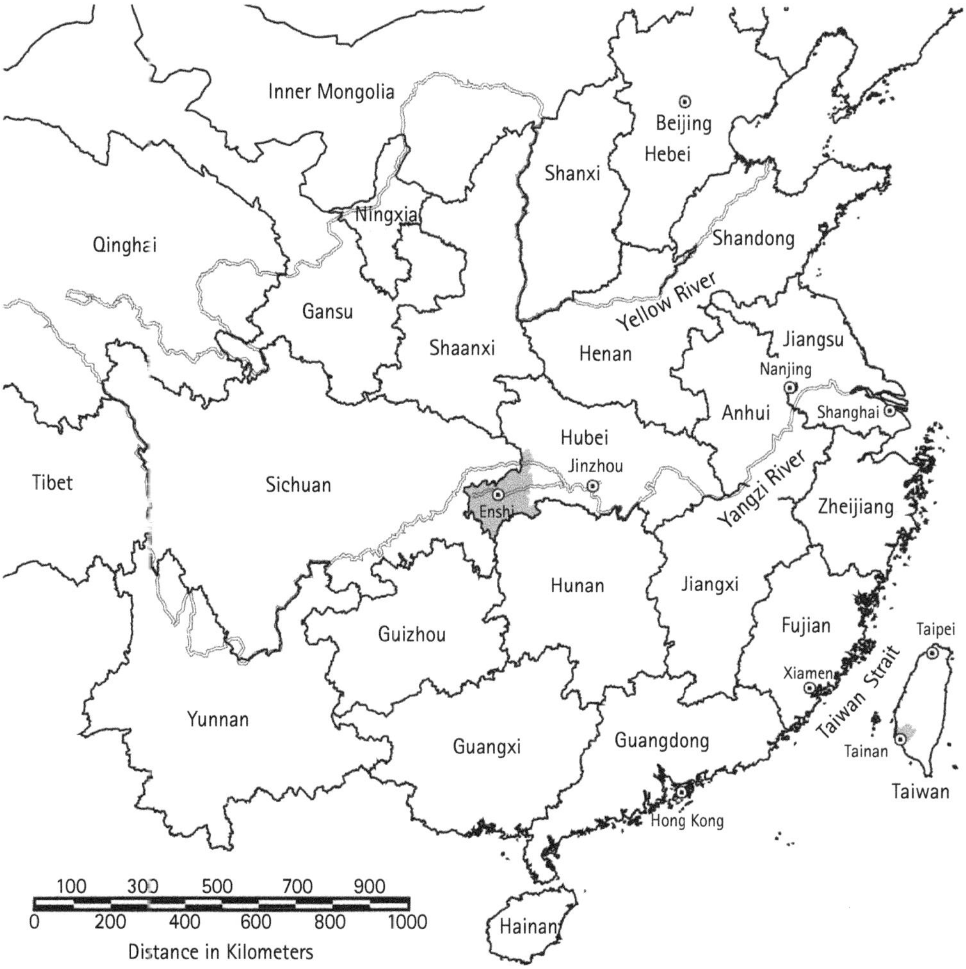

FIG. 11.1 Map of China and Taiwan.

2011).[1] This project had two distinct stages: the first identifying which ethnic groups existed, the second classifying all individuals into one of the previously identified ethnic groups (Brown 2002, 2004). Most people were ethnically classified before the Cultural Revolution (see Table 11.3), when minorities suffered under strong assimilation pressures.

Taiwan's government accepts citizens' self-identification, except that indigenous peoples targeted by affirmative action policies require certification. Three distinct Han ethnic groups dominate Taiwanese politics: Hoklo, Hakka, and Mainlanders. Hoklo and Hakka are ethnic Taiwanese, whose ancestors immigrated between 1624 and 1895. Mainlanders fled to Taiwan after WWII. Taiwan Aborigines are Austronesian peoples, linguistically and genetically related to Polynesians.[2] The 'mountain' peoples, about 2%

[1] Minority status confers affirmative action benefits which vary regionally.

[2] 'Aborigines' is the preferred English appellation of the Alliance of Taiwan Aborigines; 'Indigènes' and 'First Nations' are also accepted.

Table 11.1 Ethnic (M. *minzu*, 民族) groups in China[a]

Han 汉族, 91.5% of PRC population (estimated % of PRC population)[b]	**non-Han minority groups** (M. *shaoshu minzu* 少数民族) (% of PRC population, population size)[c]
Mandarin 官话 (all Mandarin speakers) (71.5%)	Zhuang 壮族 (1.30%, 16.2 million)
Wu 吴 (including Shanghainese) (8.5%)	Manchu 满族 (0.86%,10.7 million)
Yue 粤 (Cantonese) (5.0%)	Hui 回族 (0.79%, 9.8 million)
Xiang 湘 (4.8 %)	Miao 苗族 (0.72%, 8.9 million)
Min 闽 (Fujianese) (4.1%)	Uyghur 维吾尔族 (0.68%, 8.4 million)
Hakka 客家 (3.7%)	Tujia 土家族 (0.65%, 8.0 million)
Gan 赣 (2.4%)	Yi 彝族 (0.62%, 7.8 million)
	Mongols 蒙古族 (0.47%, 5.8 million)
	Tibetans 藏族 (0.44%, 5.4 million)
	and more

[a] *Source languages are:* M. Mandarin Chinese; T. Minnan Chinese, also called Taiwanese; J. Japanese.

[b] *Source:* Ramsey (1987).

[c] *Source:* Population Census of the PRC (2000).

Table 11.2 Ethnic (M. *zuqun* 族群) groups in Taiwan[a]

Han 漢 (98% of Taiwan's population)			**Aborigines** (M. *Yuanzhumin* 原住民)	
Taiwanese (M. *Taiwan ren* 臺灣人 or M. *Bensheng ren* 本省人)		Mainlanders (M. *Waisheng ren* 外省人) (13%)	Plains Aborigines (M. *Pingpu zu* 平鋪族) (current percentage of population unknown)[b]	Mountain Aborigines (M. *Gaoshan zu* 高山族) (2%)
Hoklo (T. *Hok-lo lang* 福佬人 or M. *Taiwan ren* 臺灣人) (73%)	Hakka (M. *Kejia ren* 客家人) (12%)	many groups	many groups, including Siraya	many groups, including Taroko

[a] *Source:* Taiwan Census (2008).

[b] Plains Aborigine affiliation is not included on the 2008 census.

of Taiwan's population, successfully defended their territories from the seventeenth century until 1930 when the Japanese colonial government used modern artillery to 'pacify' them. Plains Aborigines are descendants of Austronesians who adopted Han languages and customs over centuries of interactions (Brown 2004). What proportion of Taiwan's

Table 11.3 Historical regimes

Taiwan		China	
Dutch	1624–1661	Ming	1368–1644
Zheng	1662–1683	Qing	1644–1911
Qing	1683–1895		
Japanese	1895–1945	Republican	1911–1949
Nationalist	1945–1996	War with Japan	1937–1945
Martial-law period	1947–1987	Civil War	1945–1949
		Communist (PRC)	1949– present
		Maoist era	1949–1976
Transitional period	1987–1996	Cultural Revolution	1967–1976
		Reform era	1978–present
Full electoral democracy	1996–present		

population the 'plains' peoples constitute is unclear, because by the mid-twentieth century, many had acquired a Taiwanese identity (a Han category). In the mid-1990s, a national recasting of Aborigines as laudable ancestors led many to reclaim plains Aborigine identity and continues to fuel problematic assertions of Taiwanese as genetically Austronesian (see Chen 2008 for an even-handed evaluation of Taiwanese genetic ancestry).

11.2 Macrostructure Can Drive Population Membership Shifts

Historical evidence indicates that regime changes and mass migrations, both in concert and singly, drove shifts in the membership of ethnic populations. These macrostructural factors influenced population membership most importantly by changing individuals' *social experience*—the lived social interactions each individual has with other individuals *plus* the broad range of political-economic and historic influences that shape those interactions (Brown 2004, 2010; cf. Appiah 2006: 16). Ethnic identity formation for children in a society, and for immigrants new to a society, begins with assignment of a potentially negotiable ethnic label which in turn influences social interactions. Ethnic identities are *collective* and thus, by definition, *shared*: the social experiences of individuals in an accepted population (associated with an ethnic label) are similar enough to each other to be considered versions of one another, both by the individuals themselves and by outsiders (Brown 2004, 2010; cf. Sperber 1996: 81). Ethnic identities are often *claimed* in terms of shared cultural ideas or practices, but the effects of culture are subordinate to those of social experience, especially the political economic context dictating subsistence and employment (Brown 2004, 2008b, 2010; cf. Appiah 1994, 2006; Barth 1969, 1994; Harrell 1995, 1996a, 2001). Historically, regime changes and mass migrations changed which individuals had similar social experiences. In this way, these macrostructural factors shifted membership of the population sharing a collective identity with a specific ethnic label.

11.2.1 Migration-Driven Short Routes

Because identities are so rooted in social experience, migrations that bring a disproportionate influx of population have the potential to shift identities. Such mass migrations not only create new relations between immigrants and the receiving population but can also shift previous relations within subdivisions of the receiving population. These relations are strongly influenced by political dominance, which itself can be affected by such large population influx.

Seventeenth-Century Southwestern Taiwan

To control a thriving trade between Austronesians, Chinese, Japanese, and Europeans, in 1624 the Dutch established the first state-level political presence, centred near present-day Tainan City, over an estimated 65,000 Austronesians island-wide (see Figure 11.1, Table 11.3; Brown 2004: 36–40, ch. 4). Dutch colonial strategy included using Taiwan to produce a rice surplus for its Indonesian colony and to sell in China and Japan. However, the Dutch never managed to pressure even their primary Austronesian allies (the Siraya) into producing a surplus of rice. Consequently, the Dutch encouraged large-scale immigration from China. By 1650, the Dutch recorded 50,000 Han Chinese men in southwestern Taiwan (few women migrated to Taiwan). Regime change combined with mass migration occurred in 1661, when Zheng forces, which had been battling the incoming Qing regime on the Chinese mainland, retreated to Taiwan, bringing about 30,000 people, mostly Han Chinese men, including 20,000 troops. In 1683, the Qing regime defeated Zheng forces and took control of Taiwan.

Under the Dutch and Zheng regimes, Han immigrant men were largely forced to seek wives among the indigenous peoples or mixed Han-Austronesian population. Comparison of 1650 and 1685 population records for southwestern Taiwan suggests that the mixed-ancestry population maintained an Austronesian identity while under Dutch control, but shifted to Han identity with the Zheng or Qing takeover.[3] Assignment to these ethnic categories affected social experience. The Dutch valued Aborigines as military allies and viewed Han with suspicion. The Zheng and Qing regimes required Aborigines to provide corvée labour service not required of Han and viewed Han as their main constituency. The Qing statute that all unmarried Han men must return to

[3] Despite a Manchu ruling family, the Qing regime used Han language, customs, and administrative practices.

China meant that the population remaining in southwestern Taiwan in 1685 (the most populous area) would have had a significant proportion—an estimated 40–50%—of mixed-ancestry households.

Migrations and intermarriage combined with changes in political regimes and economic interests to shift the ethnic identity of the mixed-ancestry population. The population sex ratio in seventeenth-century Taiwan was so male-biased that natural selection may have contributed influence (Brown 2004: 227), for Han men who refused indigenous or mixed wives were unlikely to have heirs (genetic or economic). Whether this 'short route' to Han identity was initiated by mixed-ancestry individuals' own tactical negotiations for a more prestigious ethnic status or by the ascriptions of Zheng or Qing officials based on patrilineal ancestry, subsequent social experience, including government taxation and neighbours' treatment, would have changed individuals' sense of themselves over time, authenticating the Han identity.

Two Millennia in Enshi (Hubei Province, China)

Acquisition of Han identity by people in the mountains of Enshi, which separate the Hubei plains from the Sichuan basin, was shaped by mass migrations (see Figure 11.1; Brown 2004: ch. 5). Waves of forced migration—to occupy a strategically important border region, provide agricultural and other labour, and promote stability by diluting a rebellious population with more compliant Han—occurred over two millennia. The earliest known immigrants were part of a third-century BCE expansion strategy, but the mountainous terrain kept it on the fringes of dynastic struggles (and recorded history) until the Yuan (Mongol) dynasty (1276–1368) brought it back into the tribute-paying realm. Oral histories and genealogies report forced migrations during the Ming and Qing dynasties (see Table 11.3) and remember the order of different patrilineal clans' arrival.

Each wave of migration merged earlier immigrants with 'locals' (M. *tujia*), not only through intermarriage but also in shifting group membership. One clan could be considered 'local' by a second clan and an 'outsider' by a third. Over time, people in Enshi came to share major characteristics with Han elsewhere in China: language (Southwest Mandarin), marriage practices, footbinding—as well as some aspects of ancestor worship, funerary rites, and patrilineal inheritance (other aspects differed noticeably). Since identity as local or outsider was relative, neither status was strongly linked to any cultural marker. By the early twentieth century, both locals and outsiders, including wealthy Confucian-educated landowners, practised some customs such as placing graves within villages that were noticeably different from Han elsewhere.

Political control did not constitute regime change at the local level for centuries. Despite imperial soldiers (forced migrants) stationed there, Enshi remained under 'local' or indigenous rule until the Qing period, although a 1395 imperial order did require indigenous rulers to learn Confucian principles and Han writing. Effective regime change at the local level occurred in 1735, when Enshi came under the same Qing bureaucratic administration as Han areas elsewhere. Mass Han immigration again followed, constituting another wave that merged all earlier immigrants as 'locals'. Subsequent regime changes followed those elsewhere in China: the founding of the Republic and later the People's Republic (see Table 11.3).

The process of mass immigration driving ethnic identity change of former 'outsiders' to 'locals' continued up until PRC officials assigned ethnic identity labels to individuals in the late twentieth century. The last large-scale Han immigration prior to that time occurred in the late 1930s, when Japan's invasion of China forced the removal of Hubei's provincial capital to mountainous Enshi City. This wave of immigrants again lumped together the descendants of all previous waves as local Han—poorer and less sophisticated than these cosmopolitan outsiders, but Han nonetheless. Some local men conscripted in the 1940s and taken to Taiwan (as Mainlanders) were shocked, upon their first allowed return to Enshi in the 1980s, to discover that, in the interim, their relatives had been classified by the PRC government as members of a non-Han ethnic minority, called Tujia (discussed in section 11.2.2). Before 1949, 'locals' were accepted as Han, the 'short-route' process of shifting population membership via mass migrations unquestioned.

Post-WWII Taiwan

The impact of regime change and mass migration on the creation of ethnic populations—assigning membership to new categories—occurred rather dramatically in the context of China's civil war. During Taiwan's time as a Japanese colony, China, which had nominally become a republic, was torn apart by feuding warlords, Western imperialism, Japanese invasion, and armed conflict between Nationalist and Communist troops (see Table 11.3). After WWII, Chinese Nationalists took control of Taiwan, bringing one to two million people there before their defeat on the mainland in 1949. This combination of regime change and mass migration led to the formation of two new ethnic identities within the Han category: Taiwanese and Mainlanders (see Table 11.2; Brown 2008; cf. Gates 1981; Corcuff 2000).

The receiving Han population in Taiwan had previously been subdivided into Hoklo and Hakka ethnic groups, but the Nationalist government merged them into a single 'Taiwanese' population (M. *bensheng ren*, literally, people from this province, i.e. Taiwan), with membership defined by their historic presence in Taiwan under Japanese rule. People accompanying the Nationalists' retreat came from all over China, but once in Taiwan, the government combined them into a single category, in opposition to Taiwanese: Mainlanders (M. *waisheng ren*, literally, people from outside this province). Membership was determined by the shared social experience of fleeing the Communist unification of China. For example, refugees from Fujian were Mainlanders, even though they shared (Minnan) language and many customs with Hoklo, who were Taiwanese. Mainlanders, like many politically dominant ethnic populations, did not recognize themselves as an *ethnic* population (until democratization loosened their hold on political dominance).

Taiwanese–Mainlander distinctions were ubiquitous social experience throughout the martial-law years. At the beginning of each school year, every child was required to state for the class from which province his or her father came, marking Taiwanese from Mainlanders. Language was a frequent basis for discrimination, especially in employment, since most Taiwanese had a recognizable accent in speaking Mandarin, because they had to learn it quickly after it was declared the national language virtually overnight. (Japanese had been the national language during the colonial period.) Individuals caught speaking their native Hoklo or Hakka languages in public places were subject to fines. Forty years of such government-sanctioned distinctions between Taiwanese and Mainlanders led to self-identification along these lines. Here, we see mass migration accompanied by regime change leading to the formation of new ethnic identities by merging previously distinct populations. We also see the ability of the ruling elite to shape social experience such that it reinforced the identity labels assigned by the government, despite the fact that the Nationalist government wanted to instil Chinese identity among the Taiwanese.

11.2.2 Regime-Tipped Long Routes

Because identities are so strongly rooted in social experience, political regimes' implementation of social and economic policies have the potential to shape and shift identities. When such policies act in concert with mass migration, as in the short-route examples already discussed, the shifts happen very rapidly and dramatically. However, states also have the potential to affect changes with the implementation of new policies, as is often done by regimes that have relatively recently come to power.

Early Twentieth-Century Southwestern Taiwan

After two hundred years of Qing rule, plains Aborigines in southwestern Taiwan who had not assimilated (shifted identity) had largely acculturated—i.e. adopted many of the customs and cultural ideas of the dominant Han in their area. At the turn of the twentieth century, both villagers and neighbouring Hoklo classified Toushe, Jibeishua, and Longtian villages in Tainan County as plains Aborigine communities, despite much acculturation and some intermarriage with Hoklo (see Figure 11.1; Brown 2004: ch. 3). Villagers had spoken Taiwanese as their native language for decades, and grew the same cash crops—rice and sugarcane—under the same rental terms as their Hoklo neighbours. They dressed like neighbouring Hoklo of the same economic class. They used the same forms of marriage, though in different proportions. Except for the Christian minority, villagers buried their dead in the same way, with a Daoist priest presiding, and differences in tending graves and worship went largely unnoticed. Most villagers linked surnames with property inheritance, as Hoklo do, but some expressed ambivalence about it. Marriage forms did not always carry the same meanings. In Hoklo villages, poor men who married uxorilocally (to their wife's home) would have been scorned, but they did not face such stigma in Toushe, Jibeishua, and Longtian; uxorilocal marriage rates were higher here than in neighbouring Hoklo villages. Local culture moved toward a Hoklo model, but ideas changed after practices were introduced, and some ideas spread more quickly than others (Brown 2007).

It is significant that the type of Han living near these villages were Hoklo, because Hoklo bound their daughters' feet while Hakka—the other major category of Han in Taiwan at the time—did not. Despite all the shared cultural practices and ideas by 1900, and even some Hoklo ancestry, Toushe, Jibeishua, and Longtian were still considered plains Aborigine. In interviews I conducted in 1991–1992, elderly villagers repeatedly spoke of footbinding as the single cultural practice whose absence had marked them as non-Han in the early twentieth century: 'We savages didn't bind feet.' Ironically, this comment shows the degree to which villagers were culturally Hoklo, because they had internalized Hoklo cultural ideas that lack of footbinding proved them less than civilized, less than Han. Being culturally Hoklo, however, had not won them Hoklo identity.

Ironically, it was footbinding that finally allowed long-route acquisition of Han identity, due to the custom's interaction with a new regime's development plans. In 1894, Japan forced the Qing government to cede Taiwan. Colonial strategy included supplying Japan with Taiwan-produced rice and capitalizing on Taiwan's already prominent global exports of sugar and camphor. Achieving these goals required development of Taiwan's industry, infrastructure (including railroads and large irrigation systems), and human capital. Upon gaining control of Taiwan, Japan cut off new immigration from China and reformed the land tax system, reducing overall taxes and removing the remnants of ethnic differentiation in taxation. Over time, the colonial government also instituted public education and health measures, and later began promoting loyalty to Japan (albeit as second-class citizens; Brown 2010). Until about 1915, Japan's expenditures to control and develop Taiwan exceeded its colonial returns, but development investment continued well beyond that time, including the provision of seed money for the Jianan irrigation system, built 1919–1930, which allowed three rice crops per year in southwestern Taiwan.

In 1905, the same year that the first colonial census revealed that 68% of Taiwan's female Hoklo population had bound feet (Brown 2004: 156), the colonial government announced a ban on footbinding, which it considered barbaric, but did not enforce the ban. By 1915, however, the colonial government realized that footbinding interfered with its goals to expand Taiwanese women's labour in the agricultural sector. Hoklo women with bound feet did not work in paddy fields, since the paddy mud ruined the binding cloth, which was expensive to replace. To further economic development, enforcement of the footbinding prohibition was implemented through the colonial household registration system, under which police inspected every household two or three times a year (see Wolf and Huang 1980: 16–33; Brown 2004: 8–9, 54). Starting in 1915, footbinding was added to the inspection. Women who could were required to unbind their feet,[4] and household heads were ordered not to allow (re)binding. Elderly Han villagers told me that Japanese police officers who oversaw the unbinding explained it would expand women's agricultural work. It did, moving Hoklo social experience toward that of plains Aborigines: women worked in paddy fields.

[4] There was a range of footbinding forms; the 'lotus' form broke the arch and could not be unbound (Brown et al. 2012).

The footbinding ban inadvertently affected marriage and identity because the custom of footbinding had marked an ethnic, endogamous boundary. In other words, the custom marked *role expectations* for brides—normative expectations about which category of women were acceptable holders of the role of bride (Brown and Feldman 2009). Before 1915, Hoklo would not bring in (never-bound) plains Aborigine brides for their sons, nor marry their (usually bound) daughters to plains Aborigine men (virilocally). Only very poor Hoklo men—who could get a wife no other way—married into plains Aborigine households (uxorilocally). By forcibly removing the marker, so that no brides had bound feet, the Japanese government created the potential for negotiation over which women could subsequently become brides.

Change in which women became virilocal brides occurred almost immediately, and can be seen in the frequency of uxorilocal marriages among plains Aborigines (where a husband married to his wife's home: Table 11.4). (Most marriages in that time and place were virilocal, where a wife married to her husband's home, or uxorilocal; Lipatov et al. 2011: 906.) The colonial household registers provide information on marriages and ethnic identity for every person alive in Taiwan between 1906 and 1945 (see Lipatov et al. 2011: 905 for a description of the database which is the source of these calculations). The database for predominantly Hoklo Danei Township in southwestern Taiwan includes 21,151 people over those forty years. Toushe village, with 7,332 people, the only Danei village in which plains Aborigines were a majority, has been distinguished for more direct comparison with Jibeishua village, with 3,268 people, from Dongshan Township. Using only women's first marriages (remarriage after widowhood or divorce are excluded), a chi-square test of the shift in marriage associated with the 1915 footbinding ban shows significant association between the frequency of uxorilocal marriages among plains Aborigines and the presence or absence of footbinding among the Han (see Table 11.4; Brown and Feldman 2009: 22140). Further evidence of changes in which women became virilocal brides comes from a comparison of the rates of uxorilocal marriages between plains Aborigines and Han by decade (see Figure 11.2; Lipatov et al. 2011). Through 1915, the rates of uxorilocal marriage were significantly higher for plains Aborigines than for Han, but after 1916, the rates for plains Aborigines and Han are *not* significantly different. In other words, the 1915 ban marks a turning point at which there was both a

Table 11.4 Uxorilocal first marriages by plains Aborigines in Danei Township, Taiwan

	1906–1915	1916–1945
Virilocal	62	232
Uxorilocal	33	37

$p = 0.0001$, using χ_1^2

Data source: Computer database of the Taiwan Household Registers.

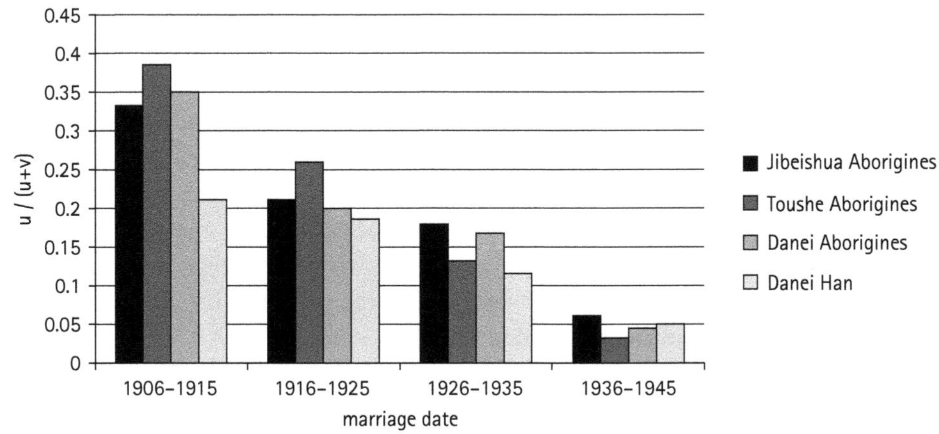

FIG. 11.2 Uxorilocal marriage rates of plains Aborigines and Han in Danei Township and Jibeishua village, Taiwan.

significant increase in the number of plains Aborigine women who became virilocal brides *and* when the proportion of plains Aborigine women who became virilocal brides was no longer significantly different from the proportion of Han virilocal brides.

These quantitative assessments show that plains Aborigines in the household registers shifted to virilocal marriages rapidly after the 1915 footbinding ban. The only way that this shift was possible was for ethnic endogamy to end. When all first-time brides had natural feet, Toushe, Jibeishua, and Longtian exchanged brides with neighbouring Hoklo villages. Ethnographic interviewing found that it was at this point that Hoklo stopped calling their new affines by the only Hoklo-language ethnonym for plains Aborigines: 'savages' (T. *hoan-a*). As marital exchanges between Toushe village and its Hoklo neighbours increased in frequency, the ethnic boundary disappeared. The people of Toushe acquired membership in the Han population.

Through the mid-1990s (across two regime changes; see Table 11.3), Toushe, Jibeishua, and Longtian villagers shared similar lived experiences and suffered the same political-economic discrimination as their Hoklo neighbours. Economic distinctions between Hoklo, Hakka, and plains Aborigines had disappeared at the beginning of Japanese rule, when colonial restructuring of taxation and land tenure gave these groups the same rates. They were in the same ethnic category for day wages, educational quotas, limitations on businesses, constraints on political involvement, and (once WWII began) conscription for corvée and military service (Japanese and mountain Aborigines were in different categories). Under Nationalist martial-law rule, these villagers—like other Taiwanese—had to learn Mandarin, lost jobs, and found themselves subordinated to Mainlanders. During democratization, they were as moved by campaign speeches made in the Taiwanese language as neighbours from villages that had always been considered Hoklo. Those moved *as Hoklo* included young people—many of whom had no idea of their Aborigine heritage when I interviewed their grandparents in 1991–1992—*and* the senior generation, who remembered being taunted as 'Toushe savages' in their youth. Elderly villagers who spoke about once being Aborigine made it clear to me that identity existed long ago and they were Hoklo at the time of the interview.

Even though the Japanese colonial government's aim in eliminating footbinding among the Hoklo was economic—expanding women's agricultural labour contributions—implementation had the unexpected consequence of shifting ethnic membership. Because the custom marked an ethnic, endogamous boundary, the colonial regime's removal of it tipped the balance of social negotiations about which category of women were acceptable holders of the role of virilocal bride and led to the shift in ethnic population membership.

Enshi Locals as Non-Han

Post-WWII China provides other examples of a recently instituted regime enacting policies that change ethnic population membership. While government policy in colonial Taiwan finally tipped long-route ethnic change into assimilation with the ethnic majority, in Enshi we see the potential of state policies to tip long-route change back to minority status. We also see complications of policy implementation in a large state bureaucracy. The PRC government's ethnic identification project, begun in the 1950s, had a goal of fixing an ethnic label for every citizen. This project required two stages of identification and classification, whose distinct implementation relied on people with different network positions and roles in the state's hierarchical bureaucratic structure. The individuals who held these roles also had a non-random differentiation in the referent populations from which they derived their normative expectations. The dynamics of social negotiation within the context of these structural conditions allowed complications in policy implementation: in Enshi, the identification and classification stages occurred decades apart (Brown 2004: ch. 5). This time lag affected local people's social experience, their government-assigned ethnic identity, and their attitude to that assignment.

The first stage brought ethnologists, linguists, officials, and Party members from the central government into Enshi and surrounding areas in the 1950s to identify the different ethnic groups living there. Those with the authority to identify ethnic categories were both outsiders and mostly Han. Having Han outsiders fill this identifier role is notable because outsiders, being ignorant of local social experience, generally rely on cultural ideas and customs to mark ethnic identity (Brown 2002). What struck these authorities were regional customs—shared across clans variously considered 'local' (M. *tujia*) or 'outsider' (M. *kejia*)—that indicated weak parental authority and ran strongly counter to widespread Han beliefs and practices largely standardized across Han areas of China in the Ming and Qing periods (Brown 2007; cf. Watson 1988). Faced with local customs that violated the cultural ideas and social expectations derived from their external Han referent population, the central government's team initially identified the population in this region—including Enshi in Hubei and adjoining areas in Hunan, Sichuan, and Guizhou provinces—as Miao (another non-Han category in the region, which includes a range of Tibeto-Burman peoples such as Hmong).

A period of social negotiation followed. Locals objected to the region's identification as primarily Miao, for they saw themselves as Han and disdained Miao as 'backward'. Rather than argue the central government team of experts wrong, they took the politically expedient path of lobbying for status as 'Tujia'—which means 'local families' or 'locals' in Mandarin Chinese. Hunan officials who were familiar with local social experience did contest identification of the area as a non-Han region, eligible for ethnic minority affirmative action benefits (Brown 2002). Nevertheless, in 1957, the central government recognized Tujia as an ethnic minority identity, distinct from Miao. Here, the first stage ended, with 'Tujia' an officially accepted ethnic label but no population as yet assigned to it.

The second stage should have followed immediately, but one month after Tujia recognition, a policy shift placed a national moratorium on the entire ethnic identification project. The timing of this moratorium meant that, in areas with significant Tujia population, the second stage of classifying individuals into one of the officially recognized groups waited until local officials got orders to have the ethnic classification of the population ready for the 1982 census. Now in the Reform era, provincial officials were eager to capitalize on new state policies granting extra economic development funds from the central government to areas with significant minority populations, yet they were leery of mandates for Han immigration—evident from forced mass migration of Han to Xinjiang and Tibet—where the non-Han population was considered too large. In this context, local officials struggled to classify the population. As locals themselves, with local social experience and a local referent population, they understood that none of the official criteria they were supposed to use to distinguish minorities from Han (language, culture, territory, economy) actually differentiated locals from outsiders (Brown 2002; Wu 2005). Initially, local officials used surnames to distinguish Tujia and Han, but when this method yielded 60% of the Enshi population as Tujia, their provincial supervisors told them to redo the classification to obtain a non-Han population between 25 and 40%. Eventually, local officials used the year that a family's patrilineal ancestors migrated to Enshi to determine official ethnic identity. Families whose ancestors arrived before 1735 (when the area shifted to Qing bureaucratic rule) were Tujia; those who came after were Han.[5] This process resulted in an acceptable 37% of Enshi's population as Tujia. The time lag in implementation of the second stage allowed provincial officials to recognize that the mandate for Han immigration was based on ethnic population size and spurred their strategic management of the classification process, thus influencing ethnic membership for many people.

The time lag between stages had further consequences for ethnic membership because social experience did not become ethnically marked in Enshi until the mid-1980s. Most locals, who had not lobbied for Tujia recognition in the 1950s, maintained their pre-1949 Han identity throughout the Maoist era. Individuals' continued acceptance as Han meant that they escaped ethnic persecution, which was severe

[5] Because local officials did not finish classifying the entire population in time for the 1982 census, the Tujia population figures jump dramatically from the 1982 to the 1990 census.

during the Cultural Revolution (e.g. Khan 1996: 141). Thus, older locals—who had spent most of their lives as Han—did not accept minority identity when I interviewed them in 1996. Several reported that, when they first received their national identity card (in the mid-1980s) and discovered it said Tujia, they went to the police station to request a change (to no avail). The middle generations varied more, in their reactions and their experience of Cultural Revolution assimilationist propaganda. One woman petitioned to change her Han status to Tujia when she found out her parents' village was Tujia, but although she repeatedly urged her adult son to consider the affirmative action benefits, he thought it a stigma to be non-Han and refused to change his registration. The youngest people, of course, experienced most of their formative years as Tujia, in a Reform era that promoted Tujia heritage as honourable, granted affirmative action benefits (e.g. number of children allowed, points spotted on the national college entrance examination), and made ethnic tourism lucrative. Most youths I spoke with in 1996 had no reservations about their Tujia identity, a trend which appears to continue (Brown 2010). These generational differences in attitude to minority identity correspond to social experience. Older people who spent most of their lives as Han resisted the non-Han identity, while younger people who spent most of their lives as non-Han embraced it. Officials in Enshi shrewdly managed the classification process and appear to continue managing identities to maintain 25–40% of the population as minority (thus avoiding centrally mandated Han immigration) yet without riling local sentiments (Brown 2002, 2010). Here, local and provincial government officials have been successful in manipulating identities, a point made bluntly by an ethnologist in Beijing in 1995: 'We tell them what they are, and after a while they get used to it.'

Prmi Acceptance of Externally Assigned Identities

The hubris of this Beijing ethnologist raises questions of *why* an externally assigned identity need not provoke resistance. A key factor is well illustrated by the results of the PRC ethnic identification project among people who call themselves Prmi in their own language (Harrell 1996b, 2001: ch. 10). Prmi constitute a single population spanning the Sichuan–Yunnan provincial border in southwestern China; they share a language and religion (Tibetan Buddhism) and intermarry without regard to province of origin. However, in the classification stage, officials labelled Yunnanese Prmi as Pumi and Sichuanese Prmi as Tibetans. These labels reflect social experience. The Pumi are a small ethnic population. Their current small voice in Yunnan provincial affairs continues their pre-1949 political experience as one of many minorities, and so excited no protest. In Sichuan, having a political voice joined with other Tibetan Buddhists also predates 1949. The region was long controlled by the lama at the local monastery, requiring Prmi in Sichuan to work with other Tibetan Buddhists—many of whom are linguistically and culturally distinct from the Prmi—to have a political voice. When interviewed in the early 1990s, Sichuanese Prmi saw no contradiction in referring to themselves as Prmi in their own language and Tibetans (Zang) in Mandarin, as both labels accommodate their experience of religion as the most salient aspect of their identity. At the same time, Yunnanese Prmi saw no need to be considered Tibetan, as their religion was not the means to political-economic power in Yunnan. Prmi on both sides of the border accepted their (differing) imposed identities because their respective classifications matched how their social experience differed by provincial residence. The key, then, to not provoking resistance to an imposed identity lies in social experience: people accept an imposed identity *if* it corresponds to their actual social experience.

The crucial importance of social experience allows regimes so much potential for manipulating collective identities because regimes can affect the macrostructural factors which determine social experience. Recognition of this power, and the PRC's successes among the Prmi and Tujia (among other cases), fuelled the Beijing ethnologist's claim that people eventually accept government-dictated membership in an ethnic population. But such assignment does *not* always succeed. PRC failures in shaping the identity of Tibetans, many Muslim peoples (e.g. Uyghur), and Taiwanese as Chinese, are internationally known. States are not monolithic, so conflicting or shifting policies as well as competing interests within a state's bureaucratic administration influence outcomes. In Enshi, the outcome went in the PRC government's favour, but in Taiwan, under both Nationalist martial-law rule and Japanese colonial rule, such complicating factors worked against government-desired outcomes. Because individuals have multiple and dynamic connections across the social structure in which they live, they hold multiple roles and cultural beliefs, fill hierarchical positions relative to other similarly embedded individuals, and enact many practices. Structural associations among roles, beliefs, or practices thus create what may be called associational skewing, i.e. unexpected consequences for political regimes attempting deliberate manipulation of identities (Brown and Feldman 2009).

Martial-Law Taiwanese Are Not Chinese

After its defeat in the Chinese civil war, the Nationalist government in Taiwan *wanted* Taiwanese to accept membership in a national Chinese identity (in order to further its long-term goal of wresting control of China from the Communists). Given ideology and beliefs conflating Chinese national identity and Han ethnic identity (Brown 2004: ch. 1), the Nationalist regime assumed Taiwanese assimilation into a national Chinese population would follow directly from their clear membership in the ethnic Han population. The regime reinforced this expected result, for example, treating Taiwan as only one of the many provinces purportedly part of the Republic of China (the country name still used in Taiwan).

The Nationalists' promotion of Chinese national identity failed among the Taiwanese. Not only was an illegal Taiwan independence movement formed, but quotidian acts of resistance were legion, including the maintenance of Hoklo and Hakka languages and folk religion. Discriminatory policies in employment and education distinguished Taiwanese from Mainlanders in social experience and ethnic identity (discussed earlier; Brown 2008b). Taiwanese remained a largely endogamous ethnic population. Although many Mainlander soldiers needed wives, they had more success among the small population of mountain Aborigines than among the much larger Taiwanese population. Although both shared a larger Han ethnic identity, it was the lower-order ethnic categories of Taiwanese and Mainlanders that were more salient to their daily lives and thus took precedence.

Mainlanders remained strongly associated with Chinese national identity. Virtually all had some connection—military, industrial, or educational—to the national government that had earned their removal to Taiwan, a connection easily marked by ability to speak Mandarin Chinese (the national language). Moreover, because Mainlanders during the martial-law period did not recognize 'Mainlander' as an ethnic identity, it was their national identity as Chinese (M. *zhongguo ren*) which best captured their social experience as refugees and reliance upon Nationalist government connections. The close association of Mainlanders with Chinese national identity led to an extrapolation from the discriminatory distinction between Taiwanese and Mainlanders. If Taiwanese were different from Mainlanders, and Mainlanders were Chinese, then Taiwanese were also different from Chinese.

Colonial Taiwanese Are Not Japanese

Japanese colonial authorities also failed to instil a desired Japanese identity in Hoklo, Hakka, and plains Aborigines (here called 'Taiwanese' for expediency) due to Taiwanese social experience of discrimination and an associational skewing effect (Brown 2010). In fact, by emphasizing cultural practices as a means to assimilation, colonial authorities amplified the associational effect.

Despite an assimilation (J. *dōka*) policy aimed primarily at the Taiwanese elite from 1919, enforcement of sweeping assimilationist changes in the Taiwanese population (under a policy of 'imperialization', J. *kōminka*) did not occur until WWII (which began in East Asia in 1937 with Japan's invasion of China). Colonial authorities promoted Japanese language and literacy, slogans praising the Japanese emperor, and wearing kimono. They converted Chinese surnames to Japanese, imposed Shinto religious practices, and recruited 'volunteers' for military service. They used such positive incentives toward Japanese identity as expanding (Japanese-language) education. Most poor, rural Taiwanese had at least two years of primary education, sufficient to give them basic Japanese language skills, and many middle- and upper-class Taiwanese pursued higher-level education, a select few even attending college in Japan. Colonial authorities also attempted to remove Chinese cultural influence, in ways that Taiwanese experienced as discrimination: levying fines for speaking Hoklo or Hakka languages in schools and government offices; prohibiting use of classical Chinese language in theatres, publications, or even private lessons; and suppressing the Taiwanese folk religion.

Taiwanese interviewed in the 1980s and 1990s—including farmers, working-class urbanites, and the highly educated elite—expressed self-identification as part of a not-Japanese, Taiwanese we-group, and repeatedly discussed anti-Taiwanese discrimination which had made them second-class to Japanese (Brown 2010; cf. Gates 1987). Most commonly mentioned were the regular house inspections and inequalities in employment and education. Taiwanese were fined—the equivalent of a month's wages for a factory worker—if they did not meet Japanese standards of cleanliness, which required daily attention to sweeping, washing, and hauling waste, both at home and at food markets. Employment inequalities were institutionalized in ethnically calculated day wages for skilled workers (Japanese earned more than Taiwanese) and in limiting Taiwanese employment opportunities to agriculture, medicine, and low- to middle-level factory managerial positions and government jobs (e.g. in the water-canals office). Limitations on how much capital Taiwanese but not Japanese could have in a single business kept the Taiwanese elite second-class. Educational opportunities—with much potential for influencing social experience—were limited by strict ethnic quotas for students beyond elementary school and teachers at all levels. Japanese constituted about 5% of Taiwan's colonial population, but half the student enrolment in the top three girls' high schools was reserved for Japanese. Such discriminatory policies worked against assimilation policies by creating an accumulated social experience of ethnic difference for Taiwanese across all economic classes.

In sum, Japanese assimilation efforts focused on cultural aspects of identity, notably aesthetic and religious practices, proved to be an unconvincing means of invoking Japanese identity. The possible exception here is language change, which, however, had clearer consequences for employment. The differential ability of cultural versus social changes to affect ethnic population membership indicates a differential underlying dynamic of causality (Brown 2004, 2007, 2008b; Brown and Feldman 2009). In terms of Japanese colonial strategy, the misplaced focus on *cultural* change meant that, for all their efforts, authorities could not build a network with which people could identify.

Moreover, by focusing on cultural practices and ignoring social experience—especially its economic aspects—Japanese colonial authorities amplified associational skewing. First, the focus on cultural practices made it very difficult to achieve a socially accepted fluency. The 'authenticity' of non-natives' execution of such practices and grasp of the beliefs behind them is always open to question by a native (cf. Lindholm 2008). By association, the cultural beliefs and practices of individuals who hold the social role of native are unassailably authentic. Thus, the colonial government's assimilation goals were inherently discriminatory (cf. Ching 2001). The success of even the most highly educated Taiwanese could potentially be found wanting by the lowest-ranking Japanese. Second, Taiwanese experienced daily discrimination that made them socially—and especially economically—subordinate to Japanese. Moreover, the unquestioned membership of Japanese residents in Taiwan as ethnic Japanese served as the basis for these employment and educational inequalities. Thus, in both institutionalized discrimination and the cultural focus of assimilation efforts, Japanese were associated with a level of civilization that Taiwanese could never achieve (cf. Harrell 1995).

Footbinding in China

Associational effects resulting in failure to manipulate collective identity, as well as the primacy of socio-economic conditions over aesthetics and beliefs, are not limited to ethnic or national identities. In China, unlike Taiwan, binding of daughters' feet—which was integral to Han gender identity before 1949 (Ko 2005)—ended not due to political campaigns targeting beliefs related to gender identity, but due to changes in economic conditions (Brown et al. 2012; Brown and Satterthwaite-Phillips, under review; cf. Gates 2001, forthcoming 2014; Bossen et al. 2011; contra Mackie 1996, Appiah 2010).

Footbinding, which was widespread from the Ming through Republican periods, ended at different times in different locales of China between the late nineteenth and mid-twentieth centuries, with the PRC government finally halting its last remnants in the 1950s. From the late nineteenth century, the Chinese elite and Christians opposed it, setting up anti-footbinding societies; the Qing and Republican governments banned it outright, connecting it to national honour. These political movements explicitly sought to counter beliefs that women who did not meet the aesthetic standards of beauty and sexuality dictated by footbinding either did not marry or married poorly. Indeed, rural women *believed* they were footbound for marriage, but quantitative assessment of actual marital mobility shows most footbound women were *not* significantly more likely to marry up than not-bound women. Mistakenly touted as an exemplar of the primacy of cultural ideas over the economic value of labour (e.g. Ko 2005) and even over natural selection (Richerson and Boyd 2008), footbinding, for rural Han commoners at least, was a form of labour control that boosted the contribution of young girls to handcraft production, most often spinning, by disciplining them to sit for long hours.

The cultural belief in footbinding's association with marriage made mothers believe that they must bind their daughters' feet, but changing economic conditions ended the custom, not any shift in cultural ideas related to gender identity. When industrially produced thread and textiles sold for less money than handicrafts in rural markets, families shifted their daughters to other, usually more physically demanding, labour and ceased footbinding. Elderly rural women interviewed in the twenty-first century recalled that, even as their fathers allowed them to have not-bound feet to work in the fields, instead of spinning or doing other commercial handwork, their mothers continued to mistakenly insist they must bind or end as spinsters; many had never heard of the pre-1949 political campaigns to end footbinding. For rural Chinese, who constituted the majority of the population, economic conditions succeeded in removing footbinding from gender identity, where political manipulations failed.

11.3 Deriving Macrostructures From Micropopulations

Taiwanese and Chinese history demonstrate the ability of macrostructural changes arising from mass migrations, government policies, and economic conditions to shift ethnic population membership by reshaping the social experience of individuals and thereby recasting their identities. Causation, as discussed so far, follows the generally accepted principle of the social construction of individuals (e.g. Bourdieu 1977; cf. Appiah 2006; Johnson-Hanks, this volume). However, such a macro-level view makes it difficult to see individuals' agency (a common criticism of Bourdieu, e.g. de

Certeau 1988: 59; Comaroff 1985: 5). Consequently, these ethnic population changes may appear transformational—the result of periodic external shocks that shift the system from one state to another. However, such macro-scale transformations are not simply a state-dependent transformative process.[6] Rather, they occur because individuals act, and their actions have consequences that aggregate to constitute social structures.

Specialized data are needed to understand how variation in actions—i.e. of different individuals' social negotiations of their identity—results in broader, population-scale changes. The necessary evidence includes data: on the roles, relative positioning, and economic circumstances of individual agents in the hierarchical, networked structures in which they act; on individuals' machinations within those structures; and on the frequency and distributions of social role expectations and cultural beliefs in the agents' own population and an appropriate referent population. The first of the 'long-route' case studies, given earlier, provides a clear case in point.

11.3.1 The End of Hoklo Footbinding: Causality and Negotiations by Individuals

In enforcing its footbinding ban, the Japanese colonial government removed the last major, visible marker of plains Aborigine identity for Toushe, Jibeishua, and Longtian villagers. An external observer can see removal of the boundary between Hoklo and plains Aborigines as a logical, semiotic implication of the ban, but this implication is not sufficient causal explanation for the process of change. Hoklo did not collectively discuss the semiotics and so agree to grant Hoklo ethnic membership to these communities. How, then, was the ethnic boundary actually removed? The suddenly elevated rate of bride exchange verifies the removal. This population-level shift is the aggregated result of many micro-level decisions about marriage partners. In Taiwan at that time, parents generally arranged marriages without consulting the bride and groom, so the shift represents decisions by many parents. Individuals making these decisions had multiple potential partners, with Hoklo parents having a wider range to choose from than did most villagers of Toushe, Jibeishua, and Longtian.

The new, dominantly virilocal marriage pattern (see Figure 11.2) did not emerge without individual effort and social negotiation. One woman, whom I will call Ms Kou, said she was the first Hoklo woman to marry into Toushe as a first-time bride (never having previously married). She reported that it took three years for her husband's father—Mr Ong—to negotiate the marriage agreement with her parents, even though he was an uxorilocally married Hoklo man. Hoklo himself, Mr Ong would have understood the Hoklo Kou family with whom he was negotiating by virtue of his fluency in the normative role expectations and cultural beliefs of the Hoklo referent population. He would have understood their probable suspicion of his morals,

[6] An example of a state-dependent transformative process is songbirds that sing in the morning if they are in a state with sufficient energy from having enough forage.

due to associated expectations and beliefs about any man married uxorilocally (which Hoklo considered an affront to parents and patriline); their certain concern about their daughter's safety in a 'savage' village; and yet their probable appreciation of his status in Toushe village—the household Mr Ong headed was one of the wealthiest in Toushe at the time. Nevertheless, it required the intervention of Mr Ong's friend—the bride's brother—to finally win the elder Mr Kou's agreement to the marriage. This friend, returning from an extended work-related sojourn, could vouch for Mr Ong personally and for Toushe's acculturation, allowing Mr Ong to negotiate the marriage agreement in terms of the normative expectations of the dominant Hoklo referent population and not as a 'savage' or an uxorilocally married man.

We cannot know whether the change to Hoklo identity would have occurred if Mr Ong had not been so persistent, but if he had not negotiated this marital agreement, I think it possible that few others in Toushe would have attempted to bring in a Hoklo daughter-in-law. In pursuing the marriage, Mr Ong contributed significantly to the ethnic change that occurred. He set a social precedent by having a Hoklo woman in the role of virilocal bride in Toushe village. This precedent brought Toushe marriage practices further into line with Hoklo role expectations, created virilocal, affinal social-network ties between Toushe and Ms Kou's natal village, and spread (via these connections) knowledge of how acculturated (i.e. Hoklo) Toushe actually was. These consequences interacted to improve incrementally Toushe's connection to a Hoklo referent population and to promote omission of Toushe's plains Aborigine heritage. Mr Ong took advantage of a social opportunity to solidify Hoklo identity for his son, an opportunity created unintentionally by the footbinding ban, but it was not a foregone conclusion that his negotiations would be successful.

To effect macro-level social change requires that many individuals across the population take up the new practice and that some individuals initiate the change by adopting the new practice before it is common (Brown 2004, 2007). Sociological studies discuss the importance of such initiators (and identify an S-shaped adoption curve across a population; e.g. Rogers 2003), but ethnographic analysis demonstrates that the spread of practices requires much more than a decision on the part of a single individual to adopt the practice. A process of social negotiation—such as negotiations about brideprice and dowry and other pre-existing, locally relevant marital contract issues—is often required for a potential initiator to be able to enact the desired practice. It is not simply a matter of who wants to innovate but more importantly a matter of which of those would-be initiators are socially able to carry out the innovation. Thus, understanding how macrostructural changes derive from micropopulations requires identifying the systemic conditions under which the tactical manipulations of specifically positioned individuals are likely to succeed.

Social negotiations are tactical or manipulative social interactions, occurring within a context of the constant social interactions that constitute the human social environment. Social interactions form ties between individuals, maintaining a dynamic network which shifts as ties are created, maintained (whether strongly or weakly), and cut off (Brown 2008b; Brown and Feldman 2009). The hierarchical character of interactions influences their aggregation and thereby shapes the network into a structure.

Each individual's position in the networked structure influences his or her interactions (and his or her perception of them) as well as what tactical manipulations are possible (cf. Bourdieu's concept of social capital—e.g. Bourdieu 1977: 184). Tactical manipulations by oneself and others move individuals within the structure, sometimes to their advantage, sometimes not—for example, the manipulation of Tujia identity by local government officials. Hence, individuals are constantly involved in maintaining their position—constantly (re)constructing their social niche.

11.4 Theorizing Human Population Dynamics

These Taiwanese and Chinese cases of shifting ethnic membership raise questions about how to theorize empirically based units of human population which are subject to the dynamics of individuals' tactical manipulations. Such social negotiations are individuals' means of regulating the social niche in which they live, but these negotiations are also subject to social and cultural selection pressures from normative role expectations and cultural beliefs, and subject to associational skewing that causally modifies outcomes (Brown and Feldman 2009; cf. Runciman 2009). A modified niche construction theory approach can explain these dynamics and offer predictions about the conditions under which a population's membership is likely to change.

11.4.1 Niche Construction Theory (NCT) on Social Inheritance

NCT argues that evolutionary causal processes are not just the impact of environmental selection pressures on individual organisms but necessarily also include the feedback of individual organisms' impact on the environment because their niche construction changes the environment that the organisms pass on to subsequent generations, which in turn alters selection pressures. (For further discussion of NCT, see Odling-Smee et al. 2003; Odling-Smee 2011, this volume.) What we have seen in these case studies is how social environmental selection pressures impact on individual humans as well as how individuals' social negotiations feed back to modify the social environment that they pass on to subsequent generations. There is a strong analogy between the environmental inheritance posited by NCT and social inheritance, at least up to a point.[7]

[7] My view of social inheritance begins, not with NCT's view of social inheritance, but with its view of ecological inheritance.

Following Odling-Smee's (2011: 181) description of the NCT view of ecological inheritance, we can describe social inheritance as transmitted through the external environment, not reproduction. It seldom depends on transmission of discrete replicators. Rather, the altered selective environment comes from individuals choosing or perturbing components of their social environment. Social inheritance is continuously transmitted by multiple individuals to multiple individuals within and between generations, throughout the lifetimes of individuals. It is not always transmitted by relatives (genetic, social, or cultural). In each generation, each individual inherits not only a society (which modifies his or her local selective environment), but an initial individual–society relationship built up by immediate social ancestors—the position or status from which that individual begins in a given hierarchically structured social network. This relationship is an individual's social niche, an 'address' in social space and time, which is influenced by parental choices and which yields a 'resource package' of reliable, verifiable expectations for creating, achieving, and holding roles. To this point, the analogy works well. However, in my view, the analogy between social and ecological inheritance breaks down over culture, because social inheritance is not reducible to cultural units in the way that NCT posits ecological inheritance is ultimately reducible to genetic units.

11.4.2 Units and Verifiability

Units of culture are notoriously difficult to specify (Brown 2008a), despite their famous label as 'memes' (Dawkins 1976), seminal definition as 'socially learned information' (Cavalli-Sforza and Feldman 1981; Boyd and Richerson 1985), and recent inclusion (with genetic units) in 'semantic information [which] is anything that reduces uncertainty about selective environments relative to the fitness interests of [individual] organisms' (Odling-Smee 2011: 184). These definitions all assume that society is ultimately reducible to cultural units. Based on empirical evidence of different social and cultural dynamics, and in agreement with sociologist G. W. Runciman (2009), I view society and culture as distinct, parallel inheritance tracks (see Figure 11.3; Brown 1995, 2004, 2008; Brown and Feldman 2009; cf. Durham 2008). Thus, I define distinct units for culture (shared meaningful ideas or beliefs) and society (role expectations). Runciman (2009: 3, 92) also identifies roles as fundamental social units and discusses cultural units as beliefs (though he defines them as 'memes' where I do not). Runciman, however, sees culture as preceding society, which suggests that, like those who do not distinguish social and cultural units, he sees social inheritance as reducible, at least phylogenetically, to cultural inheritance.

I suggest that the key difference between social and cultural units is their verifiability (Brown and Feldman 2009; cf. Brown 1995), and this difference means that society cannot be reduced to culture. Social role expectations may be unverified but it is possible to verify them, and each individual throws them out if they prove inaccurate or if he cannot put them into operation. Cultural beliefs cannot be verified as accurate or inaccurate, whether because people in general cannot verify the belief (e.g. the

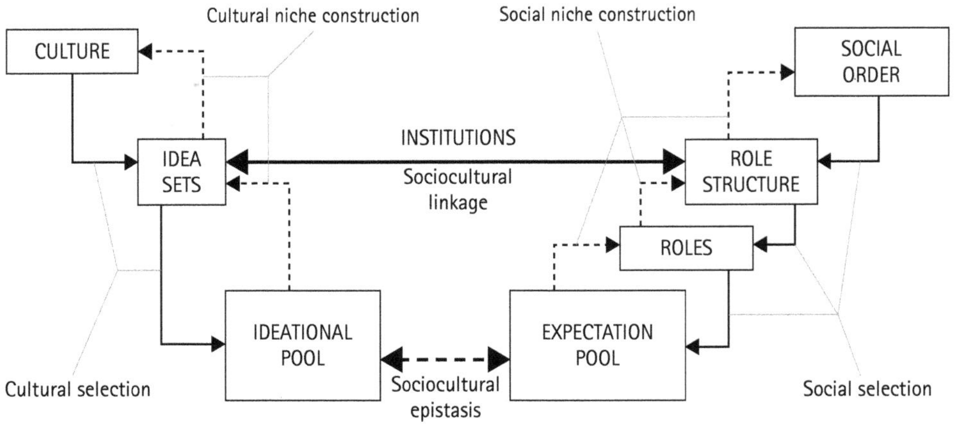

FIG. 11.3 Cultural inheritance track and social inheritance track with cross-track associations.

existence of God) or because a specific individual or population cannot verify the belief due to knowledge limitations (e.g. whether the earth was flat, for thirteenth-century Europeans). A belief may become verifiable with the expansion of knowledge and thus opens to social negotiation and the possibility of becoming an expectation. The reverse process is also possible. The ability of non-human primates to abstract role expectations (e.g. Worden 1996; Tomasello et al. 2005) suggests that cultural units may have derived phylogenetically from social units (Brown 1995, 2008a). Whether that means unverifiable units are reducible to verifiable units in humans, however, is an open question.

11.4.3 Selection Pressures and Institutional Association

The distinction of verifiability is important for understanding social and cultural selective pressures as well as the possibility of associational modifications of those pressures. Individuals bring to every social interaction specific expectations about the behaviour of others, and the expectations people already have will influence the behaviour that they accept from others. An expectation's accuracy is thus tested with each relevant social interaction. What behaviour actually ensues in an interaction, however, is negotiated via speech and actions; this negotiation process constitutes social selection (Figure 11.3). Individuals benefit by enforcing their expectations as accurate because this accuracy maintains their social competence and may improve or defend their social rank. Consequently, people resist unexpected behaviour. Ease, cost, and degree of resistance are based on the relative rank of the interacting individuals. Nevertheless, individuals can successfully negotiate the unexpected achievement of a role or an unexpected behaviour for a role holder. Such social precedents constitute social niche construction because they feed back to change shared expectations of roles through the verification of these new expectations as accurate.

An instance of such social selection pressures occurred when Mr Ong from Toushe initially approached Mr Kou to negotiate a marriage contract. Mr Ong was in a difficult position for social negotiation, due to one or both sets of Mr Kou's (probably unverified) normative Hoklo role expectations—those expectations predicting uncivilized behaviour from a 'Toushe savage' and those expectations predicting unscrupulous behaviour from a Hoklo man who married uxorilocally. Mr Ong's ability to demonstrate 'civilized' behaviour (by Hoklo standards) could have caused Mr Kou to discard his expectations about a Toushe man as inaccurate, to convert these expectations into (unverifiable) beliefs precluding contradiction, or to shift his assumption of Mr Ong's ethnic membership from 'savage' to Hoklo. In the latter response, Mr Ong still faced institutionally associated role expectations and cultural beliefs based on his uxorilocal marriage.

The institutional association would amplify social and cultural selection effects on Mr Kou's prejudice against him, with verifiability a factor in Mr Ong's ability to respond. Mr Kou's beliefs that an uxorilocally married (Hoklo) man is untrustworthy stemmed from a set of Han cultural ideas about a man's perpetual obligations to his parents and the semiotic implication that a man who marries to his wife's home has turned his back on his obligations to care for his parents in old age and provide offerings to their spirits after death. In one sense, these beliefs might be verifiable—by marrying his daughter to Mr Ong's son, Mr Kou could find out whether Mr Ong was trustworthy—but unverifiable in the short term of wanting to know whether such an act would subject his family to risk.

Mr Kou's cultural beliefs would influence other information that he received. This is cultural selection: when the (unverifiable) ideas that an individual already has internalized as meaningful—that she believes or thinks that others generally believe—influence her broadcasting, reception, or internalization of subsequent ideas (Figure 11.3). There is a feedback from individuals to the larger culture—cultural niche construction—via the back influence of individuals' belief (or disbelief) on whether particular ideas are accepted by others as meaningful. Thus Mr Kou's belief only added to the general perception of these beliefs as true.

Mr Kou's beliefs about uxorilocal men were further strengthened by their institutional association with (verifiable) expectations in kinship (a networked structure of roles). An institution is the integration of a cultural idea set into a social role structure, such that the cultural beliefs and role expectations reinforce each other. Institutional associations exemplify sociocultural linkage (Figure 11.3), because beliefs and expectations are transmitted together. Mr Kou's (verifiable) role expectations about uxorilocally married men stemmed from observations or reports of the difficulties that underlie the documented higher divorce rates for uxorilocal than for virilocal marriages for Hoklo in colonial Taiwan (Wolf and Huang 1980). Institutional linkage to cultural beliefs confers moral weight on role expectations (the movement from 'is' to 'ought' against which Hume warned), which would have made Mr Kou's suspicions more intransigent.

Against such institutionalized prejudice, Mr Ong himself apparently could make no headway, even over several years of negotiations. Behaviour cannot overcome unverifiable belief. It required someone strategically socially placed relative to Mr Kou—his son, the bride's brother—to break the impasse. The younger Mr Kou, trusted by both his father and Mr Ong, could vouch that the role expectations for an uxorilocally married man were inaccurate in Mr Ong's case, *or*—and this might have been easier—he could shift the most salient role that the elder Mr Kou assigned to Mr Ong from that of an uxorilocally married Hoklo man to that of his son's friend. Thus, the younger Mr Kou was in a position, which Mr Ong was not, to persuade the elder Mr Kou to enter into a marital contract with Mr Ong.

11.4.4 Associations: Institutional and Epistatic

Footbinding shows more aspects of associational modification of selection pressures (Brown and Feldman 2009).[8] For plains Aborigines and Hoklo, footbinding was institutionally associated with the social role of the bride because it defined an ethnic endogamous boundary. Hoklo cultural ideas regarding what constituted civilization were integrated with the kinship role structure in a way that restricted the category of woman who could be recruited as a virilocal (first-time) bride for a Hoklo groom. The woman had to be Hoklo, as indicated by having bound feet. Interestingly, footbinding may have remained the final ethnic marker for epistatic reasons because holders of parental roles influenced whether binding occurred. Han mothers bound their daughters' feet, and as a consequence, it was Han women who held the knowledge of how to enact the practice; but most intermarriage between Han men and Aborigine women occurred where neither elder generation of women knew the binding procedure.

Role expectations related to a single role are associated by definition, but the association of two roles can lead to unexpected associations between specific role expectations. Social linkage occurs when expectations for two (or more) roles are 'transmitted' together—because they share at least one role expectation explicitly linking them (Figure 11.4). For example, the roles can be simultaneously recruited and filled, as in the roles of bride and groom. In social epistasis, expectations for two or more roles overlap for the structural reason that one individual holds multiple roles. For example, a young woman who is both a bride and a hand-spinner is subject to expectations for both roles.

Whether the association between footbinding and marriage for the Han population is seen as institutional or epistatic divides the current debate in China studies over footbinding. Those who view footbinding as driven by gender identity essentially view this set of cultural ideas as integrated into the kinship role structure—an institutional association which conferred moral weight on bridal role expectations of footbinding. Those, including myself, who view footbinding as institutionally associated not with marriage but with handicraft production, essentially view footbinding's association with marriage as

[8] Footbinding also illustrates sociocultural epistasis (association between beliefs and expectations due to structural overlap), cultural epistasis (association between beliefs due to semiotic implications), and cultural exaptation (where a custom originates prior to its association with a cultural belief). See Brown and Feldman (2009) for further discussion.

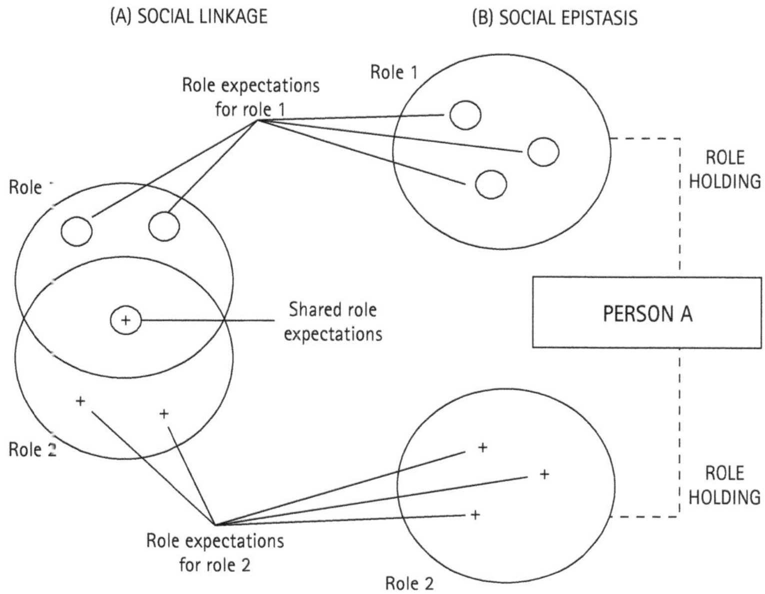

FIG. 11.4 Social linkage and social epistasis.

epistatic because footbound, unmarried girls who constituted the handcraft labour force were also recruited for the role of (virilocal) bride. The overlap created by the same category of person—footbound, unmarried girls—holding crucial roles in both role structures (the kin structure and the handwork production structure) allowed social epistasis.

11.4.5 Explaining and Predicting Changes in Population Membership

The modified NCT approach used here allows prediction of conditions for changing population membership at both the macro and micro levels. Regime changes and mass migrations proved so powerful in shifting ethnic membership because these events radically altered the social selective environment by changing macrostructural relations. Suddenly, under conditions of regime change and mass migration, established social niches held by different ethnic populations required renegotiation, yielding changes in social experience. When renegotiation resulted in ethnic-based discrimination, these macrostructural changes were particularly powerful in shifting ethnic membership. This NCT approach does not suggest that regime change and mass migrations are the only conditions which yield population membership change,[9] but it explains why these conditions render such change likely.

[9] Intermarriage contributes to changing population membership, but it is not sufficient in itself, as it may lead to acculturation (the adoption of some cultural ideas and practices) without assimilation (adoption of a whole new identity).

At the micro level, the verifiability distinction between role expectations and beliefs predicts that expectations are more likely to change population membership, and also explains why: their verification is subject to social negotiation and influenced by precedent. The transition from plains Aborigines to Hoklo for Toushe, Jibeishua, and Longtian villagers was not due to a change in cultural beliefs, but to a change in role expectations. When the expectation of having a footbound bride became forcibly inaccurate for Hoklo, due to colonial enforcement of the footbinding ban, it allowed individual-level social negotiations about marriage partners that accumulated to remove the ethnic border and shift ethnic population membership.

An empirical human population, which is recognized by its members and allows shifting membership, is defined by the accumulated outcomes of social negotiations (social selection pressures) and social precedents (social niche construction). Like the biological concept of species, or the sociological concept of civilization, the concept of population is a moving target with indistinct edges. As with other evolutionary concepts, population must be defined relative to a particular set of selection pressures and niche construction feedback in a specific time and place. There is still much work to be done to fully theorize human population dynamics, but niche construction theory is a robust basis for exploration and modification.

References

Appiah, K. A. (1994). 'Identity, Authenticity, Survival: Multicultural Societies and Social Reproduction', in A. Gutmann (ed.), *Multiculturalism: Examining the Politics of Recognition*. Princeton, NJ: Princeton University Press, 149–163.

Appiah, K. A. (2006). 'The Politics of Identity', *Daedalus* 135/4: 15–22.

Appiah, K. A. (2010). *The Honor Code: How Moral Revolutions Happen*. New York: W. W. Norton.

Barth, F. (ed.) (1969). *Ethnic Groups and Boundaries*. Boston, MA: Little, Brown and Company.

Barth, F. (1994). 'Enduring and Emerging Issues in the Analysis of Ethnicity', in H. Vermeulen and C. Govers (eds), *The Anthropology of Ethnicity: Beyond 'Ethnic Groups and Boundaries'*. Amsterdam: Het Spinhuis, 11–32.

Bossen, L., X. Wang, M. J. Brown, and H. Gates (2011). 'Feet and Fabrication: Footbinding and Early Twentieth-Century Rural Women's Labor in Shaanxi', *Modern China* 34/4: 347–383 (doi: 10.1177/0097700411403265).

Bourdieu, P. (1977). *Outline of a Theory of Practice*. Cambridge: Cambridge University Press.

Boyd, R. and P. J. Richerson (1985). *Culture and the Evolutionary Process*. Chicago, IL: University of Chicago Press.

Brown, M. J. (1995). '"We Savages Didn't Bind Feet": The Implications of Cultural Contact and Change in Southwestern Taiwan for an Evolutionary Anthropology', PhD dissertation, Department of Anthropology, University of Washington, Seattle, WA.

Brown, M. J. (2002). 'Local Government Agency: Manipulating Tujia Identity', *Modern China* 28/3: 362–395.

Brown, M. J. (2004). *Is Taiwan Chinese? The Impact of Culture, Power and Migration on Changing Identities*. Berkeley, CA: University of California Press.

Brown, M. J. (2007). 'Ethnic Identity, Cultural Variation and Processes of Change: Rethinking the Insights of Standardization and Orthopraxy', *Modern China* (special issue on Orthopraxy) 33/1: 91–124 (doi: 10.1177/0097700406294701).

Brown, M. J. (2008a). 'Introduction: Developing a Scientific Paradigm for Understanding Culture', in M. J. Brown (ed.), *Examining Culture Scientifically*. Seattle, WA: University of Washington Press, 3–16.

Brown, M. J. (2008b). 'When Culture Does Not Affect Behavior: The Structural Basis of Ethnic Identity', in M. J. Brown (ed.), *Examining Culture Scientifically*. Seattle, WA: University of Washington Press, 162–183.

Brown, M. J. (2010). 'Changing Authentic Identities: Evidence from Taiwan and China', *Journal of the Royal Anthropological Institute* 16: 459–479 (doi: 10.1111/j.1467-9655.2010.01634.x).

Brown, M. J. and M. W. Feldman (2009). 'Sociocultural Epistasis and Cultural Exaptation in Footbinding, Marriage Form, and Religious Practices in Early 20th-Century Taiwan', *Proceedings of the National Academy of Sciences of the USA* 106/52: 22139–22144 (doi: 10.1073_pnas.0907520106).

Brown, M. J., L. Bossen, H. Gates, and D. Satterthwaite-Phillips (2012). 'Marriage Mobility and Footbinding in Pre-1949 Rural China: A Reconsideration of Gender, Economics, and Meaning in Social Causation', *Journal of Asian Studies* 71/4: 1035–1067.

Brown, M. J. and D. Satterthwaite-Phillips (under review). 'Economic Correlations to "Beauty" Custom: Footbinding in Pre-1949 China'.

Cavalli-Sforza, L. L. and M. W. Feldman (1981). *Cultural Transmission and Evolution: A Quantitative Approach*. Princeton, NJ: Princeton University Press.

Certeau, M. de (1988). *The Practice of Everyday Life*. Berkeley, CA: University of California Press.

Chen, S.-J. (2008). 'How Han Are Taiwanese Han? Genetic Inference of Plains Indigenous Ancestry among Taiwanese Han and its Implications for Taiwan Identity', PhD dissertation, Department of Anthropological Sciences, Stanford University, Stanford, CA.

Ching, L. T. S. (2001). *Becoming 'Japanese': Colonial Taiwan and the Politics of Identity Formation*. Berkeley, CA: University of California Press.

Comaroff, J. (1985). *Body of Power, Spirit of Resistance: The Culture and History of a South African People*. Chicago, IL: University of Chicago Press.

Corcuff, S. (2000). 'Taiwan's "Mainlanders": A New Ethnic Category', *China Perspectives* 28 (March–April): 71–81.

Dawkins, R. (1976). *The Selfish Gene*. Oxford: Oxford University Press.

Durham, W. H. (1991). *Coevolution: Genes, Culture and Human Diversity*. Stanford, CA: Stanford University Press.

Durham, W. H. (2008). 'When Culture Affects Behavior: A New Look at Kuru', in M. J. Brown (ed.), *Examining Culture Scientifically*. Seattle, WA: University of Washington Press, 139–161.

Fox, R. G. and B. J. King (eds) (2002). *Anthropology Beyond Culture*. New York: Berg.

Gates, H. (1981). 'Ethnicity and Social Class', in E. M. Ahern and H. Gates (eds), *The Anthropology of Taiwanese Society*. Stanford, CA: Stanford University Press, 241–281.

Gates, H. (1987). *Chinese Working-Class Lives: Getting By in Taiwan*. Ithaca, NY: Cornell University Press.

Gates, H. (2001). 'Footloose in Fujian: Economic Correlates of Footbinding', *Comparative Studies in Society and History* 43/1: 130–148.

Gates, H. (forthcoming 2014). *Hand and Foot: Footbinding and Female Labor in Sichuan*. Oxford: Routledge.

Harrell, S. (1995). 'Introduction: China's Civilizing Projects', in S. Harrell (ed.), *Cultural Encounters on China's Ethnic Frontiers*. Seattle, WA: University of Washington Press, 3–36.
Harrell, S. (1996a). 'Introduction', in M. J. Brown (ed.), *Negotiating Ethnicities in China and Taiwan*. Berkeley, CA: University of California Press, 1–18.
Harrell, S. (1996b). 'The Nationalities Question and the Prmi Problem', in M. J. Brown (ed.), *Negotiating Ethnicities in China and Taiwan*. Berkeley, CA: University of California Press, 274–296.
Harrell, S. (2001). *Ways of Being Ethnic in Southwest China*. Seattle, WA: University of Washington Press.
Johnson-Hanks, J. A., C. A. Bachrach, S. P. Morgan, and H.-P. Kohler (2011). *Understanding Family Change and Variation: Toward a Theory of Conjunctural Action*. Dordrecht: Springer.
Khan, A. (1996). 'Who Are the Mongols? State, Ethnicity, and the Politics of Representation in the PRC', in M. J. Brown (ed.), *Negotiating Ethnicities in China and Taiwan*. Berkeley, CA: Institute of East Asian Studies, University of California, 125–159.
Ko, D. (2005). *Cinderella's Sisters: A Revisionist History of Footbinding*. Berkeley, CA: University of California Press.
Lindholm, C. (2008). *Culture and Authenticity*. Oxford: Blackwell.
Lipatov, M., M. J. Brown, and M. W. Feldman (2011). 'The Influence of Social Niche on Cultural Niche Construction: Modelling Changes in Belief About Marriage Form in Taiwan', *Philosophical Transactions of the Royal Society B* (Special Issue: Human Niche Construction) 366: 889–900 (doi:10.1098/rstb.2010.0247).
Mackie, G. (1996). 'Ending Footbinding and Infibulation: A Convention Account', *American Sociological Review* 61/6: 999–1017.
Mullaney, T. S. (2011). *Coming to Terms with the Nation: Ethnic Classification in Modern China*. Berkeley, CA: University of California Press.
Odling-Smee, F. J. (2011). 'Niche Inheritance', in M. Pigliucci and G. B. Müller (eds), *Evolution: The Extended Synthesis*. Cambridge, MA: MIT Press, 175–207.
Odling-Smee, F. J., K. N. Laland, and M. W. Feldman (2003). *Niche Construction: The Neglected Process in Evolution*. Princeton, NJ: Princeton University Press.
Richerson, P. J. and R. Boyd (2008). 'Cultural Evolution: Accomplishments and Prospects', in M. J. Brown (ed.), *Explaining Culture Scientifically*. Seattle, WA: University of Washington Press, 75–99.
Rogers, E. M. (2003). *Diffusion of Innovations*, 5th edn. New York: Free Press.
Runciman, W. G. (2009). *The Theory of Cultural and Social Selection*. Cambridge: Cambridge University Press.
Sperber, D. (1996). *Explaining Culture*. Oxford: Blackwell Publishers.
Tomasello, M., M. Carpenter, J. Call, T. Behne, and H. Moll (2005). 'Understanding and Sharing Intentions: The Origins of Cultural Cognition', *Behavioral and Brain Sciences* 28: 675–735.
Watson, J. L. (1988). 'The Structure of Chinese Funerary Rites: Elementary Forms, Ritual Sequence, and the Primacy of Performance', in J. L. Watson and E. S. Rawski (eds), *Death Ritual in Late Imperial and Modern China*. Berkeley, CA: University of California Press, 3–19.
Wolf, A. P. and C.-S. Huang (1980). *Marriage and Adoption in China, 1845–1945*. Stanford, CA: Stanford University Press.
Worden, R. P. (1996). 'Primate Social Intelligence', *Cognitive Science* 20/4: 579–616.
Wu, X. (2005). 'The New Year's Eve Dinner and Wormwood Meal: Festival Foodways as Ethnic Markers in Enshi', *Modern China* 31: 353–380.

CHAPTER 12

THE CAUSAL RELATIONSHIP BETWEEN FERTILITY AND INFANT MORTALITY

prospective analyses of a population in transition

HILLARD KAPLAN, PAUL L. HOOPER, JONATHAN STIEGLITZ, AND MICHAEL GURVEN

12.1 Introduction

Understanding the relationship between fertility rates and infant and juvenile mortality risks has been a long-standing interest of demographers and biologists. Standard demographic transition theory is one classic approach (Coale 1973; Coale and Watkins 1987; Davis 1945, 1963). This theory posits that mortality rates for infants and children drive fertility rates, mediated through cultural, social, and psychological processes. Those social processes have the effect of keeping population size stable. Under conditions of high mortality, high fertility rates are required to keep populations from shrinking. When there is a transition from high to low mortality rates, cultural processes, albeit with some lag, will result in reduced fertility rates. After a period of rapid growth, the population returns to stable equilibrium.

Life history theory derived from evolutionary biology posits a bidirectional relationship between fertility and juvenile mortality risks (Blurton Jones 1986; Cole 1954; Kaplan 1996; Smith and Fretwell 1974). According to the theory, organisms have limited budgets and face a trade-off between quantity and quality of offspring. Increasing fertility reduces available time and energy to invest in each individual offspring, increasing their likelihood of dying. At the same time, for organisms that engage in postnatal investment in offspring (including all birds and mammals), the death of an existing offspring releases time and energy for the mother to invest in new replacement offspring. The theory proposes that natural selection results in organisms possessing characteristics that will tend to maximize biological fitness, often measured by the number of surviving offspring they produce. According to the theory, the effect of an exogenous reduction in mortality risks to offspring will depend on how it affects the trade-off between quantity and quality of offspring. By increasing the number of existing offspring demands on the mother, reductions in mortality will tend to reduce fertility, but if the cause of reduced mortality is due to an increase in available resources, fitness-maximizing organisms may respond by increasing fertility. For most organisms, populations will eventually return to equilibrium due to resource constraints.

Many traditional native South Americans living in the lowland areas of Amazonia are currently undergoing a transition in mortality rates. The demographic history of those groups reveals large swings in population size, due to the effects of contact with Europeans. During the first centuries of contact, there were large demographic collapses throughout the Americas, due to both introduced diseases and economic, social, and political disruption (see Livi Bacci 2007 for a review). The rubber boom in the late nineteenth century caused further demographic disorder. The effects of these disruptions were greater for settled groups than for small-scale nomadic societies, living on small water courses in the interior of the forest. Many of those small-scale societies continued to live traditional lifestyles until the middle or latter part of the twentieth century. As they make increasing contact with global forces (including change in access to land and resources, schooling, and public health), they are now experiencing demographic change; in many cases, the result is rapid population growth as high as 3–5% annually (Perz et al. 2008).

The aim of this paper is to examine the causal processes underlying recent change in demographic outcomes in one traditional native South American society, the Tsimane of lowland Bolivia, with the specific goal of shedding light on the relationship between fertility and mortality. While the vast majority of Tsimane remain subsistence-level forager-farmers, people and communities vary in distance from the nearest town, their ability to speak Spanish (from none to fluent), their access to medical care, their educational level (from none to completion of secondary school), and their degree of economic exchange with outsiders (Gurven et al. 2007). Using prospective data collected between 2002 and 2010, we first present descriptive data on age-specific fertility and total fertility rates (TFR), and how they vary by distance from the town of San Borja (population approximately 23,000). We then examine the determinants of parity progressions among parous women, using a proportional hazards approach. Next, a comparable analysis is given regarding the transition to first birth. The subsequent section examines the determinants of mortality risks for infants, first for higher-order births and then for firstborn children. Finally, we compare secular change in mortality and fertility rates among the Tsimane to Bolivia as a whole, using historical demographic data.

12.2 Methods

Longitudinal data on health status, fertility, and mortality were collected from 2002 to 2010 under the aegis of the Tsimane Health and Life History Project (www.unm.edu/~tsimane/). Census data were maintained throughout this period, documenting the age, sex, community membership, and parents of each individual born or living in the sample villages. Retrospective reproductive histories were collected from a subset of adults in this sample according to methods described in Gurven et al. (2007). Yearly medical exams and interviews were conducted by a mobile team of Bolivian physicians and biochemists. These exams recorded height, weight, pregnancy status, and parity. White blood cell count was quantified by biochemists using a microscope or a portable dry hematology analysis machine (QBC Diagnostics). Spanish-speaking ability and completed years of schooling were also recorded.

The subsequent analyses of age at first birth, parity progressions, and infant mortality draw on the census data collected during the sample period (2002–2010), supplemented by retrospective demographic interview data where available. In order to understand the patterning of variance in demographic outcomes across individuals and communities, multilevel models (Gelman and Hill 2007; Hox 2002) were estimated, including individual- and community-level predictors and random effects. Mixed-effects Cox proportional hazards models (Therneau 2012) were employed to predict time to next birth following a birth that occurred during or immediately preceding the sample period, and to predict age at first birth. Mixed-effects logistic regression models (Bates et al. 2012) were employed to predict the likelihood of each child born dying within one year of birth, with separate models for higher-order births and first births. Total fertility and infant mortality rates before 2002 were drawn exclusively from retrospective demographic interviews. Nation-wide Bolivian demographic data were obtained from the UN Common Database, UNICEF.

12.3 Results

The total fertility rate for the entire sample of Tsimane from 2002 to 2010, based on 5,730 risk years for births, is 8.8. Figure 12.1 shows the age-specific fertility rates from which the TFR is derived. There is some evidence of regional variation in age-specific fertility. If we divide the population into the more remote Tsimane villages and those that are either near the town of San Borja or at a large Catholic mission settlement, fertility rates are somewhat higher for women in more remote communities during peak reproductive years (Figure 12.2), resulting in TFRs of 9.3 and 8.4 for remote and near villages, respectively.

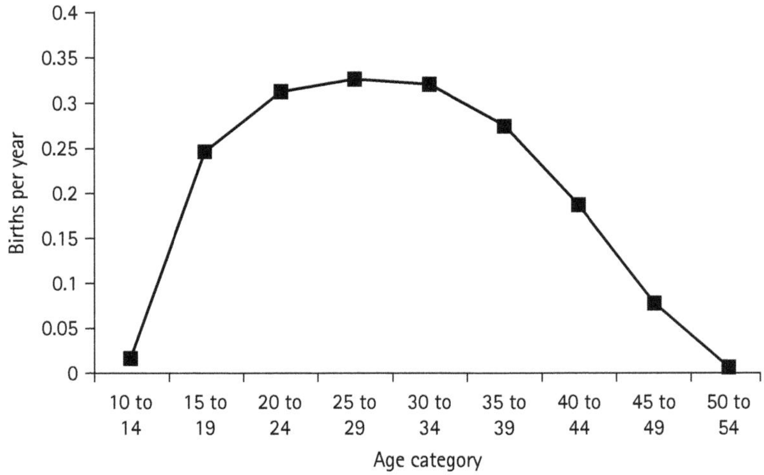

FIG. 12.1 Age-specific fertility rate among Tsimane women.

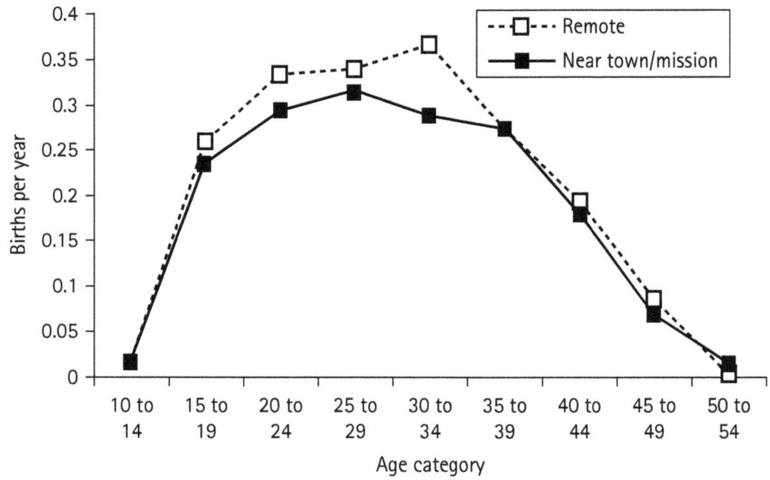

FIG. 12.2 Regional variation in age-specific fertility rate.

12.3.1 Parity Progressions for Parous Women

Table 12.1 shows the results of a Cox proportional hazards analysis of 1,159 risk intervals for the progression from one birth to the next birth among parous women in 14 communities (of which 659 intervals terminated with a birth). Model A examines the variance captured by the individual- and community-level random-effect terms with no predictor variables, and shows that the individual-level variance is much greater than the community-level variance. Model B examines the ability of individual-level variables to predict fertility. It shows that (a) rates of progression decrease with a woman's age, radically after age 35 (see Figure 12.3a for the corresponding functions); (b) the death of

Table 12.1 Multilevel Cox proportional hazard models of probability of giving birth among parous women

	Model A	Model B				Model C				Model D			
Fixed effects		B	exp(B)	SE	p	B	exp(B)	SE	p	B	exp(B)	SE	p
Age 20–24 (baseline)	–	0	1	–	–	–	–	–	–	0	1	–	–
Age 25–29	–	–0.220	0.802	0.121	0.069	–	–	–	–	–0.221	0.802	0.121	0.069
Age 30–34	–	–0.431	0.650	0.142	0.002	–	–	–	–	–0.439	0.645	0.142	0.002
Age 35–39	–	–1.072	0.342	0.160	<0.001	–	–	–	–	–1.078	0.340	0.160	<0.001
Age 40–44	–	–2.338	0.097	0.228	<0.001	–	–	–	–	–2.347	0.096	0.229	<0.001
Age 45–49	–	–2.131	0.119	0.407	<0.001	–	–	–	–	–2.156	0.116	0.409	<0.001
Prev. infant survives (baseline)	–	0	1	–	–	–	–	–	–	0	1	–	–
Prev. infant does not survive	–	1.020	2.774	0.225	<0.001	–	–	–	–	1.023	2.781	0.226	<0.001
BMI < 20.5 (baseline)	–	0	1	–	–	–	–	–	–	0	1	–	–
BMI ≥ 20.5	–	0.352	1.422	0.159	0.027	–	–	–	–	0.346	1.414	0.159	0.030
Spanish fluency: none (baseline)	–	0	1	–	–	–	–	–	–	0	1	–	–
Spanish fluency: moderate	–	–0.042	0.959	0.124	0.740	–	–	–	–	–0.035	0.966	0.130	0.790
Spanish fluency: high	–	–0.558	0.573	0.183	0.002	–	–	–	–	–0.561	0.571	0.198	0.005
Community mean age	–	–	–	–	–	–0.026	0.974	0.095	0.780	0.057	1.058	0.096	0.550
Community mean BMI	–	–	–	–	–	0.234	1.263	0.268	0.380	0.187	1.206	0.270	0.490
Community mean Spanish fluency	–	–	–	–	–	0.047	1.048	0.299	0.880	0.173	1.189	0.311	0.580
Community distance from town (km)	–	–	–	–	–	0.004	1.004	0.004	0.400	0.003	1.003	0.004	0.500
Random effects	Variance	Variance				Variance				Variance			
Individual	0.760	0.571				0.745				0.571			
Community	0.044	0.050				0.039				0.042			

Notes: N = 1159 risk intervals belonging to 575 females aged 20–49 in 14 communities.

the infant whose birth began the interval increases the likelihood of progression to the next birth (Figure 12.3b); (c) women with Body Mass Index (BMI) less than 20.5 (one standard deviation below the mean) also have lower progression rates (Figure 12.3c); and (d) women who speak Spanish fluently have significantly lower rates of progression to next birth than do women who speak some or no Spanish (Figure 12.3d). Model C examines community-level predictors, and shows them to be generally weak. Model D combines individual- and community-level effects simultaneously; this model yields no substantial improvement over the individual-level Model B, suggesting that the regional effects in Figure 12.2 are due to individual factors that vary by region.

12.3.2 Progression to First Birth for Nulliparous Women

Table 12.2 shows a Cox proportional hazards analysis of the predictors of first birth. Again, Model A shows much larger individual variance than community variance. Model B, with individual-level predictors, shows that (a) there is a general trend to earlier first birth over time; (b) BMI higher than the early adolescent female mean of 18.1 is associated with faster transition to first birth (Figure 12.4a); and (c) education has a large effect on the transition to first birth (although the direction of causality may be unclear). Those with no or one year of education undergo transition first; those with two or three years of education experience transition slightly later; while those with four or more years of education have their transition last (Figure 12.4b). While Model C shows some association between mean community values (particularly mean birth year) and age of first birth, Model D indicates that these effects are better attributed to individual-level rather than community-level characteristics.

12.3.3 Infant Mortality Rates: Higher-Order Births

Table 12.3 presents an analysis of the predictors of infant mortality in the first year of life using mixed-effects logistic regression (N=747 births of which 4.6% died in the first year of life). It shows that (a) children born to mothers over age 35 have about twice the mortality risk of children born to younger mothers; (b) mother's BMI negatively predicts infant death, with the odds ratio decreasing by 14% for each additional unit of BMI; and (c) there is a strong interaction between the length of the preceding birth interval and geographical region of residence in determining mortality: short intervals of less than two years increase mortality hazards by more than fourfold in remote regions with no access to medical treatment, but do not affect mortality rates in regions with greater access. Neither maternal education nor the ability to speak Spanish are significantly related to mortality rates (results not shown). The multilevel random-effect terms in model A though D capture effectively zero variance in mortality, while the community-level predictors included in models C and D show little predictive ability.

Table 12.2 Multilevel Cox proportional hazard models of probability of first birth among nulliparous females

	Model A	Model B				Model C				Model D			
Fixed effects		B	exp(B)	SE	p	B	exp(B)	SE	p	B	exp(B)	SE	p
Year of birth	–	0.118	1.125	0.027	<0.001	–	–	–	–	0.112	1.118	0.028	<0.001
BMI < 18.1 (baseline)	–	0	1	–	–	–	–	–	–	0	1	–	–
BMI ≥ 18.1	–	0.770	2.159	0.194	<0.001	–	–	–	–	0.673	1.959	0.218	0.002
Education: ≤ 1 year (baseline)	–	0	1	–	–	–	–	–	–	0	1	–	–
Education: 2–3 years	–	–0.016	0.984	0.170	0.930	–	–	–	–	–0.089	0.915	0.185	0.630
Education: ≥ 4 years	–	–0.615	0.541	0.192	0.001	–	–	–	–	–0.670	0.512	0.226	0.003
Community mean year of birth	–	–	–	–	–	0.247	1.280	0.108	0.023	0.063	1.065	0.102	0.530
Community mean BMI	–	–	–	–	–	0.332	1.393	0.304	0.270	0.070	1.073	0.268	0.790
Community mean education (years)	–	–	–	–	–	–0.129	0.879	0.088	0.140	–0.013	0.987	0.084	0.880
Community distance from town (km)	–	–	–	–	–	0.000	1.000	0.005	0.970	–0.002	0.998	0.004	0.640
Random effects	Variance	Variance				Variance				Variance			
Community	0.119	0.019				0.039				0.005			

Notes: N = 590 females aged 10–20 in 14 communities.

Table 12.3 Multilevel logistic regression models of infant mortality in the first year of life

	Model A				Model B				Model C				Model D			
Fixed effects	B	exp(B)	SE	p	B	exp(B)	SE	p	B	exp(B)	SE	p	B	exp(B)	SE	p
(Intercept)	−3.074	0.046	0.178	<0.001	0.006	1.006	1.570	0.997	−13.116	0.000	12.310	0.287	−10.719	0.000	12.088	0.375
Maternal age ≤ 35	–	–	–	–	0	1	–	–	–	–	–	–	0	1	–	–
Maternal age > 35	–	–	–	–	0.825	2.281	0.394	0.036	–	–	–	–	0.825	2.283	0.396	0.037
Maternal BMI	–	–	–	–	−0.152	0.859	0.070	0.030	–	–	–	–	−0.161	0.851	0.071	0.024
Near town/mission × long IBI[a] (baseline)	–	–	–	–	0	1	–	–	–	–	–	–	0	1	–	–
Near town/mission × short IBI[b]	–	–	–	–	−0.508	0.602	0.641	0.428	–	–	–	–	−0.556	0.574	0.646	0.390
Remote × long IBI[a]	–	–	–	–	−0.019	0.981	0.525	0.971	–	–	–	–	−0.128	0.880	0.633	0.839
Remote × short IBI[b]	–	–	–	–	1.455	4.284	0.450	0.001	–	–	–	–	1.348	3.850	0.611	0.027
Community mean age	–	–	–	–	–	–	–	–	0.092	1.096	0.168	0.586	0.082	1.085	0.172	0.635
Community mean BMI	–	–	–	–	–	–	–	–	0.434	1.544	0.499	0.384	0.379	1.460	0.531	0.476
Community mean IBI (months)	–	–	–	–	–	–	–	–	−0.090	0.914	0.063	0.155	−0.013	0.987	0.081	0.869
Community distance from town (km)	–	–	–	–	–	–	–	–	0.002	1.002	0.006	0.796	0.003	1.003	0.006	0.574
Random effects	Variance				Variance				Variance				Variance			
Mother	0.000				7.55E-15				0.000				1.33E-12			
Community	2.11E-14				6.49E-11				0.000				1.05E-11			

Notes: N = 747 later-born infants belonging to 461 mothers in 14 communities.

[a] long IBI = preceding interbirth interval ≥ 24 months.

[b] short IBI = preceding interbirth interval < 24 months.

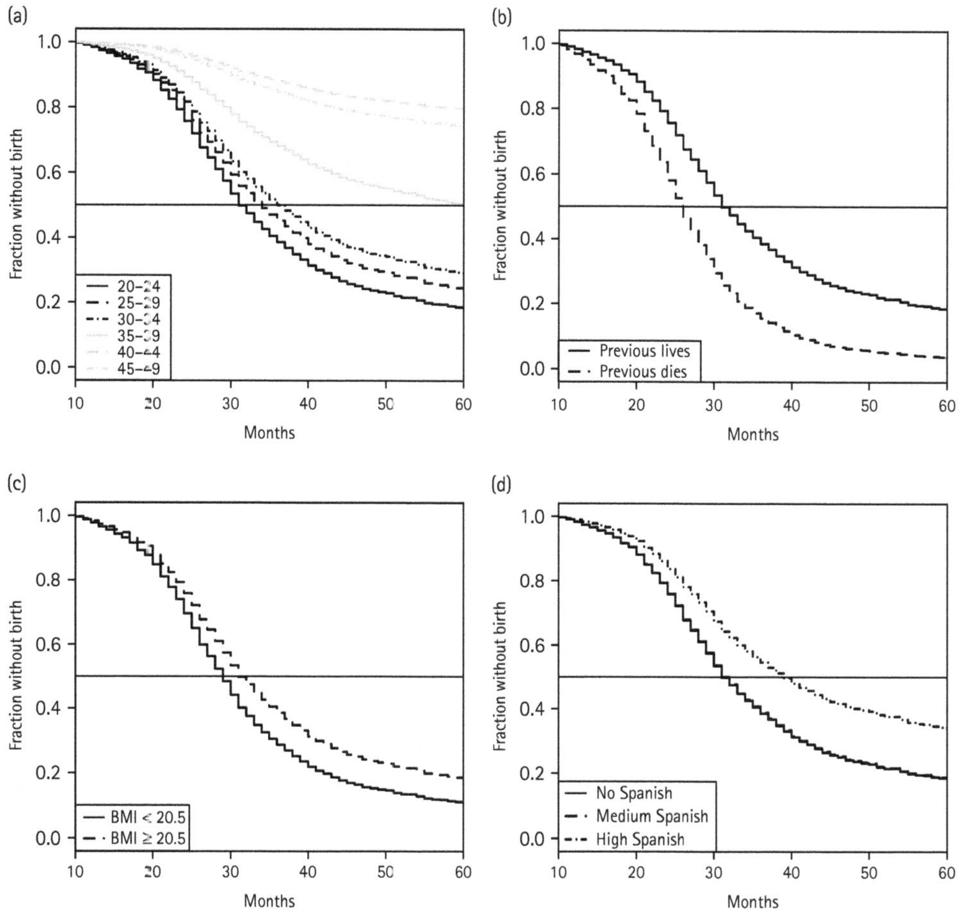

FIG. 12.3 Progression to next birth for parous women as a function of (a) age, (b) survival of previous infant, (c) BMI, and (d) Spanish fluency.

12.3.4 Mortality Risks for Firstborn Infants

Firstborn children die at almost twice the rate of later-born children (7.9% vs 4.6%). Table 12.4 shows the predictors of infant mortality for 198 firstborn children. It shows (a) a strong effect of mother's age at birth on mortality risk, with the odds ratio decreasing by 24% with each additional year of age; (b) a counter-intuitive impact of maternal BMI (unlike older mothers), where heavier mothers have higher infant mortality rates; and (c) the infants of mothers with high white blood cell counts are more likely to die, with an odds ratio of 1.15 for each additional 1,000 cells per cubic millimeter of blood. Neither maternal education nor the ability to speak Spanish were significantly related to mortality rates (results not shown). The variance attributed to the community-level random effect in Model A is obviated with inclusion of individual- and community-level predictors, although none of the community-level fixed effects is statistically significant.

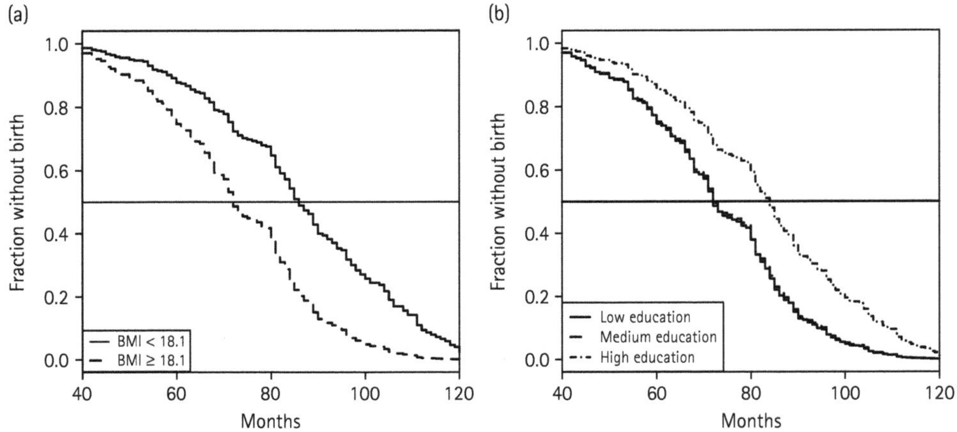

FIG. 12.4 Progression to first birth for nulliparous women as a function of (a) BMI and (b) education.

12.3.5 Changes in Fertility and Infant Mortality over Time: Tsimane and Bolivia Compared

Figure 12.5 shows a comparison between mortality and fertility rates of the Tsimane and the rest of Bolivia, using historical demographic data. There is a general similarity in the downward trajectory of infant mortality rates in both populations. Tsimane infant mortality increased slightly in the 1970s, when roads were built connecting La Paz to San Borja, potentially increasing pathogen exposure due to an influx of highland migrants to the region. The steady decrease in Tsimane infant mortality appears only after 1990, when the first health clinic 'Horeb' was created 6 kilometers from San Borja. By 2009, we find that the infant mortality rate has dropped to about 4–6% for both the Tsimane and for Bolivia as a whole.

The historical trajectories of fertility rates in the two populations, while moving in the same direction, are more dissimilar. From 1975 to 2009 the TFR in Bolivia as a whole dropped by 46% (from 6.6 to 3.5), while there has been relatively little change in the Tsimane TFR over time. By 2009, the gap in TFR between Tsimane and Bolivia is almost six births.

12.4 Discussion

These results provide some fundamental insights into the determinants of fertility and mortality in natural fertility populations undergoing socio-economic and epidemiologic transition. Physiological energy stores appear to play an important role in determining fertility. Women with greater energetic reserves (as indexed by BMI) had an

Table 12.1 Multilevel logistic regression models of firstborn infant mortality in the first year of life

	Model A				Model B				Model C				Model D			
Fixed effects	B	exp(B)	SE	p	B	exp(B)	SE	p	B	exp(B)	SE	p	B	exp(B)	SE	p
(Intercept)	−2.541	0.079	0.294	<2e−16	−5.559	0.004	3.076	0.071	32.814	1.8E+14	20.438	0.108	31.407	4.4E+13	20.413	0.124
Maternal age	–	–	–	–	−0.276	0.759	0.145	0.058	–	–	–	–	−0.278	0.757	0.158	0.078
Maternal BMI	–	–	–	–	0.269	1.309	0.098	0.006	–	–	–	–	0.286	1.332	0.099	0.004
Maternal WBC[a] ($\times 10^3/mm^3$)	–	–	–	–	0.140	1.151	0.076	0.064	–	–	–	–	0.127	1.136	0.087	0.145
Community mean age	–	–	–	–	–	–	–	–	−0.611	0.543	0.533	0.252	−0.308	0.735	0.558	0.581
Community mean BMI	–	–	–	–	–	–	–	–	−0.913	0.401	0.637	0.152	−1.177	0.308	0.663	0.076
Community mean WBC[a] ($\times 10^3/mm^3$)	–	–	–	–	–	–	–	–	−0.463	0.630	0.401	0.249	−0.578	0.561	0.438	0.187
Community distance to town (km)	–	–	–	–	–	–	–	–	0.015	1.015	0.013	0.239	0.015	1.015	0.014	0.283
Random effects	Variance				Variance				Variance				Variance			
Community	0.144				0.053				0.000				0.000			

Notes: N = 198 firstborn infants in 14 communities.

[a] WBC = white blood cell count.

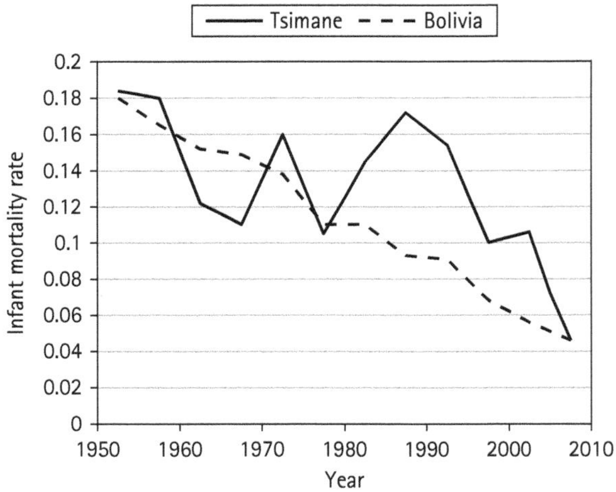

FIG. 12.5 Infant mortality rates of Tsimane and Bolivia by year.

earlier first birth and shorter intervals between subsequent births. As expected from models positing a causal relationship from mortality to fertility, the loss of the previous infant hastens the transition to the next birth, presumably due to the termination of lactation and faster resumption of ovulation (Wood 1994), as expected by models positing a causal relationship from mortality to fertility.

There is also evidence that fertility affects mortality. In remote areas with minimal access to medical services, a short interval between births increases the mortality risk to the subsequent infant about fourfold. Age of first birth also has a significant effect on mortality rates, reducing the mortality risk of the earlier-born infant by a quarter for each additional year that a young women delays reproduction. Thus, with respect to production of surviving offspring, there is a clear trade-off between fertility (both its initiation and its resumption following a birth) and infant survival.

Physiological energy stores also appear to affect mortality. Greater maternal BMI is associated with a lower risk of infant mortality for higher-order births. For first births, the results were counter-intuitive, since maternal BMI is associated with a higher risk of infant mortality. One possible explanation of this result is that there is more competition between maternal and infant energy stores when mothers are adolescents and still increasing in body weight (which continues past age 20 among Tsimane girls). Early maternal physiology may be sequestering more resources for itself at a cost to the infant. It is possible that some of the increased risk of mortality for firstborn children is due to this energy competition, and not just due to the inexperience of young mothers. This counter-intuitive finding does have precedent in developed societies: in the US, obese women experience more neonatal mortality, regardless of weight gain during pregnancy, due to more pregnancy complications related to short gestation and low birth weight, pre-eclampsia, and prolonged labour (Chen et al. 2009).

Taken together, these results provide support for the view that natural selection has acted on the physiological mechanisms governing human fertility to maximize the

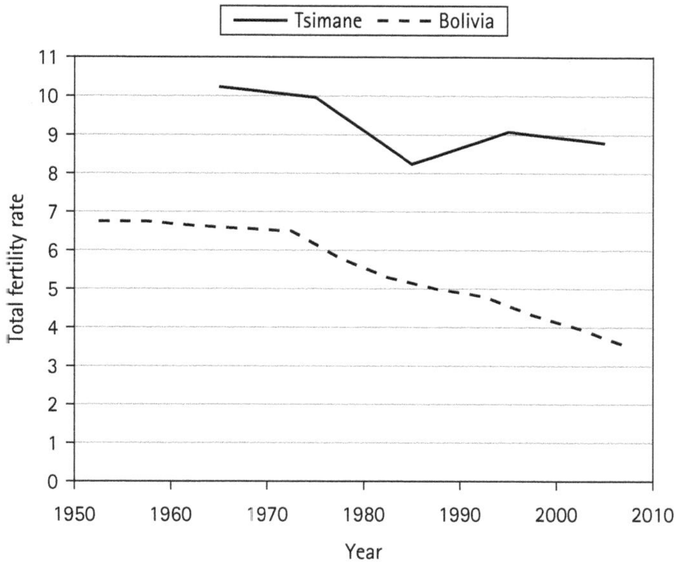

FIG. 12.6 Total fertility rates of Tsimane and Bolivia by year.

production of *surviving* offspring through the optimal combination of fertility and parental investment. The transition to the next birth is affected by both energy stores and whether there is an existing living infant to invest in. Most birth intervals are longer than two years; holding energy stores constant, having a shorter interval results in increased mortality risk.

The effects of secular change also provide support for this view. Mortality has been decreasing rapidly over time. Given the observed regional effects on mortality, it appears that much of the secular trend in mortality rates is due to increased access to Western medical treatment, especially antibiotics and rehydration therapy. One remote area that is served by a medical post with itinerant teams visiting the villages had similar infant mortality rates to villages near the town of San Borja, where there is a hospital.

The multilevel models developed here provide additional insights into the processes determining fertility and mortality in natural fertility populations. Results show that the effects of community-level variables are quite weak. Most determinants of fertility and mortality modelled here operate on the individual level, with the exception of regional effects on mortality. Both infant and child mortality from 1950–2000 increased with distance from San Borja. This does not mean that there is little scope for social effects on demographic rates among the Tsimane (and similar groups). We have shown in other papers that kin—especially post-reproductive kin—play a major role in providing the food energy to support reproduction (Hooper 2011; Kaplan et al. 2010). It appears to be the case that, in the present setting, the social effects on fertility and mortality are achieved through social assistance, with impacts on energy and care, rather than through cultural effects based on conformity or other process of value transmission.

The multilevel models also suggest that unobserved heterogeneity is greater at the individual than group level. In relation to the terminology used by Wachter (this volume), it may not be possible to distinguish frailty from life-luck, even with the random effects terms for individual heterogeneity. With few births per woman during the sample period, the differences in women in hazards of fertility and the likelihood of their baby dying could be due to unobserved differences in health, tendency to conceive, parenting abilities, etc.; or alternatively, to differences in life-luck, such as flu epidemics and random variation in chromosomal segregation during meiosis. There does seem to be large scope for unobserved heterogeneity in both the initiation and nature of reproduction. The counter-intuitive relationship between maternal BMI and infant mortality for first births may reflect unobserved heterogeneity in adolescent growth patterns, maternal feeding practices, and perhaps more subtle differences in family assistance. The relatively high risks of fertility for some women in their late 40s may also represent differences among women in health, ovarian follicle number, ovum quality, and perhaps other factors.

The secular change in infant mortality rates among the Tsimane is dramatic and keeping pace with the rest of Bolivia. Nevertheless, infant mortality rates still show large yearly fluctuations, probably reflective of temporal variance in respiratory and diarrheal epidemics (cf. Omran's characterization of the 'Age of Pestilence' where mortality is high and fluctuating; Omran 2005).

In contrast with mortality, there has been little or no change in total fertility rates over time among the Tsimane, whereas TFRs have dropped from 6.6 to 3.6 for Bolivia as a whole. This is likely due to the fact that lowering family size in order to increase parental investment in education still has little pay-off in the context of their isolation and rural lifestyle (McAllister et al. 2012). At the same time, modernization is beginning to affect personal values regarding fertility. McAllister et al. (2012) showed that values about ideal family size are changing among some women, and are affected by education and Spanish-speaking ability. About 10% of parous women in our sample speak Spanish fluently. Their rates of transition to next birth were lower than for women who speak little or some Spanish. Presumably an increasing number of women will speak Spanish as time progresses. Schooling will also be more common for current cohorts; Tsimane girls and boys aged 10–19 now have an average of 3.2 and 3.5 years of schooling, respectively, compared with 0.4 and 1.3 years of schooling among women and men aged 40+. For the transition to first birth, there is also evidence that young women with more education (≥ 4 years) initiate reproduction later. In some cases, however, pregnancy may have terminated schooling rather than vice versa.

In sum, however, the effects of education and Spanish speaking on actual fertility as of yet are quite weak, and do not overshadow the increasing nutritional status of women. For this reason, hazards of births are increasing slightly over time, after controlling for the death of the previous child. This may be reflective of a 'pre-decline rise' in fertility (Dyson and Murphy 1985). It is also noteworthy that over the course of the study, progression rates to first birth increased by about 12% per year (i.e. mothers' average age at first birth decreased). This finding is consistent with the biological view that in natural fertility populations, the physiological mechanisms underlying fertility regulation should tend to maximize population growth. It is less consistent with the cultural homeostasis view proposed by demographic transition theories (e.g. Coale and Watkins 1987), which argued that cultural institutions will act to maintain population equilibrium by lowering fertility when mortality is low and vice versa. Although the rise in fertility following mortality decline has been observed now among the Tsimane, other transitioning groups in South America, and elsewhere (Dyson and Murphy 1985; McSweeney and Arps 2005), classic demographic transition theories generate the expectation of stable fertility levels that gradually decline following drops in mortality. Our findings highlight the kinds of mechanisms at the individual level that can lead to growing fertility rates as a response to changes in mortality. As the process of modernization continues, we expect greater scope for community and social network effects on fertility as people begin to model and preferentially interact with others sharing similar behavioural strategies (Casterline 2001).

To conclude, we propose that under traditional, natural fertility conditions: (1) fertility and mortality are tightly linked; (2) the physiological processes underlying fertility outputs are tuned to individual condition (in terms of energy inputs and stores), existing vessels of investment, and the environmentally variable connection between birth spacing and subsequent infant mortality; and (3) fertility outputs will tend to maximize descendant production over the long run. With modernization, as populations become more internally structured by socio-economic status and differing strategies for parental investment, the scope for cultural and network effects may be much greater, with less control being exercised by a fitness-maximizing physiology.

References

Bates, D., M. Maechler, and B. Bolker (2012). 'lme4: Linear Mixed-Effects Models Using S4 Classes', R package, <http://CRAN.R-project.org/package=lme4>.

Blurton Jones, N. B. (1986). 'Bushman Birth Spacing: A Test for Optimal Interbirth Intervals', *Ethology and Sociobiology* 4: 145–147.

Casterline, J. E. (ed.) (2001). *Diffusion Processes and Fertility Transition: Selected Perspectives.* Washington, DC: National Academy Press.

Chen A, S. A. Feresu, C. Fernandez, and W. J. Rogan (2009). 'Maternal Obesity and the Risk of Infant Death in the United States', *Epidemiology* 20/1: 74.

Coale, A. J. (1973). 'The Demographic Transition', in *Proceedings of the IUSSP Liege International Population Conference.* Liege: IUSSP, i.53–72.

Coale, A. J., and S. C. Watkins (eds) (1987). *The Decline of Fertility in Europe.* Princeton, NJ: Princeton University Press.

Cole, L. C. (1954). 'The Population Consequences of Life History Phenomena', *Quarterly Review of Biology* 29: 103–137.

Davis, K. (1945). 'The World Demographic Transition', *Annals of the American Academy of Political and Social Science* 237: 1–11.

Davis, K. (1963). 'The Theory of Change and Response in Modern Demographic History', *Population Index* 29: 345–366.

Dyson, T. and M. Murphy (1985). 'The Onset of Fertility Transition', *Population and Development Review* 11/3: 399–440.

Gelman, A. and J. Hill (2007). *Data Analysis Using Regression and Multilevel/Hierarchical Models*. Cambridge: Cambridge University Press.

Gurven, M., H. Kaplan, and A. Zelada Supa (2007). 'Mortality Experience of Tsimane Amerindians of Bolivia: Regional Variation and Temporal Trends', *American Journal of Human Biology* 19: 376–398.

Hooper, P. L. (2011). 'The Structure of Energy Production and Redistribution among Tsimane Forager-Horticulturalists', PhD thesis, University of New Mexico, Albuquerque.

Hox, J. 2002. *Multilevel Analysis: Techniques and Applications*. New York: Psychology Press.

Kaplan, H. S. (1996). 'A Theory of Fertility and Parental Investment in Traditional and Modern Human Societies', *Yearbook of Physical Anthropology* 39: 91–135.

Kaplan, H. S., M. Gurven, J. Winking, P. L. Hooper, and J. Stieglitz (2010). 'Learning, Menopause, and the Human Adaptive Complex', *Annals of the New York Academy of Sciences* 1204: 30–42.

Livi Bacci, M. (2007). *Conquest: The Destruction of the American Indios*. Cambridge: Polity Press.

McAllister, L., M. Gurven, H. Kaplan, and J. Stieglitz (2012). 'Why Do Women Have More Children Than They Want? Understanding Differences in Women's Ideal and Actual Family Size in a Natural Fertility Population', *American Journal of Human Biology* 24/6: 786–799.

McSweeney, K. and S. Arps (2005). 'A "Demographic Turnaround": The Rapid Growth of Indigenous Populations in Lowland Latin America', *Latin American Research Review* 40: 4–29.

Omran, A. R. (2005). 'The Epidemiological Transition: A Theory of the Epidemiology of Population Change', *The Milbank Quarterly* 83: 731–757.

Perz, S., J. Warren, and D. Kennedy (2008). 'Contributions of Racial-Ethnic Reclassification and Demographic Processes to Indigenous Population Resurgence: The Case of Brazil', *Latin American Research Review* 43: 7–33.

Smith, C. C. and S. D. Fretwell (1974). 'The Optimal Balance between Size and Number of Offspring', *American Naturalist* 108: 499–506.

Therneau, T. (2012). 'coxme: Mixed effects Cox models', R package version 2.2-1, <http://CRAN.R-project.org/package=coxme>.

Wood, J. W. (1994). *Dynamics of Human Reproduction: Biology, Biometry and Demography*. New York: Aldine de Gruyter.

PART V

CONSTRUCTING POPULATIONS IN THE LONG TERM

CHAPTER 13

GENETICS AND THE RECONSTRUCTION OF AFRICAN POPULATION HISTORY

FRANCESC CALAFELL AND DAVID COMAS

THE study of the extant genetic variation provides a powerful tool both to test hypotheses on population history and to generate new ones. Recently, the analytical framework of human population genetics has completed a cycle: the lineage-based description of mitochondrial DNA and Y chromosome diversity, where often populations vanished only to be replaced by marauding hordes of disembodied haplogroups, is giving way to the multivariant analysis of SNP allele frequencies. This cycle, in effect, has returned the field to the toolkit pioneered by Cavalli-Sforza in the late 1980s. There is now, however, an important difference: population definition can be left to Bayesian algorithms, creating a tension between the individual, the population as socially defined, and the reified Bayesian population constructs.

In the present paper, we will review the recent literature on African population genetics produced by both approaches, with emphasis on a few selected issues. East Africa shows an amazing array of cultural, linguistic, and genetic diversity, and can be imagined as a source or a sink of population movements. A similar connection between genetic and linguistic diversity has been made for Southern Africa, and we will discuss its recent candidacy to humankind's cradle and the relevance of Khoe-San groups. Contrary to the latter view, the Bantu expansion appears to have homogenized a vast swathe of Africa. We will therefore briefly review the population dynamics of that expansion and the meaning of the haplogroups that have been proposed as *Bantu markers*. An important case in point is the multifaceted nature of Pygmy populations, resulting from a complex interplay between identity, phenotype, adaptation, and admixture. Finally, we will touch on populations north of the Sahara with emphasis on contributions from the North, East, and South.

13.1 Populations: The Object of Population Genetics?

Population genetics, the description and interpretation of genetic diversity within populations, operates within a well-defined theoretical framework that models very precisely how different processes shape genetic variation and allows the reverse inference, namely, that from the observed genetic diversity one may be able to glimpse the processes that generated that diversity. Classically, some forces impinging on genetic diversity have been related strictly to the genomic machinery, such as mutation and recombination (the latter, though, has plenty of room for individual and population variation, as recently discovered; Baudat et al. 2010; Berg et al. 2010; Myers et al. 2010). Other processes, such as changes in population size (founding events, bottlenecks, expansions) and gene flow (migration, both as a collective and an individual phenomenon) affect populations and fit squarely into the domain of this volume. Finally, natural selection is a multifarious process that can sieve out diversity that is detrimental to the survival or reproduction of the individual, or reward with better survival odds the carriers of some variants, with the major qualification that the fate of an allele (genetic variant) may be linked to the physical, biological, and cultural environment. Genetic variation that allows adaptation to high altitude has recently been discovered (Yi et al. 2010). Natural selection has also been triggered by the changes in diet brought by the cultivation of starchy foods (Perry et al. 2007), dairy husbandry (Bersaglieri et al. 2004; Enattah et al. 2002), or by local pathogens, such as the malaria parasite (Verra et al. 2009).

Before diversity-shaping processes can be investigated, population genetics requires a methodological simplification of one classical definition of a population. Population, as understood by classic authors like Darwin, is a group bound by reproductive links, that is, composed by individuals who exchange genes. This strict definition is clearly unworkable and in contradiction with the models that admit gene flow. One would struggle to find cases in which a clear border can be drawn that separates groups of humans that absolutely do not exchange genes. As this chapter shows, many current methods in population genetics allow us to infer how much different source populations have contributed to the admixed make-up of an individual or of a population sample. This strict population definition is usually watered down to a group of individuals in which *most* reproductive mates are found. In practice, human population geneticists do not seek the assistance of demographers to tally up marriages (assuming marriage as a proxy for a reproductive unit, a tenet that is adhered to with various strength in the global cultural landscape). Rather, they define, by some sort of cluster analysis, which are the actual groups of endogamous individuals, and then use them as sampling units. This is a key point: given the costs associated with recruiting individuals to volunteer for genetic studies and with the actual laboratory and data analysis, human population geneticists always study population *samples*, not the entire population. In section 13.2, sample sizes range from one to a few thousands. Sampling strategies tend to be simple and aseptic: geneticists resort to self-identification ('a Yoruba is whoever says he is a Yoruba') or to a geographical/pedigree approach ('individuals with all four grandparents born in Tunisia'). It is often impossible to know the consequences that these approaches may have when it comes to interpreting results. Moreover, as we shall discuss in the next section, in some cases the sampled population is made to disappear by considering certain lineages only, or by replacing it with statistical constructs.

13.2 A Short Primer on Human Population Genetics Methods

The raw material of population genetics analysis is genetic variation, which comes in a few varieties with specific properties (for an extended explanation on this subject, see Jobling et al. 2004). The human genome is composed of three billion basic units ('bases', or 'nucleotides', that come in four possible states), organized in 23 pairs of chromosomes plus an extra bit (also known as mitochondrial DNA). We inherit each chromosome in a pair from each parent, with two exceptions: (i) the 23rd pair, the sexual chromosomes, which in women are XX and men are XY—the Y chromosome is found only in men, who inherit it from their fathers; and (ii) mitochondrial DNA (mtDNA), found in both sexes but inherited solely from the mother. These are also the two exceptions that escape recombination, the process in which chromosomes within each pair (and the lineages they represent) are shuffled when producing gametes (sperm or eggs).

Genetic variation can be as simple as the alternate presence of one of two nucleotides at a given position in the genome. These are the Single Nucleotide Polymorphisms or SNPs, which are created by mutation at the extremely slow rate of once every hundred million conceptions for a particular site on a given chromosome. This rate implies that most SNPs pre-date the inception of current populations and, particularly for those in which the least frequent allele is not that rare (for instance, with a frequency above 10%), both alleles in most cases are present in most global populations (Gravel et al. 2011). Combinations of alleles in neighbouring sites are called haplotypes, which are (with the important exceptions noted above), shuffled by recombination.

Insertion/deletion polymorphisms (or indels) have properties similar to those of SNPs. They represent the single gain or loss of one or a few bases, and tend to be rare events. In contrast, short tandem repeats (STRs, also called microsatellites), in which polymorphism is also created by the gain or loss of DNA bases, mutate much faster. In STRs, a short (one to six nucleotides) motif is repeated a variable number of times. STRs mutate much faster than SNPs, at a rate of up to once every thousand conceptions. This property is used to date particular events, such as the foundation of a lineage, drawing on evidence of the accumulation of STR variation within that lineage.

Human population genetics has recently completed a circle. It started with the so-called *classical* polymorphisms (i.e. blood groups and other protein polymorphisms) that were analysed on the basis of their allele frequencies using potent statistical instruments such as Principal Component Analysis (PCA). This era culminated with the publication of the magnificent book by Cavalli-Sforza et al. (1994). However, a few shortcomings of classical polymorphisms can be pointed out: they are relatively few, their connection to actual DNA variation was mostly unknown, and they could have been subjected to confounding natural selection (that is, two populations could be genetically similar not because of a shared history but because they lived in similar environments; think of skin colour in Africans and Australian Aborigines). The polymerase chain reaction technique (PCR) and automated sequencing heralded the next, uniparental marker, era: mtDNA and the non-recombining region of the Y chromosome (NRY) could be routinely analysed and a firm phylogeography could be established for both genomic regions, allowing the dissection of population structure with unprecedented precision and reliability. Nonetheless, mtDNA and the NRY have their own quirks: they each behave as a single locus and can be swayed by natural selection, and their variation may reflect the peculiarities associated with their sex-specific transmission.

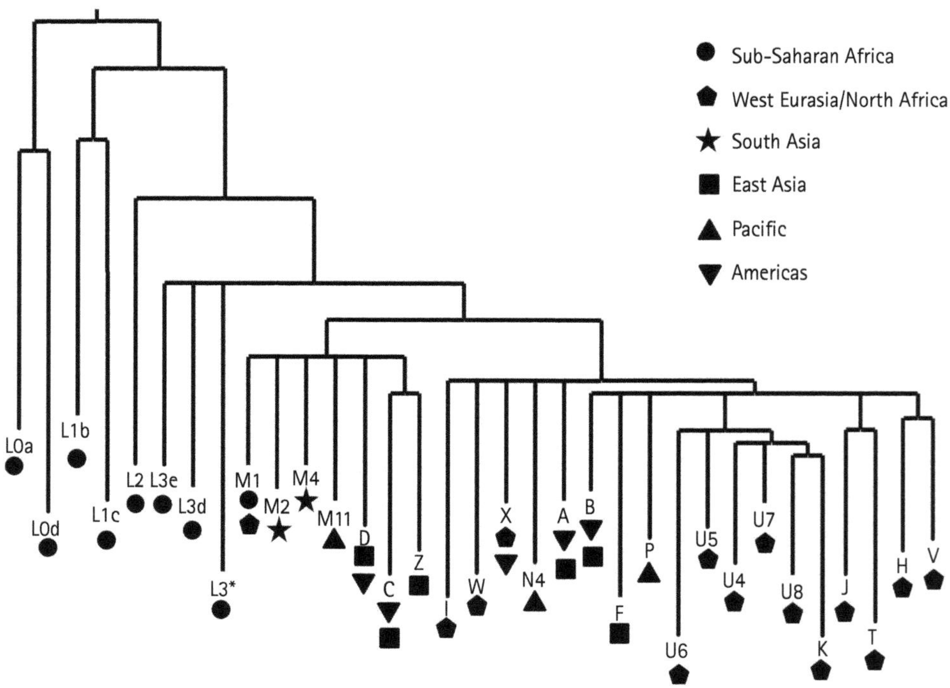

FIG. 13.1 Phylogenetic tree of mitochondrial DNA haplogroups, with the geographic regions where they are more frequent.

Often, haplogroups—namely, the more stable, deeper branches of the mtDNA and NRY phylogeny (see Figures 13.1 and 13.2)—are treated as autonomous entities and confused with the entire population. For example, one can often read expressions such as 'the Bantu haplogroup', which can be erroneously interpreted to mean that the Bantu are monomorphic for that particular haplogroup or that this haplogroup is exclusively found in the Bantu. Often papers are devoted to the geography and history of single haplogroups, masking the more complex diversity existing in each population.

Another technological development spearheaded a new breakthrough in human population genetics: SNP array genotyping platforms have made it affordable to genotype hundreds of thousands of polymorphisms. Now the whole genome is covered, and the action of selection is masked by a vast majority of putatively neutral markers. Demographic events such as migration and bottlenecks have effects on the whole genome, whereas natural selection affects genetic variation at the few genes that are relevant for the traits that are under natural selection; considering the whole genome is more likely to result in inferences referring to demographic history than to natural selection. The results are again treated in terms of allele frequencies and subjected to PCA, to which newer techniques may also be added, such as the Bayesian classification algorithms implemented in STRUCTURE (Pritchard et al. 2000) or ADMIXTURE (Alexander et al. 2009); see also Lawson (this

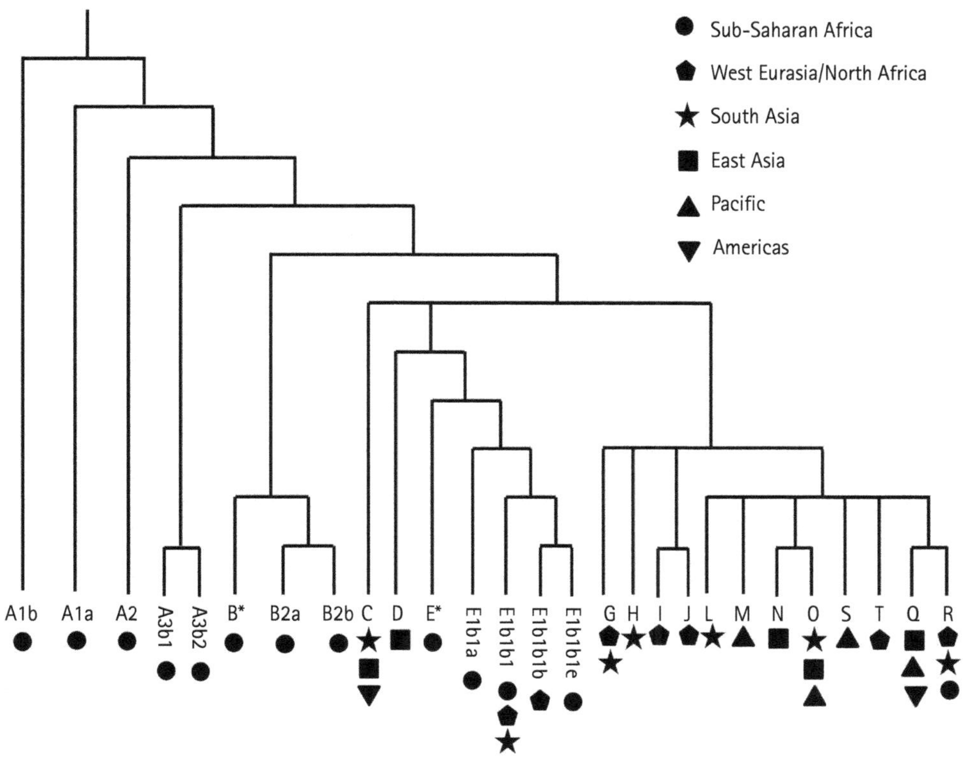

FIG. 13.2 Phylogenetic tree of Y-chromosome haplogroups, with the geographic regions where they are more frequent.

book). In these algorithms, a given number of ancestral components is pre-set and the algorithm assigns a probability to each individual of belonging to each component. These components are often reified and treated as if they were actual populations.

At this point, we have seen a number of ways in which a population is constructed in human population genetics: as an idealized reproductive unit; as a sampling unit that relies on self-identification, geography, or language; as a phylogenetical lineage in mtDNA or NRY; or as a statistical construct resulting from a Bayesian analysis. In sections 13.3 through 13.8, we shall see how these different constructs are utilized when studying the genetics of African populations, and how their use impinges on the conclusions that can be drawn.

13.3 The African Genetic Landscape: An Introduction

Africa is a continent of great cultural and environmental diversity. Traditional subsistence models include pastoralism, farming, and a sizable number of hunter-gatherer populations (Pygmies, the Hadza in Tanzania, and Khoisans in Southern Africa; see Figure 13.3 for the location of these populations, and of others mentioned below). Roughly one-third of current world languages are spoken in Africa (Scheinfeldt et al. 2010), and belong to four families: Niger-Kordofanian languages are spoken throughout Western, Eastern, Central, and Southern Africa, and include the common Bantu languages. The Afro-Asiatic family includes languages spoken in Northern, Central, and Eastern Africa (as well as in western Asia), belonging to the Cushitic, Chadic, and Semitic branches. The Nilo-Saharan language family is spoken predominantly in Central and Eastern Africa and includes the Sudanic and Nilotic languages. Finally, the Khoisan language family contains languages with click sounds in Eastern (the hunter-gatherer Hadza and Sandawe) and Southern Africa (such as the also hunter-gatherer San). This is the most controversial language family in Africa, since the Eastern and Southern branches are rather divergent. However, a systematic study of the phonemic diversity of world languages (Atkinson 2011) pointed to Southern Africa as the richest part of the globe and a probable place of origin for human linguistic diversity.

Anatomically modern humans originated in Africa and spread to the rest of the world, where they probably met and exchanged a few genes with local hominids such as Neanderthals (Green et al. 2010) and Denisovans (Reich et al. 2010). Africa has been the continent with the largest population for the longest time; in population genetics terms, it has the largest effective population size, probably three times larger than that of the other continental populations (Melé et al. 2012; Tenesa et al. 2007). This particular history has had multiple genetic consequences: genetic diversity is greater in Africa than in any other continent (Henn et al. 2011; Tishkoff et al. 2009), even in small hunter-gatherer populations. For instance, Calafell et al. (1998) found that all of the multiple alleles at two STRs ever described in samples from worldwide distributed samples were present in one Pygmy sample. At many loci, haplotypes found elsewhere are a subset of those found in African populations (Melé et al. 2010; Tishkoff et al. 1996; Tishkoff et al. 1998). And when the phylogenetic trees

FIG. 13.3 Populations mentioned in the text. Positions are given as a general reference rather than an exact location. The grey area represents the extent of the Bantu languages. The dashed and dotted lines encircle the West and the East Pygmies. The numbers indicate approximate locations as follows: 1. !Kung San; 2. Juu San; 3. Ju|hoansi; 4. ≠Komani; 5. Xhosa; 6. Zulu; 7. Mbugu; 8. Sandawe; 9. Maasai; 10. Hadza; 11. Luhyia; 12. Luo; 13. Beja; 14. Yoruba; 15. Mandenka; 16. Mozabite; 17. Karreitje People; 18. Nama; 19. /Gui and //Gana; 20. Khwe; 21. Herero; 22. Bantu speakers.

of the unilinearly transmitted portions of the genome (that is, mtDNA and the NRY) are reconstructed, sequences found outside of Africa branch out of an African root.

The general patterns of genetic diversity in Africa have been studied by different means. The first continent-wide synthesis was presented by Salas et al. (2002) using mtDNA control region sequences. The control region is the segment of the mtDNA molecule that is most polymorphic and that optimizes the sequencing efforts. By defining the phylogeography of African mtDNA in this way, haplogroups (groups of sequences defined by a common set of mutations) could be linked to some particular geographical locations and historical events. For instance, Eastern and Western African populations carried distinct sets of haplogroups; Western Pygmies and the San also had specific haplogroups. Other haplogroups had geographical extents that matched the expansion of the Bantu-speaking populations, to the point that their presence in non-Bantu populations could be attributed to admixture from a Bantu source (Plaza et al. 2004). In a more recent study, Behar et al. (2008) used the entire mtDNA molecule and could refine many of the conclusions in Salas et al. (2002).

A comparable, continent-wide study of the Y chromosome is not available. However, a global survey of the global Y-chromosome variation (Underhill et al. 2000) revealed that the deepest branches were found only in Africa, and, in particular, among Pygmies and Khoisan speakers.[1] Again, a more recent analysis confirms and refines this general landscape (Cruciani et al. 2011). And also in parallel with mtDNA, a set of *Bantu markers* has been detected (Cruciani et al. 2002); however, the differentiation between Eastern and Western Africa is less pronounced for the Y chromosome than it is for mtDNA. This is contrary to the usual pattern in which the Y chromosome is more differentiated than mtDNA, which is usually attributed to patrilocality (Pérez-Lezaun et al. 1999; Seielstad et al. 1998). This is definitely an area where a more fluent dialogue between population geneticists and demographers is likely to be fruitful.

Recently, it has become possible to genotype large sets of polymorphisms that cover most of the genome, reaching unprecedented precision. In a seminal paper, Tishkoff et al. (2009) analysed 1,327 STRs and insertion/deletion polymorphisms (indels) in 2,527 individuals from 120 populations. Given the breathtaking scope of that work, many of its results will be discussed in the following sections of this paper. To summarize very briefly: the genetic landscape of Africa appears dominated by the homogeneity of Niger-Kordofanian speakers; it is quite complex in Eastern Africa and the Sahel; and a (probably ancestral) connection between Pygmies and Khoisan speakers is revealed. In a global study, Jakobsson et al. (2008) analysed 525,910 SNPs in a much more limited sampling of African populations and recovered the same Pygmy-Khoisan background (although, given the sparse sampling, in that case it could also have been called 'non-Bantu'; the awareness of the implications of limited population sampling is crucial to the adequate interpretation of genetic results). Sikora et al. (2010) contributed further sampling in South East Africa and emphasized the role of geographical distance in genetic differentiation. Finally, some studies report very different findings: Henn et al. (2011) focused on the Khoisan speakers, finding no evidence of a Pygmy-Khoisan connection, and suggesting Southern Africa as the cradle of humankind.

[1] As time estimates on phylogenetic trees depend on mutation rate estimates, and since these are in flux lately, we would not dare venturing dates for these branches that are more precise than the 50,000–150,000 year bracket (Hawks 2012; Lynch 2010; Nachman and Crowell 2000; Roach et al. 2010).

13.4 South Africa: The Latest Candidate to Humankind Cradledom

In 1968, *Man the Hunter* (Lee and DeVore 1968) presented a novel premise for a research programme: the study of the contemporary hunter-gatherer populations could open a window into the lives of our Palaeolithic ancestors. Among the populations studied by the contributors to that book, the !Kung San of Southern Africa had a starring role. It was realized that their social structures (and, by implication, those of the Palaeolithic groups) were much richer than previously imagined. As we shall see in part of this section, their genetic structure is also much more diverse than could be imagined for a small, largely endogamic population.

As mentioned in section 13.3, the Khoisan language family comprises click languages spoken in SW Africa and by the Hadza and Sandawe in Tanzania. Their phonemic diversity (and that of the neigbouring Bantu languages, isiZulu and isiXhosa, which have borrowed 15 click consonants) anchors the correlation between linguistic diversity and distance from Africa that has led to the suggestion that Southern Africa is the cradle of human languages (Atkinson 2011).

The genetic structure of the Southern African Khoisan speakers has been studied in relation to their putative place as the most direct descendant of the ancestors of humankind, or in relation to their special connections to other African populations, or for their own interest, which is probably at least as important as the other two aspects. Most (about 70%) of mitochondrial DNA lineages in the Southern African Khoisans belong to haplogroups L0d and L0k, which branch directly from the root of all human mtDNA sequences (Behar et al. 2008). The male-mediated counterpart of the mtDNA tree, namely the NRY, showed haplogroups A and B branching successively from the root (Karafet et al. 2008); both are well represented (by subgroups A2, A3b1, and B2b) in Southern African Khoisan speakers. However, a recent study (Cruciani et al. 2011) has shaken the NRY tree and now it is haplogroups A1a and A1b, both absent in Southern Africa and found at low frequencies in a few Northern and West-Central African populations, that branch off first and are presumed to be more related to the human ancestral Y chromosome. Finally, geographic patterns of linkage disequilibrium and population differentiations, based on 580,000 genowide SNPs, are consistent with an origin of modern humans in Southern Africa rather than Eastern Africa (Henn et al. 2011) although the recent analysis of seven groups of Khoisans shows a more complex pattern of admixture that blurs the exact localization of the origin of humankind (Schlebusch et al. 2012). Other types of evidence, beyond genetics and linguistics, such as the archaeology at Pinnacle Point (Marean et al. 2007) appear also to favour an early presence of anatomically modern humans in South Africa.

The Southern African Khoisan speakers are not alone in the ancestral branches of human diversity. Mitochondrial haplogroup L0d can be further split into sub-haplogroups L0d1, L0d2, and L0d3; the former two are found indeed in Southern Africa, while the latter is found solely, albeit at low frequencies, in Eastern Africa (Gonder et al. 2007). This can be interpreted as pointing to an ancestral connection between Eastern and Southern Africa; note that the recent Henn et al. (2011) study has reversed the assumed direction of that connection. In the NRY tree, subhaplogroup A3b2 (close kin of Southern African A3b1) is also found in Eastern Africa, while B2b is shared between Southern African Khoisan speakers and western Pygmies. Note the inferential rationale in this approach: if two populations share identical lineages (that are found at some frequency, among other lineages, in these populations), or lineages that are close to each other in the mtDNA or NRY, then these populations are themselves related. Obviously, other interpretations are possible, which are not always taken into account: for instance, these shared lineages could have been acquired independently from a third population, or could have disappeared by random phenomena in other populations whose links then appear to be more tenuous than they really are.

Tishkoff et al. (2009), while genotyping 1,327 STRs and indels in 2,527 individuals from 120 populations, found it difficult to extricate the Southern African Khoisan speakers from the Eastern Mbuti Pygmies; note that this component was found also at lesser frequencies in the Southern African Bantu speakers and in the Cape Coloured, but not in the much more densely sampled Eastern Pygmies. This connection (based on a putative ancestral population as inferred by a Bayesian statistical method from the allele frequencies) disappeared in the studies by Henn et al. (2011) and Schlebusch et al. (2012). Two key points could explain this difference: the number of markers, larger in Henn et al. (2011) and Schlebusch et al. (2012); and, more importantly, the number and diversity of the Southern African Khoisan sample. The seven !Kung San in the sample used by Tishkoff et al. (2009) contrast to the thirty-five ≠Khomani Bushman, five Ju|'hoansi Bushman, and twelve Juu San in Henn et al. (2011), and the even more diverse sampling in Schlebusch et al. (2012), including twelve Karretje People, seven Nama, seven /Gui and //Gana, seventeen Ju|'hoansi, thirteen !Xun, seventeen Khwe, seventeen ≠Khomani, eight Herero, and nineteen Bantu speakers. It may be the case that low sample sizes prevented Tishkoff et al. (2009) from differentiating San from Mbuti components.

At present, when we are marching towards the mass production of whole human genome sequences, one of the first individuals to be fully sequenced was !Gubi, a speaker of Tuu (a Khoisan language) from the southern Kalahari (Schuster et al. 2010); four other Khoisan genomes were investigated in the same paper. A striking result was found: any two Khoisan individuals were more divergent than a European from an East Asian. Much of the genetic diversity that was lost when humans left Africa to colonize the rest of the world has been preserved in these small, besieged populations. On average, a nucleotide chromosome position was 20% more probable to be polymorphic in a Khoisan speaker than in other humans. Schuster and colleagues mention that the Khoisan speakers analysed contained some SNP alleles that are known to increase bone mineral density, others that destroy ingested natural toxic substances faster, or that increase sprint performance, or that decrease the amount of water needed. Given that no statistical analysis is provided for the prevalence of these superpower genes in the Khoisan speakers (actually, given the low sample size, it would have been meaningless), we are left with the impression that the Khoisan speakers need these superpowers to survive in the Kalahari desert. There is a further twist: the five individuals analysed are the elders (and probably the oldest) of their villages; they all have reached the ripe age of 80. Whether these SNPs, or others among the 3.761 million SNPs discovered, are key to longevity, remains to be known.

Superpower genes are also among the highlighted discoveries in the recent work by Lachance et al. (2012); they have sequenced the entire genomes of five additional hunter-gatherer individuals (Pygmies, Hadza, and Sandawe) and identify numerous loci that harbour signatures of local adaptation, including genes involved in immunity, metabolism, olfactory and taste perception, reproduction, and wound healing.

A non-Khoisan South African has also been sequenced: none other than Archbishop Desmond Tutu, also by Schuster et al. (2010). The authors discovered that he (a Xhosa) is closer to the other Bantu speaker sequenced so far, a Nigerian Yoruba (Bentley et al. 2008), than to the Khoisan speakers. However, his mtDNA sequence belongs to L0d, a haplogroup much more frequent in the Khoisan, as discussed already, which would imply maternal gene flow from the Khoisan to the Xhosa; conversely, one of the Khoisan speakers carried an NRY haplogroup (E1b1b1) that is much more frequent in Bantu speakers. The Xhosa have also been studied with sample sizes larger than one (Patterson et al. 2009), and they also found them much closer to the Nigerian Yoruba than to the San. This is an example of a choice population geneticists face: for a given budget, resources can be allocated either to sampling width (more individuals from more populations) or to genotyping depth (more polymorphisms, up to the entire genome). Currently, genotyping costs are falling and the ultimate frontier has been reached: it is possible to genotype the entire genome; but while the financial costs of producing whole genomes is decreasing, the infrastructure and the know-how to deal with massive amounts of data remain difficult to acquire. In the same paper, Patterson et al. (2009) study the Cape Coloured by means of genome-wide SNPs and can estimate the contributions to this mixed population as 23% European, 22% South Asian, 18% Indonesian, and 37% Xhosa. This dissection points to a flagrant violation of the biological concept of population; though, by many definitions (including self-identification) the Cape Coloured are indeed a population, they are not a self-contained mating unit.

Southeastern Africa has a distinct genetic landscape. The populations of Mozambique carry a particular set of mtDNA haplogroups (L0a1b, L0a2*, L2a1a2, L2a1b1, and L3d1a1), which together constitute 50% of the Mozambican mtDNA sequences, while they are found at most 17% elsewhere (in Angola) (Salas et al. 2002). The autosomal SNPs also reveal a South-East African specific component (Sikora et al. 2010). This, as discussed in section 13.7, may have some implications for understanding the Bantu diaspora.

13.5 East Africa: Source or Sink?

East Africa (which, in this section, we will limit to modern-day Kenya and Tanzania) is the most linguistically diverse region in Africa and one of the richest in the world. The four language families spoken in Africa converge there: Khoisan (although the linguistic affiliation of the two click languages spoken in Tanzania, namely Hadza and Sandawe, is hotly debated), the East Sudanic branch of Nilo-Saharan (with languages like Maasai and Samburu), the Cushitic branch of Afro-Asiatic (Iraqw, Burunge), and the Niger-Kordofanian Bantu languages (Sukuma, Gogo). It is understandable, then, that the Bantu (but heavily influenced by Arabic) Swahili is used as a lingua franca. As we shall see, this linguistic diversity is paralleled by a high degree of genetic differentiation among and within populations, which has been interpreted both as a source of genetic diversity for non-Africans, or, more recently, as the result of migrants from elsewhere contributing their genetic (and cultural) heritage to the region.

The oldest anatomically modern human remains, such as Omo and Hertho, have been found in East Africa; this fact, together with the rich fossil record for ancestral hominins, led the proponents of the Out of Africa hypothesis to suggest that East Africa was the birthplace of modern humans (Stringer and Andrews 1988).

In a detailed analysis of mtDNA sequences in Tanzanian populations, Gonder et al. (2007) found that local East African lineages such as L0d, L0f, and L5, are among the oldest in Africa, and retrieved also sequences from branches of the mitochondrial tree, such as M1, N1, and J, which are much more frequent outside of Africa. In the NRY tree, most African chromosomes fall in the E1b1b1 (M35) clade; within that, a particular branch is found mostly in Africa, namely E1b1b1e (M293) (Henn et al. 2008). The geographic distribution of the E1b1b1e haplogroup, in association with the microsatellite diversity estimates for populations, is consistent with an expansion through Tanzania to Southern-Central Africa, although prior to the historical Bantu expansion. This type of reasoning, usual in human population genetics, may lead to the misunderstanding that the population that expanded E1b1b1e carried only this lineage. Even worse, it often seems that haplogroups expanded of their own accord, with no reference to the demographic patterns involved: how many individuals moved, where did they move from and to, why did they move and who they were are considerations that are often left unexplained.

The autosomal analysis by Tishkoff et al. (2009) contained a large number of samples from East Africa that allowed a dissection of the genetic make-up of that region in exquisite detail. Overall, the region in Africa with the smallest correlation between genetic and geographic distance was, precisely, East Africa, which may be a reflection of the variegated origins of populations living in close contact there. A Bayesian analysis revealed a number of components within East Africa. The Khoisan-speaking Hadza and Sandawe each had their specific components; neither was related to the specific components of southern African Khoisan speakers. Afro-Asiatic and Nilo-Saharan speakers shared a component that was also found in Ethiopian Afro-Asiatic speakers, the Cushitic Beja in the Red Sea coast of Sudan, and even further afield, by the Mozabite Berbers in Algeria. This can be interpreted as the genetic correlate of the introduction of Cushitic languages into East Africa, while the spread of Sudanic languages would not seem to be accompanied by a concomitant gene flow. A component that is spread in Africa from the Senegalese Mandinka to the South African Xhosa, and that is associated with Niger-Kordofanian speakers, is also found in East Africa, where it reaches its maximum frequency in the Bantu speakers. Surprisingly, it is also prevalent in the Sudanic Luo, which is further testimony to the facility with which Sudanic languages were adopted. Finally, the Cushitic Mbugu carry a component not found elsewhere.

Samples from the Sudanic Maasai and the Bantu Luhya were included in the Hapmap phase III project (Altshuler et al. 2010), and while the latter were quite close to other African samples such as the Nigerian Yoruba, the former were surprisingly close to European and Middle Eastern populations (Melé et al. 2010; Sikora et al. 2010). However, the much denser population sampling in Tishkoff et al. (2009) provides a higher-resolution analysis in which the Maasai seem connected exclusively to other Eastern and Northern African populations, indicating that non-Bantu populations in East Africa seem to be the descendants of relatively recent immigrants rather than the direct kin of the ancestors of all non-Africans.

All the genetic studies that focus on East Africa recognize external gene flow into this region, mostly from the West and the North. However, when African and other populations are analysed together, a pattern of decreasing genetic diversity from East Africa often emerges, which has often been interpreted as the hallmark of an early expansion out of East Africa (Jakobsson et al. 2008). It is assumed that the geographic path that best fits this decline is more likely to have been the path out of Africa, and Melé et al. (2012) found that this path went through the Bab el Mandeb strait rather than further north along the Nile and across the Sinai.

13.6 The Pygmies: Identity and Phenotype

The two main groups of hunter-gatherers in Africa are some Khoisan speakers in Southern and Eastern Africa, and Pygmies in Central Africa. They are scattered through the equatorial rainforest and two main groups of Pygmies: Eastern (located in the Democratic Republic of Congo and Rwanda) and Western (Cameroon, Republic of Congo, Gabon, and the Central African Republic). However, other groups, called 'pygmoids', might be the result of admixture between Pygmy and neighbouring populations (Cavalli-Sforza 1986). African Pygmies are defined by their geographic, cultural, and phenotypic features. Linguistically, they speak languages borrowed from their neighbours that replaced their ancestral languages (Ruhlen 1994). Their shorter stature represents one extreme of the height spectrum in humans. Although this phenotypic trait has attracted the attention of human biologists during the last century, the adaptive process and the genetic variants related to the phenotype remain elusive. The putative adaptive advantage of the Pygmy phenotype is still under debate (Perry and Dominy 2009) and different hypotheses have been proposed to explain the reproductive advantages of the Pygmy stature: thermoregulatory adaptation to the rainforest (Cavalli-Sforza 1986), adaptation to a food-limited environment (Hart and Hart 1986), improvement of mobility in the dense tropical forest (Diamond 1991), and adaptation for an earlier reproductive age due to shortened lifespan (Migliano et al. 2007).

Genetic studies have provided useful insights into the origin and relationships of Pygmies with surrounding populations. Genetic autosomal markers have shown that Pygmies, as well as Khoisan-speakers, are located near the root of African diversity, suggesting a common origin for hunter-gatherers with ancient divergence (Jakobsson et al. 2008; Li et al. 2008; Tishkoff et al. 2009), although recently it has been suggested that the origin and deepest diversity of our species should be located within Khoisan-speakers (Henn et al. 2011; Schlebusch et al. 2012)), as described earlier.

Analyses of non-recombinant markers (Y-chromosome and mtDNA lineages) have shown differential patterns in Eastern and Western Pygmies. Both Pygmy groups share a high frequency of the Y-chromosome B2b lineage (Berniell-Lee et al. 2009; Wood et al. 2005), which is also found in Khoisan-speaking populations, suggesting a possible common root among African hunter-gatherers or recent contact (Batini et al. 2011b). Substantial differences are found in maternal (mtDNA) lineages: Western Pygmies exhibit lineages within the L1c mtDNA haplogroup (Batini et al. 2007; Batini et al. 2011a; Quintana-Murci et al. 2008), whereas most of Eastern Pygmy lineages fall into subclades of L0a, L2a, and L5 (Batini et al. 2011a), suggesting a possible ancient maternal separation between the two groups of Pygmies (Destro-Bisol et al. 2004). Recently, based on autosomal and uniparental markers, it has been estimated that African Pygmy populations diverged from the ancestor of the non-Pygmy Africans around 60,000 years ago (Kya) (Batini et al. 2011a; Patin et al. 2009; Verdu et al. 2009); while the split between the two Pygmy populations (Western and Eastern) was placed at approximately 25 Kya (Batini et al. 2011a; Patin et al. 2009) with little gene flow since then. Besides the divergence which has occurred since this ancient split, differential migration between neighbouring Bantu populations and Pygmies has taken place: Western Pygmies have experienced substantial gene flow with Bantu speakers, whereas the gene flow with Eastern Pygmies has been reduced. Male lineages that are at high frequency in Bantu speakers, such as E1b1a or B2a, are found in Western Pygmies (Berniell-Lee et al. 2009); and L1c lineages are found in Western Bantu speakers, suggesting an asymmetrical gene flow between Western Pygmy females and Bantu males. Thus, it seems that the genetic and adaptative features of Pygmy groups (Eastern and Western) have a common origin and have not evolved independently as a result of convergent evolution.

Beyond their phenotype, Pygmies, along with the Khoisan speakers, may represent the descendants of the ancestral populations living in the southern half of African before the Bantu expansion. While for the Khoisan speakers their connection to an ancestral population seems clear, the isolation of the Pygmies and their genetic adaptation to the tropical rainforest may have made this link more tenuous and difficult to recognize.

13.7 The Bantu and the March to the South

The Bantu branch of the Niger-Kordofanian family includes some 500 languages spoken by more than 400 million people throughout sub-Saharan Africa. The wide distribution of Bantu languages in the African continent has been linked to one of the largest human cultural movements, the so-called *Bantu expansion*, which has been described as a complex phenomenon involving the transmission of language, technology, and probably genes (Diamond 1997). This process took place around 5,000 years ago when Bantu-speaking people in North-West Cameroon and Southern Nigeria spread southwards, probably related to the spread of agriculture and iron technologies (Newman 1995; Phillipson 1993; Vansina 1995), and isolated, erased, or admixed with hunter-gatherers present in the territory. The expansion followed two major paths, one to the south-east and one to the south-west, avoiding the equatorial rainforest. However, some authors have highlighted the reductionism of this population migration linked to the spread of languages, and have underlined the relevance of local migration processes (Ehret 2001; Lwanga-Lunyiigo 1976; Schoenbrun 2001).

Genetic studies have revealed a high degree of homogeneity among Bantu-speaking populations as a result of their common and recent origin. The frequency and distribution of several mitochondrial lineages in Bantu populations (such as L0a, L2a, L3b, and L3e) and of Y-chromosome haplogroups (such as E1b1a and B2a), have suggested a genetic footprint of the Bantu expansion (Beleza et al. 2005; Berniell-Lee et al. 2009; Cruciani et al. 2002; Pereira et al. 2001; Plaza et al. 2004; Salas et al. 2002; Wood et al. 2005). However, differences in the genetic composition of the different Bantu areas have been found, especially in terms of the degree of assimilation of hunter-gatherers and drift constraints (Beleza et al. 2005; Pereira et al. 2002; Plaza et al. 2004; Salas et al. 2002). Recently, a deeper look at paternal lineages using a large battery of Y-chromosomal markers has shown certain heterogeneity within Bantu speakers (Montano et al. 2011), again highlighting how, either by virtue of their maternal or paternal transmission, or just by chance, mitochondrial DNA and the Y chromosome may provide different views and need to be integrated if one aspires to reconstruct population history.

13.8 North Africa: Between the Sand and the Sea

Despite being part of the African continent, the human population history of North Africa has been different to, although not totally independent of, the sub-Saharan area. Its demographic history has been conditioned in an east-west direction by the barriers imposed by the Sahara Desert to the south and the Mediterranean Sea to the north. North Africans are often grouped with Near Eastern populations because populations in both regions speak primarily Afro-Asiatic languages, like Arabic, and phenotypic attributes, like lighter skin pigmentation, differentiate many North Africans from sub-Saharan African populations. Archaeological records point to an ancient occupation of the territory going even beyond 120,000 years ago, and different cultures arose and vanished (Aterian, Ibero-Maurusian, Capsian) until the arrival of agriculture from the Near East. A complex demographic scenario where autochthonous people have been identified with Berber ancestors, later Arab arrivals from the Near East, influence of European peoples, and sub-Saharan migrations through the desert, has been posited (Bosch et al. 2001).

Classical genetic markers have grouped North Africans with non-sub-Saharan populations (Bosch et al. 1997; Cavalli-Sforza et al. 1994) having some traces of gene flow with sub-Saharan groups. The east-west clinal gradient in the area, and the absence of major genetic differences between Berber- and Arab-speaking groups suggest that the Neolithic and Arabization processes were mainly cultural, with little demographic impact. Uniparental markers have not found differences between Arabs and Berbers: mtDNA shows an autochthonous group of lineages (U6) and a clinal distribution of certain haplogroups with influence of European Middle Eastern and sub-Saharan haplogroups (Fadhlaoui-Zid et al. 2011). The interpretation of Y-chromosome results remains controversial since the presence of specific lineages in North Africa (such as E1b1b1b) has been attributed to an ancient origin (Bosch et al. 2001) or to the Neolithic expansion from the Middle East (Arredi et al. 2004). Genome-wide analyses in North African populations were, until recently, limited to a sample of the isolated Mozabites (Algerian Berbers) that, when compared to a worldwide sample of populations, show a shared background with European and Near Eastern populations, but also traces of sub-Saharan gene flow (Li et al. 2008; Tishkoff et al. 2009). Their characteristics, peculiar as they might be, were extrapolated to North Africa as a whole. This has now changed with the publication of a study (Henn et al. 2012) comprising 125 individuals from seven North African populations (North and South Morocco, the Western Sahara, Algeria, Tunisia, Libya, and Egypt). Two distinct, opposite gradients of ancestry were observed: an east-to-west increase in likely autochthonous North African ancestry and an east-to-west decrease in likely Near Eastern Arabic ancestry. The indigenous North African ancestry may have been more common in Berber populations and appears most closely related to populations outside of Africa, but divergence between Maghrebi peoples and Near Eastern/Europeans likely precedes the Holocene (>12,000 ya). Signatures of sub-Saharan African ancestry vary substantially among populations. These sub-Saharan ancestries appear to be a recent introduction into North African populations, dating to about 1,200 years ago in southern Morocco and about 750 years ago into Egypt, possibly reflecting the patterns of the trans-Saharan slave trade that occurred during this period. Even with a sample that is relatively small and that could be more geographically precise, the genotyping depth has allowed population history to be reconstructed with unprecedented precision.

13.9 Conclusion

We have presented a review of the population genetics of African regions that are closest to our own expertise; many interesting questions have been left out, such as the complexity around Lake Chad, or the interest of West Africa, both intrinsically and as a main source of the African slave diaspora. And many questions remain untapped. For instance, two key points define a time window for which little is known with certainty: the origin of anatomically modern humans, and the Bantu expansion. Did the Bantu expansion erase all traces of past populations south of the equator? Certainly, it did not north of it. Better population sampling, with regard both to which areas are sampled and how populations are defined, is needed to address these and other questions.

Population sampling is a thorny issue in human population genetics. Geographical, linguistic, and cultural aspects are taken into account in various degrees when deciding what is a population unit that can be sampled; often, real or perceived isolation, linguistic particularity, or some historical link are reasons that lead to one population being sampled over others. However, sampling is often opportunistic: one has the right connections that allow a given population to be accessed or cajoled into participation (the ethical implications of sampling non-Western populations is a field in itself which we will not further develop here). Sample sizes are also often opportunistic, reflecting the intersection of how many individuals can be sampled and how many genotypes can be paid for. Such critical arguments are easy enough to make; still, how population sampling impinges on genetic results is a facet of research that should not be forgotten when interpreting results in terms of genetic histories.

Populations, obviously, are sets of human individuals. Mitochondrial and Y-chromosome lineages are not, in that sense, populations. They might be used to tag populations, or as proxies, but this implies, as we have seen for the Bantu, that a population may be reduced to one or a few lineages, leading to an oversimplified picture of the demographic history of the population. And an ancestral population inferred with methods such as STRUCTURE or ADMIXTURE remains a statistical construct with an at most tenuous connection to actual, past populations. Thus, one should be cautious not to conflate lineages and statistical constructs with actual populations made of real humans and their marvellous biological, social, and historical complexity.

We have also reviewed the march of the genotyping technologies, and we have glimpsed the future: whole genome resequencing, in which every single base in a human genome is determined. In the near future, the expectation is that whole genomes will be resequenced in population samples. While this would be a magnificent resource for population genetics, the analytical techniques needed to extract meaning from genetic sequences should not lag behind in their development. Research in population genetics should be driven by competing hypotheses to be tested and resolved, not by the techniques that are available.

REFERENCES

Alexander, D. H., J. Novembre, and K. Lange (2009). 'Fast Model-Based Estimation of Ancestry in Unrelated Individuals', *Genome Research* 19/9: 1655–1664.

Altshuler, D. M. et al. (2010). 'Integrating Common and Rare Genetic Variation in Diverse Human Populations', *Nature* 467/7311: 52–58.

Arredi, B. et al. (2004). 'A Predominantly Neolithic Origin for Y-Chromosomal DNA Variation in North Africa', *American Journal of Human Genetics* 75/2: 338–345.

Atkinson, Q. D. (2011). 'Phonemic Diversity Supports a Serial Founder Effect Model of Language Expansion from Africa', *Science* 332/6027: 346–349.

Batini, C., V. Coia, et al. (2007). 'Phylogeography of the Human Mitochondrial L1c Haplogroup: Genetic Signatures of the Prehistory of Central Africa', *Molecular Phylogenetics and Evolution* 43/2: 635–644.

Batini, C., J. Lopes, et al. (2011a). 'Insights into the Demographic History of African Pygmies from Complete Mitochondrial Genomes', *Molecular Biology and Evolution* 28/2: 1099–1110.

Batini, C., G. Ferri, et al. (2011b). 'Signatures of the Preagricultural Peopling Processes in Sub-Saharan Africa as Revealed by the Phylogeography of Early Y Chromosome Lineages', *Molecular Biology and Evolution* 28/9: 2603–2613.

Baudat, F. et al. (2010). 'PRDM9 Is a Major Determinant of Meiotic Recombination Hotspots in Humans and Mice', *Science* 327/5967: 836–840.

Behar, D. M. et al. (2008). 'The Dawn of Human Matrilineal Diversity', *American Journal of Human Genetics* 82/5: 1130–1140.

Beleza, S. et al. (2005). 'The Genetic Legacy of Western Bantu Migrations', *Human Genetics* 117/4: 366–375.

Bentley, D. R. et al. (2008). 'Accurate Whole Human Genome Sequencing Using Reversible Terminator Chemistry', *Nature* 456/7218: 53–59.

Berg, I. L. et al. (2010). 'PRDM9 Variation Strongly Influences Recombination Hot-Spot Activity and Meiotic Instability in Humans', *Nature Genetics* 42/10: 859–863.

Berniell-Lee, G. et al. (2009). 'Genetic and Demographic Implications of the Bantu Expansion: Insights from Human Paternal Lineages', *Molecular Biology and Evolution* 26/7: 1581–1589.

Bersaglieri, T. et al. (2004). 'Genetic Signatures of Strong Recent Positive Selection at the Lactase Gene', *American Journal of Human Genetics* 74/6: 1111–1120.

Bosch, E., F. Calafell, D. Comas, et al. (2001). 'High-Resolution Analysis of Human Y-Chromosome Variation Shows a Sharp Discontinuity and Limited Gene Flow between

Northwestern Africa and the Iberian Peninsula', *American Journal of Human Genetics* 68/4: 1019–1029.
Bosch, E., F. Calafell, A. Pérez-Lezaun, et al. (1997). 'Population History of North Africa: Evidence from Classical Genetic Markers', *Human Biology* 69/3: 295–311.
Calafell, F. et al. (1998). 'Short Tandem Repeat Polymorphism Evolution in Humans', *European Journal of Human Genetics* 6/1: 38–49.
Cavalli-Sforza, L. L. (1986). *African Pygmies*. Orlando, FL: Academic Press.
Cavalli-Sforza, L. L., P. Menozzi, and A. Piazza (1994). *History and Geography of Human Genes*. Princeton, NJ: Princeton University Press.
Cruciani, F., P. Santolamazza, et al. (2002). 'A Back Migration from Asia to Sub-Saharan Africa Is Supported by High-Resolution Analysis of Human Y-Chromosome Haplotypes', *American Journal of Human Genetics* 70/5: 1197–1214.
Cruciani, F., B. Trombetta, et al. (2011). 'A Revised Root for the Human Y Chromosomal Phylogenetic Tree: The Origin of Patrilineal Diversity in Africa', *American Journal of Human Genetics* 88/6: 814–818.
Destro-Bisol, G. et al. (2004). 'The Analysis of Variation of mtDNA Hypervariable Region 1 Suggests that Eastern and Western Pygmies Diverged before the Bantu Expansion', *American Naturalist* 163/2: 212–226.
Diamond, J. D. (1991). 'Anthropology: Why Are Pygmies Small?' *Nature* 354: 111–112.
Diamond, J. D. (1997), *Guns, Germs, and Steel: The Fates of Human Societies*. New York: W. W. Norton & Company.
Ehret, C. (2001). 'Bantu Expansions: Re-Envisioning a Central Problem of Early African History', *International Journal of African Historical Studies* 34: 5–41.
Enattah, N. S. et al. (2002). 'Identification of a Variant Associated with Adult-Type Hypolactasia', *Nature Genetics* 30/2: 233–237.
Fadhlaoui-Zid, K. et al. (2011). 'Mitochondrial DNA Structure in North Africa Reveals a Genetic Discontinuity in the Nile Valley', *American Journal of Physical Anthropology* 145/1: 107–117.
Gonder, M. K. et al. (2007). 'Whole-mtDNA Genome Sequence Analysis of Ancient African Lineages', *Molecular Biology and Evolution* 24/3: 757–768.
Gravel, S. et al. (2011). 'Demographic History and Rare Allele Sharing among Human Populations', *Proceedings of the National Academy of Sciences of the USA* 108/29: 11983–11988.
Green, R. E. et al. (2010). 'A Draft Sequence of the Neandertal Genome', *Science* 328/5979: 710–722.
Hart, T. B. and J. A. Hart (1986). 'The Ecological Basis of Hunter-Gatherer Subsistence in African Rain Forest: The Mbuti of Eastern Zaire', *Human Ecology* 14: 29–55.
Hawks, J. (2012). 'Longer Time Scale for Human Evolution', *Proceedings of the National Academy of Sciences of the USA* 109/39: 15531–15532.
Henn, B. M., L. R. Botigué, et al. (2012). 'Genomic Ancestry of North Africans Supports Back-to-Africa Migrations', *PLoS Genetics* 8/1: e1002397.
Henn, B. M., C. Gignoux, M. Jobin, et al. (2011). 'Hunter-Gatherer Genomic Diversity Suggests a Southern African Origin for Modern Humans', *Proceedings of the National Academy of Sciences of the USA* 108/13: 5154–5162.
Henn, B. M., C. Gignoux, A. A. Lin, et al. (2008). 'Y-Chromosomal Evidence of a Pastoralist Migration through Tanzania to Southern Africa', *Proceedings of the National Academy of Sciences of the USA* 105/31: 10693–10698.
Jakobsson, M. et al. (2008). 'Genotype, Haplotype and Copy-Number Variation in Worldwide Human Populations', *Nature* 451/7181: 998–1003.

Jobling, M., M. E. Hurles, and C. Tyler-Smith (2004). *Human Evolutionary Genetics: Origins, Peoples and Disease.* New York: Garland Science.

Karafet, T. M. et al. (2008). 'New Binary Polymorphisms Reshape and Increase Resolution of the Human Y Chromosomal Haplogroup Tree', *Genome Research* 18/5: 830–838.

Lachance, J. et al. (2012). 'Evolutionary History and Adaptation from High-Coverage Whole-Genome Sequences of Diverse African Hunter-Gatherers', *Cell* 150/3: 457–469.

Lee, R. B. and I. DeVore (1968). *Man the Hunter.* Chicago, IL: Aldine.

Li, J. Z. et al. (2008). 'Worldwide Human Relationships Inferred from Genome-Wide Patterns of Variation', *Science* 319/5866: 1100–1104.

Lwanga-Lunyiigo, S. (1976). 'The Bantu Problem Reconsidered', *Current Anthropology* 17: 282–286.

Lynch, M. (2010). 'Rate, Molecular Spectrum, and Consequences of Human Mutation', *Proceedings of the National Academy of Sciences of the USA* 107/3: 961–968.

Marean, C. W. et al. (2007). 'Early Human Use of Marine Resources and Pigment in South Africa during the Middle Pleistocene', *Nature* 449/7164: 905–908.

Melé, M., A. Javed, M. Pybus, F. Calafell, et al. (2010). 'A New Method to Reconstruct Recombination Events at a Genomic Scale', *PLoS Computational Biology* 6/11: e1001010.

Melé, M., A. Javed, M. Pybus, P. Zalloua, et al. (2012). 'Recombination Gives a New Insight in the Effective Population Size and the History of the Old World Human Populations', *Molecular Biology and Evolution* 29/1: 25–30.

Migliano, A. B., L. Vinicius, and M. M. Lahr (2007). 'Life History Trade-Offs Explain the Evolution of Human Pygmies', *Proceedings of the National Academy of Sciences of the USA* 104/51: 20216–20219.

Montano, V. et al. (2011). 'The Bantu Expansion Revisited: A New Analysis of Y Chromosome Variation in Central Western Africa', *Molecular Ecology* 20/13: 2693–2708.

Myers, S. et al. (2010). 'Drive against Hotspot Motifs in Primates Implicates the PRDM9 Gene in Meiotic Recombination', *Science* 327/5967: 876–879.

Nachman, M. W. and S. L. Crowell (2000). 'Estimate of the Mutation Rate per Nucleotide in Humans', *Genetics* 156/1: 297–304.

Newman, J. (1995). *The Peopling of Africa: A Geographic Interpretation.* New Haven, CT: Yale University Press.

Patin, E. et al. (2009). 'Inferring the Demographic History of African Farmers and Pygmy Hunter-Gatherers Using a Multilocus Resequencing Data Set', *PLoS Genetics* 5/4: e1000448.

Patterson, N. et al. (2009). 'Genetic Structure of a Unique Admixed Population: Implications for Medical Research', *Human Molecular Genetics* 19/3: 411–419.

Pereira, L., L. Gusmão, et al. (2002). 'Bantu and European Y-Lineages in Sub-Saharan Africa', *Annals of Human Genetics* 66/5–6: 369–378.

Pereira, L., V. Macaulay, et al. (2001). 'Prehistoric and Historic Traces in the mtDNA of Mozambique: Insights into the Bantu Expansions and the Slave Trade', *Annals of Human Genetics* 65/5: 439–458.

Pérez-Lezaun, A. et al. (1999). 'Sex-Specific Migration Patterns in Central Asian Populations, Revealed by Analysis of Y-Chromosome Short Tandem Repeats and mtDNA', *American Journal of Human Genetics* 65/1: 208–219.

Perry, G. H. and N. J. Dominy (2009). 'Evolution of the Human Pygmy Phenotype', *Trends in Ecology and Evolution* 24/4: 218–225.

Perry, G. H. et al. (2007). 'Diet and the Evolution of Human Amylase Gene Copy Number Variation', *Nature Genetics* 39/10: 1256–1260.

Phillipson, D. W. (1993). *African Archaeology*. Cambridge: Cambridge University Press.
Plaza, S. et al. (2004). 'Insights into the Western Bantu Dispersal: mtDNA Lineage Analysis in Angola', *Human Genetics* 115/5: 439–447.
Pritchard, J. K., M. Stephens, and P. Donnelly (2000). 'Inference of Population Structure Using Multilocus Genotype Data', *Genetics* 155/2: 945–959.
Quintana-Murci, L. et al. (2008). 'Maternal Traces of Deep Common Ancestry and Asymmetric Gene Flow between Pygmy Hunter-Gatherers and Bantu-Speaking Farmers', *Proceedings of the National Academy of Sciences of the USA* 105/5: 1596–1601.
Reich, D. et al. (2010). 'Genetic History of an Archaic Hominin Group from Denisova Cave in Siberia', *Nature* 468/7327: 1053–1060.
Roach, J. C. et al. (2010). 'Analysis of Genetic Inheritance in a Family Quartet by Whole-Genome Sequencing', *Science* 328/5978: 636–639.
Ruhlen, M. (1994). *The Origin of Language: Tracing the Evolution of the Mother Tongue*. New York: John Wiley and Sons.
Salas, A. et al. (2002). 'The Making of the African mtDNA Landscape', *American Journal of Human Genetics* 71/5: 1082–1111.
Scheinfeldt, L. B., S. Soi, and S. A. Tishkoff (2010). 'Colloquium Paper: Working toward a Synthesis of Archaeological, Linguistic, and Genetic Data for Inferring African Population History', *Proceedings of the National Academy of Sciences of the USA* 107/2 (supplement): 8931–8938.
Schlebusch, C. M. et al. (2012). 'Genomic Variation in Seven Khoe-San Groups Reveals Adaptation and Complex African History', *Science* 338/6105: 374–379.
Schoenbrun, D. (2001). 'Representing the Bantu Expansions: What's at Stake?', *International Journal of African Historical Studies* 34: 1–4.
Schuster, S. C. et al. (2010). 'Complete Khoisan and Bantu Genomes from Southern Africa', *Nature* 463/7283: 943–947.
Seielstad, M. T., E. Minch, and L. L. Cavalli-Sforza (1998). 'Genetic Evidence for a Higher Female Migration Rate in Humans', 20: 278–280.
Sikora, M. et al. (2010). 'A Genomic Analysis Identifies a Novel Component in the Genetic Structure of Sub-Saharan African Populations', *European Journal of Human Genetics* 107: 786–791.
Stringer, C. B. and P. Andrews (1988). 'Genetic and Fossil Evidence for the Origin of Modern Humans', *Science* 239: 1263–1268.
Tenesa, A. et al. (2007). 'Recent Human Effective Population Size Estimated from Linkage Disequilibrium', *Genome Research* 17/4: 520–526.
Tishkoff, S. A., E. Dietzsch, et al. (1996). 'Global Patterns of Linkage Disequilibrium at the CD4 Locus and Modern Human Origins', *Science* 271/5254: 1380–1387.
Tishkoff, S. A., A. Goldman, et al. (1998). 'A Global Haplotype Analysis of the Myotonic Dystrophy Locus: Implications for the Evolution of Modern Humans and for the Origin of Myotonic Dystrophy Mutations', *American Journal of Human Genetics* 62/6: 1389–1402.
Tishkoff, S. A., F. A. Reed, et al. (2009). 'The Genetic Structure and History of Africans and African Americans', *Science* 324/5930: 1035–1044.
Underhill, P. A. et al. (2000). 'Y Chromosome Sequence Variation and the History of Human Populations', 26: 358–361.
Vansina, J. (1995). 'New Linguistic Evidence and the "Bantu Expansion"', *Journal of African History* 36: 173–195.
Verdu, P. et al. (2009). 'Origins and Genetic Diversity of Pygmy Hunter-Gatherers from Western Central Africa', *Current Biology* 19/4: 312–318.

Verra, F., V. D. Mangano, and D. Modiano (2009). 'Genetics of Susceptibility to Plasmodium falciparum: From Classical Malaria Resistance Genes towards Genome-Wide Association Studies', *Parasite Immunology* 31/5: 234–253.

Wood, E. T. et al. (2005). 'Contrasting Patterns of Y Chromosome and mtDNA Variation in Africa: Evidence for Sex-Biased Demographic Processes', *European Journal of Human Genetics* 13/7: 867–876.

Yi, X. et al. (2010). 'Sequencing of 50 Human Exomes Reveals Adaptation to High Altitude', *Science* 329/5987: 75–78.

CHAPTER 14

SPECIES, POPULATIONS, AND GROUPS IN HOMININ EVOLUTION

SARAH ELTON AND JASON DUNN

14.1 Introduction

Hominins underwent an extensive adaptive radiation in the late Miocene, Pliocene, and Pleistocene. Over twenty species have been identified in the human fossil record (Wood and Lonergan 2008), some of which were contemporaneous and possibly sympatric. Although extremely well studied, most hominins are not particularly abundant in the fossil record, and some are known from only a small number of sites. Thus, much attention has been paid to the diagnosis of and differences between species, and fewer studies have examined variation and populations below species level. Similarly, the pressures that may cause speciation to occur, such as global climatic fluctuation (DeMenocal 2004), have been much better studied than the factors that may have caused intraspecific divergence and hence the formation and dynamics of populations or subspecies and subpopulations or groups. The paucity of the fossil record and the patchy distribution of individuals geographically and over time present numerous challenges when considering hominin populations and subpopulations. Thus, we use a referential modelling approach in this chapter, taking data from modern African primates to investigate the likelihood of hominin subspecies in the Pliocene and Pleistocene, and some of the environmental and behavioural conditions under which population and subpopulation differentiation may be most likely.

African primates, particularly chimpanzees and baboons, are often used as models for hominins (Elton 2006). Baboons and some other Old World monkeys are in many ways likely to be ecologically similar to hominins, being predominantly terrestrial yet exploiting a wide range of habitats, behaviourally flexible and able to respond to changing and changeable environments, and feeding eclectically off a wide range of resources, including C4 (tropical grass) derived foodstuffs (Elton 2006; Ulijaszek et al. 2012). Chimpanzees, on the other hand, are probably more cognitively similar to hominins, more similar in body plan, and likely to share important aspects of social structure, such as male kin bonding (Foley and Lee 1989; Elton 2006). Thus, we use both African monkeys and apes in this chapter to provide context for hominin subspecies and groups. Along with examining modern primate geographic ranges and relationships, if any, with patterns of intraspecific variation, we consider the processes that contribute to creating this variation. This includes using Sewall Wright's concept of the 'fitness landscape' (Wright 1932) to explore how shifts in plastic features such as body weight may contribute to differentiation. Even small behavioural and biological differences can alter an organism's ecological niche, and abiotic and biotic environmental factors also influence niches and the settlement patterns and dispersals of populations and subpopulations in their own right. Thus, we also touch on these to help shed light on what shapes the composition and structure of distinctive hominin subpopulations.

14.2 Subspecific Diversity: A Brief Review

Botany and zoology have very different conventions when considering variation within species, with botanists often recognizing varieties and forms as well as subspecies and species (Mallet 2007); only zoological conventions and examples are considered here. In zoology, a population is commonly defined as a set of organisms in which any pair of members can breed together. Implicit in this is that populations comprise individuals of the same species, living within the same geographic area. In this paper, the concept of 'population' is closely aligned to that of subspecies, albeit with the caveat that it is unlikely that every possible breeding pair actually has the opportunity to breed. Subspecific designations can be an important means of describing diagnosable features of non-reproductively isolated populations, an approach favoured by Mayr and Wright, the originators of the biological species concept (Isaac et al. 2004). Dividing species into subspecies assumes structured variation. Variation is a fundamental of biology, as species obviously do not comprise a set of homogenous individuals. Even if the genotype is identical, variation is expressed in the phenotype through the influence of external agents—climate, nutrition, illness, and so on. Variation can be stochastic—as might commonly arise through developmental insults to an individual—but for it to be seen as meaningful for demarcation of subspecies it must be structured, with members sharing recognizable characteristics. Variation within species is often spatially structured (Thorpe 1987; Winker 2010). This was recognized taxonomically over a century ago by Sundevall who was first to use trinomial terms for geographic varieties (Winker 2010). Polytypy—ordered subdivision within a taxonomic unit—including subspecific differentiation is extensive across the animal world (Thorpe 1987). This may be clinal variation related to a changing environmental variable, such as the size of deer in the Americas related to temperature (Millien et al. 2006). Other variation is categorical: for example, where distinct lineages 'remeet' at a hybrid zone, such as the hooded and carrion crow (Saino 1998). Indeed, true panmixia (the potential for random mating across the whole of the geographic range of a species) is rare and thus homogeneity is almost non-existent, except in species that breed in only one locality (Thorpe 1987).

Just as the concept of 'a species' is open to debate, defining a subspecies is not straightforward (Winker 2010). Quantitative criteria such as the '75% rule' have been proposed (Amadon 1949), which states that for a group to be considered a subspecies it must be 75% different in some trait or traits from other putative subspecies. However, this is not widely employed—often subspecies are defined on the basis of some arbitrary level of diagnosable phenotypic differentiation, especially when it is geographically based (Winker 2010). While objectivity can be difficult, being able to categorize at subspecies level is highly useful, allowing intraspecific variation to be quantified and helping to detect reduction in gene flow between populations. The highly polytypic baboons (*Papio hamadryas* subspecies), for example, have extensive, spatially structured biological and behavioural variation across their geographic range, including hybrid zones between populations (Jolly 1993). So extensive is the variation in *Papio hamadryas* that many researchers view the six commonly recognized subspecies as separate species (see Jolly 1993 for a review, and see also Figure 14.1). Within these major divisions, subforms are also recognized, such as the division of chacma baboons into the grey-footed and southern/Cape forms (Grubb et al. 2003). Among primates as a whole, structured variation and hence subspecific designation is very common (e.g. Grubb et al. 2003; Brandon-Jones et al. 1994), although variants may be viewed as species to promote conservation (Mace 2004).

Subspecific differentiation occurs in ways similar to specific differentiation, with geographic and hence reproductive isolation being highly influential. Within a wide

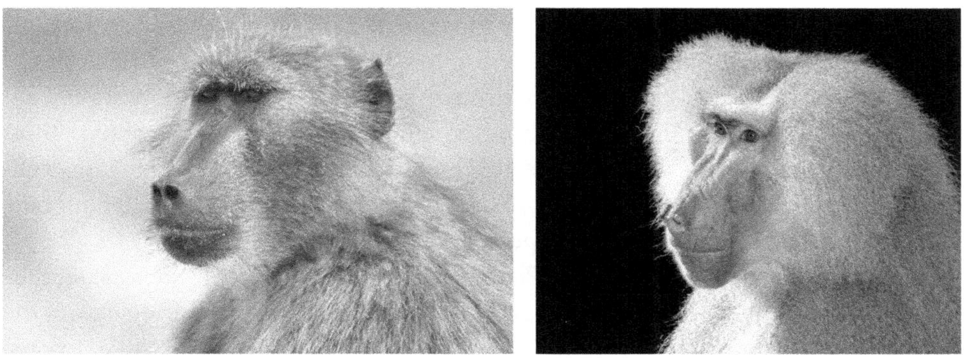

FIG. 14.1 The many faces of *Papio hamadryas*. Left: *P. h. ursinus* from southern Africa (http://commons.wikimedia.org/wiki/File:Papio_ursinus_2.jpg). Right: *P. h. hamadryas* from East Africa (http://commons.wikimedia.org/wiki/File:Papio_hamadryas_(aka).jpg).

geographic distribution, there may be insular populations (cut off by sea or areas of unsuitable habitat) where divergence is pronounced and a subspecific designation is warranted. There are excellent examples of this in primates, including the morphologically distinctive Bale monkey (*Chlorocebus aethiops djamjamensis*, part of the vervet group), which is found within a relatively small area of Ethiopia, isolated from other vervet subspecies, and, unusually for *C. aethiops*, is dependent on highland bamboo forest (Elton et al. 2010). Even if clinal variation is present, the systematic, population-based differences between forms at geographical extremes of the whole species range (in which populations are contiguous) may be sufficient to make subspecific designations. This is again evident in vervet monkeys, where populations in western, eastern, and southern Africa are morphologically distinct even when size differences are taken into account—and hence seem to be 'good' subspecies (Elton et al. 2010). This notwithstanding, there are some classic zoological examples (especially in birds) of geographically unstructured variation (and hence few subspecies) even in species with very large geographic distributions. The European robin (*Erithacus rubecula*) is a case in point. Although some insular forms (for example, on the Canary Islands and Great Britain) are recognized as distinct subspecies, most forms across mainland Europe, from Spain to Siberia, are not given subspecific designation (Dietzen et al. 2003).

There are a number of possible explanations for the relative lack of subspecific diversity in European robins. It has been suggested (Phillimore and Owens 2006) that the modest phylogenetic divergence, which results in fewer good subspecies, in Palearctic (and Nearctic) birds may be because the taxa in these areas are evolutionarily 'young' due to the effects of post-glacial recolonization, and so may not have had time to diverge. In general, taxonomic diversity tends to be higher in the tropics than in temperate regions, which may be linked to the fact that tropical environments are more ancient than temperate ones (and hence have had the opportunity to 'build' diversity), or even because greater biotic interaction (as a result of existing species diversity) may promote taxonomic diversification (see Mittelbach et al. 2007 for an extensive review). Even in African catarrhines, species diversity appears to be greater at the equator than in the subtropics (Eeley and Foley 1999). Patterns of subspecific diversity are also likely to be shaped by movement patterns and hence opportunities for gene flow within a species. In British birds, for example, increased dispersal distance is related to lower subspecific diversity (Belliure et al. 2000). A significant proportion of mainland European robins are migratory (Dietzen et al. 2003). Thus, chances of gene flow across the whole geographic range, creating spatially unstructured patterns of variation, are higher than in species that do not habitually migrate, working against the formation of subspecies.

So what might be expected for Plio-Pleistocene hominins, found in tropical and subtropical Africa? Inhabiting the tropics, the probability of taxonomic diversity—at the subspecific as well as the specific level—is relatively high. Prior to two million years, there is no good palaeontological reason to assume that hominin movement patterns would differ greatly to those seen in modern sub-Saharan African primates, as hominin fossils from this period are not found outside Africa. Extrapolating from modern humans and primates, there is indeed no compelling reason to believe that any hominin practised the type of migration seen commonly in birds, even though the dispersal ability of *Homo* is very well documented, albeit with movement probably occurring over many generations (Antón et al. 2002). On these bases, therefore, tropical mammals, especially primates, seem much more appropriate models when considering subspecific diversity than do temperate birds.

Widespread modern primates show considerable spatially structured variation (Cardini et al. 2010). Given that genetic diversity is also spatially structured in humans (Lawson Handley et al. 2007), with observable phenotypic differences in different regions, it can be assumed broadly that variation in hominins would also follow the patterns seen in modern human and non-human primates, at least some of the time. However, assuming that some subspecific diversity was present in hominin species is a long way from actually being able to identify such diversity securely or even predict the extent to which that diversity occurred.

Subspecies have been recognized in the hominin fossil record: several authorities demarcate Neanderthals and modern humans only at the subspecific level (see review in Harvati et al. 2004), for instance. As might be expected, determining hominin subspecies is fraught with difficulty. Given the nature of the terrestrial fossil record, with a paucity of specimens patchily distributed in time and space, it is inevitable that palaeo-anthropologists view hominins primarily as either species or individuals rather than subspecies and groups. Necessarily, hominin species are described on the basis of very few, often fragmentary, specimens, and their palaeobiology is reconstructed from the characteristics of a small number of individuals and extrapolated to the species. Quantifying intraspecific variation and demonstrating that it is structured rather than stochastic can be challenging even if working with large samples of extant organisms that can be mapped and observed behaviourally as well as studied morphologically and genetically (*sensu* Cardini et al. 2007). In hominins (and indeed, most other extinct terrestrial animals), there are insufficient fossils to robustly quantify intraspecific variation, even in species that appear abundant and are relatively well represented in the fossil record. For example, fossils of *Paranthropus boisei*, a Plio-Pleistocene hominin from East Africa, are numerous but in a study of its dental microwear, only seven out of a possible 53 specimens preserved the features necessary for analysis (Ungar et al. 2008). Theoretically, however, given patterns of variation in modern primates, subspecies may have occurred in hominins with a relatively wide geographic distribution. It is possible, for example, that *Australopithecus bahrelghazali*, dated to around 3–3.5 Ma and found only in Chad in Central Africa, is a regional variant of the more-or-less contemporaneous *Australopithecus afarensis*, found in East Africa (Wood and Lonergan 2008).

Examining populations within hominin species is made even more challenging by the fact that their fossils are not evenly distributed in space or time, with the majority of specimens coming from a small number of highly productive fossiliferous regions such as the East African Rift Valley or the Cradle of Humankind in South Africa. Given that subspecific variation is often spatially structured (Thorpe 1987; Winker 2010), the lack of knowledge about actual geographic ranges (or even lack of hypothesized ranges) makes it difficult to construct a framework for exploring population-level variation in hominins. So why bother about populations and subpopulations in hominins at all? Darwin (1874) noted that species could arise as the result of subspecific diversity and differentiation. One compelling reason for considering hominin populations is thus the potential for them to provide the 'raw material' for speciation. It is a logical extension to assume that populations would be further subdivided, again based on analogy with modern primates, into smaller breeding groups with a distinct home range. This links to another reason why it is useful to think about populations and also subpopulations: behavioural and ecological differences. Even on such a small scale, these are evident between groups, and may help reduce competition and subtly partition niches. Considering populations and subpopulations therefore helps to understand palaeobiology and adaptation.

14.3 Hominin Subspecies: Using Modern Primates as Models

The first aim of this chapter is to model the possible subspecific diversity within African Plio-Pleistocene hominins. We reconstruct the possible geographic ranges of three African Plio-Pleistocene hominin species (*A. afarensis, A. africanus*, and *P. boisei*) using simple GIS distribution models, and also calculate the extent of their ranges (plus that of *P. robustus*) based solely on recorded fossil localities. The links between geographic range and subspecific diversity are explored in modern primates and used alongside the hypothesized species ranges in a simple referential model to assess the likelihood of subspecific diversity in each hominin species. In passing, we also consider potential subspecific differentiation in Pleistocene *Homo neanderthalensis*, although it is not the focus of this chapter.

14.3.1 Methods

Modern primate species ranges were calculated in ArcGIS 9.1 from distribution polygons based on observational data, downloaded from the IUCN Red List (IUCN 2011). The IUCN Red List favours a 'splitting' taxonomy whereby many taxa recognized by most authorities as subspecies are inflated to species, so we manually reassigned species to subspecies where appropriate, based on Grubb et al. (2003), one of the most widely used and accepted taxonomic authorities, and which represents a consensus view. Informed by our research (Dunn et al. 2013), Jolly (1993) was used as the authority for subspecific differentiation in the *Papio* baboons (*Papio hamadryas* subspecies). These subspecies were then grouped together as species and the area of each species distribution calculated from polygons using the ArcGIS *geometry* tools. Spearman's rank correlation in SPSS 19.0 was used to explore the relationship between geographic species range and number of subspecies. Linear regression, also in SPSS 19.0, was used to examine the predictive value of the relationship between log10 species range and number of subspecies. To account for lack of independence of data due to phylogeny, independent contrasts (Felsenstein 1985), using a consensus ultra-metric tree downloaded from 10K Trees (Arnold et al. 2010), were also calculated for these data and used in regression via the R (R Core Team 2013) package 'caper' (Orme et al. 2012). As body mass is a fundamental ecological variable that is both available in hominins and often correlated with animal life history variables that, in turn, may influence speciation rates (Cardillo et al. 2003), we also explored the relationships between species range, subspecies number, and log10 male and female body masses using Spearman's rank correlation, linear regression, and independent contrasts. It has been suggested widely that forest fragmentation promotes speciation (see, for example, Cardini and Elton 2009), so a Mann Whitney U in SPSS 19.0 was used to test whether there was a significant difference in subspecies number between arboreal and terrestrial taxa.

Species ranges for the hominins were first reconstructed on the basis of recorded fossil finds (Table 14.1). For *A. afarensis, A. africanus*, and *P. boisei*, that have a spread of localities, simple polygons based on the findspots at the geographic extremes of the range were constructed in QGIS Application (QGIS Development Team 2013) using the World Equidistant Cylindrical (Sphere) projection. Given the small numbers of localities for each species, this meant practically that most findspots were used to define the edges of the polygon, and all the localities for each species, regardless of their date, were included. For *A. afarensis* a second polygon was drawn, incorporating the one findspot, Bahr-el-Ghazal in Central Africa, of *A. bahrelghazali*. The area of each polygon was calculated using the geometry 'area' function in QGIS. No polygon based on finds was constructed for *P. robustus*, as it is found in only five localities (listed in Table 14.1) situated in a highly restricted geographic area. Instead, its geographic species range was calculated from the published area (<47,000 hectares) for the Cradle of Humankind (Berger 2005). For *H. neanderthalensis*, the range was calculated on the basis of a polygon adapted from the distribution given in Krause et al. (2007).

Fossil (and for later hominins, archaeological) finds are the only secure evidence that a species was present in an area. However, it is possible to employ species distribution models, developed for modern biodiversity assessment and based on known occurrence in conjunction with environmental data, such as temperature, precipitation, vegetation, and elevation, for the area of interest (in this case Africa) to explore potential ranges of extinct organisms (Maguire and Stigall 2009). Distributions were thus modelled for *A. afarensis* (excluding *A. bahrelghazali*), *A. africanus*, and *P. boisei* (but not *P. robustus*) using the BIOCLIM/DOMAIN 'Predict' tool in DIVA-GIS (Hijmans et al. 2012). Considerable amounts of data on Plio-Pleistocene palaeoclimates and environments are available, based on many different lines of evidence (Elton 2008). Nonetheless, past climates and environments were highly variable spatially as well as temporally (Trauth et al. 2009), and there are significant challenges extrapolating from global to regional to local scales of evidence (Elton 2008). We therefore chose to keep the environmental

Table 14.1 Geographic coordinates for Plio-Pleistocene hominin localities

Hominin locality	Region	Species[a]	Longitude (decimal degrees)	Latitude (decimal degrees)	Date (millions of years ago)
Taung	Southern Africa	*A. africanus*	25	–28	2.6–2.8
Makapansgat	Southern Africa	*A. africanus*	29	–24	2.85–2.58
Sterkfontein	Southern Africa	*A. africanus*	28	–26	2.6–2.0
Swartkrans	Southern Africa	*P. robustus*	28	–26	1.8–1.0
Kromdraai	Southern Africa	*P. robustus*	28	–26	2.0–1.5
Coopers	Southern Africa	*P. robustus*	28	–26	1.9–1.4
Drimolen	Southern Africa	*P. robustus*	28	–26	2.0–1.5
Gondolin	Southern Africa	*P. robustus*	28	–26	1.9–1.5
Malapa	Southern Africa	*A. sediba*	28	–26	1.99
Bahr-el-Ghazal	Central Africa	*A. bahrelghazali*	19	16	3.6
Hadar	East Africa	*A. afarensis*	41	11	3.4–3.0
Dikika	East Africa	*A. afarensis*	41	11	3.4
Laetoli	East Africa	*A. afarensis*	31	–4	3.85–3.63
Omo Shungura	East Africa	*A. afarensis*	36	5	3.4
Maka	East Africa	*A. afarensis*	42	11	3.4
Fejej	East Africa	*A. afarensis*	36	5	4.0–4.18
Belohdelie	East Africa	*A. afarensis*	42	11	3.4
Koobi Fora	East Africa	*A. afarensis*	36	4	3.3
Lothagam	East Africa	*A. afarensis*	36	3	3.9–3.0
Konso	East Africa	*P. boisei*	37	5	1.43
Omo Shungura	East Africa	*P. boisei*	36	5	2.6–2.2
Koobi For a	East Africa	*P. boisei*	36	4	1.9–1.52
Chesowanja	East Africa	*P. boisei*	36	1	1.4
Peninj	East Africa	*P. boisei*	36	–2	1.4
Nachukui	East Africa	*P. boisei*	35	3	2.6–1.78
Olduvai	East Africa	*P. boisei*	35	–3	1.8–1.7
Chiwondo	East Africa	*P. boisei*	33	–9	2.3

Note: [a] Target species for this analysis: other hominin species may also have been recovered from the same locality. Longitude and latitude from Holmes et al. (2005); dates from Wood (2011).

data used in the distribution modelling as simple as possible, to offset error caused by over-interpretation/-extrapolation. Hence, we used only elevation and biome (vegetation) information in the *A. afarensis* and *A. africanus* models. Africa has experienced significant topographic changes since the Plio-Pleistocene, including the uplift of the East African Rift (Trauth et al. 2007; Sohl et al. 2009). Thus, the PRISM3 Pliocene topographic dataset (Sohl et al. 2009) was used to approximate elevation across Africa. The biome dataset for Africa, based on palaeobotanical and palaeoclimatic reconstructions, was also taken from PRISM3/BAS (Salzmann et al. 2008). Although the *A. afarensis* sites span one million years (4–3 Ma; Table 14.1) and *A. africanus* sites have been dated to between 2.85 Ma and 2.0 Ma, the paucity of data meant that all fossil localities for each species (as listed in Table 14.1) were included for each species model. Hence, the PRISM3 dataset, developed to represent the period between 3.3 and 3.0 Ma, was used without further amendment for both species, as our exercise is at best an approximation.

The earliest *P. boisei* fossils are dated to 2.6 Ma (Table 14.1), around the time of the onset of glaciation in the Northern Hemisphere (Trauth et al. 2007) which marks the shift from Pliocene to Pleistocene climatic patterns. *P. boisei*, present in East Africa for over a million years (Table 1), therefore experienced profound changes in global climate over its tenure (Wood et al. 1994; Elton 2008). By 1.5 Ma, the topography of the East African Rift stabilized into more or less the formation seen today (Trauth et al. 2007). To reflect the shift in climate and topography from Pliocene to Pleistocene patterns during the time of *P. boisei*, two distribution models were constructed for this species, again pooling all localities regardless of date because of paucity of sites. The first used the PRISM3 datasets as described above to model possible African distribution. To account for the stabilizing of the East African Rift, the second model used modern elevation data drawn from the WORLDCLIM (Hijmans et al. 2005) altitude dataset (resolution 2.5 by 2.5 arc minutes, which is approximately 25 km^2 in area at the equator). To account for climatic shifts between the Pliocene and Pleistocene, modern annual average precipitation and temperature data (which have a close functional relationship with biome) for Africa were also taken from WORLDCLIM (BIO12 and BIO1, 2.5 by 2.5 arc minutes) for use in the second *P. boisei* model. These values are obviously not identical to those that would have prevailed in the past: in the early Pleistocene, African climate was wetter than today and extremely spatially and temporally variable because of changes in solar intensity due to the orbital precession of the Earth (Trauth et al. 2009). Some precipitation estimates for hominin localities do exist (e.g. Hernández Fernández and Vrba 2006), but they have their own limitations and could compound error, especially as the climate data required for distribution models necessarily must extend beyond the occurrence sites themselves. Temperature estimates based on larger-scale continental or global data (that could be extended to the wider area of interest) would necessitate all points being altered by the same amount (for example, raising or lowering the whole dataset by 1°C), which would negate any effect in a single model. It was thus assumed that because of the temporal averaging of localities and the likelihood of localized climatic variability, not only in time but also over small areas, internal model consistency should take precedence over absolute palaeoclimatic accuracy, if such a phenomenon is even possible, and the modern data were used without adjustment.

DIVA-GIS distribution models report probability of occurrence on a sliding scale from 'excellent' to 'not suitable' (Hijmans et al. 2012). To calculate the areas of potential species ranges, polygons were constructed in QGIS quantifying the continuous distribution of 'excellent' and 'very good' probabilities for each species, radiating from the fossil findspots. Some areas were deemed to have 'excellent' probability of occupation but were separated from the continuous distribution and were thus not included in the polygons. Again, the World Equidistant Cylindrical (Sphere) projection was used and the area of each polygon calculated using the geometry 'area' function in QGIS.

14.3.2 Results and Discussion

The modern primate data in Table 14.2 show that there is a positive trend in geographic range versus subspecific diversity: pooling arboreal and terrestrial species, those with larger ranges tend to have greater numbers of subspecies (Spearman's rho, one-tailed significance test: $N = 47$, $r = 0.48$, $p < 0.001$). Extremely similar trends (results not shown) occur when arboreal and terrestrial samples are analysed separately. In the pooled sample, the predictive value of the relationship between subspecies number and log10 range is fairly small but significant (Table 14.3). When independent contrasts for subspecies number and log10 range are used in linear regression, the adjusted r^2 increases (Table 14.3). There was no significant correlation between male body mass (Spearman's rho, two-tailed, $N = 46$) and species range ($r = 0.09$) or number of subspecies ($r = 0.10$). A similar result was found for analysis of female body mass ($N = 45$) and species range ($r = 0.05$) or number of subspecies ($r = 0.11$). Results of the linear regression and independent contrasts linear regression for log10 species range/subspecies number and log10 body masses were also non-significant (Tables 14.3 and 14.4). There was no significant difference between numbers of arboreal and terrestrial subspecies (Mann Whitney U, $P > 0.05$; arboreal $N = 31, \bar{x} = 2 \pm 3$, terrestrial $N = 16, \bar{x}\ 2 \pm 2$).

Although there were significant relationships with and without phylogeny between geographic range and subspecific diversity, there were many deviations from the overall trend. The terrestrial *Chlorocebus preussi* has one of the smallest geographic ranges of sub-Saharan African catarrhines, but includes two subspecies, one of which is found on Bioko Island (Oates 2008). *Mandrillus leucophaeus*, also terrestrial, also has two subspecies in a relatively small range; again, one subspecies is found on Bioko Island (Oates and Butynski 2008). Insularity can accelerate morphological evolution (Nowak et al. 2008), which is a likely cause of the diversity evident in *C. preussi* and *M. leucophaeus*, as well as contributing to divergence in the arboreal *Piliocolobus pennantii*, which has a small range and four subspecies, one of which is found on Bioko (Oates et al. 2008). *Gorilla beringei*, another terrestrial catarrhine with a relatively small species range, is currently divided into two subspecies, the highland and lowland forms (Robbins and Williamson 2008). This geographic difference, combined with the small population size of the highland form, is likely to have promoted subspecific diversity and is a reminder that the exploitation of different types of terrain can promote reproductive isolation

Table 14.2 Calculated geographic ranges and number of subspecies for modern African catarrhine primate species

Species	Geographic range (km²)	Number of subspecies	Primary habitat classification	Male body mass (kg)	Female body mass (kg)
Piliocolobus rufomitratus	165	0	Arboreal	9.67	7.21
Piliocolobus kirkii	1,300	0	Arboreal	5.8	5.46
Piliocolobus gordonorum	4,806	0	Arboreal	No data	No data
Chlorocebus solatus	9,933	0	Terrestrial	6.89	3.92
Chlorocebus preussi	14,172	2	Terrestrial	4.5	4.5
Piliocolobus pennantii	19,937	4	Arboreal	9.0	5.8
Cercopithecus sclateri	29,645	0	Arboreal	4.0	2.5
Mandrillus leucophaeus	36,210	2	Terrestrial	17.5	12.5
Cercopithecus erythrotis	45,808	2	Arboreal	3.6	2.9
Gorilla beringei	53,399	2	Terrestrial	162.5	97.5
Cercopithecus erythrogaster	60,369	2	Arboreal	4.1	2.4
Theropithecus gelada	86,234	2	Terrestrial	19.0	11.7
Chlorocebus lhoesti	176,931	0	Terrestrial	5.97	3.45
Cercopithecus hamlyni	195,235	2	Arboreal	5.49	3.36
Cercocebus torquatus	235,465	0	Terrestrial	9.47	5.5
Mandrillus sphinx	266,234	0	Terrestrial	31.6	12.9
Colobus satanas	269,108	2	Arboreal	10.4	7.42
Colobus polykomos	281,386	0	Arboreal	9.9	8.3
Cercopithecus diana	285,617	2	Arboreal	5.2	3.9
Miopithecus talapoin	320,577	0	Arboreal	2.5	2.0
Procolobus verus	330,697	0	Arboreal	4.7	4.2
Pan paniscus	341,643	0	Terrestrial	45.0	33.2
Cercocebus atys	360,297	2	Terrestrial	11.0	6.2
Piliocolobus badius	360,568	3	Arboreal	8.36	8.21
Colobus vellerosus	392,658	0	Arboreal	8.5	6.9
Miopithecus ogouensis[a]	398,813	0	Arboreal	1.38	1.12
Allenopithecus nigroviridis	414,829	0	Arboreal	6.13	3.18
Cercopithecus petaurista	437,286	2	Arboreal	4.4	2.9
Cercopithecus mona	508,489	0	Arboreal	5.1	No data
Cercopithecus campbelli	508,774	2	Arboreal	4.5	2.7
Gorilla gorilla	570,056	2	Terrestrial	170.4	71.5
Lophocebus aterrimus	590,644	2	Arboreal	7.84	5.76
Cercopithecus cephus	638,393	3	Arboreal	4.29	2.88

(*continued*)

Table 14.2 Continued

Species	Geographic range (km^2)	Number of subspecies	Primary habitat classification	Male body mass (kg)	Female body mass (kg)
Cercocebus galeritus	1,052,221	4	Terrestrial	9.61	5.26
Piliocolobus (unnamed species from central assemblage)	1,093,445	8	Arboreal	9.72[b]	7.13[b]
Cercopithecus nictitans	1,097,211	2	Arboreal	6.67	4.26
Lophocebus albigena	1,197,865	0	Arboreal	8.25	6.02
Cercopithecus pogonias	1,728,454	7	Arboreal	4.26	2.90
Cercopithecus neglectus	1,877,687	0	Arboreal	7.35	4.13
Colobus angolensis	1,888,514	7	Arboreal	9.68	7.57
Cercopithecus mitis	1,925,484	18	Arboreal	7.70[c]	4.36[c]
Pan troglodytes	2,045,671	4	Terrestrial	49.57[d]	40.37[d]
Cercopithecus ascanius	2,149,685	5	Arboreal	3.7	2.92
Colobus guereza	2,507,400	8	Arboreal	11.70[e]	8.55[e]
Chlorocebus patas	5,920,182	0	Terrestrial	12.4	6.5
Chlorocebus aethiops	11,931,093	6	Terrestrial	4.89[f]	3.54[f]
Papio hamadryas	13,818,390	6	Terrestrial	21.86[g]	11.91[g]

Notes: [a] Sequences unavailable in GenBank and hence species removed from independent contrasts analysis. All body mass data taken from Smith and Jungers (1997) except for *Piliocolous pennantii* from Groves (2007) based on the Bioko Island form.
[b] Mean of *P. sp. parmentieri, P. sp. langi, P. sp. oustaleti*
[c] Mean of *C. mitis, C. m. stuhlmanni, C. m. erythrarchus*
[d] Mean of *P. t. troglodytes, P. t. verus, P. t. schweinfurthii*
[e] Mean of *C. g. matschiei* and *C. g. guereza*
[f] Mean of *C. aethiops* populations from Kenya and Botswana
[h] Mean of all subspecies except *P. h. kindae* for females and all except *P. h. kindae* and *P. h. papio* for males.
For records marked as 'no data', no precise body mass data for indigenous African populations were available.

and hence diversification (Winder et al. 2013). *Theropithecus gelada*, found only in the Ethiopian and Eritrean highlands, has two subspecies, *T. g. obscurus* (distributed east of the Rift Valley) and *T. g. gelada* (west of the Rift) (Groves 2001). The Rift may be a natural geographic boundary, inducing vicariance.

Towards the upper end of geographic range size, *Cercopithecus mitis* has the highest number of subspecies (eighteen). *C. mitis* is an arboreal primate, found in diverse habitats (Lawes 1990), which may have encouraged divergence. The *C. mitis* range is also discontinuous, which probably reflects climate change and forest history since its evolution (Lawes 1990) around 2.2 Ma (Tosi et al. 2005). This forest fragmentation was probably a key factor in monkey subspecific and specific differentiation in the Pleistocene,

Table 14.3 Results of linear regressions of number of subspecies (dependent variable, y) on species range and body mass

Model	Predictor variable (x)	N	Adjusted r^2	F	P
Linear regression	Log10 range	47	0.155	9.466	<0.001
Linear regression	Log10 male body mass	46	-0.021	0.063	Ns
Linear regression	Log10 female body mass	45	-0.023	0.028	Ns
IC linear regression	Log10 range	45	0.272	17.46	<0.0001
IC linear regression	Log10 male body mass	44	0.009	1.371	Ns
IC linear regression	Log10 female body mass	43	0.011	1.467	Ns

Note: IC = independent contrasts. Ns = non significant ($p > 0.05$).

Table 14.4 Results of linear regressions of log10 species range (dependent variable, y) on body mass

Model	Predictor variable (x)	N	Adjusted r^2	F	P
Linear regression	Log10 male body mass	46	-0.020	0.055	Ns
Linear regression	Log10 female body mass	45	-0.023	0.005	Ns
IC linear regression	Log10 male body mass	44	-0.004	0.850	Ns
IC linear regression	Log10 female body mass	43	-0.022	0.082	Ns

Note: IC = independent contrasts. Ns = non significant ($p > 0.05$).

creating refugial islands and restricting gene flow (Cardini and Elton 2009). Although neither arboreal nor terrestrial primates showed greater subspecific diversity across the sample as a whole, fragmentation may help to explain why some arboreal primates in Africa (such as *P. pennentii*, *Colobus guereza*, *C. angolensis*, *Cercopithecus pogonius*, and the central assemblage of *Piliocolobus*, as well as *C. mitis*), have relatively large numbers of subspecies (Cardini et al. 2010). However, not all arboreal primates have high levels of subspecific diversity. In contrast to its sister taxa *Piliocolobus* (the red colobus monkeys), *Procolobus verus* (the olive colobus) has been described as monotypic (Nowak et al. 2008) and has no diagnosed subspecies (Grubb et al. 2003). Even more strikingly, *Cercopithecus neglectus* has a very large range but no diagnosed subspecies (Grubb et al. 2003). This lack of subspecific differentiation may be because their range is continuous across Central Africa (Aghokeng et al. 2010). However, the observation that they host geographically distinct strains of simian lentivirus (Aghokeng et al. 2010) indicates the potential for population divergence. Detailed geometric morphometric analyses of *C. neglectus* skulls across their range, shown in studies of other African primates to quantify very small but evolutionarily meaningful variation (Cardini and Elton 2009), may help to identify incipient or realized subspecific divergence.

Hypothesized geographic species range areas for the hominins are given in Table 14.5, with polygons of the distribution models shown in Figure 14.2. Although African Plio-Pleistocene hominins are the focus of this chapter, we briefly consider subspeciation in Neaderthals as an example of a temperate latitude hominin. The Eurasian robin, as discussed above, is not an ideal comparator for any primate (least of all one belonging to the genus *Homo*), but it does provide a classic example of a widespread temperate species with very few subspecies. This prompts the question about whether subspecies should be expected in Neanderthals. Ancient DNA supports the division of Neanderthals into three distinct geographic groups—in Western Europe, Western Asia, and in the south of their range, between which there was nonetheless some gene flow (Fabre et al. 2009). It is not clear whether these genetic differences map onto subspecies, especially as most morphological work on Neanderthals stresses their similarities rather than differences (Hawks 2012). The estimated Neanderthal geographic range lies towards the upper end of species range sizes for African catarrhines. Eeley and Foley (1999) noted a strong skew towards smaller ranges in African catarrhines,

Table 14.5 Areas of hypothesized ranges for fossil hominin species

Hominin species	Distance between two furthest fossil sites (km)	Area of hypothesized species range (km^2)	First and last appearance dates (Ma)
A. afarensis	1847	A: 270,558 B: 8,677,224 C: 2,542,155	3.9–2.9
A. africanus	668	A: 15,737 B: 1,614,269	2–3 Ma
P. boisei	1618	A: 181,155 B: 3,859,234 D: 7,869,705	2.4–1.2 Ma
P. robustus	~25	E: 470	2–1 Ma
H. neanderthalensis	N/A	F: 8,013,426	0.2–0.028 Ma

Note: The number of kilometers between the two most distant points was calculated on the basis of longitude and latitude coordinates (Table 14.1) using the US government National Hurricane Center longitude/latitude distance calculator (http://www.nhc.noaa.gov/gccalc.shtml). For hypothesized species ranges: A = polygon area based on sites; B = polygon area based on DIVA-GIS model using PRISM3 data; C = polygon based on sites including *A. bahrelghazali*; D = polygon area based on DIVA-GIS model using WORLDCLIM data; E = based on published hectare value (Berger 2005); F = polygon area constructed from the distribution given in Krause et al. (2007).

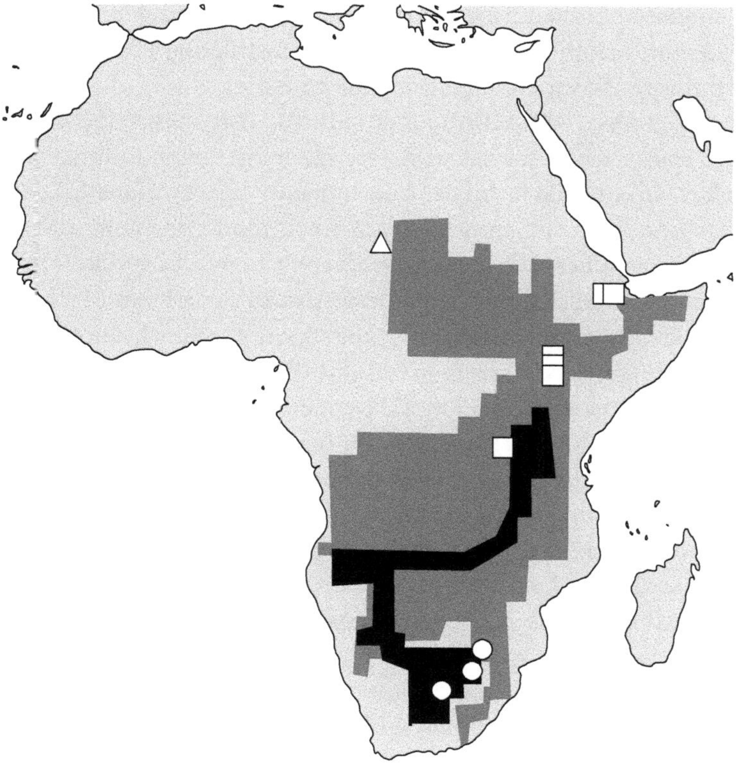

FIG. 14.2 Fossil occurrences and hypothesized geographic ranges for *Australopithecus* species. *A. bahrelghazali* site represented by white triangle, *A. afarensis* by white squares, *A. africanus* by white circles. The hypothesized range for *A. afarensis* is shown in dark grey shading and hypothesized range of *A. africanus* shown in black.

replicated in our data, and also observed that species ranges were smaller nearer the equator. Given the temperate, Eurasian distribution of Neanderthals, therefore, a more appropriate comparator might be the most geographically widespread macaque, *Macaca mulatta*, found in tropical, subtropical, and temperate Asia at more northerly latitudes than any sub-Saharan African primate. *M. mulatta* has a species distribution (extracted from IUCN data as described earlier) of 6,647,129km^2, somewhat less than that estimated for Neanderthals. Like Neanderthals, distinct molecular clades (in west and east) have been reported (Brandon-Jones et al. 2004). However, a highly comprehensive and detailed analysis of *M. mulatta* morphology suggests that the species is monotypic, with no 'good' subspecies (Fooden 2000), even though multiple subspecies have been identified in the past (Brandon-Jones et al. 2004). Subspecies can be defined on the basis of molecular evidence but the available data for *M. mulatta* do not appear to identify clades consistently (Brandon-Jones et al. 2004). Hence, for the time being, influential authorities do not attribute any subspecies to *M. mulatta* (Brandon-Jones et al. 2004), providing important context for interpreting Neanderthal intraspecific diversity. In particular, inferences about hominin subspecies made on the basis of molecular data alone should be treated with caution, and further research on subspecific divergence in temperate mammals would be useful in constructing a framework in which to investigate intraspecific variation in temperate hominins.

Reference to modern African primates indicates that subspecific diversity is possible in the tropics even in species with very small distributions, influenced by modern and past environmental factors and evolutionary history. Nonetheless, at 470 km^2 the tiny *P. robustus* range in Southern Africa seems more consistent with a substantial human home range rather than the total geographic range of a species. Only one of the geographic ranges calculated for the modern African catarrhines (Table 14.2), that of *Piliocolobus rufomitratus*, is smaller, as is the range of the poorly known and relatively newly discovered *Rungwecebus kipunjii* (insufficiently studied to include in the analysis here), currently estimated at 20.6 km^2 (Davenport and Jones 2008). The range of the insular Zanzibar red colobus (*P. kirkii*) is over twice the size of the apparent *P. robustus* range and the highly endangered *P. gordonorum* has a range nearly eight times larger. Given that the *P. robustus* geographic range is at the very bottom of the distribution of modern African catarrhine primates, it is very possible that its actual distribution was much larger than the fossil record suggests. This is supported by the observation that *P. robustus* was likely to have been a highly flexible feeder (Lee-Thorp et al. 2003), a generalist rather than a specialist, which would imply a much larger geographic distribution than is evident from the fossil record. Unfortunately, the clumping of *P. robustus* fossil localities makes it impossible to employ GIS prediction models.

The question of whether *P. robustus* would have diverged into subspecies is also vexed. There has been much debate (reviewed in Cofran and Thackeray 2010) over whether its fossil record is too variable to be contained in a single species, with some assigning the Swartkrans specimens to *Australopithecus crassidens*. The variability present in the *P. robustus* fossil record may be attributed to change over time, with different forms representing chrono-species or chrono-subspecies, but given the imprecise dating of cave sites this is very difficult to test. It is difficult to see that subspecific diversity could be maintained contemporaneously if the geographic range of *P. robustus* was as small as the fossil locality distribution suggests, as subspecies would be living sympatrically. Sympatric subspecies are found in plants and invertebrates, and may coexist because of reproductive isolation (caused, for example, by different flowering or breeding seasons; Arista et al. 1997; Binks et al. 2012) or differential resource/habitat exploitation (Forbes et al. 2009). Sympatric subspeciation is highly uncommon in mammals. It is possible—but by no means probable—that significant morphological divergence, possibly resulting in the formation of a second subspecies, could have occurred in sympatric groups of *P. robustus* because of differential resource exploitation and hence morphological character displacement. However, dietary variation is reasonably well documented in *P. robustus* (Lee-Thorp et al. 2003) and although seasonal variations within individuals are evident (Sponheimer et al. 2006), there is nothing to suggest systematic differences between different groups that may cause divergence. This notwithstanding, even with such a small range, *P. robustus* would have been divided into subpopulations (such as mating groups) which may have adopted localized behaviours and traditions (see section 14.4).

The other southern African hominin considered here, *Australopithecus africanus*, has a geographic range based on fossil localities that also falls around the bottom of the African catarrhine distribution. When interpreting the likelihood of subspecific diversity in hominins, there is the added complexity that the localities are drawn from the whole known tenure in the fossil record and may be widely separated in time, with the consequent possibility that the range may have shifted over the lifetime of the species. Thus, even a range based on known localities may be artificially large for the species at a particular point in time. Predicated on the notion that the distribution of fossil sites accurately reflects the range of the species, which is obviously far from certain, subspecific diversity would be unlikely in *A. africanus*, even though it has significant cranial (Ahern 1998) and postcranial (Harmon 2009) variation. All the modern catarrhines with similar ranges either have no subspecies or subspecies due to insularity (specifically, on Bioko). Although *P. pennantii* has a range only slightly larger than the estimated range based on localities for *A. africanus*, its subspeciosity is probably the product not only of 'the Bioko effect' but also because of intense fragmentation of its forest habitat (Cardini and Elton 2009). *A. africanus*, found in central southern Africa and inhabiting woodland and grassland as well as forested areas (Elton 2008), was unlikely to have experienced vicariant subspeciation due to isolation on an oceanic or forest island. It must be considered, however, that because of the uneven nature of the terrestrial fossil record and consequent sampling bias, the known range captures only a fraction of the actual range. Hence, a second hypothesized range, based on GIS distribution modelling, predicts a much larger geographic spread for *A. africanus*, towards the upper end for modern catarrhines. Given that there was a statistically significant positive relationship between geographic range size and subspecific diversity, structured variation (translating to subspecies) would be much more likely in *A. africanus* with this range. For example, *Pan troglodytes*, which has a slightly larger geographic distribution, has four subspecies. Nonetheless, another appropriate comparator for *A. africanus*, the terrestrial guenon *C. patas*, has a very large range but no valid subspecies recognized by Grubb et al. (2003). Other authorities (e.g. Kingdon 1997) do recognize subspecies—for example, subdividing eastern and western forms—but further research is required before subspecific differentiation can be securely identified (Grubb et al. 2003). This indicates that although diversity can be present in a widespread primate, it does not automatically or easily equate to the structured variation necessary to allocate groups to subspecies. Thus, it cannot be assumed automatically that given a large range *A. africanus* would subspeciate, although the general trend within modern catarrhines points to increased probability of subspecific differentiation as geographic range size gets larger.

The estimated ranges based on fossil localities for *A. afarensis sensu stricto* (i.e. without *A. bahrelghazali*) and *P. boisei* could incorporate subspecific diversity, even though they both fall in the lower half of the modern catarrhine distribution. *T. gelada* and *G. berengei* provide examples of terrestrial primates with smaller ranges that each have two subspecies, although *C. patas, Mandrillus sphinx, Cercocebus torquatus*, and other terrestrial primates with larger ranges have no recognized subspecies. In Plio-Pleistocene East African hominins, the presence of the Rift and geologically active environments may have facilitated structured population differentiation due to the presence of natural barriers to gene flow. It has been argued that *P. boisei* is polytypic (Suwa et al. 1997), but interestingly its intraspecific variation does not appear particularly high compared to that seen in modern humans, apes, and colobus monkeys (Wood and Lieberman 2001) and aspects of its morphology show stasis over time (Wood et al. 1994) and by implication from locality to locality. This relative homogeneity is mirrored in diet, at least on the basis of stable isotope analysis (Cerling et al. 2011). According to studies of dental microwear, the diet of *A. afarensis* was also not very variable, either over time or across different environments (Grine et al. 2006). However, *A. afarensis* does show significant cranial variation over time and also between the two sites where it is best sampled, Hadar and Laetoli (Lockwood et al. 2000). These morphological differences could be evidence for structured differentiation at the subspecific level, either temporally (chrono-subspecies), or spatially, or both. Indeed, the variation within *A. afarensis* is substantial, even though care must be taken when determining whether the variation is caused by sexual dimorphism, geography, or time (Kimbel and Delezene 2009). Adding the locality of Bahr-el-Ghazal to the *A. afarensis* range estimation not unexpectedly increased it considerably, bringing it to the higher end of the modern catarrhine range, similar to *P. troglodytes* but not yet approaching the very large distributions of *C. aethiops* and *P. hamadryas*. *A. bahrelghazali* is known only from a small amount of mandibular material with associated dentition (Brunet et al. 1995) and relatively little has been published on it. The material that is known appears to fit within *A. afarensis* (Kimbel and Delezene 2009) and these simple comparisons of geographic range indicate the feasibility of treating *A. bahrelghazali* as a subspecies of *A. afarensis*.

The GIS model for *A. afarensis* (Figure 14.2), which used only *A. afarensis s.s.*, predicts an even wider species range than is seen based on localities, approaching that of the extremely widespread *P. hamadryas* and *C. aethiops*. Unlike in these modern taxa, however, the *A. afarensis* predicted range does not extend into West Africa. Associated with woodland and bushland habitats, it is uncertain whether *A. afarensis* exploited tropical forest, even though earlier hominins probably did (Elton 2008). Thus, rainforest regions may have been a barrier to dispersal across Africa and pockets of rainforest may have acted to isolate populations and hence promoted diversity. In the context of large, mainly terrestrial and eurytopic primate ranges—descriptors that can each be applied to *A. afarensis* and indeed most Plio-Pleistocene hominins—all the species ranges predicted here using GIS-based modelling seem appropriate, especially given that many modern primate distributions have been negatively affected by anthropogenic factors. Again, *C. patas* serves to provide a cautionary note when extrapolating from size of range to the probability of structured diversity, but it stands out as an exception compared to other primates with substantial ranges that all contain subspecies. Thus, given the models and comparison with modern primates, the spatial patterning of its sites (including Bahr-el-Ghazal), and the considerable variation within the species, subspecific diversity in *A. afarensis* seems highly likely.

Incorporating *A. bahrelghazali* into *A. afarensis* substantially widens an east-west distribution in Africa that was already larger than that seen in other Plio-Pleistocene hominins, including *P. boisei* (Table 14.1). Although its specimens are found as far apart as Malawi and Ethiopia (a distance of over 1600 km), *P. boisei* localities occupy a very restricted longitudinal range (37–33°E). However, its GIS models (Figure 14.3), using both Pliocene and modern data, predict an extensive westward distribution. Much spatially structured variation in widespread modern African primates is longitudinal rather than latitudinal (vervets: Cardini et al. 2007; red colobus: Cardini and Elton 2009; blue monkeys: Cardini et al. 2010). In baboons, significant west–east variation also exists, and the north–south variation that is observed occurs over a large area that covers most of sub-Saharan Africa (Dunn et al. 2013), considerably more than that occupied by *P. boisei*. Morphological differentiation in widespread African primates is driven in part by allometry caused by environmentally induced body size divergence

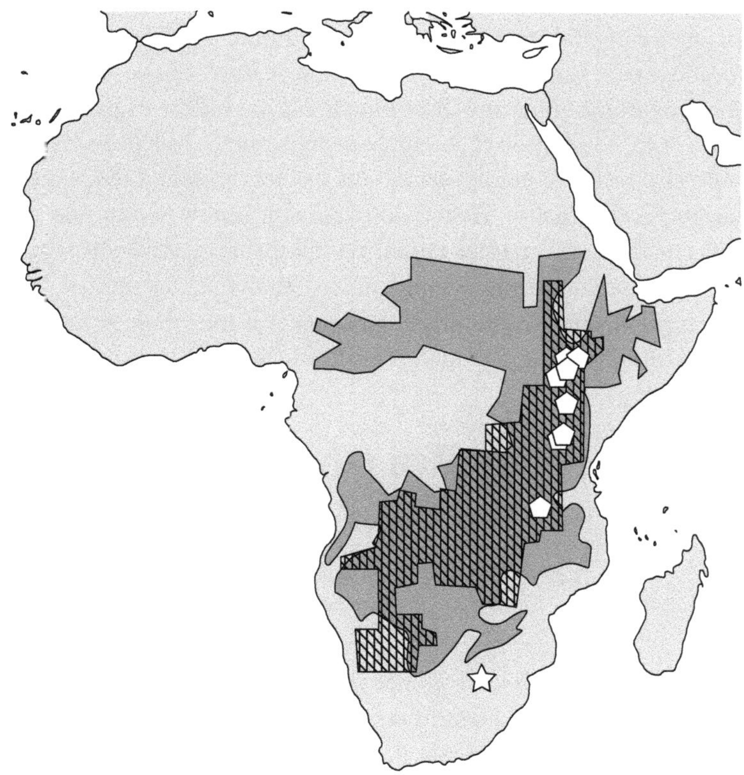

FIG. 14.3 Fossil occurrences and hypothesized geographic range for *Paranthropus boisei*. *P. boisei* sites represented by white pentagons, *P. robustus* by white star. Hypothesized ranges for *P. boisei* based on WORLDCLIM and PRISM3 data shown in dark grey shading and black hatching respectively.

(Cardini et al. 2010). Although *P. boisei* occupied heterogeneous environments, with habitat shifts in time as well as space (Elton 2008), the known north–south fossil distribution, at least on present evidence, shows little spatial patterning. Environmental gradients significant enough to promote subspecific divergence might only be experienced on greater longitudinal or latitudinal scales. If east–west distributions in African Plio-Pleistocene hominins, including *P. boisei*, were as wide as predicted here, they may well have promoted structured geographic variation, but as fossils are sampled only from a more restricted area, this may be difficult to detect with current specimens.

To sum up this section, using modern African catarrhines as referents in combination with species range estimates for Plio-Pleistocene hominins indicates that subspecific diversity was probable in the East African species *A. afarensis* and *P. boisei* and possible for the southern African *A. africanus*, with fossil data for *P. robustus* too clumped to make any inference. Better-resolution palaeo-environmental data would help to improve our understanding of potential hominin occupation areas and hence geographic ranges. Distributions are also influenced by abiotic and biotic environmental factors, including biome, food availability, competitors, and geographic barriers such as mountain ranges and large rivers, which may also 'feed' genetic and morphological divergence. Although it may be possible to include some of these features in a model, even apparently 'fixed' features such as mountain ranges and rivers can appear, disappear, or move over the course of three million years. A classic example is the River Nile, a major geographic entity in Africa which did not follow its present course until the late Pleistocene (Goudie 2005). Accurate geomorphological and ecological data for the past would also improve predictions about species range and the features that may promote subspecific divergence. Hence, one challenge is now to improve upon and test the predictions made in this chapter. Quantifying subspecific diversity would also benefit from the development of more sophisticated analytical techniques to allow variance due to temporal, spatial, and other influences on morphology to be partitioned meaningfully within the relatively small Plio-Pleistocene hominin sample available.

14.4 The Role of Hominin Subpopulations and Groups in the Adaptive Landscape

In the final part of this chapter, we consider briefly the role of hominin subpopulations and groups in providing the 'raw material' for structured intraspecific divergence, which may in turn lead to full speciation. Wright (1932) envisaged species as occupying positions on an imagined and topographically varied evolutionary 'landscape'. In his conceptual model, elevated locations equate to adaptive states and depressed areas maladaptive ones. Natural selection is the driving force that pushes a species 'up the hill' in the model, or adapts it to its surroundings in reality. Different species inhabit different adaptive landscapes and these vary with environment. The varied topography of the landscape results in discrete rather than continuous variation, producing the discrete species observed in nature, each notionally hovering over some adaptive peak. The evolutionary importance of subspecific variation can be viewed within this context. Wright (1932) recognized that species were not homogenous and thus considered a species to be a cloud rather than a point on an adaptive landscape. This has profound implications. A species spread widely over an adaptive landscape is likely to constitute subspecies occupying different but adjacent peaks. As such, the species is less likely to go extinct when certain local conditions change either by stochastic effects such as drought, famine, or disease, or more long-term environmental change. Additionally, a wide distribution over this theoretical landscape increases the catchment area for the adaptive peaks essential to evolutionary survival if the environment should change.

The cohesiveness of a species is maintained by reproduction (Wright 1932). A reduction in gene flux facilitates the actions of differential selection and genetic drift, enabling divergence. Using Wright's model, a breakdown in reproduction leads to constriction in the species 'cloud' on the adaptive landscape. Each resulting discrete entity may occupy a separate adaptive peak, which over time may lead to further constriction and eventual splitting of the groups. For this reason some authors have considered subspecific differentiation as representing incipient speciation. However, low levels of gene flow can impede speciation, which requires divergent selection on several traits (Rice and Hostert 1993). As such, the notion that subspecies are incipient species can be challenged. Jolly (1993) argued that subspecies may be stable and long-standing features of a species rather than on the verge of budding off. While evolutionary trajectories may begin to diverge, secondary contact between recently separated taxa ('reticulation') may result in hybrids that are adaptively suited to the environment and change the adaptive trajectory of a lineage (Hey et al. 2003). Nevertheless, the forces that sculpt subspecies, whatever their destiny, are the same as the forces that sculpt species, and heterogeneity within a species thus presents an important insight into natural selection (Winker 2010).

Several hominin species survived in the fossil record for a million years or more, around five times as long as our own species, *Homo sapiens sensu stricto*, has existed. These species, including *A. afarensis* and *P. boisei*, experienced profound climatic change that required responses, which may have been ecological, behavioural, genetic, morphological, and possibly even cultural, and may have led to biological adaptation. Although not truly formalized in Linnaean taxonomy, species may have subspecies or populations that in turn may have subordinate groups or subpopulations (Figure 14.4). In a species that has a reasonably large range, responses to environmental change may differ from population to population and group to group, partly because environments tend to differ over large areas. In the previous section, we argued that *P. boisei* and *A. afarensis* were both likely to have had extensive geographic ranges. Based on modern African primates, we also put forward the hypothesis that they would have been divided into subspecies. But how might those subspecies have formed, and how might individuals—on which natural selection generally acts—have been grouped and behaved within those subspecies?

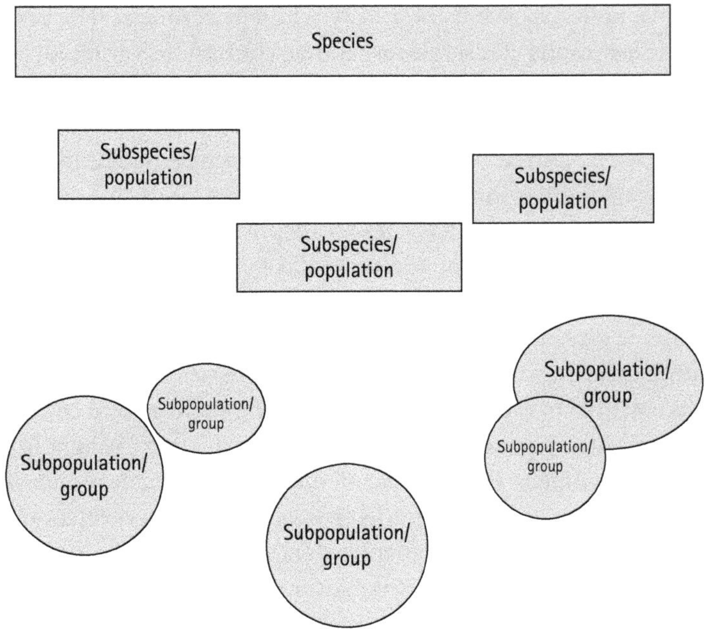

FIG. 14.4 Schematic of species subdivisions.

Primates, like humans, are social and live in groups that form subsets of populations, below the subspecific level. These are not defined on the basis of structured morphological or genetic variation but are instead identified usually through observation of association, competition, cooperation, and breeding patterns. As is the case in humans, these groups often contain related individuals: chimpanzees are male-kin bonded for example, whereas most Old World monkeys are female-kin bonded. We can reasonably assume that extinct hominins also lived in groups, but these are considered only rarely. Hominins were probably male-kin bonded, a parsiminous assumption based on the mating systems of African apes and most humans (Foley and Lee 1989). Recently, independent palaeobiological data have been collected that support this for at least one Plio-Pleistocene hominin species. Strontium isotope analysis of the teeth of *P. robustus* indicates that larger individuals, inferred to be male, came from a much smaller geographic area than their smaller, probably female, conspecifics (Copeland et al. 2011). This suggests male philopatry and female dispersal, as is seen in chimpanzees (Copeland et al. 2011). Further, it is possible that some hominins lived in multi-female groups monopolized by a single male, suggested by evidence for extended male growth (continuing into adulthood) in the skulls of *P. robustus* (Lockwood et al. 2007). Although this gorilla-like pattern is somewhat at odds with the isotope data and it is highly likely that the social system of *P. robustus* does not have a direct analogue in the modern world (Copeland et al. 2011), these lines of evidence shed important light on the structure of hominin groups and possible population genetics in at least one species.

Individual movement between subgroups is important for gene flow, which in turn maintains both genetic heterozygosity in a given mating group and a level of genetic homozygosity over a given area. Dispersing only relatively small distances to breed, as occurs in catarrhines, can help to structure subspecific variation, especially if natural barriers prevent extensive gene flow between distinct populations. Depending on circumstances, however, localized individual transfer out of a group may not happen routinely or even at all. Female chimpanzees may take up to two years to move fully into a new group, dividing their time between that and the group in which they were born (Goodall 1985). In some chimpanzee subpopulations, such as Gombe, female transfer rates are lower than in nearby groups such as Mahale, possibly because of isolation, and behavioural measures are taken to avoid inbreeding (Goodall 1986). This indicates how groups within a subspecies may become isolated and yet avoid the deleterious consequences of inbreeding. It further suggests how different groups have the opportunity to exploit their own adaptive peak and diversify even over very small spatial scales. A good example of this diversification—albeit cultural—is the different discrete traditions, maintained for generations, observed in nearby chimpanzee groups. For example, in Tanzania, Gombe chimpanzees use levers and pound objects whereas those at Mahale do not (Whiten et al. 1999). Further, based on research in captive chimpanzees, it appears that small groups living in close proximity can adopt and maintain their own distinct cultural norms in a short period of time (Whiten et al. 2005).

Behavioural and biological traits, including some chimpanzee traditions, can emerge stochastically but may then go on to produce evolutionarily meaningful differentiation. However, differentiation can also be more ordered, as occurs under selection. The environment—broadly defined—often provides the impetus for this selection, and acts as an important interacting factor when considering group composition, ecology, and behaviour. Although group sizes vary from species to species and place to place, there is often an ecologically optimal group size, and groups can split permanently to facilitate feeding or reproduction when they exceed that (Dunbar 1988). This could give realistic opportunities to exploit varying adaptive landscapes. Such 'budding' also promotes wider-scale dispersal, which has occurred many times during primate and hominin evolution and may, like environmental change, fuel evolutionary novelty. Thus, as groups or populations come across novel competitors, predators, habitats, and foodstuffs, and exploit their potential for phenotypic variation, they may diversify along an environmental gradient.

When considering subspecific differentiation in hominins, it is possible that small differences in the local environment, reinforced by competition for food and differential reproductive success, may lead to morphological shifts that are initially plastic but later genetic. Research into the widespread African vervet monkey has suggested that differentiation of body size (a labile feature that profoundly influences ecology and behaviour) could be the first step in population divergence, providing a way of moving from one adaptive peak to another (Elton et al. 2010). Indeed, size is so plastic that it has been shown to differ considerably in Kenyan vervets even over a few kilometres in response to environmental changes, specifically the presence of human food discards (Turner et al. 1997). Such a response requires morphological plasticity—the ability to gain weight, which if it occurs in juveniles can lead to structural skeletal change caused by increased growth—but also behavioural plasticity. Dietary flexibility, whereby conspecifics are able to exploit (and indeed may choose to exploit) different foods, is a good example of behavioural plasticity (Ulijaszek et al. 2012), which is probably the result of interaction between learned and preprogrammed (i.e. genetically determined) behaviours (Mery and Burns 2010).

This vervet scenario—of dietary and body mass shifts that allow the potential for movement between peaks in an adaptive landscape—might seem difficult to tie up with what is known from the hominin fossil record. For example, *P. boisei* shows no evidence of body size increase over time (Wood et al. 1994), and dietary reconstructions for both *A. afarensis* (Grine et al. 2006) and *P. boisei* (Cerling et al. 2011) suggest that these species had little temporal or spatial change in diets. However, a single line of evidence, such as dental microwear or stable isotope analysis, only provides one facet of a complex entity such as diet (Ulijaszek et al. 2012). Given the variability in feeding behaviour in many modern non-human primates and humans, in which food preferences may differ by population, by group, by season, by individual, by age (Ulijaszek et al. 2012), and the palaeobiological evidence for individual dietary variability in certain hominin species (e.g. Sponheimer et al. 2006), it is reasonable to assume that all hominins had some degree of behavioural plasticity that extended to diet and enabled a response to novel environments or circumstances. Just as vervet exploitation of cultivated foods emerged opportunistically and may be missed in a snapshot observation of their feeding behaviour, small alterations to behaviour or feeding preferences in individual hominin groups might not be identified palaeontologically but nonetheless could provide the raw material for future significant change.

Plastic behavioural responses to challenges in slightly different environments may cause two sister groups (or, on a larger scale, populations) to diverge. Initially, this behavioural plasticity may influence morphology plastically, such as exploitation of slightly different foodstuffs leading to different bone deformation during ontogeny of the masticatory system (Ulijaszek et al. 2012). Plasticity could then play a facilitating role in evolutionary change if individuals exhibiting those morphological shifts had a competitive and hence reproductive advantage and the trait was reinforced by selection (Ulijaszek et al. 2012). Essentially, behavioural plasticity at the individual level may be one of the elements needed for groups and hence higher organizational units to move between adaptive peaks. Despite apparent dietary homogeneity, there is evidence for morphological change over time in *A. afarensis*, with more recent individuals having a larger mandibular corpus (Lockwood et al. 2000), a plastic or even selected trait shift that could be related to dietary diversification within at least some of the population or species. In *P. boisei*, more recent specimens exhibit a larger P4 crown area (Wood et al. 1994). Since tooth morphology is genetically determined and not remodelled in the way that bone is, this could indicate either genetic adaptation—presumably to diet—or genetic drift, both of which are evolutionary forces. Although relatively modest morphological changes such as these seem a long way from the appearance of a new species, the process of intraspecific divergence, which may lead ultimately to speciation, is dependent on them.

14.5 Conclusion

In this chapter, we have attempted to explore the potential for subspecific differentiation in African hominins and identify some of its possible drivers. Our approach is no substitute for study of fossil and genetic material but instead has generated some predictions about intraspecific diversity that may be empirically testable. Gaining a deeper understanding of pattern and process in human evolution will require better knowledge about intraspecific diversity in hominins. This needs not only sophisticated morphological and molecular tools but also a better integration of palaeo-anthropological discoveries into a wider theoretical evolutionary biology framework.

References

Aghokeng, A. F., A. Ayouba, S. Ahuka, F. Liegoies, P. Mbala, J.-J. Muyembe, E. Mpoudi-Ngole, E. Delaporte, and M. Peeters (2010). 'Genetic Diversity of Simian Lentivirus in Wild De Brazza's Monkeys (*Cercopithecus neglectus*) in Equatorial Africa', *Journal of General Virology* 91: 1810–1816.

Ahern, J. C. M. (1998). 'Underestimating Intraspecific Variation: The Problem with Excluding STS 19 from *Australopithecus africanus*', *American Journal of Physical Anthropology* 105: 461–480.

Amadon, D. (1949). 'The Seventy-Five Per Cent Rule for Subspecies', *Condor* 51: 250–258.

Antón, S. C., W. R. Leonard, and M. L. Robertson (2002). 'An Ecomorphological Model of the Initial Hominid Dispersal from Africa', *Journal of Human Evolution* 43: 773–785.

Arista, M., P. L. Ortiz, and S. Talavera (1997). 'Reproductive Isolation of Two Sympatric Subspecies of *Juniperus phoenicea* (Cupressaceae) in Southern Spain', *Plant Systematics and Evolution* 208: 225–237.

Arnold, C., L. J. Matthews, and C. L. Nunn (2010). 'The 10kTrees Website: A New Online Resource for Primate Phylogeny', *Evolutionary Anthropology* 19: 114–118.

Belliure, J., G. Sorci, A. P. Møller, and J. Clobert (2000). 'Dispersal Distances Predict Subspecies Richness in Birds', *Journal of Evolutionary Biology* 13: 480–487.

Berger, L. R. (2005). *Working and Guiding in the Cradle of Humankind*. Johannesburg: Prime Origins Publishing.

Binks, R. M., J. Prince, J. P. Evans, and W. J. Kennington (2012). 'More than Binding Divergence: Reproductive Isolation between Sympatric Subspecies of a Sea Urchin by Asynchronous Spawning', *Evolution* 66: 3545–3557.

Brandon-Jones, D., A. A. Eudey, T. Geissmann, C. P. Groves, D. J. Melnick, J. C. Morales, M. Shekelle, C.-B. Stewart (2004). 'Asian Primate Classification', *International Journal of Primatology* 25: 97–164.

Brunet, M., A. Beauvilain, Y. Coppens, E. Heintz, A. H. E. Moutaye, and D. Pilbeam (1995). 'The First Australopithecine 2,500 Kilometres West of the Rift Valley (Chad)', *Nature* 378: 273–275.

Cardillo, M., J. S. Huxtable, and L. Bromham (2003). 'Geographic Range Size, Life History and Rates of Diversification in Australian Mammals', *Journal of Evolutionary Biology* 16: 282–288.

Cardini, A., J. A. F. Diniz Filho, P. D. Polly, and S. Elton (2010). 'Biogeographic Analysis Using Geometric Morphometrics: Clines in Skull Size and Shape in a Widespread African

Arboreal Monkey', in A. M. T. Elewa (ed.), *Morphometrics for Nonmorphometricians*. Heidelberg: Springer, 191–217.

Cardini, A. and S. Elton (2009). 'Geographic and Taxonomic Influences on Cranial Variation in Red Colobus Monkeys (Primates, Colobinae): Introducing a New Approach to "Morph" Monkeys', *Global Ecology and Biogeography* 18: 248–263.

Cardini, A., A.-U. Jansson, and S. Elton (2007). 'A Geometric Morphometric Approach to the Study of Ecogeographical and Clinal Variation in Vervet Monkeys', *Journal of Biogeography* 34: 1663–1678.

Cerling, T. E., E. Mbua, F. M. Kirera, F. K. Manthi, F. E. Grine, M. G. Leakey, M. Sponheimer, and K. T. Uno (2011). 'Diet of Paranthropus *boisei* in the Early Pleistocene of East Africa', *Proceedings of the National Academy of Sciences* 108: 9337–9341.

Cofran, Z. C. and J. F. Thackeray (2010). 'One or Two Species? A Morphometric Comparison between Robust Australopithecines from Kromdraai and Swartkrans', *South African Journal of Science* 106: 40–43.

Copeland, S. R., M. Sponheimer, D. J. de Ruiter, J. A. Lee-Thorp, D. Codron, P. J. le Roux, V. Grimes, and M. P. Richards (2011). 'Strontium Isotope Evidence for Landscape Use by Early Hominins', *Nature* 474: 76–78.

Darwin, C. (1874). *The Descent of Man*. New York: Appleton and Company.

Davenport, T. R. B. and T. Jones (2008). *Rungwecebus kipunji*, in *IUCN 2012: IUCN Red List of Threatened Species*, Version 2012.2. Available online at <www.iucnredlist.org>.

DeMenocal, P. B. (2004). 'African Climate Change and Faunal Evolution during the Pliocene-Pleistocene', *Earth and Planetary Science Letters* 220: 3–24.

Dietzen, C., H.-H. Witt, and M. Wink (2003). 'The Phylogeographic Differentiation of the Robin *Erithacus rubecula* on the Canary Islands Revealed by Mitochondrial DNA Sequence Data and Morphometrics: Evidence for a New Robin Taxon on Gran Canaria?', *Avian Science* 3: 115–131.

Dunbar, R. I. M. (1988). *Primate Social Systems*. London: Croom Helm.

Dunn, J. D., A. Cardini, and S. Elton (2013). 'Biogeographic Variation in the Baboon: Dissecting the Cline', *Journal of Anatomy* 223/4: 337–352.

Eeley, H. and R. A. Foley (1999). 'Species Richness, Species Range Size and Ecological Specialisation among African Primates: Geographical Patterns and Conservation Implications', *Biodiversity and Conservation* 8: 1033–1056.

Elton, S. (2006). 'Forty Years On and Still Going Strong: The Use of the Hominin-Cercopithecid Comparison in Palaeoanthropology', *Journal of the Royal Anthropological Institute* 12: 19–38.

Elton, S. (2008). 'The Environmental Context of Human Evolutionary History in Eurasia and Africa', *Journal of Anatomy* 212: 377–393.

Elton, S., J. Dunn, and A. Cardini (2010). 'Size Variation Facilitates Population Divergence But Does Not Explain It All: An Example Study from a Widespread African Monkey', *Biological Journal of the Linnean Society* 101: 823–843.

Fabre, V., S. Condemi, and A. Degioanni (2009). 'Genetic Evidence of Geographical Groups among Neanderthals', *PLoS ONE* 4: e5151.

Felsenstein, J. (1985). 'Phylogenies and the Comparative Method', *American Naturalist* 125: 1–15.

Foley, R. A. and P. C. Lee (1989). 'Finite Social Space, Evolutionary Pathways, and Reconstructing Hominid Behavior', *Science* 243: 901–906.

Fooden, J. (2000). 'Systematic Review of the Rhesus Macaque, *Macaca mulatta* (Zimmermann, 1780)', *Fieldiana Zoologica* 96: 1–180.

Forbes, A. A., L. L. Stelinski, T. H. Q. Powell, J. J. Smith, and J. L. Feder (2009). 'Sequential Sympatric Speciation across Trophic Levels', *Science* 323: 776–779.

Goodall, J. (1986). *The Chimpanzees of Gombe*. Cambridge, MA: Belknap Press.

Goudie, A. S. (2005). 'The Drainage of Africa since the Cretaceous', *Geomorphology* 67: 437–456.

Grine, F. E., P. S. Ungar, M. F. Teaford, and S. El-Zaatari (2006). 'Molar Microwear in *Praeanthropus afarensis*: Evidence for Dietary Stasis through Time and under Diverse Paleoecological Conditions', *Journal of Human Evolution* 51: 297–319.

Groves, C. P. (2001). *Primate Taxonomy*. Washington, DC: Smithsonian Institution Press.

Groves, C. P. (2007). 'The Taxonomic Diversity of the Colobinae of Africa', *Journal of Anthropological Sciences* 85: 7–34.

Grubb, P., T. M. Butynski, J. F. Oates, S. K. Bearder, T. R. Disotell, C. P. Groves, and T. T. Struhsaker (2003). 'Assessment of the Diversity of African Primates', *International Journal of Primatology* 24: 1301–1357.

Harmon, E. (2009). 'Size and Shape Variation in the Proximal Femur of *Australopithecus africanus*', *Journal of Human Evolution* 56: 551–559.

Harvati, K., S. R. Frost, and K. P. McNulty (2004). 'Neanderthal Taxonomy Reconsidered: Implications of 3D Primate Models of Intra- and Interspecific Differences', *Proceedings of the National Academy of Sciences* 101: 1147–1152.

Hawks, J. (2012). 'Dynamics of Genetic and Morphological Variability within Neandertals', *Journal of Anthropological Science* 90: 81–87.

Hernández Fernández, M. and E. S. Vrba (2006). 'Plio-Pleistocene Climatic Change in the Turkana Basin (East Africa): Evidence from Large Mammal Faunas', *Journal of Human Evolution* 50: 595–626.

Hey, J., R. S. Waples, M. L. Arnold, R. K. Butlin, and R. G. Harrison (2003). 'Understanding and Confronting Species Uncertainty in Biology and Conservation', *Trends in Ecology and Evolution* 18: 597–603.

Hijmans, R. J., S. E. Cameron, J. L. Parra, P. G. Jones, and A. Jarvis (2005). 'Very High Resolution Interpolated Climate Surfaces for Global Land Areas', *International Journal of Climatology* 25: 1965–1978.

Hijmans, R. J., L. Guarino, and P. Mathur (2012). *DIVA-GIS Version 7.5 Manual*. Available online at <http://www.diva-gis.org/docs/DIVA-GIS_manual_7.pdf>, accessed June 2013.

Holmes, K. M., K. A. Robson Brown, W. P. Oates, and M. J. Collins (2005). 'Assessing the Distribution of African Palaeolithic Sites: A Predictive Model of Collagen Degradation', *Journal of Archaeological Science* 32: 157–166.

Isaac, N. J. B., J. Mallet, and G. M. Mace (2004). 'Taxonomic Inflation: Its Influence on Macroecology and Conservation', *Trends in Ecology and Evolution* 19: 464–469.

IUCN (2011). The IUCN Red List of Threatened Species, Version 2011. Available online at <http://www.iucnredlist.org>, accessed 30 September 2011.

Jolly, C. J. (1993). 'Species, Subspecies, and Baboon Systematics', in W. H. Kimbel and L. B. Martin (eds), *Species, Species Concepts, and Primate Evolution*. New York: Plenum Press, 67–107.

Kimbel, W. H. and L. Delezene (2009). '"Lucy" Redux: A Review of Research on *Australopithecus afarensis*', *Yearbook of Physical Anthropology* 52: 2–48.

Kingdon, J. (1997). *The Kingdon Guide to African Mammals*. London: Academic Press.

Krause, J., L. Orlando, D. Serre, B. Viola, K. Prufer, M. P. Richards, J. J. Hublin, C. Hanni, A. P. Derevianko, and S. Paabo (2007). 'Neanderthals in Central Asia and Siberia', *Nature* 449: 902–904.

Lawes, M. J. (1990). 'The Distribution of the Samango Monkey (*Cercopithecus mitis erythrarchus* Peters 1852 and *Cercopithecus mitis labiatus* I. Geoffroy, 1843) and Forest History in Southern Africa', *Journal of Biogeography* 17: 669–680.

Lawson Handley, L. J., J. Goudet, A. Manica, and F. Balloux (2007). 'Going the Distance: Human Population Genetics in a Clinal World', *Trends in Genetics* 23: 432–439.

Lee-Thorp, J. A., M. Sponheimer, and N. J. van der Merwe (2003). 'What Do Stable Isotopes Tell Us about Hominid Dietary and Ecological Niches in the Pliocene?', *International Journal of Osteoarchaeology* 13: 104–113.

Lockwood, C. A., W. H. Kimbel, and D. C. Johanson (2000). 'Temporal Trends and Metric Variation in the Mandibles and Teeth of *Australopithecus afarensis*', *Journal of Human Evolution* 39: 23–55.

Lockwood, C. A., C. G. Menter, J. Moggi-Cecchi, and A. W. Keyser (2007). 'Extended Male Growth in a Fossil Hominin Species', *Science* 318: 1443–1446.

Mace, G. M. (2004). 'The Role of Taxonomy in Species Conservation', *Philosophical Transactions of the Royal Society London B: Biological Sciences* 359: 711–719.

Maguire, K. C. and A. L. Stigall (2009). 'Distribution of Fossil Horses in the Great Plains during the Miocene and Pliocene: An Ecological Niche Modeling Approach', *Paleobiology* 35: 587–611.

Mallet, J. (2007). 'Subspecies, Semispecies, Superspecies', in S. A. Levin (ed.), *Encyclopedia of Biodiversity*. Oxford: Elsevier, 5 pp. Online update, available at <http://www.sciencedirect.com/science/article/pii/B9780123847195001386>.

Mery, F. and J. Burns (2010). 'Behavioural Plasticity: An Interaction between Evolution and Experience', *Evolutionary Ecology* 24: 571–583.

Millien, V., S. K. Lyons, L. Olson, F. A. Smith, A. B. Wilson, and Y. Yom-Tov (2006). 'Ecotypic Variation in the Context of Global Climate Change: Revisiting the Rules', *Ecology Letters* 9: 853–869.

Mittelbach, G. G., D. W. Schemske, H. V. Cornell, A. P. Allen, J. M. Brown, M. B. Bush, S. P. Harrison, A. H. Hurlbert, N. Knowlton, H. A. Lessios, C. M. McCain, A. R. McCune, L. A. McDade, M. A. McPeek, T. J. Near, T. D. Price, R. E. Ricklefs, K. Roy, D. F. Sax, D. Schluter, J. M. Sobel, and M. Turelli (2007). 'Evolution and the Latitudinal Diversity Gradient: Speciation, Extinction and Biogeography', *Ecology Letters* 10: 315–331.

Nowak, K., A. Cardini, and S. Elton (2008). 'Evolutionary Acceleration in an Endangered African Primate: Speciation and Divergence in the Zanzibar Red Colobus Monkey', *International Journal of Primatology* 29: 1313–1339.

Oates, J. F. (2008). *Cercopithecus preussi*, in *IUCN 2013: IUCN Red List of Threatened Species*, Version 2013.1. Available online at <www.iucnredlist.org>.

Oates, J. F. and T. M. Butynski (2008). *Mandrillus leucophaeus*, in *IUCN 2013: IUCN Red List of Threatened Species*, Version 2013.1. Available online at <www.iucnredlist.org>.

Oates, J. F., T. Struhsaker, and N. Ting (2008). *Procolobus pennantii*, in *IUCN 2013: IUCN Red List of Threatened Species*, Version 2013.1. Available online at <www.iucnredlist.org>.

Orme, D., R. Freckleton, G. Thomas, T. Petzoldt, S. Fritz, N. Isaac, and W. Pearse (2012). caper: Comparative Analyses of Phylogenetics and Evolution in R, R package version 0.5. Available online at <http://CRAN.R-project.org/package=caper>.

Phillimore, A. B. and I. P. F. Owens (2006). 'Are Subspecies Useful in Evolutionary and Conservation Biology?', *Proceedings of the Royal Society B: Biological Sciences* 273: 1049–1053.

QGIS Development Team (2013). QGIS Geographic Information System, Open Source Geospatial Foundation Project. Available online at <http://qgis.osgeo.org>.

R Core Team (2013). R: A Language and Environment for Statistical Computing, R Foundation for Statistical Computing, Vienna. Available online at <http://www.R-project.org/>.

Rice, W. R. and E. E. Hostert (1993). 'Laboratory Experiments on Speciation: What Have We Learned in 40 Years?', *Evolution* 47: 1637–1653.

Robbins, M. and L. Williamson (2008). *Gorilla beringei*, in *IUCN 2013: IUCN Red List of Threatened Species*, Version 2013.1. Available online at <www.iucnredlist.org>.

Saino, N. (1998). 'Congruence between Morphological Variation and Altitudinal Gradient across a Hybrid Zone between Carrion and Hooded Crows', *Italian Journal of Zoology* 65: 407–412.

Salzmann, U., A. M. Haywood, D. J. Lunt, P. J. Valdes, and D. J. Hill (2008). 'A New Global Biome Reconstruction and Data-Model Comparison for the Middle Pliocene', *Global Ecology and Biogeography* 17: 432–447.

Smith, R. J. and W. L. Jungers (1997). 'Body Mass in Comparative Primatology', *Journal of Human Evolution* 32: 523–559.

Sohl, L. E., M. A. Chandler, R. B. Schmunk, K. Mankoff, J. A. Jonas, K. M. Foley, and H. J. Dowsett (2009). PRISM3/GISS Topographic Reconstruction, *U.S. Geological Survey Data Series 419*.

Sponheimer, M., B. Passey, D. de Ruiter, D. Guatelli-Sternberg, T. Cerling, J. Lee-Thorp (2006). 'Isotopic Evidence for Dietary Flexibility in the Early Hominin *Paranthropus robustus*', *Science* 314: 980–982.

Suwa, G., B. Asfaw, Y. Beyene, T. White, S. Katoh, S. Nagaoka, H. Nakaya, K. Uzawa, P. Renne, and G. WoldeGabriel (1997). 'The first skull of *Australopithecus boisei*', *Nature* 389: 489–492.

Thorpe, R. S. (1987). 'Geographic Variation: A Synthesis of Cause, Data, Pattern and Congruence, in Relation to Subspecies, Multivariate Analysis and Phylogenesis', *Bolletino di Zoologia* 54: 3–11.

Tosi, A. J., K. M. Detwiler, and T. R. Disotell (2005). 'X-Chromosomal Window into the Evolutionary History of the Guenons (Primates: Cercopithecini)', *Molecular Phylogenetics and Evolution* 36: 58–66.

Trauth, M. H., J. C. Larrasoan, and M. Mudelsee (2009). 'Trends, Rhythms and Events in Plio-Pleistocene African Climate', *Quaternary Science Reviews* 28: 399–411.

Trauth, M. H., M. A. Maslin, A. Deino, A. G. N. Bergner, M. Duehnforth, and M. R. Strecker (2007). 'High- and Low-Latitude Forcing of Plio-Pleistocene East African Climate and Human Evolution', *Journal of Human Evolution* 53: 475–486.

Turner, T. R., F. Anapol, and C. J. Jolly (1997). 'Growth, Development and Sexual Dimorphism in Free Ranging Vervet Monkeys in Kenya', *American Journal of Physical Anthropology* 103: 19–35.

Ulijaszek, S., N. Mann, and S. Elton (2012). *Evolving Human Nutrition: Implications for Public Health*. Cambridge: Cambridge University Press.

Ungar, P. S., F. E. Grine, and M. F. Teaford (2008). 'Dental Microwear and Diet of the Plio-Pleistocene Hominin *Paranthropus boisei*', *PLoS ONE* 3: e2044

Whiten, A., J. Goodall, W. C. McGrew, T. Nishida, V. Reynolds, Y. Sugiyama, C. E. G. Tutin, R. W. Wrangham, and C. Boesch (1999). 'Cultures in Chimpanzees', *Nature* 399: 682–685.

Whiten, A., V. Horner, and F. B. M. de Waal (2005). 'Conformity to Cultural Norms of Tool Use in Chimpanzees', *Nature* 437: 737–740.

Winder, I. C., G. C. P. King, M. Deves, and G. N. Bailey (2013). 'Complex Topography and Human Evolution: The Missing Link', *Antiquity* 87: 333–349.

Winker, K. (2010). 'Subspecies Represent Geographically Partitioned Variation, a Goldmine of Evolutionary Biology, and a Challenge for Conservation', *Ornithological Monographs* 67: 6–23.

Wood, B. (ed.) (2011). *Wiley-Blackwell Encyclopedia of Human Evolution*. Hoboken, NJ: Wiley Blackwell.

Wood, B. and D. E. Lieberman (2001). 'Craniodental Variation in *Paranthropus boisei*: A Developmental and Functional Perspective', *American Journal of Physical Anthropology* 116: 13–25.

Wood, B. and N. Lonergan (2008). 'The Hominin Fossil Record: Taxa, Grades and Clades', *Journal of Anatomy* 212: 354–376.

Wood, B., C. Wood, and L. Konigsberg (1994). '*Paranthropus boisei*: An Example of Evolutionary Stasis?', *American Journal of Physical Anthropology* 95: 117–136.

Wright, S. (1932). 'The Roles of Mutation, Inbreeding, Crossbreeding, and Selection in Evolution', in D. F. Jones (ed.), *Proceedings of the Sixth International Congress on Genetics*: Ithaca, New York. Austin, TX: Genetics Society of America, i.355–366.

CHAPTER 15

RESIDENCE PATTERNS AND THE HUMAN-ECOLOGICAL SETTING IN HISTORICAL EASTERN EUROPE

a challenge of compositional (re)analysis

MIKOŁAJ SZOŁTYSEK

15.1 INTRODUCTION

AT least since 1970s, family historians, demographers, and sociologists alike have been seeking to brand major areas of Europe with a particular type of family system. The idea that some European regions have distinct household and family patterns has been asserted with a high degree of confidence by scholars studying the living arrangements of Eastern European societies. One of the central tenets of sociological and historical studies of the family has long been the existence of a specific, peculiar 'Eastern European type' of domestic group organization. Its inherent features—i.e. complex family organization, universal nuptiality, and familistic societal values—were presumed to transcend physical, political, and cultural boundaries in sustaining a common family model across the vast territory of the European 'East'. The popularity of this reductionist view of historical family behaviour in Eastern Europe has persisted among Western scholars, even though it is incompatible with the general acknowledgement that the region has a wide variety of languages, confessions, and cultures, as well as heterogeneous ecological and institutional settings. This perspective appears to persist largely because Western family historians know little about Eastern Europe, and are unfamiliar with 'native' Eastern European literature on family and demography.

While the growing interest among Western scholars in studying macroregional differences has led to a series of heuristically interesting attempts to map the history of European families, it has also blocked their awareness of variation within these regions, especially with regard to Europe's eastern territories. This essay seeks to place the study of population and its heterogeneity in Eastern Europe on a very different footing. Based on a thorough reading of the 'native' literature on family and demography of the eastern part of the continent, and drawing on an unprecedented collection of census microdata from historical Eastern Europe, our goal is to identify and differentiate the composition and behaviour of multiple subpopulations in the area. The compositional approach adopted here not only calls into question the existence of a demographically uniform Eastern Europe, it also demonstrates how and why identifying and sorting subpopulations really is necessary for gaining an understanding of the continent's family history. The new history of families suggested here entails probing beneath aggregated trends of demographic patterns in order to observe and evaluate the diverse familial responses to polymorphic demographic, institutional, and environmental factors. A detailed investigation of this kind challenges the conventional wisdom that standard units of population are homogeneous.

This paper has three major sections. We open with a brief account of how the perception of the divergent 'Eastern European family pattern' developed, including a review of some counterfactual, albeit thoroughly ignored, evidence from Western mainstream family history. In the following section, we present in detail selected contributions of Eastern European scholars. The concepts of familial characteristics and developments outlined in these studies reveal a striking diversity of familial behaviour in the eastern part of the continent. We pay special attention to patterns in Poland–Lithuania that can be discerned with the use of recently digitized historical census microdata and refined methodologies, as well as (in section 15.4) to possible explanations of the polymorphic character of familial behaviour in the several regional subpopulations of this area. We close with a general discussion and conclusion.

15.2 The 'Great European Divide': A Western Misconception?

In capturing the Eastern European familial 'specificity', Western scholars have tended to rely on two approaches in particular: (1) using single dispersed case studies to infer the existence of spatially bounded 'ideal' types of family structures, and (2) treating nation-level aggregate statistics as if these stood for a dominant familial behaviour. The former approach has led to a variety of conceptual problems, which were caused when isolated demographic facts were used to make broad assumptions about the demography of large groups. The latter approach discounted information about regional patterns and local-level variation by tacitly assuming that the individuals who composed a given aggregate were homogenous (cf. Kreager 2009). Both approaches paved the way for a regionalization of family systems at a very high level of generality, in which Eastern Europe was treated as a more or less undifferentiated whole. Scholars using these research strategies tended to assign certain demographic and familial characteristics to various societies of the eastern part of the continent regardless of context, which led them to conclude that the region's population was fixed and undifferentiated.

Eastern European 'specificity' was first 'captured' in the works of Le Play. His mid-nineteenth-century mapping of family types (Le Play 1877–1879, vol. I, facing p. 683) was the result of his intensive household sample surveys, a method he called '*sociographie microscopique*' (Micheli 2000). This mapping was based on the direct observation of 57 families scattered all over Europe and the Mediterranean. However, Le Play's conclusions about family structures in Eastern Europe derived from the study of only seven families, who were concentrated largely in two highly dispersed regions, from the Urals to Hungary and Slovakia (Le Play 1877–1879, vols 2 and 4).[1] Despite a stunning dearth of evidence, this highly respected scholar of family sociology drew bold conclusions, claiming that patriarchal, patrilocal, and multigenerational households could be found among the 'Eastern nomads, Russian peasants, and the Slavs of Central Europe', and also among the Hungarian population (Le Play 1871: 94; Le Play 1982[1872]: 259; also Le Play 1867: 69).

Le Play's nineteenth-century assessment of Eastern European difference penetrated deep into the collective consciousness, and was later repeated in modern historical demography and family history. The enduring myth of a demographically uniform Eastern Europe in which people marry young and live in patriarchal households can be traced back to J. Hajnal's 1965 article on marriage patterns in Europe (Hajnal 1965). Based on aggregate national statistics from around 1900, Hajnal posited the existence of a 'European pattern', the distinctive features of which were a high age at marriage and a high proportion of people who never marry at all. According to Hajnal, this pattern pervaded 'the whole of Europe except for the eastern and south-eastern portion'. Hajnal significantly hardened Le Play's initial distinctions between Eastern Europe and the rest of the continent, and was keen to equate the marriage pattern of several countries located east of an imaginary line from St Petersburg to Trieste with marriage characteristics of 'non-European civilizations' (1965: 104). To reveal the presence of the 'Eastern European pattern', Hajnal used indirect measures of nuptiality from census aggregates of Bulgaria, Serbia, Hungary, Romania, Bosnia, and Russia/USSR (1965: 102). Importantly, however, he overlooked a large group of Eastern European territories between Germany and Russia that later became Poland, Czechoslovakia, and the Baltic states of Lithuania, Latvia, and Estonia. All of these territories had been part of Western

[1] These were four families (one of the semi-nomad Bachkirs) from the Ural area; another one from central Russia, between Ryazan and Nizniy Novgorod; one family from the town of Hatvan in north-central Hungary, and one from the village of Schemnitz in the Slovakian sub-Carpathian area within the Hungarian state. One additional family was located in the Bulgarian territories of the Ottoman Empire, but is treated here as belonging geographically to the Balkans.

European civilization throughout the Middle Ages and the early modern era (Halecki 1952). This selection bias in Hajnal's mode of presentation very likely led him to exaggerate differences between Western and Eastern European marriage behavioural patterns (Sklar 1971).

Long after Le Play, wild extrapolations from single cases have continued to foster tacit assumptions about Eastern European families. For example, Czap's study of a single Russian community of Mishino (some 170 kilometers south-east of Moscow; 128 households in 1814, with a population of 1,173 persons) posited the existence of a specific 'Eastern European family type' (as opposed to that of the West), which was characterized by a high proportion of multiple-family households, a mean household size of significantly more than five persons, and early and universal marriage within the population (Czap 1982: 18).[2] Czap had no reservations about calling marriage behaviour on the Mishino estate 'a robust non-European marriage pattern' (without inverted commas; see Czap 1978: 114, 116), and he also seemed to believe that his findings were representative of large areas of pre-modern Russia (Czap 1982: 6). A subsequent study of another isolated locality in the European part of Russia (64 households and some 500 'souls' in 1813) even suggested that a 'Mishino pattern' may have existed across the Eastern territories (Hoch 1986; see Dennison 2003 for criticism). Partly because of the inclination to brand major areas of Europe as having a particular type of household system, and partly because of the lack of knowledge about Russia and Eastern Europe in general among Western family historians, Czap's observation has regularly been assumed to be representative of the whole continent to the east of Hajnal's line (Sovič 2008: 144).

In the early 1980s, Hajnal referred to Mishino (alongside examples from fifteenth-century Tuscany) when illustrating the differences between two kinds of household formation system in pre-industrial Europe: the simple household system found in the northwestern part of continent and the joint household system. He also suggested that the two supranational large-scale family systems he posited could be spatially conceptualized as referring to territories to the west and to the east of his line (Hajnal 1982: 452). He then used Mishino's joint family system as if it represented the situation in all of the European East, or at least among all Eastern European serf populations. 'Indeed,' he argued, 'it seems likely that something like the Mishino household formation system prevailed among populations numbering in the millions' (1982: 468). He then jumped thousands of kilometers south-west from Mishino to use circumstantial data from Hungary and Croatia to supplement his case for the prevalence of joint family systems in other Eastern European regions (Hajnal 1982: 469).

Laslett departed from Hajnal's dichotomous East–West model of European familial differences, arguing instead that there were two additional sets of tendencies, which he saw as bridging the most extreme contrasts in domestic group organization on the continent (Laslett 1983). Like Hajnal, however, Laslett used for his ambitious mapping

[2] Both Hajnal's and Laslett's perceptions of family type in Eastern Europe seem to have been deeply influenced by Czap's findings; see Laslett (1983: 517, 520–521, 549); Hajnal (1982: 467–469, 473).

exercise data from several single localities, which he thought were illustrative of the various regions of Europe with distinguishable family and household forms ('West', 'West/Central or middle', 'Eastern', and 'Mediterranean'; see Laslett 1983: 516, 526–527). Laslett took the 'eastern tendencies' illustrated by the Mishino data as 'justifiably associated with the domestic group structure of European Russia as a whole and some of its surrounding areas' (1983: 529). Even greater geographical and definitional uncertainties surround his notion of the West/Central or middle domestic groups, as explicated by the population of Grossenmeer in 1785, a small village of some 150 houses near Oldenbourg in northwestern Germany (Laslett 1983: 522–523). Whole populations in this immense area stretching from Oldenbourg in German-speaking lands to Ryazan province south of Moscow in the east (a distance of approximately 2,500 kilometers) were said to have had either 'Western/Central' or 'Eastern' familial traits, or to have displayed an unspecified mixture of the two.

Laslett's and Hajnal's tentative generalizations have long been held in high regard, and the long-term scarcity of research material available for Eastern Europe made them particularly durable. Two French historical demographers stated, for example, that 'it is possible to find a few features common to the area between southern Serbia and northern Estonia, passing through Poland and Russia . . . , the average size of the household (eight to nine persons), which was much higher than in western Europe, and a strong propensity towards multiple households, both being extreme forms of tendencies in evidence in central Europe' (Burguière and Lebrun 1986: 36; similar judgements can be found in Alderson and Sanderson 1991; Kaser 2001: 34; Fauve-Chamoux 2001: 221; Thornton 2005: 52).

Laslett's and Hajnal's tentative generalizations also provided a ready-made framework for scholars from other fields hoping to understand more recent developments. Therborn, for example, referred to Hajnal's nuptiality hypothesis in his global history and sociology of the family, observing that 'the classical European family divide, running from Trieste to St Petersburg . . . is still visible in 2000' (2004: 305). In a similar spirit, demographers used Hajnal's bipolar division of the continent from around 1900 to explain European differentials in fertility and nuptiality behaviour since the Second World War (Grassland 1990: 186–187; Monnier 2004; Blum and Rallu 1993; Monnier and Rychtarikova 1992; Sobotka 2003: 475).

Several scholars have suggested that there is a need to move towards a more diverse approach to understanding the prevailing forms of familial organization in both parts of the continent (Todorova 2006; Sovič 2008; Plakans and Wetherell 2001, 2005; Kertzer 1991). Still others have shown that mapping residence patterns empirically with the systematic use of archival sources can produce a more variegated picture of regional and national populations (Szołtysek 2008a, 2008b). Wall tackled the topic of intra- and interregional variation in familial organization in Western Europe, and suggested that there was considerable variation in domestic group organization and nuptiality (Wall 1979, 1991, 2001).

Although an East–West polarized distinction between family patterns in Europe seems to have been pervasive and largely resistant to criticism, the orthodoxy proclaiming the East–West familial dichotomy has never been entirely uncontested. In the early 1970s, Sklar carefully collected census data for every political entity of the Eastern European region from around 1900, which she then minutely decomposed into smaller administrative units (Sklar 1971, 1974). Having far surpassed Hajnal's earlier contributions, both in terms of data collection and geospatial awareness, Sklar argued that '. . . the East European regions that were to become Latvia, Lithuania, Estonia, Czechoslovakia, and Poland after World War I did not exhibit the Eastern European pattern of marriage behaviour, but were actually closer to the West European pattern'. By referring to values of the singulate mean age at marriage, she argued that 'nuptiality in these regions at around 1900 followed the West European late marriage pattern', with the female mean age at first marriage fluctuating between 24 and 27 (see Figure 15.1), and the mean age for men fluctuating between 25 and 30. Sklar observed commonalities across Estonia, Latvia, Lithuania, Poland, and Czechoslovakia in the proportions single in different age groups, finding only a slight departure from this general tendency in the territories that later became Poland. She concluded that people in all of these areas 'married rather late, and moderate proportions never married at all' (1974: 232–234). In light of this abundant evidence, Sklar felt comfortable concluding that, in Eastern Europe around 1900, both 'Western' and 'Eastern' European marriage patterns prevailed, with the latter being followed by the Balkan countries. She substantiated her claims using some basic data on household size that showed only negligible differences at around 1900 between Sweden and the Baltic, Czech, and Polish provinces (with the mean

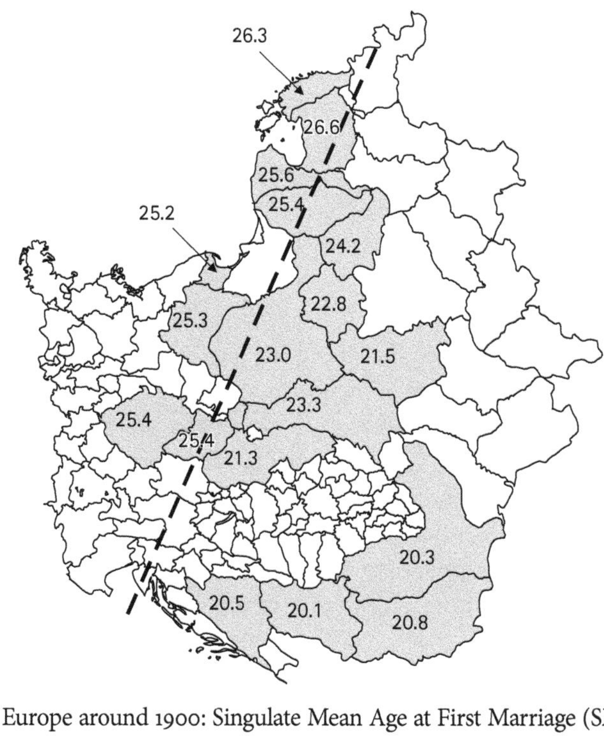

FIG. 15.1 Eastern Europe around 1900: Singulate Mean Age at First Marriage (SMAM) for females.

household size (MHS) fluctuating between 4.7 and 5.2 persons), but more significant differences between the West and Bulgaria and Serbia (MHS 5.8 and 7.2, respectively).

Not surprisingly, Sklar had a very critical stance on Hajnal's assessment of Eastern European nuptiality. She argued that he not only exaggerated the difference between 'European' and 'East European' marriage patterns, but that he also purposely left out of his analysis those Eastern European countries which exhibited one or more characteristics of the 'Western European' pattern of marriage (Sklar 1971: 36–50). Sklar seemed to want to relocate the dividing line suggested by Hajnal more towards the east, thereby moving countries such as Latvia, Lithuania, Estonia, Poland, and Czechoslovakia into the zone of 'Western' marriage and household characteristics (Sklar 1974: 232, 234). However, Sklar's approach seems to have been based on a broader definition of the Eastern European population than many scholars before and after her (cf. Kreager 2009). By suggesting that there were at least two distinct marriage patterns in the region, Sklar's work was the first in modern population history to dispel notions of historical Eastern Europe as a demographic monolith. Her research not only uncovered a transition zone along the north–south axis, which seemed to distinguish the marriage patterns of East-Central Europe from those of the Balkans; it also revealed the presence of noteworthy differences within East-Central Europe itself.

Sklar's substantial contribution to the historical demography of Eastern Europe generally went unnoticed by mainstream scholars working on the geography of family forms. In fact, the positions of 'revisionists' remain rather obscure within the mainstream discourse even today (see Heady 2010; De Moor and Van Zanden 2010; cf. Szołtysek 2012a, 2012b), and further efforts to persuade scholars to accept a less stereotypical image of the families from outside of 'Western Europe' are clearly needed. In the sections that follow, we will seek to broaden the intellectual horizons of the debate opened up by Sklar by offering an overview of almost entirely unknown scholarly contributions of Eastern Europeanists on historical family and demography. Tracing the lineage of work from east of the ostensible Hajnal Line reveals sharp differences between the findings of Eastern European researchers and the dominant assumptions of Western scholars.

15.3 Eastern European Studies on Family, Household, and Marriage

15.3.1 The Czech, Slovak, and Hungarian Literature

Sklar's agenda for conducting more detailed studies into the structure of the family in Eastern Europe resurfaced most extensively in the Czech and Slovak literature. Ever since the historical population listing known as 'The 1651 Register of Subjects According to Their Religion' (*Soupis poddaných podle víry*), which covered almost all of the lands of historical Bohemia, was first examined with the use of quantitative methods (Čaňová et. al. 1987; Grulich and Zeitlhofer 1999: 36–40),[3] one of the basic premises of Czech and Slovak scholars has been that there may have been an intermediary marriage and household formation pattern in Central Europe (Horska 1989; Čaňová and Horska 1992; Langer 1994: 44). Horska was the first to introduce the concept of the 'Central European model of the family' (Horska 1989; also Čaňová and Horska 1992), by which she meant a nuptiality pattern that was intermediate between the Northwestern and Eastern European models. She also asserted that, from the seventeenth through the nineteenth centuries, family households in the Czech countries were almost never of the patriarchal type; instead, they were most frequently composed of parents and children, as was the case 'elsewhere in Western Europe' (Horska 1989: 142; Horska 1994: 101, 104). Several studies confirmed that picture, pointing to an overwhelming dominance of nuclear households in Bohemia (up to 79%). In addition, a significant fraction of the young unmarried population in Bohemia were found to have worked as unmarried servants in the households of non-kin (Čaňová 1989, 1992; Čaňová and Horska 1992: 102; Horský and Maur 1993: 13).[4] All of these features made it possible to treat the Bohemian variant of the 'Central European' pattern of the family as being more or less compatible with patterns observed in Western Europe.

At the same time, however, Czech and Slovak scholars were very sensitive to historical population differentials that have existed on the territories later to become Czechoslavakia, and warned that an important demographic fault-line passed across the Czech lands during the early modern period. Whereas in Bohemia more complex family types could have been more widespread only before the seventeenth century, the 'great family' was much more usual in Moravian Carpathians and Slovakia where it frequently involved patterns of co-residence resembling joint-property systems of *zadruga* type[5] (Horska 1989: 142, 1994: 101–104; Horský and Maur 1993: 14–15; Čaňová and Horska 1992: 94–95; Langer 1994: 44–45; also Švecová 1989: 215). A non-negligible geographical pattern was also believed to have existed in the Czech lands with regard to nuptiality, as age at first marriage declines as we proceed from the northwestern to the southeastern parts of the region (Čaňová and Horska 1992: 90–94; Horska 1994: 102; Švecová 1989: 211). Švecová drew on ethnographic literature to link these two different family and demographic regimes in the area of the former Czechoslovakia with two types of property devolution: the one-heir system known as '*rodina jednonástupnická*'

[3] *Soupis* was drawn up in 1651 by the Habsburg monarchy in the form of register of households.

[4] In works of Laslett, Hajnal, and their followers, life-cycle service became a constitutive centrepiece of the eccentric North-West European marriage and household formation pattern, ensuring late marriage and the formation of a new production and consumption unit at marriage. Domestic service facilitated the accumulation of savings for the establishment of new households, and was also believed to promote economic growth in Western Europe, where it was supposed to strengthen 'acquisitory impulses' along with 'individualistic' behaviour (Laslett 1977; Hajnal 1982; also Macfarlane 1978).

[5] *Zadruga* refers to the Balkan joint family, characterized by patrilineality, patrilocality, patricentrism, ancestor worship, and celebration of the household patron saint.

(an equivalent of Le Play's *famille souche*[6]), and joint-property systems known as '*rodina nedielová*' (Švecová 1989: 215–216; also 1986, 1966). In the latter, the development of nuclear or stem family arrangements was prevented by the joint-property system, the tradition of equal partible inheritance among sons, and the strictly agrarian environment (Švecová 1966: 85; 1986: 204; also Langer 1994: 44). Instead of nuclear families, three- or four-generation families with partilocal marriage and patriarchal power relations were quite prevalent, a pattern which often persisted well into the twentieth century (Švecová 1989: 214–217). While these complex family collectives did not completely dominate the region of Slovakia, their incidence was connected with a variety of local ecotypes. According to Švecová, within Central Europe the Western Carpathians have long represented the border between the two different family models suggested by Hajnal (Švecová 1986: 204).

A similar variegation of family forms was also found in eighteenth- and nineteenth-century Hungary. Andorka disproved the notion that polynuclear households represented a general pattern in Hungary. Although they were fairly widespread in the Transdanubian region, places where the share of nuclear households was much greater and extended and multiple families were much less prevalent could easily be found in other areas of the country (Andorka 1976: 344). A later study of seven localities (Andorka and Farago 1983) suggested that household structure in Hungary 'seems to have been intermediate between western Europe on the one hand and Serbia and Russia on the other', but allowed that important differences might have existed within the country itself.

Farago (1986) pinpointed those differences more precisely by allocating the marriage and household organization patterns in Hungary to three specific categories: the 'western European', the 'eastern European', and finally the 'east-central European' family model (*Ostmitteleuropäische Familienmodell*) (Farago 1986: 135–136; also Farago 1998). The latter category was supposed to encompass behavioural patterns between those Hajnal had contrasted as 'Northwest European', and Russian or Eastern European. Őri (2009) also found very patchy patterns of marriage and household formation across pre-industrial Hungary that defy classification in a simple dichotomous model, and argued that at the end of the eighteenth century the country had both 'Eastern' and 'Western' modes of marriage and household formation with several intermediate patterns. Adaptive variation seems to have existed in a range of scales within the truly polymorphous population of Hungary, including among regional groups and among the local populations of a single region. This huge mosaic of patterns cannot be explained simply by ethnocultural (or, rather, ethno-religious) models, and instead calls for a much more complex set of explanations that takes into account a wide variety of demographic, socio-economic, and ecological factors.

[6] In Le Play's *famille souche* (now usually translated as 'stem family') the father selects one child to remain on or near the parental homestead, to work the farm and eventually inherit it, thus continuing the family line. All other children leave the parental family. According to Le Play, the stem family was the historically dominant form of peasant household in most parts of Europe.

15.3.2 Family and Household in Lithuania and Belarus: Time Invariance and Time Awareness

Although Hajnal's distinction between two supranational, large-scale family systems in pre-industrial Europe was based on data from disparate countries with very different conditions and from very different points in time, his conclusions have been commonly viewed as describing 'primordial' features of Eastern European populations, and as representing the whole continent to the east of the St Petersburg–Trieste 'line' in all historical periods (e.g. De Moor and Van Zanden 2010; Therborn 2004). By contrast, studies by Lithuanian and Belarusian scholars have taken a more time-variant approach to historical Eastern European populations. These works clearly found that, in some historical periods, the differences between the East and the West in terms of the composition of residential groups were much less pronounced than would be expected, or were even negligible.

Višniauskaitė demonstrated that the 'grand indissoluble family' (an equivalent to the term 'joint family' commonly used in Western terminology) never constituted a dominant household form in ethnic Lithuania between the sixteenth and the end of the nineteenth centuries.[7] The nuclearized family system of Lithuanians was a direct consequence of the decomposition of lineage relationships, which affected the Baltic countries starting in the thirteenth and fourteenth centuries, and which led to a marked decline in family communes. Both of these processes were additionally strengthened by the agrarian reforms of the mid-sixteenth century (Višniauskaitė 1971: 4). Increases in peasant obligations due to manorialism and demands for compulsory labour inflicted upon the peasants in eighteenth-century Lithuania, led to an accumulation of family labour on the holding, which resulted in a sharp rise in the number of multiple family households (33% of all domestic units in the years 1700–1800) (Table 15.1). Paradoxically, however, the only point in time when really complex multifocal families formed in some parts of Lithuania was during the 1930s and 1940s; that is, when capitalism was already a fact of life (Višniauskaitė 1971: 7).

Kapyski and Kapyski's analysis of household structure in 252 settlements in various Belarusian territories between the end of sixteenth century and the mid-seventeenth century revealed that, on average, a household was comprised of no more than 1.2 conjugal family units, and that more than 85% of all domestic groups had only one such unit. Most of the remaining multiple-family domestic groups contained two small co-residing families (Figure 15.2). The transition from the sixteenth to the seventeenth centuries was marked by an increasing simplification of peasant residential patterns, and one-family households made up the majority of domestic units across the Belarusian territories (Kapyski and Kapyski 1993: 43). Golubev obtained similar results (1,700

[7] Aggregated data for 15 estates with 791 households; see Višniauskaite (1971: 8–12). By transposing the data from 1594–1700 onto Laslett's typology, we find that 81% were simple households, and only 6.9% were multiple-family domestic groups.

Table 15.1 Development of family patterns in Lithuania (16th–19th centuries)

Period	Households (Freq. = 100%)	Simple (%)	Multiple-family (%)	Complex (multiple-family + extended) (%)
1594–1600	300	79	9.3	21
1616–1700	491	82.9	5.5	17.1
1775–1790	130	54.6	33.8	45.4
1806–1880	162	47.5	30.2	52.5

Source: Višniauskaite (1964); own calculations.

peasant domestic units were analysed), finding that 73% of all households consisted of individual families. The share of the latter was, however, shown to decline on a trajectory of movement towards eastern Belarus (Golubev 1992: 88).

Referring to his study of several communities from central Belarus, Nosevich demonstrated that nuclear family households were completely dominant in Belarus in the second half of the sixteenth century (between 70 and 89% of total households) (2004: 81–87). At the same time, however, he pointed to the emergence of a more complex family pattern during the eighteenth and nineteenth centuries, which he linked to the gradual increase in feudal obligations imposed on the peasantry by Eastern

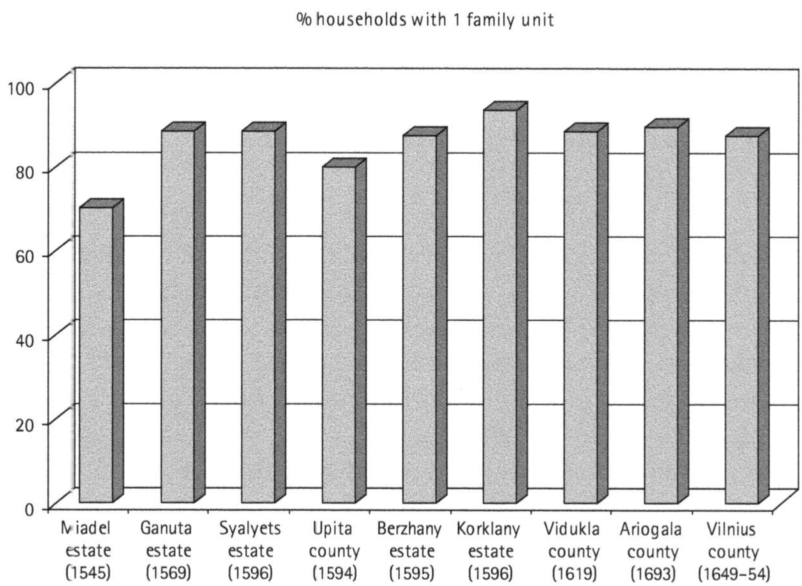

FIG. 15.2 The 'prehistory' of Lithuanian and Belarusian family pattern (16th–17th centuries).

Table 15.2 Development of family forms in two Belarusian parishes

Parish	Census year	Households (Freq. = 100%)	Mean household size	Conjugal family units/ household	% simple households	% extended households	% multiple households
Koren (Catholic)	1740	69	7	1.4	56.5	7.3	36.2
	1793	156	7.2	1.7	30.1	11.5	58.3
	1811	177	7.9	1.8	29.4	10.7	59.9
	1834	171	7.9	1.7	23.4	17	59.6
	1850	171	8.2	1.8	17.1	15.3	67.6
Zembin (Greek Catholic)	1795	79	6.7	1.7	46	14	39
	1811	72	9.2	2.1	32	8	60
	1834	59	7.9	1.6	30.5	20.3	49.2
	1850	48	9.5	1.9	22.9	12.5	64.6

Source: Nosevich (2004).

European landlords (Nosevich 2004: 157–176) (Table 15.2). Nonetheless, throughout almost the entire eighteenth century, the rural population in Belarus followed a pattern of rather moderate household complexity, which stood in marked contrast to the patterns found in Russia. Towards the end of the eighteenth century, the family patterns in Belarus gradually transformed into more communal forms already typical of the vast regions of Russia, with a share of complex families rising significantly above 50%. According to Nosevich, it was this nineteenth-century phenomenon, but not its various antecedents, that appeared so attractive to Western scholars (Nosevich 2007).

15.3.3 Historical Poland–Lithuania: A Sink of Familial Diversity

A more precise identification of the actual variation of familial behaviour across the large territory of Eastern Europe was attempted by Szołtysek in his studies of residence patterns in historic Poland–Lithuania (Szołtysek 2008a, 2008b, 2009).[8] Using measures of household structure and composition, he revealed that, at the end of the eighteenth century, three household and family patterns with substantial numerical and qualitative differences existed in the Polish territories. The structural progression within larger regions was shown to have nearly always moved in the same direction: from less

[8] Throughout the early modern era, the Polish–Lithuanian Commonwealth was one of the largest countries in Europe, covering most of the immense territory between the German-speaking areas and Tsarist Russia.

kin-centred, more nucleated, and neo-local households in the west, to much higher levels of household complexity on Poland's eastern edges.

The present analysis builds on this research, and expands it by focusing on the living arrangements of elderly people (aged 65 or older) in multigenerational households. The residential patterns of the aged are one of the key indicators of differentials in domestic group organization and welfare functions of the family (Szołtysek 2012c). Also, looking at the residence patterns from the perspective of the elderly reduces the effects of demographic conditions on the potential for intergenerational co-residence (Ruggles 2010).[9] This analysis evaluates the case for regional variation in the living arrangements of the aged by exploiting a vast collection of historical data on 21,132 rural households from late eighteenth-century Poland–Lithuania (*CEURFAMFORM* (CEFFD) database). These domestic groups belonged to 236 parishes with 900 settlements, with an overall population of nearly 130,000 persons (Figure 15.3). The data were derived from various types of population listings, which were painstakingly assembled from regional historical archives in Poland (Warszawa, Kraków, Jelenia Góra), Belarus (Minsk), Ukraine (Kiev), and Lithuania (Vilnius), and subsequently transcribed into a computer data file.[10] These enumerations listed individuals by residential units, providing a standard set of demographic identifiers (age, sex, marital status) and kinship relationship pointers. More than 90% of these listings come from the period 1766–1799, and the rest are scattered between 1666 and 1751.

The goal was to use elderly-focused variables to capture the existing interregional variation in family system organization in historical Polish lands. In addition, using the ordinary least squares regression, the intention was to investigate whether the observed diversity could be accounted for by referring to measures of demographic structures and economic conditions. Recently, Ruggles put forward a set of influential hypotheses suggesting that variation in rates of intergenerational co-residence can be successfully explained by demographic rates and agricultural activities alone (Ruggles 2009, 2010).

Multigenerational cohabitation of the elderly was subdivided into stem-family and joint-family types in order to capture two distinct manifestations of the complexity of family co-residence. The elderly were classified as living in stem-family type arrangements if they were co-residing with no more than one married child, and in joint-family type arrangements if they were cohabitating with two or more married children (Ruggles 2010; Gruber and Szołtysek 2012).[11] In order to account for regional variations in demographic conditions, simple control variables for mortality, fertility, and nuptiality were constructed (see Table 15.3). These variables were identical to those used by Ruggles (2009, 2010) and subsequently by Gruber and Szołtysek (2012); the detailed descriptions and rationale for each of them are documented in those studies.

[9] Even in populations where few households have the potential to include three generations, most persons over the age of 65 have the demographic potential to reside with children and grandchildren.

[10] The database development was supported by the Marie Curie Intra-European Fellowship project (FP6-2002-Mobility-5, Proposal No. 515065) at the Cambridge Group for the History of Population and Social Structure, Cambridge, UK, 2006–2008.

[11] Married couples in which both partners were aged 65 or older were treated as single observations, as they shared a single living arrangement.

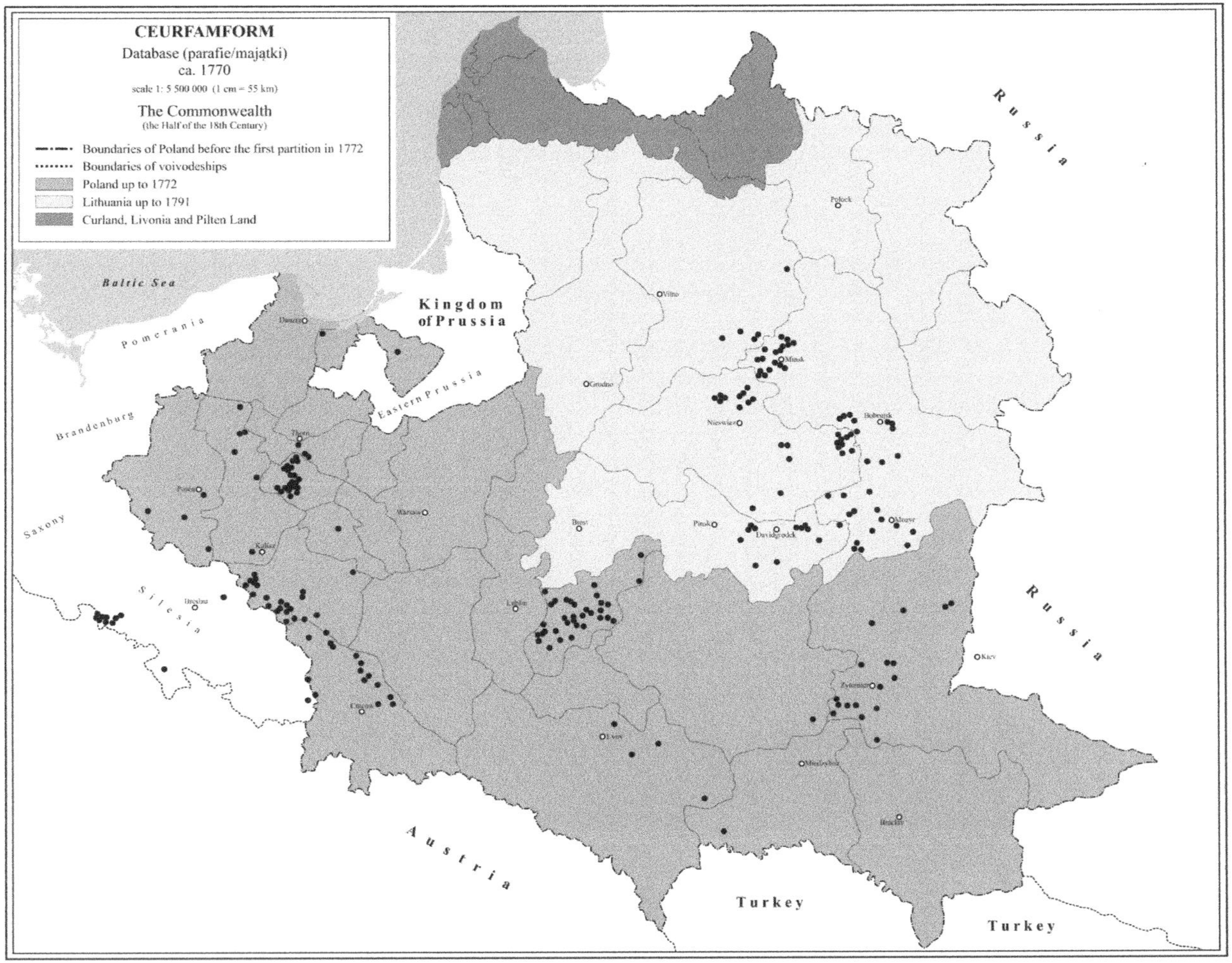

FIG. 15.3 Data spatial distribution of the CEURFAMFORM database (parishes).

Table 15.3 Characteristics of census samples and values of variables used in the analysis

Region	Percent elderly residing in		N
	stem family	joint family	
Poland–Lithuania 18th c.	44.3	15.7	3,419
PL West	36.8	0.8	1,254
PL1	6.3	–	80
PL2	35.2	2.1	145
PL3	37.4	0.9	348
PL4	41.2	0.8	262
PL5	44.6	0.5	184
PL6 + 7	36.6	0.4	235
PL East	48.5	24.3	2,165
PL8	53.9	14.5	572
PL10	44.4	31.3	374
PL11N	48.0	18.8	542
PL11S	46.8	33.2	677

Source: CEFFD database.

We have grouped the 236 parishes studied into regions, either on the basis of their administrative affiliation or geographical proximity (Figure 15.4). These cover the western and southern fringes of the Polish–Lithuanian Commonwealth (regions 1 to 5, 6 + 7); extend eastwards into the borderland between present-day Poland, Belarus, and Ukraine (region 8); and then move even farther east into Belarus, south from Minsk (regions 11N and 11S, respectively), and into western Ukraine (region 10). While the spatial coverage is rather dispersed, the 'samples' cover major socio-economic and cultural regions of historical Poland–Lithuania. The database comprises exclusively rural societies engaged in small- and medium-scale farming. An overwhelming majority of the population of all of the regions were serfs living in personal and hereditary subjugation.

Part of the regression analysis reported here comes from a larger study in which the Polish data were inspected along with 100 samples drawn from countries around the world since the mid-twentieth century, and from Western Europe and North America since the mid-nineteenth century (Gruber and Szołtysek 2012).[12] Data for the 1918 census of Albania and nineteenth-century Serbia were also included. While this additional material was important both in structuring the overall argument, and also in guiding the statistical

[12] These were computerized by IPUMS International and the North Atlantic Project, and researched by Ruggles (in the text also as 'Ruggles' collection'; see more in Gruber and Szołtysek 2012).

analysis, in the present context it is important only insofar as it creates a background against which Polish–Lithuanian peculiarities can be explained.

Table 15.3 summarizes the regional variations for the two measures of living arrangements within Poland–Lithuania. It should be noted that regions 1–7 represent the western part of the country (Figure 15.4), whereas the rest of the regions (8–11S) make up the eastern territories.[13]

The proportion of the population living in stem families did not vary significantly in different regions of the country. While the proportion of the elderly in stem families in western Poland was, on average, 12 percentage points higher than in the east, some of the western territories (regions 4–5) displayed rates of co-residence that were similar to those in the east. However, apart from a marked departure from the overall trend in region PL 1 in the very north, there was a general trend among regions in the western part of Poland–Lithuania towards somewhat lower rates of co-residence of the elderly in stem families than in the eastern parts of the country.

When we look at the spatial distribution of joint-family co-residence among the aged within Poland–Lithuania, sharp contrasts become apparent. The Polish–Lithuanian data reveal a striking difference between the country's western and eastern halves.[14] In the western part of Poland (regions 1–7) we find only negligible traces of elderly people in co-residence in joint families (hardly above the 1% level), or rates that are very similar to those characteristic of nineteenth-century Western Europe (0.2–0.6%; see Gruber and Szołtysek 2012: 112). In contrast, the rate of joint-family co-residence in eastern Poland (at 24.3%) is not only higher than in any country in Ruggles' collection, it also surpasses the corresponding figures for Albania (Gruber and Szołtysek 2012: 112). The highest rates can be found for regions 11S and 10 in the eastern part of historic Poland–Lithuania (today southern Belarus and northern Ukraine; see Figure 15.4), where more than 30% of the elderly were living with at least two married children.

In order to control for the effects of economic and demographic conditions on the residential patterns of the elderly, we used regression analysis to predict the living arrangements of the aged in each of the census samples, and then to compare this predicted family structure with the observed values. As the true contrast within Poland–Lithuania was in the prevalence of joint-family arrangements, here we focus only on the region-specific proportions of the aged living with at least two married children. This dependent variable was regressed on the nine predictors. Regional variations in their values are reported in Table 15.4.

There is a great deal of random variation in the regional values of all of the independent variables, but some regularities can be discerned. Although there seems to be no general east–west contrast across the territories of historical Poland, there were rather dramatic differences in the age at marriage. For the most part, males in the western regions of

[13] To secure a sufficient number of cases, regions 6 + 7 were combined (the two regions were demographically and structurally similar). Region 9, unlike all of the other eastern territories, was omitted from the analysis because it provided too few cases for the age-specific analysis of residence patterns.

[14] Basic demographic variation between Polish eastern and western territories was by no means a crucial factor responsible for these observed differences in family arrangements of the aged. The east–west gradient in joint family co-residence across Poland–Lithuania persists when regional demographic differences as measured by differences in nuptiality, the share of elderly, and marital fertility, are held constant.

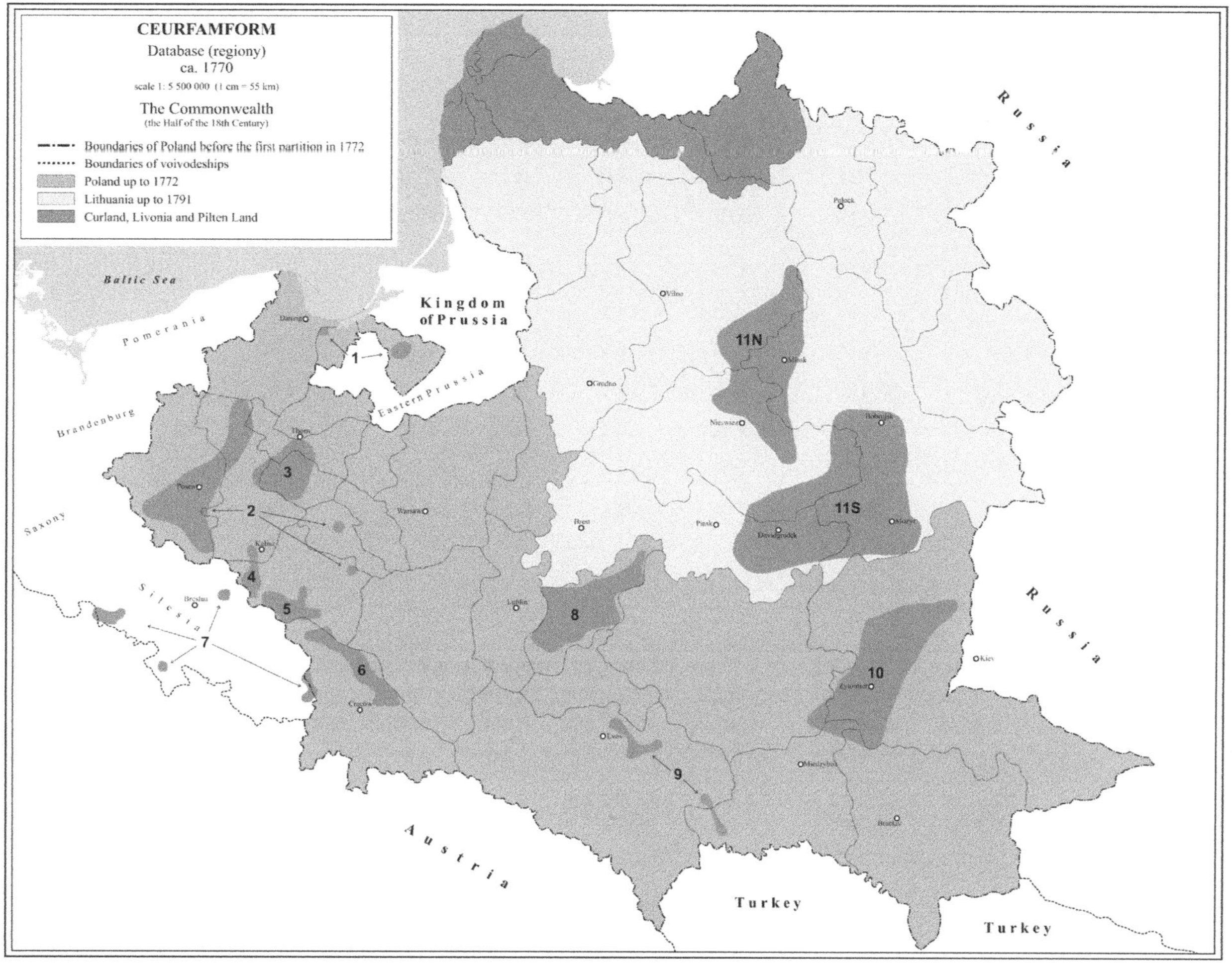

FIG. 15.4 Data spatial distribution of the CEURFAMFORM database (regions).

Table 15.4 Census samples by percent of stem and joint families among the aged

	Agricultural employment	Percent elderly	Marital fertility	Female marriage (SMAM)	Male marriage (SMAM)	Non-marriage	Unmarried elderly women	Elderly couples	*De jure* census		
	Log of % men aged 18–64 employed in agriculture	% of population aged 65 or older	Age-standardized marital fertility ratio	Female singulate age at marriage	Male singulate age at marriage	% never married at ages 45–54 (both sexes)	% pop 65 + who are women without spouses	% pop 65 + who are residing with spouse	*De jure* census enumeration rule	Population	Households
Poland–Lithuania 18th c.	**4.5**	**3.0**	**64.3**	**20.0**	**24.0**	**1.2**	**27.6**	**53.5**	**0**	**127,847**	**21,132**
PL West	**4.5**	**2.9**	**69.4**	**22.5**	**27.3**	**1.8**	**30.8**	**56.8**	**0**	**47,938**	**7,921**
PL1	4.5	3.5	83.9	25.1	28.5	0.9	25.8	62.9	0	2,543	545
PL2	4.5	3.6	78.1	24.0	28.9	2.5	34.6	51.5	0	4,559	646
PL3	4.5	3.2	71.1	21.5	27.8	2.8	29.9	58.0	0	12,893	1,984
PL4	4.5	3.7	75.1	22.3	26.4	1.3	32.1	53.1	0	8,358	1,317
PL5	4.5	2.2	57.1	22.1	26.6	1.6	25.7	64.2	0	8,320	1,562
PL6 + 7	4.5	2.3	71.6	21.4	27.8	1.2	34.2	54.4	0	11,265	1,867
PL East	**4.6**	**3.0**	**61.5**	**18.4**	**22.0**	**0.7**	**25.6**	**51.5**	**0**	**79,909**	**13,211**
PL8	4.6	2.5	58.1	18.2	23.6	0.8	24.7	45.2	0	25,193	4,520
PL10	4.6	3.0	60.5	19.1	23.1	1.2	26.6	52.0	0	14,026	2,100
PL11N	4.6	3.7	63.1	20.3	24.2	0.6	23.3	53.3	0	16,285	2,855
PL11S	4.6	3.2	64.4	16.7	19.9	0.4	27.3	53.3	0	24,405	3,736

Source: CEFF database.

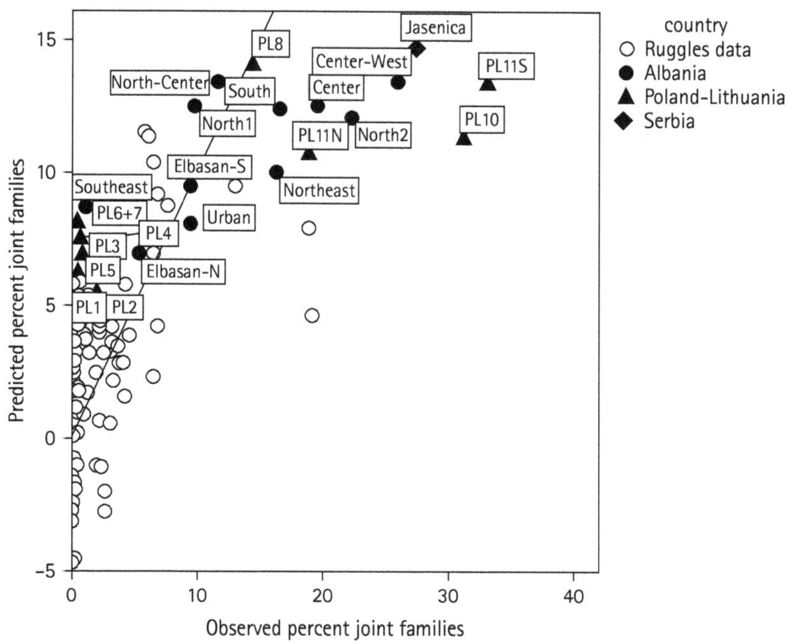

FIG. 15.5 Predicted and observed percentage of elderly in joint families (Poland–Lithuania against 112 census samples).

Sources: Szołtysek (2011); Ruggles et al. (2009); North Atlantic Population Project (2009); Halpern and Gruber (2012a, 2012b); Kaser et al. (2011).

historical Poland married very late, well above the age of 25. By contrast, male Ukrainians and Belarusians tended to marry at much younger ages. The female ages at marriage in Poland's eastern regions in the east (as low as under age 19) were, meanwhile, generally consistent with the characteristics of Hajnal's 'joint household formation systems'. The average female age at marriage rose above 21–22 years only in the western Polish regions.

Figure 15.5 plots the observed against the predicted values for the model discussed above (the model adjusted R^2 was 0.41). All of the eastern Polish values except for one are dispersed well below the prediction line, along with those of some other Eastern European regions (e.g. Albania). The predictive power of the model was found to be very weak: the proportions of the elderly living with at least two married children were much higher than expected from the model in all of the areas studied. On the other hand, the majority of the Polish western areas are strongly clustered near the left axis and display values of joint-family co-residence that are much lower than predicted. The gap between the western and eastern territories of Poland seems to be as large as is possible.

Figure 15.6 shows the goodness of fit of a further model, in which only the countries with at least 2% of the elderly living in joint co-residence arrangements were included (the number of observations thus dropped to 59).[15] This regression model is more

[15] In practice, this led to a circumvention of the comparison of the historical Eastern European data

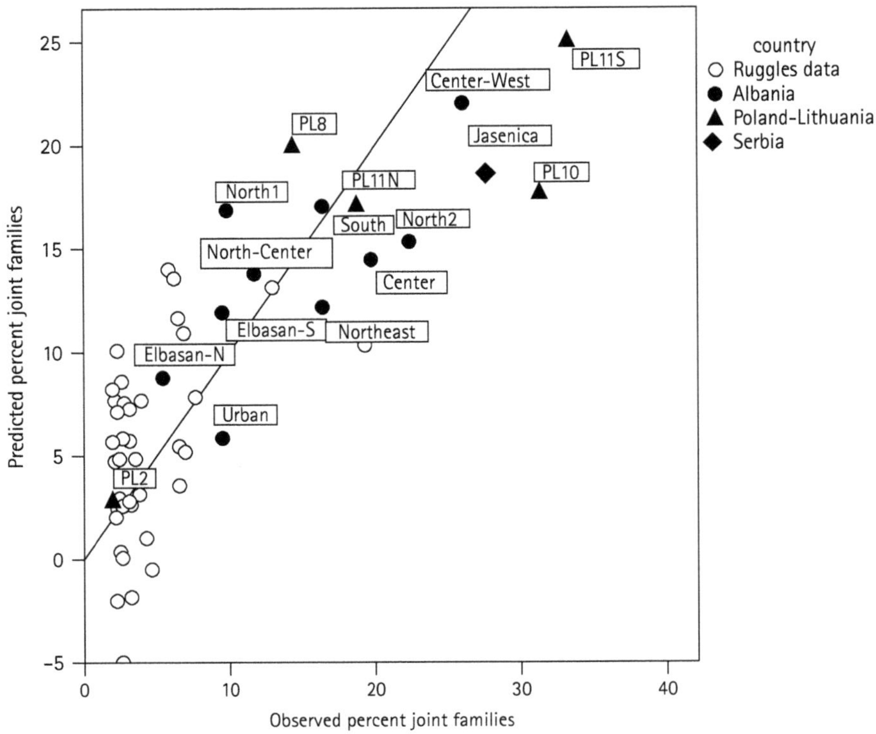

FIG. 15.6 Predicted and observed percentage of elderly in joint families (Poland–Lithuania against 54 census samples).

Sources: Szołtysek (2011); Ruggles et al. (2009); North Atlantic Population Project (2009); Halpern and Gruber (2012a, 2012b); Kaser et al. (2011).

powerful (a 40% increase of adjusted R^2, from 0.41 to 0.57). However, for the two Polish eastern regions with the highest propensity for joint-family co-residence (regions 10 and 11N), it still shows a rather weak association between the predicted and the observed values. This suggests that a simple structural model does not provide a full explanation of elderly co-residence patterns in these regions of Poland–Lithuania (43% of the observed variation remains unexplained). It may well be that, for these regions to fit into the predicted pattern, an additional factor not covered by the control variables, such as the ecotype or a cultural disposition to live together with relatives in the same household, should be taken into account.

What do all of these exercises tell us about the familial nature of various subpopulations of historical Poland? First, they point to a striking diversity in familial behaviour at the 'meso' level. This variation calls into question the validity of efforts to homogenize Eastern European family patterns, and shows that it is dangerous to assume that we can extrapolate directly from the micro or family level to macro levels (regional or national). This research shows that the notion of a homogenous family system in Eastern Europe is at best superficial, and likely to be seriously misleading.

Secondly, it appears that the measure of joint family structure in Poland–Lithuania is not closely tied to basic measures of demography. The regions with the highest proportions of joint-family arrangements in our datasets have higher observed percentages than those that were predicted. This suggests that we should think about adding more variables to the model. The obvious temptation is to turn to the supposed cultural, rather than structural, factors that might be responsible for the variations in the frequency of this type of co-residence. However, while it is likely that cultural influences played some role in fostering the complexity of residential arrangements, we believe that differences in the observed patterns of joint-family structure across Poland–Lithuania may be more satisfactorily explained by a mixture of economic, ecological, and institutional factors—or, better, constraints (cf. Verdon 1998). Cultural differences, ecotype specificities, and political economic and demographic factors might be the pieces of a puzzle that have yet to be put together in studies of family systems. The use of multilevel interpretations that attempt to sort out the social, ecological, economic, and cultural factors that influence the observed differences is probably the only way to provide an accurate picture of the polymorphic nature of family patterns across Poland–Lithuania. In subsequent paragraphs, we will try to 'bite that bullet'.

15.4 Towards Multilevel Explanations of Polish–Lithuanian Diversity

In the following section, we discuss a number of contextual influences on the observed variation in family patterns in Poland–Lithuania. Although we draw upon knowledge about one particular area, the observations below might explain a wider Eastern European 'puzzle'; that is, they may account for patterns observed in other territories of the region where similar variations in familial forms existed. The provisional set of associations discussed in the following paragraphs involve a combination of demographic, political-economic, and cultural factors. Their complex relationship, as manifested in just one region of Eastern Central Europe, reaffirms how important it is to provide a more nuanced explanation of the aggregate demographic properties of populations, using knowledge of localized contexts.

15.4.1 Population Density

Population density is a powerful demographic indicator which, under certain circumstances, may exert a profound influence on the working of a family system, including patterns of living arrangements (Cain and McNicoll 1988). At the end of the eighteenth century, there were significant differences within Poland–Lithuania which suggested a more or less clear-cut west–east division in population spatial distribution.

Regions 1–7 (this and all subsequent spatial references refer to Figure 15.4) generally represented the most populated areas of the country (22–33 persons/km^2). In some, like Silesia (region 7), population density rose as high as 50 persons/km^2 during the late eighteenth century (Kuklo 2009: 217).

Moving east, we observe a gradual decrease in population density. Region 8 represented a transitional zone (15–18 persons/km^2) between the more densely inhabited western Poland and Belarusian and Ukrainian provinces, where during the last decades of the eighteenth century population density was decidedly lower. A majority of settlements from region 11N were located in the zone of moderately low population density (11–13 persons/km^2). The core areas of region 11S sharply departed from those patterns, and thus stood out even in relation to the generally sparsely populated eastern areas (5–8 persons/km^2) (Karpachev and Kozlovskij 1972). The population density in right-bank Ukraine (region 10) was very similar to that of central Belarus (11–13.5 persons/km^2) (Krykun 1977).

The findings of agricultural economists suggest that there were direct links between low population density, simple farming technologies, and land abundance on the one hand, and the non-existence of landless labouring classes and of hiring and labour exchanges on the other (Boserup 1965). In Polish–Lithuanian regions where population (like labour) was scarce, land was abundant and the level of agricultural advancement was often very low, population was the greatest form of wealth, which left no room for the development of non-cultivating labouring classes. These objective economic circumstances influenced the individual behaviour of peasants, creating, for example, strong disincentives for the development of life-cycle service. Whereas in the 'West', considerable land mobility offered alternatives to staying on the family farm and provided incentives for entering the ranks of hired labour (e.g. life-cycle 'service'), at the edges of the Polish eastern territories the more complex domestic groups provided various insurance substitutes to younger household members, thus discouraging them from leaving home. Whether socio-economic conditions in the 'Eastern' clusters facilitated the development of cultural attitudes which devalued hired labour, or whether these values were already present in the family living environment of Belarusian peasants (Obrębski 2007: 139), is difficult to judge (see more in Szołtysek 2009). Last but not least, differences in population density seem to have acted as an ecological sustainability test (Szołtysek and Zuber-Goldstein 2009: 24) through which localized landowners' strategies were filtered and adjusted to prevailing demographic and agricultural conditions (see section 15.4.4).

15.4.2 Ethnic and Religious Differentiation

The Polish–Lithuanian population under investigation was strongly differentiated by ethnic and religious affiliation. Polish-speaking communities of the Catholic faith predominated in the western territories (regions 2–6) (sometimes intermixed with some Protestant German-speaking settlers). However, the northern part of that cluster (region 1) and the western part of region 7 had retained their non-Polish (mostly German) and predominantly Protestant character since the sixteenth century (Litak 2006: 168; Szady 2010: 228–250).

Since the late Middle Ages, the division between western and eastern Slavdom, along with the dividing line between the Western and Eastern Churches, had run southward through region 8 (Szady 2010: 228–250). At the end of the eighteenth century, the latter was dominated by Uniates (Greco-Catholics), who constituted 75% of the population, and were mainly Ruthenians (proto-Ukrainians) (Budzyński 2008: 142–144). Moving east from region 8, the preponderance of Uniates over Catholics was becoming even more overwhelming.

Regions 11N and 11S consisted almost exclusively of Greco-Catholics of Ruthenian/Belarusian origins, with only some minor Polish and Lithuanian influences. Larger Polish and Catholic populations could be found only at the northern fringes of region 11N (Litak 2006: 168). Region 11S was inhabited by the Greek Catholic population of Poleshuks, whose linguistic and ethnic identity was an amalgamate of Belarusian and Ukrainian elements (Obrębski 2007). In western Ukraine—west of the Dnieper River (also region 10)—the Ukrainian population predominated (Litvin 2006: 269–270). At the end of eighteenth century, the Uniate/Orthodox Church was much more dominant than the Catholic Church (Litak 2006: 168).

Although contemporary demographic research has posited that some specific racial and ethnic groups may differ in their propensity to include non-nuclear units in their domestic groups (e.g. Goldscheider and Goldscheider 1993), a comparable hypothesis regarding ethnic differences in household composition and household formation patterns in the past so far does not seem to exist. However, much better suited to the current analysis is a set of hypotheses related to the distinction between Western and Eastern Christianity. Following Goody (1983), Mitterauer argued that Christianity played a critical role in eroding the kinship principles underlying the social organization of early medieval European societies. At the same time, he clearly distinguished between different effects of Western and Eastern Christianity, and saw this difference as a crucial marker of divergent family behaviour in historic Eastern and Southeastern Europe. According to Mitterauer, the penetration of the specific agrarian regime that he saw as facilitating the nuclearization of family structure in Eastern Europe, the so-called 'hide system' (see section 15.4.4), 'never went beyond the dividing line between the Western and Eastern Church' (Mitterauer 1994: 3). It was only through the values of Western Christendom that a high marriage age and the rejection of patrilineal principles of household formation eventually became established within the seigniorial framework (Mitterauer 1999: 220). This diverging effect of Western and Eastern Christendom is explained not so much by differences between the two churches in family and marriage rules, as by the weaker institutional power of the Orthodox Church to prevail against the long-term effects of the kinship customs and practices of the pre-Christian substratum (Mitterauer 1994: 11–12; Mitterauer 2003: 42–43).

The Polish data seem to fit nicely into the set of hypotheses postulated by Mitterauer, particularly because the observed difference between the truly divergent family system organization clearly overlapped with the historical boundary between the two types of Christianity in spatial terms. Again, however, the influence of religion appears to have worked not through the aspect of faith or the observance of doctrine, but because of the more effective institutional organization in mitigating the persistent effects of the historical kinship structures.

15.4.3 Divergences in Historical Kinship Organization

Social anthropologists have long believed that household structures were a by-product of underlying kinship structures, or at least were closely intertwined with them (e.g. Yanagisako 1979). Such kinship structures were said to have been particularly resistant to change and capable of placing their imprints on various social and demographic phenomena over the long run. It is therefore interesting to note that a wealth of information from nineteenth- and early twentieth-century kinship studies can be linked to familial behaviour observed among the subpopulations in question.[16] Among the critical issues in the present context are the longevity of archaic forms of communal social organization based on ancestral kinship in the territories in question, and the prevailing patterns of inheritance and devolution of property.

Following Bogišić (1884), nearly all Southern Slavic literature has deemed the so-called *zadruga*-type family forms a relic of ancient all-Slavic forms of ancestral organization. Several Eastern-Central European authors have also signed on to this theory (for review see Szołtysek and Zuber-Goldstein 2009). However, some of them also provide striking acknowledgment of the diversity of family patterns in Eastern-Central Europe. In particular, they suggest that it is necessary to recognize differences in the way family forms develop in the territories that would later become the western and eastern lands of historic Poland–Lithuania.

Kadlec's discussion of the Central European variant of a *zadruga* form known as *niedział* (literally 'something undivided')—a commune of people bound by ancestral kinship who jointly manage a shared estate under the guidance of a single leader—is here particularly informative (Kadlec 1898). He pointed out that such communal forms of patrilineal organization survived for long periods only in southern Slavdom and in Russian countries, but vanished much earlier in regions inhabited by western Slavs (similarly Gimbutas 1971: 136). Among Poles, these communes disappeared earlier than among the Czechs—most likely before the end of the sixteenth century. The more rapid process of individualization of family life and property laws in western Slavdom is mainly attributable to the influence of Western concepts, especially the terminology of German law (*Ius Theutonicum*) (Kadlec 1898: 1–2, 10, 49, 53, 75, 100–107, 125, 130).

Balzer (1899) found large family communes in archival documents pertaining to legal aspects of property organization and devolution in medieval Bohemia, Poland proper (western regions in CEFFD), as well as in the Polish eastern borderlands (including regions 8–11S). However, these *zadrugal* forms varied in durability. They disappeared most rapidly from the territories of the Polish Crown and Bohemia, and, if they lasted longer, then usually as relatively simple and small two-generational communes (Balzer 1899: 185, 193, 241–242). On the western fringes of Ukraine (including

with the Ruggles data from Middle Eastern, African, and Asian countries. Five out of six Polish western regions were now also excluded from the modelling.

[16] Most of the authors mentioned below were not anthropologists or even ethnographers. Most of them were amateur or professional historians (historians of law in particular) working primarily with a range of documentary materials pertaining to legal aspects of family and property.

region 10), family communes lasted well into the sixteenth century and in some minor regions they even survived until the eighteenth century (Balzer 1899: 191–199). Central European family communes also differed in their life-cycle characteristics. In western Poland and Bohemia, they took the form of temporary joint-property groups, which usually split either immediately or shortly after the demise of the head. Farther to the east in Poland, 'undivided family units' were more durable: in these remote areas, communes among brothers sometimes lasted their entire lifespan (Balzer 1899: 193–199).

According to Łowmiański, the disintegration of lineage groups into small families was already occurring among Slavs during the period of intense settlement between the seventh and tenth centuries. However, this process did not result in only the conjugal family gaining primacy. In early modern Poland proper (regions 1–7), family communes of the scale of extended Balkan *zadrugas* did not occur; households of more than one married couple occurred, but only exceptionally (Łowmiański 1967: 357–358). However, in some regions of Slavdom, a strong lineage system survived until very recent times (Łowmiański 1967: 346–350). For example, Łowmiański, drawing upon substantial archival material on medieval and early modern settlement processes, and court cases on property disputes and transfers, but only sporadically on quantitative data on population, juxtaposed small and nuclearized families of the Polish Crown (western regions in CEFFD) with residential communes from Belarus (regions 11N and 11S), where the population used familial property communes extensively (Łowmiański 1967: 360–362). According to Łowmiański, the grand Belarusian families from the late early modern period were the continuation of a previously vanished institution prevalent on the western lands of the Polish country.

What the above studies suggest is that the character of the Eastern European population was multifaceted. When we look at the dynamics of kinship organization over the medieval and early modern times, we see that there was not one, but a variety of 'populations' across the vast expanse of Eastern Europe that could be distinguished not only by the ways in which the kinship idiom structured principles of inheritance, property, and family organization, but also by the divergent rhythms of the development of these principles. The observer is particularly struck by the congruence of differences in late early modern living arrangements across various parts of Poland–Lithuania and the differences in the pace at which old ancestral forms of social organization were gradually vanishing in different areas. However, while it seems plausible to interpret internal east–west dichotomies in Poland in family organization as resulting from the dynamics of kinship organization, additional research seems to be required before we can convincingly posit causal and temporal links between phenomena observed at the household level and historical trajectories of kinship. Given the prominence of economic explanations in recent analyses of variation in historical patterns of co-residence, it seems advisable to consider other types of influences (Ruggles 2009; also Alderson and Sanderson 1991).

15.4.4 The Manorial System and its Local Variants in Poland–Lithuania

While the economic and social life of all of early modern East-Central Europe was dominated by the rise of a 'new' serfdom and of a manorial–serf economy, noble landlordism was most pronounced in the Polish Commonwealth (e.g. Hagen 1998). The massive growth in the landlord's powers over the rural population led to an expansion of previously modest familial manor farms into large-scale domainal economies that had the goal of producing surpluses for sale in the urban markets of Western Europe. This type of seigneurialism prompted landlords to claim from their peasant subjects not only rents in cash and kind, but, above all, the labour services which were essential to the very functioning of demesne farms (Kula 1976). However, the new system, which arose within the framework of manorialization, based on methodical village structures and a standardized unit of arable known as the 'hide' (*Hufe*), had two essential features, which interacted to a large extent with family arrangements of the rural subjects: (1) the principle of single-heir impartible farm succession, which meant that only one of the sons could inherit and marry; and (2) a rule which dictated that only one married couple with children could live off a particular hide. The uniform populating of *Hufes* with nuclear families and the simultaneous prevention of a numerical accretion of farming families on them resulted from a systematic policy of seigneury devised in order to facilitate the most beneficial collection of tribute (Mitterauer 1999). As this policy encouraged the formation of conjugal families, bilateral kinship, and the loosening of genealogical ties, it worked against the sustainability of complex families, favouring instead the neo-local formation of the family.

The centuries-long process of the diffusion of the manorial system over Poland–Lithuania was, however, marked by strong regional disparities. In general, in the western parts of the historical Kingdom of Poland (regions 1–7 in CEFFD), pre-existing Polish–Slavic legal and economic arrangements had been restructured into the hide system starting in the thirteenth and fourteenth centuries (Hagen 1998: 154–156). During the two subsequent centuries, this legal and agrarian regime was further extended into the southeastern areas of the Polish Crown largely inhabited by the Ruthenian population (including region 8 and 9) (e.g. Budzyński 2008: 85–94). Only by the end of the 1630s, the new order arrived in historic Volhynia and the northern shores of the right-bank Ukraine (region 10) (Litvin 2006: 122, 135, 144–145). The final and much delayed manifestation of these processes was the introduction of the 'hide constitution' among rural populations of Belarusians and Lithuanians in the Grand Duchy of Lithuania from the sixteenth century onwards (*pomiara włóczna*)[17] (regions 11N and 11S) (French 1969).

Ideally, a direct consequence of *pomiara* was the advanced standardization of the material living conditions of the peasants, which was reflected in the normalization of their arable land, their feudal obligations, and the material resources essential to their performance. Under normal conditions, a successful implementation of the reform accelerated the progressive dissolution of old forms of social-territorial and familial organization, especially complex forms, and stimulated the already ongoing process of the individualization of families (e.g. Kernazhȳtski 1931: 123–125, 128–129; Golubev

[17] Literally: apportioning grounds into 'włóka'—hide.

1992: 88; Litvin 2006: 154–155). However, instead of a complete replacement of the previous rules and arrangements, institutional and settlement hybrids sometimes emerged in some of these eastern territories (for example in region 8). In some other locations, the popularization of a unified land structure, and especially of individual land management, was long hindered by traditions of collective landowning (e.g. Litvin 2006: 152; Rosdolsky 1954). This is why, in some parts of the Ukrainian and Belarusian lands (regions 10–11N), traditional property, economic, and family relations prevailed well into the sixteenth century, but were ended by the intensive development of manorialism in the seventeenth and eighteenth centuries (Gurbik 2011). In other areas of the Polish east (particularly in Polessya; region 11S), the implementation of the hide system and its associated features was severely hampered by environmental difficulties (harsh ecological conditions due to extensive swamps). In these areas, primitive forms of agriculture and archaic family arrangements survived well into the 1920s (French 1969: 131; Kernazhŷtski 1931: 73–78; Obrębski 2007).

Since the peasant populations in historic Poland did not, as a rule, have the status of independent farmers (both legally and economically), a great number of key issues related to family formation processes should be viewed from the perspective of landlords' strategies and 'policies'. Indeed, there is considerable evidence that Polish–Lithuanian landlords were customarily concerned with their peasants' property transfers and residential arrangements (e.g. Kula 1976). Estate instructions from Poland–Lithuania suggest that, in most parts of the territory, the maintenance (or, if necessary, the restoration) of tax- or labour-capable family units was among the landlords' most explicit economic interests. This orientation in seigniorial authority suggests that there may have been strong interventions by landlords in the process of property transfer, which provided strong incentives for neo-local household formation among subject farmers (Szołtysek and Zuber-Goldstein 2009).

Wishing to have the maximum possible number of peasant families ready to perform duties for the demesne, feudal lords in the western parts of Poland–Lithuania (regions 1–7 in CEFFD) often separated co-residing couples of different or the same generations, or they employed other strategies to encourage the formation of households in a neo-local manner (Kula 1976; Woźniak 1987: 93–94; Szołtysek and Zuber-Goldstein 2009: 23–24).[18] Similar tendencies can also be found in the eastern parts of Poland–Lithuania and Ukraine, where the serf-owner's perennial desire was to increase the population, and thus to increase the number of arable cultivators. These goals were usually achieved through marrying the serfs off early and universally, but also by preventing the co-residence of too many potential labour-capable units (Pochilevich 1952: 406; Golubev 1992: 57–76; Kapyski and Kapyski 1993: 44–45). When necessary to make neo-localism feasible, special provisions and

[18] These neo-local principles seemed to be common knowledge even to contemporary observers, as it is easy to find descriptions similar this one from 1767: 'A serf, having no more property besides the clothes he gained while being in service, is usually forced to take over a holding together with an acreage just after his marriage' ('Monitor' 1767; quoted from Woźniak 1987: 108).

support were available to the landlord: '[. . .] and where two household heads are to be found in one house,' one of the *Instructions* stipulated, 'a holding from which two days of service can be fixed must be immediately apportioned, building new premises has to be prescribed, support in wood must be given, as well as one year of freedom of all obligations' (Pawlik 1915: 24).

Taken together, these actions seemed to have created favourable conditions for neo-local family formation and to have constrained intergenerational co-residence in Polish eastern lands (Szołtysek and Zuber-Goldstein 2009). However, in many of these eastern territories, especially in Belarus and Ukraine (regions 10–11S), which suffered severe population losses in the seventeenth and early eighteenth centuries due to wars, the 'neo-local incentives from above' were strongly inhibited by general ecological constraints, as well as by place-specific agricultural and demographic conditions.

Despite the abundance of land suitable for cultivation by the rural classes until the end of the eighteenth century, the scarcity of labour and the almost complete lack of a market for hired labour, coupled with low levels of agricultural development, made effective multiplication of the numbers of labour-capable nuclear household units unlikely in the east (Szołtysek 2009: 81; also French 1969). Lithuanian-Belarusian landlords seemed to have been well aware that certain socio-economic and ecological conditions imposed constraints on their otherwise more-or-less 'Western' economic orientation. The *Instructions* suggest that Belarusian seigneurs understood quite well why, given the poor agricultural conditions of their lands and the often limited resources available for supporting individual families, a temporary co-residence of several family units might help to prevent the creation of economically unviable households. 'The estate manager should not allow family households to split,' one of the *Instructions* stipulated, 'unless there are two male adults in the subunit wishing to stay where it was before, and at least one adult son in the branch is intending to become independent [. . .].' This is 'because singletons [single householders] split between two households are likely to fall into poverty due to the lack of sufficient manpower' (Grodzienska Crown estate, 1777, west from region 11N, after Pawlik 1915: 134; see also 47, 53, 167, 277).

Although economic considerations seemed to have played a decisive role in preventing multiple-family groups from splitting unconditionally, the resulting accumulation of family members in one dwelling might have corresponded well with the Belarusian and Ukrainian 'ancient' common-law practices (Obrębski 2007). To the extent that these were not fiercely and successfully opposing the seigniorial policies, manorial regulations might have been effective enough to have created a relatively high number of the extended and multiple-family households actually observed in the Polish eastern regions in the CEFF database. This means that the cultural or economic preferences of the peasants, combined with institutional pressures, formed a complex web of interactions. It is essential that we seek to disentangle and understand this complex web of interactions among the various populations by linking demography, political economy, and culture.

15.5 Summary and Conclusion

Western scholars of Eastern Europe typically see the region as the locus of complex family organization and familistic societal values, a view that has attained the status of dogma in Western social sciences and demography. A few scholars have criticized these mainstream perspectives on the topic from an empirical, conceptual, or epistemological point of view, and have called for moving beyond the stereotypical and artificial divisions of Europe into 'Western' and 'Eastern'. In the 1990s, Kertzer argued that '[. . .] eastern Europe, like western Europe, displayed a diversity of household systems in preindustrial times' which were linked to 'regional differences in political economic arrangements and ecological conditions' (1991: 163). However, as the persistent usage by demographers of the division proposed by Hajnal to explain European demographic differentials since the Second World War suggests, the position of the 'revisionists' is still largely absent from mainstream discourse (see, however, Szołtysek 2012a, 2012b). This essay represents an attempt to persuade scholars to accept a less stereotypical image of families outside of Western Europe.

Drawing on a reading of the literature long screened off from the main current of European thought, as well as a collection of census microdata from historical Poland, Belarus, and Ukraine, we have identified in this paper multiple subpopulations in the area of Eastern-Central Europe separated by significant variations in household formation, marriage, and residence patterns. Thus, the challenge of the (re)compositional analysis of Eastern Europe's familial history primarily involves questioning the existence of a demographically uniform population of the region. Continuing to retain a dichotomous division into two zones of familial behaviours defined across some imagined 'line' no longer appears to be defensible. These findings need to be integrated into mainstream research and discussion so that scholars can gain a better understanding of marriage, family, household, and community patterns in Europe and elsewhere.

It also follows that all propositions that link the contemporary separateness of the Eastern European populations *en masse* according to a specific feature of their social, political, and economic life with their supposedly historically based peculiarity in terms of marital behaviour or familial forms (e.g. Todd 1985; Hartman 2004) will never reflect historical reality on the ground, and as such can never be validated. There were many Eastern European populations in the past, and the notion of a homogenous family system in the region can be maintained only at the price of a level of generalization that is so superficial that it is likely to be seriously misleading.

However, the compositional approach adopted here not only makes the kind of simplifications that Hajnal, Laslett, and others fostered untenable, it also puts the study of Eastern European populations and their heterogeneity on a very different footing. This approach calls into question the existence of a demographically uniform population of Eastern Europe. It goes beyond older homogenizing models of

Eastern European family patterns by adopting a new methodological outlook that relies on an inherently variegated understanding of the population, and highlights the several ways in which different populations were nested in their respective local contexts.

In terms of methodology, our approach to understanding past populations advocates the systematic use of archival sources to document the contours and variability of households, nuptiality, and life course patterns, thus enabling Eastern European regions to be defined by spatially organized historical statistics. Instead of relying on conventional and simplistic notions of dividing lines and 'ideal family systems', this new scientific programme takes a more flexible view of the nature and permeability of frontiers and transition zones, and the ways in which familial and demographic borders have been crossed and diffused, both across space and over time. Such critical investigations of the differences in European historical family systems over space and time should replace simplistic claims and tentative hypotheses based on isolated demographic facts or national aggregates of individuals. As we have shown, this shift in thinking about the populations of historical Eastern Europe does not preclude meaningful generalizations. Generalizations need not be universally applicable. Rather, to deal adequately with the real world they must first be shown applicable to local regions.

Finally, this approach highlights the importance of the interaction of a variety of human populations with their environment in the broader sense. When assigning certain familial characteristics to a given population, it is essential to take into account the population's interaction with the localized but potentially powerful forces of ecology, demography, and political economy. The same demographic manifestations may in fact result from different pressures coming from otherwise unobserved effects of a specific ecotype or institutional setting, while different familial outcomes can be produced by seemingly similar contexts in one or several domains. A careful sorting of social, ecological, economic, and cultural factors is particularly advisable for future studies of Eastern European populations, in which we can expect the several explanatory factors to be strongly intermingled.

References

Alderson, A. S. and S. K. Sanderson (1991). 'Historic European Household Structures and the Capitalist World-Economy', *Journal of Family History* 16/4: 419–432.

Andorka, R. (1976). 'The Peasant Family Structure in the 18th and 19th Centuries (Data from Alsónyék and Kölked in International Comparison)', *Acta Ethnographica Academiae Scientiarum Hungaricae* 25/3–4: 321–348.

Andorka, R., and T. Farago, (1983). 'Pre-industrial Household Structure in Hungary', in R. Wall and J. Robin (eds), *Family Forms in Historic Europe*. Cambridge: Cambridge University Press, 281–307.

Balzer, O. (1899). 'O zadrudze słowiańskiej: Uwagi i polemika'. *Kwartalnik Historyczny* 13/2: 183–256.
Blum, A. and J.-L. Rallu (1993). 'European Population', in A. Blum and J.-L. Rallu (eds), *European Population*, ii: *Demographic Dynamics*. Montrouge: John Libbey Eurotext, 1–48.
Bogišić, V. (1884). 'D'une forme particulière de la famille chez les Serbes et les Croates', *Revue de droit internationale et de législation comparée* 16: 374–409.
Boserup, E. (1965). *The Conditions of Agricultural Growth: The Economics of Agrarian Change under Population Pressure*. Chicago: Aldine.
Budzyński, Z. (2008). *Kresy południowo-wschodnie w drugiej połowie XVIII wieku*. Studia z dziejów społecznych 3. Przemyśl-Rzeszów: Wydawnictwo Naukowe TPN.
Burguière, A. and F. Lebrun (1996). 'The One Hundred and One Families of Europe', in A. Burguiere, C. Klapisch-Zuber, M. Segalen, and F. Zonabend (eds), *A History of the Family*. Cambridge: Polity Press, ii.11–94.
Cain, M. and G. McNicoll (1988). 'Population Growth and Agrarian Outcomes', in R. D. Lee et al. (eds), *Population, Food and Rural Development*. Oxford: Oxford University Press, 101–117
Čaňová, E. (1989). 'Population of the Třeboň dominion' [An analysis of the List of Subjects of 1586], *Historicka demografie* 13: 33–58.
Čaňová, E. (1992). 'Slození domácností v Cechách v roce 1651', *Historicka demografie* 16: 63–66.
Čaňová, E. and P. Horska (1992). 'Existuje stredoevropsky model rodiny pro predstatisticke obdobi?', in Z. Pavlik (ed.), *Snatecnost a Rodina: Brachnost i Semia*. Prague: Academia, 90–104.
Čaňová, E., P. Horska, and E. Maur (1987). 'Les listes nominatives de la Boheme, source de donnees pour l'histoire sociale et la demographie historique', *Annales de demographie historique* 24: 295–312
Cerman, M. (2001). 'Central Europe and the European Marriage Pattern: Marriage Patterns and Family Structure in Central Europe, 16th–19th Centuries', in R. Wall, T. K. Hareven, J. Ehmer, and M. Cerman (eds), *Family History Revisited: Comparative Perspectives*. Newark: University of Delaware Press, 282–307.
Chojnacka, H. (1976). 'Nuptiality Patterns in an Agrarian Society', *Population Studies* 30/2: 203–226.
Czap, P. (1978). 'Marriage and the Peasant Joint Family in Russia', in D. Ransel (ed.), *The Family in Imperial Russia*. Urbana: Illinois University Press, 103–123.
Czap, P. (1982). 'The Perennial Multiple Family Household, Mishino, Russia, 1782–1858', *Journal of Family History* 7: 5–26.
De Moor, T. and J. L. van Zanden (2010). 'Girl Power: The European Marriage Pattern and Labour Markets in the North Sea Region in the Late Medieval and Early Modern Period', *The Economic History Review* 63: 1–33.
Dennison, T. K. (2003). 'Serfdom and Household Structure in Central Russia: Voshchaznikovo, 1816–1858', *Continuity and Change* 18/3: 395–429.
Farago, T. (1986). 'Formen bäuerlicher Haushalts- und Arbeitsorganization in Ungarn um die Mitte des 18. Jahrhunderts', in J. Ehmer and M. Mitterauer (eds), *Familienstruktur und Arbeitsorganisation im ländlichen Gesellschaften*. Graz: Böhlau, 103–183.
Farago, T. (1998). 'Different Household Formation Systems in Hungary at the End of the Eighteenth Century: Variations on John Hajnal's Thesis', *Historical Social Research* 23/1–2: 83–111.

Fauve-Chamoux, A. (2001). 'Marriage,Widowhood, and Divorce', in D. I. Kertzer and M. Barbagli (eds), *Family Life in Early Modern Times 1500–1789*, i: *The History of the European Family*. New Haven and London: Yale University Press, 221–256.

French, R. A. (1969). 'Field Patterns and the Three-Field System: The Case of Sixteenth-Century Lithuania', *Transactions of the Institute of British Geographers* 48: 121–134.

Gimbutas, M. (1971). *The Slavs*. New York: Praeger.

Goldscheider, F. K. and C. Goldscheider (1993). *Leaving Home before Marriage: Ethnicity, Familism, and Generational Relationships*. Madison: University of Wisconsin Press.

Golubev, V. F. (1992). *Sialianskaie zemlevladanne i zemlekarystannne na Belarusi XVI–XVIII ctct*. Minsk: Nauka i Technika.

Goody, J. (1983). *Development of the Family and Marriage in Europe*. Cambridge: Cambridge University Press.

Goško, J. (1976). *Naselennja ukrainśkich Karpat XV–XVIII st: Zaselennja, Mihracii, Pobut*. Kiev: Naukova Dumka.

Grandits, H. (2010). 'Introduction: The Reshaping of Family and Kin Relations in European Welfare Systems', in H. Grandits (ed.), *Kinship and Social Security in Contemporary Europe*, i: *Family, Kinship and State during the Century of Welfare: Eight Countries*. Frankfurt and New York: Campus, 23–46.

Grassland, C. (1990). 'Systèmes démographiques et systèmes supra-nationaux: La fécondité européenne de 1952 à 1982', *Revue Européenne de Démographie* 2: 163–191.

Gruber, S. (2009). 'Household Formation and Marriage: Different Patterns in Serbia and Albania?', in A. Fauve-Chamoux and I. Bolovan (eds), *Families in Europe between the 19th and 21st Centuries: From the Traditional Model to Contemporary PACS*. Cluj-Napoca: Cluj University Press, 229–248.

Gruber, S. and M. Szołtysek (2012). 'Stem Families, Joint Families and the "European Pattern": How Much of a Reconsideration Do We Need?', *Journal of Family History* 37/1: 105–125.

Grulich, J. and H. Zeitlhofer (1999). 'Struktura jihočeských venkovských a městských domácností v 16. a 17. století. (Příspěvek k dějinám sociální každodennosti poddaných v období raného novověku)', *Historicka demografie* 23: 31–84.

Gurbik A. O. (2006). 'Seljans'kij simejnij lad', in V. A. Smolij (ed.), *Istorija ukrains'kogo seljanstva: Narisi*, 2 vols. Kiev: Naukova Dumka, i.152–159.

Gurbik, A. O. (2011). 'Ekonomichne stanovishche ukraïns'kikh zemel' u skladi Rechi Pospolitoï (1569–1647)', in V. M. Litvin et al. (eds), *Ekonomichna istoriia Ukraïni: Istoriko-ekonomichne doslidzhennia*. Kiev: Institut istoriï Ukraïni NAN Ukraïni, i.375–410.

Hagen, W. W. (1998). 'Village Life in East-Elbian Germany and Poland, 1400–1800: Subjection, Self-defense, Survival', in T. Scott (ed.), *The Peasantries of Europe from the Fourteenth to the Eighteenth Centuries*. London: Longman, 145–190.

Hajnal, J. (1965). 'European Marriage Patterns in Perspective', in D. V. Glass and D. E. C. Eversley (eds), *Population in History: Essays in Historical Demography*. London: Edward Arnold, 101–143.

Hajnal, J. (1982). 'Two Kinds of Preindustrial Household Formation System', *Population and Development Review* 8: 449–494.

Hajnal, J. (1983). 'Two Kinds of Preindustrial Household Formation System', in R. Wall and J. Robin (eds), *Family Forms in Historic Europe*. Cambridge: Cambridge University Press, 65–104.

Halecki, O. (1952). *Borderlands of Western Civilization: A History of East Central Europe*. New York: Ronald Press.

Hartman, M. S. (2004). *The Household and the Making of History: A Subversive View of the Western Past*. Cambridge: Cambridge University Press.

Heady P. (2010). 'Introduction: Care, Kinship and Community', in P. Heady and P. Schweitzer (eds), *Kinship and Social Security in Contemporary Europe*, ii: *Family, Kinship and Community at the Start of the 21st Century: Nineteen Localities*. Frankfurt and New York: Campus, 13–59.

Heady, P., S. Gruber, and T. Bircan (2010). 'The Quantitative Background', in P. Heady and P. Schweitzer (eds), *Kinship and Social Security in Contemporary Europe*, ii: *Family, Kinship and Community at the Start of the 21st Century: Nineteen Localities*. Frankfurt and New York: Campus, 61–90.

Hendrickx, F. (2005). 'West of the Hajnal Line: North-Western Europe', in T. Engelen and A. P. Wolf (eds), *Marriage and the Family in Eurasia: Perspectives on the Hajnal Hypothesis*. Amsterdam: Aksant, 73–103.

Hoch, S. L. (1986). *Serfdom and Social Control in Russia: Petrovskoe, a Village in Tambov*. Chicago, IL: University of Chicago Press.

Horska, P. (1989). 'K historickemu modelu stredoevropske rodiny', *Demografie* 31/2: 137–143.

Horska, P. (1994). 'Historical Models of the Central European Family: Czech and Slovak Examples', *Journal of Family History* 19/2: 99–106.

Horský, J. and E. Maur (1993). 'Die Familie, Familienstrukturen und Typologie der Familien in der böhmischen Historiographie', *Historicka demografie* 17: 7–35.

Kadlec, K. (1898). *Rodinný nedíl čili zádruha v právu slovanském*. Praha: Nákl. vlastním v komisi Bursíka & Kohouta.

Kapyski, Z. and B. Kapyski (1993). 'Belaruskaia veska i iae nasel'nitstva u kantsy XVI-pershai palove XVII st. Vopyt demagrafichnai kharakterystyki', *The Belarusian Historical Journal* 2: 42–45.

Karpachev, A. M. and P. G. Kozlovskij (1972). 'Dinamika chislennosti naselenia Belorussii vo vtoroi polovine XVII –XVIII vv.', *Ezhegodnik po agrarnoj istorii Vostochnoj Evropy za 1968 god*, 81–94.

Kaser, K. (2001). 'Serfdom in Eastern Europe', in D. I. Kertzer and M. Barbagli (eds), *The History of the European Family*, i: *The History of the European Family*. New Haven and London: Yale University Press, 25–62.

Kernazhŷtski, K. I. (1931). *Agrarnaya réforma ŷ Babruĭskim starostve i ékonomichnae stanovishcha yago nasel'nitstva z XVII da palovŷ KhIKh stal. (Farmavan'ne kapitalizmu)*. Minsk: Belaruskaia͡ akadėmiia͡ navuk.

Kertzer, D. I. (1991). 'Household History and Sociological Theory', *Annual Review of Sociology* 17: 155–179.

Kreager, P. (2009). 'Darwin and Lotka: Two Concepts of Population', *Demographic Research* 21/16: 469–502.

Krykun, M. (1977). 'Nasielenie prawobrzeznoj Ukrainy w 1789 g.', *Problemŷ istoricheskoĭ demografii SSSR*, 92–103.

Kuklo, C. (2009). *Demografia Rzeczypospolitej przedrozbiorowej*. Warszawa: DiG.

Kula, W. (1976). 'The Seigneury and the Peasant Family in Eighteenth-Century Poland', in R. Forster and O. Ranum (eds), *Family and Society. Selection from the 'Annales. Economies, Societes, Civilisations'*. Baltimore: Johns Hopkins University Press, 192–203.

Langer, J. (1994). 'Household—Social Environment—Ecotypes', *Ethnologia Europe Centralis* 2: 43–55.
Laslett, P. (1977). 'Characteristics of the Western Family Considered over Time', in *Family Life and Illicit Love in Earlier Generations: Essays in Historical Sociology*. Cambridge: Cambridge University Press, 12–49.
Laslett, P. (1983). 'Family and Household as Work Group and Kin Group: Areas of Traditional Europe Compared', in R. Wall and J. Robin (eds), *Family Forms in Historic Europe*. Cambridge: Cambridge University Press, 513–563.
Le Play, F. (1867). *La Réforme Sociale en France déduite de l'observation comparée des peuples Européens*, i. Paris: Dentu.
Le Play, F. (1871). *L'organisation de la famille selon le vrai modele signale par l'histoire de toutes les races et de tous les temps*, 3rd edn. Tours: Alfred Mame et fils.
Le Play, F. (1982[1872]). 'Le Réforme Sociale', in C. Bodard Silver (ed.), *Frederic Le Play on Family, Work, and Social Change*. Chicago: Chicago University Press, 259–262.
Le Play, F. (1877–1879). *Les Ouvriers Européens*, 6 vols. Tours: Alfred Mame et fils.
Litak, S. (2006). *Atlas kościoła łacińskiego w Rzeczpospolitej Obojga Narodów w XVIII wieku*. Lublin: Towarzystwo Naukowe KUL.
Litvin, V. M. (ed.) (2006). *Istorija ukrains'kogo seljanstva: Narisi*, 2 vols, i. Kiev: Naukova dumka.
Łowmiański, H. (1967). *Z dziejów Słowian w I tysiącleciu n.e.* Warszawa: Państwowe Wydawn, Naukowe.
Macfarlane, A. (1978). *The Origins of English Individualism: The Family, Property and Social Transition*. Oxford: Basil Blackwell.
Micheli, G. (2000). 'Kinship, Family and Social Network: The Anthropological Embedment of Fertility Change in Southern Europe', *Demographic Research* 3, article 13, DOI: 10.4054/DemRes.2000.3.13.
Mitterauer, M. (1994). 'Medieval Roots of Family Developments', unpublished conference paper (available in the library of the Cambridge Group for the History of Population and Social Structure).
Mitterauer, M. (1999). 'Ostkolonisation und Familienverfassung: Zur Diskussion um die Hajnal-Linie', in V. Rajšp and E. Bruckmüller (eds), *Vilfanov zbornik: Pravo-zgodovina-narod. In memoriam Sergij Vilfan*. Ljubljana: ZRC, 203–221.
Mitterauer, M. (2003). 'European Kinship Systems and Household Structures: Medieval Origins', in P. Heady and H. Grandits (eds), *Distinct Inheritances: Property, Family and Community in a Changing Europe*. Münster: Lit Verlag, 35–52.
Mitterauer, M. and A. Kagan (1982). 'Russian and Central European Family Structures: A Comparative View', *Journal of Family History* 7: 103–131.
Monnier, A. (2004). 'The European Union at the Time of Enlargement', *Population* 59/2: 315–336.
Monnier, A. and J. Rychtarikova (1992). 'The Division of Europe into East and West', *Population: An English Selection* 4: 129–159.
Nauck, B. (2007). 'Value of Children and the Framing of Fertility: Results from a Cross-Cultural Comparative Survey in 10 Societies', *European Sociological Review* 23: 615–629.
Nosevich, V. (2004). *Tradicionnaja belorusskaja derevnja v evropejskoj perspektive*. Minsk: Tèxnalogija.

Nosevich, V. (2007). 'The Multiple-Family Household: Relic of a Patriarchal Past or More Recent Phenomenon?', paper presented at Cambridge Group for the History of Population and Social Structure, August.

Obraniak, W. (1968). 'Oblicze demograficzne wsi wielunskiej w epoce Sejmu Wielkiego', *Studia Demograficzne* 16: 109–122.

Obrębski, J. (2007). *Polesie: Studia etnosocjologiczne*, ed. A. Engelking. Warszawa: Oficyna Naukowa.

Őri, P. (2009). 'Marriage Customs and Household Structure in Hungary at the End of the 18th Century: The Case of County Pest-Pilis-Solt (1774–1785)', in A. Fauve-Chamoux and I. Bolovan (eds), *Families in Europe between the 19th and 21st Centuries: From the the Traditional Model to the Contemporary PACS*, Supplement of the *Romanian Journal of Population Studies*. Cluj-Napoca: Cluj University Press, 167–192.

Pamporov, A. (2008). 'Patterns of Family Formation: Marriage and Fertility Timing in Bulgaria at the Turn of the Twenty-First Century: A Case-Study of Sofia', *History of the Family* 13/2: 210–221.

Pawlik, S. (1915). *Polskie instruktarze ekonomiczne z końca XVII i z XVIII wieku*, i. Kraków: Akademia Umiejętności.

Perier, P. (2002). 'Le Play and his Followers: Over a Century of Achievement', *International Social Science Journal* 50/157: 343–348.

Perkovskij, A. L. (1979). 'Evoljucija simi i gospodarstva na Ukraini v XVII: Peršij polovini XIX st', *Demohrafični doslidžennja* 4: 37–46.

Plakans, A. (1973). 'Peasant Families East and West: A Comment on Lutz K. Berkner's Rural Family Organization in Europe', *Peasant Studies Newsletter* 2: 11–16.

Plakans, A. (1987). 'Interaction between the Household and the Kin Group in the Eastern European Past: Posing the Problem', *Journal of Family History* 12: 163–175.

Plakans, A. (2002). 'Agrarian Reform in the Family in Eastern Europe', in D. I. Kertzer and M. Barbagli (eds), *Family Life in the Long Nineteenth Century 1789–1913*, i: *The History of the European Family*. New Haven and London: Yale University Press, 73–108.

Plakans, A. and C. Wetherell (2001). 'The Search for Place: East European Family History, 1800–2000', in R. Wall, T. K. Hareven, J. Ehmer, and M. Cerman (eds), *Family History Revisited: Comparative Perspectives*. Newark: University of Delaware Press, 257–281.

Plakans, A. and C. Wetherell (2005). 'The Hajnal Line and Eastern Europe', in T. Engelen and A. P. Wolf (eds), *Marriage and the Family in Eurasia: Perspectives on the Hajnal Hypothesis*. Amsterdam: Aksant, 105–126.

Pochilevich, D. L. (1952). 'Zemleustrojstvo i pozemel'nyj kadastr v Belorussii, Litve i Ukraine v XVI-XVII vv.', in B. D. Grekov (ed.), *Materialy po istorii zemledelija SSSR*. Moscow: Izd-vo AN SSSR, i.322–410.

Polaszewski, L. (1978). 'Struktura społeczna ludnoici w parafii Szubin w 1766 roku', *Przeszłoie Demograficzna Polski* 10: 157–175.

Polla, M. (2006). 'Family Systems in Central Russia in the 1830s and 1890s', *History of the Family* 11: 27–44.

Polla, M. (2007). 'Peasant Families in Northern Russia: Nineteenth-Century Regional Patterns', *Historical Social Research* 32/3: 270–298.

Reher, D. S. (1998). 'Family Ties in Western Europe: Persistent Contrasts', *Population and Development Review* 4/2: 203–234.

Rosdolsky, R. (1954). 'Die Ostgalizische Dorfgemeinschaft und ihre Auflösung', *Vierteljahrschrift für Sozial- und Wirtschaftsgeschichte* 41/2: 97–145.
Ruggles, S. (2009). 'Reconsidering the Northwest European Family System: Living Arrangements of the Aged in Comparative Historical Perspective', *Population and Development Review* 35: 249–273.
Ruggles, S. (2010). 'Stem Families and Joint Families in Comparative Historical Perspective', *Population and Development Review* 36: 563–577.
Rumlova, E. (1993). 'Demograficka a socialni struktura obyvatelstva panstvi Dymokury v polovine 17. stoleti', *Historicka demografie* 17: 153–200.
Rychtarikova, J. (1993). 'Nuptialité comparée en Europe de l'Est et en Europe de l'Ouest', in A. Blum and J.-L. Rallu (eds), *European Population*, ii: *Demographic Dynamics*. Montrouge: John Libbey Eurotext, 191–210.
Sakalo, O. (2008). 'Domogospodarstva sil's'kogo naselennja Get'manščini drugoi polovini XVIII st.: Dejaki istoriko-demografični aspekti (na prikladi sela Vedmeže Romens'koi sotni Lubens'kogo polku)', *Kraeznavstvo* 1–4: 168–174.
Seligová, M. (1993). 'Příspěvek ke studiu rodinných struktur v Čechách v 17. století. Panství Děčín—sonda', *Historická demografie* 17: 111–130.
Sklar, J. (1971). 'East European Nuptiality: A Comparative Historical Study of Patterns and Causes', unpublished PhD dissertation. Ann Arbor, University Microfilms Library Services (71-9925).
Sklar, J. (1974). 'The Role of Marriage Behaviour in the Demographic Transition: The Case of Eastern Europe around 1900', *Population Studies* 28: 231–247.
Sliž, N. (2004). 'Dasledavanne gistoryi sjam'i: njavykarystanyja magčymasci belaruskaj gistaryjagrafii', *Gistaryčny Al'manach* 10, <http://kamunikat.fontel.net/www/czasopisy/almanach/10/07.htm>.
Smith, D. S. 1993. 'The Curious History of Theorizing about the History of the Western Nuclear Family', *Social Science History* 17: 325–353.
Sobotka, T. (2003). 'Re-emerging Diversity: Rapid Fertility Changes in Central and Eastern Europe After the Collapse of the Communist Regimes', *Population-E* 58/4–5: 451–486.
Sobotka, T. (2008). 'The Diverse Faces of the Second Demographic Transition in Europe', in T. Frejka, T. Sobotka, and J. M. Hoem (eds), *Childbearing Trends and Policies in Europe* (Special Collection), *Demographic Research* 19, article 8: 171–224.
Sovič, S. (2008). 'Moving Beyond Stereotypes of "East" and "West"', *Cultural and Social History* 5/2: 141–163.
Švecová, S. (1966). 'Klasifikácia rodinných foriem na slovenskom materiáli', *Ceský lid* 53: 85–89.
Švecová, S. (1986). 'Slovenská a ceská rodina', *Ceský lid* 73: 203–205.
Švecová, S. (1989). 'Dva typy tradicnej ceskoslovenskej rodiny v Ceskoslovensku', *Ceský lid* 76: 210–222.
Szady, B. (2010). *Geografia struktur religijnych i wyznaniowych w Koronie w II połowie XVIII w.* Lublin: RW KUL.
Szołtysek, M. (2003). 'Ludność parafii bujakowskiej w XVIII i XIX wieku: Między "unikalnym" systemem formowania się gospodarstw a swoistością pogranicza', unpublished PhD dissertation, University of Wrocław.
Szołtysek, M. (2004). 'Astride the Hajnal Line—Household and Family in the Upper Silesian Parish of Bujakow, 1766–1803', *Polish Population Review* 11: 59–93.

Szołtysek, M. (2007). 'Central European Household and Family Systems, and the "Hajnal-Mitterauer" Line: The Parish of Bujakow (18th–19th Centuries)', *The History of the Family* 1: 19–42.

Szołtysek, M. (2008a). 'Three Kinds of Preindustrial Household Formation System in Historical Eastern Europe: A Challenge to Spatial Patterns of the European Family', *The History of the Family* 13/3: 223–257.

Szołtysek, M. (2008b). 'Rethinking Eastern Europe: Household Formation Patterns in the Polish-Lithuanian Commonwealth and European Family Systems', *Continuity and Change* 23: 389–427.

Szołtysek, M. (2009). 'Life Cycle Service and Family Systems in the Rural Countryside: A Lesson from Historical East-Central Europe', *Annales de Démographie Historique* 1: 53–94.

Szołtysek, M. (2012a). 'Spatial Construction of European Family and Household Systems: Promising Path or Blind Alley? An Eastern European Perspective', *Continuity and Change* 27/1: 11–52.

Szołtysek, M. (2012b). 'The Genealogy of Eastern European Difference: An Insider's View', *Journal of Comparative Family Studies* 43/3: 335–371.

Szołtysek, M. (2012c). 'Family Systems and Welfare Provision in Poland–Lithuania: Discrepancies and Similarities', MPIDR Working Paper WP-2012-016.

Szołtysek, M. and D. Biskup (2008). 'Różnorodność czy tożsamość? Chłopskie gospodarstwo domowe na ziemiach Rzeczypospolitej i Śląska pod koniec XVIII wieku', in C. Kuklo (ed.), *Rodzina i gospodarstwo domowe na ziemiach polskich w XV–XX wieku: Struktury demograficzne, społeczne i gospodarcze*. Warszawa: DiG, 363–390.

Szołtysek, M. and B. Zuber-Goldstein (2009). 'Historical Family Systems and the Great European Divide: The Invention of the Slavic East', *Demográfia: English Edition* 52/5: 5–47.

Tchmelyk, R. (1992). 'Dejaki mirkuvannja pro viniknennja maloi sim'i na ukaini', *Narodna tvorcist' ta etnografija* 3: 41–45.

Tchmelyk, R. (1999). *Mala ukrains'ka seljans'ka simja druhoi polovyny XIX–počatku XX st.* Lviv: Institut narodoznavstva.

Therborn, G. (2004). *Between Sex and Power: Family in the World 1900–2000*. London: Routledge.

Thornton, A. (2005). *Reading History Sideways: The Fallacy and Enduring Impact of the Developmental Paradigm on Family Life*. Chicago: University of Chicago Press.

Todd, E. (1985). *The Explanation of Ideology: Family Structures and Social Systems*. Oxford: Basic Blackwell.

Todorova, M. (2006). 'On the Epistemological Value of Family Models: The Balkans within the European Pattern', in *Balkan Family Structure and the European Pattern: Demographic Developments in Ottoman Bulgaria*. Washington, DC: The American University Press, 199–211.

Verdon, M. (1998). *Rethinking Households: An Atomistic Perspective on European Living Arrangements*. London: Routledge.

Višniauskaitė, A. I. (1971). 'Razvitie litovskoj krest'ianskoj sem'i', in *Proceedings of the VII International Congress of Anthropological and Ethnological Sciences*, Moscow. The Hague: Mouton, 1–12.

Wall, R. (1979). 'Regional and Temporal Variations in English Household Structure from 1650', in J. Hobcraft and P. Rees (eds), *Regional Demographic Development*. London: Croom Helm, 89–113.

Wall, R. (1991). 'European Family and Household Systems', in *Historiens et populations: Liber Amicorum Etienne Helin*. Louvain-la-Neuve: Academia, 617–636.

Wall, R. (2001). 'Transformation of the European Family across the Centuries', in R. Wall, T. K. Hareven, J. Ehmer, and M. Cerman (eds), *Family History Revisited: Comparative Perspectives*. Newark: University of Delaware Press, 217–241.

Woźniak, A. (1987). *Kultura mazowieckiej wsi pańszczyźnianej XVIII i początku XIX wieku*. Wrocław: Zakład Narodowy im. Ossolińskich.

Yanagisako, S. J. (1979). 'Family and Household: The Analysis of Domestic Groups', *Annual Review of Anthropology* 8: 161–205.

CHAPTER 16

LINKING LATE-IMPERIAL AND EARLY MODERN POPULATION DYNAMICS IN THE LOWER YANGZI VALLEY

an analysis of Xiaoji township

MARK ELVIN

16.1 Summary

DR Josephine Fox and I recently published a reconstruction of consistent interlocking age-specific rates of mortality, nuptiality, and fertility in 13 areas of the late-imperial Lower Yangzi region (hereafter LYV) (Elvin and Fox 2008, 2009). These were in effect long-term means, with a chronological focus of 1814. We did not attempt to identify trends over time. The sample size was almost 20,000 cases, but the great majority of individual cases lacked at least one or other of the main parameters for the fields in which we were primarily interested. Overall, though, coverage of these parameters was adequate for the region as a whole; and this was used where necessary as a basis for estimating or adjusting rates for areas where the dataset for the area coverage of a particular measure was too small. The main conceptual framework for the technical demography was the two-parameter logit system pioneered by the late Bill Brass, with additions from work by Heather Booth and Basia Zaba for fertility.

The present study follows this up by focusing on a small area lying in the heart of the same region at a later time, over four 12-month periods that fall within 1931–1935.[1] This is the rural township of Xiaoji zhen 峭岐鎮, using the local pronunciation; the standard modern form is 'Qiaoqi zhen'. It is located in Jiangyin county in the province of Jiangsu.

[1] Abbreviated dates are given hereafter by the *second* of a calendar year partially covered.

The data employed come from a little book, *An Experiment in the Registration of Vital Statistics in China*, compiled by Qiao Qiming 喬啟明, Warren Thompson, and Chen Caizhang 陳彩章, and published in 1938 (hereafter referred to as *QTC*). Its quality of information is unusually high, and I am extremely grateful to Professor Arthur Wolf for recommending it to my attention. The main objective of the analysis here is to see if the demographic patterns that we identified for the early nineteenth century were still visible in the fourth decade of the twentieth, at least for *rural* society outside new conurbations like Shanghai, and regardless of the effects of migration, which *QTC* covers in detail but our earlier work could not.

A central issue here is how to compare the more realistic year-by-year *variations* shown by *QTC*, often startling, with our *average* values, drawn from cases covering a much wider span of years (approximately a century, ignoring outliers). Interestingly the general shapes and the mean values of the key measures in the four 12-month periods in the early 1930s are remarkably close to those of the mid-Qing dynasty LYV standard life table that was one of the most important products of our early work. (Termed 'standard' since relatively small area-specific shifts in the two parameters, when applied to the standard, can relatively accurately generate—via the appropriate two-parameter logit-based modification for the area—each of the direct-data sections of all 13 area life tables from it.) Much the same observation applies to fertility, and, above all for the shapes of the frequency curves, to nuptiality.

It needs to be remembered that between the time covered by these late-imperial regional measurements and the early modern one there lay a period of economic, social, and demographic disruption caused by the Taiping Uprising in the 1850s and early 1860s, and its aftermath. The apparent continuity is thus to some degree illusory, being in fact between an earlier mid-dynasty averaged 'norm' (to some extent an artefact) and a subsequently somewhat restabilized situation under the early Republic. It should also be remembered that our earlier study showed marked variations between the 13 component areas we studied in the region, at the extreme a gap of 13.0 years between the highest and the lowest area expectancies of life at birth, and that in the countryside these may well have continued (Elvin and Fox 2006, 2008). So far we don't know about other rural localities as precisely as we do about Xiaoji.

The most important finding as regards methodology has been the demonstration from the Xiaoji data of the importance of the *omission of high-level perinatal mortality* from household information gathered census-fashion on a single occasion, and the determination of the level of such omissions, which could even have been slightly higher if the Xiaoji investigators had had the resources to call on families weekly rather than monthly. This emerges clearly because of the difference that appears between the data they gathered to compile an age distribution pyramid, in which mortality over the first year of life was *apparently* almost non-existent, and the annual quasi-life tables based on monthly visits to families, in which it is prominent and severe. (Contrast ages 0–5 in Figure 16.3 and Figure 16.8, below.) I suspect that this omission has been a common phenomenon in much Chinese demographic data used for the timespan covered here. Significantly, this omission affected our LYV standard in the earlier project relatively little as the data for it were gathered almost exclusively from the situation of widows, including living male children (with numbers and ages in a usable percentage of cases), at the exact moment that their husbands died, and almost invariably with a record of the widow's age in the traditional style of the number of calendar 'years' through which she had at least partly lived. Reiterated exploratory modelling based on the need for internal consistency pointed here to serious infant mortality (of the order of > 300‰ for age < 1 for females) in the LYV. This pattern was later provisionally confirmed by Xiaoji monthly data. But Xiaoji is of course only a sample of one sub-area. All that can be said definitely on this new factual foundation is that there are now clearly questions to be asked and answered of all previous estimates regarding this point.

16.2 The Township

Xiaoji early in the 1930s was a township of 221 villages and four market towns. It had 21,864 inhabitants as of 1 March 1932. Mean family size was 4.7 persons; mean household size was 4.8 persons. Numbers varied directly with the economic status of a family or household. The total area of the township was 13.9 square miles, just under 3,000 hectares. Population density was 1,573 people per square mile, about 7.3 per hectare. There were 2,673 per square miles of *cultivated* land, or 12.3 per cultivated hectare, reflecting intensive farming based on multicropping, rice in the summer with its monsoon rains, followed by wheat in the drier winter. There was population pressure on the land, causing a large number of people to leave the farms each year, and according to the authors of the survey 'the land is more intensively cultivated year by year. For the same reason local disturbances have frequently occurred and the number of wandering people has been increasing' (*QTC* 3–32). There are passing references to irrigation systems (*QTC* 43 and 50), but no reference to the work their maintenance would have entailed. Fertilizer collection, crucial in the absence of fallowing, and with only a limited range of rotations and crop variation, must also have been a burden, though it, too, is not mentioned. There were numerous mulberry trees growing leaves for silkworms, and this implies a need for labour to gather them. No mention is made of silk-reeling, however, and this activity may by this date have migrated to the city of Wuxi where it had been industrialized. This phenomenon is recorded as attracting some local out-migrating labour (*QTC* 58). Most households are said to have had 'at least one loom' for weaving cotton, the yarn for which was probably imported as nothing is said about cotton being grown locally or even its being spun.

Much of the rice was sold at once after the harvest, and a large proportion of the proceeds used to pay debts and taxes. Many people did not have enough reserves to see their families through to the next rice harvest. Recurring indebtedness was widespread. Just under 14% of the people had an 'unknown' occupation, a category excluding those who had 'no occupation', children at school, and those under 7, so presumably transitory and opportunistic (*QTC* 3–32). There was thus a pervasive economic instability. What we don't know is if this was anything new historically, or, if it was, when and how it developed.

Diseases took a heavy toll. *QTC* mentions epidemics of cholera (1932, 1933), measles (1932), and malaria (1932, 1934), and regularly occurring infection by dysentery, diphtheria, tuberculosis, and meningitis. Their prevalence was increased by the lack of personal and public hygiene, the absence of sanitary precautions regarding drinking water, little reliable medical knowledge, and quackery and 'superstition' (*QTC* 6–7, 40–52).

After the devastation caused by the Taiping Uprising and government suppression in 1864, the population recovered, and is said to have 'doubled' in the 60 or so years before the project. This would imply on the order of 1% per year during this period.

Given this level of irregularity, continuities from the past have to be sought in terms of underlying *patterns* rather than absolute numbers. This theme is developed in the later parts of this chapter.

16.3 The Project

The organizers recruited a staff of 180 local volunteer assistants under the direction of Chen Caizhang, who lived in the area of study and spoke the local dialect. The work had the permission of the local government authorities, and a network of community support was built up, though with considerable difficulty in the face of initial local hostility, intentional misinformation, misunderstandings, and fears such as that mentioning death brought bad luck. Great effort was devoted to cross-checking the results. Chen found that the cooperation of the women and children was the most important single factor in whether the work of registration proceeded effectively in a locality or not. 'Women and children', he reported, '. . . have a better memory than men for such matters of purely village interest as births and deaths' (*QTC* 6–10).

The project recorded its results in terms of *mid-year points* defined as the 1 March *between* 1 September in the preceding Western calendar year and the subsequent 31 August in the same year. Thus when *QTC* gives a year as the date of a figure, it usually means either this midpoint or the two six-month periods flanking it equally on each side.

16.4 Was the Population Rising or Falling?

Over the period 1932–1935, defined as just described, Xiaoji's female and male populations appear to have had *natural* intrinsic rates of per-mil annual *growth* of about 6.1‰ and 6.45‰ respectively, under two-thirds of a percent a year. The *registered population* appears however to have *decreased* because recorded out-migration exceeded recorded in-migration. The net rates of annual migration varied from −0.5% (1932) to −3.1% (1935) for women, and from −0.8% (1933) to −2.4% (1935) for men. Over the four-year period, there were 1,998 migratory movements in or out by women, and 1,651 such migratory movements by males. These can all be calculated from Tables 16.1, 16.2, and 16.3 below.

We still know little about the phenomenon, however, because 'migration' took a variety of forms. More than half was 'local' in the sense that it was seasonal, typically in search of work in the silk industry of nearby Wuxi, or else 'short-range', this being defined as within the local county. Apart from females who secured jobs in Shanghai as amahs,[2] there was a significant rate of eventual return from the big cities. It was observed that 'after finding [Shanghai and Wuxi] also intolerable, or having saved money after working there some years, [some of] the migrants may return' (*QTC* 59).

These results are derived from calculations based on data in Table XXI[3] and Table LIV in *QTC* (32 and 70), which do not intermesh entirely accurately in one respect discussed later, so there is an element of approximation. These data are laid out in Table 16.1, consisting of 16.1a on births and 16.1b on deaths:

Table 16.1 Annual changes in the total population of Xiaoji calculated from data on annual births and births per thousand, and on annual deaths and deaths per thousand

16.1a Births

	Females			Males		
	Births	rate ‰	Total	Births	rate ‰	Total
1932	505	49.0	10,306	552	47.8	*11,548**
1933	460	45.3	*10,155*	487	43.0	*11,325*
1934	420	42.1	*9,976*	428	38.2	*11,204*
1935	440	44.6	*9,865*	572	51.1	*11,194*
All	*1,825*			*2,039*		

* 11,558 in Tables XVII and XXI. Compiler's error?

16.1b Deaths

	Females			Males		
	Deaths	rate ‰	Total	Deaths	rate ‰	Total
1932	427	41.4	*10,314*	508	44.0	*11,546*
1933	378	37.2	*10,161*	398	35.2	*11,307*
1934	549	55.0	*9,982*	552	49.3	*11,197*
1935	225	22.8	*9,868*	277	24.8	*11,169*
All	*1579*			*1735*		

Note: Figures in roman type are in the source; those calculated by me are in *italics*.

[2] Resident female domestic servants who typically would function as nurses, wet-nurses, or nannies as well as in other ways.

[3] Table numbers shown in capital roman form are from *QTC* to distinguish them from the tables produced for the present chapter.

The problem has now emerged. If we judge on the basis of a comparison between the birth *rates* and the death *rates* in the third and sixth columns of 16.1a and 16.1b, the population was *rising* in two out of the three annual transitions, 1934 to 1935 being the exception. If we look at *numerical totals*, though, it was slowly *falling* across all three of them. Note that the relationship between totals and rates is given in 16.1a by the formula 505/0.049 ≈ 10,306, etc. The total 10,306 is *not* italicized as, uniquely, it appears explicitly in Table XXI (*QTC* 32), and so can be linked directly to the other two roman-type numbers in the first line of the female section of Table 16.1a, which come from Table LIV. So far as can be told, the research team were working in terms of mid-year *points* rather than periods of time. The underlying logic can thus be applied to the 'outputs' of deaths and births, and calculations made on them taken together. The point-total 'output' *Tb* or *d* for a given 'year' of a given sex thus equals the complete point-total *Tb* or *d* multiplied by the year's rate *rb* or *d* as appropriate. There seems no other plausible way to interpret the data as presented.

The female births can be seen from Table 16.1 to exceed the female deaths by 61.5 per year on average; the male births likewise can be seen to exceed the male deaths by 76.0 per year on average. The first impression is thus of population *growth*. This impression has to be adjusted, however, for average annual net *out-migration* figures of 169.5 females and 166.3 males (Table 16.2). Unfortunately we lack the depth in time of data that would allow us to attempt this analysis properly. It does rather suggest that net out-migration had recently been rising. There was certainly a terminal surge in 1935, but there are no data that allow us to show how this may have fitted into longer-term patterns.

It is thus natural, since here we can get an answer, to ask what the impact of in-migration and out-migration were year-by-year on the size of the population of Xiaoji. Table 16.2 gives a summary of the migration figures in Tables XLV and XLIX (*QTC* 60 and 65).

There is a mismatch between the decrease calculated from Table 16.1's records of the actual *numbers* of births and deaths and the *larger decrease* that emerges in Table 16.2 showing the net out-migration from Xiaoji. The most plausible hypothesis is that this mismatch was largely due to the effects of migration. However, using the female data on annual birth rates in Table 16.1a with its directly recorded initial female population of 10,306, we find a mean intrinsic increase rate of 6.07% over 4 years. This gives a (notional) final total female population of 10,5558.5, that is, 10,306 times 1.00607. This total is 10,306 + 252.5. This second figure thus gives us an approximation[4] to the difference of 234.5 between the decline calculated via data on net migration and the decline implicit in the numbers of births and deaths actually recorded. The same procedure for males, using an averaged initial total (from 16.1a and 16.1b), determines a mean annual intrinsic rate of increase of 6.68%. The difference in numbers of 'people' that this gives is 11,864.4 − 11,547 = 317.4, as compared to 365.5 in Table 16.2. This is at least tolerable if one is inclined to tolerance.

[4] More precise calculation gives an annual rate of increase of 6.07% and a final total female population of 10,5585.5. Given the roughness of the data, nothing is gained from this.

Table 16.2 Gross and net in-migration (+) and out-migration (–) in Xiaoji 1931–1935

	Females				Males			
	Gross			Net	Gross			Net
1932	+138	–187	=	–49	+50	–158	=	–100
1933	+156	–309	=	–153	+85	–281	=	–196
1934	+136	–309	=	–173	+152	–247	=	–95
1935	+215	–518	=	–303	+202	–476	=	–274
Total				–678				–665

	Females		Males
Decreases from Tables 1a and 1b	–441 and –446	:	–354 and –377
Net decreases from migrations	–678	:	–665
Mean differences between decreases	234.5	:	365.5

16.5 Marriages

Nuptiality offers the most straightforward comparisons of patterns between mid-Qing times and early Republican ones, so it is useful to begin here. Coverage of types of marriage in the Xiaoji survey is, however, limited to those in the so-called 'major' mode, a term introduced by Arthur Wolf and Chieh-shan Huang (Wolf and Huang 1980). The term refers to an arranged marriage between young *adults*, mostly in their late teens, who are ready for sexual cohabitation and the bearing of children. Marriage in some form was very close to universal for women both in late-imperial times and in the early decades of modern China. This was not true of males, as there was normally an insufficiency of marriageable-age women to meet the demand. The shortage was a result of the higher female 'mortality' at early ages, a term that here necessarily includes female infanticide and quasi-infanticide resulting from gender-differentiated care. Remarriage by widows took place, although it was ideologically disapproved of, but the rate can be calculated to have fallen quite sharply with female age (Elvin and Fox 2009: 96–98). There are no data for this in the Xiaoji study though there are for early Republican China in the Buck land survey section on population (Buck 1964 [1937]), and these—*faute de mieux*—formed a part of the calculation of the LYV estimate.

'Minor-mode' marriage was the practice of transferring *young* girls before the ages of sexual maturity to the families of their designated husbands-to-be, with sexual relations being entered into later either after a very reduced marriage ceremony, or merely parental approval and an announcement of marriage. This transition to reproduction took place at ages seemingly not greatly different from those for major-mode marriages. There was usually a financial consideration involved, which could include straightforward purchase (*QTC* 34). Acquiring a little daughter-in-law in this way could both reduce the costs of any marriage ceremonies and present-giving incurred by the husband's family, provide it with additional labour, and make it easier for the mother-in-law to establish her control. It was thus probably a commoner practice among poorer families. Qiao and his colleagues were of the opinion that the custom had been increasing 'in recent years', that is, just prior to the 1930s when they wrote.

Figure 16.1 shows the distribution of the ages of women at major-mode marriage at the two periods more than a century apart. Since the patterns are relatively alike, it is clear that, locally at least, the mean age has risen in this time by over a year and a half. Since we are here comparing a township with a region, this does not of course prove that this shift was a region-wide one. What is most significant is that the sharply peaked *shape* has persisted.

Figure 16.2 shows that different calendar 'years' varied quite noticeably in the numbers of women marrying. The 'year' 1932 is, in particular, conspicuous for only being surpassed in numbers over the span of ages 14 through 23, at two years of age, namely 17 and 22. There would thus seem to have been years that were, so-to-speak, 'good' for

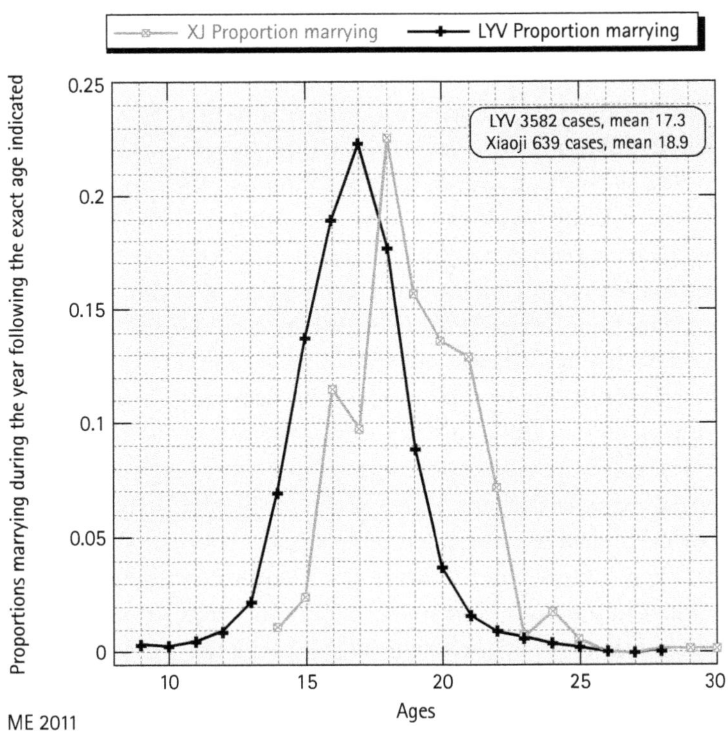

FIG. 16.1 Distribution of the ages of women in Xiaoji 1932–1935 and the Lower Yangzi Valley (*circa* 1814) at major-mode marriage.

Sources: Elvin and Fox (2009: 103); Chiao, Thompson, and Chen (1938: 36–37).

marriages, and others that were less good. This sample size is too small to say anything much more definite than this.

One other feature of Xiaoji marriages that deserves mention is their seasonal asymmetry. The six months from November through April accounted for 81.2% of total marriages, and the remaining six months for 18.8%. There were good practical reasons for this relating to family finances, climatic conditions, and the work schedule, but also astrological (*QTC* 33–34). Just as the short range of ages into which most of female marriage was compressed shows a strong sense of what behaviour was thought appropriate for what particular age, so the sharpness of the dividing lines between boundary months hints at a probable local cultural predilection for actions deemed seasonally suitable.

16.6 Deaths

As mentioned earlier, Dr Fox and I have produced a 'standard' life table for the Lower Yangzi Valley (LYV) around a focal point of 1814, but stabilized against the effects of short-term fluctuations by having been drawn from cases some of which occurred in probably every year during the century from 1750 to 1850 (Elvin and Fox 2008, 2009). It is 'standard', as

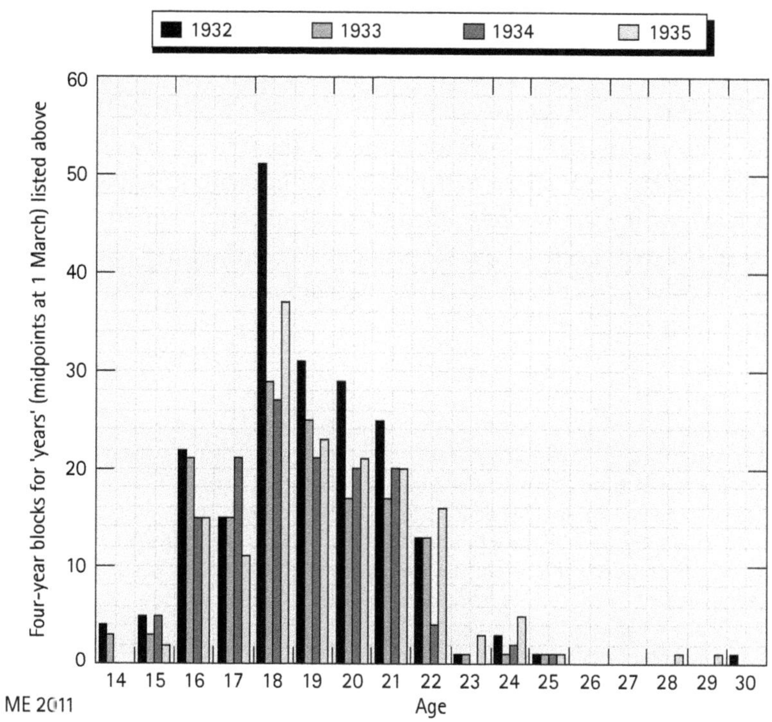

FIG. 16.2 Annual variations in numbers of women entering major-mode marriage in Xiaoji, by age and by calendar 'year' (midpoints).

already noted, in the sense that life tables fitting well with the recorded data and estimates derived from these, for each of its selected prefecture-level and county-level components can, for all of those looked at to date, be easily and precisely generated from it via the Brass two-parameter logit method.[5] By 'easily' is meant that to transform the standard values into a close approximation of the mortality directly recorded in or estimable from the data[6] in one of the localities in the region studied, one only has (1) to add or subtract to the standard logit for each age a single value for the parameter 'alpha' (roughly, the general 'height' of the logit values associated with the overall rate of survival) that differs in the LYV from 0.0 by less than an absolute value of 0.3 in all cases, and under the absolute value of 0.2 in all but three cases; and (2), likewise, to multiply the logit for each value by a single value of the parameter 'beta' (roughly the degree of 'twist' in the logit curve at high and low ages) that lie on either side of 1.0 in the narrow band between 0.952 and 0.110 (Elvin and Fox 2009: 347). There is a coherent mortality 'family' as regards patterns of age-specific mortalities in most of the LYV region, and this extends to some places quite a long distance away from the LYV itself, such as Dingzhou to the north, Zunhua to the north-east, and Guiyang to the south-west.

How though, for the purpose of comparison, do we generate at least a quasi-life table[7] for early modern Xiaoji from data most of which only cover a period of four years? Of the two ways available from the data to hand, the most obvious is fairly certainly the worse. The *QTC* survey contains a *pyramid* of ages in Xiaoji as of 1 March 1932. By smoothing it—in order to avoid the problem of the occasional apparent 'reincarnation'—a sort of life table can be generated.

The results of this procedure are shown in Figure 16.3, a clockwise rotated half-pyramid, restricted to females only in the interests of clarity. The expectation of life at birth is 28.5 years, and 35.9 years at age 10. These levels are quite reasonable in the light of the earlier LYV study. What then is wrong with it? (Apart from, that is, a certain clumsiness that derives from the fact that our basic data other than for ages 1 and 2–5 are in five-year packets, and the possibility that the particular smoothing process chosen may be in a certain sense demographically capricious, even though not mathematically so.)

[5] For age x, logit $x = 0.5\ ln\ ((1-l(x))/l(x))$ where $l(x)$ indicates the proportion of a cohort still alive at x, and *ln* refers to the natural logarithm. In words, the logit for age x is half the natural logarithm of the ratio of the proportion of those who have died to those still alive. Except for very young and very old ages this system yields a roughly linear equivalent for the age-specific death curve. Hence its convenience.

[6] This is an oversimplification in the interests of brevity. The reconstructed life table is only one of a number of parts of the model that is being built into a coherent stable whole that matches a collection of somewhat fragmented items of *five* basic types of data. For a summary view see the linkage diagram (Figure 2.1 in Elvin and Fox 2008: 371).

[7] A 'quasi-life table' may be defined as a set of age-specific rates of survival from age 0 to age ω (omega, the highest age of any demographic interest) that formally resembles a normal life table, but is drawn from a set of data too limited in time to be plausibly regarded as *in itself* generally representative of the population from which it comes. Such tables may still be useful, nonetheless, in helping to identify aspects of the pattern of age-specific survival that are relatively resistant to variation, even when the samples are small, and for sketching the probable range of variations.

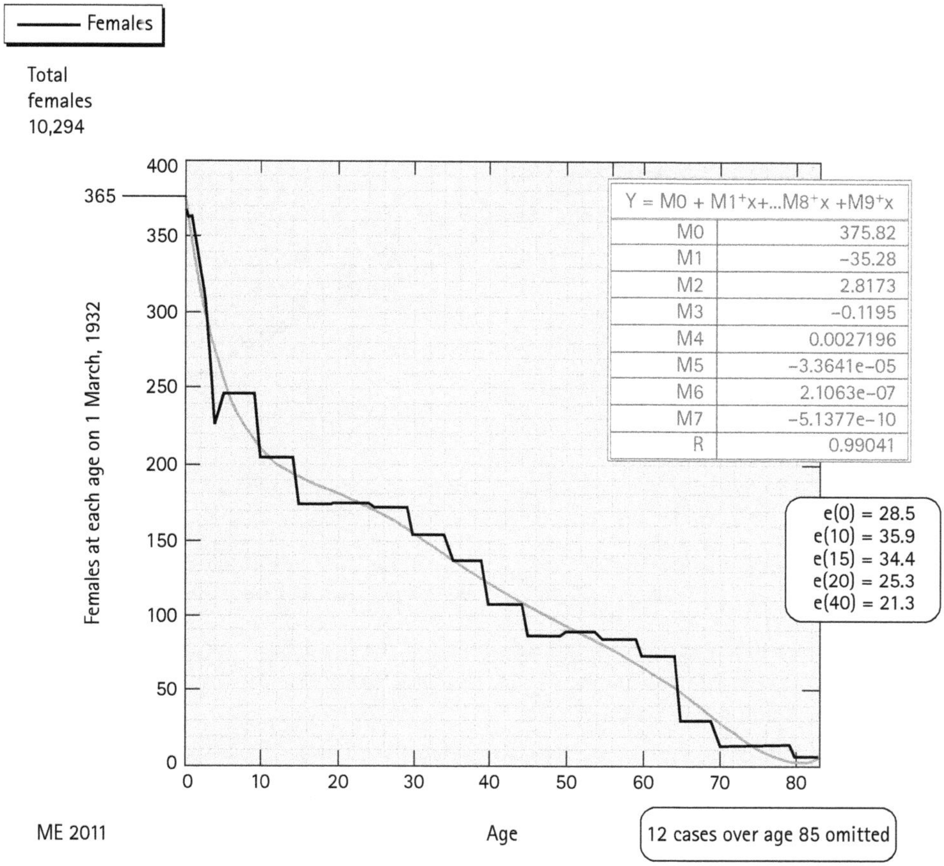

FIG. 16.3 Quasi-life table generated from the age distribution pyramid for Xiaoji, 1 March 1932.

Source: Chiao, Thompson, and Chen (1938: 22).

To repeat, what is wrong with this construction? The answer is that a pyramid of this sort is likely to have omitted *infant deaths* around the time of birth. If these are substantial, as seems probable in this case, then the entire structure is to some degree awry, and it can be misguided to try to integrate it with other patterns, above all fertility, which has to supply the shortfall. The argument falls into two parts, one logical and the other empirical:

(1) Let us assume that the census-style survey for the age pyramid was done accurately, and that the question asked by the investigators was something like, 'What were the ages of all the people living in this household on March 1, 1932?' (*QTC* 113). We also know that there was only *one* such age distribution survey incorporated into the *QTC* project, namely that shown in Figure 16.3, and that it was conducted over about four weeks for the midpoint of the survey 'year' 1932. The total was 21,864 (*QTC* 10 and 22). If we simplify the model

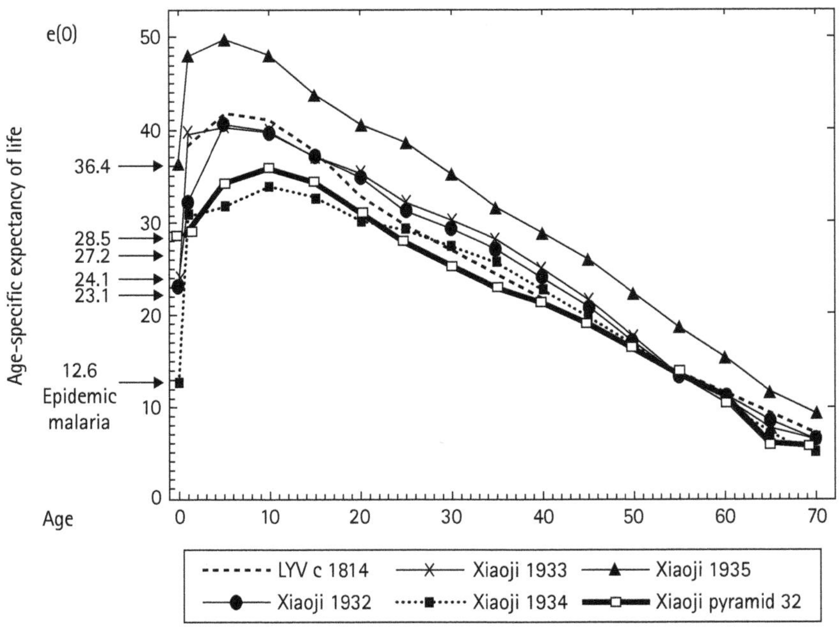

FIG. 16.4 Expectations of life in female annual quasi-life tables for Xiaoji 1931–1932 to 1934–1935 compared with the Lower Yangzi standard (*circa* 1814) and the life table derived from the 1932 Xiaoji age pyramid.

Sources: Elvin and Fox (2009); Chiao, Thompson, and Chen (1938).

by assuming that births are equally distributed around the year, which was in fact far from being the case, and consider only neonate deaths *within the first month of life*, then *most* of them will be missed in such a census, since the infants will either have been born and died *before* the survey date, or have been born and then died *after* the recording date but before the start of the next survey 'year'. In fact, if there is only a single recording operation, as here, then over 11/12ths of the total of such fleeting lives will not be recorded, *even if the question is answered accurately*. Figure 16.4, again for women only, enables us to compare the first two years of a graph of the *expectancies* of further life at various ages derived from this quasi-life table (i.e. based on a single-survey age pyramid, and shown by the heavier solid line with white markers), to the LYV standard and the four *QTC* quasi-life tables derived by the second method. This reveals almost *non-existent first-year mortality* as shown in the quasi-life table for Xiaoji in 1932. One has only to examine the almost vertical upwards slopes of the other graphs from age 0 to age 1, as mortality rapidly declined, with the barely altering one for the pyramid. Expectation of life rises, at early ages, as future years of survival rise. The standard format for the life table, using the declining *proportion* of the initial cohort that *survives* at each age, going from 1 at age 0 to 0 at age ω, is shown in Figure 16.5 which, for balance, covers only males. In both Figures 16.4 and 16.5 the unmodified LYV standard

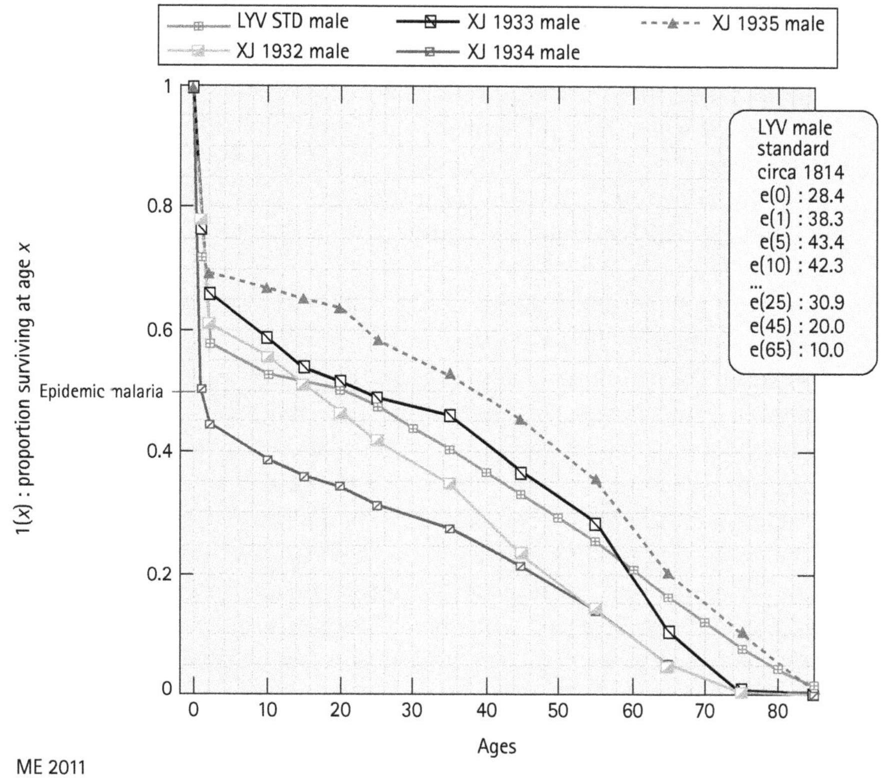

FIG. 16.5 Male annual quasi-life tables for Xiaoji from 1931–1932 to 1934–1935, with the male standard for the Lower Yangzi Valley *circa* 1814 for comparison.

Sources: Elvin and Fox (2009); Chiao, Thompson, and Chen (1938).

can be seen situated very roughly towards the mean of the various quasi-life tables, and with a rather similar shape, of which the most striking feature is the sharp inflection around age 2, which will be discussed towards the end of this chapter.

(2) The second approach converts the age-specific *death rates* per thousand for a given sex in four separate years recorded for Xiaoji in *An Experiment in the Registration of Vital Statistics* (*QTC* Table XXXVII, 53) into the *proportions surviving* at each age x, usually written as $l(x)$, from the initial 1.0 to a final 0.0. In Figure 16.4 these values are expressed in the alternative form of an individual female's 'expectation' or 'expectancy' of further years to live at each age (calculated as areas under the relevant section of the $l(x)$ curve and bounded below by the x-axis). What appears in these curves is what *could* be regarded as a reflection[8] of a proper life table *if* the pattern for that year were to be

[8] 'Reflecting' in a colloquial sense, not a technical mathematical one.

> maintained steadily across a longer period. Palpably this is what did *not* happen. The curves reflecting the age-specific survivorship pattern in the graph constantly shift about from one transient statistical incarnation to another. In 'height' at least, these patterns for successive years differ from each other, sometimes only slightly, as in the case of 1932 and 1933, but at other times markedly, as for 1935. The year 1933, it should be noted, saw a severe epidemic of malaria in which mothers and children were the most severely hit.

Three important points can now be made. The first is that the main pattern of the population dynamics of the LYV in late-imperial times clearly persevered into the early 1930s in at least some rural areas, most notably as regards the *general shape* of the age curves of marriage and death, and those derived from and so reflecting the latter. The second is that the methods used for asking questions from written data and living informants for past demographic surveys of China before very recent times need to be re-examined to determine whether or not there are likely to have been omissions of severe very early infant mortality analogous to those demonstrated here as having been lost from the reconstruction based on the age pyramid due to insufficiently time-dense reporting. The third is that, in accordance with Philip Kreager's demographic insights, standard methods applied to historical data tend to miss or smooth out the frequently severe demographic instabilities from one year to the next, and that these instabilities are often significant events. They are exemplified in the foregoing by the quasi-life tables for the four survey years, but the study of Xiaoji fertility will later turn up some equally striking examples. There are also the tricks that can be played on population models by *migration*. These have been known to some degree for some time, but the disruption that they can cause has perhaps been a little easily forgotten. In the present case, this has involved converting apparent slight population *growth* into apparent slight *decline*. Handling the instabilities effectively, however, and above all how they *moved over time* from one level to another, presents what is probably the most interesting immediate technical challenge for the future.

16.7 Births

When data on births by the age of the mother are only available for a few years, as is the case for the present study of the early 1930s in Xiaoji township, it is not easy to decide which of the various measures available for fertility is the most relevant for study. Phenomena like slump, catch-up, and overshoot do not lend themselves easily to models formulated in terms of years, yet they are evident enough even in the short span of time open to our scrutiny. What mixes of positive and negative feedbacks control the factors that create constraints and stimuli to reproduction, and how do they become significant and then fade away again?

Seeing that we have limited data, it seems arguable that the most important initial consideration is the straightforward one of how population replacement is, or can be, achieved within a context determined by (1) a broadly given pattern of age-specific *deaths*, but of a markedly varying overall *level* across time, and (2) a broadly given pattern of age-specific female *marriages*, generally a stabler factor than (1) but still subject to the kind of variation shown in Figure 16.2. This replacement can then be defined in terms of how many live births are needed *on average*, over the years, to mothers of which specific ages, consistent with the approximately known underlying biologically constrained potentials that vary with age. The individual years in the *QTC* study varied significantly in fertility in the period that we are considering. Thus total live births of both sexes in Xiaoji to women between the ages of 15 and 44 ranged from a high of 1,055 in 1932 to a low of 846 in 1934. This was a decrease of almost 20% over, roughly, two years. The next year, 1935, it jumped up more than 19% again (*QTC* 45). This yoyo-ing is probably in part a small-sample effect. But it is similarly probable that it cannot be dismissed as just that. How, though, does one find one's way securely through this?

How best to determine a historically meaningful average is a tricky issue when the number of years for which there is good information is a mere four. It is a reasonable assumption, however, but still only an assumption, that since the provisional standard for the Lower Yangzi Valley *circa* 1814 has been shown in the foregoing sections to give a basic indication of the nature of rural patterns of mortality and nuptiality that were still present more than a century later, it will also offer a guide to the approximate nature of the later average fertility pattern, at least in little-modernized areas like Xiaoji. We therefore digress a moment to look back.

The period of years around 1814 can be summarized as regards fertility in the Lower Yangzi Valley region as follows: for the *exact* replication of the female population, married women with living husbands had to bear on average female children in such numbers that, *if* they, the mothers, survived the full childbearing span, and remained unwidowed, or at least remarried following any widowhood, *then* they would end up with a statistical mean of 2.45 girls each. The number of boys can be determined from this by identifying the most plausible sex ratio at birth—not an easy matter in practice for the Chinese society of this time—and using it to multiply the figure for girls. Thus at 105.5, the average LYV woman who, with her husband, married at the mean age for matrimony, and lasted the full childbearing span, as limited by the Yangzi Valley late-imperial pattern of nuptiality, probably bore a mean total of just over five children of *both* sexes (Elvin and Fox 2009: 104). The range for the smaller county- and prefectural-level units of this time ran from a low of 4.0 children of both sexes in Shangrao up to a high of 6.6 in Jiaxing (Elvin and Fox 2008: 355). Spatial variation is thus a crucial variable to pursue further as we don't have wide enough spatial coverage to know really accurately where on this continuum an approximate regional mean value lay. But the grail on the far side of getting the picture clear is the prospect of gaining an understanding of the likely causes behind such a mean difference—between two areas in the same region—of more than two-and-a-half children born per woman completing her age of childbearing.

According to the late-imperial Lower Yangzi Valley standard life table the number of females surviving out of every 1,000 born fell by the end of the first year of life to 688.

In comparison, out of every 1,000 females born in Xiaoji, averaged over our four years, the number of those surviving after the first year of life fell to 608 (*QTC* 53). We would thus expect that, given the broad similarity of demographic structure in other respects, early Republican Xiaoji would have needed a higher mean female birthrate over the years than the late-imperial Lower Yangzi Valley to achieve replacement. (Of course the socio-economic composition of the sample studied in Xiaoji was probably significantly lower.) It is not possible to conjure up any precise relevant figures, but a reasonable case can, and will, be made that they are within a band of substantial probability lying on either side of a female reproduction ratio of just over 2.8 girls per woman living the full childbearing span. At the sex ratio at birth used for the LYV study, namely 105.5, this indicates a total reproduction ratio for both sexes together in Xiaoji of a little over 5.75 children per mother.

There is, though, a data problem with the sex ratio. *QTC* records a mean value of 105.9 for the first three years, which obviously raises no major issues. But the survey has 130.0 (*sic*) for the fourth year, 1935, which the authors think, almost certainly rightly, is likely to be an error; but they cannot, they report, find what might have gone wrong in the recording process. By concentrating where possible mainly on the female line, from mother to daughter to grand-daughter, and so on, we can largely sidestep this dilemma for our present purposes.

Figure 16.6 shows the variations in the number of births of *both* sexes (no gender-divided record being available) per thousand married women. The year with the highest birthrate is 1935, and that with the lowest is 1934, the year in which malaria was particularly devastating for women and neonates. The peak ages for childbearing lie—predictably—in the 20s; and it is also here that the divergences between the years are the greatest. These local records are juxtaposed in the graph with the empirically based general schedule of Heather Booth and Basia Zaba (Zaba 1994) which shows the proportions of a single female child that would be (notionally) born each year to a sexually active woman, unconstrained from 13 through 49 either by being unmarried, or widowed but unremarried, or by mortality, effecting a 1:1 replacement of herself by the end of her age of childbearing. (The programme we used to simulate this incorporates a variable mix of these limitations based on our data and entered—originally for fit-finding exploration—from the keyboard.) If we ignore 1934, a year badly battered by a malaria epidemic, and some 'kinks' that are probably the result of an unduly small sample, there is no persuasive evidence here for either the late starting or the early stopping that has sometimes been suggested as a characteristic of late-imperial Chinese women's reproductive behaviour. Rather the contrary. There is also no sign of birth limitation within marriage, though in our earlier work we thought that there were interesting indications that this might well have happened in the LYV with families that already had at least one son who had reached the age of 5 or more (Elvin and Fox 2008). This third possibility is not ruled out by the data given here, just not supported. The generally good fit of the Booth-Zaba (BZ) schedule with the pattern of the Xiaoji empirical data is important, as they are, so far as I know, independent of each other. This similarity increases the plausibility of Figure 16.7, where the modelling is done on the basis of the BZ pattern.

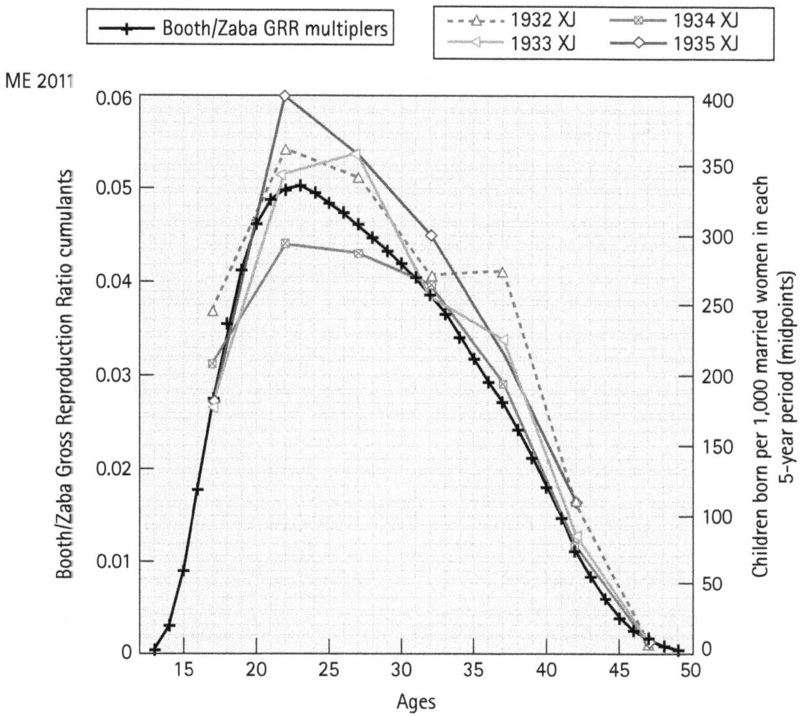

FIG. 16.6 Births of children of both sexes per year per 1,000 married women, Xiaoji between 1932 and 1935 (right-hand scale), compared with the Gross Reproduction Ratio (females only) of Booth and Zaba (left-hand scale).

Sources: Chiao, Thompson, and Chen (1938); Zaba (1994).

In order to find what the age-specific birth rates for females *would have had to have been*—note the conditional, since there is no guarantee that it happened, and in one case it certainly did not—to *replace* the loss of females from deaths during a given year, two factors need to be kept in mind. The first is reductions imposed by the known schedules of nuptiality and of widowhood due to male mortality. Second, the latter need to be offset by the rate of widow remarriage, based on rates found for the LYV. The task, then, is to find a *multiplier* for the proportions of a birth assigned to each childbearing year by the basic BZ fractional age-specific birth cumulants, so that the sum cumulates them exactly to unity.[9] This multiplier determines the number of female children an average mother needs to have, under the conditions stated, for the potentially breeding female population to replicate itself if this set of conditions is conceived of as permanent. Of course the real conditions were shifting constantly, as

[9] We use the same programme as was used by the LYV project for this purpose. It, and a record of a sample run, can be found at <http://gis.sinica.edu.tw/QingDemography/>. It might make the operation more intelligible to say that the aim is to have the female cohort, considered *as a whole*, replace itself exactly. Those still alive and reproductively active have, so to speak, to make up for those dead or, after some age, reproductively inactive (as through widowhood).

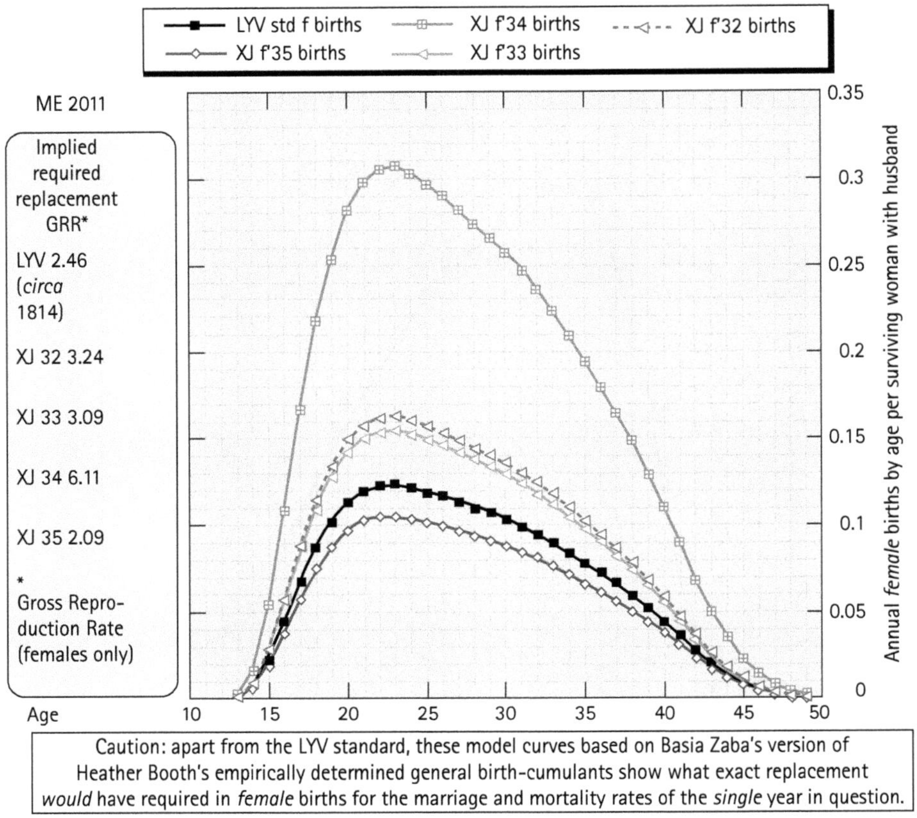

FIG. 16.7 The female births per year required per surviving female at age (x) with living husband, under given schedules of nuptiality and mortality, needed to replace the same year's female mortality in Xiaoji, by year, for 1931–1932 through 1934–1935.

has been seen, but this multiplier offers an interesting approximate measure as long as it is used with the appropriate caution. Just as we earlier came close to ignoring the males, even if not quite, we now note that we can ignore the old. Or, we can leave out the old at least until we can quantify the probably significant contribution that surviving grandmothers made to the survival and well-being of their grandchildren.

The results of the modelling exercise are shown in Figure 16.7 for the four years of the Xiaoji study, along with the same for the LYV in the years *circa* 1814. The first thing apparent is that the years are ranked in an order that is, as might be expected, *approximately the reverse* of their order in Figure 16.6. Thus the year recorded as actually having the highest fertility, 1935, now appears as the year that requires the lowest fertility, that is, the lowest Gross Reproductive Ratio, to replace itself. This inversion implies the obvious point that the net losses of one year have in fact often to be made up by the net gains of another year, or the net gains absorbed by later net losses, if a longer-term constancy of numbers is to be maintained. Otherwise there is long-term population growth or long-term population decline. Nowhere is this point more striking than in the case of the year with the lowest recorded actual births per thousand married women: 1934. It not only requires the highest rate of births to reproduce itself, but the figure itself is surreal in a single-year context. If we assume a sex ratio of 105.5, a mean two-sex total of more than 12.5 children per childbearing woman is needed. Thus if there is any sort of middle- or long-term near-equilibrium in a population of the kind we have been discussing, the likelihood is that it will be a *dynamic* one in which the temporary surpluses and deficits of individual years are being constantly shunted forward over time through processes that are modified by both damping and intensifying under appropriate circumstances. What is not clear for the moment is how that trick is done. What sort of signalling is at work? Signalling what? How fast? And how controlled when it goes malfunctional?

16.8 Further Considerations

Some supplementary remarks may be added here on topics that are related, but not central, to the main theme of the continuity of late pre-modern LYV demographic patterns. First, the way curves other than the regional standard were generated needs explanation.

16.8.1 Constructing Life Table Values for Regional Subunits: Spatial Variation in the LYV

The provisional female and male *standard* life tables for the Lower Yangzi Valley *region* were based on finding age-specific survival rates that would, as nearly as we could manage, make it possible to combine a range of a variety of empirical LYV data on marriages, births, and deaths into a self-consistent interlocking whole, assuming a stationary population.[10] To construct the life tables of the subunits, *per contra*, the only practical way, given relatively small samples for all but three of them, was to use the most systematic and almost certainly the most accurate *partial* data, namely those on the ages at death of faithful widows over the ages of 48.5 Western years, as a basis from which to derive (with some routine additional manoeuvring as indicated in Elvin and Fox 2008: 338–341) the 'alpha' and 'beta' parameters that were needed in each case to transform the whole of the LYV regional life table so that the derived $l(x)$, that is the proportion of a cohort surviving at age x, now matched as closely as possible the subunit data for $l(x)$ and other associated measures over the span of years of age these latter covered. (An example is the frequency of the survival of one or both of a widow's parents-in-law when she was bereaved at a given age.) For this we used, besides Brass's logit system, our own computer search

[10] This process is summarized in a flow diagram on p. 371 of Elvin and Fox (2008).

programme. In one case the fit, measured by the sums of the squares of errors over 18 age groups (namely, those shown along the horizontal axis of Figure 16.8, plus the cohort survival ratio for age 1) was 'exact' over three decimal places. In six further cases the fit was 'exact' in this same sense but over two decimal places; in four more cases over one decimal place; and in the last the sum of squared errors was marginally just over 0.5. Two very small subunit datasets, those from the counties of Poyang and Le'an, were omitted.

There is no justification for assuming that these subunit patterns, albeit with variations in absolute level, *always* had the same *relationship* between deaths below and deaths over 48.5 years of age as that established for the regional standard which we used as a template. It is unlikely that in fact they did, regardless of the deceptive appearance inevitably given by the graph that this was so. What does seem plausible is the more limited conclusion that there were very significant *differences* between subunits, as flagged in Figure 16.8 by the gap of 13 years already noted in the expectation of life at birth between the highest and the lowest. The highest figure, that from Shangrao county, is based on a small sample, and even when this is combined with that from the next highest, which is for Rugao county, the sum of their cases is still only a quarter of that for the lowest, namely Jiaxing prefecture. What is useful about the independent Xiaoji study is that it suggests, as in Figure 16.4 and Figure 16.5, that template applicability at least sometimes still held to a reasonable approximation in the countryside in the early 1930s, even if this single swallow hardly makes a summer.

As already noted, the LYV standard was drawn from a range of groups of people that were on average probably above the socio-economic level that would have been typical of the mid-Qing dynasty. Most importantly, 13.0% of the women either came from, or married into, a 'gentry family' defined in the Chinese sense of having at least one male member holding or having held an imperial office, examination degree, or title, however modest. This parameter was probably fairly (though not very) close to being systematically listed, and membership of this upper group was roughly five times what would have been the case for a nationally representative sample, remembering that *two* families were involved in each marriage. About another 6% of cases came from families in categories that also suggest a higher than average socio-economic status, such as sub-bureaucrats, army officers, or merchants. The economic status of the majority (approximately 80%) was not systematically indicated in the LYV data (Elvin and Fox 2009: 74–76). Qing society, with only a small proportion of families in inherited statuses, also had a high pre-modern rate of *social mobility* because of partible inheritance among male heirs, some distribution of parental wealth to daughters (notably in the costs of marriage), and the formally largely open access to most high office through competitive examinations. While some degree of social visibility was needed for inclusion in the faithful-widow dataset, it should also be remembered that faithful widows were a special group. Many of them lived in straitened circumstances because they lacked a male rice-winner, and while they were spared the hazards of childbirth they were often not spared those of poverty. The Xiaoji study's finding, more than a century later, that the average crude death rate for all ages in the township was notably lower for those who came from relatively richer families (*QTC* 55) cannot be automatically assumed to have

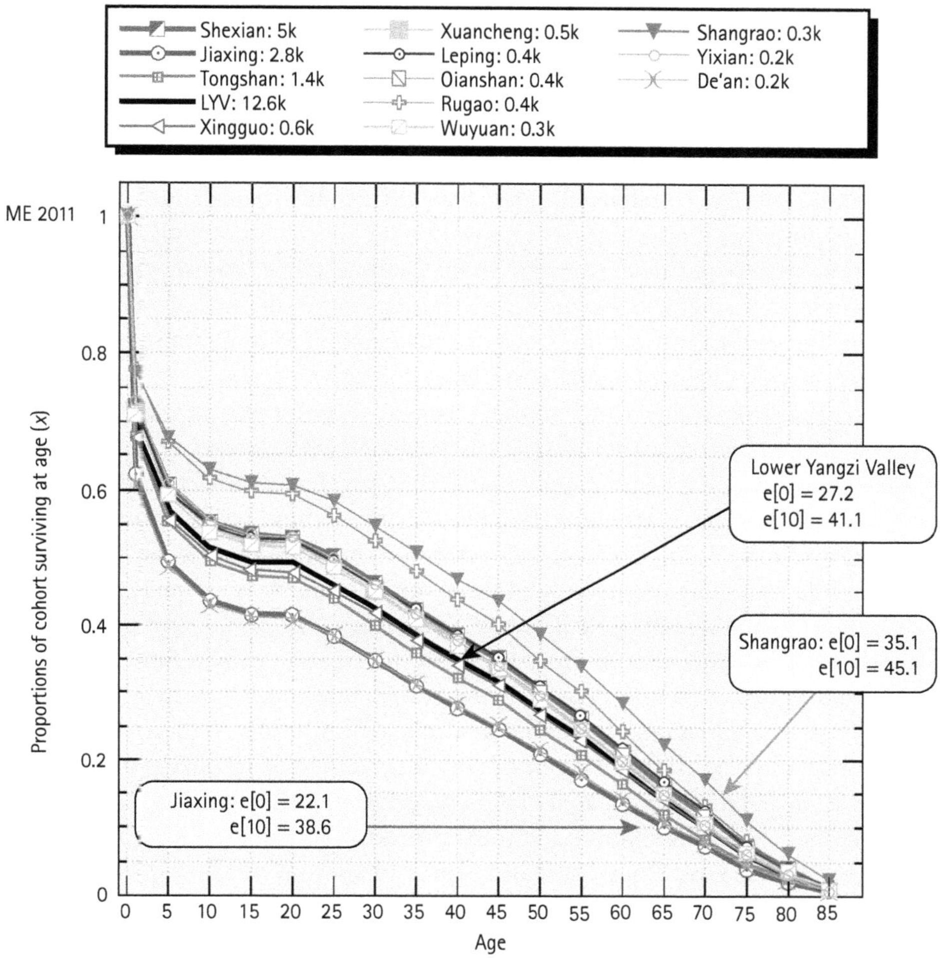

FIG. 16.8 Life tables of females in 12 regions of the Lower Yangzi Valley *circa* 1814 by administrative subdivisions.

applied to them or at least to have applied in the same measure. It is also interesting in passing that the *infant* mortality rates in Xiaoji were virtually the *same* for *all economic classes* (*QTC* 57). Overall, membership of a specific *spatial group* probably correlated more strongly with the level of faithful-widow mortality than did differences in wealth and income, though this is only an impressionistic judgement, explicable, if real, by as yet unidentified differences in local environments. We do not yet know very well which subunits had relatively richer and which relatively poorer local economies.

We can now propound what seems a strong hypothesis but one that is, as indicated, strictly speaking not proven on account of our formally arbitrary modelling assumption of the exact applicability of the general regional life table template to subunits in the region.[11] This hypothesis is that the difference in life expectancy at various ages between

[11] The subunits, as discussed, were modified for the alpha-beta parameter on the basis of what was required for *local* older female mortality, on which we have *very* good data.

any two given subunits was likely to be most striking between their values for *perinatal plus early childhood mortality*, rather than between their subunit age-specific mortalities after about age 10. The gap in life expectancy between the highest and lowest subunits in the LYV sample, for example, already *halves* between birth and age 10, more specifically from one of 13 to one of 6.5 years. This is where the main focus of relevant environmental and medical historical research will surely have to be in the immediate future. For our immediate purposes it also suggests that we should be wary about talking too easily, especially as regards pre-modern times, about *simply* 'high' or 'low' mortality over the full range of ages. It might, for example, be more appropriate in the present case to characterize the general LYV situation as one of a *combination* of 'comparatively high' early childhood mortality with a 'moderate' mortality during later years.

We have thus to combine in our thinking the year-to-year *time* variations, illustrated by Xiaoji, with the variations across *space*, as illustrated in the foregoing reconstruction of a selection of the LYV subunit populations. The Qing dynasty prefectural and county subunits were, moreover, *administrative* divisions. They thus had a somewhat *variable* relationship to more 'natural' units, such as those we might have preferred to have had, based on hill, mountain, or river boundaries, or soils and water supply. They sometimes followed, but sometimes cut across, these latter. The same is probably true of their relation to socially or economically closely linked units such as those that shared a sub-ethnicity or a marketing area. Each particular pattern will have to be determined case by case. The variations shown by data collected in terms of these middle-level administrative divisions, mainly counties of *circa* 0.2 to 0.25 m inhabitants, are thus likely to be *understated* as compared to those in 'natural' or closely linked communities of these sorts. Nor did the Chinese administrative areas, unlike some European administrative systems, in most cases distinguish urban zones from rural ones.

16.8.2 The Interface of Demography with Other Disciplines

Links with the related aspects of economic history, social psychology, and anthropology are important but hard to establish firmly. An example of this may be seen from Figure 16.9 which presents the fragmentary data for female minor-mode marriages in the LYV. These have been labelled 'virilocation', namely moving by the future bride-to-be to the home of the parents of the agreed husband-to-be, but with sexual relations deferred until a later date decided by the husband's parents. These data are fragmentary because minor-mode marriages were fairly certainly substantially under-recorded in our sample, and the relative frequencies of minor and major modes cannot be reliably inferred from their relative numbers. (These are placed inside each distribution curve in heavy type.) Other modes, such as uxorilocal marriage by a male, are also extremely rarely mentioned, presumably not only because they were in fact rare, but also because they were felt to be only marginally eligible for inclusion in a roll of honour for widowed fidelity.

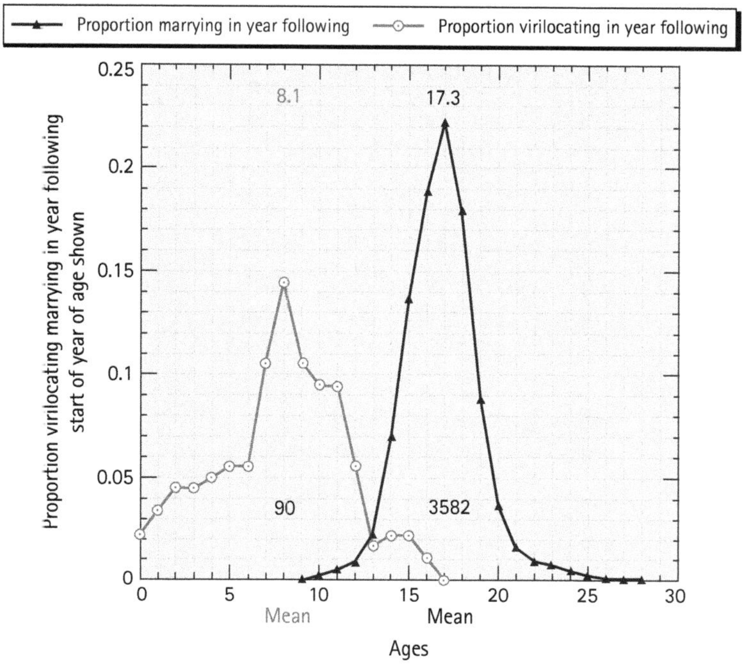

FIG. 16.9 Ages of females at virilocation ('minor-mode marriage') and major-mode marriage in the LYV *circa* 1814.

The distribution curve of female age-specific virilocation also bears the stigmata of the irregularities, like brief reversals of trend, usually associated with an inappropriately small sample.

The most interesting feature is the mean age for female virilocation, which is effectively also the modal age, namely 8 years of age. This fits quite well with the small amount of mid-Qing anecdotal evidence on the subject. An example is the ferocious polemical poem by Zheng Xie 鄭燮 against the practice of 'little daughters-in-law' (translated in my 'Unseen Lives', in Ames et al. 1998: 158–160), where the chaste sister-like wife-to-be is 11 years old, and, so far as can be told, has just moved, since there is quarrelling between her in-laws as to how she should be treated. Because it is a polemic, there is a good chance that the poet presented what he thought his readers would regard as a reasonably typical case. The peak frequency was the ages between 7 and 11 for females, as shown in Figure 16.9, and so between 10 and 14 for their husbands-to-be, since the mean gap in ages between spouses in the LYV was + 3.2 years at this time. This makes good demographic sense. The girl at this age would have had a high probability of surviving to bear children, as shown earlier in Figure 16.4—on the order of 40 years' expectancy of life—and the boy would have been at close to his lifetime lowest level of mortality. An arrangement made at these ages would have had a good chance of lasting, which it would not have had five to six years earlier. What puzzles me is that Arthur Wolf, rightly acknowledged as the foremost expert on this institution, should have found from his fieldwork that girls in

Table 16.3 Differential female mortality in the first year of life

16.3a. Gender-related 'excess' of first year of life female deaths

1932	(309.6 - 222.7)/309.6	≈	28.1%
1933	(409.4 - 237.3)/409.4	≈	42.1%
1934	(605.7 - 496.9)/605.7	≈	18.0%
1935	(258.6 - 225.6)/258.6	≈	12.8%

16.3b. Gender-related 'excess' of conjectured perinatal female deaths

1932	17.5%
1933	29.4%
1934	5.9%
1935	0.4%

16.3c: Gender-specific infant sensitivity to unidentified changes in their microbial-supportative environment as shown by percentage increases and decreases in first-year deaths compared to the preceding year in Xiaoji (from QTC Table XXXVII)

	females	males
1932–1933	+ 32.2% :	+ 6.6%
1933–1934	+ 47.9% :	+ 109.4%
1934–1935	- 57.3% :	– 54.6%.

early modern Taiwan, his area of most intensive study, seemingly moved homes predominantly as virtual babies. The bride was 'seldom more than a year old' (Wolf and Huang 1980: 82).

Qiao and his colleagues do not give any ages for minor-mode marriage in Xiaoji in the 1930s, only observing that 'many parents buy small girls from the very poor families and rear them as fiancées for their sons' (*QTC* 35). Elsewhere they note that 'most girls under 14 who have come to the area' had come as child-brides (*QTC* 61). This places an upper bound on the ages thought appropriate for this practice very close to that in Figure 16.9, but not a mean or modal age. It looks as if there could be a puzzle here awaiting future resolution. I have no reason to doubt Wolf's data. So, if Taiwan was different, why?

16.8.3 Infanticide, Abortion, and Differential Neglect

Neither the LYV nor the Xiaoji data give us any direct quantitative information on either infanticide or abortion. The Xiaoji survey notes that 'infanticide is a face-losing thing' (*QTC* 56), and by implication difficult to record adequately. In the LYV project, we thought that we might determine a rough measure for a plausible minimum level of the *combined effect* of female infanticide and seriously inadequate parental first-year-of-life care for girls by taking the *difference* between first-year female deaths and first-year male deaths, and dividing it by the *total* female first-year deaths (all expressed per thousand). For the LYV standard life table, based on a large sample stabilized by drawing on a wide range of localities over a long period of time, this gave us a value of 10.3%.[12] That, as we then saw it, suggested that just over 10% of first-year female deaths might be thought to have been caused by infanticide or first-year-of-life differential neglect. A key assumption here was that, over the long run, given *no* infanticide and *equal* care, girls would have had at least the *same* first-year-of-life deaths per thousand as the boys. The rates for both would have been rising or falling broadly together from one calendar year to the next as the consequences of predominantly similar causes—for example, transient famines or epidemics. In fact, most evidence from elsewhere in the world suggests that girls would probably do slightly better under such conditions (Lazarus 2002: 288). We therefore regarded this procedure as fixing a likely *minimum*.

The Xiaoji data in Table XXXVII (*QTC* 53) can be subjected to the same calculation, but for a relatively small sample for each of four individual years. This produces four strikingly different annual percentages, only one of which is close to the LYV value.

The later Table XLII (*QTC* 57) in the Xiaoji data states that it refers to 'deaths under 1 year of age', but its smaller numbers of cases suggests that they almost certainly relate *only to perinatal* deaths, that is to say those at and immediately around birth.

It should be emphasized that this is only a guess. Comparison gives different absolute levels but an almost similar *proportionate* year-by-year rise and fall, as can be seen from Figure 16.10, which, it should be noted, has two different scales.

The year 1933 was in gender-specific terms by far the worst year for baby girls, according to both measures. Why remains a problem, unless it was some peculiar gender-weighted effect of the cholera, or perhaps of another cholera-like disease, in that year. Something important, if ill-understood, appears to have been at work here.

[12] Working in life table per-mil terms, this was calculated as (312.13 – 280.0) / 312.13 = 0.1029.

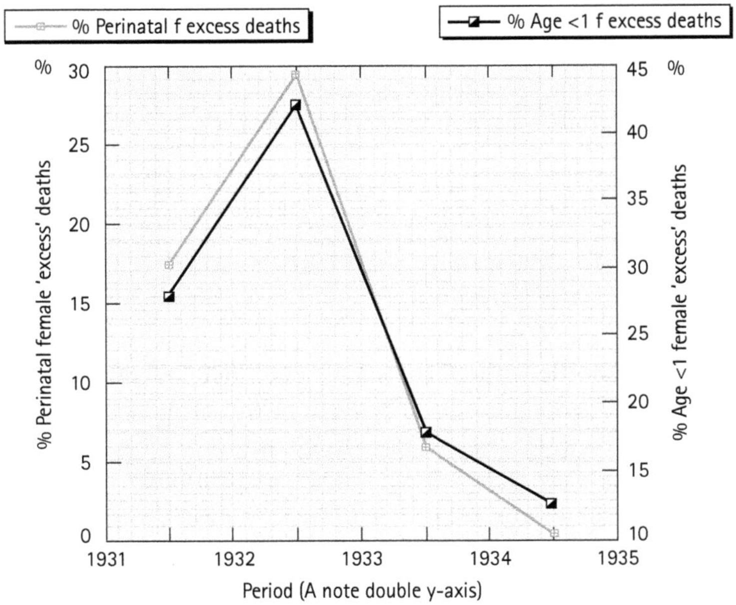

FIG. 16.10 Gender-related 'excess' female deaths in Xiaoji in the first year of life, compared with conjectured female perinatal deaths.

For the moment we can merely note three points that may be relevant. (1) The *QTC* report on Xiaoji at several places stresses 'the lower esteem in which female babies are held in China' (e.g. *QTC* 52 and 57). (2) The lower levels in 1934 and 1935 may reflect the fact that the proportion of baby girls still alive who did not have, or had not by this time, acquired resistance to whatever disease was so fatal in these years, was probably much lower than two years earlier. (3) *Neither gender* was necessarily more sensitive to changes in what might be provisionally called, for want of better terms, the multiple interacting complex of microbial-therapeutic and nursing-nutritional environments. Table 16.3c shows the three percentage annual increases and decreases in infant deaths at age < 1 of each of the two genders between the four recorded years for females and males respectively. The surge in infant *male* deaths in 1933–1934 was by far the most abrupt change, and it fell away with great, even if not comparable, rapidity in the following year.

16.8.4 General Changes in Background Factors since the Mid-Nineteenth Century

Technical modernization of the Chinese-owned sector of the most advanced part of the Lower Yangzi Valley regional economy had already started by the later 1860s in Shanghai (Elvin 1993; and the rougher but wider-ranging Elvin 1981). The impact on the *rural* economy more than a modest distance from the treaty-port city was, however, mostly limited to areas with relatively special local conditions. Shanghai's Chinese-run sector was racing ahead economically under the stimulus of foreign firms from a wide range of nations, which provided contracts, foreign examples and experts, capacity to purchase key components simply from abroad, and opportunities for training, plus the easy availability of working capital and a relatively skilled workforce. These included such atypical characteristics as: (i) the dependence of commercial demand for kerosene-powered and other types of engines on special circumstances, such as the need to lift irrigation water twice; (ii) the unusually high local use of wage labour as opposed to tenant family labour; or (iii) the presence of large-scale commercial agriculture. Machinery, apart from that for pumping, was not normally substituted easily for careful manual labour in labour-intensive multicropping farming. High yields per unit area were required to provide sustenance and taxes, given the limited supply of arable land per person. It does not seem that Xiaoji had any of the important atypical characteristics listed above.

By the early 1930s, Shanghai, long linked into the world economy, was pumping migrants of both sexes in and out at annual rates in the low hundreds of thousands, but with a consistent net increase over time (Elvin 1974: Introduction). What is interesting about the demographic picture sketched in the present essay is that up to the early 1930s this dynamic development relatively nearby had only a minimal detectable impact on the basic population dynamics of a rural township in Jiangyin county about 150 km higher up the Yangzi, in spite of improved transportation and generally increasing migration both in and out (*QTC* 60, 65).

16.9 Discussion

The structure of the provisional LYV template is in some respects counter-intuitive at first sight, whether for the years around 1814 or in its near-reincarnation in Xiaoji 1932–1935. It is also relevant to note at this point that LYV-reconstructed *year-by-year* mortality needed smoothing before it ran in a fully consistent fashion across the teenage years (Elvin and Fox 2008: 372, figure 2.2). It is thus unlikely to be the last word. But is it closer than what we have so far had to hand on the demography of the lower Yangzi Valley in late-imperial and early modern rural China? In my personal view, probably yes, but let us consider what is still probably its closest competitor. Figure 16.11 gives the well-known 1976 reconstruction, by George Barclay and colleagues at Princeton, of the demography of Chinese farmers in 'South China' during 1929–1931. Their work, which was based on data collected from just over 50,000 women in the context of the organization of the Buck land utilization survey, presents a curve for females that, at first sight, would seem to most people a rather *more* plausible pattern. Nonetheless, for reasons not satisfactorily explained, but perhaps embodying a wary caution, the Princeton Group reconstruction *failed to provide a value for l(1)*—i.e. the survivorship proportion of a cohort at the end of the first year of life—and only started at the age of 5. I have therefore supplied one of 0.7 calculated on the basis of the Princeton *e*(0) of 21.0, the expectation of life at birth, so that

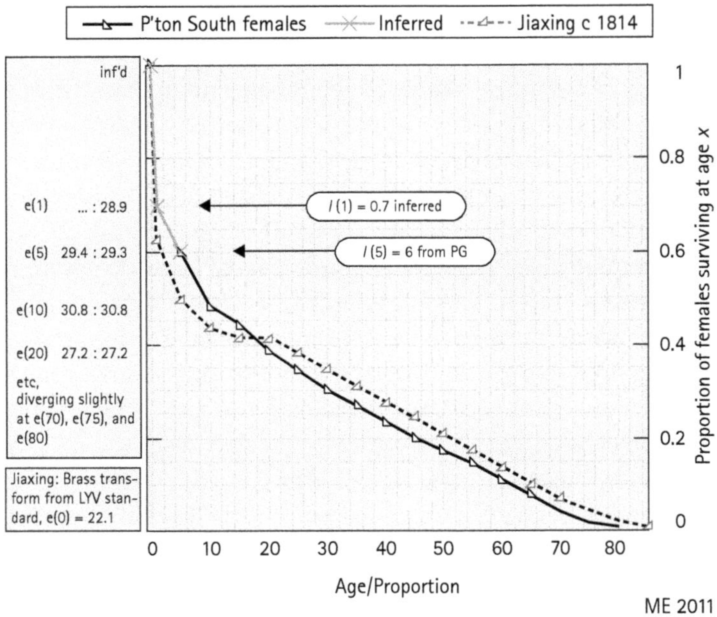

FIG. 16.11 A comparison between the female life table for 'South China' *circa* 1930 prepared by the Princeton Group in 1976 and the female life table for Jiaxing prefecture in the Lower Yangzi Valley *circa* 1814.

Source: for S. China: Barclay, Coale, Stoto, and Trussell (1976).

the *other* *e*(x) values come out as near matching as seems possible.[13] (See the larger left-hand panel in Figure 16.11.) This is in general a dubious procedure, but there is no obvious other way here of approximating the pattern of the proportions of the cohort surviving in such a way as to at least broadly match another across the span of ages.

This result is compared in Figure 16.11 with the LYV reconstruction of Jiaxing prefecture circa 1814, since the female expectation of life at birth there (22.1 years) only differs from the Princeton reconstituted value for females in South China (21.0) by about one year. The Jiaxing pattern can be derived from the LYV standard by setting alpha at 0.160 and beta at 1.049, and using a standard Brass logit transformation. The Jiaxing female expectation of life at age 10 is 38.6 as opposed to Princeton's South China value of 30.8. What is of primary interest, though, is the difference in the *patterns of inflection* around about the ages of 10 to 25.

Apart from the massive loss of life in the first year evident in Jiaxing, the serious differences as regards shape and slope are two. The first is the sharp inflection that starts in Jiaxing at around the age of 2, but does not appear in the hypothesized Princeton

[13] With minor, but probably unavoidable, divergences most likely due to slight differences between the Princeton algorithms and mine, and perhaps slightly differing inputs since their published values are only rounded to one decimal place, as opposed almost certainly to their operationally used ones. They also stop at age '75+', with '+' undefined (Barclay et al. 1976: 620).

Table 16.4 Female survivorship proportions l(x) from the LYV Jiaxing prefecture and from Xiaoji

Age	Jiaxing f circa 1814	Xiaoji f 4-year mean 1932–1935	Xiaoji f: 1932	1935
10	0.44	0.30	0.44	0.62
15	0.42	0.28	0.42	0.61
20	0.41	0.26	0.39	0.58

Note: 1933 and 1934 can probably be classed as somewhat unusual years because of the severity of epidemics, and have therefore been omitted in the table. Their values were, respectively (i) 0.47, 0.44, 0.40; and (ii) 0.27, 0.24, 0.22. Remember the LYV, from which the Jiaxing curve is derived, is an average across around a century.

version. The second major difference with the Princeton life table is what might be called the 'shelf' between ages of about 10 to somewhere between 15 and 20, which seems to suggest that for a brief decade the advance of mortality has almost been brought to a halt. It is, however, as we have seen, very nearly matched by the values of the recorded, not reconstructed, survivorship proportions of the Xiaoji quasi-life tables (Table 16.4).

Our original self-critical conjecture was that this 'shelf' feature in the LYV and its transforms had to be accepted due to our need, in order to cover this decade of life, to splice together two sections of our data that drew upon different sorts of records, and so presumably we had only been able to combine them relatively imperfectly. It now looks as if this concern, while it still probably has an element of truth, is too severe on our process of reconstruction. The direct Xiaoji picture from the data, where no such splicing is required, suggests that something like the 'shelf' through ages 10–15–20 was not inherently demographically impossible, but reflected some aspect of 'reality'.

References

Barclay, G., A. Coale, M. Stoto, and J. Trussell (1976). 'A Reassessment of the Demography of Traditional Rural China', *Population Index* 42/4: 606–635.

Buck, J. (ed.) (1964 [1937]). *Land Utilization in China*. Nanking: University of Nanking; repr. New York: Paragon.

Chiao, C. M., W. S. Thompson, and D. T. Chen (1938). *An Experiment on the Registration of Vital Statistics in China*. Oxford, OH: Scripps Foundation for Research in Population Problems.

Elvin, M. (1974). 'Introduction', in M. Elvin and G. W. Skinner (eds), *The Chinese City between Two Worlds*. Stanford, CA: Stanford University Press, 1–17.

Elvin, M. (1981). 'Making Progress Pay: A Basic Problem in China's Early Modernization', in Chin Hsiaoyi (ed.), *Symposium on the History of the Republic of China*. Taipei: KMT History Committee, iii.254–280, at 281–286.

Elvin, M. (1993). 'Le transfert des technologies', *Nouveaux Mondes* 2: 73–85. Geneva: CRES.

Elvin, M. (1998). 'Unseen Lives: The Emotions of Everyday Existence Mirrored in Chinese Popular Poetry of the Mid-Seventeenth to the Mid-Nineteenth Century', in R. T. Ames (ed.), *Self as Image in Asian Theory and Practice*. Albany, NY: State University of New York Press, 113–199.

Elvin, M. and J. Fox (2006). Qing Demography Project website, Academia Sinica, Taipei. Created by Tzai-Hung Wen. Available online at <http://gis.sinica.edu.tw/QingDemography>. Contains data, PERL computer programmes, and commentary.

Elvin, M. and J. Fox (2008). 'Local Demographic Variations in the Lower Yangzi Valley during Mid-Qing Times', in T. Hirzel and N. Kim (eds), *Monies, Markets, and Finance in Early Modern Societies: East Asian and Global Perspectives*. Berlin: LIT Verlag, 333–376.

Elvin, M. and J. Fox (2009). 'Marriages, Births, and Deaths in the Lower Yangzi Valley during the Later Eighteenth Century', in C. Wing-Chung Ho (ed.), *Windows on the Chinese World: Reflections by Five Historians*. Lanham, MD: Lexington, 67–112.

Lazarus, J. (2002). 'Human Sex Ratios: Adaptations and Mechanisms, Problems and Prospects', in I. Hardy (ed.), *Sex Ratios: Concepts and Research Methods*. Cambridge: Cambridge University Press, 287–311.

Wolf, A. and C.-S. Huang (1980). *Marriage and Adoption in China, 1845–1945*. Stanford, CA: Stanford University Press.

Zaba, B. (1994). *The Demographic Impact of AIDS: Some Stable Population Simulation Results*. London: Centre for Population Studies, London School of Hygiene and Tropical Medicine.

PART VI

IDENTIFYING SUBPOPULATIONS FOR DISEASE TREATMENT AND CONTROL

CHAPTER 17

FROM POPULATIONS TO CLINES IN MODERN STATISTICAL GENETICS

CHRIS SPENCER

17.1 Background

THE concept of a population is important in statistics. It is often a term used to describe a large, possibly infinite collection of events from which a typically small sample is taken. The assumption that the population is large and homogeneous allows a helpful simplification: that observations are 'independent and identically distributed, or i.i.d.'. And this turns out to be useful for conducting statistical inference—the formal process of comparing mathematical models of the world in order to make decisions or draw conclusions. In allowing observations to be described in this way, we can test for differences in the parameters of the distributions from which the observations are assumed to be drawn. Explicitly we are assuming that, conditional on the population label, no other relevant information determines the outcome of draws from it. Here, relevant information is deliberately vague; in a data-rich world there is often a long list of measurable factors whose relative importance is typically subjective. Choosing to allow for them comes at a cost: as more and more populations are required for the i.i.d. assumption to hold, we typically lose the ability, or statistical power, to detect the differences in parameter estimates between populations, because the samples from each become smaller.

In studies of organisms like humans, there are an almost uncountable number of potentially relevant quantities that one could measure. In this sense, every individual is unique, and therefore arguably could be in a *population* all of their own. However, many of these measurement variables may be irrelevant for any given scientific question and therefore it is acceptable to ignore them in order to make the statistical model of the world simpler. An old adage in statistics from Professor George Box—that 'all models are wrong but some are helpful'—encapsulates the necessity to simplify the complexity of the world in order to make inferences. The unease we have at describing the continuums of diversity of organisms like humans as discrete groups is at odds with the fact that it is necessary to define populations for the purposes of scholarly analysis. These issues go to the heart of statistical analysis and more broadly what it means to describe a population, which is the theme of this book.

Population genetics is the study of differences in the DNA sequence which, when packed into chromosomes, comprise the genome of an organism. As humans, we inherit two of these genomes, one from our mother, and one from our father. Globally there are more than 20 million places in the 3-billion-letter-long genome that commonly differ between one genome and next (The 1000 Genomes Project Consortium 2010). And these variants can occur in a vast number of combinations; in fact, aside from monozygotic twins, every individual inherits a unique genome. Only relatively recently have we been able to measure so many genetic variants in large numbers of individuals. For some problems in statistical genetics, defining discrete populations is unsatisfactory because it makes assumptions about which components of the complex structure of genetic variation are important in any given context: how do we know which of these factors are important and therefore how do we define a population?

The aim of this chapter is to describe how populations can be conceptualized and inference performed. It is based on the idea that all individuals are unique, but that genetic variation is structured, with individuals more similar to some than others. We introduce the approach by describing recent increases in large-scale genetic data, why old models do not exploit the richness of the data, and what can be learnt from the new approaches.

17.2 An Explosion of Genetic Data

In recent years there has been an explosion in the amount of so-called 'genome-wide' genetic data. The fuel for the revolution has come from massive advances in the technology used to assay genetic variation, and a scientific drive to understand how these polymorphisms relate to phenotypes, particularly common diseases (Wellcome Trust Case Control Consortium 2007). The motivation is twofold: first, by associating genetic markers with disease phenotypes we can use the data to inform us of underlying biological mechanisms by trying to infer the molecular consequence of DNA changes; secondly, we can use these markers to identify individuals that have a predisposition to clinically relevant phenotypes. The scientific method of choice for identifying these variants is called an association study. In theory it is a powerfully simple idea: the frequencies of DNA variants in cases of individuals with a disease are compared to a sample from the population at large. When a genetic variant has a statistically significant increase in frequency in the cases, it is implicated in disease predisposition.

Powered by large sample sizes, these genome-wide association experiments have been highly successful. In 2005 we only knew of a handful of predisposing variants for several diseases. In 2011 there are over 2,000 genetic variants identified across hundreds of clinical phenotypes. There is an open debate as to how best to use these clues to disease biology in order to improve health and health care, but they represent a huge leap forward on the road to developing new approaches to prevention, diagnosis, and treatment.

One example is a study of the genetic basis of the disease multiple sclerosis (Sawcer et al. 2011), published in the journal *Nature*. The genome-wide association study involved DNA analysis of over 25,000 individuals of European ancestry. Multiple sclerosis (MS) is characterized by damage to nerve cells and is accompanied by a localized inflammatory or immune response, which accumulates over time and leads to a degenerative neurological condition. A key question for scientists trying to understand the disease was whether an errant biological process damaged the nerves *leading to* an inflammatory or immune response, or whether it was a dysregulated inflammatory or immune response *leading to* damaging of the nerves. The study pinpointed 57 parts of the genome where single DNA changes increased MS susceptibility. Far more of these regions contained genes that are strongly implicated to be involved in the immune system than you would expect by chance. As you are born with your genome (and it's not influenced by the environment), genetics helps to establish causality: these observations strongly suggest that multiple sclerosis is an auto-immune disease in which the body's normal immune and inflammatory process are dysregulated and cause damage to normal (primarily nerve) tissue. Furthermore, due to the success of similar studies in other auto-immune diseases, the project was able to quantify the overlap of susceptibility variants. It was substantial, underlining immune pathways as key to disease aetiology. Observations of this kind help scientists to understand disease mechanisms and, importantly, how best to target further research.

17.3 Back to Basics

Population geneticists, like other scientists, tend to work with models of the world that have lots of simplifying assumptions. If you make as many simplifying assumptions as possible about how genetic variation in a population evolves through time, it reduces to a simple stochastic process. To make things easy, assume for example that we are haploid (only have one genome) rather than diploid (two genomes, one from your mother and another from your father). Further, consider only a single part of the human genome (often referred to as a locus) with two possible forms to the genetic variant (called alleles). To simulate evolution from one generation to the next we can imagine taking the following approach: first, sample a parent at random from the population of individuals, and let their offspring inherit the allele that their parent had. Repeat this procedure independently for each individual until the next generation is populated. And so on for successive generations. Figure 17.1 shows a small cartoon of this process. We can follow the frequency of the two forms of alleles over time in the top panel: for example, in the generation in which the mutation arose (indicated in black) it just so happens that this individual had two offspring (and likewise in subsequent generations), such that the mutation spread through the population; the change in frequency of a mutant allele is due to a set of chance events which ultimately determine its fate. As the probability that each descendant derives from a given parent is equal, it is perhaps natural to describe the collection of individuals as a population. Note that in a diploid population the notion of random mating, which follows from equal parental probability, is important for determining the frequency of pairs of alleles within individuals (this is discussed further in the chapter by Bodmer and Winney in this book).

A simple extension to this model starts with one population, which then splits into two (bottom panel of Figure 17.1). The modelling assumption dictates that once individuals are in separate populations they are reproductively isolated. Therefore, individuals can only leave descendants in the population in which they exist. At the point of the split, the two populations start with the same allele frequency, but the fact that these two populations have been isolated means that subsequent changes of allele frequencies in each population become independent stochastic processes. Over time, by chance alone, we expect the allele frequencies in the two populations to diverge. Population geneticists call this stochastic process genetic drift.

The population-genetics model we have described so far is discrete, both in terms of time (discrete generations) and in terms of the representation of individuals. However, were we to imagine very many generations and very many individuals, you might start to view the whole process as continuous; the frequency of an allele in each population now can be thought of as continuous number (between zero and one) and time can be measured as a continuous quantity. A schematic is shown in Figure 17.2. The extent of the

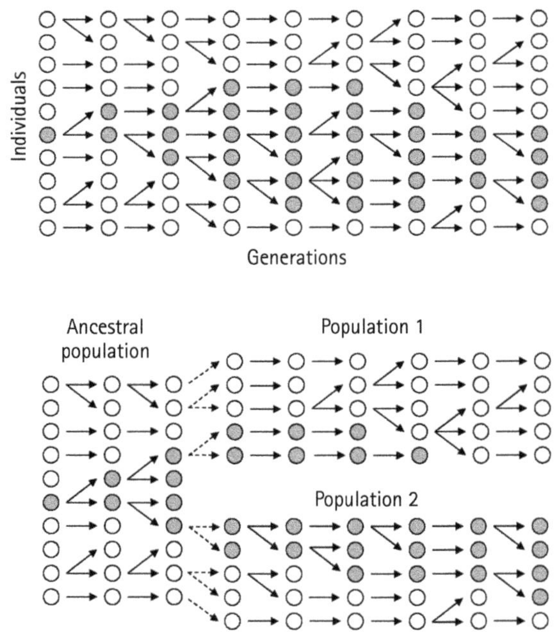

FIG. 17.1 A simple model for the evolution of the frequency of an allele. In each panel a circle represents a haploid individual. Moving from left to right, the next generation is formed by a sample parent and inheriting (arrows) the allele they carried. In the first generation there is only one individual with the grey allele. In the bottom panel, after three generations the populations split, decoupling the change in the frequency of the allele in the two resulting populations.

stochastic changes in the ancestral allele frequency, α, to the present-day allele frequency, θ, is captured by the parameter π. It turns out that the natural scaling for time is in units of population size. With some thought, you can convince yourself that the bigger the population, the less variability there can be in the change of frequency from one unit of time to the next, so a longer period is needed for big changes in frequency (i.e. when compared to changes in a small population). The parameter π therefore describes the rate at which allele frequencies change relative to a fixed population size: the amount of genetic drift.

As time goes on, and the longer this process continues, the more the allele frequencies in these two subpopulations will differ from the ancestral population. There are then two factors which determine the difference in allele frequency: the time since the split and the size of the population. We can now model the process of sampling a relatively small number of individuals from the population, and look at the counts of the alleles they carry. Within a population we imagine this process to be a set of i.i.d. random variables; each sample of an individual we take has the same probability of being one form of the allele which is simply determined by its current-day allele frequency in that population.

By comparing models where we assume individuals to be in different populations (with different allele frequencies) to those where they are in the same population we can start to make inferences about the evidence for, or against, population structure. Moreover, once we are convinced that two or more populations are different, we can begin to measure how big the differences are by estimating π, as this tells us something about the size of populations and the time since they split.

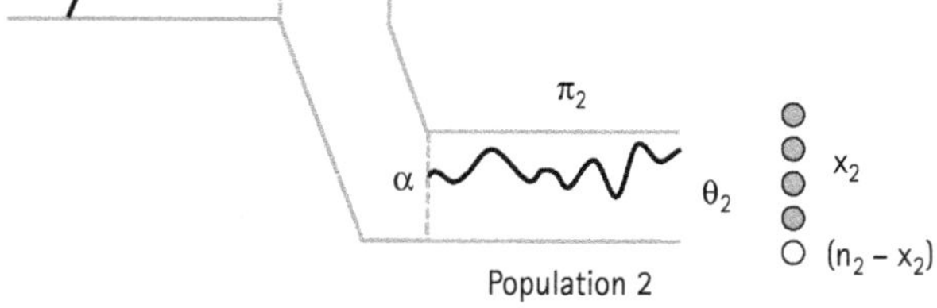

FIG. 17.2 A diagram of a statistical model for the continuous change in frequency of an allele through time. The frequency of the allele, represented by the wiggly line, changes in the ancestral population until the population splits. At this time the frequency is α. In each daughter population the frequency then evolves independently at a rate, π, which depends on the time since the split and the size of the populations. The frequency at the present day is denoted θ. The process of sampling x of the grey alleles from a sample of size n, in each of the population is also independent in the two populations. In population 2, for example, $x=4$ and $n=5$.

17.4 Identifying Populations Using More and More Data

In 2002, a paper by George Nicholson, Peter Donnelly and colleagues used models of this type to analyse a dataset containing 30 chromosomes from France, 60 from Utah in the United States of America (i.e. European Americans), and 70 chromosomes from Iceland. They use molecular methods to obtain the alleles carried by study individuals at 83 bi-allelic markers (single-nucleotide polymorphism or SNPs) to assess whether there is genetic evidence that these represent separate populations, and if so, how different they are. To allow statistical inference to be informed by all 83 SNPs, they made the assumption that allele frequencies were the result of independent stochastic processes from the model.

The results of the inference are shown in Figure 17.3. The x-axis gives a measure of genetic drift since the populations split some time in the past. The y-axis indicates how probable a given level of drift is, conditional on the data (i.e. counts of the two alleles at each of the SNPs in each collection). The level of genetic drift from the ancestral population for both the French samples and those collected from Utah is most likely (i.e. has

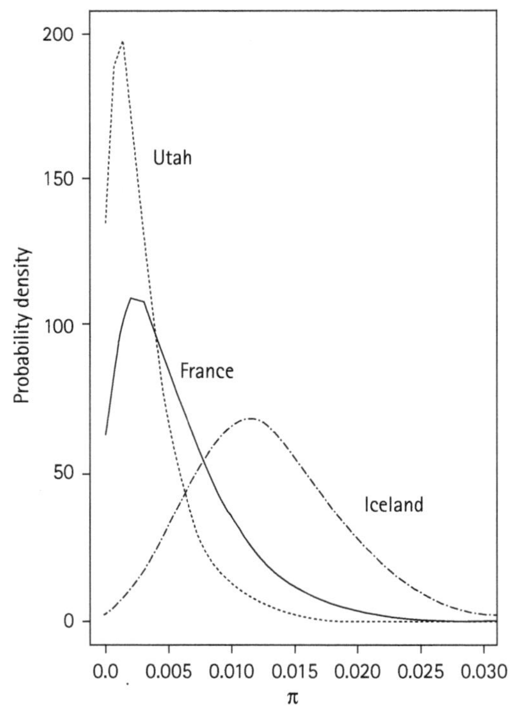

FIG. 17.3 The posterior probability of the extent of genetic drift, π, estimated from samples of three populations typed at 83 SNPs. The x-axis gives the estimated amount of genetic drift, or population differentiation, and the y-axis shows the probability given the data and the model assumptions. The figure is adapted from Nicholson et al. (2002).

highest probability) close to, but not at, zero; there is relatively little evidence that these samples are from different populations. In contrast, for the Icelandic population, most of the probability is away from zero, suggesting that this population is differentiated from the other two populations. The likely explanation is that Iceland was founded by a relatively small number of individuals who existed until recently in a relatively small isolated population leading to high levels of genetic drift.

Since 2002, there have been staggering improvements in DNA genotyping technology that have allowed scientists to collect a lot more data, both in terms of numbers of individuals and numbers of SNPs. The population-genetic study described above collected a couple of hundred chromosomes at under a hundred SNPs. However, the more recent study into multiple sclerosis involved over 25,000 individuals at over 500,000 SNPs. Studies of this size are now fairly routine, and by leveraging new sequencing data, some projects are looking at 100,000 individuals at over 20 million SNPs across the genome.

Going from 1,000 or so data points (here genotypes, the combination of two alleles that a human carries at a locus), to over 500 billion is quite an increase in scale. In statistics, more data usually mean more powerful inference, and in theory this means identifying more and more populations with ever more subtle differences in allele frequency. This raises the important question of which level of resolution is most appropriate: how should we define populations when potentially so many could exist?

17.5 Modelling Population Differentiation

To help us think through these issues we can return to the multiple sclerosis genome-wide association study. The study collected a set of cases, individuals diagnosed with multiple sclerosis, and sets of controls that are sampled at random from the population. It included about 10,000 cases and about 15,000 controls sampled from different countries, but all with European ancestry. The large samples are good news in terms of statistical power because they allow subtle differences to be detected in allele frequencies between cases and controls. As part of this process, good practice in statistical analysis is to think of ways to check that the data make sense, thereby ensuring that these differences represent real genetic associations. As humans are very good at spotting visual anomalies, a useful rule is to plot the data in as many ways as possible, and then check whether its properties are as expected. However, a downside of having all this data is that it is difficult to spot problems by eye. For example, it is time-consuming to check all 500,000 SNPs included in the study, so we have to teach a computer to do the checking for us. One example of a necessary check is to make sure that the technology that infers genotypes from DNA samples does so accurately. For the vast majority of SNPs it does, but every now and then it makes a mistake, which can lead to the wrong allele frequency being estimated for a sample and potentially a false positive association with MS susceptibility.

How can we use population genetics to help us spot these errors? One solution exploits the fact that most SNPs have accurate allele frequency estimates, and therefore can be used to estimate the level of genetic drift between the samples. By fitting a population-genetic model we can make predictions about how much we expect allele frequencies to differ between collections and use it to help spot errors. For example, if we infer two populations to have very little genetic drift between them, then we would be very surprised to see a large difference in frequency, and when we do, then the data should perhaps not be used for association analysis. An example is shown in Figure 17.4, which overlays estimated SNP allele frequencies in each of the samples as inferred from the clustering of a sample of individuals into three genotype groups based on the experimental assay.

The data shows that the samples from Norway and Italy have very high allele frequencies relative to the rest. By eye we can see that the difference between these two plots and the others creates the suspicion that their genotypes may have been incorrectly inferred

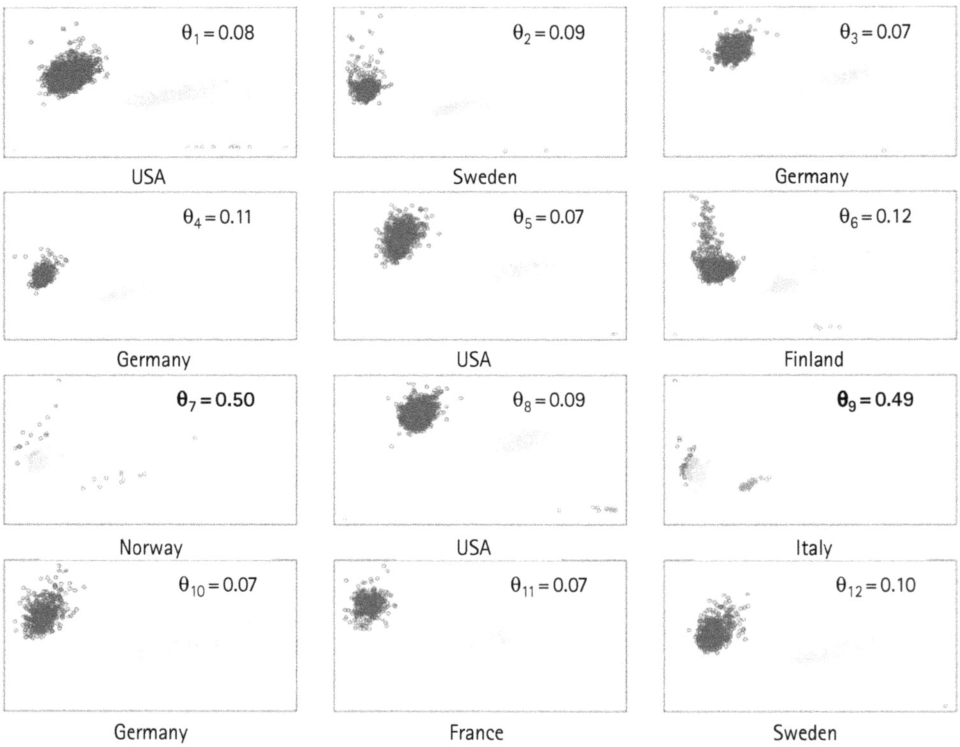

FIG. 17.4 An example of using population allele frequencies to spot genotyping errors. In each panel the x-axis and the y-axis represent the intensity of molecular probes aimed at detecting the two alleles (referred to as 'A' and 'B') at a SNP. The colours indicate the inferred genotype with AA, AB, and BB genotypes shaded dark grey, grey, and light grey respectively. Populations 7 (Norway) and 9 (Italy) have incorrectly called genotypes and this leads to very different inferred allele frequencies, θ_7 and θ_9. The level of genetic drift, π, estimated for each population can be found in the supplementary information (Coop et al. 2010) from which the figure is adapted.

from the data. A simple way to try to automatically weed out problems of this kind is to compare the allele frequencies between these groups and to identify the sample collections which have significantly different allele frequencies. But the general strategy is problematic in this setting, because we know that allele frequencies do genuinely vary between Italy and Norway, due to their population history. So there is a conundrum: if we knew which the bad SNPs were, then we could use the good SNPs to estimate the degree of genetic drift. But we can't work out which are the good (and bad) SNPs until we know how much genetic drift to expect between samples. Luckily we can solve the problem iteratively, by alternatively inferring bad SNPs and estimating the level of genetic drift, and in the end, the statistical inferences lead to the information we are after. This strategy is a powerful approach for spotting subtle aberrant allele frequency estimates, which would otherwise have been missed.

In this setting, the large amount of genome-wide data and the problem in hand—detecting errors in the inferred genotypes—dictate the correct definition of a population. As each of the sample collections were genotyped separately, we know that the features (in this case, errors) we are looking for are most readily identified by treating each experimental cohort as its own 'population'. In thinking about the most useful definition of a population in genetics, the answer typically depends on modelling as little complexity as possible, whilst grouping individuals in a way which is most informative given the model assumptions.

17.6 When is a Population Not a Population?

A key idea in the approach to ensuring accurate genotypes, as described earlier, is to use the genome-wide data to learn about the average influence of population history on allele frequencies, and then to look for SNPs where this model doesn't appear to fit well. The same concept has other uses in genetics. Of particular interest is using allele frequency differences to look for regions of the genome which have been subject to natural selection. Directional (also called positive, or Darwinian) selection in one population (for example, the ability of adults to digest the sugar lactose in populations which have adopted agriculture) can rapidly increase the frequency of the beneficial allele, faster than genetic drift would be expected to act. Thus, when compared to another population in which directional selection is absent, or reduced, large allele frequency differences can arise. In theory the same population-genetics models can be applied in this setting to detect these regions, but there is an additional complication: we don't know which populations (or subpopulations) have experienced the natural selection, and therefore we don't know how best to define a population in order detect the signal.

Figure 17.5 is from a paper by Graham Coop and colleagues (2010), which neatly demonstrates the problem. It shows the allele frequency in a set of different populations of a SNP that is known to be associated with human ability to uptake sodium, a trait which is potentially under natural selection to differing degrees in different populations. On one axis there is a measure of distance from the equator and the points are shaded according

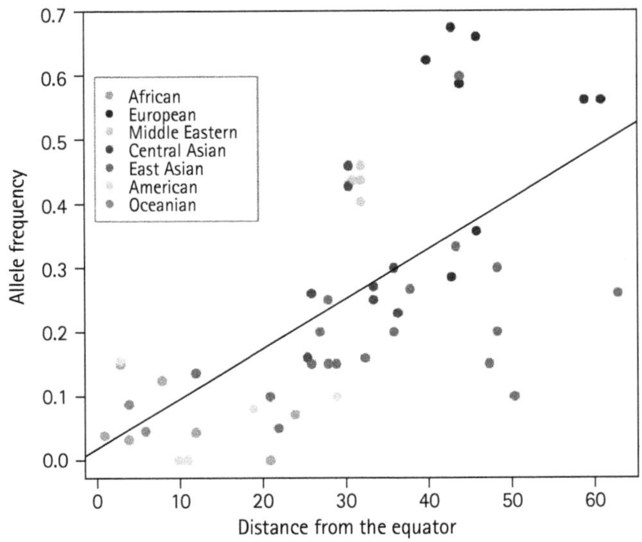

FIG. 17.5 The distance from the equator for each of 52 human populations, plotted against sample allele frequencies for the SNP AGT M235T in each population. The points are coloured according to the geographic region each population belongs to. Note that similar populations cluster together and there appears to be a correlation with distance from equator. The plot and legend are adapted from Sawcer et al. (2011), which includes a full colour version.

to where in the world the collections were sampled. Note that there is a clustering of points with similar symbols, suggesting geographically proximal populations have more similar allele frequencies. Questions we may want to ask are: Is the distribution of allele frequencies sufficiently explained by what we know about population structures? Are these patterns consistent with genetic drift alone? Or do we need to invoke some other process, perhaps natural selection, to explain these observations?

An important message of the paper is to explain why, in order to make accurate inferences (for example to decide whether we need to implicate natural selection), we need to add some complexity to our model of population history. Thus far, we have assumed a simple model in which all populations from which we have taken samples split at the same time in the past, and then evolved completely independently to the current day. Of course, in reality, especially when considering geographically proximal populations, the real history is much more complex. Firstly, population splits may happen sequentially: for example, a split when modern humans left Africa, and then a subsequent split into the populations that peopled Europe and Asia. Secondly, migration between populations can occur, either as one-off events, or as a continual process over time. These ancestral processes are represented graphically in Figure 17.6.

The effect of either of these complications is to introduce some shared variability in allele frequency between populations. It is possible to capture the essence of these processes by introducing correlation, or covariance, in the statistical model. For example, in the middle panel of Figure 17.6, the covariance matrix, Σ, includes off-diagonal elements, π_{12}, to capture the fact that populations one and two share more ancestral history by splitting from population 3 before splitting from another. We can see that this might make sense; if migration

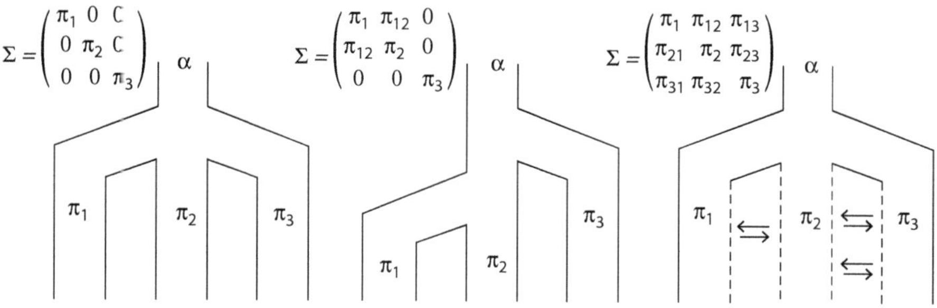

FIG. 17.6 Schematic representation of the effect of different models of population history on the correlation of allele frequencies across populations. Each of the three panels shows a model where the ancestral population, with allele frequency, α, splits into three populations. On the left the split occurs at the same time creating three daughter populations; in the middle, the splits occur sequentially, with population three separating first; on the right, the populations split at the same time but there are migration events at a higher rate between population two and three than that between one and two (no direct migration between one and three). The matrix, Σ, shows how the model induces covariance, or non-zero off-diagonal elements, in the expected allele frequencies between pairs of populations.

rates are very high, or if two of the populations have split only very recently, their allele frequencies are expected to be very similar, and in the limit case they will behave as if a single population—i.e. their expected allele frequencies will correlate perfectly after we account for sampling variability. Modelling these processes is beneficial because the covariance structure provides additional information about population history, and importantly, because of the improved model fit, it is easier to identify subtle departures from genome-wide patterns.

Their data included samples from 52 populations from across the world for which genome-wide markers were available. The empirical correlations in allele frequencies between the populations are shown in Figure 17.7. The patterns of correlation show there

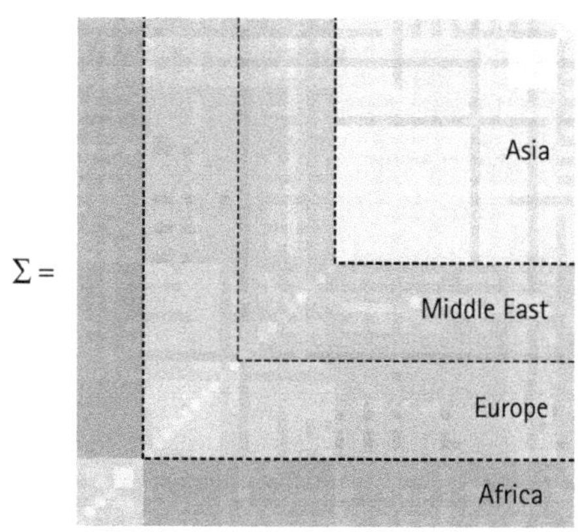

FIG. 17.7 A heat plot of the empirical correlation matrix between 52 worldwide populations. Lighter colours denote high correlation. Dotted black lines group major geographic divides in the data. As in Figure 6, each element of the matrix is a pixel denoting the correlation in allele frequency between pairs of observations. The figure is adapted from Sawcer et al. (2011).

are groups of populations, which correspond to major geographical regions and tend to have similar frequencies. It is therefore tempting perhaps to reduce these groups to a single population, but keeping the full richness of this data, in terms of all pairwise correlations, allows us to retain as much information as possible. Why might that be important? Well, if natural selection occurs in only one population, lumping it together with others could dilute the signal we are looking for.

Modelling the populations in this way allows us to test whether the allele frequency differences, which were observed at the SNP determining sodium uptake, are compatible with general patterns of variance and covariance across the genome. The way this is achieved in practice is to first estimate the covariance matrix, Σ (which is like estimating the degree of genetic drift in previous examples), and then to ask whether the observed frequencies are explained by this population structure, or whether another term or component to the model is required. This additional term is used to make a prediction about which populations might have more extreme allele frequencies, and it can be used to test specific hypotheses. Coop and his colleagues tested several additional predictors to see whether they better explained variation in particular regions of the genome. These included measures of environmental changes, such as levels of sunlight and pathogen density, as well as testing for specific changes in certain geographically defined groups. They identified several regions of the genome, including variation near both pathogen defence genes and those which determine skin pigmentation, which stood out as having undergone natural selection (Coop et al. 2010).

17.7 Every Man Is an Island (or at Least a Population)

The concept of modelling correlations in allele frequencies across populations and the interest in not grouping populations a priori lends itself to an interesting extreme. Thus far, samples are initially grouped into sets based on where they were collected, typically referred to by country. But in fact, genotype data is available for each individual within the sample, so why not model each individual as a 'population' and let the covariance in alleles between individuals capture the population structure?

The idea of using a matrix of correlation in alleles between individuals has already been employed in the genome-wide association study setting, notably in the multiple sclerosis (MS) analysis described earlier. Its purpose is to adjust for the fact that individuals who are genetically related tend more often to share alleles than those who are related more distantly. If weakly related individuals cluster as either cases or controls in a study, this can lead to spurious associations between the alleles that study individuals carry and the phenotype they exhibit. In essence, it violates the i.i.d. assumption. The motivation for modelling pairwise relatedness, rather than stratifying individuals into populations, is that we don't know the extent to which population structure influences the allele frequency of an as yet undetected disease-associated SNP. So instead of each element of the matrix Σ representing correlations between populations, they now describe correlations of alleles at SNPs across the genome between all individuals considered in pairs.

In practice, fitting a population-genetic model with parameters which relate to shared genetic drift at the individual level is extremely intensive from a computational point of view. For analysis of the 52 populations described above, the process of fitting the model is doable, but extending the same model, without approximations, to the individual level in studies of tens of thousands of samples is currently unfeasible (the matrix Σ is too large). The solution is twofold: firstly to use the empirical allele shared between individuals as an estimate of the variance/covariance matrix, and secondly, to decompose this matrix into its principle component to allow the model's fit to be assessed efficiently. These steps allow association studies to test for association at hundreds of thousands of SNPs, whilst taking into account the full pairwise correlation structure between individuals.

As well as speeding up the efficiency of statistical inference, this way of decomposing the correlation matrix is a common tool for revealing subtle structure in large datasets. Its role in population genetics extends back to the very earliest genetic studies and the resolution it can bring when applied to contemporary datasets is impressive (see the chapter by Bodmer and Winney). Principal components can be defined as a set of linear combinations of SNPs such that individuals who tend to share alleles across the genome tend to have similar scores. It is commonly the case that the first two principal components, those that explain most of the variability in the dataset, correspond to geographic clines, for example from north to south and east to west. Successive components explain other patterns in the data, which make up successively smaller fractions of the pairwise correlation structure. In this sense, rather than defining discrete populations, these continuous measures of the position of individuals in the transformed principal component space provide an alternative way of describing differences and similarities. Again, these ideas have been used in genome-wide association studies, since including the leading principal components as covariate in standard regression models can be used to protect against population structure in tests for genotype–phenotype relationships.

17.8 Natural Selection and Genetic Clines

So why might this be useful for population-genetic inference? A potentially informative application again occurs in the context of identifying natural selection, because the use of the individual-level correlation matrix avoids having to define populations. A test for increased allele frequencies between groups of individuals can be described as one in which the principal components have a larger influence than the weight assigned by the decomposition of the genome-wide matrix predicts. To investigate the potential for such an approach to spotting natural selection at each SNP along the genome, we can compare a model where the pairs of alleles, or genotypes, carried by study individuals are explained by the inferred pair-wide correlation matrix, to one where the influence of one of the leading components is increased. A statistical test can then be used to formally assess how much better the latter model fits the observed data (Pirinen 2012). As the principal components capture gradients of allele frequency change, the test can

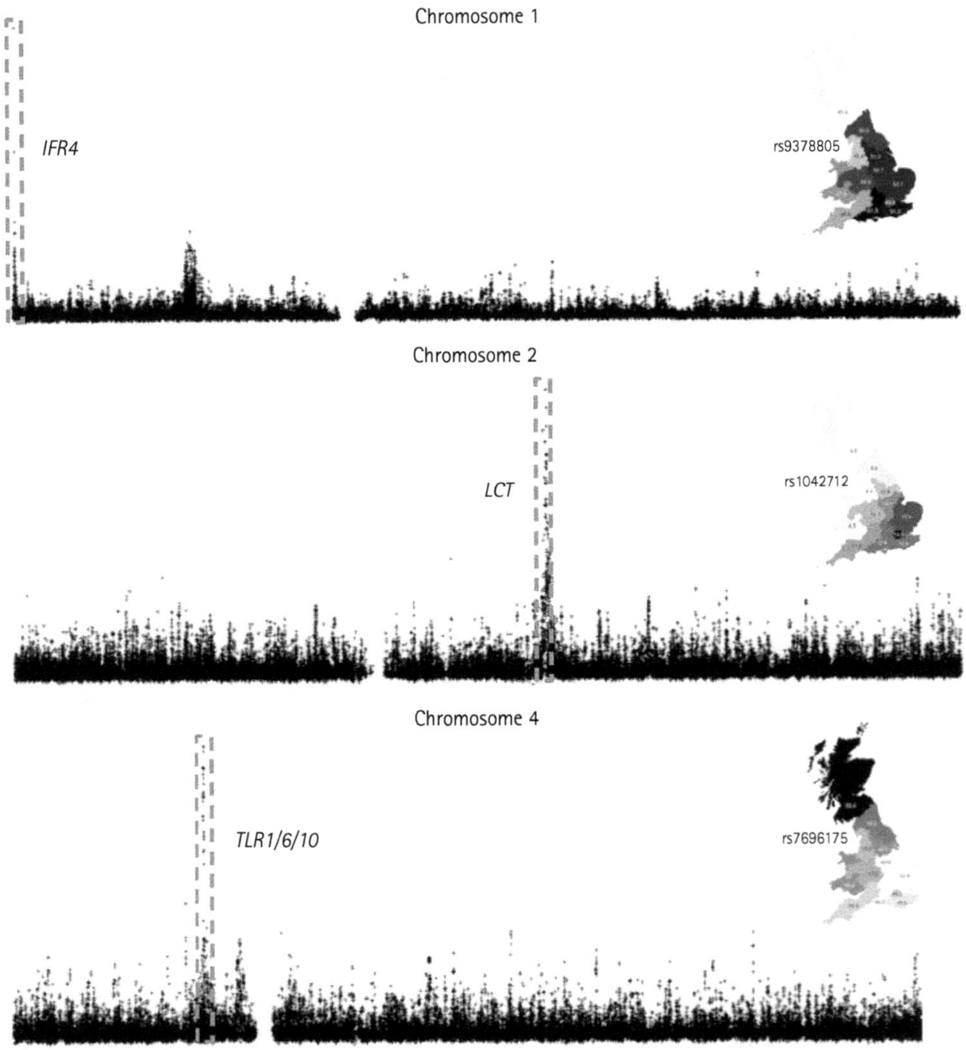

FIG. 17.8 Signals of natural selection on three human chromosomes as inferred from population analysis (Pirinen 2012) of 5,000 individuals from the UK. In each of the 3 panels the x-axis is a measure of the evidence in favour of a model where one of the principal components, which describes the relatedness between individuals is increased, relative to a model where it is at its estimated genome-wide average. On each panel a locus with strong evidence in favour of natural selection is highlighted and the allele frequency of the top SNP is plotted on a map of the UK, with dark colours denoting higher frequency. The name of the gene of interest within the region is labelled on the plot. The results are based on work by Matti Pirinen (2012).

identify which individuals are overdifferentiated, and this can be related back to their geographical relationship. Population genetics calls these patterns clines.

The results of applying this approach to all 500,000 SNPs across the 22 chromosomes of the human genome in the 5,000 UK control datasets used in the MS study are shown in Figure 17.8.

The figure highlights signals on three chromosomes by plotting the statistical evidence in favour of natural selection at each SNP against its physical position in the genome. On chromosome 2 it picks out a well-known region of the genome that is home to the lactose (*LCT*) gene. Genetic variants at the lactose locus determine ability to digest lactose into adulthood, which is thought to have been beneficial due to the spread of agriculture. By plotting which principal component is responsible for the increased allele frequency change we can infer that the influence of natural selection shows a cline from Scotland to the South East. Other regions of the genome showing strong evidence for association include a set of 'Toll-like receptor' (*TLR*) genes on chromsome 4 which are known to be involved in innate immunity. In this case, these loci are thought to have harboured beneficial alleles which increased in frequency in parts of the country most heavily affected by bubonic plague; individuals without the protective allele will have had a slightly decreased chance of survival, and this has a disproportionally larger effect where the epidemic was more severe. In fact, the part of the genome that shows the strongest evidence for natural selection by this measure is at the beginning of chromosome 1 in a gene called *IRF4*. *IRF4* is thought to be involved in a range of processes including immune system response, but also eye colour, hair colour, skin pigmentation, freckling, and skin cancer risk. It is therefore unclear which of these phenotypic consequences drove strong allele frequency difference across the UK, or whether they occurred in ancestral populations which differentially populated geographic regions. Identifying these loci and understanding the mechanism of natural selection provides insight into key recent events in evolution, and points to important biological processes.

17.9 Final Remarks

The work described in this chapter covers but a tiny fraction of the substantial theoretical and empirical research that underlies modern population genetics. It aims to highlight one interesting concept: that when you collect data on individuals directly then it is possible to do away with the idea of defining populations altogether. In settings where the scientific hypothesis directly relates to groups of individuals, due to an assumption of where features of interest might be clearest, for example looking for genotyping errors between groups of individuals assayed with different technologies, or in comparing coastal populations to inland populations to identify genetic markers of adaptation, then the a priori assumption of who belongs to which population may help to identify such effects. However, defining populations upfront can be problematic when there is little or no information on which sets of individuals might show the differences that are of scientific interest. In this case removing the grouping of individuals into populations might help either to identify features that would otherwise have been missed, or to control for patterns of similarity not captured by populations labels. Two examples in the field of statistical genetics include: genome-wide association studies, where the use of a pairwise correlation matrix, rather than population labels, helps to protect against spurious results due to population structure or cryptic relatedness; and the use of a correlation matrix to identify parts of the genome where allele frequencies are unusually differentiated along one or more of the principal components.

As with many areas of statistics, and science more generally, it is common for new approaches to take advantage of novel analytical methods which exploit the power of modern computers. The use of correlation matrices to describe patterns between thousands of samples (here individuals) with millions of observations (here SNPs) has benefited from these advances. Therefore, the concept of a population may, in more and more settings, be replaced by a set of principal components, onto which observations are mapped by transformations of the data, which successively capture key patterns of variability. In population genetics, these principal components represent clines of genetic variation and allow a more complete description of the complex human history that has led to the global patterns of genetic diversity we see today. Outside of genetics, the choice of which observations to include, and how to weight them, becomes important as it influences the principal components. Perhaps this helps to make clear that the definition of a population is nearly always subjective, influenced by both the things you choose to measure and how you choose to compare them.

References

Coop, G., D. Witonsky, A. Di Rienzo, and J. K. Pritchard (2010). 'Using Environmental Correlations to Identify Loci Underlying Local Adaptation', *Genetics* 185: 1411–1423, doi:10.1534/genetics.110.114819.

Nicholson, G., A. V. Smith, F. Jónsson, Ó. Gústafsson, K. Stefánsson, and P. Donnelly (2002). 'Assessing Population Differentiation and Isolation from Single-Nucleotide Polymorphism Data', *Journal of the Royal Statistical Society: Series B (Statistical Methodology)* 64: 695–715, doi: 10.1111/1467-9868.00357.

Pirinen, M. (2012). 'Detecting Natural Selection by Mixed-Model Analysis of Genome-Wide Human Genetic Diversity', ms.

Sawcer, S., G. Hellenthal, M. Pirinen, C. C. A. Spencer, P. Donnelly, and A. Compston (2011). 'Genetic Risk and a Primary Role for Cell-Mediated Immune Mechanisms in Multiple Sclerosis', *Nature* 476: 214–219, doi: 10.1038/nature10251.

The 1000 Genomes Project Consortium (2010). 'A Map of Human Genome Variation from Population-Scale Sequencing', *Nature* 467: 1061–1073, doi: 10.1038/nature09534.

Wellcome Trust Case Control Consortium (2007). 'Genome-Wide Association Study of 14,000 Cases of Seven Common Diseases and 3,000 Shared Controls', *Nature* 447: 661–678, doi: 10.1038/nature05911.

CHAPTER 18

POPULATION STRUCTURE AND PUBLIC HEALTH RESEARCH ON HIV CONTROL IN SUB-SAHARAN AFRICA

SIMON GREGSON AND TIMOTHY B. HALLETT

18.1 Introduction

Tremendous progress has been made in recent years in controlling the spread and impact of HIV epidemics in sub-Saharan Africa. During the last decade, the number of new ('incident') infections amongst adults and children fell substantially (from a peak of 2.8 m [2.6 m–3.0 m] in 1997 to 1.5 m [1.3 m–1.6 m] in 2010) and the proportion of adults currently infected ('HIV prevalence') started to decline (from 6.0% [5.7%–6.4%] to 4.7% [4.4%–4.9%]) (UNAIDS 2014). By the end of 2012, 7.75 m people in sub-Saharan Africa were on antiretroviral treatment, representing an estimated 63% [57%–67%] of those in need (WHO 2013). As a consequence of these trends, the number of deaths due to AIDS in the region fell from 1.9 m [1.7 m–2.0 m] in 2004 to 1.1 m [1.0 m–1.3 m] in 2013 (UNAIDS 2014; Jahn et al. 2008). During the same period, a number of promising new approaches to HIV prevention have been identified, including male circumcision (Auvert et al. 2005; Bailey et al. 2007; Gray et al. 2007), oral and topical pre-exposure prophylaxis (Abdool Karim et al. 2010; Baeten et al. 2012), and 'treatment as prevention' (Cohen et al. 2011). Nevertheless, HIV prevalence remains very high in many countries in sub-Saharan Africa and there is still a long way to go before the epidemic in the region can be said to be under control.

From a public health perspective, the primary area of interest in HIV research has been in establishing the scale and severity of epidemics, in identifying strategies that are effective in reducing their spread, and in controlling their impact on mortality and quality of life at the level of the general population. However, to answer these questions, it has proved necessary to investigate, understand, and take into account the composition or structure of national and local populations and the characteristics, roles, and patterns of interplay between many different forms of subpopulation. In some cases, these are groups of individuals who are known to each other, such as groups of sex workers and people who participate together in a community group; in other cases, they are individuals who simply share a common sociodemographic or behavioural characteristic.

In this chapter, we examine the importance of population structure in relation to four critical areas of public health research on HIV, drawing particularly on examples from our work in Zimbabwe over the past 20 years. We begin, in the first section, by describing the importance of population structure in understanding the temporal dynamics of HIV epidemics and in explaining variation in the spread and demographic impact of HIV within sub-Saharan Africa. In the second section, we discuss the 'sentinel surveillance' approach taken by UNAIDS and national HIV control programmes to track trends in HIV epidemics by using data from convenient population subgroups, the biases that have arisen from the selective nature of these subgroups, and the methods that have been developed to correct for these biases. In the third section, we turn to discussion of the role of community trials for measuring the effectiveness of interventions and some of their limitations. These trials are conducted in geographic and demographic subpopulations and provide a context in which mathematical models and observational data can be combined effectively to identify differences with these subpopulation that account for why results in some trials have not been replicated elsewhere. In the final section, we consider the challenges that population structure, particularly through its influence on the underlying natural dynamics of HIV epidemics, has posed for designing and evaluating HIV control programmes. These challenges point to the interaction between population structures including a set of community groups which together shape the incidence and spread of HIV. Again, a combination of mathematical modelling and observation appears to indicate the most fruitful way forward.

18.2 The Spread and Demographic Impact of HIV Epidemics

18.2.1 Early Estimates of the Scale and Impact of HIV Epidemics

In the 1980s, it soon became clear that HIV epidemics had the potential to extend over periods of several decades (Anderson and May 1987) and, largely due to the long period of HIV infection prior to the development of AIDS, to have severe demographic consequences, particularly in developing countries (Anderson et al. 1988). However, early mathematical models differed greatly in their predictions of the likely speed, scale, and impact of HIV epidemics. In a workshop convened by the United Nations and the World Health Organization in December 1989, leading groups of modellers were provided with a common set of input assumptions and asked to predict the peak in HIV prevalence and the maximum impact on the rate of natural population increase. For the worst case scenarios, the predictions of peak HIV prevalence (estimated for all ages) varied from 30% to 58% while population growth was reduced from an initial 3.5% per annum to between −2.5% and + 2.3% per annum (United Nations 1991). With hindsight, it became clear that there were flaws both in the demographic cohort projection models and in those that built from epidemiological foundations (Gregson et al. 2007b). Many of the discrepancies arose from inadequate consideration given to the effects of heterogeneity in behaviour within populations as well as in infectiousness over the infection period. In particular, it was not recognized at the time that the high rates of transmission seen early in epidemics would be reduced not only because of saturation of infection within high-risk groups but also because they resulted from overlapping concurrent partnerships formed by individuals in core group subpopulations, many of whom were in the early acute phase of infection and therefore highly infectious (Morris and Kretzschmar 1997). Nonetheless, these early models played a critical role in mobilizing the international response to the AIDS pandemic and current models, in which these initial limitations have been addressed, now provide a reasonable representation of behavioural, epidemiological, and demographic data (Ghys et al. 2010; Walker et al. 2003; Dorrington et al. 2001).

18.2.2 National and Within-Country Variation in the Scale of HIV Epidemics

There has been great heterogeneity in the HIV epidemics in mainland sub-Saharan Africa (Figure 18.1). In UNAIDS national estimates for 2009, HIV prevalence levels varied from under 1% in Eritrea, Mauritania, Niger, and Senegal to 26% in Swaziland (UNAIDS 2010). Within countries, the epidemic has also differed in the extent of its spread, with higher prevalence rates generally being seen in more urban areas. In Zimbabwe, where the epidemic has been more evenly spread than in most countries, HIV prevalence in antenatal surveillance sites still varied from 8% in Binga to 26% in Victoria Falls (Zimbabwe Ministry of Health and Child Welfare 2010).

Ecological studies describing factors associated with differences in HIV prevalence in national and urban or rural subpopulations have been useful in generating hypotheses about the reasons for variation in HIV spread. Early studies of African countries by Bongaarts and Moses in the late 1980s showed higher HIV prevalence in countries with lower levels of male circumcision, highlighting the possibility that circumcision might reduce male susceptibility to infection through sexual intercourse (Bongaarts et al. 1989; Moses et al. 1990; Williams et al. 2006) (Figure 18.2).

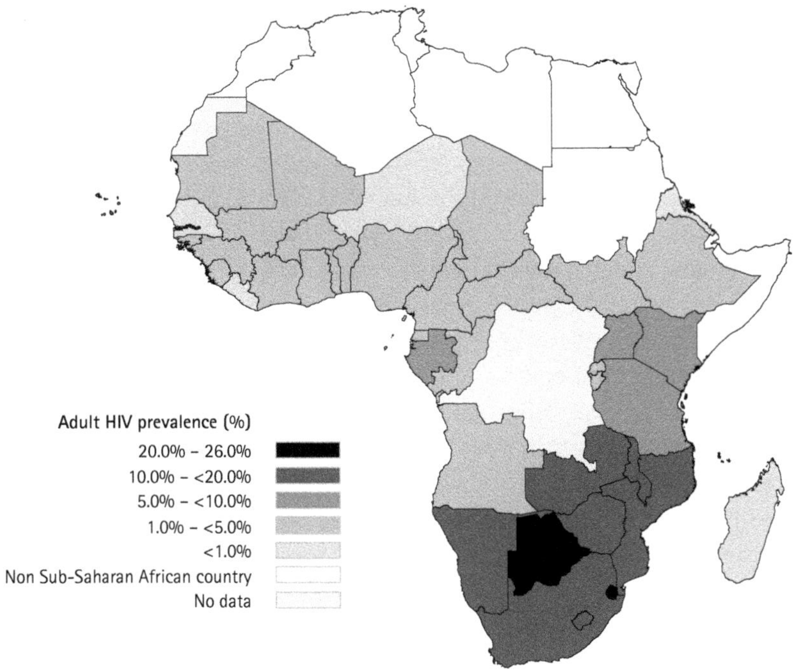

FIG. 18.1 HIV prevalence in sub-Saharan Africa, 2011.

Source: UNAIDS (2012).

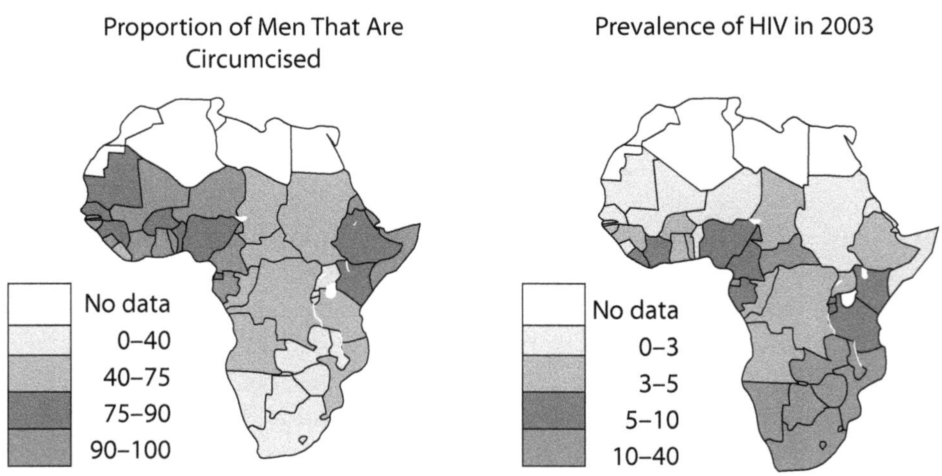

FIG. 18.2 Comparison of proportions of men circumcised and HIV prevalence in adults by country in 2003.

A similar negative association with HIV prevalence was found in a study of 89 countries by Drain and colleagues, in which greater female adult literacy, higher levels of child immunization against diphtheria, tetanus, and pertussis, and higher numbers of doctors also were found to be independently associated with lower HIV prevalence after controlling for geographic region and age of epidemic (Drain et al. 2004). In another study, Voeten and colleagues found that differences in in-migration accounted for 57% of the variation in HIV prevalence amongst pregnant women in urban areas in 28 countries in sub-Saharan Africa (Voeten et al. 2010).

In the late 1990s, an in-depth study (the 'Four Cities Study') was conducted to identify factors that might explain the higher prevalence of HIV infection in two cities in Eastern and Southern Africa (Kisumu, Kenya and Ndola, Zambia) compared to two cities in Western Africa (Cotonou, Benin and Yaoundé, Cameroon) (Buve et al. 2001b). The study found evidence that factors associated with increased risk of sexual transmission of HIV infection were more common in the high-prevalence cities. In particular, lack of male circumcision and certain STDs (herpes simplex virus type 2 (HSV-2) and trichomoniasis in women) were more common in Kisumu and Ndola than in Cotonou and Yaoundé. Younger age at first sexual intercourse (for women), younger age at first marriage, and large age differences between spouses were more common in Kisumu and Ndola; however, behavioural factors that increase exposure to HIV infection, including high rate of sexual partner change, sex with sex workers, and concurrent sexual partnerships were no more common in the high-prevalence cities (Buve et al. 2001a). The findings from this study have been important in prompting individual-level trials of male circumcision (see final section for more details) which, in turn, have led now to the implementation of male circumcision programmes in several countries (World Health Organization and UNAIDS 2011).

Early studies were also conducted in a number of countries to identify specific characteristics ('risk factors') that place individuals at increased risk of being infected with HIV to help explain this heterogeneity and in the hope that risk factors would be found that could be modified so as to slow the rate at which new individuals were acquiring HIV infection. Heterosexual intercourse was identified as the principal mode of transmission of HIV infection in sub-Saharan Africa, with the risk of acquisition, upon exposure to a previously infected partner, being accentuated in the presence of other sexually transmitted infections and particularly those such as chancroid and syphilis that cause genital ulcers (Cameron et al. 1989; Latif et al. 1989; Rottingen et al. 2001).

Drawing on findings from these studies, a proximate determinants framework (Figure 18.3) was developed (Gregson 1994; Boerma and Weir 2005), similar to those used in demography in the study of fertility (Bongaarts 1978) and early child mortality (Mosley and Chen 1984), to explain ways in which differences in underlying socio-economic, sociocultural and demographic contexts and characteristics can lead to differences in the risk of HIV infection at the individual and population levels by increasing or reducing exposure to proximate and biological determinants.

Amongst the proximate determinants, it is possible to distinguish factors that increase the risk of HIV transmission from an infected individual. For example, multiple overlapping partnerships, or short gap lengths between partners, increase the opportunity for transmission (Watts and May 1992; Morris and Kretzschmar 1997), which may be accentuated by the elevation in infectiousness that comes in the weeks following infection (UNAIDS Reference Group on Estimates Modelling and Projections 2010). Having

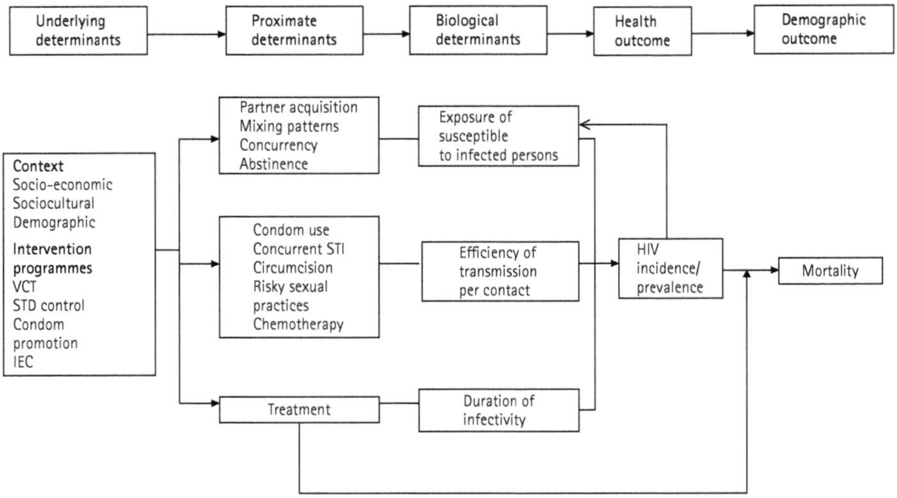

FIG. 18.3 Proximate determinants conceptual framework for factors affecting the risk of sexual transmission of HIV.

Source: Boerma and Weir (2005).

a sexually transmitted infection has been found to increase both the risk of transmission (i.e. by increasing viraemia) and the risk of acquisition (by increasing susceptibility) (Royce et al. 1997), while antiretroviral treatment has recently been shown to reduce transmission (Cohen et al. 2011).

Epidemiological mathematical models have been useful in developing understanding of the roles of the various proximate and biological factors (individually and in combination) in determining the spread of HIV infection and the impact of AIDS over time. For example, in a study in the early 1990s, a baseline model scenario, for a country with the demographic profile of Zimbabwe in the late 1980s, an annual rate of sexual partner acquisition of 2, negligible condom use, and limited mixing between subpopulations with high and low rates of sexual partner acquisition, generated a peak HIV prevalence amongst 15–49-year-olds of just over 20% and reduced population growth from 2.6% to 1% (Gregson et al. 1994). Increasing the rate of sexual partner acquisition to 3.5 generated a faster and larger epidemic in which prevalence peaked at 33% and population growth fell to just above zero. However, also increasing the level of mixing between subpopulations with high and low rates of sexual partner acquisition resulted in an even larger epidemic with HIV prevalence peaking at 48% and population growth falling to −0.8%. The study also highlighted how HIV epidemics were likely to cause substantial increases in orphanhood as well as shifts in the age and sex composition of affected populations.

These and other early mathematical modelling studies pointed to the importance of sexual networks as critical factors in determining the size and rate of spread of HIV epidemics (Anderson et al. 1990). Random or disassortative sexual mixing (as opposed to assortative or 'like-with-like' mixing) between spatial and socio-economic subpopulations and between subpopulations with different rates of sexual partner change was highlighted as being a possible explanation for larger epidemics (Garnett and Anderson 1993). Sexual networks characterized by strong 'core groups' (subpopulations such as commercial sex workers who have very high rates of sexual partner change) and large 'bridge populations' (individuals who form sexual partnerships with partners who have both high and low rates of sexual partner change) create high levels of concurrency and are more conducive to the rapid spread of HIV infection (Morris and Kretzschmar 1997; Morris et al. 1996).

Early studies of behavioural risk factors revealed substantial heterogeneity in rates of sexual partner change (Cleland and Ferry 1995). Nonetheless, typically, a small minority of women appear to have very high rates of sexual partner change. In contrast, particularly in the early years of HIV epidemics, substantial fractions of men engaged in commercial and transactional sex activity whilst continuing to have sex with their marital partners. Thus, male clients were hypothesized to constitute a bridge population between a core group of commercial sex workers and the majority of women who have intrinsically relatively low-risk behaviour (Gregson and Garnett 2000).

Migrant workers also can be an important bridge population (Morris et al. 1996). For example, Zimbabwe and South Africa may have had larger national epidemics than Uganda, in part, because of greater labour-related mobility instigated during the colonial period, facilitated by strong cultural attachment to the rural homelands and a good transport infrastructure (Potts and Mutambirwa 1990; Coffee et al. 2004; Coffee et al. 2007). Men working away from home in towns, mines, and commercial farming areas often became infected through contacts with sex workers in these locations and then spread the infection into rural populations when they returned home to visit their wives at the end of the month and during harvest and holiday periods. Commercial sex activity and other forms of transactional sex occurring locally contributed to a large rural epidemic (Gregson et al. 2001). Whereas HIV prevalence peaked at similar levels in the major urban centres of Harare and Kampala, Zimbabwe experienced a much larger national epidemic than Uganda because of the more extensive spread of infection into and within its rural population.

Mathematical models of HIV epidemic dynamics have been useful in establishing the likely population-level impact of factors found to be important at the individual level. Individuals with higher numbers of new sexual partners typically have a greater risk of acquiring HIV infection (Lopman et al. 2008); and, as shown in the aforementioned example, a higher proportion of the population having multiple sexual partners will, *ceteris paribus*, tend to result in a larger HIV epidemic in the general population. But factors that are important at the individual level do not always have an impact at the population level. In our field study in eastern Zimbabwe, we found that women aged 17–24 years were, on average, six years younger than their male sexual partners (Gregson et al. 2002) and that young women whose sexual partners were ten years or more older than themselves were twice as likely to have become infected with HIV than women of the same age whose partners were only 0–4 years older (Figure 18.4a).

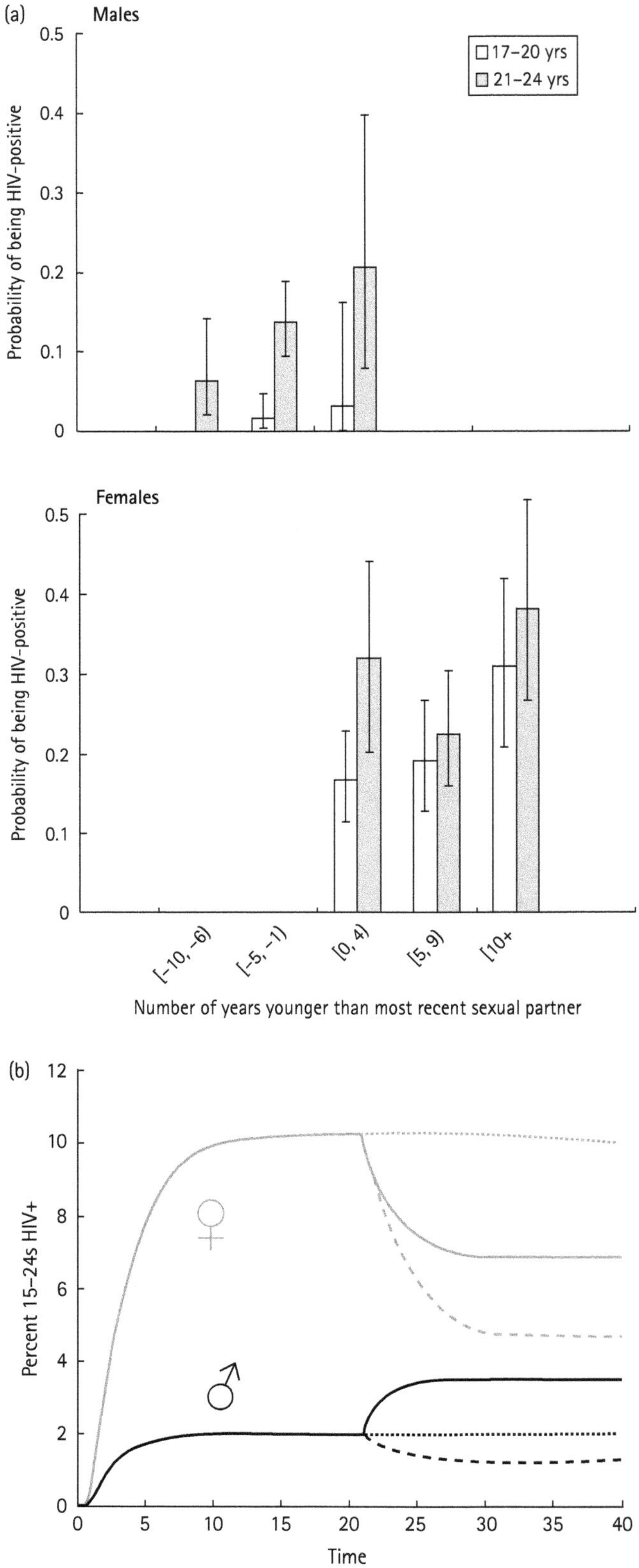

FIG. 18.4 The effect of age differences between sexual partners and risk of HIV infection.

However, in a subsequent modelling study, we found that differences in patterns of sexual mixing by age *per se* played a negligible part in explaining differences in HIV prevalence in the sexually active population (taken as aged 15–49 years) as a whole (Hallett et al. 2007b). At young ages, if the partnerships that young women would have had with older men are 'replaced' by partnerships with young men, HIV prevalence in young men increases and a sizeable fraction of young women still become infected. Whilst reduction in age differences between men and women in sexual partnerships can alter the sex ratio of HIV prevalence at young ages, these differences must be eliminated almost entirely for there to be a substantial impact on the overall size of an epidemic (Figure 18.4b).

18.2.3 Variation in Trends in HIV Prevalence over Time

Wasserheit and Aral have observed that 'Each sexually transmitted disease (STD) epidemic evolves through predictable phases, shaped by a dynamic interplay between the pathogen, the behaviour of the sub-populations in which it emerges, and the prevention efforts that are developed to limit its impact' (Wasserheit and Aral 1996). Thus, they argue that the size and character of key subpopulations and the distribution of STD infections (including HIV infection) within these groups can change over time. Certainly it seems likely that many HIV epidemics in Africa initially were concentrated within female sex workers and their male clients (Carswell 1987). Whether the epidemic was contained within these core populations, or spread into and within the general population, depended on several factors: the infectiousness of the dominant strain or subtype of the virus, the prevalence of factors that enhance the risk of HIV transmission per sexual contact (e.g. other STDs and male circumcision), the pattern of sexual contact between core and bridge subpopulations and the wider population, and the speed and effectiveness of changes in behaviour. For example, early interventions amongst sex workers may have contributed to limiting the spread of HIV into the general population in some West African populations (Boily et al. 2002; Meda et al. 1999), whilst early denial may have been a factor contributing to the larger generalized epidemics experienced in countries in Southern Africa.

Another important aspect of the temporal dynamics of HIV epidemics is that the nature and role of underlying determinants or risk factors can change over the course of an HIV epidemic due to variations between subpopulations in behavioural responses to epidemics. For example, socio-economically and educationally better-off subpopulations typically had higher levels of HIV infection in the early years of epidemics, but with greater circulation of correct information about how to avoid infection, and with increased exposure to close friends and family members dying from AIDS, HIV has become associated with poverty (Hargreaves et al. 2008). The impact of these dynamics on overall trends in the HIV epidemic in the general population is thus likely to differ between countries depending on the relative sizes of, for example, their more and less educated subpopulations.

A number of other factors may influence the temporal dynamics of HIV epidemics. These include the selective removal (through higher death rates) of subpopulations with high-risk sexual behaviour and the extent to which these groups are replaced within the population (for example, as new generations age into the sexually active population) (Garnett and Anderson 1993). Probably less important are possible feedback processes such as: sexual risk behaviour amongst children infected through mother-to-child transmission who survive into adulthood (Ferrand et al. 2009); heightened sexual risk behaviour in adolescence and later life amongst individuals who lose their parents from AIDS during childhood (Gregson et al. 2005); and new sexual partnerships formed by men and women who become widowed following the death of their spouses from AIDS (Lopman et al. 2009).

18.3 Population Structure and Surveillance of HIV Epidemics

18.3.1 Tracking Trends in HIV Prevalence Using Data from Sentinel Surveillance Populations

Since the late 1980s, the Joint United Nations Programme on HIV/AIDS (UNAIDS) has recommended HIV surveillance amongst easily accessible population subgroups (Chin 1990). In countries with concentrated epidemics, trends in HIV prevalence in high-risk/core groups were tracked, initially, through surveillance amongst patients attending STD clinics. More recently, HIV prevalence has been measured in purposive samples of these key populations and a range of methods (e.g. multiplier method, network scale-up, and respondent-driven sampling) have been used to estimate the sizes of these populations. In countries with generalized epidemics, surveillance is conducted amongst pregnant women attending for routine check-ups at selected antenatal clinics, in order to track trends in adults in the general population.

Conducting HIV surveillance amongst the subpopulation of pregnant women attending antenatal clinics proved to be convenient and feasible in most African countries. Furthermore, comparisons of HIV prevalence estimates amongst pregnant women with estimates for men and women combined in general population samples obtained in small-scale local community studies, indicated a reasonable level of agreement. However, discrepancies were found in some locations and HIV prevalence in men and women (separately) in the general population tended to be lower and higher, respectively, than amongst pregnant women (Schwartlander et al. 1999). A number of sources of bias in data from surveillance populations were identified (e.g. differences in age structure and sexual activity, and reduced fertility in infected women; Zaba and Gregson 1998), and special software packages were developed to produce national

HIV/AIDS estimates that corrected for these biases (Stover et al. 2006; Schwartlander et al. 1999).

In recent years, anonymized HIV testing has been added to larger, more nationally representative general population surveys such as the Demographic and Health Surveys. Comparisons of HIV prevalence estimates obtained from these surveys with national estimates derived from antenatal surveillance systems revealed overestimation in the national estimates. This was found to be due to over-representation of antenatal clinics in more urban high-prevalence locations (Montana et al. 2008), a problem which has been accounted for statistically in the most recent rounds of national estimates (Henry J. Kaiser Family Foundation 2008; UNAIDS 2007). Participation and acceptance of HIV testing in national general population surveys has been variable, leading to debate over whether these surveys represent a 'gold standard' for HIV estimates (Boerma et al. 2003). Indirect estimates of HIV prevalence in non-tested males and females tend to be higher than in those tested (Reniers and Eaton 2009), and there is continuing debate as to whether the overall effect of non-response on observed national estimates of HIV prevalence is sufficiently large to require adjustment (Mishra et al. 2008; Barnighausen et al. 2011).

Currently, there is interest in moving from periodic (usually biannual) antenatal clinic surveys towards utilizing routine data from pregnant women tested for HIV infection in prevention of mother-to-child transmission (PMTCT) programmes, which have been scaled up now in most African countries. However, operational challenges remain in the collection, availability, and interpretation of the PMTCT programme data (Marsh et al. 2010).

18.3.2 Limitations in HIV Prevalence as an Indicator of Changes in HIV Risk

HIV surveillance based on monitoring trends in HIV prevalence has been useful for tracking changes in the overall burden of infection within populations. However, due to the typically long duration of HIV infection (Todd et al. 2007), HIV prevalence, as an indicator, represents the accumulation of infections over periods of ten years and more. Therefore trends in HIV prevalence provide limited information on changes over time in the rate that new infections are occurring (HIV incidence) in a population. This difficulty has increased in recent years due to improvements in the survival of people living with HIV following successful introduction of antiretroviral treatment services in African countries.

Ideally, HIV incidence is estimated prospectively in representative general population cohorts, but this is expensive and rarely feasible. A number of efforts are being made to develop special biological assays and algorithms for distinguishing newly acquired infections from earlier ones using the same biomedical specimens (e.g. dried blood spots) that are used for prevalence assays (Parekh et al. 2002; Hargrove 2009; Incidence Assay Critical Path Working Group 2011). However, estimates based on

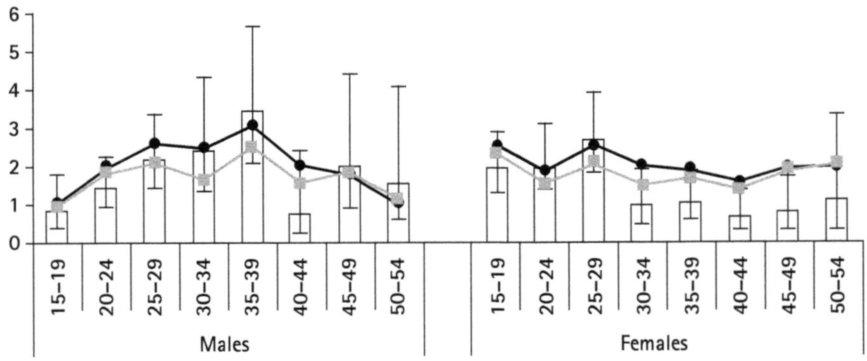

FIG. 18.5 Comparison of estimates of HIV incidence derived using two methods for estimating incidence from serial cross-sectional data on age-specific HIV prevalence (dark and light grey lines) with incidence measured in the closed cohort (bars with 95% confidence intervals), Manicaland, Zimbabwe, 1998–2003.

these types of assays currently are unreliable (UNAIDS Reference Group on Estimates 2005). In the meantime, indirect estimates of HIV incidence, and the distribution of new infection by age and sex, can be derived from age-specific HIV prevalence data collected in successive general population surveys (as in Figure 18.5) (Gregson et al. 1996; Hallett et al. 2008), or by fitting mathematical models to trends in HIV prevalence measured in surveillance data collected from pregnant women attending antenatal clinics (Hallett et al. 2009b).

18.4 The Role of Randomized Controlled Trials in Identifying Effective Interventions

Two principal forms of randomized controlled trial have been conducted to evaluate the impact of specific HIV control strategies. The first measures the individual-level efficacy of an intervention by measuring the reduction in HIV incidence amongst individuals randomly assigned to receive the intervention compared to a control group. Trials of this type were used to resolve a long-standing controversy about the possible protective effect of male circumcision. In three separate studies conducted in South Africa, Kenya, and Uganda, men in the intervention arm were found to have an almost 60% lower risk of acquiring HIV infection (Weiss et al. 2008). Another recent example is the CAPRISA 004 trial in KwaZulu-Natal, South Africa, which found that women who used an antiretroviral microbicide to protect against HIV infection during sexual intercourse had a 39% lower risk of acquiring HIV infection (Abdool Karim et al. 2010).

However, HIV prevention activities for which there is evidence of *efficacy* at the individual level do not always translate into programmes that are *effective* at lowering HIV incidence at the population level. Often this is because of factors associated with population structure, sexual networks, and the temporal dynamics of HIV epidemics. Thus, the second form of trial involves randomization at the community level (Donner and Klar 1994; Hayes et al. 1995) and is used to measure the population impact of community-level interventions or of individual-level interventions that could have an enhanced or reduced impact at the community level. The most famous example in HIV research is the Mwanza trial of improved STD treatment in north-west Tanzania (Grosskurth et al. 1995). In this trial, 12 communities, each comprising the catchment population of a health centre and its satellite clinics, were allocated to pairs based on socio-economic and spatial criteria that had been found in previous surveillance studies to influence levels of HIV transmission (i.e. roadside settlements, villages close to the lake shore, island villages, and rural villages) (Hayes et al. 1995). Within each pair, one community was allocated at random to the intervention arm of the trial and the other community was allocated to the control arm. Enhanced diagnosis and treatment services for STDs, utilizing sustainable syndromic management algorithms (rather than laboratory diagnostics), were introduced and promoted in the intervention communities. Over the subsequent two-year period, HIV incidence was found to be lower in the intervention community than in the control community in each of the six matched pairs of communities. Overall, HIV incidence was 42% lower in the intervention communities than in the control communities. This trial was implemented in the early 1990s in areas where HIV prevalence in the general population was still fairly low (4%). Subsequent trials of STD treatment and behavioural interventions (some focused on core groups and some directed to the wider population) conducted in more advanced epidemics showed reductions in risk behaviour and (sometimes) in HIV incidence amongst the population subgroups reached by the interventions, but found that these effects did not translate into reductions in HIV incidence in the general population (Wawer et al. 1999; Kamali et al. 2003; Quigley et al. 2004; Cowan et al. 2010; Gregson et al. 2007a; Ross et al. 2007). At the time of writing, several new community-randomized controlled trials were underway to test the hypothesis that interventions comprising combinations of biomedically focused approaches can reduce HIV incidence (Boily et al. 2012).

Given the importance of sexual networks in the spread of HIV infection, targeting or supporting certain population subgroups with prevention activities could lead to the diffusion of programme effects beyond the immediate target group into the wider population. As suggested above, targeting of core group populations early in HIV epidemics may be especially effective in preventing, or at least slowing, transmission into the general population. Specific approaches that were tried in the 1990s to reduce transmission within and from core groups included syndromic management of STDs (Grosskurth et al. 1995; World Health Organization 2003) and peer education programmes amongst sex workers and their clients (Wilson et al. 1990). However, in more advanced generalized epidemics, more comprehensive programmes are required. Boily and colleagues (Boily et al. 2002) have suggested that 'For epidemics with a low reproductive potential [i.e. unlikely to spread widely in the general population], modest core group interventions can significantly reduce HIV incidence and prevalence. As the epidemic matures, effective interventions should incorporate both core and non-core populations. For epidemics with a high reproductive potential, core group interventions are necessary but not sufficient to have a rapid and large-scale impact. A more general population approach is also needed early in the epidemic.'

In recent years, with the expansion of antiretroviral treatment services, increasing attention is being paid to the importance of reducing transmission from people living with HIV. Improved survival amongst this group is likely to increase its relative size within the general population in many countries. Selection for a propensity towards higher-risk sexual behaviour together with restored good health might have been expected to result in an increase in risky behaviour amongst this group, and a possible further escalation in HIV epidemics. Early studies suggest that this may not be the case (Venkatesh et al. 2010) although the reasons for this are unclear, and the impact of increased treatment availability on behaviour within the wider population has yet to be established. Potentially more important though, infectiousness levels amongst individuals on successful antiretroviral treatment have been found to be extremely low (Donnell et al. 2010; Cohen et al. 2011). As a consequence, scaled-up HIV testing services coupled with early initiation of treatment in HIV-infected individuals is being considered as an HIV prevention intervention (Granich et al. 2009; The HIV Modelling Consortium Treatment as Prevention Editorial Working Group 2012).

Community-randomized controlled trials are regarded as providing the 'gold standard' of evidence for the efficacy of an intervention and can work well when there is a straightforward replicable intervention linked to the disease outcome (Hallett et al. 2007c). However, HIV incidence trials are complex and expensive to implement and do not provide evidence for effectiveness when put to scale in 'real-world' circumstances. Furthermore, both positive and negative results can be difficult to interpret and generalize (Padian et al. 2010).

18.4.1 The Role of Mathematical Models in Evaluating HIV Control Interventions

In part because of these problems, increased use of mathematical models and observational data has been advocated in the design and evaluation of the population-level impact of alternative HIV control strategies (Garnett et al. 2011). In one early study, a mathematical model was used to identify possible reasons why the encouraging results of the Mwanza HIV/STD prevention trials were not replicated in trials of similar STD treatment interventions in the Rakai and Masaka districts of south-west Uganda (White et al. 2004). It was concluded that reductions in sexual risk behaviour prior to the trials in Uganda, lower levels of curable STDs, and the later stage of the epidemic in Uganda explained most of the contrast in HIV impact observed between the three trials. The model results indicated that 'all four simulated intervention strategies were more effective in reducing incidence of HIV infection in Mwanza than in either Rakai or Masaka' (White et al. 2004).

Similarly, in the case of the Manicaland trial in Zimbabwe, for which the intervention was focused on subpopulations at highest risk, but which did not succeed in reducing HIV incidence overall, mathematical models suggested that this could have been because the mixing between population groups was not sufficient for the impact to spread, or because there were underlying epidemiological differences between the communities in different arms that impaired an ability to detect a difference (Hallett et al. 2007a).

Theoretical investigations have also shown how an understanding of the population structure can lead to more effective or efficient use of resources in intervention. For instance, a voluntary counselling and testing intervention based on providing counselling for behaviour change should aim to reach those persons for whom behaviour change would have the greatest impact on their risk of infection, or the risk of transmission in the population. Therefore, younger and more sexually active persons should be a priority. In contrast, a testing programme that seeks to refer persons identified as HIV-infected to a treatment programme, to improve health and survival, should aim to find as many HIV-infected persons as possible that have not otherwise been tested, and should therefore prioritize middle-aged and older persons and perhaps men (who are less likely attend a health clinic for other reasons) (Hallett et al. 2009a).

Following the evidence that ART can reduce the chance of infection, many have considered how ART can be used to reduce HIV transmission in a population. As there is a strong correlation between HIV infection risk and the risk of AIDS death (Mulder et al. 1994), it follows that an immediate priority is treatment for those with most advanced disease. Once these groups are covered, however, it has been suggested that treatment should be provided at an earlier stage of infection for infected pregnant women, or for those with a long-term sexual partner who is not infected, since such persons may be amongst the most likely to transmit the virus (Delva et al. 2012).

18.5 Measuring the Effectiveness of National HIV Control Programmes

At the national level, establishing whether HIV control programmes have had an impact on rates of HIV transmission is even more problematic due to the complex natural dynamics of infectious disease epidemics and the lack of empirical counterfactuals for trends in infection levels in the absence of programmatic and behavioural responses. In this section, we start by reviewing these dynamics briefly before turning to approaches that have been developed to address them.

In epidemics of all infectious diseases, the infection spreads rapidly amongst individuals who are most exposed and susceptible to infection. After a time, infection levels saturate amongst these high-risk groups and the average rate of infection reduces to reflect the lower level experienced by those who are at less risk. Thus, declines in incidence typically occur even in the absence of effective control programmes, as illustrated for the case of SARS in Figure 18.6a (Riley et al. 2003). HIV incidence peaked in most sub-Saharan African countries within about ten years (Bongaarts et al. 2008). HIV prevalence also has fallen to varying degrees in a number of these countries. However, declines in HIV prevalence also can occur in the absence of reductions in risky behaviour, depending largely on the population structure. In the illustration shown in Figure 18.6b, *ceteris paribus*, lower heterogeneity and changes in behaviour over

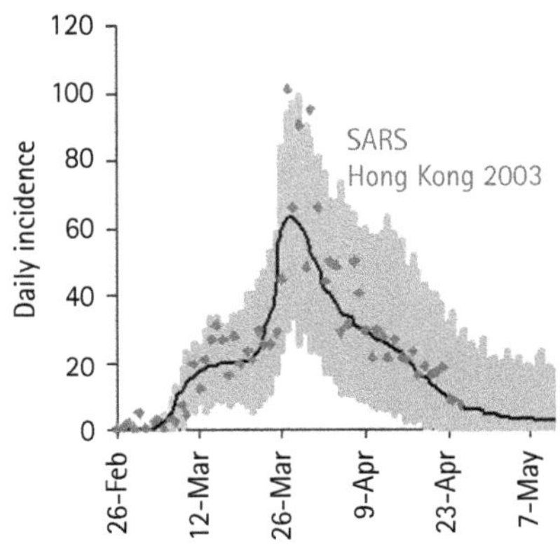

FIG. 18.6a Trends in the daily incidence of Severe Acute Respiratory Syndrome (SARS) in Hong Kong, February to May 2003.

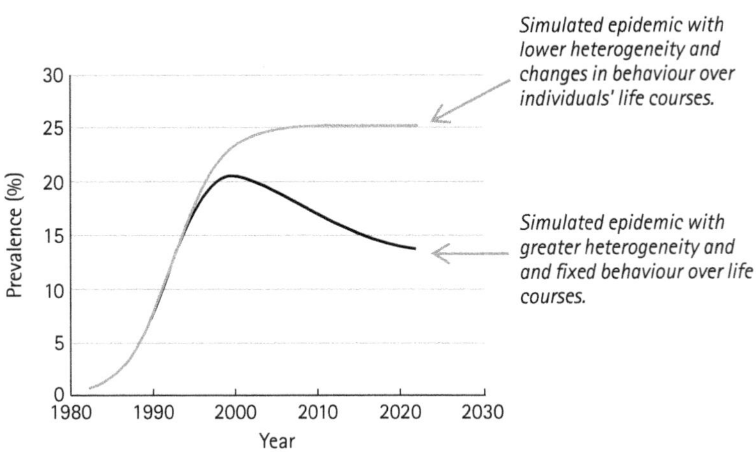

FIG. 18.6B Modelled HIV prevalence in a generalized epidemic assuming no behaviour change, or change amounting to a 50% reduction in the risk of infection.

individuals' life courses give rise to a stable level of HIV prevalence after the epidemic reaches its peak whereas greater heterogeneity but constant behaviour over the life course result in a substantial drop in HIV prevalence.

These complex temporal dynamics of infectious disease epidemics make it very difficult to determine from observed trends in HIV prevalence alone whether changes in sexual risk behaviour have occurred and, if so, whether they have had an impact on the epidemic. To address this problem amongst others, UNAIDS advocated parallel behavioural sentinel surveillance as part of its 'Second Generation Surveillance' initiative in the early 2000s (UNAIDS 2002; Amon et al. 2000). However, data on sexual behaviour can be subject to social desirability bias (Catania et al. 1990) which may change over time and particularly with increasing awareness of AIDS mortality and/or greater exposure to HIV prevention messages. Prior to HIV infection, women in sub-Saharan African populations tended to under-report sexual partnerships whilst men over-reported their numbers of sexual partnerships (Nnko et al. 2004). However, since the spread of HIV epidemics, both sexes are more likely to under-report sexual relationships. In part, due to these concerns, behavioural sentinel surveillance has been adopted systematically in only a small number of countries.

Declines in HIV incidence and HIV prevalence cannot, in themselves, be taken as evidence for reductions in risk behaviour. However, an understanding of the natural dynamics of HIV epidemics does provide a basis for generating credible conservative counterfactual model scenarios for the trends in HIV prevalence that would be expected in a given population in the absence of behaviour change (Hallett et al. 2006; Hallett et al. 2009b). Where empirical trends in HIV prevalence are lower or decline more steeply than these counterfactual trends, it can be inferred that there is evidence that reductions in risk behaviour have occurred. In the case of Zimbabwe, using Bayesian statistical methods, we found that the best explanation for trends in HIV prevalence observed at antenatal clinics followed in the national surveillance system (in both urban and rural populations, fitted separately) was reductions in sexual risk behaviour having occurred between 1999 and 2004 (Hallett et al. 2009b). This result was consistent with empirical data from a general population cohort in eastern Zimbabwe (Gregson et al. 2006) and findings from analysis of national survey data spanning this period (Gregson et al. 2010).

When population-level reductions in HIV risk are identified using modelling and statistical approaches of the form described here, a further question arises as to the extent to which these reductions can be attributed to national HIV control programmes. More specifically, the question arises as to whether there are subpopulations other than population structures defined by risk-prone sexual relations that are shaping the incidence of HIV infections and the relative success of programme interventions. These subpopulations are defined by social networks and community institutions. In rural Malawi, there have been reports of local communities responding spontaneously by adopting protective behaviours (Watkins 2005). In Southern Africa, participation in certain forms of social groupings (e.g. churches or community groups) sometimes has been found to be associated with reductions in HIV risk behaviour and could facilitate the diffusion of new ideas and adaptations to social norms through local social networks (Campbell et al. 2002; Gregson et al. 2004; Pronyk et al. 2008). There may, however, be heterogeneity in effects on health of community group participation, and negative effects sometimes have been observed (Portes and Landolt 1996; Veenstra 2000). Community groups may, for example, serve to spread incorrect information or reinforce unhelpful social norms. The characteristics of populations that participate in community groups (e.g. their levels of education), and the extent to which they provide bonding and bridging social capital (i.e., positive social spaces), are obviously important for facilitating critical dialogue about AIDS and for renegotiation of social norms. Such factors may be effective in enabling adoption of safer behaviour and uptake of services (Gregson et al. 2004; Nhamo et al. 2010).

In our own study in eastern Zimbabwe, we found that women who were members of community groups at baseline (in the late 1990s) had a lower rate of HIV infection over the subsequent three-year period than their counterparts who did not participate in these groups (Gregson et al. 2011). A similar trend was seen across a wide variety of different types of community groups, ranging from sports clubs, farmers' groups, and burial societies to more dedicated AIDS groups (Figure 18.7). However, men who participated in these types of groups tended to have *higher* levels of HIV incidence than other men. This difference may reflect differences in the individual backgrounds of women and men who join groups (more risk-averse for women; possibly less so for men) and differences in the types (e.g. more women in savings clubs; more men in sports clubs) and characteristics (e.g. more women in single-sex groups; more men in groups that

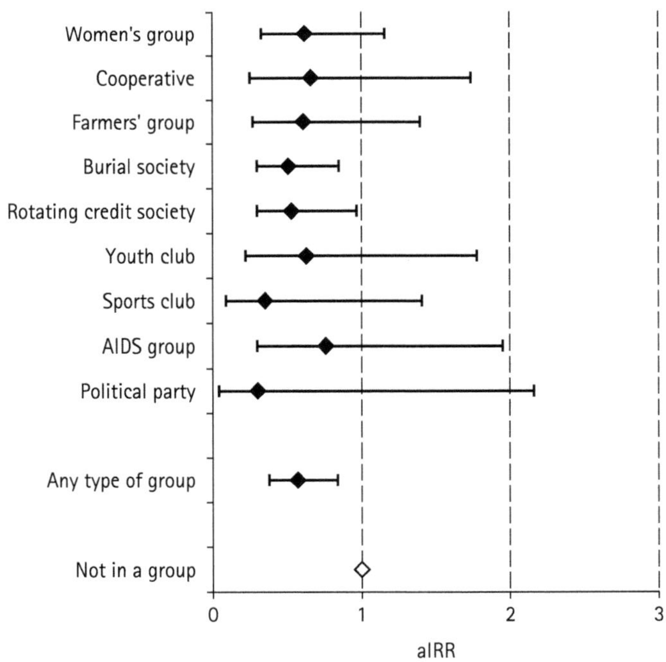

FIG. 18.7 Individual-level effect of participation in community groups on HIV incidence aIRR, incidence rate ratio for HIV infection adjusted for age. Horizontal lines show 95% confidence intervals.

consume alcohol) of the groups they join (Gregson et al. 2011). In addition, there may be important gender differences in the motivations for, and nature of, community group participation. Interestingly, the protective effect of community group membership against the risk of HIV infection seen for women in the late 1990s and early 2000s had disappeared by the mid–late 2000s, suggesting a possible transitory effect as reductions in risk behaviour took hold across the wider population and community groups became less viable in Zimbabwe during a period of severe political and economic instability (Campbell et al. 2013, Gregson et al. 2013).

These observations suggest an active response to increases in AIDS mortality from within communities but it seems unlikely that effective modifications in sexual risk behaviour would have occurred without national programmes to provide information on modes of transmission and protective behaviours. Furthermore, biomedical programmes such as routine screening of blood donations and increases in the availability of STD treatment services probably slowed the spread of HIV infection in the early stages of epidemics. Thus, where reductions in HIV risk are identified, it seems reasonable to attribute at least some of this success to national control programmes. This said, more work is needed to understand the interaction between these programmes and community groups.

A puzzle remains, however, as to why there are a number of countries in sub-Saharan Africa where declines in HIV prevalence have occurred that can be linked to reductions in sexual risk behaviour, yet there are also many other countries where declines have not yet occurred despite largely similar prevention programmes (Hallett et al. 2006). As part of a recent study on the HIV decline in Zimbabwe, ecological comparisons based on the proximate determinants framework were used to identify contextual factors unique to Zimbabwe, which could have facilitated behaviour change (Halperin et al. 2011). High AIDS mortality from the late 1990s resulting from the large epidemic was thought to have been the primary factor motivating reductions in casual sex and other multiple sexual partnerships, whilst interpersonal communication and other community-based interventions that were scaled up following the establishment of the National AIDS Council in 1999 may have assisted people in adopting protective behaviours. The severe economic decline in Zimbabwe occurred slightly after behaviour had begun to change, but reductions in incomes, which prevented men from frequenting bars and paying for transactional sex, in combination with shifting social norms, played an important secondary role in driving and sustaining the changes in behaviour. In the ecological comparisons with seven countries in Southern Africa where there was less evidence for reductions in behavioural risk, Zimbabwe stood out in having a strong marriage system and high levels of secondary school education. This unique combination may have facilitated: (i) a clearer understanding and acceptance of how HIV is sexually transmitted; and (ii) a greater ability to act upon 'be faithful' messages, reflecting the stronger marriage pattern in Zimbabwe than in neighbouring countries that also had relatively well-educated populations (e.g. Botswana and South Africa) (Halperin et al. 2011). Institutional arrangements, notably marriage and education, therefore, may contribute to HIV control directly as components of population structure and indirectly through their influence on community groups.

18.6 Conclusion

An understanding of population structure and the role it plays in the transmission dynamics of infection is critical to understanding HIV epidemics and how they can be controlled. In this chapter, we have illustrated this, using examples largely from our work, conducted at the population, subpopulation, and individual levels, in Zimbabwe over the past 20 years. This work highlights the important roles that various subpopulations have played in HIV transmission, and in HIV surveillance and research, and how good targeting and study of key subpopulations, at appropriate phases of epidemics, can result in more effective programme outcomes.

References

Abdool Karim, Q. A., S. S. Abdool Karim, J. A. Frohlich, A. C. Grobler, B. Baxter, L. E. Mansoor, A. B. M. Kharsany, S. Sibeko, K. P. Mlisana, Z. Omar, T. N. Gengiah, S. Maarschalk, N. Arulappan, M. Mlotshwa, L. Morris, D. Taylor, and C. T. Group (2010). 'Effectiveness and Safety of Tenofovir Gel, an Antiretroviral Microbicide, for the Prevention of HIV Infection in Women', *Science* 329: 1168–1174.

Amon, J., T. Brown, J. Hogle, J. MacNeil, R. Magnani, S. Mills, E. Pisani, T. Rehle, T. Saidel, and C. K. Sow (2000). *Behavioral Surveillance Surveys (BSS): Guidelines for Repeated Behavioral Surveys in Populations at Risk of HIV*. Arlington, VA: Family Health International.

Anderson, R. M., S. Gupta, and W. Ng (1990). 'The Significance of Sexual Partner Contact Networks for the Transmission Dynamics of HIV', *Journal of Acquired Immune Deficiency Syndromes* 3: 417–429.

Anderson, R. M. and R. M. May (1987). 'Transmission Dynamics of HIV', *Nature* 326: 137–142.

Anderson, R. M., R. M. May, and A. R. McLean (1988). 'Possible Demographic Consequences of AIDS in Developing Countries', *Nature* 332: 228–234.

Auvert, B., D. Taljaard, E. Lagarde, J. Sobngwi-Tambekou, R. Sitta, and A. Puren (2005). 'Randomized, Controlled Intervention Trial of Male Circumcision for Reduction of HIV Infection Risk: The ANRS 1265 Trial', *Public Library of Science Medicine* 2: e298.

Baeten, J. M., D. Donnell, P. Ndase, N. R. Mugo, J. D. Campbell, J. Wangisi, J. W. Tappero, E. A. Bukusi, C. R. Cohen, E. Katabira, A. R. Ronald, E. Tumwesigye, E. Were, K. H. Fife, J. Kiarie, C. Farquhar, G. John-Stewart, A. Kakia, J. Odoyo, A. Mucunguzi, E. Nakku-Joloba, R. Twesigye, K. Ngure, C. Apaka, H. Tamooh, F. Gabona, A. Mujugira, D. Panteleeff, K. K. Thomas, L. Kidoguchi, M. Krows, J. Revall, S. Morrison, H. Haugen, M. Emmanuel-Ogier, L. Ondrejcek, R. W. Coombs, L. Frenkel, C. Hendrix, N. N. Bumpus, D. R. Bangsberg, J. E. Haberer, W. S. Stevens, J. R. Lingappa, and C. Celum (2012). 'Antiretroviral Prophylaxis for HIV Prevention in Heterosexual Men and Women', *New England Journal of Medicine* 367: 399–410.

Bailey, R. C., S. Moses, C. B. Parker, K. Agot, I. MacLean, J. N. Krieger, C. F. M. Williams, R. T. Campbell, and J. O. Ndinya-Achola (2007). 'Male Circumcision for HIV Prevention in Young Men in Kisumu, Kenya: A Randomised Controlled Trial', *The Lancet* 369: 643–656.

Barnighausen, T., J. Bor, S. Wandira-Kazibwe, and D. Canning (2011). 'Correcting HIV Prevalence Estimates for Survey Nonparticipation Using Heckman-Type Selection Models', *Epidemiology* 22: 27–35.

Boerma, J. T., P. D. Ghys, and N. Walker (2003). 'Estimates of HIV-1 Prevalence from National Population-Based Surveys as a New Gold Standard', *The Lancet* 362: 1929–1931.

Boerma, J. T. and S. Weir (2005). 'Integrating Demographic and Epidemiological Approaches to Research on HIV/AIDS: The Proximate Determinants Framework', *Journal of Infectious Disease* 191: S61–S67.

Boily, M.-C., C. M. Lowndes, and M. Alary (2002). 'The Impact of HIV Epidemic Phases on the Effectiveness of Core Group Interventions: Insights from Mathematical Models', *Sexually Transmitted Infections* 78: 78–90.

Boily, M.-C., B. Mâsse, R. Alsallaq, N. S. Padian, J. W. Eaton, J. F. Vesga, and T. B. Hallett (2012). 'HIV Treatment as Prevention: Considerations in the Design, Conduct, and Analysis of Cluster Randomized Controlled Trials of Combination HIV Prevention', *Public Library of Science Medicine* 9/7: e1001250.

Bongaarts, J. (1978). 'A Framework for Analysing the Proximate Determinants of Fertility', *Population and Development Review* 4: 105–132.

Bongaarts, J., T. Buettner, G. Heilig, and F. Pelletier (2008). 'Has the HIV Epidemic Peaked?' *Population and Development Review* 34: 199–224.

Bongaarts, J., P. Reining, P. Way, and F. Conant (1989). 'The Relationship between Male Circumcision and HIV Infection in African Populations', *AIDS* 3: 373–377.

Buvé, A., M. Caraël, R. J. Hayes, B. Auvert, B. Ferry, N. J. Robinson, S. Anagonou, L. Kanhonou, M. Laourou, S. Abega, E. Akam, L. Zekeng, J. Chege, M. Kahindo, N. Rutenberg, F. Kaona, R. Musonda, T. Sukwa, L. Morison, H. A. Weiss, M. Laga, and Study Group on Heterogeneity of HIV Epidemics in African Cities (2001a). 'The Multicentre Study on Factors Determining the Differential Spread of HIV in Four African Cities: Summary and Conclusions', *AIDS* Supplement 4: S127–S131.

Buvé, A., M. Caraël, R. J. Hayes, B. Auvert, B. Ferry, N. J. Robinson, S. Anagonou, L. Kanhonou, M. Laourou, S. Abega, E. Akam, L. Zekeng, J. Chege, M. Kahindo, N. Rutenberg, F. Kaona, R. Musonda, T. Sukwa, L. Morison, H. A. Weiss, and M. Laga (2001b). 'Multicentre Study on Factors Determining Differences in Rate of Spread of HIV in Sub-Saharan Africa: Methods and Prevalence of HIV Infection', *AIDS* 15: S5–S14.

Cameron, D. W., J. N. Simonsen, L. J. D'Costa, A. R. Ronald, G. M. Maitha, M. N. Gakinya, M. Cheang, J. C. Ndinya-Achola, P. Piot, and R. C. Brunham (1989). 'Female to Male Transmission of HIV-1: Risk Factors for Seroconversion in Men', *The Lancet* 2/8660: 403–407.

Campbell, C., K. Scott, M. Nhamo, C. A. Nyamukapa, C. Madanhire, M. Skovdal, L. Sherr, and S. Gregson (2013). 'Social Capital and HIV Competent Communities: The Role of Community Groups in Managing HIV/AIDS in Rural Zimbabwe', *AIDS Care* 25 (Supplement 1): S114–S122.

Campbell, C., B. Williams, and D. Gilgen (2002). 'Is Social Capital a Useful Conceptual Tool for Exploring Community Level Influences in HIV Infection? An Exploratory Case Study from South Africa', *AIDS Care* 14: 41–54.

Carswell, J. W. (1987). 'HIV Infection in Healthy Persons in Uganda', *AIDS* 1: 223–227.

Catania, J. A., D. D. Chitwood, D. R. Gibson, and T. J. Coates (1990). 'Methodological Problems in AIDS Behavioural Research: Influences on Measurement Error and Participation Bias in Studies of Sexual Behaviour', *Psychological Bulletin* 108: 339–362.

Chin, J. (1990) 'Public Health Surveillance of AIDS and HIV Infections', *Bulletin of the World Health Organization* 68: 529–536.

Cleland, J. and B. Ferry (eds) (1995). *Sexual Behaviour and AIDS in the Developing World*. London: Taylor and Francis.

Coffee, M. P., G. P. Garnett, M. Mlilo, H. A. C. M. Voeten, S. K. Chandiwana, and S. Gregson (2004). 'Patterns of Movement and the Risk of HIV in Rural Zimbabwe', *Journal of Infectious Disease* 191 (Supplement 1): S159–S167.

Coffee, M. P., M. N. Lurie, and G. P. Garnett (2007). 'Modelling the Impact of Migration on the HIV Epidemic in South Africa', *AIDS* 21: 343–350.

Cohen, M. S., Y. Q. Chen, M. McCauley, T. Gamble, M. C. Hosseinipour, N. Kumarasamy, J. G. Hakim, J. Kumwenda, B. Grinsztejn, J. H. Pilotto, S. V. Godbole, S. Mehendale, S. Chariyalertsak, B. R. Santos, K. H. Mayer, I. F. Hoffman, S. H. Eshleman, E. Piwowar-Manning, L. Wang, J. Makhema, L. A. Mills, G. de Bruyn, I. Sanne, J. Eron, J. Gallant, D. Havlir, S. Swindells, H. Ribaudo, V. Elharrar, D. Burns, T. E. Taha, K. Nielsen-Saines, D. D. Celentano, M. Essex, T. R. Fleming, and HPTN 052 Study Team (2011). 'Prevention of HIV-1 Infection with Early Antiretroviral Therapy', *New England Journal of Medicine* 365: 493–505.

Cowan, F. M., S. J. Pascoe, L. Langhaug, W. M. Mavhu, S. Chidiya, S. Jaffar, M. Mbidzo, J. M. Stephenson, A. M. Johnson, R. M. Power, G. Woelk, and R. J. Hayes (2010). 'The Regai Dzive Shiri Project: Results of a Randomized Trial of an HIV Prevention Intervention for Youth', *AIDS* 24: 2541–2552.

Delva, W., J. W. Eaton, F. Meng, C. Fraser, R. G. White, P. Vickerman, M.-C. Boily, and T. B. Hallett (2012). 'HIV Treatment as Prevention: Optimising the Impact of Expanded HIV Treatment Programmes', *Public Library of Science Medicine* 9: e1001258.

Donnell, D., J. M. Baeten, J. Kiarie, K. K. Thomas, W. Stevens, C. R. Cohen, J. McIntryre, J. R. Lingappa, and C. Celum (2010). 'Heterosexual HIV-1 Transmission after Initiation of Antiretroviral Therapy: A Prospective Cohort Analysis', *The Lancet* 375: 2092–2098.

Donner, A. and N. Klar (1994). 'Methods for Comparing Event Rates in Intervention Studies When the Unit of Allocation Is a Cluster', *American Journal of Epidemiology* 140: 279–289.

Dorrington, R., D. Bourne, D. Bradshaw, R. Laubscher, and I. M. Timaeus (2001). *The Impact of HIV/AIDS on Adult Mortality in South Africa*. Cape Town: South African Medical Research Council.

Drain, P. K., J. S. Smith, J. P. Hughes, D. T. Halperin, and K. K. Holmes (2004). 'Correlates of National HIV Seroprevalence: An Ecological Analysis of 122 Developing Countries', *Journal of Acquired Immune Deficiency Syndromes* 35: 407–420.

Ferrand, R. A., E. L. Corbett, R. Wood, J. Hargrove, C. E. Ndhlovu, F. M. Cowan, E. Gouws, and B. G. Williams (2009). 'AIDS among Older Children and Adolescents in Southern Africa: Projecting the Time Course and Magnitude of the Epidemic', *AIDS* 23: 2039–2046.

Garnett, G. P. and R. M. Anderson (1993). 'Factors Controlling the Spread of HIV in Heterosexual Communities in Developing Countries: Patterns of Mixing between Different Age and Sexual Activity Classes', *Philosophical Transactions of the Royal Society of London Series B: Biological Sciences* 342/1300: 137–159.

Garnett, G. P., S. Cousens, T. B. Hallett, R. W. Steketee, and N. Walker (2011). 'Mathematical Models in the Evaluation of Health Programmes', *The Lancet* 378: 515–525.

Ghys, P. and G. P. Garnett (2010). 'The 2009 HIV and AIDS Estimates and Projections: Methods, Tools and Analyses', *Sexually Transmitted Infections* 86: ii1–ii2.

Granich, R. M., C. F. Gilks, C. Dye, K. M. De Cock, and B. Williams (2009). 'Universal Voluntary HIV Testing with Immediate Antiretroviral Therapy as a Strategy for Elimination of HIV Transmission: A Mathematical Model', *The Lancet* 373: 48–57.

Gray, R. H., G. Kigozi, D. Serwadda, F. Makumbi, S. Watya, F. Nalugoda, N. Kwanuka, L. H. Moulton, M. A. Chaudhary, M. Z. Chen, N. Sewankambo, F. Wabwire-Mangen, M. C. Bacon, C. F. M. Williams, P. Opendi, S. J. Reynolds, O. Laeyendecker, T. C. Quinn, and M. J. Wawer (2007). 'Male Circumcision for HIV Prevention in Men, Rakai, Uganda: A Randomised Trial', *The Lancet* 369: 657–666.

Gregson, S. (1994). 'Will HIV Become a Major Determinant of Fertility in Sub-Saharan Africa?', *Journal of Development Studies* 30: 650–679.

Gregson, S., S. Adamson, S. Papaya, J. Mundondo, C. A. Nyamukapa, P. R. Mason, G. P. Garnett, S. K. Chandiwana, and R. M. Anderson (2007a). 'Impact and Process Evaluation of Integrated Community and Clinic-Based HIV-1 Control in Eastern Zimbabwe', *Public Library of Science Medicine* 4: e302, doi: 10.1371/journal.pmed.0040102.

Gregson, S., C. A. Donnelly, C. P. Parker, and R. M. Anderson (1996). 'Demographic Approaches to the Estimation of Incidence of HIV-1 Infection among Adults from Age-Specific Data in Stable Endemic Conditions', *AIDS* 10: 1689–1697.

Gregson, S. and G. P. Garnett (2000). 'Contrasting Gender Differentials in HIV-1 Prevalence and Associated Mortality Increase in Eastern and Southern Africa: Artifact of Data or Natural Course of Epidemics?', *AIDS* 14: S85–S99.

Gregson, S., G. P. Garnett, and R. M. Anderson (1994). 'Assessing the Potential Impact of the HIV-1 Epidemic on Orphanhood and the Demographic Structure of Populations in Sub-Saharan Africa', *Population Studies* 48: 435–458.

Gregson, S., G. P. Garnett, C. A. Nyamukapa, T. B. Hallett, J. J. C. Lewis, P. R. Mason, S. K. Chandiwana, and R. M. Anderson (2006). 'HIV Decline Associated with Behaviour Change in Eastern Zimbabwe', *Science* 311: 664–666.

Gregson, S., E. Gonese, T. B. Hallett, N. Taruberekera, J. W. Hargrove, E. L. Corbett, R. Dorrington, S. Dube, K.-L. Dehne, and O. Mugurungi (2010). 'HIV Decline Due to Reductions in Risky Sex in Zimbabwe? Evidence from a Comprehensive Epidemiological Review', *International Journal of Epidemiology* 39: 1311–1323.

Gregson, S., P. R. Mason, G. P. Garnett, T. Zhuwau, C. Nyamukapa, R. M. Anderson, and S. K. Chandiwana (2001). 'A Rural Epidemic in Zimbabwe? Findings from a Population-Based Survey', *International Journal of STD and AIDS* 12/3: 189–196.

Gregson, S., P. Mushati, H. Grusin, M. Nhamo, C. Schumacher, M. Skovdal, C. A. Nyamukapa, and C. Campbell (2011). 'Social Capital and Reduced Female Vulnerability to HIV Infection in Rural Zimbabwe', *Population and Development Review* 37: 333–359.

Gregson, S., C. Nyamukapa, G. P. Garnett, P. R. Mason, T. Zhuwau, M. Caraël, S. K. Chandiwana, and R. M. Anderson (2002). 'Sexual Mixing Patterns and Sex-Differentials in Teenage Exposure to HIV Infection in Rural Zimbabwe', *The Lancet* 359: 1896–1903.

Gregson, S., C. A. Nyamukapa, G. P. Garnett, M. Wambe, J. J. C. Lewis, P. R. Mason, S. K. Chandiwana, and R. M. Anderson (2005). 'HIV Infection and Reproductive Health in Teenage Women Orphaned and Made Vulnerable by AIDS in Eastern Zimbabwe', *AIDS Care* 17: 785–794.

Gregson, S., C. A. Nyamukapa, B. Lopman, P. Mushati, G. P. Garnett, S. K. Chandiwana, and R. M. Anderson (2007b). 'A Critique of Early Models of the Demographic Impact of HIV/AIDS in Sub-Saharan Africa Based on Empirical Data from Zimbabwe', *Proceedings of the National Academy of Sciences* 104: 14586–14591.

Gregson, S., C. A. Nyamukapa, C. Schumacher, S. Magutshwa-Zitha, M. Skovdal, R. Yekeye, L. Sherr, and C. A. Campbell (2013). 'Evidence for a Contribution of the Community Response to HIV Decline in Eastern Zimbabwe?', *AIDS Care* 25 (Supplement 1): S88–S96.

Gregson, S., N. Terceira, P. Mushati, C. A. Nyamukapa, and C. Campbell (2004). 'Community Group Participation: Can It Help Young Women to Avoid HIV? An Exploratory Study of Social Capital and School Education in Rural Zimbabwe', *Social Science and Medicine* 58: 2119–2132.

Grosskurth, H., F. Mosha, J. Todd, E. Mwijarubi, A. Klokke, K. Senkoro, P. Mayaud, J. Changalucha, A. Nicoll, G. ka-Gina, J. Newell, K. Mugeye, D. Mabey, and R. Hayes (1995).

'Impact of Improved Treatment of Sexually Transmitted Diseases on HIV Infection in Rural Tanzania: Randomised Controlled Trial', *The Lancet* 346: 530–536.

Hallett, T. B., J. Aberle-Grasse, E. K. Alexandre, G. Bello, L.-M. Boulos, M. P. A. Cayemittes, M. Chantel, B. Cheluget, J. Chipeta, R. Dorrington, S. Dube, K. Eugenie, E. M. Gaillard, J. M., Garcia Calleja, G. P. Garnett, P. D. Ghys, S. Greby, S. Gregson, S. Hader, J. Hanson, W. Hladek, S. Ismail, G. John, S. Kassim, W. Kirungi, M. Louis, K. Lucien, A. Mahomva, L. Marum, T. Maselle, N. Monica, O. Mugurungi, T. Rehle, J. Stover, and N. Walker (2006). 'Declines in HIV Prevalence Can Be Associated with Changing Sexual Behaviour in Uganda, Urban Kenya, Zimbabwe and Urban Haiti', *Sexually Transmitted Infections* 82: 1–8.

Hallett, T. B., S. Dube, A. Mahomva, G. Ncube, O. Mugurungi, S. Gregson, and G. P. Garnett (2009a). 'The Role of Testing and Counselling for HIV Prevention and Care in the Era of Scaling-Up Antiretroviral Therapy', *Epidemics* 1: 77–82.

Hallett, T. B., G. P. Garnett, Z. Mupambireyi, and S. Gregson (2007a). 'Measuring Effectiveness in Community Randomised Trials of HIV Prevention', *International Journal of Epidemiology* 37: 77–87.

Hallett, T. B., S. Gregson, J. J. C. Lewis, B. Lopman, and G. P. Garnett (2007b). 'Behaviour Change in Generalised HIV Epidemics: Impact of Reducing Cross-Generational Sex and Delaying Age at Sexual Debut', *Sexually Transmitted Infections* 83: i50–i54.

Hallett, T. B., S. Gregson, O. Mugurungi, E. Gonese, and G. P. Garnett (2009b). 'Is There Evidence for Behaviour Change Affecting the Course of the HIV Epidemic in Zimbabwe? A New Mathematical Modelling Approach', *Epidemics* 1: 108–117.

Hallett, T. B., P. J. White, and G. P. Garnett (2007c). 'The Appropriate Evaluation of HIV Prevention Interventions: From Experiment to Full-Scale Implementation', *Sexually Transmitted Infections* 83: i55–i60.

Hallett, T. B., B. Zaba, J. Todd, B. Lopman, M. Wambura, S. Gregson, and J. T. Boerma (2008). 'Estimating Incidence from Prevalence: Methods and Validation', *Public Library of Science Medicine* 5: e80.

Halperin, D. T., O. Mugurungi, T. B. Hallett, B. Muchini, B. Campbell, T. Magure, C. Benedikt, and S. Gregson (2011). 'A Surprising Prevention Success: Why Did the HIV Epidemic Decline in Zimbabwe?', *Public Library of Science Medicine* 8: e1000414.

Hargreaves, J., C. P. Bonell, T. Boler, D. Boccia, I. Birdthistle, A. Fletcher, P. M. Pronyk, and J. R. Glynn (2008). 'Systematic Review Exploring Time Trends in the Association between Educational Attainment and Risk of HIV Infection in Sub-Saharan Africa', *AIDS* 22: 403–414.

Hargrove, J. W. (2009). 'BED Estimates of HIV Incidence Must Be Adjusted', *AIDS* 23: 2061.

Hayes, R., F. Mosha, A. Nicoll, H. Grosskurth, J. Newell, J. Todd, J. Killewo, J. Rugemalila, and D. Mabey (1995). 'A Community Trial of the Impact of Improved Sexually Transmitted Disease Treatment on the HIV Epidemic in Rural Tanzania: 1. Design', *AIDS* 9: 919–926.

Henry J. Kaiser Family Foundation (2008). 'Understanding the New UNAIDS Estimates'. Menlow Park, CA: The Henry J. Kaiser Family Foundation.

Incidence Assay Critical Path Working Group (2011). 'More and Better Information to Tackle HIV Epidemics: Towards Improved HIV Incidence Assays', *Public Library of Science Medicine* 8/6: e1001045.

Jahn, A., S. Floyd, A. C. Crampin, F. D. Mwaungulu, H. Mvula, F. Munthali, N. McGrath, J. Mwafilaso, V. Mwinuka, B. Mangongo, P. E. M. Fine, B. Zaba, and J. R. Glynn (2008). 'Population-Level Effect of HIV on Adult Mortality and Early Evidence of Reversal after Introduction of Antiretroviral Therapy in Malawi', *The Lancet* 371: 1603–1611.

Kamali, A., M. Quigley, J. S. Nakiyingi, J. Kinsman, J. Kengeya-Kayondo, R. Gopal, A. Ojwiya, P. Hughes, L. M. Carpenter, and J. A. G. Whitworth (2003). 'Syndromic Management of STIs and Behaviour Change Interventions on Transmission of HIV-1 in Rural Uganda: A Community Randomised Trial', *The Lancet* 361: 645–652.

Latif, A. S., D. A. Katzenstein, M. T. Bassett, S. Houston, J. C. Emmanuel, and E. Marowa (1989). 'Genital Ulcers and Transmission of HIV among Couples in Zimbabwe', *AIDS* 3: 519–523.

Lopman, B., C. A. Nyamukapa, T. B. Hallett, P. Mushati, N. Spark-du Preez, F. Kurwa, M. Wambe, and S. Gregson (2009). 'The Role of Widows in the Heterosexual Transmission of HIV in Manicaland, Zimbabwe, 1998 to 2003', *Sexually Transmitted Infections* 85: i41–i48.

Lopman, B., C. A. Nyamukapa, P. Mushati, M. Wambe, Z. Mupambireyi, P. R. Mason, G. P. Garnett, and S. Gregson (2008). 'Determinants of HIV Incidence after 3 Years' Follow-Up in a Cohort Recruited between 1998 and 2000 in Manicaland, Zimbabwe', *International Journal of Epidemiology* 37: 88–105.

Marsh, K. A., O. Bolu, S. Bodika, K. Seipone, S. Wonkongkathep, F. Baryarama, A. Yansaneh, C. Eure-Miller, and J. M. Garcia Calleja (2010). 'Can PMTCT Program Data Replace HIV Sentinel Surveillance among Pregnant Women in Africa?', *Journal of HIV/AIDS Surveillance and Epidemiology* 2/1: article 5.

Meda, N., I. Ndoye, S. M'Boup, A. Wade, S. Ndiaye, C. Niang, F. Sarr, I Diop, and M. Caraël (1999). 'Low and Stable HIV Infection Rates in Senegal: Natural Course of the Epidemic or Evidence for Success of Prevention?', *AIDS* 13: 1397–1405.

Mishra, V., B. Barrere, R. Hong, and S. Kong (2008). 'Evaluation of Bias in HIV Seroprevalence Estimates from National Household Surveys', *Sexually Transmitted Infections* 84: i63–i70.

Montana, L. S., Mishra, V. and Hong, R. 2008. 'Comparison of HIV Prevalence Estimates from Antenatal Care Surveillance and Population-Based Surveys in Sub-Saharan Africa', *Sexually Transmitted Infections* 84: i78–i84.

Morris, M. and M. Kretzschmar (1997). 'Concurrent Partnerships and the Spread of HIV', *AIDS* 11: 641–648.

Morris, M., C. Podhisita, M. J. Wawer, and M. S. Handcock (1996). 'Bridge Populations in the Spread of HIV/AIDS in Thailand', *AIDS* 10: 1265–1272.

Moses, S., J. E. Bradley, J. C. D. Nagelkereke, A. R. Ronald, J. O. Ndinya-Achola, and F. A. Plummer (1990). 'Geographical Patterns of Male Circumcision Practices in Africa: Association with HIV Seroprevalence', *International Journal of Epidemiology* 19: 693–697.

Mosley, W. H. and L. C. Chen (1984). 'An Analytical Framework for the Study of Child Survival in Developing Countries', *Population and Development Review* 10: 25–45.

Mulder, D. W., A. J. Nunn, A. Kamali, J. Nakiyingi, H.-U. Wagner, and J. F. Kengeya-Kayondo (1994). 'Two-Year HIV-1-Associated Mortality in a Ugandan Rural Population', *The Lancet* 343: 1021–1023.

Nhamo, M., C. Campbell, and S. Gregson (2010). 'Contextual Determinants of HIV Prevention Programme Outcomes: Obstacles to Local-Level AIDS Competence in Rural Zimbabwe', *AIDS Care* 22: 1662–1669.

Nnko, S., J. T. Boerma, M. Urassa, G. Mwaluko, and B. Zaba (2004). 'Secretive Females or Swaggering Males? An Assessment of the Quality of Sexual Partnership Reporting in Rural Tanzania', *Social Science and Medicine* 59: 299–310.

Padian, N. S., S. I. McCoy, J. E. Balkus, and J. N. Wasserheit (2010). 'Weighing the Gold in the Gold Standard: Challenges in HIV Prevention Research', *AIDS* 24: 621–635.

Parekh, B. S., M. S. Kennedy, T. Dobbs, C. P. Pau, R. Byers, T. Green, D. J. Hu, S. Vanichseni, N. L. Young, K. Choopanya, T. D. Mastro, and J. S. McDougal (2002). 'Quantitative Detection of Increasing HIV Type 1 Antibodies after Seroconversion: A Simple Assay for Detecting Recent HIV Infection and Estimating Incidence', *AIDS Research and Human Retroviruses* 18/4: 295–307.

Portes, A. and P. Landolt (1996). 'The Downside of Social Capital', *The American Prospect* 26: 18–21.

Potts, D. H. and C. C. Mutambirwa (1990). 'Rural–Urban Linkages in Contemporary Harare: Why Migrants Need their Land', *Journal of Southern African Studies* 16: 177–198.

Pronyk, P. M., T. Harpman, et al. (2008). 'Is Social Capital Associated with HIV Risk in Rural South Africa?', *Social Science and Medicine* 66: 1999–2010.

Quigley, M., A. Kamali, J. Kinsman, I. Kamulegeya, J. S. Nakiyingi, S. Kiwuwa, J. Kengeya-Kayondo, L. M. Carpenter, and J. A. G. Whitworth (2004). 'The Impact of Attending a Behavioural Intervention on HIV Incidence in Masaka, Uganda', *AIDS* 18: 2055–2063.

Reniers, G. and J. Eaton (2009). 'Refusal Bias in HIV Prevalence Estimates from Nationally Representative Seroprevalence Surveys', *AIDS* 23: 621–629.

Riley, S., C. Fraser, C. A. Donnelly, A. C. Ghani, L. J. Abu-Raddad, A. J. Hedley, G. M. Leung, L.-M. Ho, T.-H. Lam, T. Q. Thach, P. Chau, K.-P. Chan, S.-V. Lo, P.-K. Leung, T. Tsang, W. Ho, K.-H. Lee, E. M. C. Lau, N. M. Ferguson, and R. M. Anderson (2003). 'Transmission Dynamics of the Etiological Agent of SARS in Hong Kong: Impact of Public Health Interventions', *Science* 300: 1961–1966.

Ross, D. A., J. Changalucha, A. Obasi, J. Todd, M. L. Plummer, B. Cleophas-Mazige, A. Anemona, D. Everett, H. A. Weiss, D. C. Mabey, H. Grosskurth, and R. J. Hayes (2007). 'Biological and Behavioural Impact of an Adolescent Sexual Health Intervention in Tanzania: A Community-Randomized Trial', *AIDS* 21: 1943–1956.

Rottingen, J.-A., D. W. Cameron, and G. P. Garnett (2001). 'A Systematic Review of the Epidemiolocal Interactions between Classical STDs and HIV: How Much Is Really Known?' *Sexually Transmitted Diseases* 28: 579–597.

Royce, R. A., A. Sena, W. Cates, and M. S. Cohen (1997). 'Sexual Transmission of HIV', *New England Journal of Medicine* 336: 1072–1078.

Schwartlander, B., K. Stanecki, T. Brown, P. Way, R. Monasch, J. Chin, D. Tarantola, and N. Walker (1999). 'Country-Specific Estimates and Models of HIV and AIDS: Methods and Limitations', *AIDS* 13: 2445–2458.

Stover, J., N. Walker, N. C. Grassly, and M. Marston (2006). 'Projecting the Demographic Impact of AIDS and the Number of People in Need of Treatment: Updates to the Spectrum Projection Package', *Sexually Transmitted Infections* 82: 45–50.

The HIV Modelling Consortium Treatment as Prevention Editorial Working Group (2012). 'HIV Treatment as Prevention: Models, Data and Questions: Towards Evidence-Based Decision-Making', *Public Library of Science Medicine* 9: e1001259.

Todd, J., J. R. Glynn, M. Marston, T. Lutalo, S. Biraro, W. Mwita, V. Suriyanon, R. Rangsin, K. E. Nelson, P. Sonnenburg, D. Fitzgerald, E. Karita, and B. Zaba (2007). 'Time from HIV Seroconversion to Death: A Collaborative Analysis of Eight Sites in Six Low- to Middle-Income Countries before Highly Active Antiretroviral Therapy', *AIDS* 21: S55–S63.

UNAIDS (2002). *Initiating Second Generation HIV Surveillance Systems: Practical Guidelines.* Geneva: UNAIDS.

UNAIDS (2007). *AIDS Epidemic Update: December 2007.* Geneva: UNAIDS.

UNAIDS (2010). *Report on the Global HIV/AIDS Epidemic.* Geneva: UNAIDS.

UNAIDS (2012). *Report on the Global AIDS Epidemic.* Geneva: UNAIDS.

UNAIDS (2014). *The Gap Report*. Geneva: UNAIDS.

UNAIDS Reference Group on Estimates, Modelling and Projections (2005). *Statement on the Use of the BED-Assay for the Estimation of HIV-1 Incidence for Surveillance or Epidemic Monitoring*. London: Imperial College.

UNAIDS Reference Group on Estimates, Modelling and Projections (2010). 'HIV: Consensus Indicators Are Needed on Concurrency', *The Lancet* 375: 621–622.

United Nations (1991). *The AIDS Epidemic and its Demographic Consequences*. New York: United Nations.

Veenstra, G. (2000). 'Social Capital, SES, and Health: An Individual Level Analysis', *Social Science and Medicine* 50: 619–629.

Venkatesh, K. K., G. de Bruyn, M. N. Lurie, L. Mohapi, P. M. Pronyk, M. Moshabela, E. Marinda, G. E. Gray, E. W. Triche, and N. A. Martinson (2010). 'Decreased Sexual Risk Behavior in the Era of HAART among HIV-Infected Urban and Rural South Africans Attending Primary Care Clinics', *AIDS* 24: 2687–2696.

Voeten, H. A., D. C. Vissers, S. Gregson, B. Zaba, R. G. White, S. J. de Vlas, and J. D. Habbema (2010). 'Strong Association between In-migration and HIV Prevalence in Urban Sub-Saharan Africa', *Sexually Transmitted Diseases* 37: 240–243.

Walker, N., K. A. Stanecki, T. Brown, J. Stover, S. Lazzari, J. M. Garcia-Calleja, B. Schwartlander, and P. D. Ghys (2003). 'Methods and Procedures for Estimating HIV/AIDS and its Impact: The UNAIDS/WHO Estimates for the End of 2001', *AIDS* 17: 2215–2225.

Wasserheit, J. N. and S. O. Aral (1996). 'The Dynamic Typology of STD Epidemics: Implications for Prevention Strategies', *Journal of Infectious Diseases* 174: S201–S213.

Watkins, S. C. (2005). 'Navigating the AIDS Epidemic in Rural Malawi', *Population and Development Review* 30: 673–705.

Watts, C. H. and R. M. May (1992). 'The Influence of Concurrent Partnerships on the Dynamics of HIV/AIDS', *Mathematical Biosciences* 108: 89–104.

Wawer, M. J., N. K. Sewankambo, D. Serwadda, T. C. Quinn, L. A. Paxton, N. Kiwanuka, F. C. L. Wabwire-Mangen, C. Li, T. Lutalo, F. Nalugoda, C. A. Gaydos, L. H. Moulton, M. O. Meehan, S. Ahmed, and R. H. Gray (1999). 'Control of Sexually Transmitted Diseases for AIDS Prevention in Uganda: A Randomised Community Trial. Rakai Project Study Group', *The Lancet* 353: 525–535.

Weiss, H., D. T. Halperin, R. C. Bailey, R. J. Hayes, G. P. Schmid, and C. Hankins (2008). 'Male Circumcision for HIV Prevention: From Evidence to Action?', *AIDS* 22: 567–574.

White, R. G., K. K. Orroth, E. L. Korenromp, R. Bakker, M. Wambura, N. Sewankambo, R. H. Gray, A. Kamali, J. A. G. Whitworth, H. Grosskurth, D. Habbema, and R. J. Hayes (2004). 'Can Population Differences Explain the Contrasting Results of the Mwanza, Rakai, and Masaka HIV/Sexually Transmitted Disease Intervention Trials? A Modelling Study', *Journal of Acquired Immune Deficiency Syndromes* 37: 1500–1513.

Williams, B. G , J. O. Lloyd-Smith, E. Gouws, C. Hankins, W. M. Getz, J. Hargrove, I. de Zoysa, C. Dye, and B. Auvert (2006). 'The Potential Impact of Male Circumcision on HIV in Sub-Saharan Africa', *Public Library of Science Medicine* 3: 1032–1040.

Wilson, D., B. Sibanda, L. Mboyi, S. Msimanga, and G. Dube (1990). 'A Pilot Study for an HIV Prevention Programme among Commercial Sex Workers in Bulawayo, Zimbabwe', *Social Science and Medicine* 31: 609–618.

World Health Organization and UNAIDS (2011). *Progress in Scale-Up of Male Circumcision for HIV Prevention in Eastern and Southern Africa: Focus on Service Delivery*. Geneva: World Health Organization.

World Health Organization, UNICEF, and UNAIDS (2013). *Global Update on HIV Treatment 2013: Results, Impacts and Opportunities*. Geneva: World Health Organization.

World Health Organization (2003). *Guidelines for the Management of Sexually Transmitted Infections*. Geneva: World Health Organization.

Zaba, B. and S. Gregson (1998). 'Measuring the Impact of HIV on Fertility in Africa', *AIDS* 12: S41–S50.

Zimbabwe Ministry of Health and Child Welfare (2010). *ANC Surveillance Report 2009*. Harare: Zimbabwe Ministry of Health and Child Welfare.

CHAPTER 19

INTERVENTIONS IN CONTEXT

STEPHEN J. KUNITZ

19.1 Introduction

At the conference in 1970 to mark the founding of the Human Sciences Programme, Geoffrey Harrison and Anthony Boyce (1972: 3–4) observed that populations could be defined in several ways: (1) hierarchically in national populations, and populations of states, counties, cities, and towns; (2) for the purpose of studying fertility, mortality, and migration, in which a population 'need be no more than the group of people in which the demographer is interested'; (3) 'genetic populations [i.e.] defined by the extent to which individuals share a common gene pool'; (4) for sociological and anthropological purposes, in which 'population units may be solely defined in terms of point of view'; and (5) in ecology as 'the extent to which people share in exploiting and developing the same environmental resources'.

These definitions are problematic for this chapter, which deals with biomedical interventions to reduce mortality in different populations. Harrison and Boyce seem to suggest that this is the province of demography, but their concept of populations as studied by demographers seems to have been that the history and sociocultural features of populations were not especially relevant. To be sure, in their volume the demographic transition was discussed and the decline of mortality due to infectious diseases was mentioned (De Jong 1972), but there was no consideration of the implications of what was called in almost the very same year the epidemiologic transition (Omran 1971; Weisz and Olszynko-Gryn 2010). This refers to the relative, and in many instances absolute, increase of non-infectious causes of death that occurs as infectious diseases recede. In this chapter, however, I argue that, whatever the requirements for the study of fertility and infectious disease control might or might not have been, to understand interventions into chronic diseases that had already become visible as they were writing, historical and sociocultural considerations are of particular relevance.

19.1.1 Biomedical Interventions: Definitions and Distinctions

We would all agree that something procedural and concrete, like vaccines, antibiotics, and surgery would qualify as biomedical interventions. But what about cognitive-behavioural therapy? A nicotine patch to help with smoking cessation would count, but would a telephone quit-line run by a health department? Or graphic health warnings on cigarette packages? Or prevention programmes in schools? Or a change in the taxes applied to tobacco? And what about combined modalities, such as antidepressants and psychotherapy? For the purposes of this paper, I will use an expansive definition: any intervention advocated by health professionals that is aimed directly at improving health counts as a relevant intervention.

I will distinguish between types of interventions, technologies, and disease conditions. The first distinction is between pervasive and discrete interventions. Discrete interventions are simple, likely to be technologically based (like immunizations or appendectomies), time-limited, and do not require much or any change in beliefs, behaviour, and/or culture on the part of health care providers or their publics. Pervasive interventions require extensive change in beliefs and behaviour, which may be perceived as very threatening. For example, in the deaf population a unique culture has developed, and the use of cochlear implants has in the past been criticized as something close to apostasy because it has meant abandoning the culture and leaving the deaf community.

The second distinction is between conditions that are amenable to health care interventions and those that are not (Nolte and McKee 2004). Non-amenable conditions, such as homicide, are regarded as beyond the reasonable reach of health care systems, and I will not consider them further. Conditions amenable to health care interventions are those that can be reduced or even eliminated by the health care system. For example, with early screening several types of cancer can be reduced even if not eliminated entirely. Similarly, smoking-related conditions such as ischemic heart disease, chronic obstructive pulmonary disease, lung cancer, and cerebrovascular disease are also amenable to health care interventions. Not only are they treatable with more or less success, they are preventable. (Lung cancer, which has a long lag time between smoking cessation and onset, is often not included when rates of death due to amenable conditions are calculated.) Thus the incidence of deaths due to conditions amenable to health care is a useful indicator of the effectiveness of health care systems broadly construed, and when populations are found to differ and one is found to have much higher rates than another, it can reasonably be asked whether the health care system in places where rates are high is adequately serving, and responding to, the needs of its public.

Another useful distinction is between clinical and population-based modes of intervention. It speaks to the prevention paradox first described by Geoffrey Rose (1992). Rose observed that for a great many conditions, high-risk individuals are very likely to develop a condition but account for very few cases in the population. The vast majority

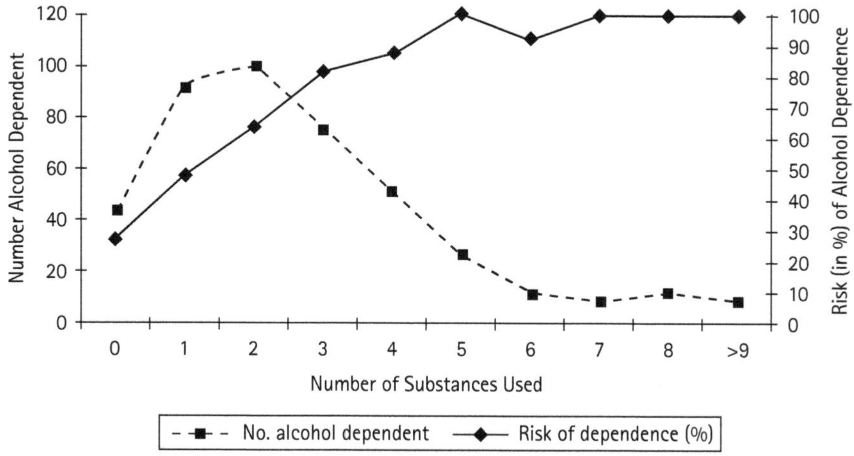

FIG. 19.1 The risk of alcohol dependence and the number of people alcohol-dependent, by number of substances used.

Source: Kunitz (2008a).

of cases occur among people at moderate risk because there are so many more of them than there are of high-risk individuals. The clinical approach focuses on high-risk patients whereas the population-based approach is aimed at reducing risks for the entire population. For instance, the risk of death from coronary heart disease increases substantially as serum cholesterol increases, but the largest absolute number of deaths due to coronary heart disease occurs at moderate elevations of serum cholesterol. To make a measurable impact on the entire population, serum cholesterol must be reduced at all risk levels.[1]

An illustration of the prevention paradox is displayed in Figure 19.1, taken from a study of substance use and misuse by Navajo Indians (Kunitz 2008a). It shows that the more mood-altering substances an individual has consumed, the higher is the risk that he/she will also be alcohol-dependent. It also shows, however, that the greatest number of alcoholics is found among people who have consumed a small number of other substances.

Another example of the same phenomenon also comes from work in the American Southwest with a mixed population of Navajos, Anglos, and Hispanics. Driving under the influence of alcohol is a cause of many motor vehicle accidents in the Mountain West. The data in Figure 19.2 indicate that the risk of an accident increased as the blood alcohol level recorded at the time of a prior arrest for DWI also increased (Woodall et al. 2004). It also shows, as Rose's work would lead us to expect, that the vast majority of accidents occurred among people whose blood alcohol level at their prior arrest was in the moderate range. The same results are observed when the total

[1] This has been interpreted to require widespread use of cholesterol-lowering drugs, which suits the pharmaceutical industry but may not be as effective as changes in diet and activity levels (Topol 2012: 19–22).

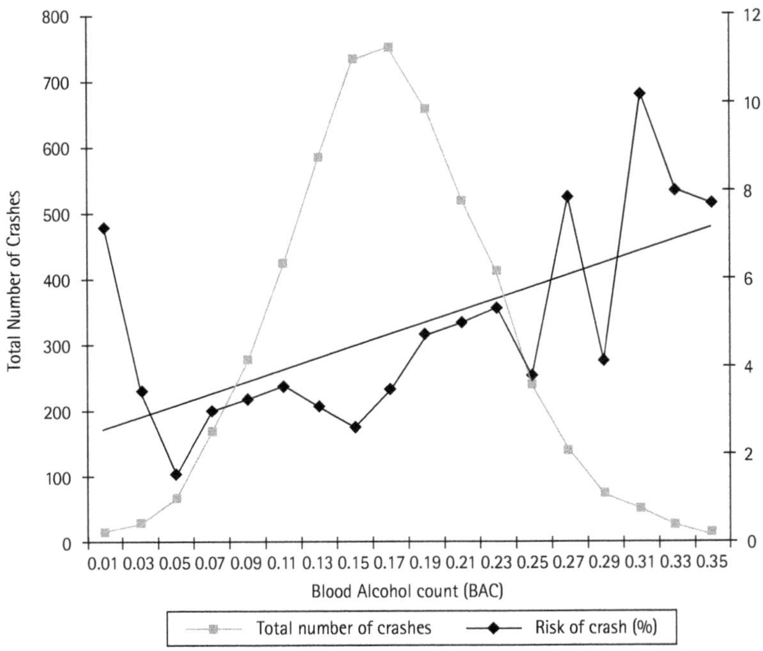

FIG. 19.2 Risk and number of alcohol-related crashes by blood alcohol level at previous arrest.

Source: Woodall et al. (2004).

number of prior arrests was substituted for blood alcohol level. The implication is that focusing on high-risk patients in a clinical setting—treating their high cholesterol, getting them to reduce their use of psychotropics, treating their irresponsible use of alcohol—may have very beneficial effects for the individuals involved but will not have a significant impact on the incidence of the relevant condition in the entire population.

A final distinction, between definitive and halfway technologies, was made by Lewis Thomas (1974) in the early 1970s. Definitive technologies are those that result in the elimination or greatly diminished incidence and severity of particular conditions, such as immunizations for vaccine-preventable diseases and antibiotics for tuberculosis, typhoid, and other infectious diseases. Halfway technologies are the kind used to manage conditions that cannot be prevented or cured: stents, renal dialysis, prosthetic limbs and joints, and Gleevec to treat chronic myelogenous leukemia. Unlike definitive technologies, these are often expensive because there are high initial costs as well as recurring costs: e.g. dialysis involves regular treatments several times a week, often for years; and Gleevec, given daily, currently costs about $29,000 per year. Indeed, avoidance of recurring costs has been one of the rationales for disease eradication programmes, perhaps the most obvious example of a definitive technological intervention (Kunitz 1987).

Cross-cutting many of these distinctions is the classification that public health workers often use of primary, secondary, and tertiary prevention. Primary prevention involves preventing a condition from developing at all. Immunization of children against a variety of infectious diseases is one example; installing systems for the delivery of clean water and for the sanitary disposal of waste would be another. Secondary prevention refers to the treatment of an existing condition, such as the use of antibiotics to treat pneumonia. And tertiary prevention refers to dealing with the late effects of an illness, such as renal dialysis for end-stage kidney disease, or physical rehabilitation after a stroke.

19.1.2 Populations

In this paper most of my examples are drawn from studies of American Indians in the northern Midwest and the Southwest, with a couple of examples, when the data are richer, from Aboriginal Australians and Canadians. These are populations that meet all of the criteria listed by Harrison and Boyce, though there are often definitional problems. For example, each American Indian tribe defines its own membership, usually by blood quantum, which may be very strict or very relaxed (Thornton 1997). In the US Census, however, people may define themselves as one or all of a number of racial groups, including American Indian. For statistical purposes, the US Public Health Service uses both census data and health care utilization data at facilities serving Indian beneficiaries. Beneficiaries are people who are enrolled members of federally recognized tribes. There is often a mismatch between race as listed in the Census and as listed on a death certificate. Because there is most consistency of these categories in places where American Indians are especially numerous, I have used data from the northern Midwest and the Southwest.

My examples come primarily from studies among American Indians in the United States, for two reasons. First, since World War II there has been a dramatic change in the most prevalent causes of death in this population, and many of the interventions that are now required involve significant changes of belief and behaviour, as well as in technology. In this regard, they are very similar to the populations of high- and middle-income countries in general (World Health Organization 2008).[2] Second, although there are important differences among them, American Indians are beneficiaries of the same

[2] According to the WHO, worldwide coronary heart disease and cardiovascular disease account for more deaths than any other condition. 'In high-income countries more than two thirds of all people live beyond the age of 70 and predominantly die of chronic diseases: cardiovascular disease, chronic obstructive lung disease, cancers, diabetes, or dementia. Lung infection remains the only leading infectious cause of death. In middle-income countries, nearly half of all people live to the age of 70 and chronic diseases are the major killers, just as they are in high-income countries. Unlike in high-income countries, however, tuberculosis and road traffic accidents also are leading causes of death. In low-income countries less than a quarter of all people reach the age of 70, and more than a third of all deaths are among children under 14. People predominantly die of infectious diseases: lung infections, diarrhoeal diseases, HIV/AIDS, tuberculosis, and malaria. Complications of pregnancy and childbirth together continue to be leading causes of death, claiming the lives of both infants and mothers' (World Health Organization 2008).

federal policies and programmes; they share similar relationships with the federal government; and they also share a history of expropriation, genocide, and poverty. These commonalities reduce some of the sources of variability often observed among populations.

19.2 Health Policy

Before proceeding, it will be useful to provide some background on health care for American Indians, particularly those west of the Mississippi River and living on or near reservations. American Indians, especially those residing west of the Mississippi, are unusual in the United States because as a result of treaties signed with the United States Congress in the first three-quarters of the nineteenth century, they have a right to health care provided by the federal government. Indian health policy has a long and tortuous history that I cannot deal with in any depth here (Kunitz 1996, 2011). It suffices to make two points.

First, in 1956, responsibility for health care, both clinical and public, was transferred from the Bureau of Indian Affairs (part of the Department of the Interior) to the US Public Health Service (part of what was then the Department of Health, Education, and Welfare). Services were provided through a regionalized bureaucracy that reported to a central office in Washington, DC. It was assumed by the congressional and professional leadership favouring the transfer that highly skilled health workers would soon reduce mortality and morbidity among American Indians to a point where they could safely and effectively integrate into the US population. This assumption regarding the power and efficacy of biomedical interventions was widely, but not universally, shared at the time (Strickland 1972; McKeown 1976; Kunitz 1991).

Though effective, the Public Health Service also often manifested disadvantages characteristic of large service organizations, including excessive paperwork and occasional non-responsiveness, and cultural insensitivity to diverse groups of clients. Consequently, beginning in the 1960s during the War on Poverty, there was increasing decentralization, with responsibility for the provision of many services devolving to Indian tribes or tribal entities.

Second, as we shall see, though federal policy with regard to the provision of services is uniform across the states with large reservation populations, regional and cultural differences as well as the devolution of services have resulted in significant variations in local conditions and actual service provision.

19.2.1 Health Consequences of Recent Policy Changes

Growth of the Indian population accelerated after World War II and coincided with, and in part resulted from, the great improvement in health services that followed assumption of responsibility for care by the US Public Health Service (Watson 2006).

The death rate declined from the time of transfer in the mid-1950s to the mid-1980s, and it was the so-called avoidable conditions—those amenable to intervention by the health care system—that declined most rapidly (Hisnanick and Coddington 1995). Then the decline stagnated and reversed slightly. Over the same period, mortality rates for the non-Indian population continued to decline (Indian Health Service 2004: 156; Kunitz 2008b).

This seems to have been due to several factors. The reduction in death rates was largely the result of the decline of infectious diseases, including tuberculosis, childhood gastroenteritis, and pneumonia. As they declined, however, non-infectious conditions began to increase in both relative and absolute importance. Among the most important were diabetes and smoking-related conditions such as lung cancer and heart disease. Diabetes accounted for about 50% of the increase from the mid-1980s to the late 1990s; smoking-related conditions for about 10 to 20% of the increase. Deaths due to diabetes appear to have declined slightly since about 2000.[3] There are, however, large regional differences in mortality that are obscured by the overall trends described above and that are examined in the next section.

19.3 Regional Differences

19.3.1 Deaths Amenable and Not Amenable to Health Care Interventions

Table 19.1 displays age-adjusted mortality rates in 1999–2001 as estimated by the Indian Health Service for several different regions (Indian Health Service 2008: 49). The rates have been adjusted for undercounting of Indian deaths, which differ among regions.[4] The Northern Area Offices include states in the northern tier from Michigan and Indiana in the east to Montana and Wyoming in the west. The Southwestern Area Offices include Arizona, New Mexico, Utah, Colorado, Nevada, and part of Texas. What is striking about these data is how consistently high the death rates are across the Northern tier of states and how low, relatively, they are in the Southwestern states.

This large regional difference is new: a study done by the Indian Health Service in the mid-1950s showed that age-adjusted death rates were essentially the same in

[3] In the most recently available data, the Indian Health Service reported a slight decline in diabetes deaths between 2003 and 2004, but it is not yet clear whether this is a random fluctuation or the start of a long-term trend. The former possibility seems most likely, but only a longer period of observation will tell (Indian Health Service 2009: 123 and Table 4.46).

[4] In general, undercounting of Indian deaths is greatest where Indians represent a small proportion of the population in places distant from reservations (Indian Health Service 2008: 49). Undercounting is lowest in the Navajo Area. See also Harwell et al. (2002) and Hahn et al. (1993).

Table 19.1 Age-adjusted death rates and life expectancies at birth, 1999–2001, in seven Indian Health Service regions

	Age-Adjusted Death Rate			Life Expectancy**	
IHS Area	*Unadjusted	**Adjusted	Unadjusted as % of Adjusted	Males	Females
North					
Aberdeen	1435.4	1470.8	97.6	63	70.6
Bemidji	1238.7	1449	85.4	65.6	70.8
Billings	1358.2	1445.1	93.9	63.8	71.4
Southwest					
Albuquerque	805.1	846.8	95.1	72.4	78.7
Navajo	923	932.3	99	71.2	77.7
Phoenix	1003	1051	95.4	68.4	75.4
Tucson	1186.3	1218.2	97.3	66	72.5

*Unadjusted for undercount of Indian deaths on state death certificates.

**Adjusted for undercount.

Source: Regional Differences in Indian Health, 2002–2003 Edition (2008).

the two regions 60 years ago, with rates in the Southwest being, if anything, slightly higher (US Public Health Service 1957).[5] In both the northern Midwest and the Southwest, deaths due to infectious diseases declined substantially, but in the northern Midwest chronic conditions have increased much more substantially than in the Southwest. Table 19.2 displays selected cause-specific age-adjusted mortality rates and all-cause mortality rates in 1999–2001 for Indians in regions across the Northern tier and Southwestern states. Deaths due to alcohol-related conditions and homicide, which are not regarded as amenable to health care interventions, do not differ between the two regions. On the other hand, heart disease and malignancies are substantially higher among Indians in the Northern tier of states than they are in the

[5] Compiled from US Public Health Service (1957):

	Southwest	Northern Plains
Age-Adjusted Death Rates 1949–53	11.15	10.15
Number of Reservations	9	9

Table 19.2 Age-adjusted cause-specific and total death rates of American Indians in the Southwest and the northern Midwest*

	Region	
Cause of Death*	Northern Midwest	Southwest
Heart disease	352.2	185.1
Cerebrovascular disease***	54.3	53.6
Malignant neoplasms	275.8	126.4
Lung cancer***	103.7	13.2
Breast cancer***	18.9	12.8
Cervical cancer***	8.5	3.7
Colorectal cancer***	27.7	13.2
Unintentional injuries	118.7	102.6
Homicide	12.2	13.2
Alcohol-related deaths	57.3	60.7
Pneumonia***	42.5	44.7
Tuberculosis***	3.2	2.4
Total all cause	1455	1012
No. IHS Areas	3	4

Notes: *Unweighted averages.

**Adjusted rates.

***Amenable to health care interventions.

Source: Regional Differences in Indian Health, 2002-2003 Edition (2008).

Southwest and account for more than 70% of the mortality difference between the two regions.

I will return to a discussion of malignancies other than lung cancer. Here I wish to point out that smoking is a risk factor for heart disease and lung cancer, and therefore it is no surprise that of all the risk factors listed in Table 19.3, smoking should be the one that differs so dramatically between the two regions (Holm et al. 2010).

Table 19.2 also displays death rates due to other causes that are amenable to health care interventions. Cervical cancer can be prevented by immunization against Human Papilloma Virus (HPV); early detection of breast, cervical, and colorectal cancer can reduce the death rate of each; and pneumonia and tuberculosis are treatable with antibiotics and supportive care. It is striking that death rates due to cervical and colorectal cancers are twice as high among Northern Plains as Southwestern tribes, the breast cancer death rates are slightly higher, and there is no difference in death rates due to pneumonia

Table 19.3 The prevalence of risk factors (in %)

	Northern Plains		Southwest	
Risk Factors	Males	Females	Males	Females
Obesity	22.4	23.7	23.7	24.1
No leisure physical activity	26.8	28.8	37.2	25.5
Binge drinking	34.3	14.2	30.4	10.8
Chronic drinking	8.8	0.8	7.4	0.2
Cigarette smoking	42.5	43.5	24.4	18.6
No seat belt use	57.6	54.3	36.5	28.4
Drinking & driving	6.7	1.8	3	1

Source: Denny and Holtzman (1999).

and tuberculosis. This suggests that health care systems on the Northern Plains perform about the same as they do in the Southwest with regard to infectious diseases but less well with regard to the prevention of deaths due to heart disease and avoidable cancers.[6]

I said at the outset that smoking-related conditions are amenable to health care interventions. However, there is no question that the high prevalence of tobacco use by a population poses a particular problem for health care providers because dealing with dependence upon tobacco is at least as difficult, time-consuming, and expensive as dealing with any other form of addiction. It requires pervasive changes in behaviour on the part of the user and changes within health care systems as well, including the investment of limited resources in activities that many health care workers may believe are beyond

[6] A study of cancer screening on the Northern Plains indicated that it was used at very low rates, and the strongest predictor of having screening for cancer at a specific site was the recommendation of a doctor or nurse (Pandhi et al. 2010). On the other hand, while the health care systems in the Southwest appear to screen patients more effectively than they do on the Northern Plains, within the Southwest these cancers are diagnosed at a later stage among Navajos Indians than among non-Hispanic whites living in the same region, indicating that the health care system serving the Navajos does not do as effective a job as the one(s) serving non-Hispanic whites. The following table shows the percent of cancers that were local at time of diagnosis (Navajo Epidemiology Center n.d.).

Site	Navajos	Non-Hispanic Whites (NM)
Prostate, men ≥50	56.2	84.6
Breast, women ≥40	52.3	64.7
Colorectal ≥50	24.3	39.1
Cervical, women ≥21	36.8	52.9

what they can reasonably be expected to do. The data in Table 19.3 indicate that current tobacco use is substantially higher among American Indians on the Northern Plains than among those in the Southwest. Other studies report similar rates (Henderson et al. 2005). Thus, the challenge to health care systems serving Northern Plains Indians is especially great,[7] but the consequences of successfully reducing tobacco use would be considerable.

A study of mortality of different regional and racial groups estimated the impact on life expectancy that would occur if several important risk factors (systolic blood pressure, body mass index, fasting blood sugar, and tobacco use) were reduced to their optimum levels. The authors combined Indians on the Northern Plains and in the Southwest and estimated that the overall effect of risk factor reduction would be an increase in life expectancy of 4.6 years for males and 5.0 years for females. The impact of reducing tobacco use to zero would be an increase of 2.4 and 2.1 years for males and females respectively (Goodarz et al. 2010). Because smoking prevalence is so different in the two regions, reducing tobacco use to zero would result in a greater improvement on the Northern Plains (closer to three years) than in the Southwest (closer to two years). The impact might be even greater if the effect of second-hand smoking could be estimated, especially on the Northern Plains where extremely cold winters that force people to remain indoors may combine with crowded housing to exacerbate the problem.

It is not obvious why health care systems on the Northern Plains deal with chronic, non-infectious conditions less effectively than do those in the Southwest. Certainly the very high rates of tobacco use are an important factor, but that does not explain the differences in death rates from cancers that are treatable if detected early, nor does it explain the somewhat lower rates of immunization observed on the Northern Plains (see Table 19.4).

It is possible that populations on the Northern Plains and in the Southwest are very different and that the former are less healthy, or more resistant to the use of services,

[7] It is not entirely clear why tobacco use is more prevalent across the North than in the Southwest. It is an indigenous American plant that followed the cultivation of corn, and there is archaeological evidence of its cultivation and use across the continent well before contact with Europeans, including among early Southwestern farmers and agricultural tribes that lived along the Missouri River in the Dakotas, as well as by Plains tribes such as the Blackfoot and Crow, who though nomadic grew it for ceremonial purposes. It appears that while the use of tobacco was widespread, it played a more central role in both spiritual and secular life among the Northern tribes than among those in the Southwest, which may have made the use of commercially sold tobacco products more acceptable among the former than the latter populations when they became increasingly available in the course of the last century (Linton 1924; Driver and Massey 1957).

Risk factors for tobacco use also differ among tribes in the two regions. Among those studied on the Northern Plains, being currently or previously married and having spent more time living off-reservation were associated with an increased likelihood of smoking. In an unnamed Southwestern tribe, young age and being male were associated with increased risk of being a smoker. There is also evidence that tobacco use starts earlier on the Northern Plains than in the Southwest (Welty et al. 2002; North et al. 2003).

Table 19.4 Immunization rates by region, 2004

Region	Pneumococcal immunizations per 100,000 pop >65	Full course of immunizations, percent of children 3–27 mos.
North		
Aberdeen	59	78.3
Bemidji	50	77.5
Billings	76	85.3
Southwest		
Albuquerque	73	85.9
Navajo	77	90.2
Phoenix	71	89.7
Tucson	86	78.6

Source: Regional Differences in Indian Health, 2002–2003 Edition.

or that the service providers themselves are less able to develop productive relationships with their publics on the Plains than in the Southwest. I consider briefly several possibilities.

19.3.2 Socio-Economic and Demographic Differences

At the regional and Area Office level there is no obvious association between mortality rates and socio-economic characteristics (see Table 19.5) except the paradoxical observation that the proportion of high school graduates in each Area is positively correlated with the death rate (Spearman's rho = 0.78, p = 0.0362). Educational attainment is consistently higher in the northern Midwest than it is in Southwest, which is just the reverse of what would be expected in light of the mortality rates displayed in the previous section.

These ecological associations are confirmed by individual-level analyses.[8] Comparisons of the probability of death of Indians in North and South Dakota with Indians from Arizona and New Mexico indicate that even adjusting for age, income, education, and sex, the probability of death is higher on the Northern Plains than in the Southwest (see Table 19.6). Older people had a higher probability of dying than young people, and women and those with higher income had a lower probability of dying than men and people with low income. Education was non-significant.[9]

[8] Data are from the National Longitudinal Mortality Study, Public Use File, Release 3, US Bureau of the Census, Department of Commerce, Washington, DC, 1 June 2008.

[9] A survey of American Indians on one Plains reservation showed that there was no association between income and education on the one hand and risk-taking behaviour and self-reported health on

Table 19.5 Income, education, and unemployment of Indians on the Northern Plains and in the Southwest

Northern Midwest	Median household income	Percent of high school graduates age ≥25 years old	Percent of males ≥16 years old who are unemployed	Percent of females ≥16 years old who are unemployed
Northern Midwest				
Aberdeen	12,310	71.6	24.7	19.2
Bemidji	19,317	75.8	14.9	12.6
Billings	14,249	75.7	26.3	17.7
Southwest				
Albuquerque	15,791	67.3	21.6	14.9
Navajo	13,984	59.5	24.8	18.6
Phoenix	16,392	64.0	18.2	15.1
Tucson	13,342	68.1	13.4	11.1

Source: Indian Health Service (May 2003), Chart 2.9.

Emigration, however, does differ in expected ways. Combining census data from 2000 (Bureau of the Census 2004) with tribal enrolment data (Bureau of Indian Affairs 2003), we observe in Table 19.7 that a higher proportion of enrolled tribal members reside on their reservations in the Southwest than in the Northern states. This is not a new pattern. Something similar was observed in the 1970 census (Bureau of the Census 1973).[10]

Income plays an important role in the emigration of Indians from their reservations. Figure 19.3 displays the association of the proportion of enrolled members living on reservations and per capita income of Indians on the same reservations. The greater the per capita income, the greater is the proportion of enrolled members on reservation. However, there is no significant correlation between income and the proportion on reservation within each region, although regions clearly differ. Indeed, in a multiple regression (data not shown), it is region rather than income that is significant. Thus, while the economic climate of each region would appear to be important, it is not the entire story.

In general, studies of American Indian demography suggest that the people who remain on reservations are poorer than those who migrate (Snipp and Sandefur 1988; Sandefur 1986); they are less well educated (Belon et al. 1983; Croy et al. 2009); and they

the other, in contrast to what had been reported from a non-Indian comparison population (Cheadle et al. 1994).

[10] However, in 1970 it was tribes, not reservations, which were the units of analysis. That is, all Sioux tribes were combined, all Apaches were combined, and Pueblos were grouped by language: Tanoan, Keresan, Hopi, etc. (Bureau of the Census 1973).

Table 19.6 Mortality regressed onto region and individual variables (odds ratios)

Variable	Model 1	Model 2
Region (Northern Plains)	2.00****	1.93***
Age	1.08****	1.08****
Sex (female)	0.36****	0.34****
Income	0.36***	0.92**
Education	1.00	–
No. obs.	1,736	2,805

Notes: * $p < 0.05$.
** $p < 0.01$.
*** $p < 0.001$.
**** $p < 0.0001$.

Table 19.7 Proportion of enrolled population residing on reservations, 2000

Region	Number of reservations	Percent of enrolled population on reservation
Southwest	32	68.7
Northern Plains	13	46.8

may be at higher risk of substance misuse (Croy et al. 2009). This is consistent with the so-called 'healthy migrant effect', and suggests the possibility that the lower the rate of migration, the healthier may be the population that remains at home: if employment and educational opportunities are available on or near reservations, the most employable and healthiest may have greater reason to remain at home (Leichenko 2003). Estimates of the effect of emigration on mortality suggest that it is real but that it cannot account for the great differences between regions.[11]

[11] It is possible that emigrants from reservations in North and South Dakota have very low mortality and that if half of them lived on their reservations and, in the extreme case, did not die during the year(s) of observation, then instead of a rate of 1,400 per 100,000 the rate would be closer to 800–900, very similar to what is observed in the Southwestern counties from which roughly half as many people proportionately emigrate. Estimates of emigrant mortality and how low it would have to be to equalize the rates in the Dakotas and the Southwest suggest that emigration may well account for some of the excess Indian deaths in Northern Plains counties compared to the Southwestern counties, but it is unlikely to account for very much. For example, if we assume that Indians who emigrate

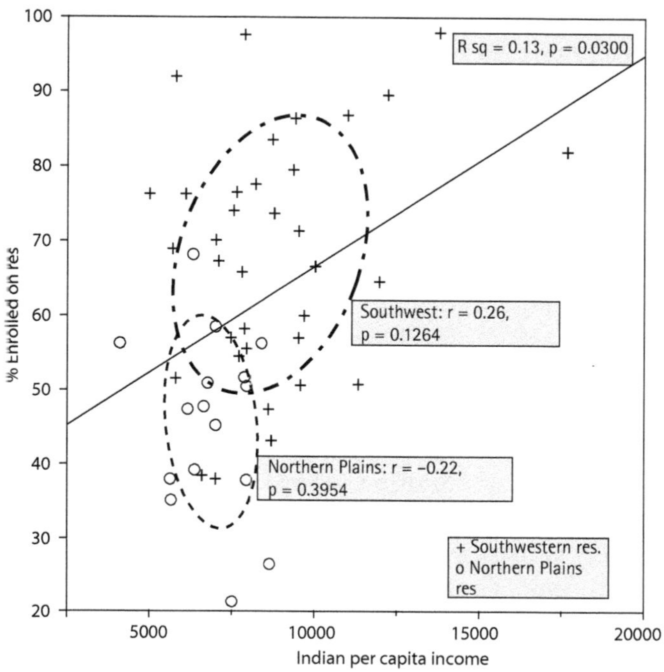

FIG. 19.3 Percent of enrolled population living on reservations in the Southwest and on the Northern Plains regressed onto per capita income of Native Americans on each reservation, 2000.

19.3.3 Indian/Non-Indian Relations

A possible reason for the differences in health care system performance may have to do with difficult relationships between the public and health care providers in each region. It may be that suspicion of government is greater among tribes on the Northern Plains than among those in the Southwest as a legacy of differences in the history of contact. No comparative surveys of attitudes of Indians and/or of providers have been done, but some proxy measures are useful. One such measure is loss of land. All Indian tribes lost access to land after the arrival of Europeans, but I focus here on loss of land after treaties were signed and reservations were established.

One way to measure loss of reservation land is by calculating how much of each of several reservations is owned by tribal entities and how much is held privately (Tiller 1996).

from reservations in these states go to cities in the same states, and if we then examine death rates of American Indians in cities in North and South Dakota, the rates are only slightly less than in reservation counties: North Dakota 1,376, South Dakota 1,237. (In North and South Dakota, these are small metropolitan areas, counties with 50,000–249,999 population: see <http://wonder.cdc.gov/wonder/help/CMF/Urbanization-Methodology.html>.) If we look farther afield, the number of Indians from the Dakotas in major metropolitan areas such as Chicago, San Francisco, and Los Angeles is said to be substantial but is diluted by a very large number of Indians from elsewhere, and the accuracy of reporting of race/ethnicity on death certificates is very problematic as well.

In the late nineteenth and early twentieth centuries, it was federal policy to encourage Indians to become private land owners. As a result, much reservation land was allotted to individual Indians, and what was left was made available for homesteading and ranching by non-Indians. Moreover, much of the land that had been allotted to Indians was ultimately acquired by non-Indians. The allotment process was carried much further on the Northern Plains than in the Southwest (Spicer 1962: 349). In the northern Midwest on average 34% of reservation land is tribally owned and the remainder individually owned. In contrast, in the Southwest, 95% of reservation land is tribally owned.

Two other measures of acculturation[12] are also useful: exogamy and English monolingualism, the latter the result of both exogamy and of having been sent to government boarding schools where use of the native language was actively discouraged. A study by the Indian Health Service in the mid-1950s (US Public Health Service 1957) revealed that in the Southwest a much higher proportion (90%) of Indians had only Indian ancestry than in the northern Midwest (<40%). Southwestern Indians at that time were also much less likely to speak English and were less likely to have attended school (see Table 19.8). Fifty years later differences persist: on average there is much greater English monolingualism among the Northern Plains (85%) than the Southwestern (53%) tribes.[13]

These various measures suggest that Northern Plains tribes have been subjected to very different experiences of contact with the federal government and Anglo-Americans than Southwestern tribes, and the result may be greater hostility to the health care system, which is very obviously an agency of the federal government. Indeed, hostility towards and distrust of the government are pervasive themes in ethnographies of Northern Plains compared with those from the Southwest (Dobie 2011: 62; Biolsi 2007; MacGregor 1946: 82; Pickering 2000: 42; Wagoner 2002: 125; Treuer 2012). Thus it should be no surprise that in response to legislation allowing tribes and tribal entities to assume responsibility for care, tribes in the northern Midwest are far more likely to have taken over management of their health care systems than are tribes in the Southwest (see Table 19.9), for previous research has shown that one of the major reasons for tribes to take over management of their health services is dissatisfaction with the responsiveness of the regional representatives of the Indian Health Service (Indian Health Service 2008; Adams 2000). In sum, greater distrust of the federal government and greater acculturation to the dominant society may conspire to have increased the likelihood that Northern tribes will manage their own health care programme.[14]

[12] Acculturation is the process by which one group acquires the attributes of another. In this case it is a measure of the acquisition of characteristics of the dominant Anglo-American society by American Indians.

[13] The Census asks questions about language knowledge and use, and these data are available at the tribal level; however, due to small samples in some populations, the same is not consistently the case at the reservation level. I have used the prevalence of English monolingualism among all people who identify themselves as members of particular Northern Plains and Southwestern tribes.

[14] It cannot be assumed that tribal management of health services among Northern tribes is responsible for the higher rates of death due to avoidable conditions because the timing of the takeovers is not known. With the publicly available data, it is difficult to determine which is cause and which effect.

Table 19.8 Comparisons of American Indians in the Northern Midwest and in the Southwest *circa* 1950

	Northern Midwest	Southwest
Number of states	9	5
Median years of school completed, AI 25 and above		
Males	7.7	3.5
Females	7.9	2.03
Median family income	1,280	744
Percent who speak English		
Ages 6–17	97.9	58.2
Ages 18–44	99.2	62.4
Ages 45 and above	89	37.1
Percent of population with 100% AI ancestry	<40	>90
Percent of Indian land tribally owned	43.2	94.8
Percent of Indian land allotted to individuals	51.9	4.5
Weighted by size of the Indian population in each state		

At this very broad level of analysis, it appears that even in the presence of the same federal policies and programmes, particular biomedical interventions such as screening for various types of cancer and preventing and treating ischemic heart disease are not applied as effectively in the northern Midwest as they are in the Southwest. Differences in income and education at both the ecological and individual level of analysis do not explain why this is the case, but it appears that the different histories of Indian–white and Indian–government relations may contribute substantially. The point is that biomedical interventions may be more or less effectively applied depending upon population and provider attributes that are very difficult to measure but that may have to do with, among other things, different historical legacies resulting in varying degrees of accommodation and trust on each side.

19.4 Discrete and Pervasive Interventions

Several examples of discrete interventions have been described here: screening for preventable and treatable cancers; influenza and pneumococcal immunizations among the elderly; and immunizations against childhood diseases among the young. Here I pay

Table 19.9 Tribal and Public Health Service management of health care programmes, 2004

	Northern Midwest	Southwest	p-value**
*Service Units**			
HIS	22	23	0.0255
Tribe	20	6	
Hospitals			
HIS	13	16	0.2530
Tribes	0	3	
Ambulatory Programmes			
HIS	28	47	0.0082
Tribes	68	51	

Notes: * A service unit is an administrative unit below the level of an Area Office in the organizational table.

Source: Regional Differences in Indian Health, 2002–2003 Edition (2008).

** Fisher's Exact Test.

attention to interventions that require pervasive changes among patients and/or providers of care, using Type II diabetes as an example. Its prevention and treatment require pervasive changes among both the public and health care systems. Primary and secondary prevention require changes in diet and exercise patterns, and secondary prevention often requires in addition the use of medication. Furthermore, among the complications of diabetes are retinopathy, gangrene, impotence, and renal insufficiency and end-stage renal disease (ESRD), all requiring tertiary prevention.

Type II diabetes did not become a major problem among indigenous peoples in the Americas and Oceania, as well as Asia and elsewhere, until the post-World War II era.[15] Since then, the problem has increased enormously. I will limit my discussion to two

[15] This was true of the Navajos, for whom it did not become a major problem until the early 1980s. Though it had been observed in the 1960s, it was thought to be a benign condition and scarcely warranted attention (Prosnitz and Mandell 1967). In my book on Navajo health and disease written in 1980–1981, I scarcely mentioned diabetes (Kunitz 1983). By 1999–2001, however, it was the fourth leading cause of death, following (1) unintentional injuries, (2) diseases of the heart, and (3) malignant neoplasms. In 1971, there were only three patients receiving treatment for ESRD; by 1985 there were 119; and by 2001 there were 636. By the end of 2010, the number of Navajos receiving dialysis was 694 (Will et al. 1997; Megill et al. 1988; Hochman et al. 2007).

issues: (1) dietary modification with an eye to primary and secondary prevention, and (2) use of renal dialysis for the treatment of end-stage renal disease.

Anthropologists have observed that in contemporary Western biomedicine and public health, as in Western cultures in general (Haidt 2012: 96–97), there is an assumption that people pursue their own self-interest and act as autonomous individuals for whom preservation of personal health is the primary goal. With respect to dietary patterns, this means that individuals are willing and able to follow the recommendations made by physicians, nurses, and diabetes educators.

As many anthropologists have also observed, however, food is not simply about nutrition, weight control, and eating five fruits and vegetables a day. It is symbolically important: what and with whom we eat are often central to definitions of the groups to which we belong. This is not only the case for Jews and Muslims who observe dietary restrictions. It is true of many groups, even if not as explicitly. Among the Innu, formerly known as the Naskapi-Montagnais, in northern Quebec, biomedical knowledge of diabetes and use of health services are both widespread, yet new cases of diabetes keep appearing at high rates and obesity is common (Roy 2006). Why?

In the 1950s new foods popular among white Canadians, such as pizza, French fries, and Coca-Cola, were viewed by Innu as markers of status and became incorporated as part of their regular diet. In subsequent decades, as political movements for autonomy became prominent, these foods began to be understood as traditional so that now there is social pressure to consume them as a manifestation of membership in the community. One informant said he would be criticized for buying a tomato. Another said she drove five hours every two weeks to do her food shopping where her choices would not be observed by her neighbours. In this setting, disregarding the dietary advice of health professionals is a political act, a way of drawing boundaries around the community and asserting its autonomy.

Similarly, among the Dakota Sioux, government provisions that were distributed in the early reservation years, when local food supplies were inadequate and undernutrition was common, have become incorporated into the diet as 'Indian' food. These 'ration' foods—beef, flour, crackers, bacon, and coffee—from which fry bread and rich stews are made, have come to be considered Indian foods that are distinguished both from older ('way-back') foods such as buffalo, cultivated corn, and a variety of berries and fruits, as well as from new foods now recommended by nutritionists.

> The food group[s] that dietitians consider under-represented in Dakota diets today are green vegetables and fresh fruits. People on diabetic diets complained that they didn't like many of the suggested selections in the diet pamphlets, saying that they were expensive, or foods that Indian people didn't like, citing in particular broccoli, spinach, asparagus, celery, and lettuce. The main components of large family meals include rich stews (preferably beef) and soups, fried potatoes, bacon and eggs, pan bread, oven-baked bread, sweet rolls, cakes, cookies, and fry bread. In criticizing diabetic diets a number of people emphasized strongly that "Indian people like to feel full" because as children they often had gone hungry. (Lang 2006a: 225)

As among the Innu, history helps to explain food preferences, but food preferences are also a vehicle for 'demarcating cultural boundaries: this disjunction of "way back" past with present, and also between Dakota–non-Dakota' (Lang 2006a: 225; Lang 2006b: 63).

In addition to these symbolic meanings, there are other considerations. When people are poor and when access to markets is limited, obtaining fresh vegetables is difficult. Furthermore, in large households, it is difficult to prepare a special diet for one person. Unless the other members are willing to change their food preferences, the diabetic person may do without. In a study of Navajo diabetics, it was found that those who lived in households in which everyone ate the same diabetic diet were more likely to follow the prescribed diet (Epple et al. 2003), suggesting that it was family support that made better control possible.

Finally, studies among the Pima of southern Arizona show that the political economy of the region (Smith-Morris 2004), including loss of land and water rights to the spreading metropolis of Phoenix and to Anglo-American farmers, has contributed to the end of farming for most people, the use of commodities and store-bought foods, and extremely high rates of Type II diabetes. All these factors have conspired to create what one observer has called a sense of 'surrender' (Kozak 1997). Because of its very high incidence and prevalence, and because of the great contemporary interest in its genetic causes, diabetes is understood by many non-diabetics as inevitable, a condition they will ultimately develop, and one more example of the destruction visited upon them by whites.

Diabetes exemplifies the many impediments to primary and secondary prevention that exist in indigenous, as well as non-indigenous, communities: the political economy of land and water loss; the symbolic importance of local food-ways; the pervasiveness of fast-food outlets; and household organization among them. Food choices are not simply an individual matter. They are also the subject of political, ideological, economic, and household considerations. Frequently these take precedence over the recommendations of physicians and health educators, even when the recommendations are based upon the latest biomedical knowledge and are understood by the population.[16]

Something similar may occur with regard to tertiary prevention. Among Aboriginal Australians, ESRD has become widespread as a result of type 2 diabetes, glomerulonephritis following streptococcal infections, and low birth weight. Many factors make it difficult for Aboriginal people to adhere to renal dialysis: language and communication barriers among and between providers, patients, and families; inability to maintain the required diet in the context of large households and traditional food preferences; the widespread problem of substance abuse by patients, their families, and friends; removal from remote communities to an urban centre where dialysis is provided; and difficulties maintaining the rigid schedule of treatments (Devitt and McMasters 1998a: 150–165).

In addition, an important feature of Aboriginal society is 'demand sharing', the obligation to share resources when they are demanded by others. Those who have resources are expected to share; it does not involve indebtedness on the part of recipients (Peterson

[16] Taking insulin or oral hypoglycemic is less problematic because their use doesn't affect others as directly as food choices do (Lang 2006a: 224).

1993). Demand sharing is the way people participate and retain membership in the community. Thus, to take one example, a man on peritoneal (home) dialysis living in a relatively remote community was very sought after because he owned a car, had a driver's license, and did not drink a great deal. Consequently he was often asked to provide rides for people, which he did willingly and frequently but at the expense of looking after himself and his dialysis equipment. The result was that he had to be hospitalized on numerous occasions when he developed serious infections (Heil n.d.).

The point of this and many similar examples (Devitt and McMaster 1998b; Joe and Young 1994) is that social obligations, the ability to meet them, and the sense of identity that results required that the patient put his responsibilities to others ahead of the need to look after himself. As in previous examples, so too are Aboriginal patients willing to utilize the health care system and understand how to use the complicated equipment involved in peritoneal dialysis. Often they do this successfully, but they often have other priorities that take precedence and that are labelled as non-compliance by health care providers.

Making effective use of the relevant biomedical interventions requires pervasive changes in patient and community behaviour and beliefs that may simply not be possible, at least in the short term. It also requires major adaptations on the part of the health care system: for example, understanding the situation of patients, their language, and the constraints they face, and providing support services close to the homes of people living in remote locations. Thus, at every level of prevention—primary, secondary, and tertiary—social processes, not lack of knowledge or willingness to use the health care system, can lead to an inability to utilize biomedical and health care interventions requiring pervasive changes in behaviour and beliefs.

19.4.1 Institutional Change

A final example of an intervention requiring pervasive change, this time at the institutional level, again comes from work with Navajo Indians. When the Public Health Service assumed responsibility for Indian health care in the mid-1950s, it arranged contracts with several universities to do projects that would lead to improved services. Perhaps the best known was a programme managed by Cornell University at Many Farms, on the Navajo Reservation (Adair and Deuschle 1970). Among other activities, the project trained bilingual Navajos as health visitors to do home visiting under the overall supervision of health professionals, providing culturally sensitive and appropriate care. When the university project ended, the health visitors were to be integrated into the existing public health nursing bureaucracy. The innovation, however, encountered overwhelming resistance from that quarter. The investigators commented that 'programmes which call for modification of bureaucratic behavior found among professions both in and out of government, call for a modification that is a threat to the establishment' (Adair and Deuschle 1970: 145). But more than that, resistance was a manifestation of 'reluctance on the part of the members of the donor society to meet the peoples of the other society in their own communities where language and behavior are strange and different' (Adair and Deuschle 1970: 149).

The point of this example is that it is not simply the public whose beliefs and behaviour must undergo change to make interventions workable. It is also the institutions that serve them. This is extremely difficult to do if it means giving up certain prerogatives and accustomed ways of doing things. As with individual patients and populations, institutions are better able to adapt to discrete rather than pervasive interventions (Coleman et al. 1966; Menzel 1960; Rogers 1995).

19.5 Discussion

Writing in the early 1970s, Lewis Thomas (1974: 115–120) claimed that science was in the same situation with respect to the non-infectious diseases as it had been with respect to the infectious diseases a hundred years earlier (Kunitz 1991). Crucial knowledge of the mechanisms of schizophrenia, cancer, stroke, and other conditions was not available, and without such knowledge, he wrote, decisive and definitive technologies to prevent and/or treat them were impossible to develop. What was available when he was writing and what are available today are the so-called halfway technologies, 'the kinds of things that must be done after the fact, in efforts to compensate for the incapacitating effects of certain diseases whose course one is unable to do very much about' (Thomas 1974: 33). These are part of tertiary prevention, like renal dialysis, organ transplants, and the iron lung. And like the iron lung, which was made obsolete by the virtual eradication of polio, it is believed by some and hoped by all that other halfway technologies will also be made obsolete by definitive technologies as simple to administer and as effective as polio vaccine.

We don't know whether the rapidly accumulating knowledge in genetics and other fields will result in definitive technologies that will have as profound an impact on the incidence, prevalence, and severity of diabetes, coronary and cerebrovascular disease, schizophrenia, and cancer as sanitation, vaccines, and antibiotics have had on the infectious diseases. It would be unwise to say that will never happen, but it is not unreasonable to say that it will not happen in our lifetime.

Prediction is difficult, especially of the future,[17] but some observers suggest that the biomedicine of the future will make possible personalized health care based upon knowledge of each individual's genetic risk profile. It has been claimed that we are entering an 'era of individualized medicine' (Topol 2012: 228), and indeed such services are already being marketed. Pursuing this topic would take me too far afield, so I will simply observe that this possibility brings us back to one of the distinctions with which I began: the difference between the clinical focus on high-risk individuals and the

[17] This profound insight has been attributed to the American philosopher and New York Yankees catcher Yogi Berra, as well as to Mark Twain, Neils Bohr, Albert Einstein, Sam Goldwyn, Will Rogers, George Bernard Shaw, and others.

public health focus on entire populations. Melzer and Zimmern (2002: 863–864) have observed that, 'With the exception of the relatively rare high penetrance, single gene disorders, genetic tests differ little from most other medical tests, providing evidence of statistical risk only.' Personalized risk profiles and the new knowledge upon which they are based will no doubt be important in the armamentarium of the clinical medicine of the future and may well become part of the public health armamentarium too (Khoury et al. 2011), but they are unlikely to be sufficient to deal with disease problems in national populations or worldwide. For that, low- and moderate-risk groups must also be the focus of attention, and for the foreseeable future this will require knowledge of a different type: knowledge of particular populations and of their history, culture, and the institutions that serve them; of the ways in which their health is shaped; and of how health can be improved by interventions that require pervasive changes at the population, provider, individual, and institutional level.

References

Adair, J. and K. W. Deuschle (1970). *The People's Health: Anthropology and Medicine in a Navajo Community*. New York: Appleton-Century-Crofts.

Adams, A. (2000). 'The Road Not Taken: How Tribes Choose between Tribal and Indian Health Service Management of Health Care Resources', *American Indian Culture and Research Journal* 24/3: 21–38.

Belon, A., M. Cattey, G. Harmon, and B. Rose (1983). 'Perception of Living Skills among Navajo High School Students in Arizona', *Journal of American Indian Education* 22/3: unpaginated.

Biolsi, T. (2007). *Deadliest Enemies: Law and Race Relations on and off Rosebud Reservation*. Minneapolis: University of Minnesota Press.

Bureau of Indian Affairs (2003). 'BIA Enrollment and Total Resident Service Area Population by Tribe: Data from BIA 2001 Labor Force Report, and Accepted Corrections', September 2003. Available online at <http://brc.arch.uiuc.edu/ihbg/negreg/July/response34_bia.pdf>, accessed 24 March 2011.

Bureau of the Census (1973). *American Indians*, 1970 Census of Population, Subject Reports, Final Report PC(2)-1F. Washington, DC: US Government Printing Office.

Bureau of the Census (2004). 'Census 2000: Data for Reservations and other American Indian and Alaska Native Areas', revised 3 March 2004. Available online at <http://factfinder.census.gov/home/aian/sf1_sf3.html>.

Cheadle, A., D. Pearson, E. Wagner, B. M. Psaty, P. Diehr, and T. Koepsell (1994). 'Relationship between Socioeconomic Status, Health Status, and Lifestyle Practices of American Indians: Evidence from a Plains Reservation Population', *Public Health Reports* 109/3: 405–413.

Coleman, J. S., E. Katz, and H. Menzel (1966). *Medical Innovation: A Diffusion Study*. New York: Bobbs-Merrill.

Croy, C. D., M. Bezdak, C. M. Mitchell, and P. Spicer (2009). 'Young Adult Migration from a Northern Plains Indian Reservation: Who Stays and Who Leaves'. *Population Research and Policy Review* 28/5: 641–660.

De Jong, G. F. (1972). 'Patterns of Human Fertility and Mortality', in G. A. Harrison and A. J. Boyce (eds), *The Structure of Human Populations*. Oxford: Oxford University Press, 32–56.

Denny, C. H. and D. Holtzman (1999). *Health Behaviors of American Indians and Alaska Natives: Findings from the Behavioral Risk Factor Surveillance System, 1993–1996*. Atlanta, GA: Centers for Disease Control and Prevention.

Devitt, J. and A. McMasters (1998a). *Living on Medicine: A Cultural Study of End-Stage Renal Disease among Aboriginal People*. Alice Springs, NT: IAD Press.

Devitt, J. and A. McMasters (1998b). *On the Machine: Aboriginal Stories about Kidney Troubles*. Alice Springs, NT: Jukurrpa Books.

Dobie, K. (2011). 'Tiny Little Laws: A Plague of Sexual Violence in Indian Country', *Harper's Magazine* 322 (February): 55–64.

Driver, H. E. and W. C. Massey (1957). *Comparative Studies of North American Indians*. Transactions of the American Philosophical Society 47, Part 2. Philadelphia, PA: The American Philosophical Society.

Epple, C., A. L. Wright, V. N. Joish, and M. Bauer (2003). 'The Role of Active Family Nutritional Support in Navajos' Type 2 Diabetes Metabolic Control', *Diabetes Care* 26/10: 2829–2834.

Goodarz, D., E. B. Rimm, S. Oza, S. C. Kulkarni, C. J. L. Murray, and M. Ezzati (2010). 'The Promise of Prevention: The Effects of Four Preventable Risk Factors on National Life Expectancy and Life Expectancy Disparities by Race and County in the United States', *PLoS Medicine* 7/3 (March): e1000248.

Hahn, R. A., J. M. Mendlein, and S. D. Helgerson (1993). 'Differential Classification of American Indian Race on Birth and Death Certificates, U.S. Reservation States, 1983–1985', *The IHS Primary Care Provider* 18/1: 8–11.

Haidt, J. (2012). *The Righteous Mind*. New York: Pantheon.

Harrison, G. A. and A. J. Boyce (1972). 'Introduction: The Framework of Population Studies', in G. A. Harrison and A. J. Boyce (eds), *The Structure of Human Populations*. Oxford: Oxford University Press, 1–16.

Harwell, T. S., D. Hanson, K. R. Moore, D. Jeanotte, D. Gohdes, and S. D. Helgerson (2002). 'Accuracy of Race Coding on American Indian Death Certificates, Montana 1996–1998', *Public Health Reports* 117/1: 44–49.

Heil, D. (n.d.). 'Mobile Bodies Keep us Healthy: Australian Aboriginal Bodies as Vehicles to Practice', unpublished ms.

Henderson, P. N., C. Jacobsen, and J. Beals (2005). 'Correlates of Cigarette Smoking among Selected Southwest and Northern Plains Tribal Groups: The AI-SUPERPFP Study', *American Journal of Public Health* 95/5: 867–872.

Hisnanick, J. J. and D. A. Coddington (1995). 'Measuring Human Betterment through Avoidable Mortality: A Case for Universal Health Care in the USA', *Health Policy* 34/1: 9–19.

Hochman, M. E., J. P. Watt, R. Reid, and K. L. O'Brien (2007). 'The Prevalence and Incidence of End-Stage Renal Disease in Native American Adults on the Navajo Reservation', *Kidney International* 71: 931–937.

Holm, J. E., N. Vogeltanz-Holm, D. Poltavski, and L. McDonald (2010). 'Assessing Health Status, Behavioral Risks, and Health Disparities in American Indians Living on the Northern Plains of the U.S.', *Public Health Reports* 125/1: 68–78.

Indian Health Service (2003). *Regional Differences in Indian Health 2000–2001*. Rockville, MD: US Department of Health and Human Services.

Indian Health Service (2004). *Trends in Indian Health 2000–2001*, US Department of Health and Human Services, Indian Health Service, Office of Public Health, Office of Program Support, Division of Program Statistics. Washington, DC: US Government Printing Office.

Indian Health Service (2008). *Regional Differences in Indian Health 2002–2003 Edition*, US Department of Health and Human Services, Indian Health Service, Office of Public Health Support, Division of Program Statistics, released March 2008.

Indian Health Service (2009). *Trends in Indian Health 2002–2003*, US Department of Health and Human Services, Indian Health Service, Office of Public Health, Office of Program Support, Division of Program Statistics. Washington, DC: US Government Printing Office.

Joe, J. R. and R. S. Young (eds) (1994). *Diabetes as a Disease of Civilization: The Impact of Culture Change on Indigenous Peoples*. Berlin and New York: Mouton de Gruyter.

Khoury, M. J., M. S. Bowen, W. Burke, R. J. Coates, N. F. Dowling, and J. P. Evans (2011). 'Current Priorities for Public Health Practice in Addressing the Role of Human Genomics in Improving Population Health', *American Journal of Preventive Medicine* 40/4: 486–93.

Kozak, D. L. (1997). 'Surrendering to Diabetes: An Embodied Response to Perceptions of Diabetes and Death in the Gila River Indian Community', *Omega* 35/4: 347–359.

Kunitz, S. J. (1983). *Disease Change and the Role of Medicine: The Navajo Experience*. Berkeley: University of California Press.

Kunitz, S. J. (1987). 'Explanations and Ideologies of Mortality Declines', *Population and Development Review* 13/3: 379–408.

Kunitz, S. J. (1991). 'The Personal Physician and the Decline of Mortality', in R. Schofield, D. Reher, and A. Bideau (eds), *The Decline of Mortality in Europe*. Oxford: Oxford University Press, 248–262.

Kunitz, S. J. (1996). 'The History and Politics of Health Care Policy for American Indians', *American Journal of Public Health* 86/10: 1464–1473.

Kunitz, S. J. (2008a). 'Polydrug Use in a Native American Population', *Substance Use and Misuse* 43/3–4: 331–339.

Kunitz, S. J. (2008b). 'Changing Patterns of American Indian Mortality', *American Journal of Public Health* 98/3: 404–411.

Kunitz, S. J. (2011). 'Health Care Policy for American Indians Since the Early 20th Century', in C. A. Bayly, V. Rao, S. Szreter, and M. Woolcock (eds), *History, Historians, and Development*. Manchester: University of Manchester Press, 146–168.

Lang, G. C. (2006a). 'Talking About a New Illness with the Dakota: Reflections on Diabetes, Foods, and Culture', in M. L. Ferreira and G. C. Lang (eds), *Indigenous Peoples and Diabetes: Community Empowerment and Wellness*. Durham, NC: Carolina Academic Press, 203–230.

Lang, G. C. (2006b). '"In their Tellings": Dakota Narratives about History and the Body', in M. L. Ferreira and G. C. Lang (eds), *Indigenous Peoples and Diabetes: Community Empowerment and Wellness*. Durham, NC: Carolina Academic Press, 53–71.

Leichenko, R. M. (2003). 'Does Place Still Matter? Accounting for Income Variation across American Indian Tribal Areas', *Economic Geography* 79/4: 365–386.

Linton, R. (1924). *Use of Tobacco among North American Indians*. Anthropology Leaflet 16. Chicago: Field Museum of Natural History.

MacGregor, G. (1946). *Warriors without Weapons: A Study of the Society and Personality Development of the Pine Ridge Sioux*. Chicago: University of Chicago Press.

McKeown, T. (1976). *The Modern Rise of Population*. London: Edward Arnold.

Megill, D. M., W. E. Hoy, and S. D. Woodruff (1988). 'Rates and Causes of End-Stage Renal Disease in Navajo Indians, 1971–1985', *Western Journal of Medicine* 149: 178–182.

Melzer, D. and R. Zemmern (2002). 'Genetics and Medicalisation', *British Medical Journal* 324/7342: 863–864.

Menzel, H. (1960). 'Innovation, Integration, and Marginality: A Survey of Physicians', *American Sociological Review* 25/5: 704–713.
Navajo Epidemiology Center (n.d.). *Cancer among the Navajo 1995–2004*. Window Rock, AZ: Navajo Division of Health, The Navajo Nation.
Nolte, E. and M. McKee (2004). *Does Healthcare Save Lives: Avoidable Mortality Revisited*. London: The Nuffield Trust.
North, K. E., B. V. Howard, T. K. Welty, L. G. Best, E. T. Lee, J. L. Yeh, R. R. Fabsitz, M. J. Roman, and J. W. MacCluer (2003). 'Genetic and Environmental Contributions to Cardiovascular Disease Risk in American Indians', *American Journal of Epidemiology* 157/4: 303–314.
Omran, A. (1971). 'The Epidemiologic Transition: A Theory of the Epidemiology of Population Change', *The Milbank Quarterly* 49/4: 509–538.
Pandhi, N., B. A. Guadagnolo, S. Kanekar, D. G. Petereit, and M. A. Smith (2010). 'Cancer Screening in Native Americans from the Northern Plains', *American Journal of Preventive Medicine* 38/4: 389–395.
Peterson, N. (1993). 'Demand Sharing: Reciprocity and the Pressure for Generosity among Foragers', *American Anthropologist* 95/4: 860–874.
Pickering, K. A. (2000). *Lakota Culture, World Economy*. Lincoln: University of Nebraska Press.
Prosnitz, L. and G. Mandell (1967). 'Diabetes mellitus among Navajo and Hopi Indians: The Lack of Vascular Complications', *American Journal of Medical Sciences* 253: 700–705.
Rogers, E. M. (1995). *Diffusion of Innovations*, 4th edn. New York: Free Press.
Rose, G. (1992). *The Strategy of Preventive Medicine*. Oxford: Oxford University Press.
Roy, B. (2006). 'Diabetes and Identity: Changes in the Food Habits of the Innu—A Critical Look at Health Professionals' Interventions Regarding Food', in M. L. Ferreira and G. C. Lang (eds), *Indigenous Peoples and Diabetes: Community Empowerment and Wellness*. Durham, NC: Carolina Academic Press, 167–186.
Sandefur, G. D. (1986). 'American Indian Migration and Economic Opportunities', *International Migration Review* 20/1: 55–68.
Smith-Morris, C. M. (2004). 'Reducing Diabetes in Indian Country: Lessons Learned from Gila River', *Human Organization* 63/1: 34–46.
Snipp, C. M. and G. D. Sandefur (1988). 'Earnings of American Indians and Alaskan Natives: The Effects of Residence and Migration', *Social Forces* 66/4: 994–1008.
Spicer, E. H. (1962). *Cycles of Conquest: The Impact of Spain, Mexico, and the United States on the Indians of the Southwest, 1533–1960*. Tucson: University of Arizona Press.
Strickland, S. (1972). *Politics, Science, and Dread Disease: A Short History of United States Medical Research Policy*. Cambridge, MA: Harvard University Press.
Thomas, L. (1974). *The Lives of a Cell: Notes of a Biology Watcher*. New York: Viking Press.
Thornton, R. (1997). 'Tribal Membership Requirements and the Demography of "Old" and "New" Native Americans', *Population Research and Policy Review* 16/1–2: 33–42.
Tiller, V. E. V. (1996). *Tiller's Guide to Indian Country: Economic Profiles of American Indian Reservations*. Albuquerque, NM: BowArrow Publishing Co.
Topol, E. (2012). *The Creative Destruction of Medicine: How the Digital Revolution Will Create Better Health Care*. New York: Basic Books.
Treuer, D. (2012). *Rez Life*. New York: Atlantic Monthly Press.
US Public Health Service (1957). *Health Services for American Indians*, PHS Publication 531. Washington, DC: US Government Printing Office.
Wagoner, P. L. (2002). *'They Treated Us Just Like Indians': The Worlds of Bennett County, South Dakota*. Lincoln: University of Nebraska Press.

Watson, T. (2006). 'Public Health Investments and the Infant Mortality Gap: Evidence from Federal Sanitation Interventions on US Indian Reservations', *Journal of Public Economics* 90/8–9: 1537–1560.

Weisz, G. and J. Olszynko-Gryn (2010). 'The Theory of the Epidemiologic Transition: The Origins of a Citation Classic', *Journal of the History of Medicine and Allied Sciences* 65/3: 287–326.

Welty, T., D. Rhoades, F. Yeh, E. Lee, L. Cowan, R. Fabsitz, D. Robbins, R. Devereux, J. Henderson, and B. Howard (2002). 'Changes in Cardiovascular Disease Risk Factors among American Indians: The Strong Heart Study', *Annals of Epidemiology* 12/2: 97–106.

Will, J. C., K. F. Strauss, J. M. Mendlein, C. Ballew, L. L. White, and D. G. Peter (1997). 'Diabetes mellitus among Navajo Indians: Findings from the Navajo Health and Nutrition Survey', *The Journal of Nutrition* 127: S2106–S2113.

Woodall, W. G., S. J. Kunitz, H. Zhao, D. R. Wheeler, V. Westerberg, and J. W. Davis (2004). 'The Prevention Paradox and Traffic Safety Outcomes from a DWI Treatment Program', *American Journal of Preventive Medicine* 27/2: 106–111.

World Health Organization (2008). *The Top Ten Leading Causes of Death by Broad Income Group, 2004*, WHO Media Centre, Fact Sheet No. 310, November.

CHAPTER 20

HORMONES AND DISEASE

contested knowledge of exogenous hormones in the evaluation of oral contraceptives and hormone replacement therapy

KLIM MCPHERSON

20.1 Introduction

Some 60 different endogenous hormones have been identified which naturally circulate in the body, moderating its physiological processes. Only small amounts of hormones are needed to affect the behaviour of cells, i.e. those exhibiting specific receptors, dramatically. A variety of given or prescribed exogenous modern chemical compounds have hormone-like effects on both humans and wildlife, in principle of a rather complicated kind. Their interference with the synthesis, secretion, transport, binding, action, or elimination of natural hormones in the body can change homeostasis, reproduction, development, and/or behaviour in the same way as endogenous hormones can.

It is clear that endogenous hormone levels affect the risk of several cancers, especially breast cancer, and possibly of cardiovascular disease. Of course synthetic or natural hormones are now used extensively to treat and to prevent disease, but also to prevent conception and to treat severe menopausal symptoms. These are used by very large numbers of women. The likely effects of exogenous hormones on disease are very difficult to predict reliably from theory and hence epidemiological study is the only method of understanding the short- and especially long-term effect of these preparations. Because short-term effects are often dramatic and advantageous (on menopausal symptoms or contraception, for example), they often have to be marketed before long-term surveillance can reliably identify the possibility of longer-term harm. This is made more complex as the dose and constituents of these products change with time. Product formulation develops either to mitigate known harm as evidence for it emerges, or to introduce a new formulation once the patent expires on an old one. Clearly the ongoing attempts to reduce the doses, commensurate with the desired effect, poses further problems of surveillance. Short-term effects usually have to be demonstrated by placebo-controlled randomized trials, while longer-term effects require observational epidemiology.

Drug development for potent hormone-like preparations happens perforce in ignorance of their long-term effects. Studying these effects is fraught, however, precisely because the beneficial short-term effects of these preparations have led to burgeoning vested interests. Epidemiology is the science of studying these effects in randomized trials and observational studies, and I will discuss four examples where the epidemiology is (or was) at risk of being inadequate, simply because proper studies could not in practice be done. Moreover, obfuscation is a dominant theme in interpreting the evidence simply because the vested interest often becomes so strong. This usually takes the form of contesting a scientific consensus on grounds of perceived bias or confounding—not usually well substantiated. It also takes the form of analysing datasets assiduously, sometimes inappropriately, to achieve desired risk estimates.

This chapter is concerned essentially with populations of women because the kind of preparations it is concerned with can, and often do, affect whole populations of women, at least in the business of deciding whether or not to take them. In making up their minds, women need to know reliably the balance of benefit to harm of so doing, and it is precisely these questions that are addressed here. The problems of integrating proper surveillance of exogenous hormones within the constraints that are intrinsic to placebo-controlled randomized trials needs to be more widely understood.

Epidemiology is not an exact science and understanding potent biases is a serious problem, hence the opportunities for designing, and analysing, a study inappropriately or interpreting it in a particular way are often too great for scientific comfort. Firstly, studies have to be representative of the population for which the results will be relevant, and this will always require the judicious choice of subjects to study. In principle epidemiological studies require careful observation of a population of ill people whose past exposure is contrasted with an observed population of healthy people (i.e. a control group). These are called case-control studies. The choice of controls is critical, whether matched individually with cases or not, and they must represent the same intrinsic possibility of exposure and of disease as did the cases before they became ill. This is difficult to arrange with certainty and much discussion will normally accompany any particular choice as to potential bias associated with that choice.

The second main method of observational epidemiology is studying cohorts of people who either are or are not exposed to the potential hazard under study. This is usually a more robust method but requires many more subjects—possibly 100 times as many for the same statistical power, and much longer follow-up. The British Doctors' Study on the hazards of tobacco involved following many thousands of doctors through their lives with regular assessment of current tobacco consumption and reliable monitoring of illness and death. Comparison gave rise to the knowledge we now have on the manifold harmful consequences of smoking. Many potential problems of interpretation can be overcome by randomizing people to a particular exposure, or not, and following them for a period to observe differences in disease rates. In this way confounding can be eliminated, even of unknown confounding variables, and selection bias is minimized so long as the people admitted into the trial can be thought to be representative of those not in the trial but for whom the results would be relevant. The main purpose of good epidemiology is to achieve sufficient statistical power to detect reliably those effects that may be small for a given individual, but could have dramatic effects on populations.

Confounding happens in observational studies, which are usually very large indeed, when people's propensity to choose the product under investigation turns out to be associated independently with the risks of plausible harms from the product. Thus people choosing hormone replacement therapy (HRT) could on average be more health-aware than those not choosing it, and hence have better outcomes. Such confounding is often very subtle, and is not ameliorated by large numbers in the studies. Hence it is often difficult to detect and to allow for. Confounding can be discounted when subjects are randomized on the simple criterion that they consume a product or not, but justifying such an experiment with large enough numbers is too often ethically and practically impossible. It is not difficult to see why. People take exogenous hormones for a reason, not randomly. This is often not compatible with random allocation. Likewise random allocation to smoke cigarettes would never be feasible—or nowadays ethical. But epidemiology is nothing if the studies it masterminds are not sufficiently representative of the populations to which the inferences it extracts are supposed to apply.

Generally in epidemiology the strength of association between an exposure and disease is measured by a relative risk (RR), the number of times higher (if RR>1) the risk (per 100,000 per annum) is among those exposed compared with those not exposed. Thus heavy lifelong cigarette smokers attract around a twentyfold (RR = 20) risk of lung cancer compared with never smokers. The safest assumption is that this RR will in general transfer from one population to another. Hence this risk applies (without evidence to the contrary) in populations far from the population where the risk was established. This is because the essential biological process whereby this risk causes this increase is thought ultimately to be the same. Hormones have a special place in this discourse, however, because endogenous hormone levels vary between people, gender especially, according to the time of the month and by age, diet, and other environmental influences—systematically. Hence exogenous hormones, since they interact with endogenous exposure, may well also have different effects according to salient characteristics of particular populations. Here we discuss only females, but much of the argument centres on the extent to which inferences made from older women may also apply to younger women. Do relative risks found among one population transfer straightforwardly to another? This is cloaked with biological uncertainty in this particular case and as such is a fertile ground for disagreement. The question is actually only resolvable by high-powered epidemiological studies, whose funding will essentially depend on the perceived importance and plausibility of the competing hypotheses.

We will discuss some examples. In many, a dominant theme is that accepting an interpretation of potential harm has commercial consequences that are profoundly important for manufacturers—and to be avoided if at all possible. Investment in drug development from the laboratory to the final testing often costs billions. The story here is not like tobacco and the wilful corruption of evidence by the tobacco industry, but there are some similarities. For example, family planners and professionals who treat menopausal women—and the women themselves—are often similarly disinclined to believe the possibility of harmful effects. Better understanding of the nature of this tendency to play down a problem will hopefully lead to better drug surveillance, and thus to a better understanding of long-term effects. Hormone-like substances have enormous potential, for instance, to improve disease treatment and prevention. A widely discussed example at present is the potential for using hormone antagonists (such as Tamoxifen for example; Descensi et al. 2009) in specific subpopulations to prevent disease. Progress on this, however, is compromised by our lack of knowledge. It may be that the effects of hormones are simply too nuanced to be picked up by epidemiological studies reliably, in which case, a much better and more lively collaboration between the laboratory, biology, physiology, social sciences, and epidemiology will be required to make significant progress.

20.2 Early Oral Contraceptive Use and Breast Cancer Risk

It has been known for a very long time that the length of exposure to endogenous hormones before first pregnancy increases the subsequent risk of breast cancer, since higher age at first pregnancy is strongly related to a greater risk in a woman's life. When oral contraceptives (OCs) were developed in the 1960s they represented a revolution in a woman's control of her own fertility, banishing furtive sexual encounters exacerbated by high risks of unwanted conception. But OCs were not readily available to unmarried women until the early 1970s. Thus the large cohort studies designed to examine unwanted long-term effects of OCs, which began in the early 1960s, recruited married women who were largely interested in child spacing. People often married partly in order to become pregnant. The numbers recruited to these studies that were nulliparous were very small indeed. So the studies were *de facto* studying the safety of oral contraceptives among parous women, when in fact the main users of oral contraceptives turned out later to be single nulliparous women. It was they who needed to know whether OCs were safe in the long term, and essentially the only reliable information was confined to the assiduous study of parous women. Exposure of nulliparous women to OCs before first pregnancy was not studied with required rigour.

Work on breast tissue samples taken from surgical specimens from women having breast enlargement, reduction, or other cosmetic operations demonstrated a quite dramatic difference in the cellular response to OCs between nulliparous and parous women (Battersby et al. 1992). The indication was in principle thus of a different effect on early-stage carcinogenesis among parous women and nulliparous women taking the pill. Cell replication rates were faster, when exposed to OCs, if women were nulliparous and if the oestrogen dose was high (as it was in early oral contraception) (Anderson et al. 1989). There was clear evidence for an oestrogen effect on cell replication rates which could translate, under plausible models of carcinogenesis, into an early-stage carcinogenic effect; this effect might be associated years later with clinical diagnosis of breast cancer (Jernström 1996). These observations did not apply at all to parous women.

These findings have given rise to the fear that OC use before first pregnancy might be more carcinogenic than OC use afterwards. Moreover, were this effect to be real it is likely that any effect on breast cancer incidence would not be observed for twenty years, since studies of the early stage of breast cancer from ionizing radiation show that changes in incidence are not manifest until at least that many years later. This impact appears particularly potent among women exposed soon after menarche. In short, to study this effect properly would require large numbers of women exposed before their first birth for several years. This kind of exposure did not happen until the mid-1970s, well after the large cohort studies had finished recruitment. Such women have still not been properly studied.

The currently inadequate epidemiological evidence is not reassuring (McPherson 1999). Simulation work has demonstrated that, were such exposure to increase the breast cancer incidence by threefold, then incidence would have risen by one-fifth by the year 2000 (McPherson 1992, 1986). This trend has indeed been documented subsequently (Cancer Research UK 2012), but there may be other plausible reasons for this finding. Resolving this uncertainty has for some time required a large case control study of breast cancer, taking detailed and validated contraception histories to disentangle the competing explanations. This has not happened, in spite of funding requests. What is also required is the development of a method for sampling breast tissue ethically and painlessly to examine pre-carcinogenic pathological indications among women of different parity and exogenous (and endogenous) exposure. Such work could then be related to exogenous hormone exposure. This also has not happened, and seems very unlikely to happen for fear of 'frightening the horses'. Reliable contraception among single women is so important that to seriously question the long-term safety of OCs—even by scientific study—is thought to be problematic and unjustified. It is quite possible that an OC-induced epidemic of breast cancer in the late twentieth century was avoided by diligent reductions in the oestrogen dose. Things might not work out so well in the future.

20.3 Third-Generation OCs and Their Relative Effect on Venous Thromboembolism (VTE)

Widely used second-generation OCs containing levanorgestrel as the progestogen were very effective, and safe except among smokers, but were no longer in patent after years of successful marketing. In 1982 various manufacturers introduced two new progestogens ostensibly to reduce the cardiotoxicity of second-generation OCs—and of course these, when licensed, were patented anew, and known as third generation OCs. So successful were these new formulations that by 1995 they had captured 50% of the (considerable) OC market.

In 1998 three studies, which were undertaken because of anecdotal reports of higher rates of venous thromboembolism (VTE), reported an elevated risk of VTE among third-generation users compared with second-generation users. The studies found a relative risk (RR) greater than two; i.e. women taking third-generation OCs suffered twice as many DVTs than those on second-generation OCs (Rosing et al. 1997). The Committee on the Safety of Medicines issued a warning to OC users to gradually change back to second-generation preparations, which of course greatly annoyed the manufacturers who stood to lose millions from their investment. As it happens, this advice caused many thousands of unwanted pregnancies (Wood et al. 1997).

The manufacturers conducted and sponsored their own studies, which perhaps unsurprisingly found much lower relative risks—indeed the pooled estimate of RR from sponsored studies was significantly lower that the pooled estimates from independent studies. Lawyers for some women with DVTs hired me as an expert against the manufacturers in a High Court action, which was a lesson in true obfuscation and pharmaceutical power. The case was lost and is amply described in a book chapter on the perils of being an expert witness, in which each bias claimed by the defence is discussed in detail (McPherson 2006).

The outcome is that third-generation OCs are still marketed even though causing higher rates of DVT and costing much more than second-generation pills. A careful study of the epidemiology of these women as a whole would very likely indicate this to be unwise for women's health.

20.4 HRT Use and Cardiovascular Disease

The composition of HRT preparations, which originated in the 1940s, has developed rapidly while being extensively marketed as a 'panacea' for menopausal women. However, oestrogen alone, which was widely prescribed from the 1960s, was shown to increase the risk of endometrial cancer (lining of the womb) by the late 1970s. The solution was to oppose the oestrogen hormonally with progesterone, which attenuated that effect while maintaining its effect on menopausal symptoms. Oestrogen and progesterone together is known as combined therapy. This was given to women with an intact uterus, while oestrogen alone continued to be prescribed to women who had had a hysterectomy. This new preparation was widely used from around 1980.

To cut a long and fascinating story short, evidence from observational (mostly cohort) epidemiological studies seemed to demonstrate a dramatic reduction by between 30% and 50% in coronary heart disease (CHD) among women taking combined HRT, compared with women who did not. Half the rate of CHD was observed in several studies among HRT takers. These studies were often of very high quality and the evidence was convincing (Barrett-Connor 1996)—so convincing that the manufacturers sought the prevention of CHD as a licensed indication for HRT from the drug regulators (the Food and Drug Administration in the US, and Medicines and Healthcare Products Regulatory Agency (MHRA) in the UK).

Part of this story involved an early, large, double-blind randomized trial (RCT), published in 1980, which was testing a cholesterol-lowering drug and oestrogen preparations for men with a history of heart disease (Coronary Drug Project 1980). By that time the role of oestrogen in preventing CHD was suggested by gender differences in incidence and mortality, and the increasing incidence among women after the menopause. This trial was enormous and, while the oestrogen given to men was a disaster (they became ill and feminine) and did not prevent CHD death, nor did the active lipid-lowering drug Clofibrate reduce cardiac risk. The investigators, troubled by this finding, decided to analyse the data according to the participants' recorded compliance with the drug, and found a big reduction in mortality among those who took more than 80% of the prescribed dose compared with those who did not. This they interpreted as a drug effect among those who took the prescribed dose. However, epidemiologists pointed out that this was an observational comparison within one arm of the randomized control trial, and could be an artefact of selection. In short, good compliers are also more likely to have more protective behaviours than those who did not comply so well.

As these findings came from a double-blind trial, neither the doctors nor the participants knew which of the several treatments being tested had been allocated to these men. One treatment in such trials is, of course, a placebo (a disguised sugar-coated pill). The effect of compliance was, moreover, almost exactly the same as the compliance effect with Clofibrate. In other words, taking a placebo reliably is highly protective for men at risk of heart attacks, while not taking it apparently is not. Adjustment for 40 baseline risk factor differences, mainly physiological measurements and not behavioural ones, made no difference to the comparison. While these subjects were men, there is no reason to suppose that the same confounding should not apply to women—i.e. that reliably taking a placebo might have a highly significant effect on subsequent mortality and hence must be a matter of the (unmeasured) characteristics of the compliers. This, in turn, has obvious implications on the interpretation of observational studies of HRT users versus non-users, since the size of the effect of the placebo between compliers and non-compliers was very similar to the putative effect of HRT on CHD. As this effect was obviously an artefact of confounding or self-selection—not pharmacology—the plausibility of a similar confounding in HRT observational studies of women would appear to be indicated.

For this reason, the licensing agencies in the UK and the US insisted that if the manufacturers of HRT wanted to license their drug for cardio-protection, they would need to do a double-blind randomized trial to demonstrate efficacy. It is essentially a ground rule of regulators that efficacy can only be properly demonstrated by a double-blind placebo-controlled randomized trial. Criteria, however, for judging long-term safety are perforce different from demonstrating efficacy for sensible regulation because drugs are widely marketed before it is possible to do observational studies. In this case the argument was uniquely persuasive. Clearly, were HRT to be licensed to prevent CHD, the market possibilities would dramatically increase. Thus, big trials were planned in the US and the UK to reliably test the hypothesis. Only the US trial, called the Women's Health Initiative (WHI), survived the scrutiny of the funders, with 8,000 women randomized between the combined HRT and the placebo, and 5,000 randomized between the placebo and oestrogen medication alone among women with no uterus. The trial was started around 1993, meaning that women might go on taking treatments for many years despite the availability of seriously conflicting evidence.

In the interim, Elina Hemminki and I noted that a different approach could be taken to resolving the controversy surrounding these treatments. Given the many small randomized comparisons between HRT and placebo treatments that companies and researchers had submitted for proof of efficacy to the regulators, could we not amalgamate these trials to test for any effect of HRT on CHD among younger women? We did this by abstracting the records from around 200 small trials for different products, obtained from existing publications. We found an increased risk of CHD among the HRT takers (Hemminki and McPherson 1997). This effect was not significantly different from no effect on risk, but was significantly different from a halving of risk. When we published this work in the *British Medical Journal*, we were completely pilloried (Naylor 1997; Sackett 2002). People just did not believe our results, partly because some trials were brought to our attention that had not been published. Would these unpublished trials make a difference?

We sought these data from the regulators. The manufacturers strongly objected, and prohibited the release of these data. We had to seek (Finnish) Supreme Court approval to override the manufacturers' objection—on grounds of legitimate public health concern. This was granted and five new trials were added to our database. Addition of these trials reinforced the previous conclusions—now with some strong evidence of publication bias (Hemminki and McPherson 2000). Two of the trials had not been published at all, each showing a large excess of cardiovascular events in the HRT group—that is, exhibiting clear evidence of an excess CHD risk in the HRT group (McPherson and Hemminki 2004). This work, including the unpublished studies, was then published in *The Lancet* in 2000. We estimated the increase in CHD events among HRT takers to be 78%. No action was taken pending the results of the WHI trial, owing presumably to the unconventional procedure followed in our analysis of synthesizing several trials.

When the results of the WHI randomized trial were published in 2002 (for combined therapy; Writing Group 2002) and in 2004 (for oestrogen alone; Writing Group 2004), the results almost exactly confirmed our results, and steps were then taken to reduce HRT use dramatically. The medical advice, however, became very different: as short a period on therapy as possible, with the lowest effective dose recommended by the regulators. Our study, however, carried a further lesson: that the synthesis of efficacy trials of widely used drugs could provide reliable information on their safety. This strategy is one that has yet to be adopted routinely.

What has happened instead is that a systematic effort has been launched to persuade the world that the Women's Health Initiative study is misleading (*Climacteric* 2012). This argument has arisen because the women randomly selected for the trial were older than most menopausal women. This selection procedure was adopted because, in order for the trials to be free of bias, they had to be both double-blind and include only non-symptomatic women—the latter criteria being crucial, since symptomatic women would know immediately which treatment they had been prescribed. Critics argued that this procedure introduced a bias because the physiological response to exogenous hormones among 50-year-old women is radically different from the 63-year-old women included in the WHI trial. A concerted effort is also being made once again to demonstrate cardiac safety among women taking HRT where evidence has previously shown that it clearly does not exist. All such attempts ignore perforce our work synthesizing efficacy trials among symptomatic menopausal women (McPherson 2012a).

20.5 Hormone Replacement Therapy and Breast Cancer

Whether or not HRT use causes an increase in breast cancer due to late-stage carcinogenic effects has always been an important question. The issue arises especially because women had been taking HRT for long periods of time to prevent menopausal bone loss (a massively effective remedy), but also to prevent coronary heart disease. Normally oestrogen is taken to be a co-carcinogen with particular effects on the breast. In 1987 we published the first epidemiological study showing a threefold increase in the risk of breast cancer among women taking HRT for more than four years (Hunt et al. 1990). If true, such a risk would be decisive for most women at that age, since the annual risk of breast cancer is 2/1000. Trebling that rate is clearly untenable. Studies continued to be published through the 1990s, most of which suggested a doubling of risk. Synthesis of these studies suggested a doubling of risk at four years' use, with quite tight confidence limits (McPherson and Mant 2005)—that is, until two studies brought the pooled risk levels down quite a bit. This revision delayed general acceptance of the risk until WHI reported a doubling of risk for five years' use in 2002. Whatever the reason for those low estimates, they successfully delayed the sensible caution associated with HRT, by then ubiquitously acknowledged, by ten or so years. Many thousands of breast cancers could have been avoided by prompt and cautious action. The Million Women Study of British women attending mammography confirms the strong association of long-term HRT use (Million Women Study Collaborators 2003).

Population rates of breast cancer have decreased importantly since the reduction in long-term use of HRT, beginning in 2002 (Ravdin et al. 2007). This could have happened in the mid-1990s if the evidence had been more carefully interpreted. What is interesting is that, since that time, HRT use has decreased rapidly, and apparently permanently, precisely because of safety considerations. Nonetheless, in the last year, two peer-reviewed journals have chosen to publish a series of lengthy articles, each taking a complete issue, and mounting a complex methodological critique of the epidemiology reviewed earlier (*Journal of Family Planning and Reproductive Health Care* 2011; *Climacteric* 2012). These studies argue that HRT is neither associated with an increased risk of CHD nor is it causal in the development of breast cancer. Vested and financial interests in favour of wide use of HRT remain, of course, profound and powerful. Their potential influence on research is such that randomized control trials in themselves are plainly unable to control it.

20.6 Conclusions

The story of attempts to develop hormonal treatments, and to make them safe to use, is part of a much larger picture of health interventions in which scientific, public, commercial, and governmental interests are all entangled. In 2011, the United Nations held a summit in New York on the burgeoning world problem of non-communicable diseases (UN General Assembly 2011). The idea was to persuade governments to take public health seriously and to implement well-tried legal and fiscal proposals to reduce the burden of chronic disease. Many of these policies require the sale and marketing of harmful products to be altered by government intervention—tobacco being an obvious example. But there are many more possibilities—especially in the food industry (Butland et al. 2007). The likelihood of commercial interests seeking to dominate the agenda for their own benefit is inevitable (McPherson 2012b).

The preceding account has outlined a state of affairs in which the reduction of female morbidity via a single set of compounds has been contested, with one consequence being the exacerbation, rather than reduction, of morbidity. This situation derives from a combination of factors:

- not knowing as much as we could about the role of hormones in early-stage carcinogenesis;
- marketing oral contraceptives that cause higher levels of serious morbidity despite strong counter-evidence;
- seemingly permanent obfuscation about the role of hormones in cardiovascular disease risk; and
- repeated attempts to deny the demonstrated role of HRT in breast cancer risk.

One clear implication is that scientific public health needs to integrate with public policy in a much more meaningful way than is happening at present. The proper and dispassionate surveillance of pharmaceutical products as currently conducted leaves too much to commercial and other interests. The problem is that public health is too often seen as a medical issue and hence beyond the responsibility of other disciplines.

The examples cited here require biology, sociology, public policy, ecology, pharmacology, epidemiology, and others to work together for the long-term health of women. Exactly the same is true for the fight against premature deaths from all chronic diseases in both sexes. There needs to be a much more sophisticated understanding among all scientists about the true threats to our health. Enlightened policies from enlightened governments could save millions of lives.

References

Anderson, T. J., S. Battersby, R. J. B. King, K. McPherson, and J. J. Going (1989). 'Oral Contraceptive Use Influences Resting Breast Proliferation', *Human Pathology* 20/12: 1139–1144.

Barrett-Connor E. (1996). 'The Menopause, Hormone Replacement, and Cardiovascular Disease: The Epidemiological Evidence', *Maturitas* 23: 227–234.

Battersby, S., B. J. Robertson, T. J. Anderson, R. J. B. King, and K. McPherson (1992). 'Influence of Menstrual Cycle, Parity and Oral Contraceptive Use on Steroid Hormone Receptors in Normal Breast', *British Journal of Cancer Research* 65: 601–607.

Butland, B., S. Jebb, P. Kopelman, K. McPherson, and S. Thomas (2007). 'Tackling Obesities: Future Choices', project report, available online at <http://www.foresight.gov.uk/Obesity/Obesity_final/Index.html>.

Cancer Research UK (2012). 'Breast Cancer Incidence Statistics: Trends over Time (Females)', available online at <http://www.cancerresearchuk.org/cancer-info/cancerstats/types/breast/incidence/#trends>, updated 7 November 2012.

Climacteric (2012). 15/3 (June): 205–293, whole issue, available online at <http://informahealthcare.com/toc/cmt/15/3>.

Coronary Drug Project (1980). 'Double-Blind RCT; Secondary Prevention of CHD', *New England Journal of Medicine* 303: 1038–1041.

Descensi, A., C. Robertson, A. Guerrieri-Gonzaga, D. Serrano, M. Cazzaniga, S. Mora, M. Gulisano, H. Johansson, V. Galimberti, E. Cassano, S. M. Moroni, F. Formelli, E. A. Lien, G. Pelosi, K. A. Johnson, and B. Bonanni (2009). 'Randomized Double-Blind 2X2 Trial of Low-Dose Tamoxifen and Fenretinide for Breast Cancer Prevention in High-Risk Premenopausal Women', *Journal of Clinical Oncology* 27: 3749–3756.

Hemminki, E. and K. McPherson (1997). 'Impact of Postmenopausal Hormone Therapy on Cardiovascular Events and Cancer: Pooled Data from Clinical Trials', *British Medical Journal* 315: 149–153.

Hemminki, E. and K. McPherson (2000). 'The Value of Drug-Licensing Documents in Studying the Effect of Postmenopausal Hormone Therapy on Cardiovascular Disease', *The Lancet* 355: 566–569.

Hunt, K., M. Vessey, and K. McPherson (1990). 'Mortality in a Cohort of Long-Term Users of Hormone Replacement Therapy: An Updated Analysis', *British Journal of Obstetrics and Gynaecology* 97: 1080–1086.

Jernström, H. (1996). 'Effects of Oral Contraceptives on Endogenous Hormones, Body Constitution and Breast Epithelium in Healthy Young Women', dissertation, Lund University, Lund.

Journal of Family Planning and Reproductive Health Care (2011). 37: 103–230 [pages contain several relevant articles].

McPherson, K. (1992). 'Modelling Latent Effects in any Association between Oral Contraceptives and Breast Cancer', in W. Morgenstern, M. Rusnak, and G. Schettler (eds), *Models of Noncommunicable Diseases: Health Status and Health Service Requirements*, Heidelburg Academy of Sciences Scientific Series. Heidelberg: Springer, 71–82.

McPherson, K. (1999). 'Type 3 Errors, Pill Scares, and the Epidemiology of Oral Contraception and Health', *Journal of Epidemiology and Community Health* 53/5: 258–260.

McPherson, K. (2006). 'One Expert's Experience', in Louis Blom-Cooper (ed.), *Experts in Civil Courts: Expert Witness Institute*. Oxford: Oxford University Press, 159–180.

McPherson, K. (2012a). 'I've Studied HRT for Years—and I'm Still Convinced it Raises Women's Risk of Cancer', *Daily Mail*, 29 May, available online at <http://www.dailymail.co.uk/health/article-2151238/Ive-studied-HRT-years-Im-convinced-raises-womens-risk-cancer.html>.

McPherson K. (2012b). 'Getting Fatter and Fatter' [book review], *The Lancet* 379: 2232.

McPherson, K. and J. O. Drife (1986). 'The Pill and Breast Cancer: Why the Uncertainty?', *British Medical Journal* 293: 709–710.

McPherson, K. and E. Hemminki (2004). 'Synthesising Licensing Data to Assess Drug Safety', *British Medical Journal* 328: 518–520.

McPherson, K. and B. Mant (2005). 'Synthesies of Studies of 4+ Years of HRT and Breast Cancer Risk', *Journal of Epidemiology and Community Health* 59/12: 1078–1079.

Million Women Study Collaborators (2003). 'Breast Cancer and HRT in the Million Women Study', *The Lancet* 362: 419–427.

Naylor, C. D. (1997). 'Meta-Analysis and the Meta-Epidemiology of Clinical Research', *British Medical Journal* 315: 617–619.

Ravdin, P. M., K. A. Cronin, and N. Howlander (2007). 'The Decrease in Breast Cancer Incidence in 2003 in the United States', *New England Journal of Medicine* 356: 1670–1674.

Rosing, J., G. Tans, and G. A. F. Nicolae (1997). 'Oral Contraceptives and Venous Thrombosis: Differing Sensitivities to Activated Protein C in Women Using Second- and Third-Generation Oral Contraceptives', *British Journal of Haematology* 97: 233–238.

Sackett, D. (2002). 'Hormone Replacement Therapy: The Arrogance of Preventive Medicine', *Canadian Medical Association Journal* 167/4: 363–364.

UN General Assembly (2011). 'Political Declaration of the High-Level Meeting of the General Assembly on the Prevention and Control of Non-communicable Diseases', A/66/L.1, 16 September.

Wood, R., B. Botting, and K. Dunnell (1997). 'Trends in Conception Before and After the 1995 Pill Scare', *Population Trends* 89: 5–12.

Writing Group (2002). 'Risks and Benefits of Estrogen plus Progestin on Health Menopausal Women: Principle Results from the Women's Health Initiative Randomized Controlled Trial', *Journal of the American Medical Association* 288: 321–333.

Writing Group (2004). 'Risks and Benefits of Estrogen on Healthy Menopausal Women after Hysterectomy from the WHI Trial', *Journal of the American Medical Association* 291: 1701–1712.

Glossary

Admixture: a term employed in population genetics to refer to the mixing of distinct populations that have been separated by genetic drift.

Age mixing: sexual partnerships formed between men and women from different age groups.

Age-specific fertility: refers to the number of births to females in a particular age category in a particular year compared to the number of females in that age category, usually expressed as births per woman or births per 1,000 women in the age category.

Allele: a particular genetic variant at a locus. A diploid individual will have one or two alleles at a locus but there may be many more present in the population.

Allometry: the change in organisms in relation to proportional changes in body size, also called biological scaling.

Allopatric speciation: the evolution of geographically isolated populations into distinct species, their divergence being facilitated by an absence of gene flow which tends to keep local populations genetically similar.

Ancestral mapping: identification of genetic relationships (parents, grandparents, cousins, etc.) between individuals. Mapping typically proceeds by constructing an ancestral tree, although it is possible (via inbreeding) that individuals will appear multiple times in an ancestral map (e.g. if two cousins have children, their grandparents would appear in both their lineages).

Ancestral recombination graph: a model showing how individuals in a sample would relate to each other if they: (a) were reproducing in a closed population without selection; and (b) have more than a single parent (i.e. they are undergoing recombination). The model is statistical in that it describes only the relationship between individuals actually sampled in the population, and not other individuals present in that population who have not been sampled.

Antiretroviral treatment (ART): therapy using anti-HIV drugs of people infected with human immunodeficiency virus (HIV). The standard treatment consists of a combination of at least three drugs (often called 'highly active antiretroviral therapy' or HAART) that suppress HIV replication, thus reducing the likelihood of the virus developing resistance. ART has the potential both to reduce mortality and morbidity rates among HIV-infected people, and to improve their quality of life.

Approximate Bayesian computation: a statistical inference procedure for inferring model parameters based only on simulating from a model, without the need to compute the likelihood (i.e. the probability of the data, given the parameters).

Binomial theorem: expansion of $(p + q)^n$. In the case of the Hardy–Weinberg law, n = 2, so the expansion is $(p + q)^2 = p^2 + 2pq + q^2$.

Biodemography: the study of commonalities across species in patterns of fertility, mortality, and life course informed by physiological parallels and evolutionary heritage.

Body mass index: an individual's body mass (in kilograms) divided by the square of their height (in metres).

Bridge population: a population at higher risk of HIV exposure whose members may have unprotected sexual relations with individuals who are otherwise at low risk of HIV exposure.

Brownian motion: an abstract mathematical process that attempts to mirror continuous jerky motion such as might be observed in random walks with tiny steps or molecules colliding in a gas.

Central European family model: type of family organization relying on a historical marriage pattern intermediate between the Northwestern and Eastern European models of low and high nuptiality.

Chi-square: a statistical distribution with *k* degrees of freedom, which is the distribution of the sum of squares for *k* independent normal random variables. It is often used in hypothesis-testing in statistics.

Chrono-species: a group of organisms that is derived from its ancestor by a process of slow, steady, evolutionary change and is not regarded as a member of the same species as its ancestor.

Clades: a group of organisms believed to compromise all the evolutionary descendants of a common ancestor; a branch of an evolutionary tree.

Classical population model: a summary term referring to the sophisticated body of ideas and procedures for reasoning about and compiling populations which prevailed from classical antiquity to the late eighteenth century. Populations were understood not, as in modern demographics, as primarily numerical phenomena composed of uniformly measured and distinct individuals that are structured in aggregate by quantitative regularities. Rather, populations are memberships built up and structured by patterns of human association, sets of relationships linking people that make them by nature constituents of local communities, networks, and other collectivities. In this view, populations are inherently open. That is: members' relationships within a population commonly involve participation in many subpopulations, as well as associations with external groups; the changing patterns of association within and between memberships may produce major alterations in individual as well as population characteristics, leading to continuing transformation of individuals and the formation of new and different populations. Populations making up society and the natural world consist of hierarchies of such memberships, structures in which many distinctive subpopulations are commonly in competition with each other. The differing associative processes by which memberships are formed and change, and consequent qualitative and quantitative differences in the composition and the nature of the ties composing a population or subpopulation, give societies differing capacities for integration and adaptation, and hence for individual and collective action. From a classical point of view, therefore, a primarily quantitative account of population structure is insufficient for purposes of explanation. The problems of integration and competition stressed in classical population thinking provided a direct stimulus to Darwin's seminal approach to population in the *Origin of Species* and to the rise of population arithmetic in the seventeenth century.

Closed population: a summary term referring to populations compiled and analysed quantitatively by formal systems of enumeration, classification, and modelling, such as censuses, surveys, clinical trials, and genomics. Quantitative methods of closed population description and analysis were pioneered in life table models in the seventeenth and eighteenth centuries, and became the prevailing administrative and scientific approach to population in modern society with the rise of social and mathematical statistics in the nineteenth century. Population genetics in the twentieth century was likewise shaped by life table and statistical sampling procedures. Statistical analysis requires populations that are closed in at least three senses. First, sampling and enumeration techniques provide exhaustive or systematically representative accounts of the aggregate of interest. Second, all individuals, events, and relationships are defined in discrete units, classifications, and variables that are decided in advance of enumeration; precise, standardized measurements are intended to ensure that analysis forecloses contextual and other influences, which thence become biases. Third, once populations are fixed in these first two senses, analysis of population structure and change is limited to purely quantitative relationships between the several units, classes, and variables.

Coalescent (multi-locus coalescent models): a tree model showing the evolution of a fixed population in the absence of selective forces. Each individual in the sample is represented as a lineage, and each lineage represented as merged back in time at a constant rate with each other lineage. This process is repeated until only a single lineage remains. In multi-locus coalescent models, a number of genetic locations are each described by a coalescent.

Commercial sex worker: an individual who exchanges sex for money or other valuable goods, such as drugs.

Communication community: theoretical concept which refers to human groups whose participant members share similar reproductive values, familial roles, institutions, and average fertility outcomes. Multiple communication communities exist within large population aggregates, such as national populations. Indeed, there may be multiple communication communities in any one location and, equally, participants may communicate virtually via institutional and other wider networks, as well as more directly in virtue of sharing the same space and life together. Individuals can participate in more than one communication community and may sometimes choose to shift allegiances between them. Observed fertility behaviour may change in consequence, or due to other internal changes amongst participants of communication communities (for example, in the perceived relative costs and benefits of childrearing). Such changes are usually a cultural response to alterations in the socio-economic or political environment.

Compositional demography: the study of population in terms of distributions of subpopulations and their members, often with reference to social, political, economic, and other inequalities. Data on composition have always been integral to demographic description and analysis, given the very different groups and processes of which societies are made up. Renewed interest by social, anthropological, and historical demographers reflects the difficulty of explaining changes in fertility and mortality trends and behaviour without systematic attention to other adjustments people make to their population memberships and the population structures to which these memberships contribute. In many environmental and social settings, the regulation of reproductive and mortality levels and trends is often not a priority, tending rather to follow strategies and institutional arrangements that serve other, more immediate, social ends. Examples include the functioning of intergenerational support, migration, and health-seeking networks; patterns of household composition in relation to marriage preferences, family life cycles, inheritance, land use, and ethnic identity; and social class differences in attitudes to gender roles and childrearing. As vital rates at national and regional levels are usually composites made up of the differing behaviour of several constituent subpopulations, differentiating them and the processes that build and change their memberships may be critical to understanding the heterogeneity of trends before, during, and after demographic transition.

Concentrated epidemic: where diseases such as HIV have spread rapidly in one or more populations but are not well established in the general population. Typically, the prevalence is over 5% in subpopulations, while remaining under 1% in the general population.

Concurrent sexual partnerships: the practice of having more than one sexual partner in a given time period, and going back and forth between them. This increases the probability of transmission of HIV and other STDs, because earlier partners can be infected by later partners.

Controls: a set of samples collected for association studies to compare against a group of individuals of a specific phenotype, such as a disease. (Genetic) differences between the two groups are likely to be associated with the phenotype.

Core groups: subpopulations which are particularly vulnerable to HIV infection, notably sex workers and their clients, men who have sex with other men, and intravenous drug users. Now referred to more often as 'key populations'.

Demes: originally an ancient Greek term for subpopulations forming principal political units in a city state, and now a conventional term in evolutionary biology referring to mating groups constituting principal subpopulations of a species and usually sharing a particular environmental niche.

Demic diffusion: a genetic model in which individuals may be located in a number of geographically located populations, or demes. According to the model, individuals mate at random within each deme and, as migration occurs between demes, mating establishes connections between them at a particular rate (i.e. individuals 'diffuse').

Demographic transition: term coined by the demographer Frank Notestein in 1945 to refer to the secular decline of fertility and mortality in Western societies from the later nineteenth century, and subsequently presumed to extend to later declines worldwide. The mechanisms behind transition were initially ascribed to general processes of economic development and social modernization, but these theories have proven insufficient to account for the diversity of declines and conditions in which they occur.

Diffusion (and social interaction theories of fertility decline): fertility theories that include diffusion and/or social interaction processes usually encompass two central elements. First, they emphasize a particular set of social interactions that are associated with the adoption of low fertility (or family planning), such as social networks, opinion leaders, public media, etc. Second, diffusion models have an explicit dynamic dimension in which the adoption of an innovation is often associated with 'contagion' or 'tipping-point' phenomena, which can accelerate social change; there may also be 'conjunctures' that represent specific short-term social contexts within which decision-making often occurs. Diffusionist accounts have appealed to demographers ever since the European Fertility Project, which indicated that fertility declines are often rapid, irreversible, and may be associated only weakly with the specific socio-economic conditions. Diffusionist approaches have been criticized for several reasons, notably: the unclear definition of what diffuses; the unclear connection to economic models and to ethnographic and historical research on individual and collective action; over-reliance on indirect empirical evidence; and insufficient knowledge about preconditions for the onset of a diffusion process.

Diploid: organisms can have multiple copies of each chromosome that makes up their DNA. Diploid organisms, such as higher animals, have two copies, one inherited from each parent.

Ecological inheritance: inheritance via an external environment of one or more natural selection pressures previously modified by niche-constructing organisms.

Effectiveness: in epidemiological research, such as addressed to HIV, 'effectiveness' describes how well a treatment works when applied in a real-life setting.

Efficacy: in epidemiological research, such as addressed to HIV, 'efficacy' describes how well a treatment works when tested in a clinical trial or laboratory study.

Eigenvalue: in principal component analysis, the eigenvalue is the change in magnitude of the associated eigenvector when it is multiplied by the data.

Eigenvector: in principal component analysis, the eigenvectors are those vectors (components) that do not change in direction when multiplied by a covariance matrix of the data. The multiplication simply makes the 'lines' larger (extend further away from the origin). The amount that they are increased by is the eigenvalue.

Epidemiological transition: a general description of the shift from infectious diseases as prevailing causes of death in a society to chronic, degenerative, and man-made diseases, which occurs in conjunction with wider improvements in social and economic conditions.

Epigenetics: study of genetic mechanisms that create phenotypes without altering the actual nucleotide sequences of genes.

Epigenetic inheritance: denotes heritable phenotypes that are not encoded in the genome.

European marriage pattern: nuptiality pattern based on late age at marriage for men and women and high proportions never marrying, supposed to have been a defining feature of North-West Europe until at least 1900.

Eurytopic: an organism able to tolerate a wide range of habitats or ecological conditions.

Evolutionary synthesis: the prevailing paradigm of evolutionary biology, articulated over the decade 1936–1947 by Ernst Mayr, Theodore Dobzhansky, Sewall Wright, R. A. Fisher, Julian Huxley, and others. By bringing together developments in several biological specialisms (such as genetics, systematics, population biology, ecology, and paleontology), developments in Mendelian genetics were reconciled with Darwinian natural selection, and broad-scale, macro-evolutionary processes were found to be accountable in terms of changes in local populations. Also often referred to as the 'modern synthesis'.

Exchangeability: a statistical concept in which a number of random variables are sampled from some probability distribution, but this probability distribution does not depend on the order in which the variables are sampled. For example, if the numbers 1 through 5 are sampled without replacement, then the result is exchangeable but not independent.

Fixed frailty model: a demographic model positing that age-specific mortality rates in a heterogeneous population differ from person to person by person-specific random factors fixed across the whole of life.

Genetic drift: the change in SNP frequency that occurs in separated populations. This occurs because actual SNPs present from one generation to the next are determined stochastically. This effect decreases with the effective population size.

Generalized epidemic: in the study of sexually transmitted diseases, an epidemic that is self-sustaining through heterosexual transmission. In a generalized epidemic, HIV prevalence usually exceeds 1% among pregnant women attending antenatal clinics.

Generative modelling: a closed or fully specified mathematical or computer model designed to imitate a physical process, from which data can be simulated.

Gene–culture coevolutionary theory: describes how human genetic and human cultural processes both contribute to human evolution, and how they interact.

Genome: the complete assembly of genes present in a given organism.

Genotype: the set of alleles (particular variants of genes) carried by an individual organism.

Gompertz mortality model: the first and most successful mathematical model for late adult mortality, proposed by Benjamin Gompertz in 1825, stipulating that the slope of age-specific rates at any age is proportional to the level of the rates, yielding an exponential curve as a function of age.

Greedy optimization: method used to find the maximum value of a function, which is 'greedy' if it tries to obtain the best possible increase of the function at every iteration.

Such a method cannot, in consequence, extrapolate from a locally maximal value to find the globally maximal value.

Haploid: organisms can have multiple copies of each chromosome that makes up their DNA. Haploid organisms, such as bacteria and archaea, have only one copy.

Haplotype: a haplotype is a single copy of the set of chromosomes for an individual. Diploid individuals consist of two haplotypes. Specifying a haplotype rather than a genotype involves knowing which mutations appear together on the same genomic copy.

Haplogroup: in humans a group of individual haplotypes that share identity by descent on the haploid chromosomes (the *Y* chromosome and the mitochondrial genome), usually defined by a specific locus with a low mutation rate.

Heterozygote in a diploid organism, an individual is heterozygous for a genetic marker if two different alleles are carried, one on each chromosome.

Heuristic model: a model of convenience, often developed in the context of attempts to isolate or explore incompletely understood phenomena, and not necessarily based on an established body of theory.

Hidden Markov model: a statistical model decomposing the probability of the data xi at locations i into two parts using a set of hidden variables yi. The probability of xi depends only on yi and on global parameters, i.e. it is conditionally independent of xi and yj for $j \neq i$. The probability of yi depends only on the previous $yi-1$ but is conditionally independent of both xj (for all j) and yi (for $j \neq I - 1$).

HIV control programme: a set of activities implemented by a national government to prevent the spread of HIV infection, provide care and treatment for those who are infected, and mitigate the impact on others (e.g. orphaned children) who are affected by an epidemic.

HIV incidence: the rate at which new people within a population or subpopulation are acquiring HIV infection over a given period of time.

HIV prevalence: the proportion of a population or subpopulation currently infected with HIV at a given point in time, usually expressed as a percentage.

Homozygote in a diploid organism, an individual is homozygous for a genetic marker if the alleles on both chromosomes are the same.

Human Genome Diversity Panel (HGDP): a genetic dataset consisting of 1,043 individuals from 51 populations around the world. 660,918 SNPs have been typed for these samples, which are available for researchers to use. For a description of the data and access information see <http://hagsc.org/hgdp/index.html>.

Inbreeding coefficient: a measure of the relatedness of individuals within a population. The coefficient is the expected level of homozygosity arising in a particular population.

Isolation-by-distance: populations undergo genetic drift separately only if there is some barrier to gene flow preventing changes in one population from appearing in the other. Isolation-by-distance models use some measure of the geographical separation between populations to describe the size of the genetic barrier, usually in terms of a migration rate that decreases with distance.

Joint families: laterally extended, multiple-family residential arrangements based on patrilocal marriage of several (if not all) children, most often sons. Young couples remain subject to the authority of the 'patriarch' (usually the senior male), and split away at the time of the husband's father's death, or after. Associated with strong patrilineal orientation, early marriage, and the corporate (joint) nature of property ownership.

Life history theory: a branch of evolutionary biology attempting to explain optimal age schedules of demographic parameters within and among species.

Life table: models of life expectation developed initially in the seventeenth and eighteenth centuries, chiefly in the context of governmental and private annuity schemes, and also as a testing ground for the calculus of probability. The demands which accurate life table analysis made on data collection provided an influential model which shaped the emergence of national and local statistical data systems in the nineteenth century.

Linkage disequilibrium: a genetic sample is in linkage equilibrium when each SNP is independent and therefore found (on diploid organisms) on the two copies of the chromosome in the proportions expected from the population-level frequency. Linkage disequilibrium is a deviation from these proportions due to correlations on the genome caused by recombination having occurred sufficiently frequently for SNPs to be correlated.

Locus: a specific bit of the genome: for example, a gene or a genetic marker.

Markov Chain Monte Carlo (MCMC): a statistical inference technique used to find the probability of parameters, given some data in which parameters cannot be directly computed. Parameter values are sampled, and previous 'good' parameters used to obtain a new 'good' sample.

Meta-population: a total population system composed of multiple local populations, separated geographically, but functionally connected through dispersal.

Natural dynamics (or natural course of infectious disease epidemics): the ways in which an infectious disease agent spreads within a population over time in the absence of control.

Natural fertility: a concept developed initially for describing and analysing reproductive levels in historical populations assumed not to practise modern or other contraceptive technologies. Natural fertility is supposed to exist in the complete absence of strategies to regulate the numbers of children. The concept has proven very helpful in stimulating exploration of the influence of breastfeeding, pathological sterility, and other influences on reproduction, and under the proximate determinants framework may be contrasted to societies practising modern contraception and abortion. The widespread existence of birth control strategies and practices not aimed at regulating numbers of children—e.g. for spacing purposes—may, however, be obscured significantly by the concept.

Natural selection: the process defined initially by Darwin in which organisms better adapted to their environment tend to survive and produce more offspring. Typically, variation exists among organisms within a reproducing population, and this variation is often, but not necessarily, heritable. When variation is causally connected to differing ability to survive and reproduce, differential reproduction becomes a key mechanism of natural selection.

Network density: the ratio of active ties existing in a social network to the ratio of all possible social ties in that network. Density is a measure of how connected a network is. Dense networks are most likely to be found in small, stable communities with few external contacts and a high degree of social cohesion. Loose social networks with low density, by contrast, are more liable to develop in larger communities subject to change that have many external contacts and a relative lack of social cohesion.

Niche: refers to both the requirements and impacts of organisms in environments. Niche construction theory defines niches evolutionarily, as the sum of all the natural selection pressures to which an organism or a population is exposed.

Niche construction: the process whereby organisms, through their metabolisms, activities, and choices modify their own and/or each other's environments. The term includes migration and habitat selection. The defining characteristic of niche construction is not the modification of environments per se, but the modification of natural selection pressures in those environments.

Non-parametric: a distribution that requires a potentially infinite number of parameters to be fully specified is described as non-parametric. The number of parameters is never infinite in practice, but instead grows with the number of data points observed. Because models of non-parametric distributions are very flexible, they are often claimed to make no assumptions about data; this, however, is not quite true, as the choice of how parameters describe the data can have a large effect on the inferred distribution.

North-West European household formation system: rules of family and household formation based on late marriage for both sexes, neo-local and nuclear household formation, and the circulation of young people between households as servants before marriage.

Nulliparous: refers to a woman who has never given birth.

Nuptiality: the frequency or incidence of marriage within a population, usually studied by measuring two components: proportions married and age at marriage.

Open population: a summary term referring to populations built up by patterns of association between individuals, such as mating or social networks. Study of processes that form, sustain, and alter population composition is based, at least initially, on direct observation, whether in the laboratory, in local ecologies, or ethnography. Observation, compilation, and measurement in local studies in turn generate variables and hypotheses that may be used to construct closed populations for sampling, simulation, and other purposes, including testing findings from local population studies in larger aggregates. Open population approaches under the classical population model generally excluded quantitative approaches, except for limited purposes; with the evolutionary synthesis, open population thinking was recognized as necessary and complementary to closed population models. Combined methodologies are now also employed increasingly in demographic research.

Panmictic: a population in which every individual has the same chance of mating with every other, and where population structure is simply random.

Parametric: a distribution that can be specified by a finite set of parameters is parametric. For example, the mean and standard deviation are parameters of normal distributions.

Parity: the number of children previously born alive to a woman.

Parity progression ratio: the proportion of women with a certain number of children who go on to have another child.

Phenotype: characteristics that are observable and are the result of genetic and/or environmental influences.

Phenotypic plasticity: the ability of individual genotypes to produce different phenotypes when exposed to different environmental conditions. Organisms may change their phenotypic state or activity (e.g. metabolism) in response to variations in environmental conditions, including the possibility of modifying their developmental trajectory by altering both their phenotypic characteristics and environmental niche.

Philopatry: the tendency of an animal or species to return to or remain near a particular site or area.

Phylogenetic analysis: investigation of the relationship between different genetic markers, individuals, or populations, based on the principle that those that are genetically most similar are the most related. The relationships are usually represented as a tree.

Polymorphic variation: genetic variation where the minor allele is found in the population at a detectable frequency (for example, 1 or 5%).

Population structure: in demography, populations are understood as structured primarily by the age and sex composition given in census and survey compilations, seen as the result of past trends in fertility, mortality, and migration. The age and sex structure and the trends that give rise to it may then be further studied in subpopulations as also usually recorded in national data, subdivided, for example, by locality, class, ethnicity, and so forth. The characteristics people report in censuses and surveys provide the basis for classifying individuals into age or other groups in the population structure. In evolutionary genetics, populations consist of members of a species, deme, or other mating group; the structure and membership of these populations is determined by mating within and between them, which determines which members form a population in virtue of their shared gene pool. The definition and characteristics of a population are also influenced by migration and genetic drift—in short, boundaries of subpopulation units are continuously subject to shift. The study of population structure, in other words, focuses on observed relations within and between subpopulations which affect human evolution. Not surprisingly, these populations seldom coincide neatly with groups identified by locality, class, ethnicity, or other classifications as reported in demographic data sources.

Prevention of mother-to-child transmission (PMTCT): refers to a four-prong strategy for stopping new HIV infections in children and keeping mothers alive and families healthy. The four prongs are: reducing HIV incidence in women; reducing unmet need for family planning; providing antiretroviral prophylaxis to prevent HIV transmission during pregnancy, labour and delivery, and breastfeeding; and providing care, treatment, and support for mothers and their families.

Principal component analysis: a method of reducing the dimensionality of multidimensional data. The first (principal) component is the line through the data that accounts for more of the variation than any other line. The second component is perpendicular to the first and explains the next greatest variation. This process can be repeated to identify a number of components, but in practice only the first few tend to be used in population genetics.

Proximate determinants: biosocial factors which influence human fertility directly, notably breastfeeding, nuptiality, coital frequency, sterility, contraception, and abortion. Demography treats economic and social influences on the timing and levels of reproduction, like education or income, as independent variables that can only alter fertility via one or another proximate determinant. Analogous proximate determinant approaches have been applied to mortality.

Recombination: when an organism reproduces, it must make copies of its DNA. Sexually reproducing organisms take DNA from both parents, 'recombining' them anew in their offspring. Non-sexual species can also recombine their DNA but this typically does not occur at every reproduction.

Second-generation surveillance: the regular, systematic collection, analysis, and interpretation of information to track and describe changes in the HIV epidemic over time. In addition to HIV surveillance and AIDS case reporting, second-generation surveillance includes behavioural surveillance to track trends in risk behaviours over time to warn of, or explain, changes in levels of infection and the monitoring of sexually transmissible infections in populations at risk of HIV. These different components achieve greater or lesser significance depending on the surveillance needs of a country, determined by the nature of the epidemic it is facing.

Self-determinants: loci within the human leukocyte antigen (HLA) system involved in determining self from non-self, as part of the regulation of the immune system in humans.

Sentinel surveillance: a system for monitoring trends in HIV prevalence in a population over time. In generalized epidemics this is done, for the general adult population, by tracking trends in pregnant women attending for routine check-ups at selected antenatal clinics. In concentrated epidemics, sentinel surveillance is conducted in samples of high-risk populations, such as men and women attending STD clinics.

Sexual network: groups of persons connected to one another sexually. The number of persons in a network, how central high-risk persons are within it, the percentage in monogamous relationships, and the number of 'links' each has to others all determine how quickly HIV/STDs can spread through a network. Sexual networks may be identified as distinct from, but often overlap with, social networks.

Single nucleotide polymorphism (SNP): a specific position in the genome where at least two alternative alleles (for example A or G) can be found in the population. The rare allele (known as the minor allele) should be detectable in the population and so will have a frequency of at least 1%.

Social desirability bias: the tendency of research respondents to answer questions in a manner that they believe will be viewed favourably by others. This can take the form of over-reporting good behaviour, or under-reporting bad behaviour.

Social influence: in the study of fertility decline, the process by which an individual's preferences, attitudes, and schema relating to parenthood are influenced by the prevailing norms and values of a social environment. The direction of social influence can differ in different stages of fertility decline. In the early stages, social influence may restrict the adoption of contraception. Once a noticeable fertility decline has set in, the direction of social influence may then shift: as low fertility becomes commonplace, the interaction among women or couples is likely to encourage a decline in the demand for children and behaviour that leads to limiting fertility, while at the same time, social interactions start to have an important reinforcing influence on the postponement of childbearing to older ages.

Social learning: the process by which individuals learn about the existence and details of an innovation, and reduce the uncertainties associated with its adoption by drawing on the experience of network partners.

Social multiplier effect: social interaction implies the presence of multiplier effects which can be important in the study of fertility change, notably in monitoring responses to family planning programmes. An increase in programme activity may increase the probability of people choosing to adopt modern contraception. A direct effect reflects an increase in the propensity to choose modern contraception, above extant contraceptive prevalence in a target population. However, because increased programme activity may influence all community members (not just those to whom a programme is directed), the prevalence of modern contraception in a wider population may increase as well. This change in the community has a feedback or social multiplier effect on individual behaviour in future periods. Hence, the total change in the prevalence of family planning exceeds the direct effect. The interaction in social networks thus enhances changes in contraceptive use that follow family planning efforts or other socio-economic changes.

Social network: a social structure made up of a set of social actors (such as individuals or organizations) and a set of dyadic ties between them. Study of these structures uses social network analysis to identify local and global patterns, locate influential entities, and examine network dynamics.

Social network data and analysis (egocentric, sociocentric): social network analyses focus on the causes and consequences of relations between people, and among sets of people, rather than on the features of individuals. The sociocentric (whole) network approach involves the quantification of relationships between people within a well-defined population, such as a classroom of children, residents of a village, or trading partners in a bloc of nations. The focus of sociocentric network analyses is on measuring structural patterns of social interaction in such populations, and how these patterns may explain outcomes, such as specific individual behaviour, the distribution of resources, or the diffusion of information. In contrast, in the egocentric (personal) network approach, a network comprises persons (what social network analysts call 'alters') that one person (or 'ego') knows: spouses, children, cousins, co-workers, friends, etc. Egocentric network analyses focus on measuring the size and characteristics of such respondent-focused networks, and the association between these networks and outcomes of interest.

Statistical phylogeography: a statistical approach to phylogenetic modelling (that is, representing relationships between individuals via trees) which accounts for geographical features such as sampling locations and physical barriers to population movement.

Stem families: multigenerational families with no more than one married child. Either the entire or bulk of family property is passed intact to one child (usually a son), while other offspring leave the family household and its property to make a living elsewhere, through whatever ecological niches are available to them. Most of the 'stem families' observed in historical Central and Western Europe took the form of *Ausgedingefamilie* ('retirement family'), under which headship was handed over to the heir at the time of his marriage.

Stochastic vitality model: a demographic model holding that some index of the health of an individual organism varies in a partly random and partly structured way across the life course, with death occurring when the index strays across some fatal threshold.

Subpopulation: constituent groups in a wider or total population, as defined by internal composition, relationships, and history, and by evolving relationships with other groups and/or the wider population in which they are found. Subpopulations are commonly identified by some selection of geographical, demographic, social, genetic, or other characteristics.

Survival curve: a curve that can be graphed as a function of age, representing probabilities or proportions of a population surviving from birth to each given age.

Sympatric speciation: refers to organisms whose ranges overlap or are even identical, so that they occur together at least in some places.

Total fertility rate: the average number of children that would be born to a woman over her lifetime if: (1) she were to experience the exact current age-specific fertility rates (ASFRs) through her lifetime, and (2) she were to survive from birth through the end of her reproductive life.

Transactional sex: transactional sexual relationships are sexual relationships where the giving of gifts or services is an important factor. Transactional sex relationships are distinct from other kinds of sex work or prostitution, in that the transactional sex provides only a portion of the income of the person providing the sex. Those offering sex may or may not feel affection for their patrons.

Index

The manufacturer's authorised representative in the EU for product safety is Oxford University Press España S.A. of el Parque Empresarial San Fernando de Henares, Avenida de Castilla, 2 – 28830 Madrid (www.oup.es/en or product.safety@oup.com). OUP España S.A. also acts as importer into Spain of products made by the manufacturer.

www.ingramcontent.com/pod-product-compliance
Ingram Content Group UK Ltd.
Pitfield, Milton Keynes, MK11 3LW, UK
UKHW051224160926
475595UK00022B/156
* 9 7 8 0 1 9 9 6 8 8 2 0 3 *